AGRICULTURE IN THE TROPICS

AGRICULTURE IN THE TROPICS
Third edition

Edited by

C.C. Webster, CMG
Formerly Director, Rubber Research Institute of Malaysia, Director-General, Palm Oil Research Institute of Malaysia and Professor of Tropical Agriculture, University of the West Indies

P.N. Wilson, CBE
Professor Emeritus of Agriculture and Rural Economy, University of Edinburgh, General Secretary, Royal Society of Edinburgh, formerly Scientific Director, Edinburgh Centre for Rural Research, Chief Agricultural Adviser, BOCM Silcock and Professor of Tropical Agriculture, University of the West Indies

Blackwell
Science

© 1998 by
Blackwell Science Ltd
Editorial Offices:
Osney Mead, Oxford OX2 0EL
25 John Street, London WC1N 2BL
23 Ainslie Place, Edinburgh EH3 6AJ
350 Main Street, Malden
  MA 02148 5018, USA
54 University Street, Carlton
  Victoria 3053, Australia
10, rue Casimir Delavigne
  75006 Paris, France

Other Editorial Offices:

Blackwell Wissenschafts-Verlag GmbH
Kurfürstendamm 57
10707 Berlin, Germany

Blackwell Science KK
MG Kodenmacho Building
7–10 Kodenmacho Nihombashi
Chuo-ku, Tokyo 104, Japan

The right of the Author to be identified as the Author of this Work has been asserted in accordance with the Copyright, Designs and Patents Act 1988.

All rights reserved. No part of this publication may be reproduced, stored in a retrieval system, or transmitted, in any form or by any means, electronic, mechanical, photocopying, recording or otherwise, except as permitted by the UK Copyright, Designs and Patents Act 1988, without the prior permission of the publisher.

First edition published by Longman Group UK Ltd 1966
Second edition 1980
Reprinted 1986, 1989
Third edition published by Blackwell Science Ltd 1998

Set in 10 on 12 pt Times
by Best set Typesetter Ltd., Hong Kong
Printed and bound in Great Britain at the
University Press, Cambridge

The Blackwell Science logo is a trade mark of Blackwell Science Ltd, registered at the United Kingdom Trade Marks Registry

DISTRIBUTORS

Marston Book Services Ltd
PO Box 269
Abingdon
Oxon OX14 4YN
(*Orders*: Tel: 01235 465500
         Fax: 01235 465555)

USA
Blackwell Science, Inc.
Commerce Place
350 Main Street
Malden, MA 02148 5018
(*Orders*: Tel: 800 759 6102
         781 388 8250
   Fax: 781 388 8255)

Canada
Login Brothers Book Company
324 Saulteaux Crescent
Winnipeg, Manitoba R3J 3T2
(*Orders*: Tel: 204 224-4068)

Australia
Blackwell Science Pty Ltd
54 University Street
Carlton, Victoria 3053
(*Orders:* Tel: 03 9347 0300
         Fax: 03 9347 5001)

A catalogue record for this title
is available from the British Library

ISBN 0-632-04054-8

Library of Congress
Cataloging-in-Publication Data is available

For further information on
Blackwell Science, visit our website:
www.blackwell-science.com

CONTENTS

	Preface	vii
	Acknowledgements	viii
	Contributors	ix
1	Socio-economic Background: The World Food Problem *J.C. Holmes*	1
2	Climate, Agriculture and Vegetation in the Tropics *C.T. Agnew*	20
3	Tropical Soils *J.K. Coulter*	48
4	Soil and Water Conservation *M. Stocking*	82
5	Land Clearing, Drainage, Tillage and Weed Control *B.D. Soane*	113
6	Rain-fed Arable Farming Systems and their Improvement *J.K. Coulter*	144
7	Rice and Rice-based Farming Systems *D.J. Greenland*	178
8	Plantation Crops *W. Stephens, A.P. Hamilton and M.K.V. Carr*	200
9	Agroforestry *P.A. Huxley*	222
10	Tropical Crops and their Improvement *N.W. Simmonds*	257
11	Production of Animal Feed *P.N. Wilson*	294
12	Tropical Grasslands Used for Livestock *P.N. Wilson*	313

13	Classes of Tropical Livestock *P.N. Wilson*	336
14	Adaptation of Livestock to Tropical Environments *P.N. Wilson*	371
15	Cattle Management in the Tropics *P.N. Wilson*	391
16	Livestock Improvement by Feeding and Nutrition *P.N. Wilson*	417
17	Livestock Improvement Through Health and Hygiene *P.N. Wilson*	447
18	Livestock Improvement Through Breeding *P.N. Wilson*	462
	References	487
	Index	528

PREFACE

This book gives a general account of the basic factors affecting agriculture in the tropics and of the application of existing knowledge of the principles of agriculture to its improvement and development. It deals with principles rather than specific details. It does not set out to give a comprehensive treatment of any given crop or farm animal nor to provide a detailed practical guide for farming operations.

As a background to the subject, the first chapter draws attention to the precarious balance between population and food supply already existing in many tropical countries. In some regions the rapid growth of the population has been accompanied by widespread degradation of land, loss of fertility and declining yields with consequent malnutrition and hunger. The rapid increase in the population will inevitably continue for some time to come, but the resources of good land and water are limited. Hence, there is a need for sustainable agricultural development and improvement on a vast scale in order to provide enough food and a better standard of living for the growing population while at the same time maintaining or, where possible, improving the productivity of the land. This task is rendered the more difficult by the constraints imposed by a number of socio-economic factors, especially the inability of most poor farmers to purchase the inputs, such as equipment, fertilisers and pesticides, needed to improve crop yields.

The ensuing chapters first deal with tropical climates, vegetation and soils and their influence on agriculture. The principles of soil and water conservation, land clearing, drainage, tillage and weed control are then outlined. This is followed by accounts of current practice and possibilities for improvement in rain-fed arable farming, rice-based farming systems and the culture of tree crops, and by a summary of the economic botany and breeding of the main tropical crops. Subsequent chapters cover the utilisation of natural and cultivated pastures, the classes of livestock important in the tropics, the adaptation of animals to tropical environments, animal husbandry and the improvement of livestock by better nutrition, hygiene and breeding.

Aspects of agriculture such as pig and poultry husbandry, where temperate practice can be modified to suit the requirements of the tropics, are not covered in any detail. However, the reader is referred in the text to standard books dealing with these latter subjects and also to books giving specialised accounts of particular crops, animals or facets of tropical agriculture.

The primary aim has been to provide an outline of the subject for students reading for degrees or diplomas in tropical agriculture and for those who proceed to work, or study, in tropical agriculture after graduating in agriculture or related subjects in the temperate zone. It is also hoped that this book will prove useful to tropical planters and farmers and to others indirectly concerned with agriculture, such as administrators, development planners, economists and veterinarians. In some sections of the book it has been assumed that the reader has an elementary knowledge of an applied science relevant to agriculture. However, such knowledge is not essential to a proper understanding of the greater part of the book.

In this third edition, all the chapters have either been extensively revised or entirely rewritten and a new chapter, providing a conspectus of all the important tropical crops and a summary of the methods used for their improvement, has been included. The review of relevant research findings incorporated in the text has been updated and references provided to recently published textbooks and scientific papers on various aspects of tropical agriculture. In consequence, the list of references will be found to be unusually long for a book of this type but, as a result of their own experience in university teaching, the authors believe that a bibliography of this nature will assist teachers in the preparation of their courses and will guide post-graduate students in their private reading.

ACKNOWLEDGEMENTS

The authors would like to thank Professor M. Gill, Dr C. Hendy, Dr G.N. Mowat, Professor J.D. Oldham, Professor M.M.H. Sewell, Dr G. Simm, and Dr D. Thomas for help in checking the manuscript, and to acknowledge the important contributions made by the individually invited authors of Chapters 1 to 10 which have brought fresh insights to the crop sections of the book.

CONTRIBUTORS

Dr Clive Agnew is an applied climatologist who has held appointments as a visiting Professor in USA, Sierra Leone and Oman, working mainly in Africa and the Middle East on topics of water resources, drought and environmental degradation. He teaches undergraduate and postgraduate courses in the fields of hydrology and water management, is currently the Director of the Masters in Research in Environmental Sciences, and senior lecturer in physical geography at University College London.

Professor Mike Carr has been involved in the tea industry for over thirty years, and is Executive Director of the newly-formed Tea Research Institute of Tanzania. He is currently Head of the Natural Resources Management Department at Silsoe College, Cranfield University.

Dr John Coulter was employed in Malaysia as a soil scientist and Director of Agricultural Research, and subsequently as Tropical Soils Adviser to the ODA and Agricultural Adviser to the World Bank. He is currently a private agricultural consultant.

Professor Dennis Greenland directed the research programmes of the International Institute of Tropical Agriculture in Ibadan, Nigeria from 1974 to 1976 and those of the International Rice Research Institute in Los Banos, Philippines from 1977 to 1987. He is now Visiting Professor in the Department of Soil Science at the University of Reading.

Andrew Hamilton is a specialist in tropical tree crops with extensive experience in commercial agribusiness and multilateral financing institutions in the Asia Pacific region. He is currently developing new palm oil and rubber investment for CDC worldwide.

Dr Peter Huxley was Director of Research at the Coffee Research Foundation, Kenya, then Professor of Horticulture at Reading University and Director of the Research Development Division of ICRAF. He is currently a consultant in agroforestry education.

Professor Norman Simmonds has more than 50 years experience of the tropics; he has been Honorary Professor at the University of Edinburgh since 1975, and is still active in the field since his retirement from the Edinburgh School of Agriculture in 1982.

Dr Brennan Soane worked on soil and irrigation studies in Zimbabwe for eleven years prior to joining the Scottish Institute of Agricultural Engineering, and from 1985–88 was President of the International Soil Tillage Research Organisation. After retirement he has been involved in undergraduate and postgraduate teaching in soil management.

Dr William Stephens has experience in plantations-related research, training and consultancy in a number of countries, covering subject areas from yield prediction to management development. He is currently Director of the International Research Centre for Plantation Studies at Cranfield University.

Professor Michael Stocking has been involved in tropical agricultural development, land resource and soil conservation since 1969, with field experience in Sub-Saharan Africa, South America and South and South-East Asia. He is currently Professor of Natural Resource Development at the University of East Anglia.

1
Socio-economic Background: The World Food Problem

J.C. Holmes

INTRODUCTION

Agriculture is carried out for three main reasons, and these apply whether the production is conducted in tropical, subtropical or temperate regions: food production; fibre production; bio-fuel production. Of these three, the production of food for human consumption is by far the most important. In the tropics farming is mostly undertaken by smallholders who, although they may grow some cash crops, practise it mainly for subsistence. It is still very largely carried out by traditional methods which employ hand, or ox-drawn, implements, make little or no use of fertilisers and other agrochemicals, and give relatively low yields per hectare and per man. Modified shifting cultivation is still the predominant system but, now that the fallow periods are shortened because of population pressure, this system is commonly wasteful, or destructive, of natural resources.

SOCIAL STRUCTURES AND CUSTOMS

Agriculture has many other attributes. Thus in some countries agricultural practices and religious beliefs are intimately interwoven. Certain foods such as eggs, pork and beef are forbidden for use by man on purely religious grounds and these 'taboos', and other proscribed practices, are important to understand in those parts of the world where they are still practised and socially relevant. Religious beliefs can have an obvious influence. For example, the fact that cattle are sacred to Hindus and cannot be killed has a profound influence on agriculture in India, where much land is burdened with unproductive cattle. Again, pig-keeping is a profitable and appropriate component of farming systems practised by the Chinese in Malaysia, but the native Malays, being Moslems, will not keep pigs. Other customs may have a marked effect on the way people use land. For example, in many African pastoral tribes it is customary for each individual, or family, to aim at keeping as many animals as possible, largely irrespective of the carrying capacity of the land or the quality of the stock. This is done partly as a matter of prestige, partly for the discharge of certain customary tribal obligations, such as the provision of the 'bride price' and partly in the mistaken hope that possession of a large number of animals will form an insurance against times of famine, drought or disease. It is a custom which commonly results in gross overstocking with consequent deterioration of pastures, erosion, poor-quality stock and very low productivity.

A circumstance which has affected agriculture over much of Africa, and in parts of Asia and Oceania, has been the existence of a rigid form of group or tribal life. Until recent times the majority of Africans lived for centuries not as individuals in the European sense, but as members of a group, clan or tribe which required them to live strictly in accordance with its customs. Initiative, or deviation from the accepted tribal pattern of life and land use, was frowned upon and commonly incurred reprisals. As a result there were relatively few agricultural innovators or experimentalists and practically no exceptionally good farmers from whom others might learn. Consequently, agricultural practices tended to change very slowly, although they were periodically modified as a result of external influence, perhaps by the adoption of a method employed by a nearby tribe, or by the introduction of a new crop. In recent times tribal customs and authority have been declining and people are living more and more as individuals. Nevertheless, the habits of centuries are not usually speedily changed; in many remoter areas, group custom and influence are still strong and farming is still carried on by traditional methods solely for subsistence. In such circumstances improved productivity cannot be achieved by merely attempting to change techniques, but requires a process of education and guided social change to persuade people to change their traditional attitude and to regard farming as a business to be undertaken for profit. For example, getting the pastoralists, mentioned above, to adopt better land use is not simply a matter of demonstrating better pasture management and animal husbandry but necessitates persuading the people to abandon their customs regarding stock, to develop a commercial attitude and to raise cattle for sale and profit.

LAND TENURE AND INHERITANCE

Land tenure systems, together with associated customs regarding inheritance and the allocation to individuals of shares in the different types of land available, commonly place obstacles in the way of agricultural development, especially in the more densely populated areas. Many variations of communal and individual tenure are found in different parts of the tropics.

Communal tenure is the commonest system, especially in Africa. It is characterised by rights in land being divided between the community and the individuals belonging to it. Among pastoral communities the use as well as the ownership of the land is usually communal, and various factors, particularly the sparse distribution of water supplies, may make the introduction of individual rights very difficult. In cultivating communities there is communal ownership of all land by the tribe or clan but, though individuals do not own land, they have user-rights in cultivated land, usually including that lying fallow, which are permanent so long as they continue to use the land. A user-right in land does not necessarily confer any right to property in established trees thereon, or even to the produce of such trees. Customs in this respect vary; probably the commonest situation is that wild, or semi-wild, trees are regarded as communal property, but if a man plants a tree it belongs to him and he may even retain the ownership of trees which he has planted after forfeiting his user-rights in the land. But custom may forbid a man who has user-rights in land from planting trees thereon, as this is regarded as an attempt to acquire permanent ownership of the land.

Where people have both crops and stock, it is customary for there to be individual user-rights in arable and fallow land, and communal user-rights in grazing land, but in some places communal grazing rights may extend to fallow land and stubbles. Inheritance most commonly involves the subdivision of the land over which a man holds user-rights among his sons; but customs vary, and in some tribes surviving wives may have at least a life user-right, while in matrilineal tribes the rights pass to the sons of the holder's sister.

While it has certain advantages, communal tenure among the cultivating communities generally places difficulties in the way of improved farming and land use. In association with group life, in which everyone tends to be regarded as equal, it militates against individual initiative. A farmer does not have the same interest in the care and development of his land as he would if it were his own property, especially if he knows that, should the fertility of his plot decline, he can move to new land. The land cannot be used as security for development loans. Customs regarding the ownership of trees may discourage people from planting profitable perennial crops. Land-use planning, farm planning and the introduction of better farming systems are rendered difficult by this form of tenure.

Forms of individual tenure include freehold and leasehold and various kinds of feudal tenure under which a landowner may grant a tenant farmer user-rights in land which are secure so long as dues are paid in the form of labour, part of the crop or some other form of tribute. Unrestricted individual ownership may have several disadvantages. Owners may use land wastefully, or hold it without developing it. Small landowners often become chronically indebted by mortgaging their land. But the most important and widespread disadvantage is the development of unsatisfactory landlord–tenant relationships. Especially in the more populous areas of the East, many small farmers rent land, commonly under customary arrangements which involve no written contracts. Competition for scarce land enables landlords to charge high rents and often they will only grant tenancies on a year-to-year basis, in the expectation of being able to raise rents frequently. In consequence, some tenants cannot afford to rent sufficient land to form an economic holding, and many may have to meet frequent demands for higher rents or face eviction at short notice without compensation for improvements.

This insecurity is naturally a strong disincentive to improving the holding and militates against any effort or investment from which benefits cannot be obtained almost immediately. Often the tenancy arrangements involve the compulsory sale of produce surplus to the farm family's requirements to a landlord who is also a trader and from whom the tenant has to purchase farm requisites and household necessities. To buy these essentials many poor farmers have to borrow advances against their crops, usually at high rates of interest, and thus become more or less continuously indebted to their landlords. Such farmers are reluctant to increase their indebtedness by buying inputs, such as fertilisers and pesticides, the benefits of which will in part depend on the vagaries of the weather, and the incidence of pests and diseases. The situation is even more unfavourable for the tenant who pays a proportion (usually one-third to a half) of his crop as rent, since under this common system of share-cropping there may be little advantage in spending money on inputs for which he has to pay the full cost while the landlord gets part of the return. One consequence of these unsatisfactory landlord–tenant relationships is to make it difficult to achieve the greatly increased yields promised by the recently developed cereal varieties, since small tenant farmers may not think it worth their while to pay for improved seed and for the fertilisers and pesticides, without which the new varieties will not attain their yield potential, when a large part of the benefit will go to the landlord.

Fragmentation, or the existence of holdings comprised of a number of scattered pieces of land, is common. Conditions which encourage fragmentation are a growing population and a law of inheritance which provides for subdivision of holdings among heirs. It is widespread under most systems of tenure because the Moslem, Buddhist, Hindu and most African tribal laws all provide for subdivision between heirs. It is often accentuated under communal ownership by the custom of allocating to each holder of user-right a share in each of the different kinds of land available. Thus, a man's holding may comprise separate plots of hilltop, hillside and valley bottom land, each of which is subdivided between his sons on his death. Fragmentation is disadvantageous in that it makes it difficult or impracticable for a farmer to use and tend all his lands efficiently. Fragments distant from the homestead cannot be given dressings of farmyard manure by people who do not possess wheeled transport, nor can they be used for the production of crops requiring frequent attention, or that are liable to be stolen. In the absence of fencing, fallow fields that are surrounded by other people's standing crops cannot be grazed, and this discourages the planting of improved pastures. The small size and irregular shape of the plots often results in waste of land and hampers cultivation by any other means than hand tools. Fragmentation also contributes to the difficulties of getting effective soil conservation measures implemented and increases the time and labour required in agricultural advisory work.

Apart from placing difficulties in the way of improving agricultural productivity, customary forms of land tenure also give rise to social problems. The bad landlord–tenant relationships often associated with individual ownership in the East result in social injustice which provokes growing dissatisfaction. There have been instances of violence accompanying attempts by small tenant farmers to force landlords to share out more equitably the extra profits obtainable from the introduction of higher-yielding cereal varieties. As population density increases, the disadvantages of communal tenure are becoming increasingly appreciated by the more progressive members of the agricultural community, especially where it is desired to grow perennial cash crops whose ownership cannot readily be divorced from that of the land itself under the traditional system. In many places a tendency to individual ownership has developed and land is bought and sold, but this often leads to a conflict of views within the community, some being strongly opposed to the change. Commonly, there is a lack of legal provision to deal satisfactorily with the changes that are occurring and a risk that the unfavourable features of certain forms of individual tenure may develop. In order to remove both social injustice and obstacles to agricultural improvement, there is a widespread need for governments to carry out land tenure reform and to deal with fragmentation by redistribution of land and consolidation of holdings.

GOVERNMENT SUPPORT AND FINANCE

Agricultural development on the scale and at the pace needed to provide enough food and a better standard of living for their rapidly expanding populations will clearly require much support from the governments of developing countries in the way of legislation, administrative action and finance for the provision or expansion of various public utilities and services. The need for land tenure reform and consolidation of holdings has already been mentioned. The provision of credit to farmers for the purchase of recurrent inputs, such as fertiliser, and for investment in machinery, buildings, fencing, etc., is likely to be essential on a considerable scale. Improvement in communications and transport will commonly be required. In some places, the development of reserves of good land will be dependent on a major road construction programme but, more generally, the need will probably be for low-cost feeder-roads to link farmers to local markets. Better organisation of marketing and, perhaps, some price control are likely to be needed to assure farmers of reasonable returns. In the absence of effective marketing arrangements, a considerable increase in food production following the adoption of improved varieties and methods could lead to a steep fall in prices which would discourage farmers from further investment and effort. More and better storage facilities will be essential, both to minimise post-harvest waste (which is currently very considerable) and to enable the produce of good harvests to be held over to alleviate scarcity in poor years. The expansion of agricultural research and advisory services is also likely to be a necessary accompaniment of improvement and development.

Apart from financing the establishment or expansion of the services mentioned above, money will be required for capital investment in land development and improvement, especially by means of irrigation, drainage and flood-control works, and for the development of local industries for the manufacture of farm inputs, such as fertilisers, or for the processing of some agricultural products. It is unlikely that more than a small part of the finance required will be forthcoming as investment, loans or aid grants from the more advanced countries, yet the extent to which capital can be found within many of the predominantly agricultural tropical countries is strictly limited. These countries must endeavour to earn, mainly by the export of agricultural produce, the foreign exchange they need to buy imported resources required to support development. This has become an increasingly difficult problem in times when inflation has very substantially raised the cost of these imports, while the prices obtainable for many tropical agricultural products have not increased in proportion or have even relatively decreased.

It is technically possible to achieve a very large

increase in the scale and intensity of agricultural production in the tropics and there is no reason to believe that limitations of technology or availability of cultivable land invariably need prevent sufficient food being produced in the developing countries as a whole. But the scale of the administrative action required, and the social and economic problems outlined above, make the achievement of the food production targets far less certain and place very great difficulties in the way of providing employment and a better standard of living for the expanding population.

POPULATION PRESSURE

These social considerations lead naturally to a consideration of population pressure, and the precarious balance between population and food supply in tropical countries. At the present time (in the late 1990s), the world population lies somewhere between 5 billion and 6 billion in size, and current estimates are that this population will double, or nearly double, in the course of the next two generations.

Formerly the world population was relatively stable, sometimes increasing slightly and sometimes diminishing, at a figure of approximately 1 billion. However, over of the last 200 years, due to the control of human disease resulting in more people reaching an adult reproductive age, the population has not merely increased but, in certain countries, has increased exponentially.

The first person to highlight this problem and to describe it in elementary mathematical terms was Malthus (1798). In this historic essay, still widely quoted, he stated various postulates. First, that food was necessary for the existence of man and second that 'the passion between the sexes' was necessary and was likely to remain so. Given these two simple hypotheses, Malthus argued that the increase in the human population was infinitely greater than the potential production of subsistence for man. He argued that, when unchecked, population increases at a geometrical ratio but that subsistence (i.e. food production) only increases in an arithmetical progression. In the late eighteenth century, science had very little effect on the production of food and yields, of both crops and animals, had remained fairly constant for many thousands of years. It was then assumed that extra food could only be produced by tilling extra land and clearly there was a finite limitation to the ability to find new land for agricultural production.

Again, at the time Malthus was writing, the cause and cure of most major contagious and infectious diseases were unknown and birth control was little, if ever, practised. Average family size (judged by the number of children born) was large – nearer 10 than 2. Malthus therefore reasoned that the check on the potential growth of the world population was by means of war, famine or disease, or to put it in his own words 'misery and vice'. If a population was unchecked (by misery and vice) Malthus calculated that it would double about every 25 years and it is obvious that, at this rate of increase, the world would rapidly run out of the required natural resources to support human life. Figure 1.1 shows the actual rates of population increase in certain tropical countries, and compares these with the very much reduced rates of increase in the developed countries where birth control is widely practised and where population numbers are relatively stable. It will be noted from Fig. 1.1 that already many tropical countries have shortened the 25-year doubling-time.

Future estimates of world population size are extremely difficult to make and are very inaccurate (Clark and Turner, 1974). Even in well-documented countries that regularly carry out a population census, such as the UK, the estimates of population size in 20 or 40 years' time vary according to when the estimate is made. This point is well illustrated in Fig. 1.2 which shows the estimate of the UK population size in the year 2001 made at various times between 1960 and 1980.

The results of a major study by the Food and Agriculture Organization of the United Nations (FAO) in collaboration with the United Nations Fund for Population Activities (UNFPA) and the International Institute for Applied Systems Analysis (IIASA) was published in 1982. This classic study assessed the land and climate resources, the potential food production and

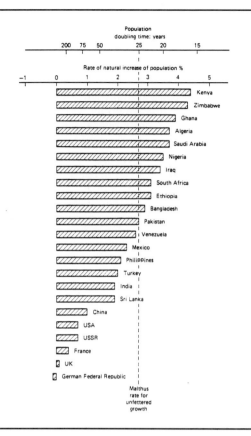

Fig. 1.1 The rate of natural increase of populations in selected countries of the world. The vertical line represents that rate of natural increase which Malthus regarded as 'unfettered' by resource limitation. (Source: Blaxter, 1986.)

population-supporting capacity of developing countries. It was based on more than 20 years of work by the staff of the soil resources, management and conservation services of the FAO and numerous co-operating institutions throughout the world. In essence, the methodology employed laid down the soil and climatic requirements for the major food crops and matched these with a soil and climatic inventory of each country and each tropical region. These data were then used to assess the 'potential population supporting capacities' of the different countries investigated.

The results of this comprehensive study should be treated with due caution. For instance, no

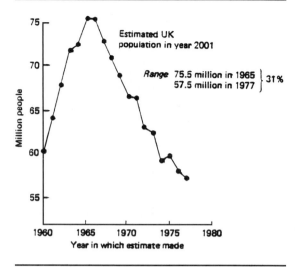

Fig. 1.2 Estimates made in different years by the Registrar General of the United Kingdom's population in the year 2001. This illustrates the considerable difficulties that there are in predicting future population. (Source: Blaxter, 1986.)

account was taken of climatic change nor of human migrations. Similarly, no full account was taken of the degrading of resources (for instance by erosion and pollution) nor was any attempt made to estimate the effect of further scientific advances in enhancing agricultural output, other than to optimise the use of current technology in the 'high input' estimates. The data reported in this chapter rely mainly on this major 1982 stock-taking exercise, and the various projections and estimates, based on these data, in later FAO publications.

LAND AND WATER USE: 1988

The estimated pattern of land and water use, consequent on the considerations outlined above, is summarised in Table 1.1.

In 1988 the tropics had 38% of the world's land and 44% of its population. Some 11% of the land was in arable or permanent crops and of this, 15% was estimated to be irrigated to some extent or other – a figure which was close to the world average. Forest and woodland occupied 38% of the land and permanent pasture 24% and

Table 1.1 Land use and populations of the world in 1988 (area in Mha). (Source: FAO, 1989a.)

Region	Land area	% total land	Arable crops area	Permanent crops area	Permanent pasture area	Forest + woodland area	Other land area	% of cropped land irrigated	Nitrogen use (kg/ha)	Cereal yield (t/ha)	Population (M people)	% population engaged in agriculture	Population per ha
World	13069	100	1373	102	3212	4049	4333	15.5	54	2.6	5205	47	0.4
Developed world	5482	42	652	22	1250	1867	1691	9.4	59	3.0	1243	9	0.2
Developing world	7587	68	722	80	1962	2182	2641	20.6	50	2.3	3962	60	0.5
Tropical world	4963	38	488	58	1168	1905	1344	14.6	28	1.8	2294	58	0.5
Tropical Africa	2237	17	133	15	636	667	786	3.6	4	1.0	468	72	0.2
Sub-saharan	890	7	36	1	254	121	478	6.6	3	0.7	101	74	0.1
West	778	6	66	9	120	392	190	1.7	3	1.0	238	66	0.3
East	171	1	13	4	45	52	57	1.7	6	1.4	81	82	0.5
Southern	398	3	18	1	217	102	61	6.8	10	1.3	48	75	0.1
Asia	1021	8	237	21	130	302	331	23.9	44	2.1	1429	62	1.4
India	297	2	166	4	12	67	49	24.7	43	1.7	836	67	2.8
Other Asia	724	6	71	18	118	235	282	22.5	45	2.6	593	55	0.8
Central America (tropical)	265	2	33	5	94	66	67	18.1	51	2.1	148	32	0.6
South America (tropical)	1386	10	85	16	307	827	151	5.6	16	2.0	243	26	0.2
Brazil	846	6	67	12	169	556	43	3.3	10	1.9	147	25	0.2
Other south America	540	4	18	4	138	271	108	13.9	35	2.1	96	28	0.2
Oceania (tropical)	54	0.4	4	0.8	0.5	43	9	0.2	14	2.3	6	54	0.1

with other (non-agricultural) land occupying 27%. It can be seen that the forests and woodlands of the tropics constituted approximately 47% of the world's total forest and woodland area.

A point emerges very clearly from the figures presented in Table 1.1. The percentage of the human population in the tropics engaged in agriculture was 58% compared with 9% in the developed world, but in spite of this extra human effort tropical countries were, and still are, typified by lower yields and by a wider ratio of human population per hectare of arable land.

There are naturally very wide differences between the various tropical countries. Thus tropical Asia had the highest density of population (1.4/ha), the highest percentage of land in cultivation (25%) and the greatest degree of dependence upon irrigation. Tropical South America, on the other hand, had a much lower population density (0.2/ha), a smaller percentage of land in cultivation (7%) and an apparent lesser use of irrigation. Africa resembled South America with 0.2 person/ha, 7% of land in cultivation and with very low agricultural yields. Tropical Oceania (excluding Australia) had a relatively small land area spread over large distances in many islands. Only 2.3% of the land area was cultivated, although with a relatively high proportion of permanent crops. In these regions, 80% of the land was under forest or woodland.

POTENTIAL LAND AND WATER USE

The potential of rain-fed crop production is dependent on many factors, such as climate, soil type, slope and soil degradation. Obtaining and combining information on all these different aspects on a wide scale was a mammoth task, made possible by 20 years of intensive work carried out prior to the publication of the 1982 report (FAO/UNFPA/IIASA, 1982). This major information base was from 117 developing countries classified by soil and climate and these were compared with the growth requirements of 15 different major crops and grassland types in order to assess the potential for food production and thus the number of people that could be supported. Because the level of agricultural inputs has a great influence on yield, the study was made for three levels of input as follows:

- *Low* – no fertiliser, pesticides or improved seeds and no conservation measures (equivalent to subsistence farming).
- *Intermediate* – some fertiliser (including about 80 kg/ha of NPK to cereals) with conservation measures and improved cropping patterns on about half the land.
- *High* – full use of inputs (including about 160 kg/ha of NPK to cereal crops), full conservation measures and the most productive theoretical mix of crops on the land equivalent to the levels of farming found in the most developed countries.

In 1988 most African countries were using little more than the low levels of inputs while tropical countries in Asia and Central America were near to the intermediate level. Tropical South America was about half way to the intermediate level.

The results produced by the study assumed that all potentially cultivable land, including that under forest and woodlands, was used to grow only staple food crops or improved pasture for livestock. In reality some land would need to be set aside for forestry, vegetables, fibres and non-food crops, and this would have the effect of reducing the total potential food production quite markedly. Thus in 1988, 38% of the land in the tropics was actually under forest and woodland. Furthermore, the average consumption per person would be increased by rising incomes and changing age structures. To allow for all these factors it was estimated that the potential population that could be supported on all cultivable land needed to be reduced by at least one-third from the potential calculated on the basis that all suitable land was used solely for optimal food production (FAO, 1984).

SOIL CONSIDERATIONS

Although a detailed treatment of soil will be given in Chapter 3, some general considerations will be dealt with here since soil is, in many cases, the main limiting factor to staple food production. Much of the land in tropical countries is unsuitable for agricultural production because of limitations of soil type and water availability. The suitability of soils for crop production depends upon their type, phase, texture and slope. Table 1.2 summarises data for Africa, Southeast Asia, Central and South America. Overall, only 23% of the land area has soils with no inherent fertility limitations but individual regions differ very markedly. Central America and Southeast Asia have the highest percentages (44% and 36%) followed by South America (20%) and Africa (19%).

The major limitations also vary. In Africa the main problems are coarse texture (20%), desert soil conditions (16%) and fertility (nutrient limitations) 15%. The main limitations in Central America are shallow soils (22%) and desert soil conditions (12%). In contrast lands in the moister Southeast Asia and South America are more limited by low fertility (26% and 41% respectively), shallow soils (each 11%) and poor drainage (8% and 10% respectively).

Table 1.2 Main problems with dominant regional soils (extent in Mha and percentage). (Source: FAO/UNFPA/IIASA, 1982.)

	Africa	Southeast Asia	Central America	South America	Total
Soils with no inherent fertility limitations (T,C,B,J,R,L,H,N,D,K)	535 (18.6)	324 (36.2)	119 (43.8)	360 (20.3)	1338 (23.0)
Soils with severe fertility limitations (A,F,P)	419 (14.6)	220 (24.5)	16 (6.0)	722 (40.8)	1377 (23.7)
Heavy cracking clay soils (V)	99 (3.4)	58 (6.5)	13 (4.9)	25 (1.4)	195 (3.4)
Salt affected soils (Z,S)	64 (2.2)	20 (2.2)	2 (0.8)	56 (3.2)	142 (2.4)
Poorly drained soils (G,O,W)	153 (5.3)	76 (8.4)	13 (4.7)	180 (10.2)	422 (7.3)
Shallow soils (I,U,E)	376 (13.1)	99 (11.0)	61 (22.3)	194 (10.9)	730 (12.5)
Coarse textured soils (Q,R)	568 (19.7)	52 (5.8)	16 (5.8)	132 (7.5)	768 (13.2)
Semi-desert and desert soils (X,Y)	459 (16.0)	43 (4.8)	32 (11.7)	94 (5.3)	628 (10.8)
Miscellaneous land units	205 (7.1)	6 (0.6)	—	7 (0.4)	218 (3.7)
Total	2878	898	272	1770	5818

KEY TO SYMBOLS
A Acrisols
B Cambisols
C Chernozems
D Podsoluvisols
E Rendzinas
F Ferralsols
G Gleysols
H Phaeozems
I Lithosols
J Fluvisols
K Kastanozems
L Luvisols
M Greyzems
N Nitosols
O Histosols
P Podzols
Q Arenosols
R Regosols
S Solonetz
T Andosols
U Rankers
V Vertisols
W Planosols
X Xerosols
Y Yermosols
Z Solonchaks

WATER LIMITATION AND IRRIGATION

Some tropical regions have potentially fertile soils but they suffer from lack of water, many regions having little if any reliable annual rainfall. In such situations yields can be dramatically improved by irrigation but this technique depends upon the availability of water for irrigation purposes and the economic cost of introducing major irrigation works. The recent and projected irrigated areas for Africa, Southeast Asia and Central and South America are presented in Table 1.3.

From the figures given in Table 1.3 it was estimated that by the year 2000 the area equipped with irrigation could be 2.5% overall but 9.1% in Southeast Asia, whilst only 1.1% in Central and South America and a very low 0.3% in Africa. These FAO predicted increases in irrigated areas to the year 2000 were based on the then current rate of expansion, and they were kept relatively conservative because of the difficulties in financing the high capital costs. In general, unfortunately the potential for low-cost irrigation expansion is greatest in countries with the greatest potential for rain-fed agriculture.

LAND AREA SUITABLE FOR RAIN-FED CROP PRODUCTION

The calculation of the land area suitable for rain-fed crop production requires an overlaying of the world soil map with climatic data, and knowledge of the growth requirements of the major food crops. This is an enormous task only recently made possible by the use of computer. The agroecological zones (AEZ) study was initiated by the FAO and considers the land area of the developing world as a mosaic of some 650 000 grid squares (each 10×10 km) each with a specific climate, growing period and soil description. A small sample of the results of such studies is given in Table 1.4, and the wide differences between regions in respect of both the percentage of suitable land for crop production, and the

Table 1.3 Irrigated area for 1975 and 1988 and projection for 2000, together with estimate of total potentially suitable area (in Mha). (Source: FAO, 1984.)

Region	1975	1988	2000 (estimated)	Total suitable (estimated)	Total land area
Africa (tropical)	3	5	6	19	2237
Asia and Oceania	60	83	98	218	1075
South and Central America	12	16	19	44	1651
Total	75	104	123	281	4963

Table 1.4 Land with suitable climate and soil for rain-fed crop production. (Source: FAO/UNFPA/IIASA, 1982; FAO, 1989a.)

	Land area (%)			Total land area actually cultivated in 1988 (%)
	with suitable soil and climate	with marginally suitable soil and/or climate	with unsuitable soil and/or climate	
Africa (less South Africa)	27	8	65	6
Southeast Asia	33	25	42	31
Central America	27	6	67	14
South America	46	8	46	8

actual use of that land, is clearly shown. Africa has a low proportion of land suitable for rain-fed production (27%) and cultivates less than 10% of the total land area. Much of the suitable land is under forest. Central America has a similar low proportion of suitable land but uses about one half of it for cropping. Southeast Asia crops nearly all of the potential rain-fed crop land (31% vs 33%) while South America has large areas of suitable land (46%) but only cultivates less than 10% of its total land area, or one-sixth of its suitable land. These differences between regions in the use of land for crops are crudely related to human population density.

The position in respect to individual tropical countries is set out in Table 1.5. The proportion of land actually cultivated, compared to the potential area available, differs widely from less than 10% in countries such as the Congo and Bolivia to 100% in India and Haiti. The variation between African countries is particularly striking. The data in Table 1.5 refer to a different year and a separate FAO study compared to the data in Table 1.4, so the two tables are not directly comparable. Also, only a few selected countries are given in Table 1.5 whereas Table 1.4 refers to whole regions of the world.

The next stage of this complicated exercise was to calculate for each cell the best food crops to maximise human nutrition by matching the ecological conditions to the growth requirements of 16 main crops and grassland communities. The average energy requirement per head of population was then used to calculate the potential population supporting capacity of the area. The study indicated that the low input farming methods in many tropical countries roughly equate to the then existing (1975) human population size and any increase in numbers would need improved methods of agricultural production to be brought into effect in order to support them.

It should again be stressed that no allowances have been made in these calculations for other essentials such as timber, fibres, vegetables and fruit, animal feed for livestock and non-food cash crops. Wood for fuel or timber was restricted to land deemed unsuitable for cropping or grazing. Most important of all, no allowance was made for inequality of food distribution, a factor of overriding importance in tropical communities.

All of these factors would, in practice, reduce the potential population supporting capacity by at least one third. Other factors would alleviate the position to some minor degree, such as in the introduction of aquaculture, or by bringing into production land currently not considered suitable for cultivation, such as by terracing steep slopes, introducing flood control measures or major drainage systems.

The most significant data are likely to be those applying to individual countries because of the difficulties of trade and distribution of food across country boundaries. Table 1.6 lists the tropical countries which were critical for food production in 1975 at the low levels of inputs, and which were predicted to be in grave difficulty

Table 1.5 Percentage of potential arable land cultivated in 1982–84. (Source: FAO, 1989a.)

Percentage	Country
<10	Angola, Congo, Gabon, Madagascar, Zaire; Bolivia, Guyana
10–19	Central African Republic, Liberia, Mozambique, Zambia; Brazil, Colombia, Peru, Venezuela
20–29	Cameroon, Sudan, Tanzania, Zimbabwe; Laos; Costa Rica, Ecuador, Nicaragua, Paraguay
30–39	—
40–49	Ghana, Mali, Uganda; Burma, Indonesia, Kampuchea, Malaysia
50–59	Guinea, Senegal; Cuba, Mexico
60–69	Kenya, Malawi, Mauritania, Nigeria; Vietnam; Dominican Republic, Jamaica
70–79	Sierra Leone, Togo; Sri Lanka; El Salvador, Trinidad and Tobago
80–89	Botswana; Philippines, Saudi Arabia, Thailand
90–99	Niger; Bangladesh, Yemen AR, Yemen PDR
100	Mauritius; India; Haiti

Table 1.6 Tropical countries critical, or near critical, at low level of inputs, for food production in the year 1975 and projected to be critical in the year 2000, assuming little or no advance in technology. (Source: FAO/UNFPA/IIASA, 1982.)

	Ratio* 1975	Ratio* 2000		Ratio* 1975	Ratio* 2000
Tropical Africa			**Tropical Asia**		
Cape Verde Is.	<0.01	<0.01	Singapore	<0.01	<0.01
Western Sahara	<0.01	<0.01	Saudi Arabia	0.21	0.17
Rwanda	0.18	0.08	Yemen PDR	0.58	0.30
Burundi	0.22	0.10	Yemen AR	0.48	0.35
Niger	0.17	0.14	Vietnam	0.90	0.74
Reunion	0.20	0.14	Sri Lanka	0.83	0.77
Kenya	0.27	0.17	Bangladesh	0.46	0.79
Comores	0.42	0.27	Philippines	1.08	0.87
Somalia	0.40	0.34	India	0.81	1.00
Ethiopia	0.59	0.36	**Central America**		
Nigeria	0.82	0.37			
Namibia	0.84	0.43	Netherlands Antilles	<0.01	<0.01
Uganda	0.97	0.45	Barbados	0.09	0.06
Mauritania	0.35	0.46	Martinique	0.26	0.23
Mauritius	0.71	0.47	Guadeloupe	0.39	0.33
Burkino Faso	0.90	0.51	Antigua	0.40	0.34
Malawi	1.27	0.61	El Salvador	0.49	0.35
Botswana	1.27	0.63	Haiti	0.51	0.40
Senegal	0.89	0.79	Windward Is.	0.58	0.49
Togo	1.69	0.82	Trinidad and Tobago	0.50	0.57
Sierra Leone	1.69	0.89	Puerto Rico	0.39	0.58
Mali	1.34	0.90	Jamaica	0.77	0.76
Benin	2.08	0.95	Bahamas	1.29	0.79
Zimbabwe	1.65	0.95	Guatemala	1.25	0.89
			Dominican Republic	1.14	0.95

*Ratio = population supporting capacity/actual or projected population. A ratio of <1 is critical.

by the year 2000. The table shows the ratios of capacity of land, at low inputs, to support a population to the actual population in 1975. A figure of 1.0 means that the potential just meets the need of the actual population. Ratios for the projected position by the year 2000 are also given. It should be remembered that the population-supporting capacity was based on all suitable land being used solely for food production, and also assumed an even distribution of the minimal food requirements. For 1975, 17 countries in tropical Africa, 8 in tropical Asia, and 11 in Central America would have been unable to produce enough food to support a reasonable standard of human nutrition if all available land had been used on a low input basis for food production (i.e. a ratio of less than 1.0) without recourse to food imports. By the year 2000, however, it is predicted that 24 countries in tropical Africa, 8 in tropical Asia and 14 in Central America would become critical. Taking a more realistic view of the land that is available and suitable for food production this would add Ghana, Tanzania and Gambia to the critical tropical Africa list, Thailand to the critical Asian list and Mexico and Honduras to the Central American list.

Some of the critical countries have oil and mineral resources whilst others have an industrial capacity or tourist industry and are, therefore, able to import some or much of the food they need from the proceeds of their national wealth. On the other hand, other countries

Table 1.7 Predicted population (in millions) for the year 2000 and predicted potential population supporting capacity with contribution of irrigation in the year 2000. (Source: FAO/UNFPA/IIASA, 1982.)

	Year 2000 population	Low inputs		Intermediate inputs		High inputs	
		Potential population	Irrigation contribution %	Potential population	Irrigation contribution %	Potential population	Irrigation contribution %
Africa							
Sub-Saharan	125	130	32	447	10	1900	3
West	315	691	4	2848	1	7538	<1
East	109	56	14	210	4	716	2
Southern	61	153	12	601	4	1992	1
Tropical Africa	610	1031	10	4105	3	12146	1
Asia							
India	1037	1298	87	1800	63	2620	43
Other tropical Asia	760	918	48	2287	18	3424	13
Tropical Asia	1797	2216	71	4087	38	6044	26
Tropical Central America	215	292	64	557	33	1293	14
South America							
Brazil	212	649	8	2995	2	7119	<1
Other tropical South America	130	554	15	1747	7	4128	3
Tropical South America	342	1203	11	4742	4	11247	2

have a very serious problem in paying for the required inputs, such as the high costs of irrigation development and fertilisers, which their country will need to increase food production from a low input level to a intermediate input level, and from the intermediate to a high input level.

The FAO and other agencies have attempted to show the theoretical increase in the supportable tropical population which could be achieved by utilising agricultural technology to a fuller extent, but assuming a stable annual cropping system. They have also calculated the extra contribution to human carrying capacity which could theoretically be achieved by more extensive use of irrigation for food production. Their estimates are presented in Table 1.7.

It is easy to be misled by the figures in the penultimate column of Table 1.7 which, misconstrued, could be taken to mean that the tropics have a theoretical potential to support a tropical population close to 30 billion. Other estimates, such as those by the World Bank (1984), paint a much more pessimistic picture. Table 1.8 provides data for some World Bank estimates for the theoretically assumed world population by the year 2000. This cruder calculation puts the maximum population sustainable at about 11 billion, roughly double the current estimate of world population size, and many people would consider this to be a much more realistic figure. The difference between what is theoretically possible (in terms of food production) and what is realistic and practical is very great and it is unwise to disregard this important fact. Estimates and predictions must be subject to large errors and should be used with discretion.

RECENT CHANGES IN LAND USE TO MEET THE NEEDS OF GROWING POPULATIONS

Tables 1.9 and 1.10 indicate how the tropical world actually changed the use of its available

Table 1.8 Estimates (in millions) made by the World Bank (1984) of the future population for different economic regions of the world.

Type of economy	Population in 1982	Projected population in 2000	Theoretical assumed maximum population
Low income (including China and India)	2267	3097	5863
Middle income (including Indonesia and Brazil)	1163	1741	3729
Oil exporters	17	33	96
Industrial market (including Europe and US)	723	718	828
Eastern Europe	384	431	523
All	4554	6082	11039

Table 1.9 Actual changes in population, arable land and agricultural inputs 1975–88. (Source: FAO, 1988; FAO, 1989a; FAO, 1990.)

	Tropical Africa	Tropical Asia	Central America	Tropical South America	Total
Population 1975 (million)	294	1048	107	177	1626
Increase to 1988	174(59%)	381(36%)	41(+38%)	66(37%)	662(41%)
Arable land 1975 (Mha)	124	231	32	70	457
Increase to 1988	9(7%)	6(3%)	1(3%)	15(21%)	31(7%)
Irrigated area 1975 (Mha)	4.0	39	47	5.6	96
Increase to 1988	1.2(32%)	18(47%)	15(31%)	1.2(22%)	35(37%)
Fertiliser use (kg N + P_2O_5 + K_2O /ha of arable + permanent crops)	5.2	21.8	54.1	38.5	22.3
Increase to 1988	2.7(52%)	46.5(213%)	24.5(45%)	13.9(36%)	27.4(123%)

Table 1.10 Actual land use in 1975 and changes (in brackets) to 1988 (Mha). (Source: FAO, 1989.)

	Tropical Africa	Tropical Asia	Central America	Tropical South America	Total
Arable land	124(+9)	231(+6)	32(+1)	70(+15)	457(+31)
Permanent crops	14(+1)	20(+1)	4(+1)	12(+4)	50(+7)
Permanent pasture	631(+5)	130(Nil)	92(+2)	285(+22)	1138(+29)
Forest and woodland	703(−36)	325(−23)	77(−11)	878(−51)	1983(−121)
Other land	765(+21)	315(+16)	60(+7)	141(+10)	1281(+54)
Total land	2237	1021	265	1386	4909

resources to meet the needs of rapidly increasing populations over the period 1975–88. As indicated elsewhere, many populations seem on course to double over the next 25 years or so.

The most striking change in land use over this period was the reduction by 121 Mha of the tropical forests and woodlands, and an increase in all other categories listed in Table 1.10. The 121 Mha which have gone from forest and woodland are balanced by an increase of 54 Mha in other land, 31 in arable, 29 in permanent pasture and 7 in permanent crops. This does not mean that all of the land taken out of forest and woodland was used in this way since it is probable that there were other movements, such as from arable to permanent pasture, and from pasture to other land through land degradation. The large increase in the 'other land' category was probably a combination of land degradation and a greater use of land for buildings and roads.

It will be seen from Table 1.9 that the increase in area of arable land of 7% was not nearly enough to provide food for the 41% increase in population. Tropical South America came nearest with an increase of 21% in arable land against a 37% rise in population. It is clear that in tropical Asia and Central America the main source of extra food supplies came from a large increase in an already vital irrigation contribution and also in fertiliser inputs. Both these areas would, in 1988, be deemed to have reached the 'intermediate input level' as used in the FAO studies already described. There must be concern that the changes in Africa did not face up to the 59% increase in population over this period, with only a 7% increase in arable land and small increases in irrigated areas (1.2 Mha) and in fertiliser use (2.7 kg/ha). South America, with vast land resources, increased the arable area by 21% and fertiliser use by 14 kg/ha while the irrigated area remained, as in Africa, relatively small.

How the population and land use has changed since 1988, or will change in future, is uncertain but it seems likely that the main thrust to feed the expected population increase will need to be through a continuing increase in the efficiency of farming (including more inputs) and a modest increase in the area of cultivated land. Unfortunately at present there is too little evidence of increased efficiency and plenty of evidence of an increase in degraded agricultural land.

HUMAN HEALTH AND NUTRITION PROBLEM

As has been shown, the current food supply is frequently inadequate to meet the needs of the present population in various parts of the tropics. Hunger and malnutrition, coupled with debilitating diseases, often reduce the energy and initiative of small farmers and are significant factors affecting agricultural productivity and development. There are numerous endemic tropical diseases; for many of these the means of prevention or cure are known but have not been fully implemented, while for others medical science has yet to find a satisfactory answer. Disease and malnutrition are often interrelated because malnutrition reduces resistance to disease, delays recovery and increases liability to relapses. Many diseases, especially those due to intestinal parasites, reduce a person's ability to produce food or to benefit from what is consumed. Disease incidence may also prevent the agricultural use of certain areas, or limit it to certain times of the year. For example, large areas are rendered unsuitable for habitation by the presence of tsetse flies, the vectors of sleeping sickness, and some malarious areas may be unusable, or usable only for the dry season grazing of cattle, unless major measures for mosquito control are undertaken.

The precise nutritional requirements of people of different races living in the various climates of the tropics have not been scientifically established, but it is generally accepted that the most important and widespread dietary deficiency is of protein. Total protein intake is very commonly inadequate and there is often a lack of proteins of high biological value derived from animal products, although this may be largely ameliorated where appreciable amounts of certain grain legumes are eaten. The degree of deficiency is influenced by the staple food crop; diets based on wheat, sorghum and millet being of better value than those based on rice or maize which, in turn, are superior to bulky diets of roots and bananas which have very low protein contents. Vitamin deficiencies and diseases resulting therefrom are also of frequent occurrence. Lack of vitamin A (the main sources of which are animal fats, yellow and red fruits and roots, fresh vegetables and palm oil) is quite common and causes night blindness and various skin troubles. Deficiency of vitamin B, producing beriberi disease, and lack of other vitamins of the B complex, causing pellagra, occurs in rice-eating countries. Skin troubles and swelling and haemorrhage of the gums, due to low vitamin C levels, are frequent.

As has been shown above, considerable agricultural improvement and development will be essential in the tropical developing countries if sufficient food of satisfactory quality is to be provided for their rapidly growing populations. Meeting any significant proportion of the requirement by imports from the more advanced countries must be regarded as impracticable for financial reasons. Most of the increased food supply must, therefore, be obtained by raising yields on land already under cultivation, which will mainly involve replacing predominantly subsistence agriculture on small farms by more intensive food production.

FARM EQUIPMENT AND REQUISITES

The implements and equipment used by the majority of small farmers are commonly limited in number and primitive in nature. This is especially so in Africa, where a farmer's tools may literally be restricted to an axe, a hoe, a knife and a box of matches. Little use is made of animal power or wheeled transport and the fact that clearing, cultivation, harvesting and transport are all carried out by manual labour naturally contributes greatly to the low productivity per man. In the East, where animal power is widely, but by no means universally, employed farmers have carts and a variety of ox-drawn implements, but many of the latter are inefficient, or made of poor materials, and there is scope for improving them. Although a high degree of mechanisation will commonly be ruled out by capital and recurrent costs, some improvement in the equipment owned by, or available on contract to, small farmers is needed to intensify farming and raise output per man. A change from shifting cultivation, in which land is cleared and rendered initially weed free by burning and is subsequently superficially hoed, to more intensive permanent cropping will require equipment for removing stumps and roots. The introduction of multiple cropping in rice-growing areas is likely to demand means of working the land more speedily than can be done by hand or with animals. In some circumstances, tractors may be the only effective means of clearing new land or of achieving the speedy and timely cultivation needed to obtain good yields in more intensive farming, but there are many difficulties in extending their economic use on small farms. There is more scope for raising production per man by introducing improved ox-drawn implements, small single-axle powered tillers and a variety of small machines, such as stumping jacks, threshers, sprayers and pumps, and this point will be dealt with in more detail in Chapter 5.

The widespread demonstration of substantial and economic responses to fertilisers with many food and cash crops had, until recently, led to some increase in their use by small farmers, although it was still on a very limited scale (see Chapter 6). Pesticides have also only been used to a relatively small extent, especially on food crops. It is difficult to see how good yields can be obtained in permanent farming systems on the majority of tropical soils without the regular and general use of appropriate fertilisers and, unless much more resistant crop varieties are bred, insecticides and fungicides will also be needed to avoid loss from pests and diseases. The recent large increases in the costs of these materials, therefore, poses a serious problem as it makes it doubtful whether small farmers will be able to afford to buy enough of them unless food prices are set, or subsidies provided, at levels unlikely to be acceptable to governments.

Correctly used, pesticides have an important part to play and without them yields would be greatly reduced or, in extreme cases, non-existent. However, it is most important to realise that all these materials have other potential toxicities over and above the target pest which they are designed to combat. In addition, agricultural fertilisers, which are important sources of plant nutrients, can also be toxic to animals and man when they find their way into potable water supplies at high concentrations. Although this is not a common problem in the tropics it nevertheless is found where intensive farming is practised in areas of very high population density.

It is for this reason that all tropical agricultural systems should utilise such potential hazardous materials with a degree of circumspection. Manufacturers' instructions should be rigorously adhered to, and care must be taken when disposing of unwanted chemicals by efficient cleaning of all receptacles and implements used to apply the materials to the target crop or animal. In temperate countries there is already a major movement aimed at either reducing or abolishing the use of such materials in agricultural systems. The so-called 'organic' or 'biological' systems are aimed to devise practices which are not reliant on agrochemical usage.

THE CONCEPT OF SUSTAINABILITY

The problem of meeting the food requirements of the increasing population demands the use of land for crop production by methods that will maintain or improve yields in the long term. Some traditional tropical farming systems have proved capable of sustaining soil fertility and crop yields more or less indefinitely, albeit at low levels of productivity per hectare and per man. For example, in some parts of Asia swamp rice-based systems have supported relatively dense populations at subsistence level for centuries, largely because fertility is imported into the system by the deposition of silt in irrigation or flood water and by nitrogen fixation. Similarly, yields of rain-fed crops on small farms in India have been sustained at modest levels by the import of fertility on to the arable land in the form of animal manure, composts and household wastes.

Recently, it has become possible to raise yields in the above systems by the use of fertilisers and improved crop varieties.

In contrast, shifting cultivation, which is basically still one of the commonest systems in the tropics, is only a sustainable system provided that the population is sparse enough to permit long restorative fallows and provided that measures are taken to prevent soil erosion. Even then there is usually a slow decline in fertility, at least in the savanna zones. Nowadays there are many places where the above provisos are no longer obtained; consequently yields are declining and large areas of land are being degraded.

In order to maintain yields under the more intensive crop production which is now widely needed to raise yields per man and per hectare and so meet the increasing demand for food, the first essentials must be to prevent runoff and erosion and to make the best possible use of animal, vegetative and household wastes as manures. Good husbandry will also be required in such matters as will timely land preparation and planting, optimum crop plant densities and good weed control. In addition, it seems that, as a rule, improved crop varieties, pesticides and, perhaps, some degree of mechanisation, will be needed to raise yields to the requisite level and that fertilisers will be necessary to sustain fertility and yields. Meeting the costs of these latter inputs on smallholdings will be a major problem.

It is for these reasons that many of the current highly intensive methods of agricultural production practised in the developed world are questionable when applied to tropical countries where problems of erosion and desertification are more acute. There is an urgent need to take a more holistic view of agricultural development in those areas which are currently being taken out of natural grassland, scrub, woodland or forest and forced into intensive systems of agricultural production. The long-term consequences of such actions should be of major concern both to the farmers and the governments of the countries concerned. In many cases legislation will be required to prevent deleterious practices from being adopted, but in many newly independent countries there seems a lack of political will to introduce more sustainable methods of production onto a reluctant rural population.

GLOBAL WARMING AND ACID DEPOSITION

There are some indications that the world climate is changing, although in what direction, and at what rate, is at present uncertain. There is currently much concern about the damage to the ozone layer. There is also much debate about pollution from industrial gases, particularly the oxides of nitrogen and sulphur. There is even greater concern about the environmental effect of felling large tracts of tropical forests. However, some contribution to the potential problems of climatic change is also made by agricultural practices in tropical countries where there is little if any industrial pollution. For instance, one of the gases concerned is methane, which is produced in large quantities from paddy fields used for swamp rice production and also from concentrations of farm animals, particularly ruminant animals, which produce methane as a breakdown product of rumination. It is, however, difficult to see what practical advice could be given to tropical agriculturalists to reduce methane production to any degree.

Indeed, it may appear that these considerations are of minor importance compared to the horrendous problems of expanding world population and food supply. However, over the next few decades the tropics will not be immune to the 'environmental considerations' now given prominence in the developed countries. Agricultural aid to tropical countries may well be increasingly dependent on their acceptance of more 'environmentally friendly' and sustainable agricultural practices, so it is important that the arguments being advanced on such matters, and the evidence that will be produced, are taken into account by those responsible for rural development strategies in the tropics.

THE USE OF AGROCHEMICALS

Many of the developments in temperate agriculture have been obtained by increasing use of agrochemicals into the farming systems, either as fertilisers (particularly fertilisers rich in nitrogen, phosphorus and potassium, as pesticides (such as fungicides, insecticides and nematicides) or as herbicides.

Steps are taken to ensure that organic produce can be correctly identified in the retail outlet in such a manner that it can command a premium for the producer, since the cost of agricultural production without the benefit of agricultural chemicals is invariably more costly. It may appear that such considerations of Western European origin are of little interest in developing countries, particularly those in the tropics, where the main thrust is to increase food production at all costs. However, it would be unwise to disregard these considerations since, in the longer term, they are bound to affect national legislation and thus tropical systems of farming just as they are influencing temperate systems of agricultural production at the present time.

SOCIO-POLITICAL CONSIDERATIONS

The point has already been made that the tropical regions where food production does not match population requirement tend to be areas where farming is difficult or inefficient and where human populations are rising. The simplest solution is to transport food from areas of oversupply to areas where food imports are sorely needed. Unfortunately the problem is not merely one of the logistics of moving quantities of food from one country to another. There are major socio-political constraints to such practices and it is important to understand some of the reasons for these.

First, most countries naturally wish to be as self-sufficient as possible for the major necessities of life, which include food, energy and water. In times of war or political unrest, countries which are net food importers are most at risk when trade is suspended. In addition, the world surpluses of food mainly take the form of the traditional crops used in the food-exporting countries and these tend to be grown in the developed world where the surpluses are in the form of maize, wheat, soya bean and meat. Unfortunately these types of food may not be those required by the underfed populations in the tropics. In addition to this, world trade is governed by various political considerations which have nothing to do with real population needs, but much more to do with the political aspirations and fiscal policies of the exporting and importing countries.

Perhaps most important of all, free movement of large quantities of food requires a very efficient transport, wholesale and retail organisation. In many developing countries these market structures are inefficient so that not only is the movement of food more costly than it should be but artificial barriers to trade are erected, often at short notice, for purely political ends.

For all these and other reasons, it is normally a good policy for any government to endeavour to plan for as high a degree of self-sufficiency in food and other essential products as is practical and economic. In many cases this must mean some degree of protection for the local producer, perhaps also matched by some kind of control on the import of 'cheap' food products dumped from the world market.

FURTHER READING

Blaxter, K.L. (1986) *People, Food and Resources*. Cambridge University Press: Cambridge.

Grigg, D. (1980) *Population Growth and Agrarian Change*. Cambridge University Press: Cambridge.

Rechcigl, M. (Jr) (ed.) (1980) *Man, Food and Nutrition*. CRC Press: Cleveland, OH.

2
Climate, Agriculture and Vegetation in the Tropics

C.T. Agnew

INTRODUCTION

The tropics are delimited by the astronomical position of the sun, i.e. the highest latitudes where the sun lies overhead which occurs at 23.5°N and 23.5°S (see Fig. 2.1). There are many alternative boundaries both geographical and meteorological (Balek, 1983) but the tropics of Cancer and Capricorn will suffice as a general demarcation. The area is vast, Strahler (1975) noted that 42% of the surface of the earth lies between latitudes 25°N and 25°S. The region contains a variety of climatological conditions and a commensurate diversity of natural vegetation and agricultural systems. The aim of this chapter is to describe the major climate characteristics, to identify the relationships between climate and plant growth in the tropics and to simplify the vegetation patterns such that they can be explained.

An appropriate starting point for any analysis of tropical vegetation is climate. It would be wrong, however, to assume that climate is the sole determinant or that the relationship between climate and plant growth is fixed. We shall return to this last point but the reader should understand that other environmental factors such as soil and the action of animals also have a significant impact on vegetation, while the activities of humans must always be taken into consideration.

At the outset it is worth summarising the key points raised below. The tropics are climatically heterogeneous but they are all characterised by large inputs of solar radiation. There are spatial variations of energy balance due to land:ocean changes and topography but rainfall differences are more significant than temperature. Hence classification of climate and vegetation follows rainfall characteristics not thermal ones. Although solar radiation inputs are large, temperature gradients are smaller than those found in higher latitudes. As a consequence atmospheric systems in the tropics function differently compared to those in temperate latitudes and care must be taken not to oversimplify the behaviour of tropical weather-forming systems as was often the case earlier in the twentieth century. This chapter commences with a description of climates and vegetation followed by a discussion of the atmospheric processes responsible. The detailed account of climatic characteristics concentrates upon precipitation and the importance of plant–soil–water relations. The resulting patterns of vegetation and agriculture are summarised and the possible impacts of climate change in the tropics are assessed.

CLIMATE AND VEGETATION CLASSIFICATION

There are several global classifications of climate all of which present climate as a static phenomenon, although the onset of global warming has raised questions over the future location of the boundaries and characteristics of tropical climates. First we need to consider the characteristics of tropical climates and the meteorological processes involved, taking note that climate refers to the average conditions experienced, normally based upon 30 years of observations.

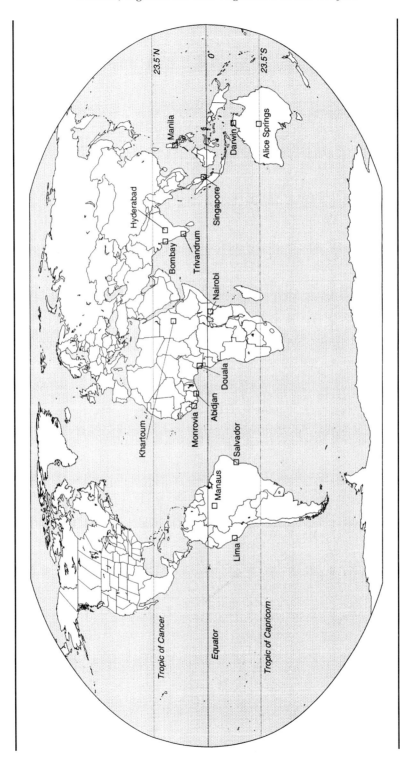

Fig. 2.1 Location of the tropics and sites of climate stations used in Fig. 2.5.

Koppen's (1931) classification identified two main climate types in these latitudes (see Fig. 2.2):

- Tropical wet climate (subdivided into rainforest with rainfall in all months, monsoon and dry winter season climates).
- Dry climates (subdivided into desert and steppe).

There are many other classifications such as Thornthwaite (1948, 1954) which was based on water balances, Strahler (1975) identified precipitation regions and developed a classification based on air mass characteristics while the United Nations Educational, Scientific and Cultural Organization (UNESCO, 1977b) placed an emphasis upon arid regions and reliable estimation of evaporation rates. The Koppen system, although highly empirical, continues to be widely used (e.g. Briggs and Smithson, 1992) because it was designed primarily to explain the distribution of vegetation. Figure 2.3 and Table 2.1 show the general distribution of natural vegetation in the tropics and Fig. 2.4 shows an idealised structure. Three major biomes (a community of similar plants and animals occupying a large area) are identified: forest, savanna and desert, with drylands accounting for over half of the area covered.

Tropical climates are characterised by high maximum temperatures and high inputs of solar radiation where water supply is the most significant variable. Unsurprisingly these biomes correspond to amounts of annual precipitation (see Table 2.2). The variation in the annual precipitation boundaries in Table 2.2 should not be of concern; distinctive vegetation regions are cartographic interpretations and Deshmukh (1986) noted that half of the tropics comprised intermediate vegetation types.

The implicit assumption of classifying these tropical biomes is that they form a *climax vegetation*, i.e. a grouping of mature vegetation where climate is the prime determinant. It has already been noted that soil and biotic factors may exert considerable local influence to alter this pattern but the climax concept can also be criticised for portraying climate:vegetation as an equilibrium relationship. Increasingly climate:vegetation interactions are seen as dynamic with constant change and adjustment. Discussions of savanna ecology are now full of the notion of 'disequilibrium'. Walker (1987), for example, argued that savanna systems were seldom at equilibrium. This has led to rejection of notions such as carrying capacity (i.e. the population which the resources of an environment can maintain) for being too static a concept (Behnke *et al.*, 1993; Scoones, 1995; Warren, 1995). Tropical rainforests are not seen as ancient, almost fossil, landscapes but as ones where structural and areal change have been constant. Writing about Africa, Meadows (1996) charted the movement of rainforest across the continent noting that contemporary distributions were determined by relatively recent events with changes during the Quaternary being particularly important. Grainger (1996, p. 173) went further and argued that, 'the biogeography of Africa is now predominately cultural not natural'. The use of

Table 2.1 Percentage of ground cover in the tropics. (After Deshmukh, 1986.)

Ground cover	%
Rainforest	14.7
Dry season forest	7.4
Wet savanna	17.6
Dry savanna	29.4
Desert	29.4
Other (bare rock)	1.5

Table 2.2 Tropical biomes and annual precipitation (mm). (Source: Whittaker, 1975 and Holdridge, 1971.)

Biome	Whittaker (1975)	Holdridge (1971)
Desert	<500	<250
Savanna	−1300	−1000
Seasonal forest	−2600	−4000
Rainforest	>2600	>4000

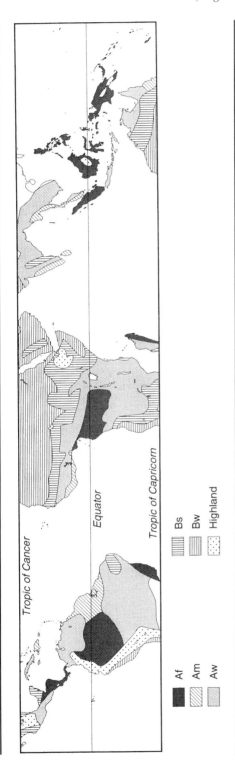

Fig. 2.2 Distribution of climates found in the tropics using the Koppen (1931) classification. (Based on Briggs and Smithson, 1992 and Strahler, 1975.) Af = tropical wet climate with precipitation all year round; Am = tropical wet climate with a brief dry season (monsoon type); Aw = tropical wet climate where winter is main dry season; Bs = semi-arid climate with annual rainfall below 750 mm and Bw = desert or arid climate.

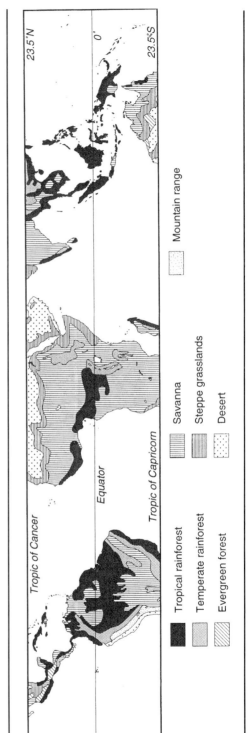

Fig. 2.3 Vegetation distribution in the tropics. (Based on Archibold, 1995 and Strahler, 1975.)

these biomes is then merely to simplify the main characteristics of tropical vegetation.

Given that precipitation is the most important climate variable influencing plant growth in the tropics rainfall regimes can be used to distinguish between various climates (see Fig. 2.5):

- *Wet climates.*
  Found in the equatorial trough (within 5° to 10° of the equator) these areas most closely conform to the perception of a tropical/equatorial climate. Heavy downpours in late afternoon, high humidities all year round and little change in temperatures. High annual precipitation in excess of 2000 mm (e.g. Douala and Manaus).
- *Wet-dry climates.*
  Influenced by the passage of the sun creating a marked seasonality, found between 10° and 20° latitude with a greater contrast at higher latitudes.
  - *Two wet seasons*
    Found closer to the equator so temperature changes are small and annual rainfall is high, between 1000 and 2000 mm (e.g. Abidjan and Salvador).
  - *Two shorter rainy seasons*
    This climate region is further from the equator but still under the influence of solar inputs during each solstice. Anticyclonic conditions are more easily established giving at least one significant dry season (e.g. Nairobi and Trivandrum) and continentality can be noticeable. There is great variation in annual rainfall but the expected range is between 600 and 1500 mm.
  - *One long rainy season (monsoonal)*
    This climate type affects large areas of Asia and parts of Australia and Africa. Annual rainfall is commonly between 750 and 1500 mm but subject to much temporal variation and the local effects of topography (e.g. Bombay and Manila).
  - *One short rainy season*
    These areas are often at the limits of cultivation with annual rainfall between 250 and 750 mm. The rain usually falls in a short rainy season of 3–5 months and there is a marked seasonality in nearly all aspects of climate including temperature and humidity (e.g. Hyderabad and Darwin).
- *Dry climates.*
  Large areas of the tropics are dry due to the influence of tropical anticyclones and reduced rainfall caused by cold offshore currents or topographical barriers. Here rainfall is sporadic and temperatures exhibit great variability (e.g. Alice Springs, Lima and Khartoum).

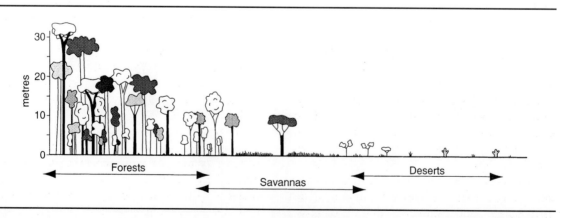

Fig. 2.4 Idealised vegetation transect from wet to dry tropics. (Based on Archibold, 1995 and Deshmukh, 1986.)

ATMOSPHERIC PROCESSES

The earth's climate system is powered primarily by inputs of solar energy. Seasonal and spatial differences of solar energy give rise to oceanic and atmospheric circulation hence the climate characteristics noted above.

Solar energy in the tropics

The amount of solar energy received at the outer edge of the earth's atmosphere is a function of latitude and season. Figure 2.6 shows the seasonal variations which produce a double maxima near the equator with increased seasonality at higher latitudes such that solar energy fluxes vary from 300 to 450 W/m^2 (units of power given as watts per square metre of the atmosphere). There is some variation in the amounts received because of turbidity (due to atmospheric particulates) and cloud cover, Reading et al. (1995) reported that in equatorial Zaire solar radiation inputs fell to around 150 W/m^2 with around 200 W/m^2 received in Amazonia and Indonesia. Nevertheless, the annual energy receipt is reasonably constant across the tropics at 12 800 MJ/m^2/yr (megajoules per square metre of the atmosphere per year), equivalent to a flux of 400 W/m^2. Much of this is lost as terrestrial re-radiation but some 100 W/m^2 is surplus at the surface of the earth and powers the major atmospheric and oceanic circulations (Sellers, 1965; McIlveen, 1992). The surface energy balance can be written as (Oke, 1978):

$$Q^* = QH + QE + \text{ or } - QS \qquad (W/m^2)$$

where Q^* is the net energy balance at the surface of the earth, QH is the energy expended on sensible heat and turbulent transfer, QE is the energy expended on evaporation and QS is the stored energy (significant for oceans and/or small time periods).

High levels of energy are released as turbulent heat transfer over the land fuelling atmospheric disturbances and rain-generating mechanisms. The moist air produced by high evaporation rates over the oceans contains high potential energy which can be released as latent heat through condensation but as more energy is used for evaporation there is correspondingly less for the vertical transfer of sensible heat. Latent heat (of vapourisation) is the energy used (or released) when a substance changes state from a liquid to a gas (or vice versa). Here it represents the energy used to evaporate water and convert it into water vapour, this energy is then released during condensation. Sensible heat is the total heat or energy content of a substance per unit mass. Here it represents the energy used to heat the atmosphere. Hence temperatures are moderated and atmospheric gradients are small. McIlveen (1992) showed that amounts of absorbed solar and emitted terrestrial radiation were approximately constant through 30°N and 30°S. As a consequence horizontal temperature gradients are small and the tropical atmosphere is relatively barotropic but the small differences in temperature can bring about significant changes in weather, for example tropical cyclones. This greater sensitivity poses acute difficulties for weather modelling and prediction in the tropics.

The main atmospheric features

The subtropical highs and trade winds

The tropics are dominated by the two persistent subtropical anticyclones that straddle the earth. These areas of high pressure, low-level divergence (air descending and flowing outwards as it reaches the earth's surface) and low humidity form part of the Hadley Cell (a large-scale, thermally driven atmospheric circulation that embraces rising air from equatorial regions and descending air around 30 to 40° latitude). Equatorial air that has been lifted aloft by solar heating begins to descend around 20 to 30° latitude. This subsiding air is adiabatically (there is no transfer of heat or mass with the surrounding atmosphere) warmed hence dry. These high pressure systems then create large arid regions that issue a reliable surface easterly wind system blowing northeasterly in the northern hemisphere and southeasterly in the southern

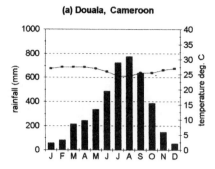
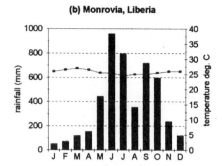
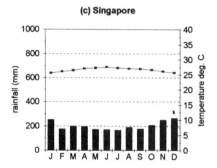
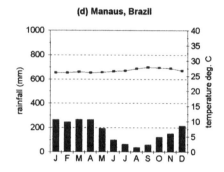
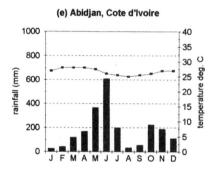
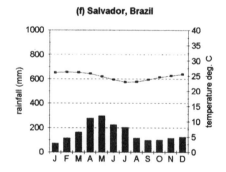
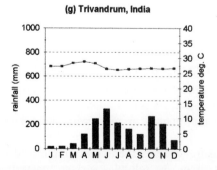
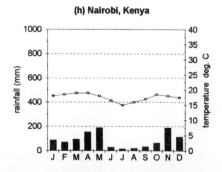

Fig. 2.5 Rainfall (■) and temperature (-■-) regimes for selected climate stations in the tropics. (Data from Griffiths, 1972; Schwerdtfeger, 1976; Takahashi and Arakawa, 1981.)
Equatorial very wet: (a) Douala, Cameroon (4150 mm); (b) Monrovia, Liberia (4624 mm). *Equatorial wet*: (c) Singapore (2430 mm); (d) Manaus, Brazil (1996 mm). *Two wet seasons*: (e) Abidjan, Côte d'Ivoire (2144 mm); (f) Salvador, Brazil (1913 mm). *Two wet seasons, marked dry*

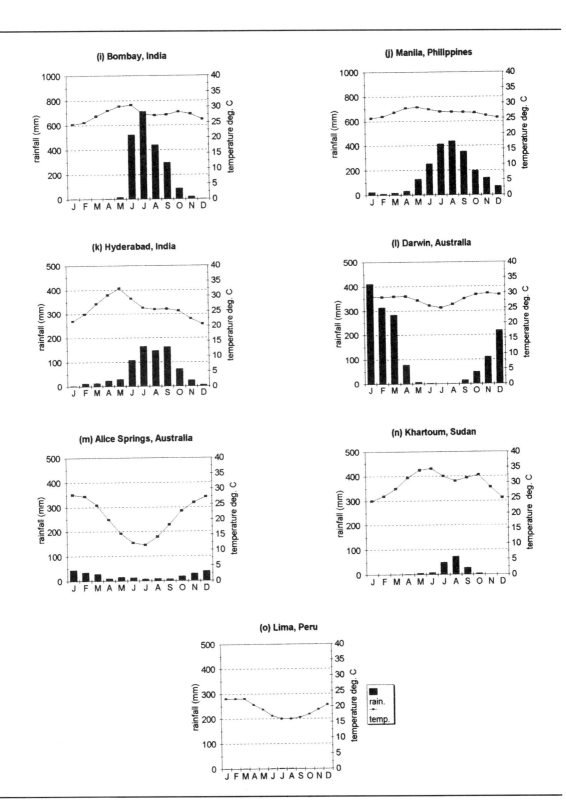

Fig. 2.5 (*Contd*) *season*: (g) Trivandrum, India (1839 mm); (h) Nairobi, Kenya (1066 mm). *One wet season, monsoonal*: (i) Bombay, India (2099 mm); (j) Manila, Philippines (2069 mm). *One wet season, semi-arid*: (k) Hyderabad, India (764 mm); (l) Darwin, Australia (490 mm). *Arid*: (m) Alice Springs, Australia (252 mm); (n) Khartoum, Sudan (164 mm); (o) Lima, Peru (10 mm).

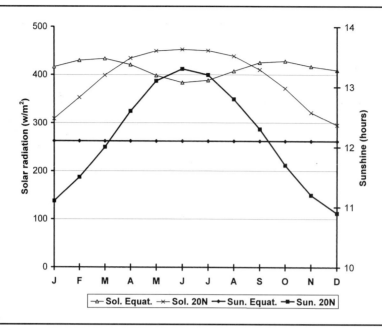

Fig. 2.6 Typical values for the northern hermisphere of daily solar radiation (W/m^2) and hours of sunshine in the tropics.

hemisphere. Balek (1983) noted that more than 30% of the earth's surface was under the influence of these easterly air flows, popularly known as the trade winds, due to their reliability over the oceans. The surface easterlies are overlain by a westerly air stream in the upper troposphere, sufficiently vigorous in places to form a fast moving jet stream (a band of rapidly moving air in the troposphere). The upper westerlies are less persistent than the surface easterlies and demonstrate greater seasonal changes associated with Asian monsoonal precipitation.

Intertropical convergence zone (ITCZ) or equatorial trough

As surface air flows out of the subtropical anticyclones in the direction of the equator it feeds the trade winds which eventually converge at the intertropical convergence zone (ITCZ). This area of low pressure is sometimes known as the equatorial trough and is associated with intense convection. A key facet of the ITCZ is the reversal of wind direction and change in cloud cover either side of the point of convergence (see Fig. 2.7). Trewartha (1981) identified a zone immediately south of the ITCZ, in the northern hemisphere, where rainfall is suppressed by an inversion (an increase of temperature with height) caused by hot dry air from the north rising over the denser moist from the south. As the height of this inversion increases southwards, so the opportunity for rain improves through convection (Kidson, 1977). Garnier (1967) identified local thunderstorms providing variable and sporadic rainfall in addition to disturbance lines (a line of thunderstorms travelling east–west). These disturbance lines (Rasmussen, 1987) occur when the stability of the overlying easterlies is disturbed by moving waveforms leading to roughly north–south belts of convergence in their lower layers.

Over land the position of the ITCZ has been related to the sun and in West Africa it has been observed to duplicate the sun's passage with a time lag of 4–6 weeks (Sumner, 1988; Farmer,

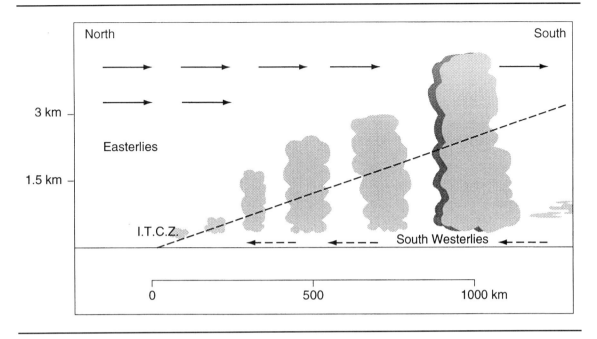

Fig. 2.7 Cross-section of the ITCZ region in West Africa. (After Trewartha, 1981 and Agnew and Anderson, 1992.)

1989), globally the pattern is more complex. The greater northward penetration in the northern hemisphere, up to 25°N (only 5 to 10°N over the oceans), is related to the larger land area and consequent higher surface temperatures. In the southern hemisphere the ITCZ only travels as far south as 15°S and remains closer to the equator over the oceans and in some areas, e.g. Peru, stays north of the equator (Reading et al., 1995). Over the oceans the ITCZ is fairly static, in the Pacific its location varies no more than 5° (Briggs and Smithson, 1992). Hess et al. (1993) investigated the mechanisms controlling the position of the ITCZ over oceans using a general circulation model (GCM) with a water covered earth which revealed the importance of an appropriate convective parameterisation. The seasonal position and movement of the ITCZ is then difficult to predict, being influenced by conditions in the upper atmosphere which in turn now appear related to oceanic circulations (see El Niño below). These surface conditions are determined by upper air flows, in particular the position and intensity of the subtropical jet streams (Kidson, 1977; Hastenrath, 1985). For example, in West Africa during winter (dry season) months when anticyclonic conditions prevail over the Sahel, the jet stream becomes convergent towards the equator (Hayward and Oguntoyinbo, 1987) producing a downward shift of air to feed high pressure at the surface. In summer months divergence aloft results in convergence of the southwesterly and northeasterly winds at the surface drawing in moist air with increased opportunities for precipitation. Low rainfall corresponds with a weak easterly jet stream associated with weaker circulation in the middle and upper troposphere. Changes in the jet stream then have consequent impacts upon precipitation.

The monsoon

The monsoon is normally described in terms of rainy season characteristics whereas it is the reversal of conditions from a marked dry period to a marked wet period that needs to be noted. This seasonal reversal in humidity and wind direction in parts of the tropics is often explained by differences in solar radiation absorbed by land and ocean. In the northern hemisphere, the heating of African and Asian land masses produces a surface low pressure that enables the penetration of warm moist air from the southwest producing a wet season. Thus the ITCZ displays considerable meridional (latitudinal) migration across the Indian sub-continent, Africa and Australia but with a comparatively weak movement over the Atlantic and Pacific Oceans and eastern Pacific region (McIlveen, 1992). However, the meridional migration of the ITCZ and onset of monsoonal precipitation appears better explained by a more complex relationship with the subtropical jet stream.

Over the Indian sub-continent in the winter dry season, surface winds are northeasterly and upper air flow is westerly and strong. The position of the jet stream, modified by the Tibetan Plateau and resultant convergence in the upper westerlies, spills subsiding air onto the region producing stable dry conditions that are only occasionally breached by traversing depressions. At the end of the northern hemisphere's winter, with a stronger north–south temperature gradient, an easterly jet stream becomes established. Thus the southwesterly surface monsoon winds are overlain by an easterly air flow in which waves develop promoting upper air convergence and monsoonal precipitation. There remains uncertainty over the precise mechanisms controlling the position and intensity of the jet stream.

El Niño southern oscillation

Discussions of tropical circulation would not be complete today without some mention of teleconnections (Flohn, 1987; Glantz *et al.*, 1991), i.e. the growing appreciation of the global interconnectedness of atmospheric and oceanic circulations. It is now widely remarked upon how changes in rainfall can be coupled to sea surface temperature anomalies and GCM modelling is leading to greater understanding of the mechanisms involved (Robertson and Frankignoul, 1990). Glantz (1987) has suggested a link between droughts in Africa and Atlantic temperatures which appear to have been successfully modelled (Folland, 1987; Owen and Folland, 1988). Mason (1995) identified sea surface temperature changes in the south Atlantic and southwest Indian Oceans that may be significant for rainfall in Southern Africa whereas Gregory (1988) examined the relationship between the El Niño and droughts over India. The most widely discussed coupling is that of the El Niño southern oscillation, or ENSO, with monsoonal rainfalls (Lockwood, 1984). Prior to an El Niño there are powerful southeasterly winds which produce a build-up of water in the western Pacific, this is later advected eastwards affecting the normal upwelling of cold water off the South American coast (Caviedes, 1988). The El Niño, or cessation of cold water upwelling, is then associated with large-scale weather oscillations in the Pacific basin. The normal, or 'Walker', circulation is reversed with a surface westerly wind and persistent high pressure subsidence in the southwestern Pacific. This in turn promotes a weakening of the monsoonal circulation and drought conditions in India and Australia. Reading *et al.* (1995) noted that during the 1983 El Niño, Australia experienced one of the worst droughts of the twentieth century, followed by unusually heavy rainfall in 1988 when the Walker circulation was re-established. Camberlin (1995) found that rainfalls in northeast Africa correlated well with the ENSO, especially droughts in Ethiopia and Uganda. However, Wuethrich (1995) reported that the El Niño phenomenon is becoming increasingly unpredictable and appears to follow a more complex pattern than previously believed.

In summary, although more than half of the earth's surface lies within tropical latitudes, the study of tropical circulations has been neglected with disproportionate attention paid to temperate regions. As a consequence many of the earlier explanations tended to use models derived from higher latitudes, e.g. frontal systems to oversimplify the motions in the tropics where changes are not so clearly related to the seasonality of solar radiation. Exacerbated by poor observing networks and large expanses of ocean, the balance is now being redressed (McIlveen, 1992). Since the 1960s there has been tremendous growth in the global network of observations enabling better computer models to be built, producing more reliable forecasts at global and regional scales.

Rainfall and atmospheric disturbances

In the tropics the seasonal changes in temperature are so small that seasonality is determined by rainfall contrasts, unlike temperate regions. There are three main precipitation forming mechanisms in the tropics: easterly waves, linear disturbances and tropical storms.

The easterly waves

The equatorial trough is associated with afternoon showers and little change throughout the year. This simple pattern is occasionally disrupted by more significant downpours linked to disturbances of the trade winds. The Coriolis force (due to the earth's rotation) is at its weakest in the tropics and the easterly atmospheric motions in this area only form a gentle series of low pressure troughs and higher pressure ridges. These occur on the equatorial side of the high pressure belts and the gentle changes in pressure would not normally be significant. In fact they are difficult to discern over land and are most clearly visible over the ocean. Although pressure gradients are small these waves are sufficient to briefly increase cloud cover and initiate precipitation. Over the oceans coherent cloud formations can be observed whereas over land movement and structure becomes confused. Although rainfall is only modest it can contribute a significant proportion of the total in areas such as the Caribbean.

Linear disturbances

These linear systems occur at higher latitudes in the tropics in northern India, south China, the Caribbean and North and West Africa where they are associated with the passage of the ITCZ. Figure 2.7 shows the relative position of the ITCZ and associated weather patterns in West Africa. In the immediate vicinity of the ITCZ, warm easterly air forms an inversion aloft suppressing precipitation but as the depth of moist air increases so cloud extent and rainfall increases (Kidson, 1977). Cloud cover is not extensive and forms a series of well-defined areas of convergence. The pattern of rainfall is then (in West Africa) a series of localised squalls or disturbance lines which travel east to west, i.e. across the direction of the surface winds. Much has been written about these disturbance lines (Gregory, 1965; Rasmussen, 1987; Jackson, 1989; Reading *et al.*, 1995) which appear to form when the stability of the overlying easterlies is disturbed by moving waveforms leading to north–south belts of convergence. The consequence is a highly localised and intensive downpour with important implications for soil conservation and plant growth.

Tropical revolving storms

Tropical storms are called hurricanes in the Caribbean and the United States, if the windspeed is above 115 km/h, and typhoons in the Pacific. At lower windspeeds of 60 km/h less intense storms are known as cyclones. Tropical storms are characterised by low pressure (down to 920 mbar), a circular pattern of the isobars with a diameter of 150–650 km, extremely high rainfall (around 200 mm/d) and steep pressure gradients that produce high windspeeds. After the storms have passed onto land they begin to weaken and the structure becomes less clear, winds decrease but rainfall can still be very high.

The precise reasons for the formation of cyclones and hurricanes are still uncertain (Briggs and Smithson, 1992; McIlveen, 1992) although they undoubtedly develop from pre-existing disturbances. The key appears to be the presence of divergence in the upper troposphere causing the development of very low pressures at the surface. In the space of a few days the sea level pressure in the centre of the cyclone can fall by 20 mbar or more. Thus the conditions are set for intensification of the storm but whether or not this happens depends upon a number of environmental factors.

Tropical storms require enormous amounts of energy to develop and to sustain their progress. The environments in which they form are hot and moist with high relative humidities. Polewards of 15° latitude sea temperatures are cooler and tropical storms diminish as they enter these cooler climes. Tropical revolving storms develop over the warmer seas but do not originate within 5° latitude of the equator and do not originate in the south Atlantic nor the southeast Pacific so some of the mechanisms generating tropical storms must be lacking in these areas. There appears to be a connection between the position of the ITCZ and regions of cyclone formation. As the ITCZ moves seasonally towards the poles, following the progress of the sun, the tropical cyclone season develops as the ITCZ reaches its most northerly position in mid-summer (northern hemisphere). But note that the ITCZ never lies to the south of 5° latitude in the south Atlantic, and always lies north of the equator in the southeast Pacific, areas where hurricanes do not originate.

Rainfall characteristics

Tropical rainfall regimes have been described above; here we are interested in the nature of precipitation and its variability. Both topics have already been reviewed for the tropics by Jackson (1989) and Reading *et al.* (1995).

Rainfall variability

Tropical regions experience the highest rainfall variability in the world, with a coefficient of variation (Cv) of 40% common, and in desert regions values in excess of 100%. UNESCO (1977a) reported, for example, both Bakel and Dakar (Senegal) lie on the same 500 mm annual isohyet yet a difference of 270 mm was recorded in 1972. Rain may fall only once every 8 years in parts of the Sahara and the Middle East, and only once every 18 years in parts of Peru. Jain (1968) characterised Rajasthan (India) as one lean year in every three and one famine in every eight while Hayward and Oguntoyinbo (1987) reported coefficients of variation in West Africa in excess of 50%.

$$Cv = (\text{standard deviation rainfall}/\text{mean rainfall})\%$$

The convective nature of the rain that does fall results in a highly variable spatial pattern. Sharon's (1972, 1981) studies of rainfall variability, for example, showed that in central Namib there was little correlation between stations 25–35 km apart and elsewhere found intense raincells with a diameter of only 5 km. Decadal rainfalls (ten day totals) at Niamey, West Africa, were found to be independent of each other, i.e. the amount falling in one period has no apparent influence upon the amount falling in the following period (Agnew and Anderson, 1992). This would not be the case where atmospheric conditions were persistent. This points to the erratic movement of the ITCZ over the region and the changeable nature of atmospheric conditions producing variable rainfalls.

The variability of precipitation in the tropics raises problems for its accurate measurement given that localised convective storms may not be recorded in standard gauges. Although there have been vast improvements in areal measurements of rainfall through radar (WMO, 1974) and satellite imagery (Barrett and Martin, 1981; Dugdale *et al.*, 1991) most rainfall data are still collected by individual observers where there are permanent settlements. One must, therefore, always approach rainfall records with a degree of caution over their reliability. This problem is further compounded by the assumption that all rainfalls follow a normal distribution that are amenable to analysis by use of their mean and standard deviations. An analysis of the same decadal data from Niamey cited above reveals that they could not be considered to follow a normal distribution. Instead this skewed decadal rainfall follows a gamma distribution (Thom, 1958) with a zero lower boundary and an infinite upper boundary. These results question the use of means to indicate the 'central tendency' of rainfall and all associated calculations such as the coefficient of variation.

Rainfall intensity

Rainfall in tropical latitudes can be very intense as well as being localised. In the tropics over a third of rainfall intensities exceed 25 mm/h (Jackson, 1989); in Malaysia, for example, intensities of 50 to 75 mm/h are common while in Uganda peak rates of 140 and 250 mm/h have been recorded. In Niger the following peak rainfalls have been reported by Hoogmoed (1986):

Niamey Ville	386 mm/h, total rainfall of 22 mm (1977)
Zinder	325 mm/h, total rainfall of 14 mm (1982)
Gaya	436 mm/h, total rainfall of 26 mm (1975)

However, such events are infrequent and when they do occur are of short duration. Delwaulle (1973), for example, noted that rainfalls in West Africa only exceeded 150 mm/h twice during the period 1966–71. High intensity is undoubtedly an important characteristic of tropical rainfall, unfortunately the data on rainfall intensities are scanty and often unreliable due to the short duration of the event.

High intensity rainfall events, while not very common, can be of great hydrological significance especially for those areas with steep slopes and only a thin surface of weathered mantle. In such conditions the surface may be quickly saturated leading to the generation of very rapid runoff producing dangerous, destructive and unpredictable flash floods. It has been said, 'More people have drowned in desert rivers than have perished from thirst,' (Nir, 1974, p. 25).

The relationships between rainfall intensity, surface infiltration and surface runoff are important for agriculture as they determine how much water enters the root zone and is potentially available for plant growth. This topic has been widely researched (Agnew and Anderson, 1992) and even modelled for water harvesting prediction (Bradley and Crout, 1995; Gowing and Young, 1996). However, there is still much uncertainty, for example whether surface crust type and raindrop kinetic energy determine infiltration rates, or whether it is surface saturation and soil moisture gradients.

CLIMATE CONDITIONS AND PLANT GROWTH IN THE TROPICS

Besides climate conditions many other factors affect the growth of plants including soil characteristics, ecological conditions such as competition and human activities. It is climate, however, that often determines the growth rate (GR) of plants which can be expressed as a function of the net assimilation rate of biomass (NAR) and the leaf area index (LAI) (Forbes and Watson, 1992) where:

$$GR = LAI \times NAR$$

Growth rate and crop biomass production should be distinguished from yield as the latter is most often a commercial definition such as 'the utilisable fraction of the plant biomass at harvest' which may or may not be directly related to above and below ground biomass. Climate influences the assimilation rate of plants through the supply of energy (sunlight and temperature) and the supply of moisture.

Energy supply to plants

Photosynthesis is vital for plant growth whereby hydrogen derived from water is combined with carbon dioxide derived from the atmosphere to produce carbohydrates using energy derived from the sun. Only the photosynthetically active region is absorbed by plant chlorophyll and changes to inputs of this part of the electromagnetic spectrum, for example through air pollution in cities or alongside roads, can have a deleterious effect on vegetation. The reverse process, respiration, whereby carbon dioxide and water vapour are released, is equally dependent upon prevailing atmospheric conditions (Monteith and Unsworth, 1990).

Table 2.3 Comparison of C3 and C4 plant growth in the tropics. (After Deshmukh, 1986, p. 25.)

	C3 plants	C4 plants
Optimum temperature for CO_2 fixation (°C)	15–25	30–45
Photosynthetic rate (mg $CO_2/dm^2/h$) under optimal conditions	15–35	40–80
Light saturation point (%)	10–25	100

Three mechanisms can be identified by which plants assimilate CO_2 during photosynthesis (Stott, 1994; Hulme, 1996):

- C3 crops such as wheat and rice (CO_2 fixed by a 3-carbon compound, phosphoglyceric acid).
- C4 crops such as maize, sugar cane, millet and the majority of savanna grasses (CO_2 fixed by a 4-carbon compound, oxaloacetic acid).
- CAM (crassualacean acid metabolism) plants such as cacti and other succulents (CO_2 fixed by crassualacean acid).

C3 crops, often found in temperate regions, are less efficient converters of CO_2 whilst C4 plants, often found in tropical climates, are more efficient (see Table 2.3) and can tolerate higher temperatures as a consequence.

The different rates of CO_2 assimilation have implications for water-use efficiency and the impacts of global warming. It is also noticeable that production of biomass (not necessarily yield) is much lower for C4 plants per kg of water consumed.

Sunlight and leaf area index

Respiration is independent of sunlight and continues through night-time and periods of low illumination. As light levels increase the rate of carbohydrate production (photosynthesis) will increase until at the point of 'light compensation' this exceeds the rate of carbohydrate destruction through respiration. Thereafter increases in light intensity will stimulate an increase in plant growth until a maximum of sunlight absorption is reached (light saturation point).

For temperate climates the plateau of light intensity is reached at about 25% of summer sunlight but this is much lower than for C4 crops in the tropics and C3 crops, such as wheat, grow well in low light conditions. As plants emerge and grow their leaf area increases but a point may be reached whereby shading and senescence of lower leaves causes the leaf area index (LAI) to decrease. Generally as plants grow so the LAI increases and light absorption increases to an maximum interception of 95% of sunlight (Forbes and Watson, 1992). For dicotyledons a LAI of 3 is taken to be an optimum for sunlight absorption whereas for monocotyledons a higher LAI of 4 to 5 is found, particularly in the tropics where there is greater sunlight penetration through the canopy as the sun is overhead. Deshmukh (1986) noted that over twice as much sunlight was intercepted by tropical crops compared to temperate crops but the net primary production was much the same due to higher rates of transpiration.

Given the high inputs of solar energy it is not surprising to find day lengths are potentially long (Fig. 2.6), a constant of 12.1 hours at the equator, rising to around 13 hours (summer solstice) and falling to 11 hours (winter solstice) at 30° latitude. Actual hours of daylight are mitigated by cloud cover and seasonal variations are evident. Hayward and Oguntoyinbo (1987) noted that for West African coasts the chances of cloud cover were two times greater (0.8–0.9) in the wet season compared to the dry season. As the levels of sunlight received in tropical latitudes are high, light is not normally a limiting factor on plant growth at the top of the canopy layer.

Photoperiodism

The initiation of flowering is linked to plant age and climate conditions, especially temperature. The flowering of some plants also responds to day length (actually the duration of night-time to be specific). Apart from flowering other processes responding to daylight include tuber formation in potatoes and bud and seed dormancy.

Unfortunately there is no simple explanation for which plants behave in this fashion although it is probably a mechanism in tropical areas with a dry season to avoid periods of desiccation. For example, most maize and soya bean are short-day plants but some (e.g. tomatoes and cucumber) are day-neutral.

Water and carbon dioxide

The variability of rainfall in the tropics means that water shortages are common. The implications for agriculture are that yields may change over short distances in response to the varying amounts of rainfall received. Such an impact is only likely, however, at the margins of cultivation in the drylands where rainfall exerts a much stronger control. Perhaps of greater significance for agriculture in the tropics is the temporal variability within seasons and interannual changes. Where there is a marked dry season, variations in the onset of the wet season can have important consequences for farmers. In areas where soils are fine textured (e.g. Vertisols in the Senegal River basin) and dry out during the dry season they are often too hard to work until after the first heavy rains. Figure 2.8 shows the structure of the rainy season can vary considerably even in a wet region such as southern Uganda. The frequency of early showers will influence the optimum planting date and in rain-fed systems will determine whether a long season or short season variety is selected. Early planting often produces the best yields and yield is reduced the longer that sowing is delayed. Alternatively even when early rains are satisfactory a dry spell later can have deleterious effects. For example, rainfalls in drought years in West Africa are more often caused by low amounts during the middle of the wet season, i.e. August, than at the beginning or cessation of the rains (Agnew, 1982, 1989). Dryland farmers have evolved a number of strategies to combat the variability of rainfall including crop selection, use of mulches and variable sowing dates.

Whether a moisture deficit constitutes a drought is open to interpretation and it is useful to distinguish between drought and desiccation (Warren and Khogali, 1992). Drought is a short-lived phenomenon whereas desiccation represents a period of increasing aridity normally the result of climate change (Agnew, 1990, 1995). Both are often assumed to be solely the result of a declining rainfall and definitions of drought are largely given in terms of precipitation (Palmer, 1965; Beran and Rodier, 1985; Druyan, 1989).

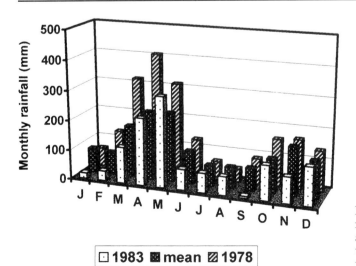

Fig. 2.8 Comparisons of rainy season regimes at Entebbe, Uganda using lowest annual total (1983, 1245 mm), highest annual total (1978, 2004 mm) and the mean for 1970 to 1990 (1516 mm).

However, agricultural drought (or ecological drought) requires consideration of the demands of vegetation and is usually defined in terms of water supplied to the crop rooting zone or meeting the needs of an idealised livestock herd (Agnew and Anderson, 1992). Hence agricultural drought can be caused by both a decline in water supplies (rainfall reduction, decreased infiltration, groundwater over-extraction) and increasing demand (planting density, intercropping, new varieties, etc.).

Photosynthesis is not limited by drought directly but by the insufficient supply of CO_2. This refers to the levels of CO_2 found in the plant stomata not in the atmosphere (there is normally sufficient CO_2 in the atmosphere, particularly for C4 plants which can photosynthesise at much lower concentrations than C3 plants). The assimilation of CO_2 is governed by the water status of the plant. Even moderate plant water deficits will affect cell enlargement and Kramer (1983) noted that plant growth and crop yield were reduced by water shortages more than any other environmental variable. The impacts of water shortages on plant growth are well known and reviewed in a number of texts such as Begg and Turner, 1976; Kozlowski, 1968a,b, 1972, 1976, 1978, 1980, 1983; Kramer, 1983 and Jones, 1992. Numerous plant processes are affected by moisture deficits from cell growth to stomatal opening but the level of tolerance varies between tropical crops. Cell enlargement is generally affected at leaf water potentials of -0.2 to -0.4 MPa with stomatal closure between -0.8 and -1.0 MPa. Kramer (1983) reported that transpiration fell at leaf water potentials of -1.2 MPa for soya bean while for maize values around -0.6 MPa were significant. The sensitivity of crops to water stress varies and Doorenbos and Pruitt (1975) provided a useful tabulation of critical periods, e.g. for tobacco (knee high to blossoming), maize (pollination period from tasselling to blister kernel stage) and sugar cane (period of maximum vegetation growth).

As comparatively little of a plant's transpiration requirements are stored in the vegetation, soil moisture is influential. The amount of soil water available to plants – available water capacity (AWC) – is normally taken to lie between field capacity (FC), the upper limit of soil water storage when free drainage has ceased, and crop wilting point (WP), around a soil water potential of -1.5 MPa, i.e.:

$$AWC = FC - WP \ (mm \ of \ water)$$

The magnitude of AWC depends upon soil characteristics such as texture, bulk density and organic matter content (Gupta and Larson, 1979). The Food and Agriculture Organization of the United Nations (FAO) (Smith, 1992) has suggested as a general guide that AWC per metre of soil is around 60 mm/m rooting depth for coarse textured soils, 100 mm/m for sandy soils, 140 mm/m for loams and 180 mm/m for clays. Not all this water is freely available. It has been known for some time (Denmead and Shaw, 1962) that the relationship between transpiration rate and soil moisture content is complex. Jackson (1989) provided a number of examples demonstrating the significance of climate, soil type and rooting system. The FAO (Smith, 1990, 1992) employed a simple relationship whereby 50% of AWC was assumed to be freely available within the rooting zone, thereafter soil water was progressively more difficult to extract to wilting point. This is similar to the Penman–Grindley 'root constant' approach (Shaw, 1994) but these simple models should not obscure the point that much still needs to be known about the relationships between water shortages and crop growth in the tropics. In particular the apportioning of energy between above ground and below ground biomass. It is known the root:shoot ratio is affected by water deficits (Monteith, 1986; Jones, 1992) but there is little supporting field data especially in the dryland tropics (Lal, 1991a). Goldstein and Sarmiento (1987) concurred that in savanna systems, where periods of desiccation are encountered, vegetation has a number of 'choices' including the development of the root system which is reflected in the root:shoot ratio, i.e. the amount of above and below ground biomass, and the control of the osmotic potential in leaves through stomatal regulation and mechanisms such as deciduous or evergreen. Tropical vegetation tends to allocate more production to above ground biomass than temperate plants; tropical forests typically have 90% above ground biomass and grasses 50% (Deshmukh, 1986). However, Goldstein and Sarmiento (1987) noted that early emerging grass species had a higher root:shoot ratio, i.e. greater development of subsurface biomass than those grasses that emerged in wetter conditions. There are, then, a number of ways in which tropical vegetation responds to constraints in water supply.

Temperature

As temperature increases so do the rates of chemical reactions in plants although growth by cell elongation is most rapid at temperatures around 20°C; above 40°C enzyme reactions are reduced (Forbes and Watson, 1992). Higher temperatures produce increased rates of both photosynthesis and respiration with the exception of the extremely hot conditions found in tropical deserts. The net growth of the plant will depend not merely upon temperature but also the extent to which other factors constrain photosynthesis. Levels of illumination and temperature are determined by solar radiation inputs, so the negative effect of high temperatures on assimilation rates causing an increase in respiration is offset by increases in sunlight absorption. The reduction in net assimilation rates due to high temperatures (below 40°C) is seldom significant except in dry conditions when water supply inhibits photosynthesis. Leaf temperature needs to be maintained at the optimum temperature through evaporative cooling and this may not be possible when water supply is constrained. For example, Goldstein and Sarmiento (1987) reported the optimum temperature for two savanna trees was found to be 25–28°C with a very narrow plateau of optimum temperatures such that at 35°C the carbon uptake was only 50% of the maximum.

Air temperature is a function of the energy budget, i.e. amounts expended on latent heat, sensible heat and advection/storage. There is the potential in the tropics for very high temperatures due to large inputs of solar energy but where evaporation is also high, air temperatures will be moderated. As a result, in the humid tropics temperatures rarely exceed the low 30s (°C) and over the oceans will be in upper 20s. In contrast hyperarid regions can experience maximum temperatures above 40°C. Further variation is created by altitude and cloud cover; as a consequence five temperature regimes can be identified (adapted from Reading et al., 1995):

- *Equatorial low altitude* (e.g. Manaus in Brazil) where temperatures are reasonably low and constant throughout the year (mean 26°C). Variations are largely a function of cloud cover.
- *Equatorial high altitude* (e.g. parts of Ecuador, Peru and Bolivia) where temperatures are also constant but are markedly below those for lower altitudes due to adiabatic cooling effects.
- *Tropical maritime* for islands (such as Hawaii and Fiji) at latitudes of 20° which are affected by the seasonal movements of the ITCZ. This introduces a seasonality to the temperature regime, with a hotter season evident but the annual variation is moderated by the maritime influence, hence the range is still only a few degrees centigrade.
- *Seasonal tropical* are typically monsoonal areas and those land masses under the influence of the ITCZ. Here the position of the maximum isotherm coincides with the migration of the ITCZ. A more marked seasonal contrast is evident and greater diurnal ranges are found. The highest and lowest temperatures are found in the dry season when energy losses are greatest. During the wet season, cloud cover reduces solar energy inputs while also reducing energy losses so that temperature extremes are mitigated. As one moves to higher latitudes and a shorter wet season so the seasonality increases and, as a consequence, the temperature variability.
- *Tropical drylands* show the greatest range of temperature. The high temperatures experienced are largely a function of low energy expenditure on evaporation coupled to significant inputs of solar radiation. Although

energy inputs are high so are potential losses and nocturnal cooling can produce a large diurnal temperature range while seasonal contrast is also noticeable (e.g. Alice Springs) as are the effects of altitude. The hot arid conditions can also be mitigated by an offshore cold ocean current (e.g. Arabia, West or Southwest Africa) where advected fog and mists become common.

Although five temperature regimes have been identified these contrasts should not obscure the general point that tropical areas are isothermal (have the same temperature) with only small temperature gradients. With the exceptions of the hyperarid and high altitude regions, temperatures in the tropics are suitable for plant growth all year round.

Other climatic conditions

Humidity and dewfall

In general relative humidities are high in tropical regions, i.e. above 80%, because of the high moisture supply, particularly in equatorial parts. At higher latitudes seasonal changes in wind direction associated with the ITCZ can encourage the introduction of low humidity northeasterlies (in the northern hemisphere) known locally as the Harmattan in West Africa. This desiccating wind can have deleterious effects on early emerging vegetation and is also believed to enhance the dispersion of meningitis among human populations. As one moves into the arid tropics humidities further reduce but it would be misleading to believe moisture is absent. Dewfall can be an important form of precipitation enhancing grazing opportunities (Lancaster and Lancaster, 1991). At some tropical locations conditions for dewfall and fog moisture are sufficiently favourable for consideration of artificial harvesting. Barrow (1987) reported that $50 m^3$ per day were collected in traps with an area of $6000 m^2$ in Peru; while polythene sheets in the Negev yielded 3631 mm per m^2 per month. Stanley-Price et al. (1986) working in Oman collected a total of 50 litres per m^2 per day and Schemenauer and Cereceda (1992) have undertaken several studies of fog water harvesting, concluding that sufficient could be collected to provide a viable water supply for a village of 330 people in Chile.

Windspeed and direction

The most pronounced feature of the tropics is the trade winds which affect nearly half the globe (Briggs and Smithson, 1992) providing a steady easterly air flow. Conversely in the equatorial trough winds are light and variable, hence their name – the doldrums. Changes to these predictable and tranquil conditions are associated with the rainfall mechanisms described above. For example, the linear disturbances within the ITCZ zone are preceded by a zone of fast moving air, which in West Africa can be seen as an advancing brown sky due to the dust transported. Tropical revolving storms contain even higher wind velocities. At a more localised scale surface heat in deserts can initiate small pockets of unstable air 'dust devils' while diurnal heating/cooling and topographical effects can create onshore/offshore breezes and mountain/valley movements.

Reliable windspeed data are notoriously difficult to obtain and tend to be found for major settlements especially at airports. Often data reported are from visual observations that will only provide a general assessment. Unfortunately data are normally averaged over 24 hours which obscures maximum gusts and changes that are perhaps significant. Vertical windspeeds are not routinely observed. The significance of wind for plant growth is on dispersal (Good, 1970), on the form of plants where exposed to high velocities and on its relationship with other climate variables. In the tropics the last point concerns the enhancement of rates of evapotranspiration through advection (the horizontal movement of air). Apart from highly localised conditions, windspeed does not appear to be a significant variable controlling plant growth with the exception of tropical storms which can cause havoc, uproot trees and devastate crops.

Crop–water models and evapotranspiration

In the tropics, water supply is the major climatic constraint on plant growth (Lal, 1991a). It was shown above that plant water deficits affect CO_2 assimilation which can lead to impaired growth or even death (wilting). Clearly the key climatological factors determining this relationship are water supply (precipitation) and water demand (evaporation).

Evaporation is a poorly understood hydrological variable. Figure 2.9 presents rates for different tropical climates but this oversimplifies the complex interplay between energy balance, vapour pressure and windspeed as well as the need to distinguish between evaporation and evapotranspiration. *Evaporation* is normally taken to be the loss of water from open bodies of water whereas the term *evapotranspiration* combines such losses from soil and vegetated surfaces with plant transpiration (Agnew, 1994). A further distinction needs to be made between potential and actual rates. Potential rates are primarily dependent upon atmospheric conditions although the surface characteristics can also play a role, e.g. water quality and depth, vegetation cover and height and soil moisture content (Ward, 1975).

The term potential (or reference) evapotranspiration (PEt) has then been defined in terms of water losses from an extensive green grass cover, of uniform height, actively growing, completely shading the ground and not short of water (Smith, 1992), but Lindroth (1993) has suggested there are several alternative definitions.

It has proved extremely difficult to measure actual evaporation (AE) or evapotranspiration (AEt) directly. Water balances, even lysimeters, are prone to errors while evaporation pans and atmometers are subject to over-estimation. Climatological formulae based on the Bowen ratio, or eddy correlation, are available (Shuttleworth, 1979) but as research tools rather than operational procedures. More recently the FAO has announced that the Penman–Monteith approach (Monteith, 1991; Smith, 1992), e.g. as used in the Meteorological Office rainfall and evaporation calculation system (MORECS) (Hough *et al.*, 1996), would be employed in future. This is based on an assessment of the energy budget above the crop, the saturation vapour pressure deficit and the atmospheric and surface resistance to water vapour transport. Paucity of data and the need for more understanding of surface resistance, limits the present

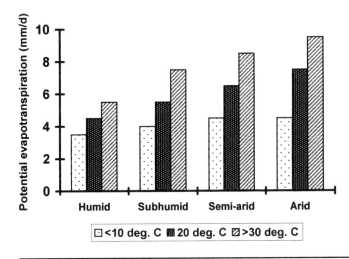

Fig. 2.9 Potential evapotranspiration (mm per day) calculated using the Penman (1948) formula for tropical climates and different thermal regimes. (Based on data from Doorenbos and Kassam, 1979.)

applicability of this approach and the Penman (1948) formula providing potential evapotranspiration (PEt) estimates is still the most widespread approach.

Most estimates of evaporation are then of potential evapotranspiration (from a grass surface) for which crop coefficients are needed (Jensen, 1974; Wright, 1981; 1982; Doorenbos and Pruitt, 1975) and some knowledge of plant–soil–water relations before actual rates can be estimated. There are two fundamental approaches: empirical or physical models. Empirical models are based on correlation of observed plant behaviour and climate conditions. Global systems such as Koppen (1931) or Holdridge (1947) employ crude measures to obtain generalised patterns of vegetation growth. More detailed regional studies are exemplified by Cocheme and Franquin (1967) who investigated climate conditions over West Africa. In the same region, Davy et al. (1976) used data from agricultural experimental stations to establish rainfall requirements which were then examined at different probabilities to determine land suitability, but they noted (p. 107) the difficulty of relating this to a largely subsistence farming system. The lack of reliable agroclimatic data and questions over the robustness of the ensuing relationships has led others to try to develop models based on the physical processes operating.

Doorenbos and Pruitt (1975) focused upon the need for reliable estimates of evaporative demand and used crop coefficients with the Penman formula to estimate crop water requirements. Although data are provided for a wide range of tropical crops the emphasis is upon irrigation need rather than the effects of water shortages. The subsequent publication by Doorenbos and Kassam (1979) set out to establish relationships between yield and evapotranspiration based on the assumption that the upper limits were set by climatic conditions and the genetic potential of the crop, i.e.:

$$[1 - (Y_a/Y_m)] = K_y [1 - (ET_a/ET_m)]$$

where Y_a is the actual yield, Y_m is the maximum yield, K_y is the yield response factor, ET_a is the actual evapotranspiration and ET_m is the maximum evapotranspiration.

A simplification of this approach was used by FAO (1986c) in order to predict crop failure from early season moisture deficits. Here soil moisture storage and the decadal differences between rainfall and potential evaporation were employed to calculate a seasonal rainfall water satisfaction index with reasonably good results. More recently the FAO have produced CROPWAT (Smith, 1990, 1992) a computer program that facilitates the computation of crop water requirements. It is based on the Penman–Monteith approach with a simplified soil moisture extraction model to calculate actual evapotranspiration. Although designed for irrigation scheduling it provides an estimate of the effects of restricted water supply on a number of tropical crops in a simplified fashion.

The water use efficiency (WUE) (see Table 2.4) is widely used to quantify the relationship between crop production and water consumption in an empirical form:

WUE = Yield per unit area / water used to produce yield.

or

$$WUE = Y/T$$

where Y is the yield and T is the crop transpiration (or WUE = Y/E where E = evapotranspiration for ecological studies).

Table 2.4 Examples of WUE for some tropical crops showing C4 plants have greater WUEs. (Adapted from Jones, 1992.)

	Y/E (×1000)
C4 plants	
Millet	2.72–3.88
Sorghum	2.63–3.65
Maize	2.67–3.34
C4 range	2.41–3.88
C3 plants	
Wheat	1.93–2.20
Rice	1.47
Alfalfa	1.09–1.60
Pulses	1.33–1.76
C3 range	0.88–2.65

A more complete analysis of WUE reveals that the rate of transpiration is dependent upon atmospheric resistance to water vapour diffusion such that (Lal, 1991a),

$$N = KT/(es - e)$$

where N is the total shoot-dry matter production, K is a constant, T is the crop transpiration and $(es - e)$ is the average vapour pressure deficit during daylight hours.

Gregory (1989) suggested an alternative form when crop evapotranspiration was employed:

$$\text{WUE} = (N/T)/[(1 + (E + R + D))/T]$$

where N/T is the transpiration efficiency, E is the evaporation, R is the surface runoff and D is the soil drainage.

For a given saturation deficit $(es - e)$ N/T is constant (Gregory, 1989) hence WUE in the tropics is amenable to improvement if transpiration increases which is a function of water supply. Many agricultural practices are found in the tropics that recognise this point. For example, water supplies can be enhanced and thus yields improved by reducing runoff losses (water harvesting for example, Gowing and Young, 1996), increasing water uptake through plant spacing and using varieties with deeper rooting systems; or reducing other losses by fallowing, weed control and mulches. It can also be seen that enrichment of atmospheric CO_2 may enhance WUE through greater transpiration efficiency, possibly allowing plants to occupy more arid tropical regions (Hulme, 1996).

CLIMATE, VEGETATION AND AGRICULTURE

Details of the crops grown and farming systems practised in the tropics can be found later in this book and elsewhere (Irvine, 1969; Arnon, 1972; Martin *et al.*, 1976; Metcalfe and Elkins, 1980; Ruthenberg, 1972, 1980; Purseglove, 1981, 1987; Barrow, 1987; Edwards *et al.*, 1990; Forbes and Watson, 1992; Jones, 1992; Loomis and Connor, 1992).

Equatorial regions: continuously wet

These regions encompass areas close to the equator where air convergence is common and rainfall persistent. Annual rainfalls are in excess of 2000 mm, there is no dry season and temperatures show little variation. They are typical of large areas in South America (e.g. Brazil, Colombia, Venezuela and Guyana) but limited to the Congo basin in Africa. These conditions are also found in parts of Malaysia, Sumatra, Borneo and many Pacific islands but seasonality starts to become noticeable through the influence of the two monsoons (e.g. Singapore). The climax vegetation is rainforest and covers 17% of the land surface of the earth (Briggs and Smithson, 1992).

Rainforest can be classified as:

- *Upper montane* less extensive, characterised by lower temperatures with mean annual values of 10°C at 3000m and 2°C at 4000m. This reduces growth rates and climate variability with frequent cloud and high humidities. Dwarf trees, shrubs, lichens, mosses and ferns are typical.
- *Swamp forests* are found in regions of inundation (e.g. coastal areas of West Africa or the Amazon River basin).
- *Lowland forests* can be distinguished into: wet and humid ecosystems. In wet forests precipitation exceeds PEt all year round (e.g. Southeast Asia, South America and Central America). In humid tropical forests (monsoon) PEt exceeds precipitation for parts of the year (e.g. Africa, Australia and India (monsoon forests)). Humid forests are the most species rich of all plant communities and contain half the world's biota (Furley, 1994). A typical structure consists of three tiers of mature trees with shrubs and herbaceous plants forming an additional two layers (Archibold, 1995). At ground level, due to low light penetration, herbaceous cover is limited but climbers and epiphytes are common.

'Traditional' occupation of the rainforest by humans involves small numbers of people operating shifting cultivation. Because the soils are inherently poor due to high rates of leaching they are soon exhausted by cultivation and need to be left fallow to recover. Thus a patchwork of areas at different stages of recovery is observed. With this level of land use the forest canopy is regenerated and soils are not eroded at excessive rates, but the regenerated forest may have a different species distribution and structure. During the twentieth century large areas of these rainforests have been cleared (Collins, 1990; Furley, 1994) although figures are often exaggerated. Grainger (1996) suggested that in Africa all forest lands are under the direct influence of human activity. The environmental impacts can be devastating including habitat destruction, loss of biodiversity, release of carbon dioxide and soil erosion on a large scale. Pearce (1996) reported that the tropics release between 1.8 and 4.7 billion tonnes of carbon each year through burning vegetation, possibly surpassing emissions by motorists in Europe and North America. As nutrients in this ecosystem are bound in the above ground biomass, removal of forest is followed by rapid depletion of soil fertility and land degradation. Overcultivation can lead to a complete removal of forest cover and the establishment of coarse grasses such as *Imperata cylindrica* (lalang) which encourage fires and the prevention of forest regeneration, although Walker (1996) suggested there may be ways of maintaining or even increasing agricultural productivity without environmental degradation.

As there are no marked seasons agriculture can continue all year round. Water is abundant but can be limiting for crops, e.g. cereals that need a dry period for grain ripening. Hence the crops that flourish in these wet equatorial regions are perennials such as rubber, oil palm, coconut, cocoa and bananas. Root crops are important for subsistence including yams (*Dioscorea* spp.), cassava (*Manihot* spp.), cocoyams (*Colocasia* spp.) and tannia (*Xanthasoma sagittifolium*). Livestock are limited to pigs and poultry due to the lack of extensive rangelands and the high incidence of disease associated with high humidities and temperatures. Individual farmers diversify by cultivating a variety of trees and shrubs that produce fruits, nuts, fuel, etc., alongside vegetable, cash and medicinal crops, e.g. tobacco. Maize can be cultivated if suitable varieties are available and weed control is effective, while rice is cultivated in parts of Southeast Asia where the change in the monsoons provides an opportunity for harvesting.

Seasonal climates: humid

These are areas of great diversity as the length of the rainy season and total precipitation reduces with latitude. This is then an area of transition between equatorial forests and tropical drylands. The region is typically known as the *savanna*.

Savannas cover 65% of Africa, 60% of Australia and 45% of South America (Deshmukh, 1986). C4 grasses and sedges dominate the landscape (Walker, 1987) with tree cover varying from 2 to 20%. The main climate characteristic is a recognisable dry season, at least one, lasting several months. Harris (1980) used this criterion to distinguish between woodland, bushland and grassland savanna (see Table 2.5). These extensive systems should be distinguished from alpine grassland which is found above the treeline on tropical mountains. Areas of high humidity and extremes of temperature produce distinctive plant communities where endemism (confinement of vegetation is one particular region) is apparent (e.g. the mountains of Papua New Guinea or the South American Andes).

Two wet seasons, high rainfall

Areas with two wet seasons and high rainfall typically receive between 1000 and 2000mm of rainfall and have a small temperature range (e.g. Entebbe, Uganda). Conditions are suboptimal for oil palms and rubber but in Africa an exten-

Table 2.5 Savanna vegetation and rainfall. (After Harris, 1980.)

Type of savanna	Annual precipitation (mm)	Dry season (months)
Woodland	>1000	2.5–5.0
Bushland	500–1000	5.0–7.5
Grassland	250–500	7.5–10

sive range of crops is found growing in the two rainy seasons (e.g. maize, various pulses, and groundnuts). Other crops include coffee, tea, bananas, sugar cane and cocoa and a variety of vegetables and fruit such as pineapples. Where soils permit, the potential agricultural productivity is impressive. In Asia rice is the main crop but livestock are limited by climate and disease.

Two wet seasons, low rainfall

Typically in areas with two wet seasons and low rainfall one of the dry seasons is more pronounced such that the annual rainfall is below 1500 mm and may be as low as 600 mm. Perennial crops can not normally tolerate the longer dry season although they may be cultivated in the wetter parts if soil moisture is conserved by mulching or the use of shade (e.g. arabica coffee region in the East Rift, Kenya). Elsewhere water deficits mean that maize, sorghum, millet, pulses and cassava are found. In the drier parts, grasslands are extensive and cattle are kept where disease and pests (e.g. the tsetse fly) can be controlled.

Monsoon: one long rainy season

In areas that experience monsoon, annual rainfall can lie between 750 and 1500 mm, and these regions are extensive in areas of Asia and large parts of Africa. The area is unsuitable for perennial crops unless they are drought resistant, such as sisal and cashew. In the wetter parts some perennials can survive and tea is an important crop. Even in the drier parts the rainy season is long and wet enough to support a variety of annual crops such as maize, millet, sorghum, groundnuts, cotton and tobacco. Rice is also widely cultivated, sometimes augmented by surface runoff or irrigation. Rangelands are productive but limited by dry season storage.

One short rainy season

Areas of one short rainy season are typically found in the Sahelian region of West Africa and the drylands of India. Drought-resistant cereals, sorghum and millet, are common along with various pulses. These plants have evolved a number of mechanisms to combat aridity (Dancette, 1980; Maiti and Bidinger, 1981; ICRISAT, 1984) including the apparent ability to remain dormant during periods of drought. In the highly variable drier climates farmers tend to be opportunistic and practice a combination of nomadic herding and cultivation (e.g. West African bush-Fulani). Soil fertility is maintained by growing leguminous crops, manuring fields with animal and human wastes and crop rotation (UNESCO, 1978). The use of fallow land to replenish both soil fertility, control weeds and maintain soil moisture is perhaps one of the key practices of dryland farmers. For instance, Gibbon and Harvey (1977), working in western Sudan, noted that on the poorer soils land was cropped for 3–4 years with millet and then left fallow for up to 20 years although sometimes groundnuts were used to reduce the fallow period down to 3 years. The food production system was further diversified by keeping livestock (e.g. over 50% of households kept livestock within the Niamey Department, Niger; 29% had poultry, 27% had cattle and 22% had sheep and goats).

The dryland farmers also have to tackle crop damage through pests and disease but perhaps the most intractable problem of all is the variable rainfall. Dregne and Willis (1983) noted that at least 180–250 mm of soil water were required to produce grain for wheat and similar crops. One strategy is to employ several sowing dates. This has the additional benefit of producing a field with variable harvest dates which reduces the risk of loosing the whole crop through pest or disease attack. However, this is a very labour intensive strategy which places limitations on the amount that can be produced (Kowal and Kassam, 1978). Gibbon and Harvey (1977) found that during the rainy season the control of weeds required 40% of all time spent on the land.

Arid climates and deserts

Deserts can be defined in terms of their climatology (annual rainfall below 250 mm) but are also considered to be areas of low or absent vegetation cover (Goudie, 1994) while McGinnies (1968) suggested that there was no universal agreement over the term desert. Whittaker (1975) provided a list of the various mechanisms by which plants have adapted to aridity:

- deep or extensive root systems
- water storage in tissues
- protective wax coating
- reduced leaf area and shedding leaves
- use of green stems for photosynthesis
- reversal of stomatal functions, i.e. C4 plants
- tolerance of desiccation
- high osmotic pressures
- rapid germination and growth.

Arid environments are characterised by pulses of water with corresponding biological pulses leading to both spatial and temporal variation in vegetation (McGinnies, 1968, 1979, 1988). Schmida *et al.* (1986) distinguished between two systems: one extensive and enduring where vegetation had adapted to arid conditions and an episodic pattern that responded rapidly to changing moisture conditions (desert winter annuals). McClearly (1968) distinguished between the flora of:

- Temperate arid areas (Patagonia, Gobi, Thar, Mojave)

and in the tropics:

- Warm arid areas (Sahara, Arabia)
- Coastal deserts (Atacama, Namib)
- Australia.

Within the classification, Australia is dominated by grasses and small shrubs, and coastal deserts contain quantities of woodland including tree cacti (*Cereus* sp.) found in the Atacama. The warm arid areas are perhaps the most difficult to characterise because of their greater diversity. The main land use is pastoralism (see Table 2.6). Pastoralists inhabit one of the earth's most hostile environments where diversity and movement are key strategies for survival. It has proved difficult to fully comprehend the manner by which pastoralists have adapted to their arid surroundings and many attempts to sedentarise

Table 2.6 Agricultural land use in the dry tropics. (Based on Ruthenberg, 1972, p. 298.)

Rainfall	Production system	Livestock
<50 mm	Pastoralism using pastures only infrequently	Camel
50–200 mm	Pastoralism with long migrations	Camel
200–600 mm	Pastoralism supplemented by dryland farming and sheep	Cattle, sheep, goats
>600 mm	Dryland farming with some herding	Cattle, sheep

and 'develop' these peoples have failed. They have evolved a number of strategies for coping with the harsh environment including:

- *Herding.* Herding animals to provide sustenance (e.g. milk, meat and blood), shelter, fuel, transport, commodities for exchange and status.
- *Herd diversification.* A herd with a variety of livestock is better able to respond to changes, and provides a greater range of products and services (Dahl and Hjort, 1976), for example camels can travel 25–30 km every day and in the wet season can live without water for up to 3 weeks by relying upon the moisture in vegetation. Louw and Seeley (1982) noted that the black Bedouin goat could withstand a 30% loss of body water, comparable to a camel, while a 12% loss was critical for most mammals. Goats are browsers as well as grazers and can therefore exploit more of the range than sheep which are essentially grazers.
- *Movement.* The essence of nomadic pastoralism is 'environmental opportunism' in that the animals and herders travel together in search of good grazing following rainfall. However, these movements are not aimless nor totally random but follow a recognisable pattern.
- *Animal sharing.* There are a variety of ways in which livestock are shared, given as an out-

right gift to a destitute household through kinship ties or loaned for, say, the period of lactation to avert hardship. The sharing of animals also distributes a household's herd thus reducing risks.
- *Storing food.* Storage is difficult because of the heat and need for movement. Camel milk lacks rennet and therefore cannot be turned into cheese, while slaughtering large livestock presents preservation problems.
- *Hunting and gathering.* Hunting of game and gathering wild fruit are relied upon in times of crisis although Nir (1974) estimated the Bushmen of the Kalahari supplied 80% of their diet by gathering.
- *Raiding and trading.* In the past if herd losses could not be recouped by loans or rapid regeneration by small livestock then pastoralists resorted to raiding. This could provide crops from dryland farmers, slaves to help with the herding and even additional animals. This practice has steadily died out during the twentieth century with trading being of more importance.

These pastoral:environmental descriptions (Agnew, 1983) are now being seen as over-simplistic and placing too much emphasis on physical environmental conditions. Lancaster and Lancaster (1992), for example, made a strong case that kinship and social organisation were better explanatory variables of pastoral behaviour. Even the long-cherished notion of rangeland carrying capacity is seen as a false concept that does not represent the non-equilibrium system operating in pastoral dryland tropics (Scoones, 1994, 1995; Warren, 1995).

CLIMATIC CHANGE

There is nothing new or startling about climatic change. The earth has been gradually warming since the end of the last ice age some 10 000 years ago. Interest has been fuelled, however, by the recent pattern of accelerated warming with an average rise of 0.5° since 1880 in the northern hemisphere (0.4°C for the globe). This warming has seen a parallel (in part only, e.g. not during the 1950s and 1960s) increase in atmospheric greenhouse gases leading to forecasts of continued warming even if emissions are reduced.

The predicted warming, however, is not accepted by all (Pearce, 1995a). Courtney (1993) pointed to inadequacies in the data record, the *Economist* (1995) suggested that there were many uncertainties while Bate and Morris (1994) reviewed the lack of scientific consensus which was illustrated in a debate held at the Royal Geographical Society in London (Olstead, 1993). Some 20 years ago Ponte (1976) was writing about the possible effects of global cooling when the onset of the next ice age was receiving consideration (Bryson, 1974, 1989) and more recently Rahmstorf (1997) has suggested changes to oceanic circulations may have the same effect. Nevertheless, the balance of opinion is that greenhouse gases have brought about accelerated warming and that this could have significant impacts on agriculture and ecosystems in the tropics. Although the greatest warming is forecast for higher latitudes, atmospheric aerosols and turbidity may offset warming for some time (Pearce, 1997) in areas such as India and the Pacific.

Global warming could cause the sea levels to rise by 20 cm in the next 25 years and 50 cm by the year 2045 (Warrick *et al.*, 1993; Semeniuk, 1994), initially through thermal expansion (Wigley and Raper, 1987). Temperatures were expected by the Inter-governmental Panel on Climate Change (IPCC) to increase by 1.4°C on average by the year 2030 (Houghton *et al.*, 1991) within a confidence interval of 0.7 to 2.0°C. More recent forecasts based on redefined greenhouse gas emissions and the effects of aerosols suggest slightly lower increases of between 1 and 3.5°C (IPCC in Pearce, 1995b) by the end of the twenty-first century as CO_2 levels double. There remains much uncertainty over changes in precipitation although general intensification of the hydrological cycle is expected to lead to higher levels of precipitation of the order 8–15% (Archibold, 1995).

These forecasts depend upon many uncertainties including future fossil fuel use and rainforest destruction, changes to cloud cover and adsorption of excess CO_2 by the world's oceans. Nevertheless, it is possible to make some forecasts based on current trends. There are various approaches: some, e.g. Wyman (1991), employ current analogues of precipitation and temperature limits on plant distribution such as Holdridge's (1959) classification to examine latitudinal shifts, while others use more detailed knowledge of growth constraints such as degree growing days used by Prentice et al., (1992) in their BIOME model employed by Hulme (1996). Cannell and Pitcairn (1993) examined the impacts of recent mild winters and hot summers in the UK as a prediction of future conditions but found it very difficult to arrive at clearly discernible changes.

Apart from climate change, enrichment of CO_2 can also lead to increased plant growth (see above) known as CO_2 fertilisation (Fakhri and Fajer, 1992). However, the C4 plants found in the tropics with their greater CO_2 assimilation mechanisms will not respond as positively to CO_2 increases compared to C3 plants in temperate areas (Warrick et al., 1991).

In the next 100 years sea level rises will affect coastal mangrove forests, coral atolls and deltas such as the Nile and the Ganges (Roberts, 1994). In tropical regions the boundary between the savannas and tropical forests is perhaps the most sensitive to change although desert margins are also highly variable. Bolin et al. (1991) suggested, based only on temperature changes with a doubling of CO_2, that the area occupied by tropical climates would increase by over 25% with a commensurate decline in subtropical and boreal forests. It is anticipated that global warming will favour an expansion of grasslands with a decrease in deserts (Archibold, 1995). There is, however, much disagreement between different GCM forecasts and uncertainty over whether ecosystems will be able to respond to the accelerated rate of change, i.e. the efficiency of migratory mechanisms (Peters, 1991). Archibold (1995) reported that changes to forest cover were far from clear with predicted changes to dry forest affecting between 0 and 71% of the present area, and moist forest changing somewhere between -10% and $+11\%$. Further uncertainty is added as within these global averages there will be marked regional shifts due to topography and soils. Stott (1994, p. 299) noted, 'there are still large uncertainties in all our estimates of global change, especially at the regional or biome levels, because our basic understanding of the relevant physics remains incomplete . . . and our actual predictions of greenhouse gas emissions remain approximate.' Sinha (1991) was also cautious because:

- CO_2 has increased over the course of the twentieth century but there is no noticeable effect on vegetation.
- Cloudiness in the tropics with higher temperatures and humidities may have a deleterious effect, e.g. on rice yields.
- Monsoonal circulation changes are uncertain.
- Too little is known about tuber crops.
- Pest and disease attacks can offset any gains.

The impacts upon tropical agriculture and food production are also far from clear. Various future scenarios have been investigated (Parry, 1990; Rosenzweig and Parry, 1993, 1994). In general terms global cereal production is expected to fall by 5% with a marked decline in developing countries of 9–11% and an increase of 10% in developed countries. Tropical agriculture is particularly vulnerable as crops are often grown near their tolerances of heat and water supply hence yields will decline under global warming unless water use efficiency is increased. This could be achieved through irrigation and fertilisers in which case world food supply could increase but only through more widespread adoption of technology. Even where climate improves in the tropics it may not result in higher production. Monsoonal precipitation is expected to increase, for example in Northern Australia, encouraging better rangeland production and improvements in cattle ranching. But as in many arid areas the major limiting soil nutrient is phosphorus; unless fertilisers are applied the improvement in climate will not necessarily lead to greater productivity (Stott, 1994).

Global warming then offers both threats and opportunities for the tropics. More widespread adoption of 'Western' technology would appear to be necessary for agriculture to benefit while the WWF (1993) proposed a range of strategies including larger nature reserves and dispersal corridors to aid spatial shifts of ecosystems in a warmer world.

CONCLUSIONS

The tropics cover an enormous area, support one-fifth of the world's human population (Stott, 1994) and contain over half of the world's biota. Major world biomes, rainforest, savanna and deserts, are to found within its boundaries under the influence of a complex and diverse climatic system. It is here that the sun's energy is converted into latent and turbulent heat transfer that powers the world's atmospheric circulation systems and where concern is growing for the global impacts of land degradation and rainforest destruction.

Climatic conditions vary from the warm, wet, equatorial regions where rain falls each month and temperatures seldom change, to the hyperarid deserts where rain is rare and temperatures fluctuate on both a seasonal and a diurnal cycle. Apart from the arid regions and areas of high elevation, temperatures are suitable for plant growth all year round and the major climatic constraint is water supply. In equatorial regions agriculture can be practised all year round but as one moves to higher latitudes greater seasonality is observed with the rainfall regime developing first into two wet seasons and then only one. As rainfall decreases perennial crops change to drought-resistant cereals and livestock become more widespread.

The region is, however, undergoing change. It is difficult to find areas that are not under the influence of human activity and patterns of vegetation are increasingly being determined by human actions not climate. The pressure on land and forest resources is enormous with expanding populations and attempts to exploit these lands through logging and agricultural development, particularly irrigation. Even climatic change appears to threaten with possible increases in pests, disease and drought. The tropics are then an area of vast climatic potential for plant growth and agriculture but a region that is in danger of being overexploited and mismanaged.

FURTHER READING

For those who require more detail there are a number of texts on the ecology (Walter, 1972; Whittaker, 1975; Golley, 1983; Loveless, 1983; Deshmukh, 1986; Evanari and Noy-Meir, 1986; Collinson, 1988; Archibold, 1995), climate and water resources (Griffiths, 1972; Schwerdtfeger, 1976; Nieuwolt, 1977; Takahashi and Arakawa, 1981; Balek, 1983; Hastenrath, 1985; Beaumont, 1989; Jackson, 1989; Agnew and Anderson, 1992; McIlveen, 1992; Reading et al., 1995) of tropical regions.

3
Tropical Soils

J.K. Coulter

SOIL FORMATION

Soils are formed by the weathering and decomposition of rocks and mineral particles, and by the addition to this weathered inorganic material of raw organic matter derived from plants and the decomposition of the latter, mainly by microbial activity. Plants contribute fresh organic material to the surface of the soil by leaf and stem fall and to the body of the soil by the death of their roots. Freely drained tropical soils, even under closed forest, do not normally show any considerable surface accumulation of raw or slightly decomposed plant remains, such as may be seen on some temperate forest zone soils. This is because the fresh organic material is speedily attacked by insects and other fauna and incorporated in the topsoil, where it is decomposed. Under aerobic conditions, and provided there is sufficient moisture, decomposition is rapid, much of the material being used in the respiration of the smaller soil animals and microorganisms and lost as carbon dioxide and water but the remainder being converted into the heterogeneous mixture of decomposition products and substances synthesised by microorganisms which is known as humus.

Rocks and rock fragments are broken down into smaller particles without change in chemical composition by a variety of physical processes. These include grinding in moving ice masses, the expansion of water freezing in cracks and crevices, the differential expansion and contraction of mineral components during diurnal temperature changes, the dissolution of cements binding mineral particles together, the abrasive action occurring between particles in flowing water (or due to wind-blown sand in dry areas) and the growth of plant roots into cracks and fissures.

Under aerobic conditions chemical weathering is mainly brought about by percolation of water containing dissolved oxygen, carbon dioxide, and organic acids derived from plant exudates and from the decomposition of organic matter. The processes of oxidation, hydration, hydrolysis and solution are involved. Ferrous and sulphide ions are oxidised. Some mineral surfaces absorb water molecules to form hydrated compounds. Hydrolysis results in the production of new mineral species with the splitting off of silica and the production of simple bases, but in the presence of dissolved carbon dioxide yields carbonates rather than hydroxides. Water containing carbon dioxide and organic acids dissolves the easily soluble components of the original minerals and also some of the products of hydrolysis. Depending mainly on the amount and rate of water movement through the soil, some of the products may be leached out of the soil and lost to the groundwater and rivers, or they may be precipitated lower down in the profile. The easily soluble cations (sodium, potassium, calcium, and magnesium) and associated anions (sulphates, chlorides, nitrates, carbonates and bicarbonates) are readily leached, although some cations may be held as exchangeable bases on the surfaces of colloidal clay and organic matter. Dissolved silica may also move out of the soil and ferric iron may be reduced to ferrous compounds and move down the profile, but is usually reoxidised and redeposited as hydrated ferric oxide.

During the development of most soils colloidal secondary clay minerals, which are more or less resistant to weathering, are formed. These are complex silicates in which silica is combined with aluminium and some other cations, notably magnesium, potassium and iron. They are formed either by physical, and relatively small chemical, alteration of the minerals of the parent material, or as a result of the decomposition of the original minerals and subsequent recombination of some of their constituents. The type of clay formed depends partly on the nature of the original minerals but largely on the intensity of weathering which is mainly governed by the amount and temperature of the water percolating through the soil.

Factors influencing soil formation

Tropical soils vary greatly owing to the wide variations in the factors of parent material, climate, vegetation and topography, which influence the soil-forming processes outlined above, and to differences in the time during which these processes have been operating.

Parent material

The parent material may influence the nutrient status, reaction, texture and structure of the soil. Under well-drained conditions, iron-rich parent materials, such as basalts and dolerites, give soils high in iron and of good structure, while parent materials low in iron but high in quartz, such as granites, give weakly structured, very erodible soils. The parent material may also affect the kind of clay that is formed, but where weathering and leaching are intense kaolinitic clays are formed irrespective of the nature of the parent material. Some of the sedimentary rocks, composed of materials which have already been through at least one cycle of weathering and transport, contain little weatherable minerals and therefore tend to give soils of poor nutrient status.

Climate

Temperature and rainfall markedly influence soil formation. In a tropical climate, soil and water temperatures are relatively high throughout the year and, where rainfall is high and reasonably well distributed, soil-forming processes can operate much faster than in a temperate climate, where chemical weathering and microbial activity are diminished by low winter temperatures and leaching is restricted by the transpiration of vegetation in summer. As rainfall has a predominant effect on the rate of chemical weathering and the amount of leaching, differences in rainfall regime have had marked effects on tropical soils. Under arid conditions, where evaporation greatly exceeds rainfall, chemical weathering is largely confined to the depth of moisture penetration where it is discontinuous and slow. Soluble substances liberated by weathering are not leached but concentrated, by evaporation, in the surface layer, which is often alkaline owing to accumulation of bases. Under a monsoon climate with moderately good rainfall through-leaching only occurs during part of the rainy season when precipitation exceeds evaporation, so that bases are not removed excessively from the profile and the soil reaction is neutral to slightly acid. Under a wet equatorial climate rainfall exceeds evapotranspiration for much of the year and soils become thoroughly leached and markedly acid. As leaching increases, silica is removed proportionately more than sesquioxides so that soils formed under hot, wet conditions have a low silica/sesquioxide ratio in the clay complex. Whenever there is any considerable leaching, kaolinitic clays, with a cation exchange capacity of only 2–4 milliequivalents (meq)/100 g are formed, but if leaching has not been very appreciable some illitic and montmorillonitic clays (with exchange capacities of 10–40 and 100–150 meq/100 g respectively) may also be formed. However, predominantly montmorillonitic clays are formed mainly in low-lying areas receiving seepage water high in bases, or in younger soils over basic rocks or limestone in lower rainfalls.

In considering the effect of climate on soil

formation, it must be borne in mind that many tropical land surfaces are very old and that the climate under which the soil was formed (palaeoclimates) may have been very different from that of today. For example, deeply weathered soils formed during past periods of much wetter climate are found in areas that are now relatively dry.

Topography

Topography can have a dominant influence on soil formation, largely overriding that of climate, in low-lying areas where water accumulates to produce permanent swamps or seasonal flooding, as described in the section dealing with soil classification later in this chapter. It also has an obvious effect on steep slopes where erosion, or movement of the surface soil by colluviation, results in the development of shallow lithosols with little root room. However, this also leads to rejuvenation of the soil by removing weathered material so that soils on steep slopes may have a higher nutrient content.

In many undulating landscapes, a regular and repetitive sequence of different soils, known as a catena, is found as one goes from ridge-top to valley bottom. This sequence of soils is due to changes in the parent material and/or differences in the site characteristics of gradient and drainage occurring down the side of the valley. If the interfluvial uplands have not been eroded, the soil on them can be deep, although where the upland area is broad and flat much of it may be seasonally wet and there may be a hard pan in the profile, as occurs over much of the 'miombo' area of Tanzania. On the other hand, on the tops of eroded ridges there are outcrops of rock and the soil on the upper part of the slope, which is likely to be shallow if the gradient is steep, is derived from the sedentary weathered products of the underlying parent rock. In the middle part of the slope the parent material may be a somewhat different colluvium which has moved down the slope under the influence of gravity and overlies the parent rock of the upper slope soil. The mid-slope soil may also receive drainage water moving laterally through the soil which carries with it dissolved salts from higher ground, thus influencing the base status and reaction of the colluvial soil.

In the valley bottom the parent material is likely to be alluvium, which may be of nearby origin or, in the case of larger river valleys, may have been transported from considerable distances and be markedly different from that derived from the local rocks.

There are commonly differences in drainage down the slope which are reflected in the colour of the soils. Good drainage and aeration on the upper slopes give rise to a red colour due to the presence of non-hydrated iron oxide, haematite. In the middle and lower slopes, which receive seepage from higher up, drainage is slower and the soils remain moist for the greater part of the year, resulting in the production of more hydrated iron oxide (goethite or limonite) and giving the soil a yellow or yellowish-brown colour. On the lower slopes and in the valley bottom, drainage may be impeded or the soil waterlogged for all or part of the year. This results in a bluish- or greenish-grey colour due to reduced iron compounds in horizons that are permanently saturated with water, or to reddish mottling due to the presence of some oxidised iron in horizons subject to seasonal fluctuations of the water table. Poor drainage on the lower slopes may be due to the formation of an iron pan (plinthite) as a result of the seeping out of water containing iron which has drained from higher up the valley side.

Vegetation

The different types of natural vegetation, which are broadly related to rainfall, influence the soil through differences in the amount of plant material they return to the soil, in their rooting habit and depth, and in the extent to which they act as a buffer against leaching by holding and circulating nutrients. Thus, the very great total weight of vegetation of closed forest, which may be as much as 375 t/ha of dry matter, results in a large contribution of organic material to the soil, in the form of leaf and stem fall and dead roots, which is reflected in the relatively high humus content of forest topsoils. The undisturbed forest ecosystem also operates on a virtually closed nutrient cycle, since the nutrients released by the decomposition of the soil organic matter are speedily reabsorbed by the mass of, mainly shallow, roots with the result that leaching losses are small and are probably offset by small amounts of nutrients drawn up from the subsoil or contributed from the atmosphere. However, on the majority of highly weathered and highly leached tropical forest soils almost all of the nutrients in the closed cycle are in the vegetation and the topsoil, the subsoil being very poor in nutrients. Once the closed cycle is broken by clearing and cultivation of the land, the fertility may be rapidly lost or used up. In contrast to the forest, the total weight of the predominantly grass vegetation of the savanna is quite small – probably not more than about 12 t/ha of dry matter – and most of this is grazed by animals or burnt in the dry season, so that the main contribution to soil organic matter comes from the dead roots. Consequently, the humus content of most savanna soils is much less than that of the undisturbed forest soils.

Time during which soil-forming processes have operated

The nature and distribution of the soils may be influenced not only by the existing topography, but also by changes that occurred during the long history of the landscape. Some geologically very old formations, such as the basement complex of Pre-Cambrian rocks which occurs over much of Africa, were long ago eroded to an almost flat surface, or peneplain. These were subsequently uplifted, or were subject to markedly increased rainfall, with the result that the rivers, which formerly flowed gently over almost flat land, were enhanced in volume and vigour and began to erode the peneplain. Eventually this led to most of the landscape consisting of young valleys with remnants of the old peneplain remaining on the flattish tops of the hills and ridges. These peneplain remains, which sometimes consist of ironstone crusts mantled with old, highly weathered soils, often form the uppermost parts of catenas. Much of the land surface of the tropics is very old and therefore the time during which parent materials have been subjected to weathering is usually longer than in the temperate zone where much of the land surface has been exposed only since the end of the last glaciation.

Because of the variation in the factors just discussed, tropical soils are very diverse, but most of them, apart from those in volcanic and alluvial areas, are highly weathered and much leached and have low or non-existent reserves of weatherable materials. They differ in nature and behaviour from temperate zone soils primarily because they have resulted from more intensive and prolonged action of the soil-forming processes.

SOIL PROPERTIES AND SOIL FERTILITY

Soil fertility may be defined as the ability to grow crops but this ability is dependent on the crop in question, the weather and thus the risk factors; economics of production have a major influence on the crops to be grown (Tinker, 1989). Soil fertility is thus part of a biological system in which soils, climate and plants interact. Of increasing importance is the concept of sustainability, that is the ability of the soil to maintain or increase productivity levels into the indefinite future. This is of particular significance at the present time when there is much arable rain-fed farming in the soil exploitation stage whereby agriculture is mining the soil without replacing plant nutrients (Jacks, 1956).

There are three important factors relating soil properties to soil fertility. The first group comprises the site factors which include the nature and depth of the profile, soil structure and soil texture; site factors are usually difficult and expensive to change. The second group consists of chemical factors which can be classified as management factors and which can be changed rather easily, though not necessarily economically, in many tropical countries. The third group relates to the biological factors that include soil organic matter and the useful and damaging soil organisms that are related to this and to the soil characteristics generally.

Site factors

Depth and nature of profile

The depth and nature of the profile have obvious effects on fertility. Very shallow soils are generally unproductive since they provide little root room for crop anchorage and extraction of nutrients and water, and they are usually either waterlogged or hold too little moisture. A good depth of soil is of particular importance in areas where the rainfall is seasonal and unreliable, since only a deep soil can hold an appreciable amount of water to supply crops in a dry spell. Annual crops rarely root below 1.5–2 m and therefore cannot use more than 2.0–2.3 m of soil, but some perennial crops and trees can root to at least 4.5–6.0 m where there is a pronounced dry season. Features of the nature, sequence, position and thickness of the soil horizons may be of importance, obvious examples being the occurrence of an anaerobic, gleyed horizon or of a layer of ironstone gravel or hardpan. The illuviation of clay to form lower horizons of clay accumulation may be moderate and beneficial by increasing water storage capacity, or excessive and thus having adverse effects on root development and the movement of air and water.

Soil texture

Apart from gravel and stones, the soil mineral particles are conventionally classified, on the basis of their equivalent spherical diameters, into size groups according to various grading systems. Thus, the system adopted by the International Society of Soil Science divides them into coarse sand (2.0–0.2 mm), fine sand (0.2–0.02 mm), silt (0.02–0.002 mm) and clay (<0.002 mm). There are several other systems including the widely used American one, which has seven grades and prescribes 0.05 mm as the upper limit for silt – a limit quite commonly used for this fraction nowadays. In the field the particles, especially the smaller ones, do not necessarily exist separately but, as described in the next section, are more usually bound together to form aggregates.

The soil texture – described in terms such as sand, sandy loam, silt, silty loam, clay loam and clay – relates to the relative proportions of sand, silt and clay in the soil. These influence the aggregate stability, permeability to air and water, drainage characteristics, water-holding capacity, ease of cultivation and nutrient status of the soil. Coarse sandy soils show little aggregation but have relatively large spaces between the particles, giving free movement of air and water but a low water-holding capacity. Fine sandy soils form aggregates which slake easily on wetting to form a rather impervious surface cap and thus they may be somewhat poorly drained and difficult to manage. Soils with much silt also form unstable aggregates and, as the pores between the fine particles are narrow, they have a high water-holding capacity but tend to suffer from impeded drainage. Clayey soils, with fine colloidal particles and narrow pores, have a high water-holding capacity, are rather impermeable to air and water and do not drain freely, but the latter disadvantages may be lessened by the development of stable aggregates and the occurrence of cracks and channels extending into the subsoil. Certain clays swell and become sticky on wetting and shrink to form hard clods when dry so that they are difficult to cultivate satisfactorily except within a limited range of moisture content.

Soil structure

The aggregation of the particles of sand, silt and clay into crumbs, clods or peds and the distribution of the pore spaces, cracks, channels or planes of weakness between them, is known as soil structure. This is of great importance for crop production because the pore space distribution largely determines whether the physical condition of the soil is favourable or unfavourable to root growth.

Roots can only ramify freely through the soil if there are pores and channels of a certain minimum size into which they can grow. Roots absorb oxygen and produce carbon dioxide, and in order that the latter may be removed and the oxygen replaced by diffusion there must be a continuous system of air-filled pores extending from the soil surface throughout the root zone. The supply of water to the roots is also effected through the pores and a productive soil must have a fairly even distribution of pore sizes capable of holding water available to crops. These requirements, as indicated by Russell (1958, 1971), mean that in a wet, well-drained soil (which is a requisite for good growth) there should be a proportion of interconnected coarse pores to permit water from heavy rainstorms to percolate into the soil instead of running off the surface, to allow adequate aeration and to provide spaces into which the roots can grow. There should also be a proportion of medium-sized pores, of diameter down to 0.003 mm, which can hold water available to the roots. In addition, there will be finer pores which, although holding water too tightly for it to be extracted by the roots, will affect the ease of cultivation of the soil.

Several mechanisms are concerned in the creation of soil structure. First, there is the shrinkage that accompanies the drying of moist soils containing a certain minimum amount of clay. An initial linking of individual soil particles may be brought about by dipolar water molecules attaching themselves to charges on the clay and associated cations to form films around the particles; as drying occurs these films become thinner, drawing the particles together until they adhere to form aggregates. In most soils the sand and silt particles are drawn into aggregates by being incorporated into a network of linked clay particles. Cracks and channels are also formed through the body of the soil when it shrinks on drying. Roots play a part in this process by drying the soil in their neighbourhood, causing shrinkage and the formation of pores which may be wide enough to admit new roots. Second, channels result from the penetration, thickening and subsequent death and decomposition of plant roots. Third, there is the activity of soil animals, both in burrowing through the soil and in passing soil through their bodies. Earthworms certainly aid in structure formation, and their casts, even if voided underground, are better structured than normal soil, probably due to the intimate mixing of mineral particles, humus, salts and, possibly, bacterial gums during passage through the worms' gut. Termites also have important effects through their mound-building activities. Finally, cultivation, especially in conjunction with weathering, can create structure. Tillage loosens compact soil, leaving large spaces between clods and crumbs. Moist clods, provided they contain sufficient clay, will shrink and crack on drying. When dry clods swell on wetting this may also produce cracks, due to uneven swelling in different parts of the clod and to the pressure exerted by absorbed air which is displaced by wetting but is entrapped by water films (Russell, 1971).

There will not usually be a system of coarser pores adequate for air, water and root movement unless a large proportion of the soil is in the form of water-stable aggregates. The formation of these aggregates and their stabilisation is not fully understood, but is largely brought about by the deposition or absorption on or between the particles of clay of certain kinds of colloidal organic matter and hydrated oxides of iron and aluminium which, on drying, cement the particles together in a fairly stable manner, so that they are not readily loosened on rewetting. The relative importance of these three kinds of material is not known since they do not act independently, being closely associated in the clay–organic complex. Many workers have found positive correlations between the degree of stable aggregation and the clay or iron oxides content of the soils. However, Deshpande *et al.* (1968) could find no evidence for an effect of iron oxide in the soils they examined and their data indicated that free aluminium oxides may play a more important role. Total soil organic matter and water-stable aggregation are seldom well correlated because aggregation is partly due to the action of the other materials mentioned, and also because only some of the components of the organic matter are active in effecting stabilisation (Griffiths, 1965). Certain polysaccharide gums, which are produced by soil organisms concerned in the early stages of decomposition of organic matter, contribute to structure and its stability in many cultivated soils. These materials may be transient intermediate products but, even if they are short-lived, it is reasonable to suppose that new polysaccharides are being continually produced provided that suitable organic material is available as a substrate for the microorganisms. Non-polysaccharide components of the organic matter evidently also contribute to structure and its stability. Fungal hyphae associated with the early stage of the decomposition of plant remains can also mechanically entangle soil particles to form fairly stable aggregates. As they are short-lived their effects are ephemeral, but they may be more important where the mycelium is being produced continually as a result of regular addition of plant remains to the soil, such as occurs under grassland and forest.

Some tropical soils possess a good and stable structure, because the particles are bound together into aggregates by strong cements of ferric and aluminium oxides. Such soils are resistant to structural degradation and erosion and can be kept under arable cropping for some years without marked deterioration in their physical properties. Other tropical soils do not contain the sesquioxide cements needed to strengthen the colloidal properties of their kaolinitic clays and are unable to form a lasting stable structure. In such soils, the effects of roots of fallow vegetation and crops (especially deep-rooting ones) and of insect and worm burrows are likely to be of particular importance in aiding water percolation and aeration. A fair to good organic matter content usually gives topsoils a crumb structure of moderate to weak stability which is difficult to maintain under arable cropping. Once the protective cover of forest or savanna vegetation is removed, the impact of high-intensity rainfall breaks down the crumbs on the soil surface. The resulting fine particles block the pores, forming a crust impeding the infiltration of water and slowing gaseous diffusion between soil and air until evaporation removes enough water to allow air to enter the pores again. Exposure to the sun, resulting in high soil temperatures, and increased aeration due to cultivation, both accelerate the decomposition of organic matter. Cultivation can also destroy structure by pulverising the soil when it is dry, and by puddling, smearing and compacting it if it is worked when too wet. Under traditional tillage with hoes and light animal-drawn implements these effects are not serious on most soils, but they can become so if intensified mechanical tillage is adopted without proper management.

Root systems assist in the maintenance of soil structure; the roots of annual crops are mostly concentrated in the top 30 cm whereas grasses like *Andropogon* have a considerable proportion of their roots at 30–60 cm (Pieri, 1992). Below ground, net production of plant material varies from about 400 to 1000 kg/ha/yr of dry matter whereas natural savanna may have over 2000 kg/ha/yr. Short-term fallows, while protecting the soil against erosion, are no more efficient than crops in improving soil conditions because their roots are shallow. Nicou and Chopart (1979) found that 10 years under fallow were necessary before there was a measurable improvement in physical properties. Chase and Boudouresque (1989) found that mulching with tree branches led to vegetative establishment on barren crusted forest soils, thus paving the way for re-establishment of trees and Mullins *et al.* (1990) reported that such hard setting soils could be improved by the addition of organic matter.

The maintenance of soil structure is thus assisted by providing a protective cover of vegetation or mulch and by minimal and timely cultivation, but under arable cropping it is commonly impracticable to avoid some tillage and impossible to give the soil a tilth strong enough to withstand the impact of tropical rainstorms. Nor is it practicable to keep the soil fully covered throughout the rainy season; the crop cover requires time to develop and the maintenance of an adequate mulch of crop or other vegetable residues is not normally practicable. The residues available on the arable land are usually insufficient, especially as some are often used for fuel or livestock feed, and transporting mulching material from uncultivated land involves much labour. Hence, under traditional annual cropping systems, a cropping period during which deterioration of the soil physical condition occurs is followed by a fallow period during which structure is partially restored. Under intensified cropping systems, however, several important structural changes take place which can be detrimental to the productivity of the soil. These include erosion and structural deterioration leading to capping and increasing bulk density. The most seriously affected areas are usually in the subhumid and semi-arid zones where the predominant soils may be *Alfisols* (q.v.) These have a weaker structure and less permeable properties, and often more fine sand and coarse silt, than the *Ultisols* and *Oxisols* of the more humid areas. Increased bulk density, measured by penetrometer, has been shown by Charreau and Nicou (1971) to increase from 150kg/m^3 in protected woodland savanna to 440 under millet in the surface soil and in the subsoil from 440 to over 1000. Soil compaction is particularly important where intensive mechanised farming has been developed. Lal and Okigbo (1990), for example, recorded a drop in equilibrium infiltration rates from more than 150 cm/h to approximately 30 cm/h with cultivation under traditional farming systems and to about 5 cm/h under intensive mechanised farming.

Chemical factors

Nitrogen

After water, lack of nitrogen is usually the greatest single cause of poor yields. Nitrogen fertiliser use in tropical agriculture is increasing but while East Asia uses 116 kg/ha of nitrogen on cropped land (Bumb, 1991) Sub-Saharan Africa uses only 8 kg/ha. Consequently most of the nitrogen used in the latter comes from mineralisation of soil organic matter, from symbiotic and non-symbiotic fixation of atmospheric nitrogen and from fixed nitrogen in rainfall.

Mineralisation of organic matter Much of the nitrogen used by tropical crops is provided by the mineralisation of soil organic matter. However, responses to fertiliser nitrogen in fertiliser trials are widespread and only exceptionally, as for example after a long forest fallow, does the soil organic matter release enough nitrogen during the cropping season for maximum growth.

Nitrogen in organic matter is made available for crop growth by mineralisation by microorganisms but the rate of release is relatively slow, that released during the growing season not usually representing more than 3% of the total amount present. In incubation tests on tropical soils, Greenland (1958) recorded 6% mineralisation of the total nitrogen content in samples from permanent grassland. Birch (1958), using a respirometer, found that his Kenya soils (0.70% N) gave about 2 ppm nitrate nitrogen on air drying after each re-wetting with an initial nitrate production of 180 ppm. In Malaya (RRI, 1967) 5–10% of the nitrogen could be mineralised by prolonged incubation.

The resistance of organic nitrogen complexes in soil to microbial decomposition is not well explained but Bremner (1968) suggests that the clay organic matter complexes in structurally stable microaggregates render some of the organic matter, within the aggregates, physically inaccessible to microorganisms. This is supported by the fact that clay soils in the wet tropics, as elsewhere, have more organic matter than sandy soils and that soils with allophane can accumulate exceptionally large amounts.

The rate of release of nitrogen from organic matter under tropical conditions is obviously extremely important. Under normal soil conditions inorganic nitrogen is continuously formed from organic nitrogen by mineralisation but some of this inorganic nitrogen is then reformed into organic forms by microbial processes. With substances having a very high carbon : nitrogen ratio inorganic nitrogen, including that added in fertiliser, will be immobilised. In temperate soils the critical ratio appears to be around 30, below which there may be net mineralisation of nitrogen. Williams (1969) stated that under an aerobic regime fresh organic matter, with an average concentration of less than 1.5% nitrogen, generally immobilised much of the available soil nitrogen. In tropical soils, however, immobilisation of nitrogen, when products of high carbon:nitrogen ratio are added to the soil, appears to be much less and Greenland and Nye (1960) found that incorporation of 10 t/ha of straw with a carbon:nitrogen ratio of about 70 caused very little immobilisation of nitrogen in a soil from Ghana. Slow decomposition, significant nitrogen fixation during decomposition, decomposition by termites and bacteria with an exceptionally low nitrogen demand, protection of the bulk of the straw from breakdown are all suggested as possible reasons. The significance of the latter should certainly not be overlooked, for the more resistant the straw to decomposition the smaller the number of flora attacking it and consequently the less the immobilisation of nitrogen. Under anaerobic conditions Williams found that the nitrogen requirement for the decomposition of rice straw in flooded soils was only one-third of that required in aerobic conditions and that such straw with a nitrogen content of 0.55% would release nitrogen during decomposition under flooded conditions.

Nitrogen is released by mineralisation of organic matter as ammonia but most plants use the nitrate form and the conversion of ammonia to nitrate, termed nitrification, is done mainly by two specialised groups of aerobic autotrophic bacteria. The first group, *Nitrosomonas*, oxidises ammonia to nitrite and the second group, *Nitrobacter*, oxidises nitrite to nitrate. Apart from the effect of the substrate, formation of nitrate is influenced by temperature and moisture levels. The optimum temperature ranges from 25 to 35°C; at 45°C, a temperature reached at the surface of bare soils in the tropics, nitrification ceases. Moisture levels have a large influence and in waterlogged soils nitrification is almost completely suppressed but ammonification is less affected and more than 100 ppm ammonia nitrogen may accumulate in such soils. Very low levels of moisture also stop nitrification. Robinson (1957) found that the critical level was about 80% of permanent wilting point in Kenya and Dommergues (1960) gave the minimum levels as pF 4.2 to 5.2.

There is much discussion in the literature on the effect of pH and liming on mineralisation. Guha and Watson (1958) incubated Malaysian soils with pH values of 4.6 to 5.4 and found that a comparatively high level of ammonification was followed by only a low level of nitrification but 4 tons per acre (10 t/ha) of calcium carbonate increased both pH and nitrification rate. The influence of air drying the soil before incubation was clearly revealed by the study as only 1.3% of the total nitrogen mineralised in one particular soil when air dried, but 5–10% when not air dried before incubation. Air drying before incubation renders many of the studies on incubation suspect, for nitrification appears to be much more influenced by this than is ammonification.

An interesting aspect of nitrification in tropical soils is the influence of certain plants in suppressing it. Wong (1964) found that water extracts of *Mikania cordata*, a Malayan compositae, could almost completely suppress nitrification. Rubber grown with this as a cover crop had depressed levels of nitrogen in the leaves, depressed rooting in the litter layer and had relatively small canopies. Meiklejohn (1953) reported almost complete absence of nitrates in soils after clearing grasses of the *Andropogonea* family, normal nitrification being suppressed for 2 to 3 years after clearing.

Nitrate flushes in tropical soils In tropical areas with marked dry seasons nitrate flushes of considerable size have been recorded in many areas. These flushes may be very large; in East Africa, for example, Semb and Robinson (1969) recorded values varying from 60 to 800 kg/ha of nitrate nitrogen in the top 40 cm. For Australia, Wetselaar (1962) quoted a much lower figure of 35 kg/ha of nitrates on a clay loam soil. The utilisation of this flush depends on the type of crop, the time of planting and the amount lost by leaching, the latter modified by soil texture since little leaching takes place until the soil has reached field capacity.

Nitrates can accumulate in the topsoil both by microbial decomposition of organic nitrogen compounds therein, early in the rainy season, and by upward movement of nitrates from the subsoil in capillary moisture during the dry season. Work by Simpson (1960) in Uganda, and

Robinson and Gacoka (1962) in Kenya suggested that some, but by no means all, of this accumulation cames from the subsoil. In Australia, Wetselaar (1962) showed that the maximum accumulation of nitrate at the end of a 6-month dry period was just below the soil crust, at a depth of about 2.5 cm. The accumulation at this depth exceeded 200 ppm nitrate nitrogen, decreasing to only a few ppm at 7.5 cm. Upward movement has also been confirmed by using chloride as a marker ion. Probably upward movement does not take place over distances exceeding about 45 cm, so that any accumulation below this depth makes little contribution to the nitrate at the surface at the end of the dry season. In East Africa, Semb and Robinson (1969), Stephens (1962) and Simpson (1960) all suggested that microbiological decomposition was the major source of the accumulation. It is generally understood that intensive drying has a partial sterilisation affect; re-wetting the soil by rain at the beginning of the wet season produces a flush of microbiological activity which releases large amounts of nitrates. This rapid increase is usually followed by an equally rapid fall as rainfall increases and the nitrates are leached into the subsoil. Leaching is normally regarded as a major source of loss though reduction to nitrous oxide and molecular nitrogen has been suggested occasionally (Simpson, 1960).

Although most reports indicate a major flush of nitrates on first re-wetting the soil after drying, Simpson (1960) could not confirm this in some Uganda soils but found a gradual accumulation of nitrates at the surface during the dry season as long as the moisture content did not fall below a critical level; Robinson (1957) suggested that this level is just below the wilting point in the Kikuyu red loam soils which he used. Simpson found that it was 88% of wilting point moisture in his Uganda soil. Griffiths and Manning in Uganda (1949) found that mulching the soil surface or shading led to much less accumulation of nitrates. Mulching reduces the surface temperature by 4–5°C and gives a much slower rate of drying and this may help explain the distinct depression of nitrate accumulation compared with that in the bare soil.

The available evidence suggests that movement of nitrates from the subsoil to the surface and nitrification close to the surface both contribute to the build-up in the top layers of the soil. During the dry season the surface 2–3 cm will become too dry for nitrification but it will continue in the lower, moister horizons and the nitrates could move close to the surface in the rising capillary moisture. This process would appear to be confirmed by the evidence of Wetselaar (1962) who showed that the accumulation just below the surface is accompanied by depletion of the subsurface at 15–25 cm; on the drying cycle there is, thus, a rearrangement of the nitrate concentration within the shallower horizons. Thus, nitrate build-up could continue for some time into the dry season if the soils are deliberately kept fallow since there are no plants to absorb the nitrate and the subsoil remains moist through lack of transpiration by plants. When moisture conditions in the subsoil are no longer suitable no nitrification will take place but redistribution could continue. Depth of sampling is thus all important, shallow sampling could indicate an increase but sampling to a 20–30 cm depth might reveal no overall change.

Rainfall Eriksson (1952) compiled worldwide data which showed that the rainfall usually contributed less than 10 kg/ha/yr of nitrogen. The average for tropical areas given by Allison (1966) was 8 kg/ha/yr, the figure for temperate areas being about 9 kg/ha/yr.

In Kenya, Bellis (1953) gave a value of about 5 g/ha/yr of nitrogen per mm of rainfall. In the monsoon areas Vialard-Gordon and Richard (1956) reported that 2.36 m of rainfall supplied 10.35 kg/ha/yr of nitrogen. An exceptionally high value of 53 kg/ha/yr was reported by Jones (1960) for northern Nigeria but subsequent checks have shown that this figure seems much too high. These figures suggest, therefore, that rainfall contributes 5–10 kg/ha/yr of nitrogen over the tropic areas generally and is thus likely to be a comparatively minor source of nitrogen supply, though it is actually about as much as the average supplied by fertilisers in Sub-Saharan Africa.

Symbiotic and non-symbiotic fixation These mechanisms undoubtedly form the major means of nitrogen acquisition by tropical soils. Meiklejohn (1955) suggested that such mechanisms must be quite efficient in tropical soils and quoted as support an example of maize being grown continuously for 16 years in Kenya without serious yield decreases. Greenland (1959), using data for tropical forest from Ghana and the Congo, calculated that the nitrogen increment in the soil and vegetation was between 50 and 150 kg/ha/yr and suggested that most of this probably came from leguminous shrubs and trees. However, the degree of nodulation in the vast majority of tropical legume species is unknown and Bonnier and Segier (1958), from their examination of a number of leguminous species in the Congo, concluded that in a tropical forest with a large return of organic matter to the soil, thus keeping up a high nitrogen level, nodulation is rare. In cultivated soils where the crop removes nitrogen and there is little return of organic matter, nodulation is considerable. This suggests that, as an area reverts to bush after cultivation, initial build-up of nitrogen may be rapid but slows down significantly as equilibrium levels are reached. Thus, figures derived from the average over a number of years cannot adequately define the capacity of the system to fix nitrogen in the initial stages.

Even with increased use of chemical fertilisers, biological fixation will continue to have a major role in supplying nitrogen for crops. In recent years much effort has gone into studying this process and promising results have been achieved at the molecular level in the manipulation of nitrogen-fixing organisms, though according to Dommergues and Ganry (1985) agriculture has yet to benefit from these advances in knowledge.

Giller and Wilson (1991) stated that there were relatively few nitrogen-fixing species shown to make a real contribution to tropical systems, the predominant group being the rhizobial association. The group of next importance is the blue-green algae, either as free-living species or in association with, for example, the aquatic fern *Azolla*. The third group is actinomycetous *Frankia* spp. which form symbiotic associations with flowering plants, particularly woody perennials of which *Casuarina* is a well-known example. The fourth group is the free-living nitrogen fixers such as *Klebsiella* and *Azotobacter* and organisms such as *Azospirillum* which live in the rhizosphere of certain plants.

Bergersen (1980) has described the methods of measuring nitrogen fixation. These include nitrogen balance studies, 15N-based techniques and the acetylene reduction method. Each method has limitations and Dommergues and Ganry (1985) stated that nitrogen-fixing estimates should be treated with caution especially when dealing with blue-green algae and rhizosphere systems. They pointed out, too, that some of the figures for *Leucaena leucocephala* of 600–1000 kg/ha/yr of nitrogen fixed were not valid.

In tropical soils most attention has been devoted to legume–rhizobium associations and to the blue-green algae which are described in Chapter 6.

There are a number of non-leguminous plants having nodules inhabited not by *Rhizobium* but by filamentous bacteria (*Frankia* spp.) which fix nitrogen and some species may be important in the tropics (Stewart, 1966; Silver, 1969). The genera with tropical distribution include *Coriaria*, *Discaria* and *Casuarina* (Kass and Drosdoff, 1970). Of these *Casuarina* is certainly very widespread, particularly on poor soils; one particular species *Casuarina papuena* is very distinctive on high nickel and cobalt soils derived from ultrabasic rocks as found in the Solomon Islands (Lee, 1969). Dommergues (1963) evaluated the fixation of nitrogen in dune sands at 760 kg/ha over 13 years, i.e. 58 kg/ha/yr. Some *Pinus* species with *mycorrhizal* associations are distributed in the tropics and Richards and Voight (1964) suggested that these fix nitrogen, a figure of 52 kg/ha/yr being given.

Fixation by free-living organisms The main genera of free-living organisms proven capable of fixing nitrogen are the blue-green algae, a number of bacteria and some fungi. Blue-green algae, better known genera being *Nostoc* and *Anabaena*, are of worldwide distribution (Stewart, 1966) and appear to be particularly abundant in the moist tropics, their contribution to the nitrogen economy of paddy soils being for long regarded as important. Stewart listed 39 species which fix nitrogen in pure culture, quite a number of which are found in the tropics. Singh (1961) reported that in India there was a profuse growth of many genera including *Anabaena*, *Calothrix*, *Nostoc* and *Tolypothrix* in paddy fields.

Blue-green algae are biologically very successful organisms and their distribution reflects their completely autotrophic nature as they are capable of synthesising all their biological requirements from carbon dioxide, water, nitrogen, and mineral salts. They appear tolerant of a wide range of temperatures, some species being found in Antarctica, others being best adapted to the 35–40°C temperatures of Indian rice fields. They are, of course, dependent on light as their source of energy for nitrogen fixation, which ceases when they are placed in the dark. Thus, they can only operate on the surface of the soil and shading by the growing crop impairs their efficiency, though too high a light intensity also impairs their fixation capabilities.

The amount of nitrogen actually contributed to crops in the field is a matter of much speculation. Measurements of their fixing capacity have been done in pure culture, in soil in flasks, pots and in the field. Stewart (1966) reported work in Japan where addition of *Tolypothrix* to the rice field gave the nitrogen equivalent of 25 kg of sulphate of ammonia/ha; a much higher figure of 900 kg/ha of nitrogen during a paddy crop was given by Singh (1961). Moore (1966) listed amounts, mostly in laboratory experiments, equivalent to 10–200 kg/ha of nitrogen over periods of 2 months to 1 year. Under cultural conditions in the laboratory the organisms are obviously capable of fixing relatively very large amounts of nitrogen. Such values appear to have little relation to what happens in the field and it is quite probable, judging by the responses which are obtained when fertilising paddy fields, that they contribute no more than 20 to 30 kg nitrogen/ha/yr. Roger and Watanabe (1986) suggested that the incorporation of *Azolla* into the soils of rice fields gave a worthwhile increase in the nitrogen status of the soil but considered that innoculation with blue-green algae was of questionable value.

In the 1970s there was much interest in rhizospheric nitrogen fixation by the bacteria occurring in the rhizospheres of many tropical grasses and cereals where they were able to derive energy from root exudates; estimates of up to 100 kg/ha/yr of nitrogen being fixed were given (Dobereiner and Campello, 1971; Dobereiner *et al.*, 1972). A similar association estimated to fix about 67 kg nitrogen/ha/yr was demonstrated for the rhizosphere of sugar cane. Subsequent investigations have shown that the rate is likely to be much lower than this and indeed the amounts may be negligible (Venkateswarlu and Rao, 1983).

Phosphorus

Most soils on old landscapes in the tropics have insufficient phosphorus for intensive agriculture and in a widespread series of trials conducted by the Food and Agriculture Organization of the United Nations (FAO) over 90% of the sites tested gave responses to phosphorus (FAO, 1968a). Total phosphorus levels are controlled by the nature of the parent materials, the degree of weathering and leaching, the amount of clay, the type of vegetation and the degree of waterlogging. Values may vary from more than 2000 ppm in soils formed on basic volcanic rocks to less than 50 ppm in sandy soils from granites. Man has had a great influence and old village sites may show 50-fold increases; on the other hand exports of groundnut and cotton seed removes large quantities and Enwezor and Moore (1966) calculated that, at that time, the export of these crops removed the equivalent of 27 000 tons of superphosphate from northern Nigeria. In temperate soils nearly half the phosphorus in the plough layer may have come from phosphate fertilisers (Cooke, 1967) but in many tropical soils virtually all of the phosphorus comes from native sources.

Soil phosphorus can be separated broadly into

inorganic and organic fractions. The latter can vary from 20 to 70% of the total (Nye and Bertheux, 1957; Stephens, 1970). Organic phosphorus cannot be used directly by plants; it has to be mineralised and this seems to be influenced by the carbon:phosphorus ratio; this is normally about 200–300 but in allophane soils it may be as high as 2000. However, a rate of mineralisation of organic matter of 3% per annum would supply only 1–10 kg/ha/yr, except where there are large amounts of organic matter as in newly cleared forest soils.

The inorganic phosphorus in the soil is in the form of minerals, usually sparingly soluble, or adsorbed on the surfaces of the clay particles. Most of the inorganic phosphorus occurs in the clay fraction. While apatite, as hydroxyapatite ($3Ca_3(PO_4)_2 \cdot Ca(OH)_2$) is the most widely occurring phosphate mineral in temperate soils, in most tropical soils the non-adsorbed inorganic phosphates are in the form of aluminium and iron compounds. Adsorption on the surfaces of iron and aluminium compounds is a source of retention in some tropical soils since such compounds have a large surface area. Thus, the nature and amount of those compounds have a profound effect on the behaviour of phosphate fertilisers added to soils. Bationo *et al.* (1989) have reported that as little as 20 kg P_2O_5/ha doubled yields of pearl millet on light sandy soils in Nigeria. By contrast dressings of up to 1200 kg/ha were required on andosols in Hawaii before maximum yields could be obtained.

There are four main factors concerned in the supply of phosphorus to plants (Williams, 1970). These are: the intensity factor, the quantity factor, the quantity–intensity relationship and one or more rate factors. The *intensity factor* may be regarded as the concentration of phosphorus in the soil solution. This is particularly important for quick-growing crops but less so for slow-growing crops or those which are not inhibited by intermittent periods of shortage, e.g. perennial grasses. The *quantity factor* is essentially the phosphate in the labile pool though there is obviously no clear-cut boundary between this and the very slowly soluble non-labile fraction. The rate of release from the non-labile to the labile form is affected by such factors as pH, aeration and waterlogging, temperature and moisture.

Quantity–intensity relationships, usually referred to as the Q/I relationships, express the buffering capacity of the soil and can be expressed in the form of sorption isotherms. The distribution of phosphate between the soil and solution phases will depend on the degree of saturation of the soil exchange complex. *Rate factors* concern the movement of phosphate ions to the root surfaces, mainly by diffusion, along a concentration gradient from a zone of higher to a zone of lower concentration. Roots can also grow towards zones of greater concentration; thus the interaction of structure, root distribution, and the effects of mycorrhiza have a strong influence on the accessibility of roots to sources of phosphorus.

As noted, many tropical soils, particularly those used for arable rain-fed agriculture, are low in phosphate and the use of phosphate fertilisers will be an essential part of improving their productivity. The fate of different kinds of fertiliser phosphates added to tropical soils is therefore of great interest. Since many such soils are acid or very acid the term 'strong phosphate fixation' is commonly used, often to imply that the soluble forms have been converted into insoluble forms. However, Wild (1950) suggested that 'phosphate fixation' be used to describe any change that the phosphate undergoes in contact with the soil which lessens the amount that plant roots can absorb, while phosphate 'adsorption' would be taken as the retention of phosphate at a surface.

In calcareous soils the fixation of phosphorus is due to the formation of a series of insoluble calcium compounds but in acid soils iron and aluminium compounds are the major reactants with phosphorus. Hsu (1965) suggested that the mechanisms involved in phosphate retention are reactions with amorphous aluminium hydroxides and iron oxides which are hydroxypolymers of aluminium and iron; these react rapidly by surface adsorption. However, strongly acid soils also have exchangeable aluminium and Al^{3+} in the soil solutions and thus precipitation as aluminium phosphate is possible. Even in the most acid soils, exchangeable Fe^{3+} does not exist so that precipitation of iron phosphate is unlikely. However, high acidity (pH < 2) is generated in the soil solution around granules of monocalcium phosphate and this can bring iron and other ions into solution (Mokwunye *et al.*, 1986).

The soils with the greatest ability to remove phosphate from solution are undoubtedly those with much amorphous material. Fox *et al.* (1968) have shown that when 44 ppm of 32 phosphate was equilibrated for 2 days with a Hawaiian andosol only 0.001 ppm phosphate remained in solution. About 1800 μg phosphate/g soil was required to adjust phosphate in the supernatant solution to 0.2 ppm. They ranked the phosphate retention capacity as: amorphous hydrated oxides > gibbsite–goethite > kaolinite > montmorillonite. While these soils need large amounts of phosphate to 'quench' the phosphate retention capacity, such large dressings have a prolonged residual effect lasting for at least 9 years after application (Kamprath, 1967).

West Africa soils have been classified by Juo and Fox (1977) on the basis of their clay mineralogy and phosphate sorption capacity (see Table 3.1).

Where precipitated aluminium phosphates control the phosphate in solution then the concentration of phosphate ions in the soil solution – the intensity – will be governed by the solubility product of the aluminium compound. From a soil management point of view, therefore, very acid soils with a small adsorption complex would respond best to higher quality rock phosphates or ammonium phosphates and light dressings of lime. Similar soils, but with a pH above 5.5, could be expected to respond well to small dressings of orthophosphates. Soils with much amorphous material will need larger amounts of phosphate initially, regardless of their pH, if they are to maintain a high phosphate intensity in the soil solution.

Because of their low solubility, phosphate fertilisers are scarcely leached and therefore tend to remain in the surface soil. In a series of lysimeter trials in Malaysia, Bolton (1968) showed that much of the applied rock phosphate or superphosphate remained in the surface 2.5 cm; some moved into the 2.5–10 cm layer but none beyond. In temperate soils, on the other hand, ploughing distributes the phosphate through at least the 15–20 cm plough layer. Concentration in the soil surface inhibits plant uptake when the soil dries out and hence the desirability of distributing the phosphorus throughout the rooting zone.

Potassium

Potassium is the nutrient cation in greatest demand by crops. The potassium in the soil solution, and that held in exchangeable form on the surface of the soil colloids, is considered to be directly available to plants. The 'fixed' potassium in interlayer spaces of certain 2:1 lattice clay minerals, such as illite and montmorillonite, is also slowly released and there is evidence that part of it is fairly readily available in tropical soils rich in these minerals because of a reversibility in the change from the 'fixed' to the exchangeable

Table 3.1 Phosphate requirements by soil type. (Source: Juo and Fox, 1977.)

Standard phosphate requirement (μg P/g soil)	Scale	Usual mineralogy
<10	Very low	Quartz, organic materials
10–100	Low	2:1 clays, 1:1 clays, quartz
100–500	Medium	1:1 clays with oxides
500–1000	High	Oxides, moderately weathered ash
>1000	Very high	Desilicated amorphous material

forms (Acquaye et al., 1967; Ahenkorah, 1970; Coulter, 1970). However, in most tropical soils the predominant clay mineral is kaolinite and the amount of 'fixed' potassium is insignificant. Non-available potassium may also be present in the crystalline matrix of certain materials such as micas and feldspars, from which it is slowly released by weathering, but prolonged weathering and leaching has left little of such weatherable minerals in most upland soils of the humid tropics.

The ease which with plants absorb exchangeable potassium depends on the ratio of exchangeable potassium to the cation exchange capacity of the soil, so that the same quantity of exchangeable potassium will be more available in a soil of low cation exchange capacity than in a soil of higher capacity. There is some evidence from investigations with a limited number of tropical crops and soils that potassium uptake is satisfactory if potassium saturation is above 2% or more of the total exchangeable bases. However, plants differ markedly both in their requirements for potassium and in the efficiency with which they can extract it, especially from soils low in this element. Boyer (1972) summarised the factors about the importance, characterisation and utilisation of potassium, especially in the humid tropical regions. He stated that exchangeable potassium was a fairly widely accepted method of evaluating availability of soil potassium to crops. A minimum level of about 0.10 meq/100g of soil has been established although this will vary with the crop and the level of productivity. The availability of potassium can also be expressed by the activity ratio:

$$\text{Activity ratio of postassium} = \frac{\text{activity of potassium}}{\text{square root of acitivty of calcium plus magnesium}}$$

Boyer (1972) suggested that there was evidence that the correlations between the potassium used by plants and the potassium activity ratio were better than those relating to exchangeable potassium. Landon (1991) quoted figures for Zimbabwe which showed various levels of potassium response for different soil types (Table 3.2). The soils of humid tropical regions usually have sufficient potassium after clearing the natural climax vegetation but some of the Oxisols may show early deficiencies; continuous cultivation over 3–5 years may give rise to potassium deficiency in many soils of the humid tropics. Leaching of potassium, which is minimal under natural vegetation, is higher on bare soil and under cultivation and potassium chloride, the most widely used potassium fertiliser, can leach readily. Bolton (1968), using a latosol in lysimeters in Malaysia, suggested that potassium chloride leached more readily than potassium sulphate and that losses of potassium could be lessened by increasing the calcium saturation of these largely aluminium-saturated soils.

Table 3.2 Potassium response levels in various soil types. (Source: Landon, 1991.)

Rating (for Central African soils)	Exchangeable K (meq/100g by ammonium acetate extraction)		
	Sands	Sandy clays	Red brown clays
Deficient (response to K likely)	<0.05	<0.1	<0.15
Marginal (some response likely)	0.05–0.1	0.1–0.2	0.15–0.3
Adequate (response unlikely but maintenance of K usually desirable)	0.1–0.25	0.2–0.3	0.3–0.5
Rich (no K required)	>0.25	>0.3	>0.5

Sulphur

Sulphur, which is required in approximately equivalent amounts to phosphate for crop nutrition, is absorbed by plants almost entirely as the sulphate ion, but most of the sulphur in surface soils is in an organic form and its accumulation and mineralisation is related to the accumulation and decomposition of the soil organic matter. Organic sulphur contents are highly correlated with organic carbon and nitrogen and the ratio is approximately 140:10:1.3 (Probert and Samosir, 1983). Smaller quantities of inorganic sulphate are also normally present, either in water-soluble form or adsorbed on kaolinites or hydrous oxides of iron and aluminium, but that adsorbed on the oxides is only sparingly available to plants. Inorganic sulphur may also occur in varying amounts in subsoils as gypsum or pyrite, depending on the drainage conditions. The sulphate ion is readily leached and both total and available sulphur tend to be low in intensely weathered and leached soils.

The pattern of sulphur behaviour is illustrated by the work of Bromfield (1972) in Nigeria which compared the results of clearing and cropping with and without the use of sulphur-containing fertilisers. None of the applied sulphur accumulated in the surface 10 cm but very large quantities of sulphate sulphur were adsorbed by the subsoils (18–32 cm). In the natural fallow virtually all of the sulphur was in the organic form. On clearing this was released by oxidation of the organic matter and sulphate sulphur accumulated in the subsoil.

In most tropical areas rainfall is the only source of additional sulphur except that provided by fertiliser use. This fertiliser source is decreasing because of the tendency to use high-analysis fertilisers which contain less impurities than before. Where there is little industry, as in most tropical countries, sulphur concentration in rainwater is closely related to distance from the sea. Within 1 km of the coast the concentration may be around 1–2 ppm. Beyond 100 km the sea has no influence. Depending on the rainfall, the amounts of sulphate sulphur deposited per year may be as low as 1.14 kg/ha, as in northern Nigeria, to 13.6 kg/ha, as on coastal areas of Hawaii (Fox et al., 1983).

Formerly, sulphur deficiencies in crops were often not identified because of the use of sulphur-containing fertilisers in experiments, such as sulphate of ammonia and single superphosphate, but deficiency symptoms and responses to sulphur have now been observed in many parts of the tropics. Thus, responses have been reported from parts of Brazil, India, West Africa, East Africa and Central Africa, where Coulter (1970) considered that sulphur deficiency was possibly as important as phosphate deficiency. Sulphur deficiency has been reported from a wide range of crops – coffee, coconuts, rice, pastures – in Papua New Guinea on several soil types, including immature volcanic ash soils, and acid sandy soils formerly under grassland (Vance et al., 1983). Sulphur deficiencies appear to be especially associated with savanna soils, presumably because they contain less organic matter than forest soils and they are associated with long periods of annual burning when the sulphur in the vegetation is volatilised as sulphur dioxide. In the West African savannas, widespread responses to sulphur have been obtained with cotton, groundnuts and forage legumes but not with cereals, whereas on some plateau soils in Zambia marked increases in maize yields have resulted from applied sulphur (Vogt, 1966). Forest soils with much organic matter are usually adequately supplied with sulphur but this is not invariably so; for example, in the tea growing areas of Malawi, where virgin soils contain 160 mg/kg of total sulphur, and 1–5 mg/kg of available sulphur (Grant and Shaxson, 1970), sulphur deficiency in tea has long been known (Storey and Leach, 1933).

Calcium, magnesium and soil reaction

Both calcium and magnesium are essential elements for higher plants and they also have an indirect effect on plant nutrition through their influence on soil pH. Soils may hold reserves of these elements in unweathered, or partly weathered, mineral particles and as cations in the inner layers of some clay minerals. These sources of calcium and magnesium may be slowly released to replenish the available supplies held as exchangeable cations or present in the soil solution. As both elements are readily leached, soils of higher rainfall areas tend to contain less of them. The highly weathered and leached Oxisols and Ultisols of the wet tropics contain only small amounts of both total and available calcium and magnesium compared with most temperate soils. However, highly acid soils are not confined to the humid tropics; extensive areas of the Sahelian region in West Africa are covered with red, sandy, strongly leached and acid soils (Wilding and Hossner, 1989). These soils are generally strongly acid, infertile weakly buffered kaolinitic systems, sands and loamy sands in texture formed on parent material that has been highly weathered and leached in wetter palaeoclimates. Base saturation is of the order of 10–20% and aluminium saturation 70–80%; pH of the top soil is in the range of 4.8–5.1.

Calcium ions are the predominant exchangeable cation in neutral and slightly alkaline soils. In strongly acid soil, exchangeable aluminium is the predominant cation. The acid nature of soils containing exchangeable aluminium results from the hydrolysis of aluminium complexes in the soil solution with the production of hydrogen ions. The hydrogen clays in acid mineral soils are unstable and decompose to give siliceous acid and aluminium ions (Kamprath, 1972). Above pH 5.6 there is relatively little exchangeable aluminium in most soils. Adverse effects of soil acidity on crop growth may thus be due to calcium or magnesium deficiency, aluminium or manganese toxicity, or to more than one of these factors, depending on the crop and soil. Many tropical crops are tolerant of considerable soil acidity but, where adverse effects have occurred, there is evidence that they are often due to a high percentage of exchangeable aluminium, resulting in aluminium concentrations in the soil solution that restrict root growth and cause aluminium to accumulate in the roots to an extent which impedes the uptake and translocation of phosphate, calcium and potassium. It has been shown that restricted root growth of cotton on many acid soils is attributable to aluminium toxicity (Adams and Pearson, 1970) and root growth of sorghum and maize is reported to be inversely related to the amount of exchangeable aluminium present (Ragland and Coleman, 1959). Kamprath (1970) found that the growth of cotton was decreased by 10% exchangeable aluminium saturation of the effective cation exchange capacity, of soya by 20%, and of maize by 45%. Abruna and Vicente-Chandler (1967) reported that sugar cane growth was severely depressed on a soil with 70% exchangeable aluminium saturation and that liming to reduce this to 30% gave a four-fold increase in growth. On the other hand, some crops which grow well on acid soils, such as pineapple, coffee, tea, rubber, cassava and upland rice, also tolerate high levels of exchangeable aluminium. Not only are there substantial differences in plant species in their tolerance of aluminium toxicity but there are also great differences within species and thus there has been considerable attention in plant breeding to select for aluminium-tolerant varieties (Laudelout, 1989). This is particularly important where the subsoils are strongly acid for it is difficult to improve these by liming as the calcium ions move relatively slowly down the profile. Where subsoil acidity is a problem the roots of susceptible species or varieties do not fully utilise these horizons and so suffer from drought.

Certain crops, such as groundnuts and sisal, have long since been known to have an appreciable demand for calcium as a nutrient and have frequently been found to respond to small dressings of lime when grown on acid soils, while other crops, such as tobacco and oil palms, are susceptible to magnesium deficiency. More recently, it has been found that, on acid soils with much exchangeable aluminium in Colombia, upland rice, cassava, cashew, mango, citrus and some forage legumes all respond to small dressings of 150–500 kg/ha of lime supplying calcium as a nutrient while heavier dressings of 2 t/ha or more depressed yields of some of the crops, notably cassava and cashew.

In temperate soils liming has long been used to ameliorate acidity and aluminium toxicity. There is much less experience on liming tropical soils.

Lathwell (1979) has described a series of liming experiments done in Oxisols and Ultisols in Puerto Rico, Brazil, Ghana and Peru to measure responses of food crops to liming. The application rates were applied to achieve pH values from less than 5.0 to more than 6.5. Some of the Oxisols in Puerto Rico, even when very acid, have low exchangeable aluminium but have rather high levels of exchangeable and easily reducible manganese. Increased calcium supply and reduced manganese availability appeared to be the reason for increased yields of maize and snap beans with levels of only 0.45 t/ha of lime. On Ultisols with levels of exchangeable aluminium of 50% or more in the subsoil, maximum yields were obtained when the aluminium saturation fell below 20%. On Brazilian Oxisols, on the other hand, with 30–80% aluminium saturation, liming rates of 1–8 t/ha ground limestone, almost doubled yields of maize and sorghum. Even at the lowest level of 1 t/ha, residual effects lasted throughout six crops.

The evidence of the adverse effects of aluminium toxicity suggests that on the more acid soils the need is to neutralise exchangeable aluminium by liming to pH 5.5 rather than to pH 6–7. Kamprath (1970) proposed that the rate of liming for acid mineral soils should be calculated as $1.5 \times$ the topsoil content of exchangeable aluminium in milliequivalents, or 1.64 t/ha of calcium carbonate equivalent/milliequivalent of exchangeable aluminium. This recommendation appears to have been adopted with some success in South America but the economic rate will depend on the tolerance of the crops concerned to aluminium and for some crops liming may only be needed to correct nutrient deficiencies of calcium or magnesium. The aim must be to determine the minimum amount of lime required for any given crop and soil situation as in many places, especially in Africa, lime has to be transported over great distances and is costly. Overliming to pH values greater than 6–7 can be harmful by decreasing phosphate availability and inducing deficiencies of zinc, boron or manganese.

There is evidence that organic matter plays an important role in ameliorating the adverse effects of aluminium in acid soils. Bationo *et al.* (1987) showed that crop residues greatly increased yields of pearl millet in Niger, particularly when used with fertilisers. Crop residues reduced the aluminium saturation from 48 to 20%. Hargrove and Thomas (1981) showed that 10% organic matter reduced the aluminium concentration at pH 4 from 1.75 cmol/kg to 0.25. It is possible that one of the causes of the decrease in yields under continuous cultivation, even when fertilisers are used, is the loss of organic matter and the consequent increase in toxicity of exchangeable aluminium.

Trace elements

Deficiencies of trace elements, or micronutrients, have seldom been reported as limiting crop production in traditional systems of subsistence farming. In these systems yields are usually low and fertility is maintained at a moderate level by fallows, by the use of animal manures and composts, or by inflow of nutrients in irrigation water or wash from higher land (Drosdoff, 1972). But there is evidence that such deficiencies can be important under more intensive and continuous rotational cropping or monoculture. It is therefore likely that if more intensive farming systems are introduced, micronutrient deficiencies will become more widespread. This tendency could be accentuated by an increasing use of high-analysis fertilisers which contain less micronutrients as impurities than the older forms.

The occurrence of trace element deficiency symptoms in the field often, but not invariably, reported to be accompanied by smaller yields, has most commonly related to plantation crops. In most cases it has been found practicable to correct the trouble by applying a small amount of the appropriate element to the soil or by spraying it directly onto the crop.

Deficiencies of copper, zinc, iron and molybdenum in citrus in various parts of the tropics and subtropics have long been known (Stiles, 1961) and in coffee widespread deficiencies of iron, molybdenum and boron have been reported (Muller, 1958). In rubber, manganese deficiency is widespread and iron deficiency of more limited occurrence in Malaysia, while zinc deficiency has been reported from Sri Lanka (Shorrocks, 1964). Other examples reported are deficiencies of boron in oil palm (Hartley, 1989) and sisal (Lock, 1969); zinc in cocoa (Greenwood and Hayfron, 1951), tung (Webster, 1950) and pineapple (Coulter, 1972); manganese in sugar cane (Lee and McHargue, 1928) and copper in pineapple (Coulter, 1972). Less commonly there have been reports of trace element deficiencies in annual crops, examples being zinc in maize in Brazil (Igue and Gallo, 1960), Puerto Rico (Drosdoff, 1972) and Zambia (Coulter, 1970); molybdenum in groundnuts in Senegal (Martin and Fourrier, 1965) and Java (Newton and Said, 1957); and boron in cotton in Brazil (McClung et al., 1961), Tanzania (Le Mare, 1970) and Zambia (Coulter, 1970).

The last two references indicate that boron deficiency is widespread in East and Central Africa where it has also been observed on wattle, pines, eucalyptus, coffee and sisal. Dramatic responses by maize, sorghum, soya bean and common beans (*Phaseolus vulgaris*) to boron application have been reported from Colombia (CIAT, 1974). There have been relatively few reports of toxicity resulting from an excessive amount of a micronutrient in an available form and these most commonly refer to manganese which may cause trouble on soils more acid than pH 5.5 under wet conditions, sometimes producing symptoms of iron deficiency by interfering with uptake or metabolism of this element. Coulter (1970) considered that manganese toxicity was potentially a serious problem on the red clays and clay loams of Central Africa which were rich in manganese readily mobilised by the temporary waterlogging that frequently occurred during the rainy season.

Chromium and nickel are commonly found in very high concentrations in soils formed on ultrabasic rocks. Even where rainfall is good or excessive, the natural vegetation is xerophytic and these well-structured soils are not suitable for agriculture. Levels of nickel of 3–70 ppm in ammonium acetate extract are recorded as toxic (Landon, 1991).

Biological factors

Organic matter

Soil organic matter refers to biological substances of plant or animal origin and consists of many different types of material with varying degrees of humification. The level in a soil depends primarily on the amount and nature of the material returned to the soil by the vegetation (which on uncultivated soils will usually be related to the climate, especially to rainfall) and on the balance between the rates of formation of humus and its subsequent decomposition. Under prolonged stable environmental conditions a state of equilibrium between these processes is achieved and the humus content remains at a fairly constant level characteristic of the soil and climate. Any marked change in environmental conditions results in a gradual change in the humus content to a new equilibrium level. Thus, if land under forest contributing, say, 10 000 kg litter/ha/yr is cleared and planted to crops providing 2000 kg residues/ha/yr, humus will continue to decompose at a rate approximately proportional to the amount in the soil, at first relatively rapidly but subsequently more slowly so that, if cropping is continued in a uniform manner for a period of years, a lower equilibrium value is eventually reached. Conversely, if cropped land is allowed to revert to a forest fallow with consequent increase in the rate of deposition of organic material, the rate of humus decomposition will at first be relatively slow, but will gradually increase as the soil humus content rises until a new, higher equilibrium level is reached.

Feller (1979) has classified soil organic matter into very coarse plant residues, coarse plant residues with part humified, much humified organic debris and organic matter bound to clay and silt. The carbon:nitrogen ratio varies from about 40:1 in the very coarse plant residues to 10:1 in the bound organic matter. Essentially soil organic matter consists of both living and non-living materials. The former contains plant roots, soil fauna such as earthworms and termites and soil microorganisms; the latter consists of macroorganic matter and humus (Floyd, 1991).

In temperate soils, earthworms have an important role in organic matter decomposition; in the tropics they are not generally found where the annual rainfall is below 800 mm. Termites are by far the most important of the macro-fauna in most tropical soils. Not only are they involved in the decomposition of plant materials but the earth-dwelling species also have an influence on the structure and permeability of the soil. They consume large quantities of the annual litter production before it is attacked by microorganisms. Furthermore, the fungus-cultivating species do not recycle the material but concentrate it in mound areas. Under certain conditions the soils from these have improved physical and chemical properties. Microorganisms, bacteria, fungi and algae play a major role in the decomposition of organic matter. As the microbial population increases following the addition of organic matter, essential nutrients are immobilised through incorporation into the biomass to be released later as the biomass decomposes. The two components of the humus are the humic and non-humic groups. Humic substances have high molecular weights and are generally classified by solubility in acid and alkali into fulvic acid, humic acid and humus (Floyd, 1991). Humic materials are bound to the clay particles by relatively weak physical bonding or by chemical bonds; polyvalent cations and oxides of iron and aluminium are involved as intermediates between the humic materials and the clay particles (Greenland, 1965). The humic substances are more resistant to microbial attack, especially when protected by the clay colloids, and thus form the stable fraction of the organic matter.

Non-humic substances are less complex and less resistant to microbial attack than the humic ones. They are specific organic compounds with clearly defined physical and chemical properties. Polysaccharides enhance aggregate stability and some of the organic acids may enhance availability of some nutrients and may complex others, e.g. aluminium.

A classification based on the stability of the organic matter fractions has been devised by Jenkinson and Rayner (1977) as shown in Table 3.3. These half-lives have been determined for Rothamsted (UK) material. In the tropics the rates are much faster and Jenkinson and Ayanaba (1977) showed them to be four times as fast in Nigeria as in the temperate soils at Rothamsted Research Station.

Table 3.3 Rates of decomposition of organic matter fractions. (Source: Jenkinson and Rayner, 1977.)

Description	Half-life (yr)
Decomposable plant material	0.16
Resistant plant material	2.31
Microbial biomass	1.69
Physically protected soil organic matter	49.5
Chemically protected soil organic matter	19.80

Other examples of the rate of decline are given by Lal and Okigbo (1990) for eastern Nigeria where the initial value of 0.9% organic carbon declined to 0.62% in the first year, 0.5% in the second year, 0.45% in the third year and 0.42% in the fourth year of continuous cultivation. Pieri (1992) gave values for Senegal of organic matter of 2.85% under forest, 2.02% after 12 years of cropping and 0.84% after 90 years of cropping. These losses in organic matter were accompanied by losses in nitrogen, cation exchange capacity (which fell from 7.8 to 2.5 meq/100 g) and loss of total exchangeable cations from 6.8 to 1.6 meq/100 g.

In summary, then, the rate of organic matter decomposition in cropped soil depends on the soil type and its initial organic matter content and on climatic and agronomic factors. High temperatures, moisture and aeration favour rapid decomposition. In general, increasing clay content and a large proportion of hydrous oxides in the clay fraction tend to decrease the rate of decomposition. Cultivation lowers the humus content at a rate and to an extent depending on the intensity of tillage, the amount of crop residues returned to the soil and the efficacy of anti-erosion measures. The loss of soil organic matter has a major impact on other factors that influence crop production; these include effects on the physical characteristics, the loss of nitrogen, calcium, magnesium and potassium and increased susceptibility to aluminium toxicity in acid soils. As a result there is usually a marked lowering of crop yields, especially if inorganic fertilisers are not used. It is therefore important, especially for arable cropping systems in which little or no use is made of fertilisers, to maintain the organic matter at as high a level as is practicable, particularly as, once the humus level has fallen appreciably, it can usually only be built up again slowly and with difficulty. Thus, Turenne and Rapair (1979) estimated that it would take 80–100 years to restore the organic matter pool in the Amazonian forest, destroyed by felling and burning. In cropped land most of the organic matter must come from crop residues but there are often competing demands for these, including livestock fodder, fuel and fencing. Furthermore the influence of termites which remove large quantities of organic material without recycling it in the soil leads to considerable losses, particularly in drier areas. However desirable the build up of organic matter in tropical soils, the problems in doing so, especially under intensive cropping, are formidable.

SOIL CLASSIFICATION

Soil classification is a way of developing and organising the search for knowledge on soils and serves as a basis for the application of soil knowledge to agriculture. However, the classification of soils presents many problems and there is not yet an agreed international classification. Many classifications, designed more specifically for tropical soils, have been published; these include those of Sys (1959, 1960) for Zaire, Aubert and Tavernier (1972) for Sub-Saharan Africa, the Commission pour Cooperation Technologique en Afrique (CCTA) Soil Maps of Africa (D'Hoore, 1965) and Dudal and Moorman (1962) for Southeast Asia. Many local classifications have been produced during the course of soil surveys but most of these are incomplete in that no attempt has been made, except in a very general way, to relate the soils at the higher categories to soils elsewhere. Some of these classifications have been made on the basis of one or more factors that specifically affect crop growth such as sodium levels, drainage characteristics, depth of peat. At such local levels these classifications are useful as they relate soil parameters to expected crop performance.

Soil surveys during the latter half of the twentieth century have provided a good general knowledge of the location of all of the important soils in the tropics. Thus much is known about the geographic distribution of tropical soils but less about their genesis and pedological relationships and even less about soil parameters in relation to crop growth.

International classifications

Of the systems designed for international use only two have been widely employed (Landon, 1991). These are Soil Taxonomy (USDA, 1975, 1982) and the FAO/UNESCO Soil Map of the World (FAO/UNESCO 1974, revised 1978). In addition the CCTA Soil Map of Africa, based on the French ORSTOM (Office de la Recherche Scientifique et Technique Outre-Mer) and Belgian INEAC (Institut National pour l'Etude Agronomique du Congo) systems, is still widely used especially in francophone Africa.

Soil Taxonomy (devised by the United States Department of Agriculture) is a hierarchical system in which the highest categories are the orders, of which there are ten. The criteria for placing soils in the same order are that they have developed by processes which are essentially similar so that horizon development and the general nature of the horizons are the same. A brief description of the orders is set out in Table 3.4 and the relationship with other older classifications in Table 3.5.

The orders are divided into suborders, great groups, subgroups, families and soil series. The terminology for each of these is based mainly on Latin and Greek roots; for example, *aridus* dry; *histos* tissue; *spodos* wood ash; *verto*, turning; *ustus*, burnt. Consequently many terms, which had been in use over a long time, have disappeared. Thus the term laterite, first used by Buchanan in 1807 to describe a ferruginous material which was soft enough *in situ* to be cut into

Table 3.4 Brief description of the orders in Soil Taxonomy.

Name of order	Characteristics
Entisols	Soils in deep regolith or earth with no horizons except perhaps a plough layer. In the tropics, soils of recently reclaimed mangrove swamps, recent volcanic ash and young river alluvium would fall in this order.
Inceptisols	Soils that have one or more diagnostic horizon, e.g. horizons with organic matter, dark colour or with some clay development, but with no significant eluviation or illuviation and with weatherable minerals present. They are most often found on young but not recent land surfaces; in the tropics they are common in volcanic ash under high rainfall and on the alluvial plains of the great river deltas and belts of coastal alluvium.
Vertisols	Soils with a high content of expanding lattice clay, high base exchange capacity (more than 30 meq %), high clay content and deep cracks on drying. The Regur, Black Cotton and Tropical Black clays which are widespread in the tropics in areas with strongly marked dry seasons are included in this order.
Aridisols	Soils of the dry areas with pale coloured surface horizons with subsurface horizons with some weathering to form clay or with high amounts of sodium, gypsum or salt. These soils are found in the dryest areas of the tropics and are, as yet, scarcely used for any form of agriculture.
Mollisols	Soils with a thick dark surface horizon (but excluding soils which, though they have this horizon, have a clay fraction dominated by allophane or a silt and sand fraction dominated by volcanic ash or soils that have features diagnositic for vertisols). Such soils are normally developed under a grass vegetation, as for example the chernozems. They are not widely represented in the tropics.
Spodosols	Soils with horizons in which there is an illuvial accumulation of free sesquioxides accompanied by appreciable amounts of organic carbon, of free iron with illuvial crystalline clay or other illuvial accumulation of organic carbon usually accompanied by an accumulation of aluminium as a non-clay form of combination. Spodosols are found only spasmodically in the tropics, normally on highly siliceous parent materials on old landscapes.
Alfisols	Soils with a horizon of clay accumulation, which sometimes has a high percentage of exchangeable sodium. They are soils of the humid regions with base saturation of >35% in the subsurface horizon. Soils of the drier areas of the tropics, with prolonged dry seasons and perhaps some of the red soils on hard limestone, commonly called terra rossas fall into this group as do some soils which have been termed red yellow podzolic.
Ultisols	Soils with a horizon of clay accumulations with <35% base saturation, formed under humid climates on old land surfaces or on pre-weathered parent materials. They have some weatherable minerals such as micas or feldspars in the silt and sand fractions. These soils, under the names of Red Yellow Podzolic and Reddish Brown Lateritic, are widespread in tropical areas with the more acidic rocks.
Oxisols	Soils with a horizon in which weathering has at some time removed or altered a large part of the silica that is combined with iron and aluminium. This results in a concentration of clay-sized minerals consisting of sesquioxides mixed with varying amounts of silicate clays with a 1:1 lattice. They may have hard or soft accumulations of sesquioxidic enriched clay. These soils occupy large areas of the humid tropics and include those soils which have been termed latosols and groundwater laterites.
Histosols	Soils with peaty accumulations. In the tropics these are of widespread occurrence in coastal swamps.

Table 3.5 Correlation of USDA Soil Taxonomy with other classification systems. (Sources: Landon (1991), Sanchez (1976), adapted from Soil Survey Staff (1960), Thorpe and Smith (1949) and Cline (1955).)

Order	Former Great Soil Groups
Entisols	Azonal soils, some low humic gleys, lithosols, regosols
Inceptisols	Andosols, hydrol humic latosols, sol brun acide, some brown and reddish brown soils, associated solonetz
Vertisols	Grumosols, tropical dark clays, regurs, black cotton soils, dark magnesium clays
Aridisols	Desert, reddish desert, sirozem, solonchak, some brown and reddish brown soils, associated solonetz
Mollisols	Chestnut, chernozem, brunizem, rendzina, some brown forest, associated humic gley and solonetz
Spodosols	Podzols, brown podzolic, groundwater podzols
Alfisols	Grey-brown podzolic, grey wooded, non-calcic brown, degraded chernozem, associated planosols and half-bog, some terra rossa estructurado and eutric red yellow podzolics, some latosols and lateritic soils
Ultisols	Red-yellow podzolic, reddish-brown lateritic, humic latosols, lateritic soils, terra roxa and groundwater laterites
Oxisols	Low humic latosols, humic ferruginous latosols, aluminous ferruginous latosols, lateritic soils, terra rossa legitima, groundwater laterites
Histosols	Bog soils, organic soils, peat, muck

building blocks but which became very hard on exposure to the weather, is now termed *plinthite*; the term would be used to describe a soil at the great group level, for example *plinthaquult* is an *Ultisol* which is wet and has a plinthic horizon; similarly most of the terms derived from *laterite*, e.g. *latosol*, *lateritic*, have disappeared.

This classification was originally designed to classify the soils of the United States into exclusive groups and has been criticised as being too closely concerned with the soils of North America, although there have been considerable modifications to accommodate the soils of the tropical regions for which Puerto Rico and Hawaii were the initial representatives. It has also been criticised for its completely new terminology and for the lengthy and complicated definitions needed to separate the classes.

The FAO/UNESCO classification, described as a legend for the soil map of the world, uses many of the descriptive terms from Soil Taxonomy but is considerably simpler. Its very broad units can accommodate virtually all tropical soils. There are 29 soil units in the legend which correspond more or less to the suborders of Soil Taxonomy but there is considerable overlap.

Some of the more important tropical soils

Although all ten orders of the USDA system of Soil Taxonomy are found in the tropics, *Spodosols* and *Mollisols* are quite limited in area. *Aridisols* cover large areas, e.g. in Africa but are of minor importance for agriculture. *Histosols* cover quite extensive areas mainly in Asia but are not widely used for agriculture. The *Oxisols* and *Ultisols* cover vast areas of the tropical landscapes, and comprise about 36% of all soils in the tropics but the *Oxisols* occur only in a few areas in Asia covering only 2% of tropical Asia (Sanchez and Salinas, 1981). The other important tropical soils are the *Alfisols* (18%), the

Entisols (12%) and the *Inceptisols* (11%). The latter two are very important for the intensive agriculture which they support in many parts of the tropics.

Oxisols (FAO classification = Ferralsols)

Oxisols are soils with an oxic horizon, a highly weathered subsurface horizon dominated by iron and aluminium hydrous oxides with some kaolinitic-type silicate minerals and sometimes quartz, but virtually no weatherable minerals. The soils have a low exchange capacity with a low level of exchangeable bases; they are usually deep, well drained, reddish or yellow coloured and with a stable granular structure, resistant to erosion and though high in clay, behave like loam soils. They are mostly of low to very low fertility and the pH varies from 4.5 to 5.0. Although strongly acid they have limited amounts of exchangeable aluminium and thus require small amounts of lime at less frequent intervals than other acid soils (Lathwell, 1979). However, they are often extremely deficient in phosphorus.

Oxisols generally occur on very old tablelands where there has been protection from erosion (Aubert and Tavernier, 1972) and are the dominant soils of continental shields in the permanently moist and wet–dry climates. There are four suborders; *aquox* are Oxisols of wet areas and are grey or mottled dark red and grey; they are seasonally saturated with water and usually have plinthite. This occurs widely over large areas of West Africa and the Katanga plateau. The *humox* suborder is an Oxisol of relatively cool moist regions with a large accumulation of organic carbon. Such soils occur in the eastern Zaire basin, central Madagascar, in South Asia and islands of the South Pacific. *Orthox* are soils of warm humid regions with less than 90 days of dry season and are largely found in West Africa, the central part of Equatorial Africa, in the humid tropics of Asia and the low altitude areas of Brazil. *Ustox* soils are dry for a substantial period and are extensive in the Brazilian plateau and East Central Africa.

Ultisols (FAO classification = Acrisols)

Ultisols are widely represented and, while less important than Oxisols in tropical America and Africa, are far more important in Asia covering 35% of the tropical areas of the latter. In China, where they are designated as red earths, they are particularly important covering 20% of the country (Felix-Henningsen *et al.*, 1989). They are soils with a subsurface horizon of clay accumulation (argillic horizon) and low base status, usually deep and well drained but lacking some of the more granular structure of Oxisols and thus more liable to erosion. Typically they are formed under forest vegetation in climates with slight to seasonal moisture deficits and surpluses. They are generally low in exchangeable bases and phosphorus though they may have some weatherable minerals. The pH is generally in the range of 4.5–5.5 but the more acid soils have an aluminium saturation as high as 80% making them suitable only for acid-tolerant crops, or otherwise requiring lime.

Suborders of the Ultisols are defined in a similar manner to those of the Oxisols. *Aquults* are soils of wet areas; *humults* are soils with a large accumulation of organic matter; *udults* are soils of humid tropical and subtropical regions; and *ustults* are soils of warm regions with long dry periods.

Because of their low level of plant nutrients many Oxisols and Ultisols are farmed under bush or grass fallow systems or with perennial crops. Successful high input systems have been developed for such crops as rubber, oil palm and tea, and large areas of these are grown on Ultisols in Southeast Asia with substantial use of fertiliser. Fertilisers, particularly lime and phosphorus, are also widely used on the Oxisols of southern and central Brazil for the growing of maize and soya bean.

Alfisols (FAO classification = Luvisols)

Alfisols are important in the transition zone to arid climates. They occur under forest or savanna vegetation in climates which have a seasonal moisture deficit. They have an argillic subsurface horizon due to silicate clay accumulation and medium to high base saturation. As such they have considerable chemical fertility but their structure is less well developed than that of Ultisols, hence the liability to structural deterioration under cultivation.

Alfisols are important over large areas of West Africa, mostly in the savanna zones, in much of Eastern and Southern Africa, in northeast Brazil and in India and the drier areas of Thailand as well as large areas in China. Like Oxisols and Ultisols, suborders of the Alfisols are defined in terms of their moisture regimes. *Aqualfs* are soils of wet places, dominantly grey throughout and saturated with water at certain seasons. *Udalfs* are soils of warmer areas in humid climates with little or no dry periods, and *ustalfs* are soils of warm places with long, dry (more than 90 cumulative dry days) in most years.

Alfisols, Oxisols and Ultisols which cover large areas of old landscapes in the tropics were formerly classified as zonal soils and comprise about 54% of tropical soils (Sanchez and Salinas, 1981). Inceptisols, Entisols and Vertisols, typical of soils formed on young landscapes, were formerly classified as interzonal and azonal soils and these comprise about 26% of tropical soils of which Vertisols total only about 3%.

Inceptisols (FAO classification = Cambisols)

Inceptisols are soils that show some profile development but are relatively immature. Because of this they are usually closely related to their parent material. Soils of the suborder *Andepts*, derived from volcanic ash, are characterised by a clay fraction that consists largely of allophane, an amorphous clay mineral. The allophane protects the humus from decomposition so that these soils have a dark-coloured humus horizon of low bulk density, highly permeable and free draining, very difficult to disperse and thus resistant to erosion. Once depleted of bases they become very acid; they have the capacity to retain large amounts of phosphorus which is only released into the soil solution when the retention capacity is nearly saturated (Younge and Plucknett, 1966). Andepts occur widely in Central and South America along the Andes, in Indonesia and in the South Pacific. In Africa they form some of the best soils in Cameroon, Kenya, Madagascar, Tanzania and Uganda. They are intensively farmed nearly everywhere in the tropics and have some of the highest population densities in rain-fed annual cropping areas.

Entisols (FAO classification = Fluvisols, Regosols)

Entisols are formed on recent deposits that are unaltered or only very slightly altered by soil-forming processes. Two suborders of the Entisols, *aquents* and *fluvents*, together with two suborders of Inceptisols, *aquepts* and *tropepts*, include many of the alluvial soils that are found in the river plains of the Indian sub-continent, Thailand, China, Indonesia and Malaysia, as well as the Amazon basin and the Nile delta. They occur under a variety of climatic conditions on flood plains, deltas and estuaries. The conditions under which they are deposited (fresh water, brackish or sea water) as well as the parent material have a major influence on their composition (Bloomfield and Coulter, 1973). Marine and brackish sediments begin as soft muds which 'ripen' gradually, a physical process in which the volume is decreased and the bulk density increased, a structure develops and permeability improves. As air penetrates the soil, chemical changes take place and oxidation causes changes in iron and sulphur compounds. These latter changes are particularly concerned with acid sulphate and potential acid sulphate soils. These occur in the suborders aquents and aquepts (Pons, 1989). *Typic sulfaquents* are the waterlogged, potential acid sulphate soils in which sulphidic material starts within 50 cm of the mineral soil surface. *Typic sulfaquepts* have a sulphuric horizon (pH < 3.5) within 50 cm of the mineral soil surface. Where the pH is <4.0 and there are jarosite mottles the soils are classified as *sulfic tropaquepts* (Sutrisno et al., 1990).

Although acid and potential acid sulphate soils cover a relatively small area, their occurrence in coastal plains and their proximity to densely populated areas make them attractive for settlement. Reclamation of these soils without proper precautions can lead to disastrous impacts on agriculture and drainage must be carefully controlled. Bos (1990) recorded that rice yields dropped from 4t/ha to 1t/ha and Bloomfield *et al.* (1968) that the yield of oil palms was greatly reduced by over-drainage. Conditions may become particularly difficult in times of drought. The 1976 drought in Indonesia led to the population of one area of Kalimantan dropping from 5300 to 1600 as rice lands were abandoned (Chairuddin *et al.*, 1990). The pH of the water in drainage canals through areas of acid sulphate and potential acid sulphate soils may fall to 2.5–3.0 with serious effects on fishing and impoverishment of the fish fauna. Many fish and shrimp ponds are being constructed in coastal areas and the presence of potential acid sulphate materials in these areas are likely to be important; furthermore many of these areas in their natural state are important as nurseries for shrimps and fish.

Vertisols (FAO classification = Vertisols)

Vertisols are the heavy cracking clay soils with a high content of shrinking and swelling (montmorillonitic) clays. They are extensive in drier areas when developed on calcium carbonate, andesitic and basaltic volcanic materials rich in ferro-magnesium minerals, or on alluvial materials derived from these. Vertisols also develop in topographic depressions in drier areas where bases are leached into the depressions. They are thus part of the catenary sequences described by Milne (1936). The surface soil is dark grey-brown to black, and this dark colour often extends into the subsoil. The colour is not usually due to a high organic matter content, which is commonly only about 1–3%, but may be due to thorough mixing of humus with the clay. A frequent surface feature is the presence of a pattern of small, rounded depressions, known as 'hog-wallows' or 'gilgai', but not all Vertisols have these and some other poorly drained soils do. The soils are neutral or somewhat alkaline, have a medium to high content of exchangeable calcium and magnesium and usually calcium carbonate concretions (and sometimes gypsum) occur at a depth of 30–70 cm where they may form a more or less hard horizon, known as 'Kankar' in India. The montmorillonitic clay minerals cause the soil to swell and become very sticky when wet and to shrink, harden and form deep vertical cracks during the dry season. In this season, a surface mulch of stable granules is formed, part of which is moved by gravity, wind, livestock and cultivation and falls down the cracks. Subsequent entry of water into the cracks at the beginning of the rains and swelling of the accumulated soil at their lower parts, produces pressures which result in some heaving of the subsoil. This process of cracking, churning and admixture of surface and subsoil, which homogenises the profile to some depth, is dependent upon the alternation of adequate wet and dry periods and may be modified under irrigation. It seems that the deep cracks do not usually completely disappear when the soil is rewetted, but that narrow fissures persist permitting some downward passage of water and constituting planes of weakness along which the cracking pattern reforms in the following dry season.

These clays are important agricultural soils because they are inherently of relatively good nutrient status. While farmers have always recognised their fertility they find them difficult to cultivate, although some have a very friable surface mulch when dry while others have a strong cloddy or blocky structure at the surface. Usually, they can only be given preparatory cultivation with hand tools or ox-drawn implements during a very limited period when the moisture content is suitable. This has greatly restricted their use for crop production by hoe farmers in Africa, although in India they have been more extensively farmed with ox ploughs. While considerable areas of these soils are now irrigated and cultivated mechanically for the production of cotton, rice, wheat, sugar cane, sorghum and various legumes, large areas are used for rain-fed agriculture. Land shaping by ridges, broad beds or cambered beds has proved useful in managing these soils for the latter. In dry areas with rainfall less than about 600 mm, moisture conservation improves yields whereas in higher rainfall areas disposal of surplus water is necessary. Thus planting in the furrows in the low veld of Zimbabwe gave better yields in years when rainfall was short or erratic; when rainfall was well distributed planting on the flat gave equally good yields (Nyamudeza et al., 1991). In Ethiopia on the other hand, with rainfall of 850–1200 mm/yr, broad bed and furrow techniques proved better than planting on the flat (Asamenew and Saleem, 1991).

Although many Vertisols occur in areas with flat to very gentle slopes, they erode quite easily on the gentle slopes and severe gullying can occur in these areas. Where these soils are used fairly intensively for the production of rain-fed arable crops, as on the undulating plains of the Deccan in India, prevention of runoff and erosion on sloping land is important, particularly as maximum infiltration of a limited rainfall is commonly essential for good crops. Kanitkar (1944) reported that, as the use of the 'regur' soils of the Deccan became intensified, because of increasing human and cattle populations, severe losses of water and soil occurred and humus content, nutrient status and crop yields all declined. The maintenance of good yields of rain-fed crops when such soils are continuously cultivated evidently demands proper soil and water conservation measures, skilful and timely tillage with ox- or tractor-drawn implements, and the use of farmyard manure or fertilisers. The broad bed and furrow technique has been used at International Crops Research Institute for the Semi-Arid Tropics (ICRISAT) to conserve both soil and moisture. Under this system annual loss of soil by erosion was 1.46 t/ha compared with 6.38 t/ha for the traditional system of cropping on the flat (Srivastava and Jangawad, 1988).

Histosols (FAO classification = Histosols)

Histosols are soils with a *histic* surface horizon, saturated with water at some season and containing a minimum of 20% organic matter if no clay is present, and at least 30% organic matter if there is 50% clay or more. Histosols develop when decomposition of plant litter fails to keep up with the rate of addition. Such soils cover an estimated 370–500 million ha from tundra to tropical environments, and about 30–40 million ha in the tropics, of which about 75% is in Southeast Asia (Maltby, 1989). The latter are predominantly swamp deposits consisting mainly of forest debris where the final stages in peat swamp development result in raised peat domes. Peatlands are fragile ecosystems which require a well-planned management strategy for their use (Gadjah Mada University, 1987). Most of the tropical peats are *dystric histosols*, i.e. they have a very low base status; while they contain considerable quantities of nitrogen they are very low in phosphorus, potassium and other cations. Crops grown on them are also liable to suffer from deficiencies of certain trace elements; for example, copper and zinc deficiencies have been reported in pineapples grown on peats in Malaysia. On deep forest peats adequate drainage is difficult and expensive because water movement is slow and the soils contain large amounts of buried timber which has to be cut through. The normal practice in Malaysia is to make large open drains, 1.5 m or more deep and widely spaced, but experience on peat soils elsewhere suggests that a shallower but more intensive drainage system might be better. Heavy machinery sinks into deep peat and therefore cannot be used for clearing, drainage and cultivation, nor can heavy lorries be used to remove produce unless considerable expenditure is incurred on road making. The roots of tree crops are either restricted by poor drainage or become exposed as the peat shrinks after draining with the result that after a few years plants such as rubber and oil palm are inadequately anchored and topple over. Pineapple is the only crop grown extensively on deep peats in Malaysia. On the other hand on the shallower peats, rubber, oil palm, Liberica coffee and, in places, rice, have been grown with some success once the nutritional problems have been identified and overcome.

Soils which contain less organic matter than the peats (such as the 'pegassy clays' of Guyana) can usually be made productive with less difficulty. Given adequate drainage and appropriate fertilisers, such soils will grow good crops of rice, sugar cane, maize, oil palm, rubber, cocoa and coconut, provided that the acidity is not too great. Not infrequently, however, these soils may have a sulphidic horizon which on drainage gives rise to high acidity and high levels of toxic aluminium.

SALINITY AND SODICITY

Salinity and sodicity are of particular interest because of the problems they cause in irrigation management. Saline soils are defined as those having a conductivity in the saturated soil extract exceeding 4 mmhos/cm (mS) and an exchangeable sodium percentage of less than 15 with a pH usually less than 8.5 (Richards, 1954). Sodic soils ('sodic' replacing the older term 'alkali', Landon, 1991) have a conductivity below 4, an exchangeable sodium percentage exceeding 15 and a pH usually above 8.5. Saline–sodic soils have a conductivity exceeding 4, an exchangeable sodium percentage exceeding 15 and a pH usually less than 8.5.

Sodic and saline soils are formed in hot arid or semi-arid regions in sites, such as the flood plains of rivers, depressions, low-lying lake margins or coastal plains, where groundwater containing salts occurs within a few metres of the surface. *Saline* soils are characterised by the accumulation of neutral salts in the surface soil resulting from the capillary rise of saline groundwater under conditions where evaporation exceeds precipitation, and in the absence of sufficient rain to provide regular percolation through the soil. During dry periods salt-charged water may be carried up to the surface where it evaporates to leave a white salt crust – hence the name 'white alkali soils' which is sometimes given to these soils. The predominant salts are usually sulphates and chlorides of sodium and calcium, but small quantities of carbonates and bicarbonates are often present and sometimes there is a considerable proportion of magnesium. Calcium, or calcium and magnesium, predominate in the exchange complex with little absorbed sodium. The surface soil is light in colour, and is 'fluffy' and permeable and not difficult to cultivate. Natural saline soils usually have a characteristic sparse vegetation and can be identified by the presence of indicator species.

Saline soils are also formed by improper management of irrigation schemes, usually through over-irrigation and lack of adequate drainage. This combination results in rising water tables and, if these are saline, salts accumulate at the surface, leading to reduction in yields and sometimes abandonment of irrigated lands. Several large irrigation schemes are suffering from salinity problems due to rising water tables, thus necessitating the installation of extensive drainage works. A major difficulty, however, is the disposal of the saline drainage water, especially from irrigation areas far from the sea; the solution to this problem in Pakistan, for example, has been the installation of a major drain, the 'Left Bank' outfall scheme, which carries saline drainage water several hundred kilometres to the sea.

Crops vary greatly in their tolerance of salinity; cotton, for example, suffers about a 10% depreciation in yield with an electrical conductivity of 10; whereas sugar cane would suffer a 50% yield loss at the same salinity level. In recent years there has been considerable attention given to breeding for crop tolerance to salinity, for example, the transfer of salt tolerance from distantly related species to wheat. Though this may eventually lead to improved crop production under relatively low saline conditions, successful cropping of more saline soils will depend on drainage and leaching out of the salts.

Sodic soils are developed from saline soils with low calcium reserves when, following a fall in the water table, percolating rainfall washes soluble salts down the profile and an appreciable proportion of the exchangeable calcium ions is replaced by sodium. Some of the sodium ions react with carbonate and bicarbonate anions resulting from the production of carbon dioxide in the soil and sufficient sodium carbonate is formed in the soil solution to raise the pH to 8.5 or above. Under waterlogged conditions sodium carbonate may also result from the reduction of sodium sulphate in the soil to sulphide which, in the absence of ions forming insoluble sulphides, is lost as hydrogen sulphide, leaving the sodium ions to react with carbonate anions. The presence of an appreciable proportion of sodium in the exchange complex and of sodium carbonate in the soil solution deflocculates the clay and humus, rendering the soil impermeable and giving it a black or dark grey colour. Clay particles move down the profile to form a clay pan in which some dispersed humus, washed down from above, may accumulate. These soils, which have been called *black alkali* or *solunetz* soils, are unproductive owing to the alkalinity and because they are very impermeable, extremely plastic when wet and form hard clods when dry. Their reclamation involves the provision of adequate drainage and the replacement of exchangeable sodium with calcium. The latter can be done by working gypsum into the surface soil which reacts with the sodium carbonate to produce calcium carbonate and soluble sodium sulphate and also replaces some of the exchangeable sodium with calcium. This treatment renders the surface layer permeable, after which irrigation water is applied to gradually wash the gypsum down the profile so that the permeability of successive layers can be improved and soluble salts washed into the groundwater.

LAND RESOURCE ASSESSMENT

It is important that both the development of new lands and the reorganisation and improvement of traditional farming in settled areas should be carefully planned in the light of a study of soil and other environmental factors and an assessment of the suitability of the land for various agriculture and forestry uses. Ideally, the planning should be based on information and quantitative data derived from detailed studies of all the relevant factors of the physical environment, but such detailed information is rarely available and it would be a lengthy and costly business to obtain it. It is, therefore, necessary to proceed by means of less detailed reconnaisance resource assessment of extensive areas followed by rather more intensive assessment of smaller areas selected as promising for development. Such surveys can thus be categorised under one of the following (Landon, 1991):

- Exploratory
- Reconnaissance
- Semi-detailed
- Detailed

The scales, aims and objectives of these levels of surveys are summarised in Table 3.6.

It will be noted from Table 3.6 that air-photo interpretation (API) plays a dominant role at the exploratory and reconnaissance stages. As the intensity of survey increases so does the amount of field work in relation to the time spent on air-photo interpretation. At the detailed level of survey, mapped boundaries will usually be drawn from the results of a field survey. The mapping units will also change. Thus at the exploratory stage air-photo interpretation will be used in the analysis of landscapes to distinguish and map areas known as land systems.

A land system is defined as an area, or group of areas, throughout which there can be recognised a recurring pattern of topography, soils and vegetation (Christian and Stewart, 1953, 1964). Since the features concerned in these patterns are indicators of potential land use, the land systems recognised can be classified into categories according to their suitability for various kinds of development. Examples of land systems might be (a) a plateau intersected by shallow stream courses and supporting grassland with intermittent trees, or (b) a sharply dissected mountain area with steep-sided, forested valleys and bare mountain tops. The recurrent units within each land system, e.g. valley bottoms and mountain tops, which differ in use potential, are referred to as *land facets* (Fig. 3.1). Land systems can be distinguished on air-photos since these not only reveal topographical features but also often show particular tones and textures characteristic of soil and vegetation types which, together with other visible features, such as the absence or extent of erosion, the amount of land surface covered by bare rocks or the area liable to seasonal flooding, give a land system a distinctive pattern (Brunt, 1967). Delineating the boundaries of land systems can readily be done on the photos as this is largely a matter of plotting the boundaries of land forms which usually coincide with easily recognisable breaks of slope. Thus, the procedure permits the delineation on

Table 3.6 Summary of types of land resource surveys. (Source: Landon, 1991.)

Type of survey	Nearest FAO equivalent nomenclature and final map scale	Aim and level	Site intensity and survey method	Approximate proportion of time input (%)			Preferred scales	
				Aer-photo unterpretation	Literature	Field work and sampling	Aerial photos	Final maps
Exploratory	Exploratory to low intensity 1 : 1 000 000 to 1 : 100 000	Resource inventory Project location Prefeasibility	Free survey of variable intensity usually much <1 per 100 ha	60 (Probable averages, very variable)	20	20	1 : 60 000 1 : 100 000	Variable
Reconnaissance	Medium intensity 1 : 100 000 to 1 : 25 000	Prefeasibility Regional planning Project location	Free survey of variable intensity usually <1 per 100 ha	50	25	25	1 : 40 000 to 1 : 20 000	1 : 50 000
Semi-detailed	High intensity 1 : 25 000 to 1 : 10 000	Feasibility Development planning	Flexible or rigid grid. Intensity 1 per 15 to 50 ha	20	20	60	1 : 25 000 to 1 : 10 000	1 : 25 000 to 1 : 10 000
Detailed	Very high intensity 1 : 10 000	Development Management Special purpose	Rigid grid 1 per 1 to 25 ha	5	20	75	1 : 10 000 to 1 : 5 000	1 : 10 000 to 1 : 5 000

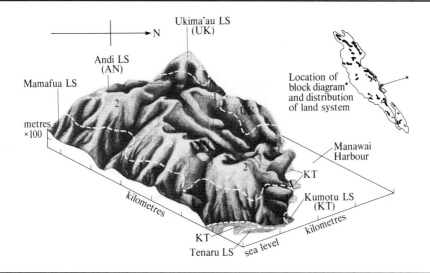

Fig. 3.1 Model illustrating a land system (the Ande system) of rounded hills and ridges with two land facets: (1) ridge summits with moderately steep crested slopes, (2) broad summits with gentle to moderate slopes. Also shown are portions of three adjacent land systems. (Source: Land Resources Division, Ministry of Overseas Development.)

the photos, and the location on the ground, of areas possessing a similar range of environmental conditions and use potential.

Air-photo interpretation can also be employed in vegetation pattern analysis for identification and mapping of different types of vegetation. This can be useful, especially in areas where little climatic and soil data are available, since each climax association, and to a lesser extent the secondary association derived from it, reflects soil and climatic conditions and represents land which, apart from topographic variations, is fairly homogeneous from the point of view of agricultural potential (agro-ecological groupings). With large-scale photos and limited ground reconnaisance quite detailed information on existing land use can be obtained and mapped. Photo-sampling techniques are also available for rapid estimation of the proportion, or area, of land under different crops or types of farming (Brunt and Rees, 1965; Brunt, 1967).

Reconnaissance surveys also use physiographic units/land systems as the final mapping unit. These may be defined in terms of soil associations and land capability units may be delineated (see below). In such surveys it is also normal to delineate those areas which are unsuitable for development, for example steep lands, flooded areas, excessively shallow soils or rocky areas.

Semi-detailed surveys may be used in feasibility studies and these should take account of the economic dimensions of the project. Soil series or soil associations are the mapping units normally used and land capability or suitability classes may also be produced at this scale.

Detailed surveys may be designed to provide data for developing projects, to provide detailed soil maps for experiment stations or to relate soil parameters to various aspects of crop growth. Such surveys would usually map the soil at very high to high intensity with map scales of 1:5000 to 1:25 000. The area represented on 1 cm^2 of a soil map at the larger scale would be 0.25 ha while that on the smaller scale would be 6.25 ha. At such scales virtually all of the soil boundaries would be checked throughout their length by ground observation and the average sampling density would be one sample per 0.5 ha at the larger scale and one sample per 8 ha at the smallest (Western, 1978).

Soil mapping units, used at the semi-detailed and detailed levels, usually consist of soil associations, series, types and phases. A soil association is a compound mapping unit consisting of a group of soils which may differ considerably in their characteristics but are geographically and topographically associated in a regular pattern which can be shown on detailed maps. An important type of soil association is the catena. A soil complex is also a compound mapping unit, but differs from an association in that it contains a complex and intimate mixture of soils which cannot be differentiated at the scale being used. The most important mapping unit is the soil series, defined as a group of soils having similar profile characteristics, except for the texture of the surface soil, and formed on the same or similar parent material. Series may be divided into soil types, according to the texture of the surface horizon, and these may be further subdivided into soil phases on the basis of some feature potentially significant for land use, such as depth, slope or stoniness.

It should be emphasised that differentiation of soils is based on observable or measurable *permanent* characters which can be used for classifying the soils and for determining their genesis. Thus, morphological characters such as texture, colour, structure, plasticity, horizons of clay accumulation, presence or absence of plinthite, calcium carbonate or salts are all important. Certain characteristics like base saturation, exchangeable sodium percentage, exchangeable aluminium can only be determined by laboratory measurements. Some of these parameters can be fairly readily related to crop performance but others may have no obvious connection. Colour may be moderately related through drainage and texture to moisture characteristics. There are other factors in tropical soils, for example the influence of termites whose influence on cropping is not well understood. However, management capacity and the economic considerations that are influenced by factors not related to agriculture, for example road systems and markets, make the interpretation of soil characteristics in terms of agricultural performance very difficult and hence present problems for land capability classification.

Land evaluation

In spite of the problems noted, it is usual and indeed necessary to interpret the findings about the land and the soils in terms of potential for agriculture. This is essential for guidance of planning decisions so that lands are developed for the most productive use. Terms for land evaluation are subject to various interpretations and there is not always a distinction between 'soil' and 'land'. Obviously the former term refers to soil *sensu stricto* but 'land' may be used to mean 'soils' or in a much wider sense to include the natural environment generally, which will include climate, topography, vegetation and possibly some of the infrastructural aspects, for example, access. Landon (1991) suggested that terminology be defined as follows:

(1) *Land evaluation.* A general term covering all forms of interpretation with no particular connotation as to final use.
(2) *Land capability classification.* A more specific term derived from the United States Department of Agriculture (USDA) (Klingebiel and Montgomery, 1961) which is based on the severity of limitations to land use.
(3) *Land suitability classification.* A term which relates to specific uses of which the United States Bureau of Reclamation (USBR) system classifying land for irrigation is an example.

Land capability

The land capability classification of the United States Soil Conservation Service (Hedge and Klingebiel, 1957; Klingebiel and Montgomery, 1961) which is widely used in the United States has also been modified for use in other countries, for example in Britain (Bibby and Mackney,

1969). In the United States system, land is graded according to its potential and the severity of the physical limitations into eight capability classes. Class I land has a minimum of physical limitations and is suitable for a wide range of crops; it has a favourable climate, is level or gently sloping, with well-drained, deep, loamy soil possessing good moisture reserves. The other seven classes suffer from increasingly severe limitations and the range of uses for which they are suitable is progressively restricted. Classes II to IV are suited to tillage crops while Classes V to VII are suited to pasture, tree crops and forestry and are not recommended for cultivation. Class VIII land is usually very steep, or broken, extremely sandy or very wet or arid, of value only for wildlife or recreational purposes.

The capability classes indicate the severity but not the nature of the physical factors limiting productivity. The latter is the basis of a division into subclasses, four kinds of limitation being recognised and denoted by letters:

w – Wetness, which may result from impermeable, or slowly permeable, soil layers, high water table, or flooding.
s – Soil limitations – shallowness, stoniness, poor structure.
e – Liability to runoff and erosion, largely dependent on slope.
c – Climatic limitations, e.g. low, ill-distributed rainfall.

Within each subclass there may be soils which suffer the same degree and kind of limitation but which differ in management requirements; for example, wetness may be due to slow infiltration of rainwater or to rising groundwater, requiring different treatments. Soils which are nearly alike in features that affect plant growth and response to management, and which can be used in the same way for the same crops are, therefore, grouped into land capability units. These units are the most detailed and specific groupings of the classification and they can only be established and mapped if a good deal of information on the soils and their management is available from research and practical experience.

Land suitability classification

This gives a more specific objective for the land use and is thus utilised where particular cropping or cultural systems are planned. The USBR system is directed towards irrigation but it is equally possible to have systems that are oriented towards tree crops (Malaysia, 1967) or rain-fed crops. There is an important economic dimension to the classification in that Class 1 lands would be expected to give the best return on the investment because they would sustain high yields of a wide range of crops at reasonable cost. Classes 2 and 3 have increasingly severe limitations in one or more of these aspects of production. Such limitations may include inferior soils, restricted drainage, uneven topography or a higher concentration of salts; all of these can be corrected but at a higher cost. Class 4 lands would only be included for irrigation where there are specific reasons for their inclusion, for example, opportunities for growing high value crops. Class 5 lands are semi-arable under present conditions and would require special investigations for their inclusion in any development. Class 6 lands would not be regarded as irrigable under present conceivable conditions. When soil mapping, classes 1, 2, 3 and 6 can be differentiated in reconnaissance surveys. Detailed surveys are usually needed to delineate classes 4 and 5. The principle of the land capability and land suitability classifications originally designed for use in the United States have been used in other systems for the tropics, sometimes with substantial modification.

The FAO (1985) has issued a set of guidelines which are complementary to those of the USBR. The basic principles state that it is necessary to evaluate land and not just soil; the main objective is to predict future conditions after development has taken place and to forecast the benefits to farmers and the national economy. This includes the need to forecast whether the proposed systems can be sustained without danger to the environment. It is emphasised that land suitability evaluation is essentially an economic concept and that assessment of physical factors alone are insufficient; they must be translated into economic terms. One way of handling this is the calculation of farm income 'without project' and 'with project' situations. The survey data needed to make this calculation will include present farming practices, the infrastructure, the economic environment, the institutional framework and the national policy framework. Although both the USBR and the FAO approach are similar in most of these aspects, the former first examines 'financial viability', i.e. the farmer's ability to repay as part of an economically justified plan of development, while the latter first considers the economic justification for the irrigation project and then the projected farm income within that economic plan.

FURTHER READING

Landon, J.R. (ed.) (1991) *Booker Tropical Soil Manual. A Handbook for Soil Survey and Agricultural Land Evaluation in the Tropics and Subtropics*. Longman Scientific and Technical: Harlow.

Pieri, C.J.M.G. (1992) *Fertility of Soils: A Future for Farming in the West African Savannah*. Springer-Verlag: Berlin.

Wild, A. (ed.) (1988) *Russell's Soil Conditions and Plant Growth*, 11th edn. Longman Scientific and Technical: Harlow.

4
Soil and Water Conservation

M. Stocking

Soil and water conservation is society's response to land degradation. Any deterioration in the quality of the land has physical symptoms: for example, gully scars in the landscape, declining crop yields, drought stress or increasing inability of the natural environment to support the land-use demands placed upon it. Nowhere are these symptoms more manifest than in the tropics, especially in semi-arid, sub-humid and savanna zones. For decades there have been accounts of the seriousness of tropical land degradation, from Jacks and Whyte's (1939) classic, *The Rape of the Earth*, to more recent but no less strident calls for action such as IUCN/UNEP/WWF (1991, p. 4 – 'we are gambling with survival'). A 'health warning' needs to be attached to some of these crisis claims (see Stocking, 1995). However, clearly there are sets of degradation problems – water and wind erosion; soil physical changes such as surface crusting and increased bulk density; reduced organic matter; increased susceptibility to drought; acidification and specific chemical deficiencies and toxicities – which are today getting progressively more serious and more difficult to deal with as human populations increase.

If the quality of our soil resources is seen to be diminishing, or if excessive amounts of water are lost to agricultural activities, then calls for action are usually instituted under the general label of 'soil and water conservation'. Most tropical, developing nations have Departments of Soil and Water Conservation in Ministries of Agriculture, Forestry, Environment, Land Use or Natural Resources – and some such as Bolivia's CODETAR (Corporación de Dessarollo de Tarija) put soil and water conservation into major regional and economic development agencies. In proper context, soil and water conservation is a component part of natural resource management for land-use activities and, as such, is inextricably linked with tillage, cropping strategies, soil fertility management, water control, grazing systems, and the whole panoply of ways of extracting a living from a seemingly reluctant, and often, delicate natural environment. Indeed, one of the major developments which will be elaborated later in this chapter, is a shift *away* from soil and water conservation as a purely technical exercise *towards* soil and water conservation as being part of an overall production strategy. The shift has been signalled, for example, in 1995 by the launch of a new journal, *Land Husbandry: the International Journal of Soil and Water Conservation*.

For the purposes of this chapter *soil conservation* will be defined as any set of measures that controls or prevents soil erosion, or maintains soil fertility, with the view to maintaining or enhancing vegetative production, while *water conservation* will refer to techniques which achieve production through increasing plant-available water. (These definitions are specifically for agricultural production systems. Soil and water conservation is also needed in urban areas, construction sites, motorway embankments and the like, in which case the principal goal of soil and water conservation would be to stabilise slopes and reduce off-site problems such as sedimentation. Similarly, soil and water conservation may be needed in catchments to protect valuable infrastructural works: for example, the Japanese *Sabo* works involving concreting whole hillsides above Kobe City are to protect the narrow coastal strip where many industries are located.) The reader must note that the two are very closely related: most measures which conserve soil, also control runoff and retain more water on-site; and all measures which increase water availability to plants reduce the soil's susceptibility to erosion or encourage sedimentation. Many measures of soil conservation (e.g. tied ridging) bring short-term benefits because of water conservation, and some measures of soil conservation (e.g. broad-base terraces) have been known to decrease yields because of waterlogging or serious physical disturbance to the topsoil.

This chapter will briefly review the nature of soil degradation and the varieties of explanations that may be put for its occurrence. Contrary to most technical treatments of the subject, a 'political–ecological' analysis will be adopted which seeks to understand not only the physical processes but also the wider causes of how and why farmers allow degradation to continue, and why they should engage in conservation. This is vital for soil and water conservation because, unless the reason why erosion occurs can rationally be explained, it is pointless trying to design programmes of conservation. For example, if small farmers in India find it more beneficial for erosion to happen on the upper parts of the landscape because they wish to harvest soil to make new paddy fields on their bottom lands, then no amount of pressure will induce them to do something (i.e. soil and water conservation) which by their perception is stupid and irrational. Indeed, soil and water conservation has been bedevilled in the past by prescriptive techniques based only on diagnosis of physical attributes of the environment. It is no wonder that it has had such a poor record of success (see Hudson, 1991). The major strategic approaches to soil and water conservation will be reviewed, followed by some examples of successful programmes of conservation. Finally, the necessary ingredients for soil and water conservation in small-farm tropical agriculture will be examined.

Because the discussion will necessarily be short, the interested reader is referred to the following texts which elaborate the essential nature of the new themes and approaches in soil and water conservation: Blaikie and Brookfield (1987) and IFAD (1992) were milestone publications in showing not only how soil degradation is as much a socio-political phenomenon as it is a physical problem but also how explanations of why erosion occurs and how people cope with it can help in designing new programmes of soil and water conservation; Hudson (1992, 1995) showed both the change in emphasis towards agronomic measures of soil and water conservation and holistic farming-system approaches to managing soil resources; Douglas (1994) and Pretty (1995) set soil and water conservation firmly into the whole debate on sustainable agriculture and gave many useful examples.

SOIL DEGRADATION

Soil degradation processes

Soil degradation is defined as a decrease in soil quality as measured by changes in soil properties and processes, and a consequent decline in productivity in terms of production now and in the foreseeable future. (*Land degradation* is normally defined more broadly than *soil conservation*, to include all aspects of physical, chemical, biological, aesthetic and human land-use qualities of the terrestrial environment – for a review of the scope of land degradation and policies to deal with it in Australia see Chisholm and Dumsday (1987).) Any use of soil will cause degradation to some extent, if only through a diminution of organic matter or export of nutrients contained in the crop. Standard practices of soil and water management will address these soil quality changes. So, for example, irrigation can help make up for a decline in the soil's intrinsic capability to supply water to plants. Inorganic fertilisers can replace nutrients attached to eroded particles of sediment. Soil degradation can, therefore, be 'hidden' by technological inputs. In so far as it is cheaper and more convenient to redress soil quality deterioration by artificial means or by compensating improvements in, say, plant genetics or reliability of water supply, soil degradation is not a problem – our social, economic, cultural and technological skills cope with the changes. However, it is when the coping mechanisms fail to accommodate the nature, and/or the rate, of change that soil degradation starts to assume problematic status. Soil degradation also becomes a problem if off-site impacts occur, such as reservoir sedimentation, increased flooding and damage to hydroelectric installations. These impacts are beyond the scope of this chapter, but it should be noted that unless society subsidises the farmer to carry out soil and water conservation, there is little incentive for the land user to undertake works from which there will be no direct benefit.

Consequently, soil degradation is partly related to socio-economic variables and to the ability of the land-use system to compensate for the inevitable changes that will occur. Take, for example, a decrease in organic matter percentage to 0.25% as is typical on the sandy soils of East Anglia in England. It is of only modest consequence to the farmer, meaning that supplementary irrigation may be needed in dry summers, additional split doses of fertiliser and lime are applied to maximise plant uptake of nutrients, and the occasional winter cover crop is planted to improve soil structure. It would be economically irrational for the East Anglian farmer directly to tackle the loss of soil and organic matter, while the alternative indirect options for maintaining production are relatively cheap and easy. For a small farmer in the humid tropics of West Africa, the same decrease in organic matter could be catastrophic, necessitating immediate soil management techniques to conserve both soil and water. The socio-economic and political environments open or close the number of options available to land users to cope with soil changes – and hence they determine whether degradation is a process worth bothering about or not. In broad general terms, tropical environments and developing countries are both physically more sensitive and socio-economically more vulnerable to degradation-induced soil quality deterioration.

Soil degradation also assumes problem levels by its interactive relationship with physical and chemical variables of the environment. In this context, it is useful to relate soil changes to the concepts of 'resilience' and 'sensitivity' of a land system (see Blaikie and Brookfield, 1987, and O'Riordan, 1995, for elaboration of these concepts). *Resilience* is a property that allows a land system to absorb and utilise change: it is the resistance to an external shock. Some soils, such as subtropical Ferralsols (Oxisols) and Cambisols, can take large amounts of erosion with little apparent impact on their productive potential. Their resistance can be compared to tropical Luvisols, where yields crash with only modest amounts of erosion. Similarly, *sensitivity* is the degree to which a land system undergoes change; or how readily change occurs with only small differences in external force. We know, for example, that in the Hill Lands of Sri Lanka, massive land slips can be triggered by only slight increases in water infiltration or terrace-construction disturbance. This sensitivity can be contrasted with the Cerrado soils of Brazil or the Nitosols of Highland Ethiopia which have open structure, excellent physical properties and ability to change very little with land use.

Therefore, soil degradation and its impact is a complex consequence of physical, chemical, social, political and economic circumstances. It is not surprising that many find the term confusing, preferring a neat technical categorisation. For them, a number of clearly defined physical processes which constitute the status of human-induced soil degradation can be recognised. The categories usually used, most recently by the *Global Assessment of Soil Degradation* (GLASOD) project (Oldeman *et al.*, 1990), are:

- *Water erosion*: Loss of topsoil (usually the finer and more fertile fraction first) by splash, sheet erosion and surface wash – this is the single, most common form of soil degradation, leading to nutrient impoverishment and reduced ability of the soil to support plant production. Also included in this category are terrain deformation such as rill and gully erosion, and mass movements such as landslides. (In many texts, rill and gully erosion are simply put alongside sheet erosion as another form of erosion. This is wrong – they are quite different. Sheet erosion and splash detachment are dynamic processes of stripping of topsoil by water, whereas gully erosion is a manifestation of a degraded catchment. By far the largest quantity of sediment that passes through a gully system is derived from sheet erosion in the gully catchment. Indeed, rills and gullies can be seen as nature's efficient way of shifting large quantities of excess water and sediment.)

- *Wind erosion*: A more or less uniform displacement of topsoil by wind, consisting of detachment by wind power and/or sand-blasting, and transport by saltation or suspension. Terrain deformation may also occur, giving hollows and dunes. Sedimentation or 'overblowing', the coverage of the land surface by wind-carried particles, is as important an impact of wind erosion as physical displacement of topsoil, damaging both structures such as roads and agricultural land.
- *Excess of salts*: The result of the accumulation of salt in the soil solution (salinisation) and of the increase of exchangeable sodium on the cation exchange of soil colloids (sodication or alkalinisation) – see Chapter 3 for descriptions of intrinsically sodic and saline soils.
- *Chemical degradation*: A variety of processes related to leaching of bases and essential nutrients and the build-up of toxic elements. Acidification is the most widespread process, induced by excessive leaching of the soil exacerbated by low organic matter, and removal of calcium in eroded sediments and crops. In turn, pH-related problems such as aluminium toxicity, micronutrient deficiencies and phosphorus fixation compound the impact of acidification on land use. A spectacular but relatively common case (e.g. Thailand, China and the Gambia) of acidification is in coastal regions and deltas with acid sulphate soils (Thionic Fluvisols) where drainage of soil containing iron pyrites decreases the pH to as low as 1.5, see Dent (1986). Pollution by mining and industrial activities and by waste accumulations give rise to location-specific chemical degradation.
- *Physical degradation*: An adverse change in properties such as porosity, permeability, bulk density, structural stability and proportion of water-stable aggregates. It is often related to a decrease in infiltration capacity and plant-water deficiency. GLASOD distinguished three types: (1) compaction, sealing and crusting – very common phenomena on most tropical land-use systems, caused by combinations of poor vegetation cover, low organic matter content of topsoil, particularly vulnerable soils containing a high silt fraction, and inappropriate tillage and machinery; (2) waterlogging – but not including paddy fields(!); and (3) subsidence of organic soils caused by drainage and/or oxidation which leads to a decrease in agricultural potential.
- *Biological degradation*: Increase in the rate of mineralisation of humus without replenishment of organic matter. This category is not separately recognised by GLASOD, but a decrease in organic matter status of soils accompanies almost all other types of degradation as well as being in its own right one of the most common processes impacting on agricultural production.

Notwithstanding the differences in severity of the various types of degradation according to societal and farming system conditions, the Food and Agriculture Organization of the United Nations (FAO) has attempted to define each process and to determine at what level each becomes serious (Table 4.1). It should, however, be noted that there are substantial interactions between the types of degradation and that usually two or more will occur together. The most obviously serious examples will form 'badlands', a descriptive name for massively degraded areas consisting of bare surfaces, sodic soils that are strongly crusted, and zero organic matter, i.e. a combination of water and wind erosion (plus often piping or tunnel erosion), excess of salts, as well as chemical, physical and biological degradation all rolled into one! On the other hand, sheet erosion by water or physical degradation by

Table 4.1 Soil degradation processes, units and definition of serious levels. (After FAO, 1979.)

Degradation process	Definition	Units	'High' severity level
Water erosion	Soil loss	t/ha/yr or mm/yr	>50 >3.3
Wind erosion	Soil loss	t/ha/yr or mm/yr	>50 >3.3
Excess of salts			
Salinization	Increase in electrical conductivity in 0–60 cm layer	mmho/cm/yr	>3
Sodication	Increase in exchangeable Na% in 0–60 cm layer	% per year	>2
Chemical degradation			
Acidification	Decrease in base saturation in 0–30 cm layer, if (a) BS less than 50%, or (b) BS more than 50%	% per year	 >2.5 >5
Toxicity	Increase in toxic elements in 0–30 cm layer	ppm per year	(not yet used)
Physical degradation			
Structure	Increase in bulk density in 0–60 cm layer, from initial level (g/cm^3): (a) <1 (b) 1–1.25 (c) 1.25–1.4 (d) 1.4–1.6	% change per year	 >10 >5 >2.5 >2

Water flow	Decrease in permeability from initial level (cm/h):
	  rapid (20)
	  moderate (5–10)
	  slow (<5)
		% change per year	
			>10
			>5
			>2

Biological degradation	Decrease in humus in 0–30 cm layer	% change per year	>2.5

Note: For all processes other than water and wind erosion, the measurements refer to annual *change* (increase or decrease) in the variable.

crusting and surface compaction could be considered more dangerous by virtue of their insidiousness, being invisible to the casual observer.

Extent of soil degradation

Spurred by international conferences such as the 1992 Rio 'Earth Summit', only recently has considerable effort been devoted to estimating the extent and global seriousness of soil degradation. Country-based estimates of land degradation were provided for the Summit and collated by UN agencies such as the FAO, Table 4.2 is an example for Asia and the Pacific. The most important large-scale assessment is the GLASOD mapping exercise (Oldeman *et al.*, 1990) which used teams of experts in 21 regions to evaluate soil degradation at a continental scale in a systematic way using the categories outlined in the previous section. One of the most detailed uses of the GLASOD estimates to assess land degradation is FAO/UNDP/UNEP's (1994) review of the severity of the process in South Asia and its effect on local people. Other studies of note include: a comparative study at national level, focusing particularly on drylands and 'desertification' using country-based studies (Dregne and Chou, 1992); a world review of erosion with

Table 4.2 Estimates of the extent of degraded land for countries in Asia and the Pacific. (Source: FAO/RAPA, 1992; Douglas, 1994.)

Country	Total land area (×1000 ha)	Arable and permanently cropped area		Estimated degraded land area	
		(×1000 ha)	(%)	(×1000 ha)	(%)
Bangladesh	13 017	9 292	71	989	7.4
China	932 641	96 115	10	280 000	30.0
India	297 391	168 990	57	148 100	49.8
Indonesia	181 157	2 126	12	43 000	24.0
Laos	23 080	901	4	8 100	35.0
Myanmar	65 754	10 034	15	210	3.2
Pakistan	77 088	20 730	27	15 500	17.3
Philippines	29 817	7 970	27	5 000	16.8
Sri Lanka	6 463	1 901	29	700	10.8
Thailand	51 089	22 126	43	17 200	33.7
Tonga	72	48	67	3	4.5
Vietnam	32 549	6 600	20	15 900	48.9
Western Samoa	283	122	43	32	11.3

no particular consistent methodology (Pimentel, 1993); and a global survey of 'hot spots', or areas and land uses which appear to be especially critical in status of soil degradation and impact on food production (Scherr et al., 1995).

GLASOD presents some alarming figures. (The 'health warning' attached at the beginning of this chapter particularly applies to these global and national statistics of the extent of soil degradation. GLASOD used 'expert' opinion. Such opinion is notoriously subjective and prone to exaggeration. Similarly, data on 'desertification' such as appears in Dregne and Chou's (1992) review have been shown to be flawed, especially where the institution involved – United Nations Environment Programme (UNEP) – has a stake in making the situation apparently critical, see Thomas and Middleton (1994).) Of the 8700 Mha of potential agricultural land, including existing cropland, pastures and forests, nearly 2000 Mha, or 22.5%, have been degraded since 1945. Half of this land is officially forest, of which 18% is degraded; 36% is rangeland, of which 21% is degraded; and 17% is cropland, of which 37% is degraded. Water erosion is the principal process identified by GLASOD, with wind erosion important in tropical drylands, and chemical degradation (including severe nutrient losses) particularly prevalent on cropland. Three hundred million hectares have been degraded so severely that rehabilitation would require extremely costly technical interventions, most of which would be uneconomic in terms of regained production. Ten per cent of the land is assessed as moderately degraded and is reversible through significant on-farm investments, and a further 9% is lightly degraded and easily reversible through good soil management.

In regional terms, GLASOD (Oldeman, 1994) shows that the degradation of cropland is most prevalent in Africa (65% cropland affected), followed by Latin America (51%) and Asia (38%), while the degradation of pasture is again more serious in Africa (31%), compared with Asia (20%) and Latin America (14%). Dregne and Chou (1992) presented an even worse picture for the drylands where the dominant land use of range was assessed at 73% degraded.

The survey of 'hot spots' (Scherr et al., 1995) gives probably a more balanced and useful picture of the extent and seriousness of soil degradation. Critical zones are identified on prime agricultural land, e.g. salinisation of irrigated areas in northeast Thailand, water erosion in humid and heavily populated southeast Nigeria. These problem scenarios are presented alongside 'bright spots' such as sustainable intensification of land use on Andosols in Indonesia and South China through multi-storey gardens, and peri-urban intensification in semi-arid Nigeria around Kano.

Soil productivity and degradation

Because of the difficulties of assessing the absolute seriousness of soil degradation, most attention recently has turned to the *impact* of the process (see reviews by Stocking, 1984, and Follett and Stewart, 1985). In the context of agricultural systems, the impact of soil degradation can be measured by the effect on different soil variables, the absolute amounts and value of nutrients connected with eroded sediments, and decrease in yields consequent upon erosion. These various measures are all encompassed by the concept of *soil productivity*, meaning the potential for production by the soil resources now and in the future. Soil productivity is very close to the concept of *sustainability*, in that it emphasises the maintenance of the intrinsic qualities of the soil *for production purposes*. With this emphasis, soil and water conservation changes from a set of technical responses with the objective of keeping soil on-site and water runoff under control to part of an integrated production strategy which meets the needs of the land user – this is a key change of focus in the current understanding of how soil and water conservation should be approached.

For South Asia specifically, FAO/UNDP/UNEP (1994) offered a useful listing of the impacts on agricultural production of soil changes which arose from land use and which had implications for long-term soil fertility. As evidence of soil fertility declined, it noted that since the early 1960s there had been a large increase in fertiliser consumption in the region – Bangladesh, India, Iran, Pakistan and Sri Lanka applied on average more than 70 kg/ha nutrients. Through the 1960s and 1970s, the result was a major increase in crop yields, but this has been attended by a large set of interrelated soil fertility problems, some of which are directly caused by soil degradation, others of which are concomitant processes arising from the need to apply ever-increasing amounts of fertiliser to make up for fertility decline. The list of problems includes:

- *Organic matter depletion*: Crop residues are used for fuel and fodder, and not returned to the soil. In Bangladesh it is reported that average organic matter has declined from 2 to 1%. This leads to degradation of soil physical properties, especially water-holding capacity; reduced ability of the soil to retain nutrients (lower cation exchange capacity and greater leaching); lower release of nutrients from the mineralisation of organic matter.
- *Negative soil nutrient balance*: Removal of nutrients in the crop harvest exceeds the combined amounts of natural and artificial replacement. Negative soil nutrient balances have been reported for all three major nutrients, and for India as a whole it is estimated that the nutrient deficit is 60 kg/ha/yr.
- *Imbalance in fertiliser application*: Initial high responses are usually reported for large doses of nitrogen, but the improved crop growth depletes phosphorus, potassium and micronutrients.
- *Secondary and micronutrient deficiencies*: Sulphur and zinc deficiencies are widely reported, and in Pakistan, with its largely alkaline soils, micronutrient deficiencies are especially serious.
- *Failure of increases in fertiliser use to be matched by increases in crop yield*: A plateau in crop yields appears to have been reached in many situations, where increased inputs are only maintaining yields not enhancing them. Pakistan, for example, has had a consistent, almost linear, increase in fertiliser usage which has not been matched by yield increases in the principal crops: wheat, rice and sugar cane.
- *Lower responses to fertilisers*: Almost no response to nitrogen fertiliser under severe phosphorus deficiency has been found, and a low response to NPK where there was zinc deficiency.

Obviously this catalogue of soil-related effects is not simply a result of soil degradation; some effects arise out of attempts to intensify production which eventually fail, having knock-on impacts on soil quality and declining productivity. Much of the change in productivity status of tropical soils arises from well-meaning but short-term attempts at increased production. FAO/UNDP/UNEP (1994) cited the case of a 33-year fertiliser experiment in Bihar, India, where, despite the use of improved varieties, wheat yields progressively declined with N, NP, and NPK fertilisation, whereas they had risen with farmyard manure. Because of soil degradation, the emphasis on *short-term production*, which tends to be promoted along with agricultural innovations such as increased use of agrochemicals, has serious implications for *long-term productivity*. Soil and water conservation has to address these complex interrelated challenges to soil quality and productivity if it is to be widely accepted by tropical farmers.

There is a variety of ways that the impact of soil degradation may be monitored, each emphasising different aspects of the process of change in soil productivity. The most common measures are:

- *Rate of erosion*: Unfortunately this is a poor proxy for actual impact in terms of yield. For example, at one site in southern Brazil (Itapiranga, Santa Catarina) a 30% decline in maize yields occurred with a cumulative erosion (over 2–3 years) of 27 t/ha for a fertile, well-structured Cambisol. The same yield decline occurred with over 900 t/ha of soil loss on a deep Oxisol.
- *Enrichment ratio* (ER): This measures an aspect of the impact of water erosion – the degree of selectivity of the removal of fine (colloidal) particles of soil which contain most nutrients and water-holding capacity of the soil. ER is simply the ratio of the concentration of a nutrient (e.g. nitrogen or phosphorus or organic carbon) in the eroded sediment to the concentration in the topsoil from whence it was derived. Typical long-term averages measured by the author in Zimbabwe on relatively sandy soils were 2.5 for nitrogen and 2.7 for phosphorus, indicating that a unit quantity of water erosion has a potential impact far greater than might be thought from the total mass of soil removed. In contrast, Tegene (1992) found ERs in the range 0.9 to 1.2 in Highland Ethiopia on relatively rich clay soils.
- *Crop yield decline*: Many studies have shown degradation-induced yield decline. A major research programme coordinated by the FAO Network on Erosion Productivity involving some 23 groups in 20 tropical countries is currently engaged in determining the precise relationships between erosion, yield and time for the main agroecologies and principal crops. Of particular importance is the translation of that yield decline into monetary terms. This can be done through:
  ○ using crop prices to assess the financial loss – typically in a single year, the value of yield loss may amount to US$15–75/ha, depending on crop type, base yield level and degree of soil degradation;
  ○ determining the cost of inputs to bring yields back to pre-erosion levels – in the Brazilian Oxisol case cited above even with an expenditure of over US$250/ha on recommended chemical fertilisers, yields never completely recovered.
- *Impact on livelihoods and farming systems*: Measures may range from increasing need and difficulty of cultivation, cost of additional land required to compensate for yield declines, need to find off-farm income, to extremely severe impacts such as disease, enforced migrations (as in the Horn of Africa), starvation and death.

An example of the impact of erosion on soil quality in the humid tropics is given in Table 4.3. The erosion 'phases' in Sierra Leone refer to varying degrees of historical erosion with the severe phase representing over 200 t/ha cumulative loss from the site since the commencement of land use. For both topsoil and subsoil, substantial changes in soil properties occur with erosion, all of which potentially could limit crop growth. A particularly common phenomenon of Ultisols and Oxisols of the humid tropics is demonstrated: the decreasing soil pH related to

Table 4.3 Influence of soil degradation on variables of soil quality – data from a deep sandy clay loam Oxisol (Ferralsol) in the transitional rainforest–savanna woodland of Sierra Leone. (Source: Sessay and Stocking, 1995.)

Soil Property	Erosion phase			
	None	Slight	Moderate	Severe
Topsoil: 0–20 cm				
pH	5.6a	5.0b	4.6c	4.4c
Exchangeable Al (meq/100 g)	1.1a	1.4b	1.8c	2.1d
Organic matter (%)	4.6a	3.9b	3.6b	2.5c
Total N (%)	0.20a	0.15b	0.12c	0.10d
Ca (meq/100 g)	2.80a	1.55b	0.32c	0.18d
Mg (meq/100 g)	1.10a	0.80b	0.40c	0.24c
K (meq/100 g)	0.14a	0.08b	0.05c	0.03d
Available P (ppm)	12.3a	8.0b	4.0c	3.2c
Subsoil: 20–40 cm				
pH	5.5a	5.0b	4.8c	4.5c
Exchangeable Al (meq/100 g)	1.47a	2.02b	2.32b	2.68c
Organic matter (%)	3.20a	2.82b	2.48b	2.10c
Total N (%)	0.15a	0.13b	0.11c	0.10c
Ca (meq/100 g)	0.43a	0.36b	0.30c	0.22d
Mg (meq/100 g)	0.33a	0.24b	0.19c	0.12d
K (meq/100 g)	0.09a	0.05b	0.04c	0.04c
Available P (ppm)	3.7a	3.0b	2.2c	2.0c

Note: Same letter within a row indicates no significant difference at 5% level of probability; all data based upon mean of six values.

increasing mobility of aluminium and fixing of phosphorus – one of the most powerful multi-factor limitations to crop root growth. In the same experiment recorded in Table 4.3, a number of soil physical properties also worsened: pF-status and related influences on plant-available water capacity, porosity and infiltration, and bulk density. Although many of these physical and chemical changes are interrelated, their combined effect is to make it extremely difficult to counteract the impact on crop yields. A corresponding trial with maize and cowpea on the erosion phases showed yield reductions of 31, 47 and 75% for maize and 15, 28 and 48% for cowpea with slight, moderate and severely eroded soils respectively. A stepwise multiple regression test on these yield losses in relation to soil properties showed that 99% of the variation in maize yield is attributable to depth of topsoil, aluminium saturation, available water and organic matter. Similarly, for cowpea yield, variations with erosion are accounted for primarily by topsoil depth, potassium and bulk density (Sessay and Stocking, 1995).

Explaining and predicting soil degradation

Why does soil degradation occur? A seemingly innocent question such as this has implications which go well beyond the physical and chemical processes in a field, and the techniques (or lack of them) employed by the farmer. Most farmers know that their practices are causing soil degradation, but either they are powerless to change without endangering their livelihoods – the typical tropical developing country situation – or it is apparently uneconomic or impractical to change – the developed country and/or intensive, mechanised farm situation. In either situation, the question 'Why does soil degradation occur?' should more realistically translate to 'Why do farmers and land users allow it?' This is the essence of the 'political economy' or 'political ecological' approach to explaining soil degradation pioneered by Blaikie (1989) in his 'chain of explanation'.

Awkward questions are often raised by the political economy approach. Instead of the land user being unambiguously the culprit for erosion, the finger of blame may also point to others. Experts may promote unwise or inappropriate recommendations. Governments may impose policies which discourage investment in land-improving techniques. International organisations may fund new schemes such as reservoirs without appreciating, for example, the displacement of marginal farmers onto steep slopes. Indeed, the whole international economic order may be blamed where developing countries cut back on their services to farmers and promote export crops without regard for the conservation implications.

To illustrate from an example quoted by Pretty (1995): in the upland farming region of the volcanic highlands of Tlaxcala, Mexico, farmers have long maintained canals, terraced fields and *cajetes* (pits at the base of the terraces) to capture soil and nutrients to compost and return to the fields. A sophisticated, indigenous, soil fertility management system provided for a wide variety of products as well as excellent conservation. With increasing work opportunities in Mexico's urban areas, many farmers are abandoning these traditional practices by substituting external inputs which demand less labour and monocrop cultivation of 'improved' varieties. Credit is only available for farmers who monocrop and use pesticides. Government subsidies are only supporting construction of terraces by tractor without *cajetes*. The impact of the loss of this traditional complex system of agriculture is that terraces are not being repaired; farmers are compensating for erosion-induced loss in soil productivity by artificial fertilisers; and soil and nutrient loss now plague downstream water storages. Inabilities to repay credit loans demand more and more off-farm wage income – a vicious cycle which impacts on the environment and further prevents land improvement. Where does the blame lie? The answer depends on one's own perspective, and the use to which the answer may be put. At the farm level, individuals have exploited the market opportunities in the city, and sought to continue land use without the original high investment in their own labour. At the level of local experts and professionals, innovations such as high-yielding monocrops and pesticides have been introduced without consideration of the wider implications to the farming system. At governmental level, credit and subsidy schemes have been implemented without appreciating the critical role of intercrops and composting in conservation. At an international level, Mexico's economy continues to be dominated by currency crises which prevent any effective assistance to rural people and which only accentuate rural–urban migration. In such an analysis of the causes of erosion, what happens in a Mexican upland field tilled by a peasant farmer, can be related to decisions taken in Geneva by a corporate banker and in New York by an international politician.

Searching for the guilty is not a fruitful exercise. However, an understanding of the complex setting and of the reasons why farmers are allowing erosion in a field is essential in the design of soil and water conservation. For example, it is pointless developing a superb technical package of bench terraces, gully control structures and tree planting for steepland marginal farmers, when the likelihood is that they will be dispossessed of their land as soon as the improvements are implemented. A campaign amongst tropical small farmers to grow cover crops which have no function other than soil improvement is doomed to failure. Therefore, how can degradation be explained and predicted at the various levels in Blaikie's 'chain':

- *Physical symptoms and economic impacts in the field*: The most important site factors for water erosion are soil type which gives an intrinsic soil erodibility, rainfall which determines the erosivity, vegetation which controls the protective cover, and slope which affects the volume and velocity of runoff. Each factor may be subdivided into a number of variables which appear to be most closely related to erosion as determined from experiments. So, for example, rainfall erosivity has for a long time been known empirically to be best determined by EI_{30}, a composite variable involving rainfall energy (E in J/m², a measure of the detaching capability of raindrops) and the maximum sustained intensity for 30 minutes (I_{30} in mm/h, a proxy for runoff volume). Similar but differing variables apply for the other factors and forms of degradation – see any standard text on soil erosion (e.g. Morgan, 1986; Hudson, 1995).

These factors and variables are employed in prediction models. The two most widely used for practical conservation design purposes are the US-derived 'universal soil loss equation' (USLE) (Wischmeier and Smith, 1978) and its most recent revision (RUSLE) (USDA/ARS, 1994) and the Zimbabwe-derived 'soil loss estimator for Southern Africa' (SLEMSA – see Fig. 4.1) which is now used in a number of tropical areas because of its simplicity and better applicability to developing country conditions (Elwell and Stocking, 1982).

Prediction models have also been developed for assessing the impact of soil erosion on crop yields. The best known is the 'erosion–productivity impact calculator (EPIC) (Williams et al., 1983), although of probably better applicability to the tropics is the 'productivity index' (PI) (Pierce et al., 1983) which is relatively simple, the 'soil life' model (Elwell and Stocking, 1984), and a composite model developed for dryland conditions in Botswana (Biot, 1990).

It is only at the level of physical symptoms in the field and their effects on productivity can explanation and prediction be made quantitative. This may partly explain why virtually all soil and water conservation research over the years, especially by the Soil Conservation Service of the United States Department of Agriculture (USDA), has been devoted to site-based physical prediction models and the determination of the values in their models. While this has an obvious function for individual fields and farmers in conservation planning for developed country agriculture, it

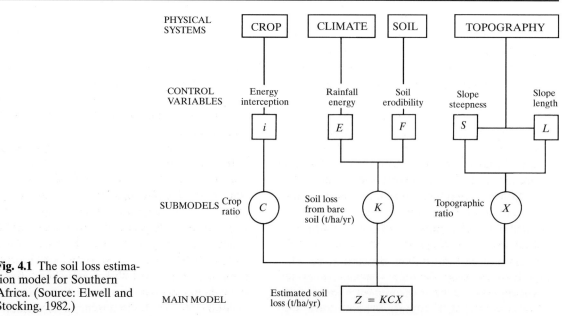

Fig. 4.1 The soil loss estimation model for Southern Africa. (Source: Elwell and Stocking, 1982.)

provides only a very small proportion of the knowledge needed to plan effectively for conservation in developing countries. There is a further danger that the use of prediction models developed mainly in temperate environments may give rise to unacceptable errors in the far more erosive tropical rainstorms and on the more erodible tropical soils.

- *Land-users resource perspective*: Soil degradation may be explained in terms of the land users' access to and employment of the basic means of production – land, labour and capital. In the tropics, soil and water conservation is crucially controlled by labour availability, the seasonal demand for family labour, gender differentiation in labour tasks and particular crops (in East Africa, for example, cash crops such as cotton, which have a high erosion hazard, are largely managed by men, while home gardens and vegetables, which are far less hazardous, are controlled by women), and the opportunity cost of labour with regard to the relative attractiveness of off-farm employment. Land is also significant: ironically, under population pressure and with only small plots available, farmers may be forced to intensify and develop more sustainable land uses. In the case of Machakos District in Kenya, erosion was far more serious in 1937, yet population density was scarcely one-third that of 55 years later (Tiffen *et al.*, 1994) – a classic example of society adapting to declining soil and land resources, and population growth spurring intensification and better land management.

    Prediction of the complex interactions of resources required for land use and their consequent impact on processes of soil degradation are very difficult. Nevertheless, the international agricultural research centres have developed various types of farming systems research which enable constraints and bottlenecks in local production systems to be identified – these potentially can have large influences on uptake of soil improvement and conservation measures, although few examples exist of actual application. Rural development practitioners have also constructed various types of rapid rural appraisal and participatory techniques, the main objective of which is to seek the land users' own analysis of the situation (see Okali *et al.*, 1994). Obvious advantages of participatory techniques are that technologies seen through farmers' eyes will be automatically analysed for their social, economic and cultural applicability. Soil and water conservation programmes that result from full participation by local people should have better chances of success, but again few examples exist (Lok, 1992).

- *National policy level and the nature of the state*: Explanations for soil degradation at the national level centre around the effects of history, current policy, legislation and incentives on the actions and attitudes of land users. Important aspects of national policy which will affect soil degradation include the distribution of rights to land, the laws of inheritance, attitudes towards support of the agrarian society, the abilities of the administration and enforcement agencies, the influence of local politicians and the relative strength of central government, educational provision, and a whole range of factors which may stem from policies such as gender divisions of labour, off-farm labour opportunities, provision of markets, and so on. While some of these factors may seem at first sight far removed from the processes of soil degradation, they can be most influential in conditioning what land users do with their resources and hence the propensity of those land-use practices to cause degradation.

    One of the most powerful examples of the effect of national policy is provided by the historical legacy of British colonialism in Africa (SADCC, 1987). Although colonial agriculturalists were well-meaning and often worked through chiefs and local governments, significant resentment was fostered by the strict enforcement of what the local people saw as repressive policies and laborious demands to dig physical conservation structures (Anderson, 1984; Stocking, 1985). As evidenced from Sukumuland in Tanzania, resistance to the 'Plough Rules' (which prevented, for example, people from planting on some of the best soils – the *mbugas* or wetland Vertisols) and the fines for failure to construct earth bunds on arable lands helped fuel protests, riots and, in at least one case, deaths. Such was the strength of feeling that several independence parties (e.g. the Tanganyika African National Union) campaigned on an anti-conservation platform, and conservation works were physically destroyed and trees uprooted.

The security to rights to use land is also an influential factor in local people's willingness to protect land resources and invest in land improvement. A change from 'open access' where there are no limits to exploitation of the resources to 'common property' where local society controls usage can have fundamental benefits for the soil. For example, the *ngitili* or village dry season grazing reserves in Sukumuland were, before the implementation of *ujamaa* (villagisation) in 1972–73, a valuable resource which helped protect the integrity of the open grazing lands especially in dry years when grazing reserves would run out long before the onset of the rains. Management of these reserves was vested in the village elders, and those who infringed the local rules had their cattle confiscated – a most effective way of ensuring compliance! Recent changes in Tanzania to private rights to land, now enshrined as national 'free-market' policy, has reportedly had a significant effect on people's willingness to plant trees and engage in agroforestry. In the Usambara Mountains, for example, the considerable land degradation caused by the illegal exploitation of the rainforest under-storey for cardamom (a valuable spice) production has been reduced because the essential shade conditions can now be reproduced on farmers' lands by planted trees.

The examples of where governments may consciously or subconsciously be promoting soil degradation by their policies are legion (see Blaikie and Brookfield, 1987). It is only through an understanding of the national policy context can suitable conservation strategies be designed which will either change national policy or work within existing policy to alter a major constraint (e.g. by providing labour subsidies for terrace construction).

- *The international economy*: Finally, the world or international economy cannot be ignored as a major explainer of why farmers allow soil erosion to occur on their land (or, conversely, why farmers practice soil conservation). The provision of aid and assistance programmes can be a powerful stimulus to governments either to ignore conservation or to promote it. The World Bank has an unhappy record of projects especially in South America, whose side-effects have been to promote deforestation. Structural adjustment programmes, enforced by agencies such as the International Monetary Fund (IMF), in countries as different as Bolivia and Uganda, are widely felt by many observers to be responsible for cut-backs in government services and neglect of the rural sector to the detriment of the protection of soil resources. Other agencies are similarly culpable, and worry about the unintended effects of donor support has led to widespread use of 'environmental and social impact assessment' (for a good review of this and the relationship with resource management see Smith, 1993). Further, the structure of the international economic order often ensures that developing countries have to follow agricultural policies which are environmentally degrading – the case of soya bean cultivation in the Brazilian *cerrado* for export to the United States in order to reduce the national debt is notorious; less well known is the huge amount of erosion caused by ploughing up grassland for the soya bean farms.

One particular case well exemplifies the international ramifications of the economic dependence of developing countries on donor aid: the marginalisation of people displaced by major dam schemes funded by the international community. For example, a consortium of donors including the British Government and the European Community provided Sri

Lanka with five major reservoirs to generate electricity and feed irrigation water to the Mahaweli Scheme. From the Victoria Dam alone, a considerable number of the 40 000 displaced families migrated into the Hill Lands, and an unknown number of these are now contributing to erosion on steep slopes of the catchments to the dams. So, a decision taken in Whitehall, London, with the persuasive support of the overseas government, can have the outcome of erosion on an isolated hillslope in South Asia.

SOIL AND WATER CONSERVATION STRATEGIES

The change in emphasis in soil and water conservation (SWC) today is well expressed by Critchley (1991, p. 51):

'The "new approach" to SWC [soil and water conservation] has come about more by necessity than design. The conventional type of SWC project – with its emphasis on building structures and reducing soil erosion – has failed so often that there has been no option but to change strategy. With the new thinking, kilometres of expensive terracing and rigid targets are out – and people's participation, flexible workplans and conservation for production take centre stage.'

In other words, a new agenda has been formulated for soil and water conservation in the tropics and for developing countries, representing a major strategic shift away from it being a package of techniques, the 'technical fix', towards soil and water conservation as an integral part of rural development. Douglas (1994) has identified a number of strategic 'dimensions' that need to be addressed in the planning and design of soil and water conservation:

(1) The *biophysical dimension*, which considers the processes of land degradation, and relates these to effects on vegetation, soil, water, productivity and other changes to the quality of the natural resources. This measures the biophysical need for soil and water conservation through techniques such as 'erosion hazard assessment', 'productivity modeling', 'land capability classification' and 'land evaluation' (see Davidson, 1992 for a useful review of the various techniques of land evaluation). These assessments, in turn, lead to the definition of technological options, or what might be termed the 'agroecological possibilities' for soil and water conservation. However, as this chapter has emphasised already, the best technological solution may not be enough. Hence, with an array of possible technologies (some perhaps technically more efficient than others – but the less efficient remaining on the option list), the soil and water conservation strategist must move quickly to the next 'dimension'.

(2) The *social and cultural dimension* takes in many of the non-quantifiable but essential ingredients for the acceptance of new technologies. This includes such aspects as the historical legacy of land use and inheritance rights. Douglas (1994) reported a case in Sukabumi District, West Java, Indonesia where in the main only farmers who owned upland plots had planted trees and engaged in terracing; the tenant farmers and share croppers who were surveyed explained that trees were considered the property of the landlord, while terraces provided too few short-term benefits commensurate with the high labour input for construction. Knowledge of land tenure rights is crucial to determining soil and water conservation strategy: they may range from owner–occupiers who will carry out private investment work; to cooperative groups such as the Kenyan women *harambee* groups who undertake major terrace construction for each other; to squatters who have to take an extremely short-term view of land use, essentially mining the soil for immediate production. The risk of land alienation is a particular problem in countries with unstable regimes and frequent conflict – the wars in the Horn of Africa continue to have impact on the management

of the natural resources, frequently to the detriment of conservation. Land fragmentation, as is common in much of South Asia, is a constraint which prevents large-scale conservation and watershed improvement works because of the need for landholders to co-operate in the siting of drainageways and terraces which would effectively mean some farmers losing much of their land to an 'unproductive' use. In Africa and India particularly, livestock have a social significance out of all proportion to their practical use in production (e.g. meat, milk or draught power). Conservation programmes of compulsory destocking have, therefore, encountered intense resistance because animals represent wealth and status in local society, notwithstanding their scrawny bodies and overgrazing of the range.

(3) The *economic and financial dimension* looks at the rationale for engaging in soil and water conservation in monetary terms. As a unit of exchange, money has dominated decision-making in developed country agriculture; it is assuming the same status in developing countries. For example, market reforms in China have radically altered the private rationality of different forms of soil and water conservation. Schemes of bench terracing whole hillsides were sensible in a centrally planned economy because the collective good was more easily realised than the private – individual peasants would not have gained if they had improved their own land, because any increase in productivity would have been shared (or taken by the state). Now it is rational for a farmer to improve private land because the gain is retained; conversely, to divert labour resources to collective activities such as terracing is irrational. The impact of the changes can be seen most spectacularly on the Loess Plateau in China; terraces, communal orchards and other large enterprises are falling into disrepair, while small-scale irrigation, composting, cover crops and other essentially 'private' forms of resource management are flourishing. Provided that monetary benefits outweigh costs, and that factors such as riskiness of investment in new technologies are also accounted for, then market forces alone have determined the success of many forms of soil and water conservation. Maize–*Mucuna pruriens* (velvet bean) cropping systems in southern Brazil have seen yields of the cereal multiply by two to three times with minimal fertiliser use, good weed control and almost no erosion – the immediate success of this system has ensured its rapid spread and local adaptation. If, however, benefits take a long time to accrue (as they often will if major earth-moving is planned), then an economic analysis must take into account that the costs are incurred well before the benefits are realised, and hence the soil and water conservation measure becomes less attractive than in a purely financial accounting of costs and benefits. Economists use discount rates and variable time horizons to incorporate these aspects into the analysis (Bishop, 1992) – inevitably they will mean that all forms of soil and water conservation must have some aspect of short-term benefit for the land user, often through providing more reliable plant water, as well as providing for long-term soil improvement.

(4) The *policy dimension* sets the political environment in which change in land-use practice might be accommodated. Farmers, governments, local officials, professionals and donors all have differing perspectives and priorities. Land users engage in soil and water conservation when soil degradation is perceived as an immediate threat to their livelihoods. Governments' policies for soil and water conservation are quite different: as Fones-Sondell (1989) discussed for Africa, short-term increases in total production are favoured and conservation *per se* is rarely a high priority, except when donors demand. Similarly, donors as represented by the professionals who work in their aid institutions, have many competing priorities, most of which have to be squeezed into the typical 3- to 5-year project cycle, an anachro-

nism in natural resource management which leads to inadequate attention to the sustainability of any improvements. At the national level, many governments have sought to raise conservation awareness through National Conservation Strategies (e.g. Zambia, Vietnam) and National Soils Policies (e.g. Syria and Uganda). A number of regional conservation policy networks now operate: e.g. the Asian Soil Conservation Network based in Jakarta (which publishes *Contour*), and the Southern African Development Community (SADC) Environment and Land Management Sector Coordination Unit in Maseru, Lesotho (with its magazine *Splash!* which since 1987 has done much to raise the awareness of policy-makers and professionals within Southern Africa to topical issues in soil and water conservation). Policy interventions can be very effective in encouraging soil and water conservation: these may range from direct incentives and subsidies (common in South Asia), food-for-work programmes (e.g. Brazil's Emergency Work Programmes for the dry northeast which include terrace and dam construction), legislation (e.g. Zimbabwe's laws prohibiting cultivation in certain hazardous zones), provision of markets and producer pricing policies, to major programmes of land reform (e.g. the Common Agrarian Reform Programme in the Philippines).

(5) The *institutional dimension*. Many land-use activities depend upon the institutions which support and service them. Institutions in developing countries can be notoriously weak, especially in the provision of agricultural extension advice and organised information for farmers. Soil and water conservation is particularly problematic for traditional government institutions organised in line ministries of agriculture, forestry, economic development, finance and so on. Multisectoral support is essential with interdisciplinary teams, including social scientists and development agents. Only a few countries have addressed the challenge: Zambia's Adaptive Research Planning Teams and Malawi's Land Husbandry Branch are good examples. In Sri Lanka in 1986, this author counted 12 ministries and 38 government agencies with a remit for some aspect of soil and water conservation – an impossible fragmentation of responsibility, leading inevitably to paralysis in institutional support with no one agency effective enough to coordinate the inputs of others. On the other hand, many community institutions can be strong and extremely effective in supporting local efforts of soil and water conservation, for example the *mobisquads* in Ghana, organised for natural resource management initiatives such as agroforestry (Dorm-Adzobu *et al.*, 1993). Non-governmental organisations (NGOs) are playing an increasingly important role in soil and water conservation: the international NGO Oxfam is a major supporter of local initiatives such as stonelines in West Africa (Atampugre, 1993), while national NGOs in India are now valued partners with government in implementing participative and integrated watershed development (e.g. PIDOW (Participative and Integrated Development of Watershed) in Karnataka which has an excellent soil and water conservation manual: Premkumar, 1994). The role of institutions at all levels is essentially to facilitate soil and water conservation: without existing groups such as women's and youth organisations, commodity associations and co-operatives, many projects have had to invent self-help groups. However, the record of imposed institutions at the local level is not good, and many have failed as soon as government and donor support is withdrawn.

How in practical terms can these 'dimensions' be integrated in order to plan and design a soil and water conservation programme? Figure 4.2 presents a conceptual framework for integrating the various strategic components. Note that the emphasis is on soil and water conservation as part of a production and land-use strategy, with ecological factors merely providing one facet of the choices that farmers have to make. The 'soil

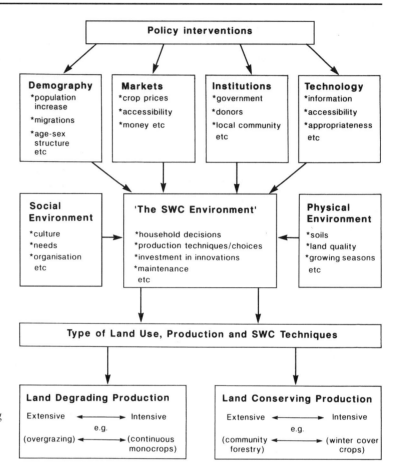

Fig. 4.2 A framework for organising the principal strategic issues in soil and water conservation (SWC). (Adapted from Scherr *et al.*, 1995.)

and water conservation environment' is thus exceedingly complex. An important implication of that complexity is that actual soil and water conservation techniques have to be extraordinarily diverse. There are no universal solutions, and techniques, if they are to be promoted by external agencies, have to be assessed for acceptability on a range of multi-factor criteria, not just technical efficiency.

SOIL AND WATER CONSERVATION TECHNIQUES

The emphasis in most standard texts on soil and water conservation is on describing and listing the range of techniques that are implementable in the major world agroecological zones. Deliberately, this chapter has avoided putting technical design issues foremost because of the necessity of seeing techniques in their holistic context. Nevertheless, we cannot entirely avoid describing the major practical approaches which have been found to work in various parts of the tropics. Texts such as Hudson (1995), general manuals of techniques such as Sheng (1986, 1990) and Critchley and Siegert (1991), and country-specific technical guidelines such as the particularly useful one for Ethiopia (CFSCDD, 1986) will give the reader ample information on dimensions of structures, design criteria, and ecological applications. This section will highlight examples of some of the major types of techniques, ranging from broadly biological approaches through to engineering structures.

Agronomic techniques

The use of plants, whether growing or dead, as a protective layer to intercept raindrops and prevent detachment of soil particles must be the first line of defence in the design of any effective soil and water conservation programme. Vegetation is nature's way of ensuring soil and water conservation, so it could be said that biological strategies merely mimic natural systems of protecting landscapes. In general terms, plants have the advantages of: low cost; technical effectiveness; relative simplicity; support to production objectives of the land user; and additional spin-off benefits. These added benefits may relate to the soil in improving organic matter levels, nutrient availability and water-holding capacity; to water availability in improving infiltration and preventing surface sealing; and to increased resistance to erosion through better surface aggregation. Indeed, agronomic or biological measures enable a 'virtuous circle' whereby both production and protection are promoted. Most modern writings in soil conservation have extolled the virtues of using vegetation as the principal means of soil and water conservation (e.g. Shaxson *et al.*, 1989; Young, 1989).

The principal agronomic measures may be divided into the following broad groups: vegetative strips and barriers; cover crops and multi-storey layers; mulching; and tillage systems which leave surface residues. A large variety of plants have been shown to have potential for soil and water conservation in tropical agriculture – a small sample only is listed in Table 4.4. In choosing a variety, first priority must go to production benefits and growth potential; then secondly, its structural characteristics and ability to maintain a dense cover throughout the rainy season.

Vegetative strips and barriers

Known generally as 'stop–wash barriers', these include grass strips of various widths up to about 1.5 m, barrier hedges, and trashlines pegged or left on the contour. Normally the barriers should be permanent, although in western Uganda it is known that farmers exploit the better fertility of the soil under the strip every few years, replanting the barrier in the old field. A dense and continuous cover is desirable so that runoff is slowed and sedimentation encouraged. In time the accumulation of sediment within and above the barrier achieves a bench-terracing effect. The usual recommendation for grass strips is that they be planted on slopes up to 15% at approximately 10 m spacing. In India, *Pennisetum pedicilatum* has been widely promoted because of its dense root structure, resistance to drought and forage value. However, other grasses with a higher local economic value could be better. For example, in parts of west Bengal and Bihar, *Palindum angustifolium* shows particular promise because of its growth characteristics and its use in specialised ropes for bed-making. As these ropes are a village industry, the added value from the activity can be great – indeed, the household return from the grass strip can exceed the crop yield from the fields that the strips are intended to protect! Support practices encouraged by vegetative barriers include contour cultivation, raised boundary bunds, strip cropping and intercropping.

Vegetative barriers are not without some disadvantages. Competition for moisture and nutrients between the barrier plants and intervening crops can be intense, depending on chosen species. 'Alley' or hedgerow intercropping systems, widely promoted by the Nigeria-based International Institute for Tropical Agriculture (IITA), have not achieved widespread acceptance because of the heavy demand for labour in pruning and the difficulties of management. Many small farmers also complain about the depredations on their crops of rats and other vermin who live in the grass strips and birds such as *Quelea quelea* who roost in hedgerows and trees.

Special mention must be made of barrier grasses which have been promoted for their unpalatability to animals. Vetiver grass (*Vetiveria zizanioides*) has received widespread attention in soil and water conservation projects, along with species of lemon grass or citronella such as *Cymbopogon citratus* and *C. nardus*.

Table 4.4 Some potentially useful species in agronomic approaches to soil conservation. Common names are primarily those used in India and Anglophone Africa. (Source: Stocking, 1993.)

Botanical name	Common name	Principal use and remarks
Grasses		
Andropogon spp.	Blue grass	Densely tufted. Average forage value
Brachiaria humidicola	Koronivia grass	Tough, resistant grazing
Brachiaria latifolia	Tanner grass	Used in East Africa for soil conservation. Dense sward
Brachiaria mutica	Para grass	Good creeping grass for erosion control. Waterways and bunds
Cenchrus ciliaris	Foxtail	Tufted. Drought resistant. Good, palatable grazing. Tolerates heavy grazing
Chloris gayana	Rhodes grass	Tufted and strongly stoloniferous. Used on improved pasture
Cynodon dactylon	Bermuda grass	Reasonable low forage. Stoloniferous. Drought resistant and hardy. Used by many farmers, gathered from bunds and waste land
Digitaria decumbens	Pangola grass	Excellent forage and erosion control
Eragrostis curvula	Weeping lovegrass	Good forage and soil binding qualities
Palindum angustifolium or *Eulaliopsis binata*	Sabai grass	Woolly tufted perennial. Used for fodder and rope-making. Highly prized but normally gathered from wild
Panicum maximum	Guinea grass	Tufted perennial. Forage, drought resistant
Paspalum notatum	Bahia grass	Perennial. Good cover, shade tolerant. Reasonable forage
Pennisetum clandestinum	Kikuyu grass	Creeping, good fodder and erosion control
Pennisetum pedicilatum	Dinanath	Annual recommended in India for soil stabilisation
Pennisetum purpureum	Napier grass Elephant grass	Excellent fodder and mulch. Good in cut-and-carry systems. Drought resistant
Saccharum munja	Sar	Good for conservation and fodder. Occurs naturally in East India
Sehima nervosum	—	Tufted perennial. Good fodder. Good on poor soils, low rain
Tripsacum laxum	Guatemala grass	Good forage. Resistant. Insecticidal properties – used on Sri Lankan tea estates against nematodes
Vetiveria zizanoides	Khus grass Vetiver grass	Strongly promoted for soil conservation. Unpalatable but good for grass strips
Perennial legumes		
Centrosema pubescens	Centrosema	Good cover, vigorous growth. Mixture with grasses possible
Desmodium intortum	Greenleaf desmodium	As centrosema
Indigofera endecaphylla	Creeping indigo	Good cover
Pueraria phaseoloides	Tropical kudzu	As centrosema
Stylonsanthes guianensis	Stylo	Mixed planting with grasses for forage
Tephrosia candida	Tephrosia	Woody hedge for fodder and fuel
Annual legumes		
Cajanus cajan	Pigeon pea Arhar dhal	Good barrier hedge, forage and food (perennial varieties available)
Canavalia ensiformis	Jack bean	Good cover, green manure and forage
Crotalaria juncea	Sunn hemp	Good green manure and forage
Lablab purpureus	Lablab bean	Cover, green manure and food
Lupinus luteus	Yellow lupin	As lablab

Table 4.4 *Continued*

Botanical name	Common name	Principal use and remarks
Macroptilium atropurpureum	Siratro	Used as legume mixture in replanting grazing land
Mucuna capitata and *M. pruriens*	Velvet bean; mucuna	Good cover, forage; good for interplanting
Vigna sinensis	Cowpea	Good cover and forage; useful in strip cropping
Other plants		
Acacia albida	—	Valued by farmers in semi-arid areas for fodder and beneficial interactions with crops
Acacia nilotica	Babul	Live fence, fodder
Azadirachta indica	Neem	Drought tolerant. Insecticidal properties of sap
Carissa spinarum	—	Thorny dwarf shrub. Use on bunds as live fence
Cassia siamea	Cassia	Drought tolerant. Good on poor soils. Hedgerow intercrop
Casuarina spp.	Casuarina	Windbreaks, erosion control, dense surface root mat
Chamaecytisus palmensis	Tagasaste; tree lucerne	Dryland leguminous fodder tree. Hardy, drought resistant. Suitable light soils on poor, eroded sites
Dendrocalmus spp.	Bamboo	Soil improver. Bund stabiliser. Barrier hedges. Gully planting
Gliricidia sepium	Gliricidia	Hedgerow intercropping. shade tree
Ipomea batatus	Sweet potato	Grown on ridges, gives good cover
Leucaena leucocephala	Subabul; Leucaena	Woody, live hedge, alley cropping, mulch, fodder supplement
Prosopis cineraria	Khejro	Said to have beneficial crop interaction. Good fodder. Dry area
Robinia pesudoacacia	Black locust	Reclamation eroded land. Steep slopes. Very resistant. Timber
Saccharum spp.	Sugar cane	Strip and ridge crop. Good barrier and soil binding
Sesbania sesban	—	Good intercrop
Zizyphus nummularia	—	Good for contour furrows. Deep rooting
Vitex negundo	Nishinda	Shrub with insecticidal properties

There are two competing views about such species. In standard technical approaches to soil and water conservation, the unpalatability of the grass is seen as a benefit in that the integrity of the strip will not be challenged. In the alternative view, the unpalatability is a disadvantage in providing no beneficial economic value to the farmer, while at the same time locking up some 15% of the land surface in unprofitable plants. It is perhaps instructive to recount an incident from Rajasthan, India, where vetiver had been planted as a conservation barrier. When one local farmer found that he could extract essential oils from the roots of vetiver and sell these at high price on the local market, the grass strips disappeared almost overnight!

Also, in respect of the need for any soil and water conservation species to be supportive of the small farm livelihood is the case of grass strips in Swaziland. The fact that virtually all of Swaziland's arable area is well protected by strips is now acknowledged to be because of the reserves of dry season grazing they contain – the foremost constraint in farming in that country is the deplorable state of the communal grazing lands and their inability to support animals through the dry season. Because the arable lands are protected from grazing through the rains, grass growth is prodigious; and only after harvest, well into the dry season, are these privately held lands opened for grazing on the stover and the grass strips. Conservation reasons for the grass strips are only mentioned by farmers as a secondary factor. Obviously, the use of any species, palatable or not, needs to be matched with farming needs and requirements, not just technical efficiency. Where land area is not a constraint, vetiver might well be a useful species amongst others to be tested and promoted (National Research Council, 1993).

Cover crops and multi-storey layers

Legume cover crops are widely used throughout the tropics and are recognised for their considerable benefit in fixing soil nitrogen (Giller and Wilson, 1991). Some of the most successful cover crops are used in plantations as under-storey forage and pasture; common species include calopo (*Calopogonium mucunoides*), the shrubby legume *Centrosema pubescens*, and tropical kudzu (*Pueraria phaseoloides*). With their creeping and trailing characteristics, they not only provide a good cover, but also smother vigorous weeds and grasses. Benefits of cover crops in plantations are usually seen in enhanced yields: Giller and Wilson (1991) reported many examples, such as rubber trees with a legume cover crop which can start to be tapped 18 months before a plantation with non-leguminous cover; and although the legumes die out after about 5 years as shade increases, the residual benefits are apparent for at least 20 years. The primary reason for the increased yields is undoubtedly fixed nitrogen – a good *Centrosema* cover fixes 150 kg/ha/yr of nitrogen – but also important is the increased organic matter, better infiltration, improved water-stable aggregates and maintenance of fertile particles *in situ*.

Cover crops have also found their place in some tropical and subtropical cropping systems. The Instituto Agronômico do Paraná in southern Brazil has been a notable leader in the development of small-farm cover cropping combinations with maize and other cash crops (Calegari, 1995). In the adjacent Brazilian state of Santa Catarina, many small farmers on nutrient-deficient Oxisols are using winter cover crops such as *Raphanus sativus* or black rye (a non legume, but good in breaking up soil compaction and building organic matter) as soil improvers and green manures, finding that their subsequent summer maize yields are two to three times greater.

Multi-storey cropping is common throughout the humid tropics, with the Sri Lankan 'Kandy' and the Tanzanian 'Chagga' home gardens being amongst the better known. Within only a fraction of one hectare, a Javanese *sondeo* was found to have up to 46 planted species (Colfer, 1991), each of which have an household economic function and each of which occupy an ecological niche giving maximum protection to the ground.

The use of vegetative cover for erosion control is not entirely without its difficulties. Some cover species, especially in drier areas, may compete aggressively with the main crop – careful experimentation, specific to the soil type and rainfall regime, needs to be undertaken because generalisations as to the utility of single species cannot be made for all conditions. While velvet bean (*Mucuna pruriens*) enhances nitrogen levels on good soils and provides better soil-water storage, on adjacent more degraded soils it can use too much water, thereby reducing main crop yields. In rubber plantations particularly, workers complain of snakes which find a ready home in the dense cover, while on some oil palm plantations on rich alluvial soils, the cover crop can grow so vigorously that it becomes difficult to control.

Mulching

This involves the use of cut grass or other vegetation residues to provide a protective layer to the ground. Mulching becomes particularly important in cropping systems where the ground is inevitably left bare for a large proportion of the rainy season. Particularly dangerous crops because of their late development in the growing season include cotton, tobacco and the Ethiopian staple cereal, *Eragrostis teff*. Teff is notorious for germinating only after the surface soil becomes puddled by early-season rainstorms. Without a good mulch cover of crop residues, erosion is immense, a fact which can be demonstrated throughout the Ethiopian Highlands.

Two basic problems exist with the widespread application of mulch in tropical farming. First, the rapid rate of mineralisation of organic matter in the heat and moisture. The actions of termites and harvester ants render much of the plant residues ineffective as a ground cover protection against erosion. While surface soil aggregation is definitely improved, there is little capacity of the dead vegetation to intercept the erosive power of rainfall. Second, in most small farming systems in Asia, Latin America and Africa, crop residues have other important functions. They are often a reserve of essential fodder: in the Mexican highlands, farmers assiduously gather every last maize stalk for stall-feeding of animals. In areas where wood fuel is deficient or expensive, residues are an important source of fuel. Throughout India, residues are fed to cattle and the manure so produced is shaped into fuel 'cakes', thus thoroughly depriving the soil of any benefit from past cropping.

Tillage systems which leave surface residues

Conservation tillage is an umbrella term which includes all methods of soil preparation that leave most residues on the surface of the soil. Included are no-till and direct drilling systems which plant immediately into undisturbed soil; they give the maximum erosion control and good soil moisture conservation, but they suffer from a dependence on the use of herbicides for weed control and surface compaction problems. Reduced tillage will usually involve a light discing which forms a seedbed without burying the plant residues; again, compaction can be a problem especially if the discing is repeated when the soil is wet. While conservation tillage has made great strides in mechanised agriculture, the investment needed in specialised equipment as well as the use of chemical inputs to counter weeds and diseases associated with organic residues renders such tillage systems inappropriate for resource-poor tropical farmers. Nevertheless, in Paraná, Brazil, the Institute Agronômico do Paraná (IAPAR) has promoted horse-drawn direct drillers which have had a slow but steady take-up by smallholders.

Mechanical soil conservation

The general principle underlying mechanical measures of soil and water conservation is the control and safe disposal of runoff water. This is accomplished by a wide array of types of earthen embankments – often called 'bunds' – and associated channels or waterways. The bunds usually function by dividing hill slopes into smaller segments, intercepting the runoff before it becomes dangerous in terms of its volume and velocity, and then leading the excess water safely off the productive land. Usually, the design of the physical measures will incorporate devices to encourage sedimentation of entrained soil particles into ditches. Because these physical measures tend to concentrate water into channelised flow, it is essential that all waterways are carefully protected. If not, the bunds will break and the waterflow will create gullies, often damaging more land than if the conservation had not been implemented. Additionally, attention has to be paid to the off-site effects of water breaking through weak or ill-maintained earth banks. For example, in the emergency drought relief work programmes in the Brazilian northeast in the 1980s, large numbers of small earth dams were built in a food-for-work campaign designed to conserve whole catchments. The banks became something of a local joke and were named 'açudes Son Risal' (literally, Alkaseltzer Dams) because, no sooner had they ponded water, the dams dissolved into the water and collapsed. The combined effect of several dams breaching was to send waves of water down the catchment, destroying all physical works and creating new gullies. It is instructive, then, to note that physical works do create additional dangers, and there must be continual maintenance to counter weaknesses and possible breakage points during heavy storms. This can place a heavy burden on local farmers and agencies, and, if there is any ambiguity as to who is responsible for maintenance, a potentially disastrous situation can occur.

According to local conditions of slope and soil type, channels and bunds are built either on the contour (where they effectively form water infiltration strips) or on a slight grade (where water flows gently across the slope and off the protected field). Contour-aligned structures are obviously more dangerous in peak floods than graded bunds, but they do have the advantage of water conservation (and possibly the disadvantage of waterlogging) and the retention of all soil within the field. Graded structures, in allowing water to flow off the site, also let the finer and more fertile particles disappear also. It is a common sight to see coarse sands retained in a field by a bund while the silts, clays and organic matter are lost. Similarly, waterways, storm drains and diversion ditches have to be designed to ensure the interception and safe disposal of water.

Grassed waterways are common on many large-scale schemes on gentle slopes; elsewhere where runoff velocity is likely to be greater, stone-pitching of the drains and protective gabions (stone-fulled bolsters wrapped in wire netting) are used. Detailed design criteria are to be found in any of the standard conservation engineering texts, for example Schwab et al. (1993).

A number of types of terrace structures designed for an FAO project in Jamaica are illustrated in Fig. 4.3. All are reverse-slope, meaning that water cannot flow over the risers: this is appropriate for steep slopes, permeable soils and/or rice paddies. Some of the terraces are suitable for smallholdings. For example, orchard

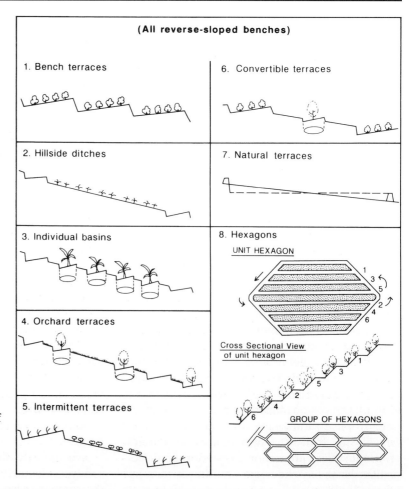

Fig. 4.3 Cross-sectional views of eight terrace structures forming reverse-slope benches for erosion control in Jamaica. (Source: Sheng, 1990.)

terraces can accommodate a few trees with arable crops on the intervening slope. Variants of this type are the Chinese 'fish-scale terraces' where each 'scale' has one or a clump of plants, and the 'eyebrow terraces' which have been used in Ethiopia on reforestation projects where each 'eyebrow' encloses a planting station for one tree. These terracettes are easy to construct by hand-hoe and can be maintained with family labour. In addition, they act as mini-water harvesting structures (see below), collecting water and maximising infiltration to the plant root zone.

A further soil and water conservation structure which has found acceptance in many parts of Africa is the 'tied ridge'. Ordinary contour ridge-and-furrow cultivation and planting has long been known to be effective in conserving moisture and reducing erosion. If the ridges are then 'tied', that is constructed with small cross ridges every 1.5–2 m, the field in effect becomes a series of basins (hence the term 'basin listing' when tied ridges are constructed by machinery). These basins not only maximise water infiltration but also provide microenvironments for different plants in intercropping. In Tanzania, for example, it is common to see drought-tolerant plants such as cassava on the top of the ridges, maize and beans on the side, and pumpkins, melons and even rice in the basin. In this way, the farmer minimises production risk and enables a range of plants to be grown with good protection. Improved no-till tied-ridging systems are now promoted in Zimbabwe, and animal-draught machinery is available for their construction (Elwell and Norton, 1988).

Gully erosion often presents a particularly difficult challenge for control. Gullies are very obvious scars on the landscape, and they tend to receive a disproportionate amount of attention, compared to sheet erosion. Their very existence indicates a degraded catchment where too much runoff and sediment is generated. Therefore, the foremost measure must be attention to the catchment by encouraging better cover, greater infiltration and dispersal of excess water flow through storm drains or into reservoirs. Normally, however, engineers will want to address the control of gully erosion directly by barriers erected within the gully. It must be emphasised that the record of such barriers is very poor when not accompanied by attention to the catchment. Rigid structures such as brick or cement small dams impound water and sediment, but they easily become undermined or by-passed in large flood flows. Temporary check dams of brushwood have been used and have the advantage of being more easily repairable. Gabions are preferred in many situations because they can flex to fill voids created by undermining; they are also better at filtering sediment from the water flow. Nevertheless, it is frustrating that, as in the extremely gullied area of central Mashonoland in Zimbabwe around St Michael's Mission and even with some of the best engineering advice backed by the local Natural Resources Board, gabions are still left high and dry in the middle of gully channels as testimony to the folly of investing solely in structures.

Water harvesting

Considered as a rudimentary form of irrigation where the farmer has no control over the timing and volume of water input, water harvesting has been used in the drylands for centuries for collecting runoff for productive agricultural use. The diversion of 'wadi' flows onto agricultural land in the Middle East has probably been practised for over 4000 years. Pacey and Cullis (1986) presented a useful account of traditional North African microcatchment techniques for growing trees (see below under 'indigenous soil and water conservation').

Critchley and Siegert (1991) gave the most comprehensive and useful manual on types of water harvesting structures, along with design criteria including the calculation of crop water demand for the various species which may be promoted in drylands by increasing effective rainfall. They distinguished between 'Negarim' microcatchments (diamond-shaped basins surrounded by small earth bunds), earthen contour bunds with minor up-slope spurs to retain water and semicircular bunds, all of which are normally found in rangeland, fodder plantings or woodlots. Cropping systems more usually have contour ridges, trapezoidal bunds, contour stone bunds, permeable rock dams or water spreading bunds. A particularly interesting account of the development and social acceptance of water harvesting structures based on stone bunds is given in Atampugre (1993). Other good sources include the review of water harvesting techniques for plant production by Reij *et al.* (1988) and Hudson's (1987) FAO review of soil and water conservation techniques for semi-arid areas which naturally dwells on techniques of rainwater harvesting and runoff concentration.

A good example of introduced water harvesting technology is the 'demi-lunes' (half-moons or semi-circular bunds) used for growing grasses, bushes, trees, and occasionally crops in the Tahoua Department of Niger, an area that was extremely overgrazed and degraded. With a mean annual rainfall of less than 300 mm, the options for rehabilitation and production are few. A local NGO started by constructing demonstration demi-lunes, each with a radius of about 2 m. With a total of about 300 of these per hectare, this gave a catchment:cultivation ratio of about 4:1. If the mini-catchments are estimated to provide about 50% of rainfall as runoff to the cultivated portion, these demi-lunes have an effective rainfall of about 600 mm. In use mainly for rehabilitating rangeland, the technique has also been shown to be effective for bulrush millet.

IMPROVING SOIL AND WATER CONSERVATION IN TROPICAL AGRICULTURE

Indigenous soil and water conservation

Farmers have been practising soil and water conservation long before the term was coined, and it is useful to review the vast range of indigenous techniques. Not only are many tried and tested over generations, but some have potential application for other areas. As Critchley *et al.* (1994) noted, the importance of the immense wealth of traditional know-how has been greatly underestimated. To ignore indigenous techniques could mean missing soil and water conservation opportunities that are adapted both to the physical environment and to local cultures. Examples are shown in Table 4.5, but there are probably many more which have yet to be found and described, especially from Asia and Central America. The following are the principal types of techniques that have been developed by local people.

Vegetative barriers and strips

The most common are narrow grass strips planted across the slope which act as water infiltration strips and sediment filters. Over the long term, the action of these strips is to 'self-terrace' the hill slope with further benefits to soil and

Table 4.5 Examples of indigenous soil and water conservation techniques. (Sources: IFAD, 1992; Critchley *et al.* 1994; Stocking, 1995.)

Country (region) – approx. annual rainfall in mm	Indigenous technique
Burkina Faso (Southwest) – 100	Stone bunds on slopes; network of earth bunds and drainage channels in lowlands
Burkina Faso (Central) – 500	Stone lines; stone terraces; planting pits (*zay*)
Cameroon (North) – 1000	Bench terraces (up to 3 m high); stone bunds
Ethiopia (Tigre) – 1900	Terraces with stone bunds
India (Rajasthan) – 800	Silt-harvesting bunds; field boundary bunds
Indonesia (Java) – 2500–3500	Rock walls; logs and grass strips across slope; rainwater diversion ditches
Kenya (Rift Valley Province) – 700	Trashlines and brushwood barriers; stone lines
Nepal (Jhiku Kola) – 2000	Silt traps; terrace systems

water conservation. Live fencing within and around fields is also traditional, and in Vietnam, for example, the shrub legume *Tephrosia candida* is planted in lines across the slope and it provides a valuable fodder for pigs, firewood for cooking, and a source of leaves for composting.

Trashlines and wooden barriers

These range from the use of banded lines of weeds (common in drier parts of Africa), crop residues, tree prunings and large logs from felled trees (as in Java) to quite sophisticated systems of pegged lines of brush which have been described from Sierra Leone and Papua New Guinea. The technical effectiveness and productivity benefits of trashlines have been amply demonstrated in Kenya (Kiome and Stocking, 1993).

Pits and basins

A large variety of small soil pits or earthen basins have been developed, ranging from small, hand-hoe, planting stations through to major structures. The *Matengo* pits in Tanzania, for example, are formed from a grid of vegetative trash which makes squares with sides of about 1.5 m; the trash is covered by soil excavated from the centre of the squares. Crops are then planted on the bunds, benefiting from the buried green manure and the moisture stored in the pits.

Earth bunds and terraces

These consist of earthen barriers dug across the slope designed to trap water and sediment, or the construction of shaped terrace systems such as the 'ladder terraces' in the Uluguru Mountains, Tanzania.

Stone bunds and terraces

These are the most common indigenous technique for drylands. Extensive areas of the Eastern Highlands of Zimbabwe and western Mozambique are covered in terraces protected by stone bunds which have been dated back to the African Middle Stone Age (*c.* fourteenth century). Throughout Africa and in parts of Nepal, Indonesia and China, versions of the same technique are operative today. Stone bunds act as simple, semi-permeable barriers, which allow the passage of excess runoff while trapping sediment. The additional benefit is that they act as a repository for unwanted field stones – though the corollary is that the technique is unlikely to be used where there are few stones locally.

Water retention and diversion ditches

The control of floodwaters and prevention of waterlogging has also exercised local farmers. A wide variety of ways of controlling excess water has been devised. In Gujarat and Rajasthan, India, for example, many of the lower slope field boundary bunds have drainage culverts which can be closed to conserve moisture or opened to allow drainage according to the needed moisture conditions within the field.

Rainwater harvesting

With the objective of increasing the effective rainfall supply to plants, rainwater harvesting involves the concentration of runoff from small catchment areas. The International Fund for Agricultural Development (IFAD, 1992) described the eastern Sudanese *teras* system which had earth bunds on three sides of a square of planted sorghum with the upslope side open to accept runoff. In this way, in a mean annual rainfall zone as low as 200 mm, cropping of cereals could still be assured.

Floodwater harvesting

On a larger scale than rainwater harvesting, this has an additional objective of capturing fertile sediment. Again, many types exist but most are similar to the *nullah* plugs described by Kerr and Sanghi (1992) throughout semi-arid India. Gullies and ravines have constructed within them plugs of rocks, earth and stones to trap sediment, and, if possible, create new bottomland rice paddies made of the soil from upslope.

These indigenous techniques are not the whole answer to improving soil and water conservation: if they were, the problem of land degradation would not still be with us – local people would have solved their own problems. Much can be learned from the indigenous application of techniques. *De facto* they are known and appreciated by local people, thus getting over the major stumbling block of most resource conservation programmes which try to introduce new techniques. They may not, however, be able to cope with the demands of increased population and lifestyle changes. They are not, therefore, a panacea but, as advocated by Critchley *et al.* (1994), they could be a useful starting point for the development of appropriate soil and water conservation technologies and programmes.

Techniques that have become accepted

Increasing population, improved farming techniques, new crops and additional market opportunities all mean that the management of the soil and water resources also has to change in keeping with a constant cycle of changing demands. Moreover, because of the hazards of soil degradation in tropical environments, the introduction of innovations, improved techniques and technologically based solutions will increasingly have to be sought. What sort of techniques have succeeded and can we learn any lessons? Figure 4.4 illustrates a number of techniques that have become widely acceptable in dryland parts of India – the major lesson is that all these techniques use the farmers' own knowledge and understanding of their difficult natural environment. Below are some more examples of techniques which have found ready acceptance.

A conservation project in the Ocoa Valley, Dominican Republic, reported by Thomas (1988), originally promoted hillside ditches protected by live barriers of citronella (*Cymbopogon nardus*). It seemed to the technicians that here was a system which was relatively simple, easy to construct and, above all, sustainable. Problems were encountered, however, with a low take-up rate of the improvements. In an informal survey, farmers revealed they liked the citronella but strongly disliked the idea of allocating scarce land for no obvious economic use for its growth. Another survey showed the desire of farmers to keep cattle, but few had enough fodder. Elephant grass (*Pennisetum purpureum*) was introduced as an alternative live barrier, and the take-up rate of the new system of hillside ditches with live barriers has increased markedly, even amongst farmers not directly involved with the project. A two-fold lesson can be drawn from this experience: first, production objectives remain the primary objective of land users; and second, conservation techniques that work must usually address a critical production constraint in the present farming system – in the Dominican case, the lack of fodder for animals.

One of the best examples of conservation structures which have become accepted in tropical small farming is the case of the Kenyan *fanya juu* terraces. *Fanya juu* is Kiswahili for 'moving upwards', which is a reference to the way soil is dug from a hillside ditch and thrown upwards to build a bund. Over time, through natural erosion processes and maintenance of the structure, the terrace builds into a series of benches. It has been found that groups of farmers, often women's associations, will cooperate together to form work groups which will tour each plot of group members. Terrace construction in this way becomes a social event, and the camaraderie and singing of such groups has been noted by many observers. *Fanya juu* terraces are now commonplace in the high potential areas of Kenya where they definitely give increased yields; they are less readily accepted in lower potential zones below about 800 mm mean annual rainfall. The lesson again is that any innovation must realise increased production, and because of the marginal crop growth in semi-arid and sub-humid Kenya, even increased moisture retention is unable to given an enhanced yield enough to pay back the investment of labour (see Kiome and Stocking, 1993, for an analysis of the economic returns of these terraces, compared to more simple conservation techniques such as trashlines, which are readily accepted). The social function of *fanya juu* terraces is also important.

Several conservation innovations, which are

related to agroforestry and have received ready acceptance by local people, are reported by Kerkhof (1990). One example concerns the Koro District of Mali where a project attempted to establish communal windbreaks of trees to combat wind erosion. When local interest waned, coercive approaches were used by village authorities and the Forest Department. Although trees were planted, the oppressive tactics ensured that no protection was provided to the saplings, leading to very poor survival rates. From 1987, a change of approach was instituted: low-cost, local initiatives were supported, rather than expensive windbreaks, and the emphasis turned to individual rather than communal actions. Ironically, windbreaks flourished because people actually wanted them, but the tree species and layouts changed. *Acacia raddiana* and *Azadirachta indica* are grown in clumps, protected by thorn branches and millet stalks, while *Acacia albida* is planted within cropped fields because the Dogon people appreciate its unique virtues of providing direct mulch and being non-competitive with crops. Since the success of these initiatives, people have spontaneously started other soil and water conservation measures, some of which are traditional (e.g. *zay* planting pits) and others which are new to the area (e.g. rock bunds). As Kerkhof (1990) noted, all this amounts to a major change from the original project, but the effect has been startling in attuning almost the whole population to conservation issues.

Some criteria for success

Soil and water conservation is perhaps the greatest challenge facing the development of sustainable agricultural systems in the tropics. For resource-poor farmers, the technological options are few and the access to sufficient resources to accomplish soil and water conservation is poor; for larger-scale commercial farming, the environmental hazards of the tropics make soil and water conservation far more demanding than it would be in temperate areas. How then can the ingredients for a successful approach to soil and water conservation be determined?

First, there is now substantial project and rural development experience, much of it emanating from NGOs and local initiatives, to build a picture of the conditions needed to maximise the chances of success of any soil and water conservation programme. It is important – and has often sadly been neglected – to learn by experience. The book edited by Hudson and Cheatle (1993) and the Swedish International Development Authority's (SIDA) review of land husbandry experiences (Lundgren *et al.*, 1993) are good examples of the type of soil and water conservation case material which is being accumulated and which must guide any future programmes.

Second, it is increasingly being recognised in soil and water conservation the crucial biophysical importance of vegetation and organic matter. While land qualities must obviously be investigated and the demands placed on them not be excessive, the role of vegetation in all its aspects crucially affects the demand–supply equation. Vegetation affects the amount and availability of soil nutrients and moisture; it affects the whole array of soil physical attributes; it controls the microenvironments in the soil, in different tillage systems and in farmers' fields. Vegetation cover, especially, is the strongest control factor in tropical environments preventing erosion and maintaining sustainable agricultural systems.

Third, this chapter has strongly advocated a realistic view of socio-economic, cultural and political aspects which control why farmers undertake land use and how they engage with the land. Of the many aspects which could be important for any one geographical area, probably the most influential relate to land tenure, rights to land and security of tenure. Without an understanding of these issues, any soil and water conservation programme is doomed to failure.

Fourth, the economic and financial situation of land users is of central importance. We are increasingly learning that small farmers in the tropics are unlikely to carry out any practices that have a negative rate of return – and many of the soil and water conservation techniques which have been promoted in the past simply fail on this count. Investments of family labour, resources and other opportunities foregone have made much that is promoted as soil and water

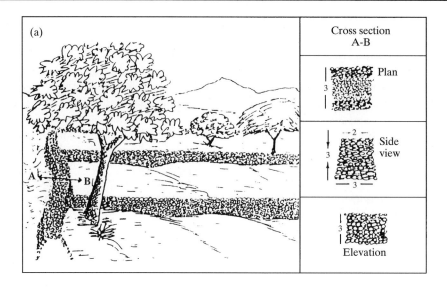

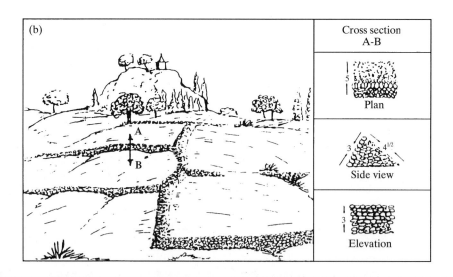

Fig. 4.4 Sketches of soil and water conservation structures which farmers utilise in dryland parts of India (a) Field boundary bund; (b) intermediate field bund; (c) silt harvesting structure; and (d) loose rock check dam. (Source: Premkumar, 1994.)

conservation economically irrational. While it is good to carry out cost-benefit analyses in assessing the viability of any set of measures, it may be sufficient in many parts of Africa and Asia to measure the labour inputs for new techniques and come to a realistic assessment of whether these can be provided by the household. Soil and water conservation has to compete against other

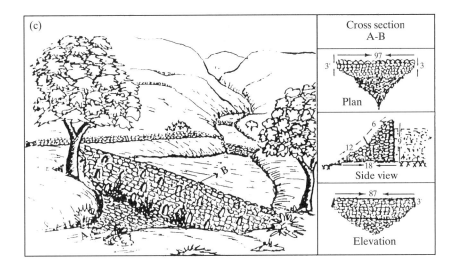

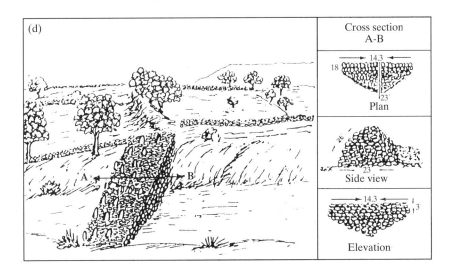

Fig. 4.4 *Continued*

farm activities such as weeding, with other rural activities such as artisanal crafts, and with the opportunities afforded off the farm, in industry and in the cities.

While the above will not guarantee success, they at least provide an outline checklist of the sorts of considerations that need to be put on the agenda of the tropical agriculturalist in assuring the sustainability of the use of the natural resources. In the final analysis, however, it can confidently be said that, if soil and water conservation can be so thoroughly integrated with the production objectives of tropical farmers that it cannot be isolated, then we will have built a surer future for tropical land, soil and water resources.

FURTHER READING

Critchley, W.R.S., Reji, C. and Willcocks, T.J. (1994) Indigenous soil and water conservation; a review of the state of knowledge and prospects for building on traditions. *Land Degrad. Rehabil.*, **5**, 293–314.

Hurni, H. and eleven international contributors (1996) *Precious Earth: From Soil and Water Conservation to Sustainable Land Management*. International Soil Conservation Organisation, and Centre for Development and Environment: Berne.

Reij, C., Scoones, I. and Toulmin, C. (eds) (1996) *Sustaining the Soil*. Earthscan: London.

Syers J.K. and Rimmer, D.L. (eds) (1994) *Soil Science and Sustainable Land Management in the Tropics*. CAB International: Wallingford.

5
Land Clearing, Drainage, Tillage and Weed Control

B.D. Soane

LAND CLEARING AND LAND PREPARATION

This section deals with the (mainly) manual clearing of land by small farmers, many of whom practise traditional, or modified, shifting cultivation. Clearing operations for the production of most crops are mainly found in the forest and savanna zones but the type and amount of vegetation to be cleared varies according to the climate, site characteristics and the time that has elapsed since previous clearing (Lal *et al.*, 1986).

Clearing in the forest zone

In the humid zone the vegetation to be cleared may be virgin growths ranging from rainforest to broad-leaved woodland, but nowadays is mostly secondary forest, or immature scrub, which has regrown since a previous period of cropping. The traditional 'slash and burn' method of clearing is undertaken during the dry season, generally using only axes, machetes and hoes although chain saws are increasingly being used to fell larger trees. As a rule, clearing is restricted to slashing shrubs and undergrowth near ground level and cutting down small- and medium-sized trees at 0.5–1.0 m above ground. Larger trees, especially if buttressed, may be cut higher but are usually left standing and are often useful in providing food, fibre or shade. As a rule, no attempt is made to eradicate the stumps and roots of felled trees unless animal or tractor-drawn ploughs are to be used.

This limited clearing is adequate for crop production with hand tools and permits rapid regeneration of the natural vegetation when the land goes out of cultivation. It also reduces the heavy labour requirement to an acceptable level. Whereas complete clearing in southwestern Nigeria required approximately 180 man-days/ha, traditional incomplete clearing needed only 57 man-days/ha (Couper *et al.*, 1981).

Clearing in the savanna zone

Vegetation in the savanna zone is variable but generally comprises a ground cover of tall grass with a variable density of scattered medium-sized trees. As in the forest zone, clearing is usually incomplete, but, although the trees are fewer and smaller, clearing in the savannas may be almost as laborious as in the forest because of the aggressive growth of grasses, the roots of which must normally be dug out (Allan, 1986). Animal and tractor-drawn ploughs are used in some savanna areas, necessitating the removal of the stumps and larger roots of felled trees.

Burning

In the past, slashed vegetation and felled trees were normally burned *in situ*, this being the only practicable way of freeing the land from debris and also having the advantage of providing a fairly weed-free seedbed for the first crop. Nowadays, the shortage of wood has led to much of the vegetation being used for fuel or building but there is still a fair amount of smaller debris left for burning.

Burning results in the loss of most of the organic carbon, nitrogen and sulphur in the felled vegetation and in the leaf and twig litter on the ground surface. The destruction of the litter is undesirable because, in the absence of thorough cultivation, it can afford some protection against raindrop impact to the surface soil and thus reduce the risk of erosion. The mineral nutrients from the vegetation are deposited in the ash, thus increasing the phosphate and exchangeable cation content, and raising the pH of the topsoil, but some of the nutrients may be lost if heavy rain falls on the bare soil after clearing (see Chapter 6).

DRAINAGE

Principles of drainage

The construction of storm drains and contour channels as components of soil conservation measures to ensure the safe removal of runoff on sloping land has already been described. This section is concerned with the need to prevent waterlogging of the roots of crops. The principles involved are the same as those well established for field drainage in the temperate zone (Van Schilfgaarde, 1974; Castle *et al.*, 1984; Smart and Herbertson, 1991; Ritzema, 1994).

Soils may be poorly drained either because there is a high water table, as in low-lying areas that receive water from surrounding higher land, or because the physical condition of the soil impedes the internal drainage of excess surface water to a depth below the root range of crops. These two causes require different ameliorative measures, but in both cases the main effect of improving drainage is to improve the aeration of the soil, thus enabling the crop to develop a deeper root system so that it can tap a larger volume of soil for nutrients and is better able to withstand periods of drought. As indicated in a later section, high or rising water tables in irrigated areas require special measures.

Soils having a water table permanently, or seasonally, high enough adversely to affect crop growth need drains to lower the water table, but the depth to which it should be lowered depends on local circumstances, especially on the nature of the soil, the rainfall and evapotranspiration regimes and the crops to be grown. Special problems arise in the drainage of acid sulphate and peat soils (see Chapter 3). Tree crops that need to develop deep root systems to anchor them firmly against high winds will require a lower water table than some annual crops or surface-rooting perennials, such as pineapple. Where rainfall is abundant and well distributed throughout the year it might be thought desirable to keep the water table below crop rooting depth at all times. But it is not certain that such thorough drainage is either necessary or economic since short periods of waterlogging of part of the root system during heavy rainfall may not be harmful. In drier areas of seasonal rainfall the chief object is to conserve water and, although it may be necessary to improve drainage during part of the wet season, it is undesirable to lower the water table more than is necessary. It is not possible to make precise recommendations regarding depth of drainage because of the lack of information on the effects of transient waterlogging of part of the root system of different tropical crops. An exception is that, wherever the groundwater contains dissolved salts in harmful amounts, it is important that drainage be deep enough to keep the water table and its capillary fringe below rooting depth.

Subsurface drains of tile, plastic pipe, rubble or brushwood are uncommon in the tropics. They are costly to install and are liable to be blocked or disrupted by the roots of perennial crops. Open surface drains are cheaper and easier to construct. They can be made by hand or mechanically with scrapers, bulldozers, drainage ploughs or dragline excavators, the latter generally being considered more economical for making main drains, while the tractor-drawn implements are better for smaller, field drains. Open drains are usually preferred, despite their disadvantages of taking up land, inconveniencing mechanised tillage operations, harbouring noxious weeds and rodents, and requiring constant maintenance. Clearing and maintenance of open drains can be largely mechanised; draglines can be used for large drains and machines have been developed for use in smaller field drains (Boa, 1958). When draining flat or gently sloping land the first step after any necessary survey is to find, or construct, an efficient outlet, the second is to cut the main drain away from the outlet, keeping it as straight as possible, and the third is to cut the field drains at intervals which will depend principally on the texture of the soil. Drains need to be closer together on heavy soils than on light soils. The gradient of the drains must be sufficient to prevent rapid silting but not so great as to cause scouring. Provided that the drains are kept in good order, a fall of from 1 in 5000 to 1 in 3000, depending on the size of the drain, is normally adequate. On hilly land natural drainage is usually satisfactory, but where an artificial drainage system is needed it should be constructed as part of the soil conservation works with field drains running across the slope on only a slight grade.

The second circumstance calling for improvement in drainage is that in which the physical condition of one or more horizons of the soil checks the downward movement of excess surface water. This may be due to lack of sufficient coarse pores and cracks in the surface soil owing to breakdown of aggregates by rainfall impact, by excessive tillage or by cultivation when the soil is too wet. In this latter case the remedy lies in protecting the soil surface with a crop cover or mulch and in avoiding untimely or excessive tillage. Drains cannot help once the structure of the surface soil has been destroyed, although they may reduce the risk of such damage occurring by keeping the surface layer drier and thus prolonging periods during which it is not too wet for mechanised tillage.

On the other hand, impeded drainage may be due to the presence of a subsurface layer of soil of low permeability below which the profile is free-draining. Such impermeable layers can occur naturally; for example as horizons of clay accumulation. They can also result from heavy equipment passing over the soil and compressing it to form a compact layer which is usually densest a few centimetres below the surface. If the soil is wet, and especially if it has a high clay content, a barrier impeding water percolation can also be formed by the smearing action of tractor wheelslip or of an implement which destroys structure and seals coarse pores and cracks. This smeared barrier may initially be very thin, but during wet periods it holds water above it and in soils of unstable structure the aggregates within the saturated layer slake and collapse to form a structureless horizon which increases in thickness with each successive wet period and, on drying, hardens to form a pan. Provided that the subsoil below is permeable, drainage can often be improved by deep ploughing or subsoiling to break up the compacted or structureless layer and provide more channels for the percolation of water. Since the object of subsoiling is to shatter the soil to produce a system of deep cracks, which should remain open as long as possible, it is best done when the soil has a low moisture content but has not dried out completely. If, however, the deeper subsoil is insufficiently permeable to allow reasonably rapid water percolation, then deep ploughing or subsoiling to break up compacted layers near the surface will only be beneficial if effective drains can also be provided.

Ridge and furrow system: cambered beds

Apart from ditches there are several ways of encouraging surplus water to run off the soil surface and improve the drainage of the topsoil. These usually take the form of mounds, ridges or broad beds. A well-known method of improving heavy soils with impeded drainage is the 'ridge and furrow' system in which the land is formed into broad ridges with a slight grade on the furrows or ditches between them.

A modification of this is the cambered bed system which has been extensively employed on heavy, ill-drained soils in parts of the West Indies for the cultivation of sugar cane, and to a lesser extent for coconuts, cocoa and citrus. It has also been used in cultivating 'vlei' soils in East Africa, as described by Robinson *et al.* (1955) and

Robinson (1959a, 1959b). These beds are constructed by first marking out the lines of the drainage ditches or furrows, which are commonly from 6 to 10 m apart. Overlap ploughing is then done, starting from the centre of the bed and working towards the drain lines. This should be effected as deeply as possible (usually to 25–35 cm on West Indian sugar estates), and forms an appreciable crown. Subsoiling is then carried out to a depth of 45–60 cm. It is essential that this be done when the clay is relatively dry as the object is to open up and shatter the soil in order to provide a deep layer of loosened, better aerated and drained soil in which roots can develop vigorously. Subsequently the drains are taken out with an implement designed to throw the soil well up towards the centre of the bed. This is usually done with the Cuthbertson drainage plough, which is a double mouldboard plough with long extension arms. Finally, the beds may be finished off by grading with a terracer blade. The result is a bed with a section as indicated in Fig. 5.1 in which some water is shed off the cambered surface of the bed (a), and some percolates through the cultivated, free-draining layer (b), to move away laterally to the drains over the cambered surface of the uncultivated, impermeable clay subsoil (c). In Trinidad the loss of the artificial structure after planting with sugar cane is rapid in the surface 5–8 cm, causing a fair proportion of the rainfall to run off the surface of the beds. Deterioration is more gradual in the remainder of the cultivated layer, in which the cane roots grow vigorously, whereas root development is much more restricted in the untilled clay immediately below.

Drainage of irrigated land

Inadequate drainage has given rise to well-known examples of failures of entire irrigation schemes in both historical and modern times. Such failures are associated with rising water tables causing accumulation of salts and/or the loss of aeration in the root zone. Even where they do not cause total failure, these injurious factors often reduce the yield of crops. Salts can accumulate in the surface soil even during fallow periods if the water table is sufficiently close to the soil surface to allow capillary rise of water and salts into the upper soil layers (Fig. 5.2).

The rise of water tables in irrigated areas may be due either to excess application of water, especially with furrow or flood application methods, or seepage losses from canals and ditches, such losses frequently amounting to 10–20% and sometimes to as much as 50%. The rate of rise of water tables shortly after irrigation schemes commence may be extremely rapid. At Bhantinda Mansa in India the water table rose at 0.4 m/yr at the beginning of the twentieth century while at West Nubaria in Egypt the water table rose at 2.5 m/yr. Unavoidable seepage from canals may be intercepted by ditches or piped drains set at sufficient depth to be below the seepage line. Such intercepted water can be used for irrigation.

The technology of drainage requirements in

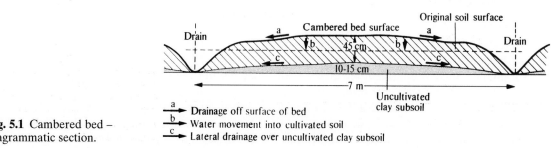

Fig. 5.1 Cambered bed – diagrammatic section.

a → Drainage off surface of bed
b → Water movement into cultivated soil
c → Lateral drainage over uncultivated clay subsoil

irrigation schemes is now well understood and documented (Rhoades, 1974; Van Hoorn and Van Alphen, 1994). The particular drainage problems in irrigation schemes in the tropics were considered by Leeds-Harrison and Rickson (1991). Drainage installation is considered to be an essential requirement at the planning stage of irrigation schemes to minimise costs and disturbance to production. However, in many schemes in the past this was not the case, necessitating the installation of drainage subsequently to rectify problems arising after cropping had commenced.

Because of the high value of irrigated land and the large investment of capital, drainage of such land is not subject to the same economic constraints which apply to non-irrigated land. Methods used include ditches, subsurface pipes and pumped wells. The disposal of drainage water in ditches should itself be subject to strict controls to minimise seepage and larger drainage canals should normally be lined for this purpose.

The placement of subsurface pipes must be carefully related to the soil and ground water characteristics. A successful scheme will often depend on the location, and placement, of pipes in an aquifer of sufficient permeability. This usually results in pipes within irrigation schemes in the tropics being much deeper (1.2–2.8 m) and further apart (50–200 m) than would be the case in non-irrigated humid temperate regions. However, the considerable depth of drains and the low gradient of much irrigated land can give rise to problems in the provision of suitable pipe drainage outfalls and pumping to the ground surface may be required. A commonly used design criterion is that the water table in irrigated areas should be not less than 1.2 m below ground level to avoid salt accumulation at the surface and loss of aeration in the root zone. The calculation of recommended spacing and depth of drains must take account of the fluctuation in the height of the water table during the interval between irrigation applications at the most critical period of the irrigation season when the interval is least. The probability of rain immediately after irrigation must also be taken into account and it is often recommended that the drainage scheme should be capable of lowering the water table in fully saturated soil to 0.3 m within 24

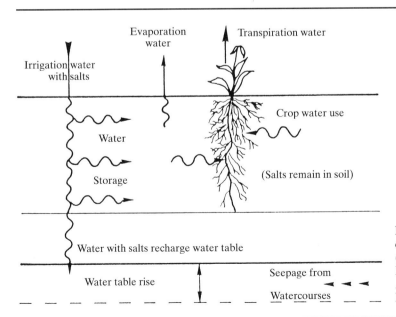

Fig. 5.2 Water and salt balance during the irrigation season. (Reproduced from Smart, P. and Herbertson, J.G. (eds) (1991) *Drainage Design*, by permission of Blackie and Son, Glasgow.)

hours and to aerate fully the root zone within 3–5 days.

Pipes used may be of sulphate-resistant cement, plastic or tile, depending on local cost and supply factors. Corrugated plastic pipe is now widely used for laterals. The use of filter envelopes of graded sand or gravel or synthetic materials around pipes is strongly recommended in unstable soils.

Where there has been a general rise in the ground water, rather than a perched water table, pumped wells are widely used (e.g. in India and Pakistan). Drain wells may be 30–100 m deep, depending on the depth of aquifers, and are usually placed at about one well per 2.5 km^2. Depending on its salinity, the pumped water may or may not be suitable for irrigation use. In some cases the use of unpumped perforated 'down' wells allows ground water near the surface to reach lower aquifers from which it can be pumped from standard wells in other locations.

The planning of drainage for irrigation projects must take into account the leaching requirement which represents the amount of applied water (adjusted for its salinity and for any rainfall) which has to be allowed to pass through the root zone to prevent the harmful accumulation of salinity (Leeds-Harrison and Rickson, 1991). The leaching requirement may well be exceeded in standard irrigation practice where surface irrigation application is employed on pervious soils. However, with overhead irrigation or with surface irrigation on low permeability soils, the leaching requirement may not be reached in normal practice, necessitating occasional prolonged application or ponding to flush out accumulated salts which must then be removed through the drainage system. The efficiency of leaching will be influenced by the presence of cracks or vertical pores which will channel much of the downward flow leaving zones between cracks relatively poorly leached.

Irrigated land is subject to surface runoff as a result of excess irrigation water or rainfall, and provision must be made for conveyance of such water with suitable precautions being taken to prevent erosion.

CULTIVATION AND SOIL MANAGEMENT SYSTEMS

Objectives

Cultivation practices serve a range of objectives which are related to the local soil and climate conditions as well as to the cropping system. The main objectives of cultivation are generally to loosen and break up the soil in order to increase aeration and water infiltration, to prepare a seedbed and to control weeds. Soil compaction arising from the passage of tractors, implements, animals or even human feet can be severe in tropical soils low in organic matter and must be ameliorated by cultivation if a satisfactory seedbed is to be produced (Soane and Van Ouwerkerk, 1994).

Small farmers in the tropics do not usually need the fine tilth required in the temperate zone for drilling small seeds. They usually broadcast small seeds, drop larger seeds into holes made with a stick or hoe and consolidate them with their feet, while many crops are established from cuttings or other vegetative means. Soils that are hard and capped at the end of the dry season require some cultivation to improve aeration and water infiltration and to prevent surface ponding or runoff, but this can be achieved by limited and quite shallow cultivation. The benefit will in any case be transient since all forms of tillage tend to aid the rainfall in breaking down soil structure and sealing the surface. Hence it is best to minimise cultivation and to leave the surface rough and cloddy where this is practicable. Unless herbicides are used, a vital function of cultivation is to destroy weeds (see later in this chapter).

It has been suggested that deeper cultivation than that customary on smallholdings could promote better root development and increase crop yields and that the use of tractors could be beneficial in this respect. But except for a few crops, such as yams, and perhaps on some structureless soils of high bulk density, the limited evidence is that there is no marked benefit from deep digging, ploughing or subsoiling unless one of the following circumstances applies:

(1) The presence of a shallow hard pan or layer of impeded drainage at fairly shallow depth with freely draining soil below.
(2) The need to eradicate deep-rooted perennial weeds.
(3) Heavy clays are being cultivated by a cambered bed, or similar, system.

Primary cultivation

In many parts of the tropics primary cultivation is traditionally carried out using a non-inverting implement which is usually locally made from wood with a simple steel share. Such an implement is known technically as an 'ard' (Krause *et al.*, 1984) and has an action similar to that of a shallow-depth chisel plough. Alternatively, an inverting implement such as a mouldboard or disc plough may be used. In India it is estimated that there are 40 million ards and 7 million mouldboard ploughs.

Non-inverting implements

The non-inverting action of an ard is essentially a low-draught operation which can be performed by animals which are too light or in too poor a condition to tackle the higher draught requirement of a mouldboard plough (Fig. 5.3). Non-inverting tillage, which is a form of minimum tillage, has the advantage of leaving most crop residues on the surface to protect the soil surface from erosion and to increase the infiltration rate. Repetitive non-inverting tillage may achieve a satisfactory seedbed with very limited resources, especially if planting is to be carried out by hand.

The use of non-inverting chisel ploughs or rippers may provide effective breaking of pans to about 35 cm provided suitable tractor power is available (Krause *et al.*, 1984). Implements with forward-inclined tines undertake a certain amount of mixing within the disturbed layer which may explain higher yields of cotton and groundnuts reported for chisel ploughs compared to heavy disc harrows working to the same depth on irrigated Vertisols in the Sudan (Dawelbeit, 1991). The change from the use of an ard to a mouldboard plough tends to proceed quite slowly and is dependent on the availability of heavier draught animals and blacksmith facilities.

Fig. 5.3 Ploughing with an ox-drawn ard in Ethiopia. (Source: Overseas Division, Silsoe Research Institute.)

Inverting implements

Inverting tillage, whether by mouldboard or disc plough, has the advantage of more effective weed control and greater soil loosening to a greater depth (Fig. 5.4). The ability to bury grass weeds is one of the most important advantages of inverting tillage. However, the higher power requirement can usually not be justified unless corresponding increases in yields are obtained. Mouldboard and disc ploughs require greater skills in operation and maintenance than do non-inverting implements (Fig. 5.5). Disc ploughs rely on their weight (or added weight) to penetrate hard dry soils, cause less inversion than mouldboard ploughs and are used primarily with tractors rather than animal power.

The need for inversion by ploughing is the source of much debate. Where soils tend to develop very high bulk densities, such as on the Luvisols of Botswana, deep mouldboard ploughing is regarded as a prerequisite for successful cropping in spite of the high demands for draught power. The use of the mouldboard plough tends to be widely held as a desirable objective in many parts of the tropics and the ploughing-in of broadcast seed is much practised in some areas (Monageng *et al.*, 1990). However, tillage research in Southern Africa is widely directed to the development of alternatives to conventional mouldboard ploughing which is recognised as a major factor in causing delayed planting, excessive soil losses through erosion and poor utilisation of early rains. Vogel (1991) reported comparisons made in Zimbabwe using non-ploughing systems, such as ripping into bare ground or through crop residues and the use of no-tillage with tied ridging. These alternatives were shown to be feasible and advantageous alternatives to ploughing for small-scale farmers.

Timeliness in primary cultivation operations

For hoe farmers, and even for those with animal-drawn ploughs, it is commonly very difficult to do preparatory cultivation during the dry season when many soils set very hard. In some areas it

Fig. 5.4 Use of tractor-drawn disc plough for primary tillage and incorporation of plant residues in Zambia. (Source: FAO, Rome.)

may be possible to complete primary cultivation at the end of the rainy season while the soil still has residual moisture, when draught animals are in good condition and crop residues can be readily incorporated. Where this can be done, it enables early planting to make full use of the first rains, but in many places it is impracticable because of late crop harvesting or weed growth early in the dry season. The usual practice is to wait until the earliest showers, or 'grass rains', soften the ground and then try to complete the work before the onset of the main rains, which usually means that much of the land is not prepared for planting in time to make full use of the early rains, part of which may be wasted in runoff from the uncultivated, uncropped land.

Early planting is important because many experiments with a variety of crops have shown that, as a rule, yields are markedly and progressively reduced the longer that planting is delayed after the onset of the rains and that yield losses from even short delays can be considerable. For example, time of planting trials in the single rainy season areas of Tanzania, covering a range of crops including maize, sorghum, soya, groundnut, sesame, cotton and castor, showed that any appreciable delay in planting after the rains had effectively started usually reduced yields considerably (Akehurst and Sreedharan, 1965). Presumably, early sown crops benefit from receiving the full season's rainfall, from less weed competition and from high soil nitrate levels at the beginning of the rains, although the last factor is probably not important because various workers have found that nitrogen fertilisers can not compensate for late planting (Turner, 1966; Jones, M.J., 1975). However, as explained in Chapter 2, uncertainty about the time of arrival of the rains and the amount and frequency of the early showers make it difficult for farmers to prepare land and plant at optimum times.

Secondary cultivation

Secondary cultivation may take place immediately after primary cultivation or be delayed for weeks, or even months, depending on the seasonal characteristics. The objectives are to break clods, level the soil surface, destroy any weed growth and, where appropriate, to firm the sur-

Fig. 5.5 A mechanic makes adjustments to the setting of a single-furrow mouldboard plough in a village workshop in Burkina Faso. (Source: FAO, Rome.)

face in preparation for sowing or planting. A variety of implements, including tined, disc and rotary cultivators or harrows, are used (Metianu, 1992). In addition, land smoothing and surface consolidation is often achieved with simple levelling bars, planks and rolls. Because secondary cultivation operations have a low power requirement, they are well suited to animal power and the low cost of such implements makes them affordable for small-scale farmers. Secondary cultivation may also include land-forming operations with ridgers or bed making equipment prior to planting. Ridges can be made with hand tools, or with animal- or tractor-drawn mouldboard or disc ridging implements.

Conservation tillage systems

Conservation tillage may be defined as a system in which soil and water conservation are paramount objectives along with the preparation of a suitable seedbed for the intended crop. There is, however, a wide spectrum of practices involving reduced and no-tillage and the management of crop residues which, in varying degrees, can be classified as having a conservation objective (Unger, 1984; Jones et al., 1990). Such practices are usually considered in distinction to 'traditional' or 'conventional' temperate zone tillage practices which generally involve a number of pre- and post-planting tillage operations in which all plant residues are buried, leaving the soil loose and bare, which would have serious erosion risks in the tropics, especially during the early life of the crop. However, as already mentioned the techniques of 'traditional' and 'conventional' tillage in most parts of the tropics involve less thorough cultivation than those in the temperate zone. The adoption of conservation tillage practices aimed at reducing erosion and runoff will not be attractive to small-scale farmers if they involve extra work or weed control problems.

In temperate regions reduced, or minimum, tillage is primarily thought of in the context of the abandonment of mouldboard ploughing and its replacement by non-inverting tined implements. However, in much of the tropics such a system, with its low draught characteristics, is frequently the traditional method of cultivation. The reduction of tillage may involve only partial disturbance in the interrow zone of row crops while retaining conventional tine cultivation within the rows. Where wheeled tool-carriers are used, strip or zone tillage can be integrated with a form of controlled traffic in which the soil in proximity to the row positions is kept permanently free of wheeled traffic (Willcocks, 1984). The use of row tillage combined with no-tillage for the interrow zone has been advocated by Ojeniyi (1991), especially for crops such as cowpeas which, in the absence of soil loosening, may show reduced yields under no-tillage because of their sensitivity to soil compaction (Kayombo and Lal, 1994). Where non-inverting tillage systems are used in the presence of plant residues, e.g. stubble mulching and mulch ripping, the partial mulch cover will provide protection against wind and water erosion but increases in soil water content may be limited if frequent tillage is needed to control weeds and if the shallow depth of tillage results in only small increases in infiltration rate (Unger, 1990).

No-tillage systems, sometimes known as no-till or direct drilling, involve the abandonment of all tillage operations and usually the use of a surface mulch of crop residues to provide suitable conditions for crop establishment and growth. The use of mulches and no-tillage systems are not necessarily inseparable. In some no-tillage systems a mulch may not be present or may be only very meagre. However, generally in the tropics 'no-tillage' implies the presence of a mulch and the benefits are usually considerably greater than using a mulch on previously ploughed land. The herbicide requirement for no-tillage systems can be greatly reduced by the maintenance of an effective mulch.

According to Lal (1985) the soil properties which favour the use of no-tillage include:

(1) Coarse-textured surface horizons or self-mulching clay soils with a high initial porosity.
(2) Resistance or reduced susceptibility to compaction.

(3) High biological activity of earthworms and other soil fauna.
(4) Good internal drainage.
(5) Friable consistency over a wide range of soil moisture contents.

Clays containing a high proportion of non-expanding clay minerals may develop a massive structure under no-tillage, especially when organic matter is low, which is unfavourable for many crops. The success of any no-tillage system in achieving anticipated levels of soil and water conservation and weed control depends fundamentally on there being enough mulch. In the semi-arid areas this can be a source of considerable difficulty but on Vertisols traditional no-tillage systems may be effective (Willcocks and Twomlow, 1992). The nutrient- and organic-rich shallow layers and the litter resulting after forest clearing can be preserved by the adoption of a no-tillage regime from the time of land-clearing (Opara-Nadi and Lal, 1987). In this way the favourable soil structure following bush fallowing can be retained and used more effectively than if cropping has been preceded by ploughing or non-inverting tillage.

No-tillage systems encourage the rapid build-up of soil fauna, especially earthworms, and their burrows, together with the undisturbed channels left by decomposing roots, improve infiltration of rainfall and reduce runoff and soil loss. Hoe farmers need less time and effort to prepare land, perhaps enabling a larger area to be planted and production per man to be increased. Where tractors are used, power requirements are lower. But savings in tractor and labour costs may be more than offset by expenditure on herbicides. Timeliness of planting should be improved since no, or reduced, tillage takes less time than normal cultivation. However, the presence of a continuous mulch of crop residues may encourage the carryover of plant disease and pests. Nutrient or acidity problems may arise which may necessitate ploughing occasionally to incorporate lime or phosphate and also the loosening of soils which have become too compact. No-tillage systems, especially where the level of mulch is insufficient, may be strongly dependent on herbicides for weed control, which may thus involve an inappropriate and costly level of technology, ill suited to local socio-economic conditions.

Mulching

Mulching with dead vegetable material, such as grass, weeds, straw or other crop residues, improves rainfall infiltration by preventing the rain from breaking down surface soil aggregates, by slowing up the movement of water on sloping ground surfaces and by improving soil structure as a result both of the addition of organic matter from the decomposing mulch and the increase of earthworm and termite activity. The value of a mulch in lessening runoff and erosion has already been mentioned. The maximum daily temperature of the topsoil under mulch is less than that of relatively dry, bare soil and this may improve the germination and establishment of heat sensitive crop seedlings. Weed growth is decreased by a heavy mulch, partly due to lower light intensity.

By lowering soil temperature, sheltering the soil from the wind and restricting weed growth, a mulch reduces evapotranspiration. The higher soil moisture content resulting from increased rainfall infiltration and lower evapotranspiration benefits crops in places where the rainfall regime may limit their growth. The improvement in infiltration has a bigger effect on soil moisture status than the reduction in evapotranspiration. Hence, it is desirable to apply mulch at the beginning of the rainy season when it can exert a maximum effect on rainfall absorption rather than in a drier time when it can only lessen evaporation.

The organic matter contributed to the soil by the decomposition of a vegetable mulch gradually releases nutrients in an available form. Nutrients may also be leached from the undecomposed mulch. Soil phosphate and potassium levels are commonly increased and acidity lessened. The large amount of potassium supplied by some mulches may induce magnesium deficiency in a crop but this can be readily corrected by the application of magnesium sulphate. Mulches usually enhance the concentration of roots in the topsoil and on some acid soils this may enable some crops to use a greater proportion of surface-applied phosphate fertiliser, especially as mulching tends to decrease the amount of applied phosphate fixed by the soil.

The effect of mulching on the available nitrogen content of the soil is variable. Nitrate levels may be lower than in unmulched soil, especially during the rainy season. This may be due to greater leaching or it may be because the mulched soil has dried out less during the dry season thus reducing the stimulating effect on nitrate production of re-wetting by early rains. In some cases a higher level of nitrate in a mulched rather than in an unmulched soil has been observed, perhaps because the more even temperature and higher moisture content of the former has permitted nitrification to proceed during the dry season (Mills, 1953). A further factor is the high carbon:nitrogen ratio of the organic matter derived from many mulching materials. This may lessen nitrification and, coupled with enhanced leaching under mulch in a wet season, can result in a crop suffering from nitrogen deficiency unless compensatory fertiliser is applied.

Mulching with vegetable materials, or sometimes with plastic as used on pineapples, has been found to increase the yields of a number of perennial crops. Hitherto, mulches have been little used by small farmers growing arable crops, except for small areas of vegetables and yams, but experiments have shown that they can give considerable increases in most years in areas of below optimum, or ill-distributed, rainfall. In many cases the yield of crops has been found to increase approximately linearly with increases in the mulch application rate. Lal (1991b) found that the yield of maize increased from 3.94 t/ha on unmulched plots to 5.27 t/ha when rice straw was applied at 8 t/ha. In Sri Lanka, Weerakoon and Seneviratne (1982) reported that the yield of cowpea was 362, 491 and 625 kg/ha when mulch was applied at 0, 4 and 8 t/ha respectively.

Problems with mulching

The production or gathering of plant, or other, residues to act as a mulch involves considerable labour and may compete with other uses of the material such as fodder, fuel, fencing or thatching. Mulch material is often suitable for stock feed and farmers will usually prefer to use it as such where animals are kept. Desiccated crop residues are readily blown around in high winds and are liable to be destroyed by termites. Because of these circumstances, the quantity of crop residues available at the beginning of the rainy season is often inadequate to provide an effective mulch. Commonly only 1–3 t/ha may be available whereas experiments have shown that much heavier applications are desirable.

Suitable material, especially grass, may be available from uncultivated areas reasonably near the farms, although this is becoming less common as increasing populations have intensified land use. In any case, small farmers without wheeled transport are usually unwilling to undertake the considerable effort of cutting, transporting and applying large quantities of material. While a full cover of mulch over the ground is usually desirable, a partial cover may be useful in some circumstances. Mulching alternate rows of some perennial crops has proved beneficial. For crops grown on ridges or mounds, such as yams, mulching on the crest of the ridge or mound can prevent excessive soil temperatures as well as retaining moist soil on the top of the ridge to avoid exposure of the tubers, which may make them unmarketable.

There are opportunities to use hand-sowing techniques in the tropics in the presence of mulch which would be considered quite unacceptable in temperate regions. The mulch can be penetrated fairly readily by a stick, hoe or spade and a hole then made in the underlying soil into which seed is dropped by hand, followed by covering with soil and firming to ensure seed/soil contact. Several designs of hand dibbers and punch-and-jab planters have been developed for use in heavy mulch and hand-pulled or pushed seeders will operate if the amounts of mulch are low. Some animal-drawn seeders are available which are usually larger versions of rolling injection planters or they may open a slot or furrow with a sweep, shoe or point but these are unlikely to be successful unless the amount of mulch is small.

Growing crops on ridges, tied ridges and mounds

In many parts of the tropics the use of ridges and mounds has been a traditional method of soil management in spite of the extra work which their preparation may entail. Experience has shown that yields of crops, particularly root crops, are often improved when planted on ridges or mounds, especially where topography or high clay content tends to result in waterlogging during the growing season.

An essential feature of ridges and mounds is the improvement in soil fertility and physical conditions within the root zone compared to conditions obtaining under flat planting (Lal, 1990). Ridging and mounding can be used to bury grass, weeds and crop residues, thus concentrating the organic matter under the root zone of the crop. There is also a concentration of the often thin topsoil within the ridge or mound, increasing the depth to infertile subsoil layers. Ash, remaining on the surface after burning cleared vegetation, will similarly be concentrated. The benefits of this process of concentration are particularly apparent where the subsoil is infertile and stoney.

On soils subject to waterlogging because of high clay content, poor structure or impervious subsoil, particularly where the topography leads to accumulation of runoff water, ridging and mounding has an important role of increasing the aeration of the root zone. Under such conditions crops grown on the flat may suffer periodic or prolonged waterlogging or even inundation, whereas the roots of crops grown on ridges or mounds remain in unsaturated soil. Root crops, such as yam, cassava, sweet potato, potato and taro, which tend to be sensitive to unfavourable aeration in the root zone, respond favourably to the improved conditions under ridges and may also be harvested more readily than when planted on the flat. Ridging facilitates weed control whether by hand, animals or tractors. Under high rainfall or topographically induced waterlogging, crops such as sugar cane will benefit from being planted on ridges even though they may tolerate limited periods of waterlogging.

On sloping land, ridges should invariably be orientated across the slope and preferably be made on the contour. The most suitable shape and size of ridges vary according to crop and soil conditions. They may be set up in permanent position, as with broad cambered ridges, or re-established annually, perhaps by moving the new ridge into the position of the previous furrow. Permanent ridges provide a virtually traffic-free zone so that soils are not degraded. The size and shape of ridges have important implications on saline soils where evaporation will lead to high salt concentrations especially on the upper extremities of ridges.

There are many variations in the planting practices on ridges (Fig. 5.6). Whereas normally crops are planted exclusively on the ridge, it is also possible to adopt multi-cropping with upland crops sensitive to waterlogging on the ridge and hydrophilic crops in the furrows. In areas of low rainfall, ridges can be used for an entirely different purpose, the gathering of sparse rainfall into the furrows where crops can be planted, leaving the ridges unplanted (Jones et al., 1989).

Crop residues and hoed weeds are often placed in the furrows where they impede rapid runoff. After harvest, ridges may be split, thus burying residues which have accumulated in the furrow and enhancing the organic matter status within the root zone of the next crop. On Vertisols in Zimbabwe, the loose highly aggregated surface soil created after drying is gathered by sweeps to build the ridges rather than by using mouldboard ridgers which would penetrate too deeply and bring up clods (Norton, 1989).

A further variant in the management of the soil surface in dry places is the use of pot-holes or non-ridged basin tillage with the crop planted in depressions where rainwater will concentrate. The creation of pot-holes with hand tools requires less energy than does ridging. Alternatively, mechanical devices, such as a chain diker, can be used to create numerous small basins, each with a capacity to hold about 25 mm of rain (Jones and Stewart, 1990).

Tied ridging

The use of tied ridging for soil and water conservation has already been mentioned. Ties

Fig. 5.6 Mixed cropping under low rainfall, with maize on the ridges showing drought symptoms whereas the groundnuts in the furrows are growing well. (Source: Overseas Division, Silsoe Research Institute.)

between ridges may be installed by hand, using hoes or shovels or by a number of mechanical implements pulled by animals or tractors. Equipment for tied ridging, also called basin tillage, has been described by Jones and Stewart (1990). A number of mechanisms have been used such as:

(1) raising shovel
(2) tripping shovel
(3) reservoir implantation
(4) 'chain' diker
(5) hydraulic motor-tripped mechanisms

A one-pass raising shovel-type machine was designed by the National Institute of Agricultural Engineering (now the Silsoe Research Institute) in the UK which ripped, ridged, planted and tied the ridges (Boa, 1966; Dagg and McCartney, 1968). However, ties are more commonly inserted as a separate mechanical operation after the ridges have been set up. The effects obtained are generally highly speed-sensitive and results may be unacceptable outside a narrow speed range. Tripping mechanisms frequently block under some conditions.

If the base of the furrows is very firm it may be necessary to cultivate them before attempting to set up tied ridges. Tied ridges may give rise to problems if mechanised cultivation for weed control becomes necessary after the ties are installed. These difficulties are partially alleviated by not installing ties in furrows to be used subsequently for wheeled traffic.

The height of ties is of critical importance with respect to the behaviour of surface water during storms. The capacity of the basin created by the ties should be large enough to accommodate the largest volume of water likely to arise in the worst storm. However, in practice, accelerated erosion may result on sloping land if ties cause water to flow suddenly across ridges and Hudson (1971) therefore recommended that ties should be somewhat lower than the ridges.

On coarse soils of high infiltration capacity, paddle-type machines can be used to insert ties at about 1 m spacing, providing a water storage capacity of about 25 mm. On fine-textured soils, or if heavy storms are expected, wheel-type raising shovel implements are more appropriate with ties inserted at a spacing of 0.75 m providing a capacity of about 50 mm. In Zimbabwe a system of no-till tied ridging has been developed (Gotora, 1991) which is applied on land already subject to erosion protection. The land is loosened by ploughing only prior to the construction of ridges; ties are then inserted with a very simple implement pulled by two or three donkeys in a single line. The technique has shown promise with maize, cotton, soya bean, sorghum and groundnuts but, although effective in reducing sheet erosion, problems of weed control and additional work requirement may render the system unattractive to small-scale framers (Vogel, 1993).

Crop responses to ridging and tied ridging

The effect of ridging on yields is influenced by the soil, the topography, the amount, distribution and seasonal variation of the rainfall, the crop grown and the time of planting. Hence comparisons of ridge- and flat-planted crops have often given variable results except that ridging usually gives reliably higher yields at sites where crops on the flat would experience waterlogging during the growing season. In various trials with a variety of crops on well-drained sites in Kenya, results varied with season but the tendency was for ridges to be beneficial on light soils in the drier areas, but often of no advantage on loams under better rainfall. Walton (1962a, 1962b) examined highly variable yield responses of cotton to ridging in Uganda and showed that during dry periods the evaporation rate was higher on ridges than on the flat and that this could adversely influence the water available to young cotton when the roots are largely confined to the ridges. However, ridged soil was shown to contain more water than flat-planted soil at 30–180 cm depth and this extra water will progressively benefit ridge-planted crops and lead to higher yields than after flat planting when dry periods occur later in the growing season. At sites where waterlogging is frequent and prolonged during the growing season, ridging will usually give reliably higher yields, the yield sometimes increasing linearly with increases in the height of the ridge.

Tieing ridges further increases the available water in the root zone and crops grown on them have been observed to have deeper, more extensive root systems than those on conventional ridges or flat planted. Hence benefits are likely to be greater in drier areas and on soils of low water-holding capacity. Early studies in East Africa by the Empire Cotton Growing Corporation (1951–56) compared the yield of a number of crops under tied ridging with that following conventional flat cultivation. The results in Table 5.1 show that much greater yield increases were obtained with tied ridging under drier areas in Tanzania than at a higher rainfall site in Uganda. At the drier sites, the less drought-sensitive crop millet showed little yield response to tied ridging compared to the large increase for cotton and sorghum. These early results have been confirmed in many studies elsewhere.

In pioneering work with mechanical installation of tied ridges in Tanzania, Dagg and McCartney (1968) found that tied ridges gave a 40% increase in the yield of maize over flat cultivation on a red clay loam as a result of greater reserves of soil water at tasselling time when plants on the flat were drought-stressed. However, on a heavy Vertisol under similar rainfall there was no shortage of available water on flat plots which yielded as well as those that were tied ridged. The advantages of tied ridging on light, poorly structured soil in Zambia was also confirmed for maize, sorghum and bulrush millet (Honisch, 1974). Hulugalle (1990) suggested that the yield benefits obtained with tied ridging in the semi-arid Sudan savanna in West Africa may not occur in the more humid Guinea savanna, where even negative yield responses may occur after tied ridging if it leads to poor weed control or to loss of aeration in the root zone.

Mounds

The gathering of topsoil into mounds with hand hoes has been practised traditionally in many parts of Africa, either for the purpose of burying grass weeds when clearing without burning in savanna areas, or for growing crops in poorly drained or seasonally swampy areas. There is a

Table 5.1 Effect of tied ridging on crop yields in relation to soil type, rainfall and crop susceptibility to drought. (Source: Empire Cotton Growing Corporation, 1951–56.)

Station	Soil (drainage)	Mean annual rainfall (mm)	Period of experiment (years)	Crop	Mean yield (kg/ha)		Increase with tied ridging (%)
					Flat	Tied ridging	
Tanzania							
Lubaga	Heavy (impeded)	790	12	Cotton	590	817	38
Lubaga	Light (free)	790	9	Sorghum	823	1173	42
Ukiriguru	Light (free)	840	6	Millet	693	745	8
Ukiriguru	Light (free)	840	7	Cotton	690	851	23
Uganda							
Namulonge	Loam (free)	1160	4	Cotton	839	877	4

widespread occurrence of depressions of moderate size with fertile soil which are swampy or flooded during the rains but gradually dry out after they cease. Making mounds permits dry season cropping in these depressions or may allow earlier planting than would otherwise be possible. In Southeast Asia mounds are sometimes used for growing other crops in paddy fields after harvesting rice, but ridges are usually preferred. The objectives and benefits of mounds are essentially similar to those of ridges but on sloping land mounds do not effectively check erosion.

POWER SOURCES AND MECHANISATION SYSTEMS FOR FIELD OPERATIONS

Power sources and power utilisation

The extent to which human, animal and tractor sources of power are used in the developing countries is shown in Table 5.2. The predominant role of human labour is evident, particularly in Sub-Saharan Africa, but also on most of the smaller farms in Asia. Animal power is more widely used in Asia than elsewhere whereas tractors only make a significant contribution in Latin America and parts of North Africa.

The figures conceal marked differences within regions in power use which are usually due to variations in size and tenure of holdings and type of crop. For example, in India relatively large holdings, employing tractors for virtually all operations, may exist side by side with very small holdings on which human labour is the only power source available (Ojha, 1987). In Kenya, farms larger than 20 ha are generally cultivated entirely by tractors whereas, on the much larger number of farms of less than 3 ha, the averages

Table 5.2 Proportional contribution (%) of total power use in 93 developing countries. (Source: FAO, 1987.)

Area	Human	Animal	Tractor
North Africa	69	17	14
Sub-Saharan Africa	89	10	1
Asia (excluding China)	68	28	4
Latin America	59	19	22
Overall	71	23	6

are 84% of cultivation by hand, 12% by oxen and 3.5% by tractor.

Changing levels of mechanisation

Because of the immense drudgery and low productivity of human labour in the tropics, there has been a widely held view that opportunities for increasing animal power, together with a progressive adoption of simple mechanisation, should be encouraged (Matthews, 1990). However, the success of any such changes is dependent on the availability of appropriate resources and skills. The adoption of even simple levels of mechanisation requires changes in management and economic aspects which have important implications for soil, water and crop relationships as well as for the farmers and their communities. The rapid adoption of mechanisation is invariably unsuccessful unless a clear cost-effectiveness can be perceived by the farmers concerned, regardless of the opinion of scientists or economists (Pingali *et al.*, 1987). Any attempts to increase mechanisation must be preceded by changes in land tenure to provide unfragmented holdings of suitable size and also depend on the availability of suitable training, technical skills, credit facilities and workshops for the repair of tractors and implements.

The future importance of mechanisation opportunities and problems is much debated in the tropics and widely differing views are held concerning the extent to which mechanisation should be encouraged. Youdeowei *et al.* (1986) wrote that 'For tropical agriculture to become as efficient and profitable as the agricultural systems in the developed countries...the shift to mechanical power is imperative'. Okigbo (1989) also believed that 'mechanisation is vital in reducing drudgery, getting more young people into farming, achieving timeliness in operations, and addressing the shortage of labour at peak periods of demand'. Mechanisation has the technical capability to reduce drudgery, to increase productivity per man-hour, to improve timeliness, to increase efficiency in handling and processing farm produce and to promote the development of newly cropped areas. In northern India, owning or hiring a tractor has become widely recognised as a symbol of prosperity and many farmers utilise private hire and contract services (Yadav and Suryanto, 1990). However, elsewhere the socio-economic circumstances may exert a powerful brake on the uptake of even the simplest levels of mechanisation. The enlargement of cultivated areas through mechanisation, even when technically feasible, often brings with it exhorbitant costs and soil management problems which farmers had not encountered when only hand or animal-drawn tools were used. In particular, mechanisation must be accompanied by skills and resources to prevent erosion and maintain fertility and an understanding of existing technical, managerial, financial and social resources available to local farmers.

Human power

Hand labour is likely to remain the dominant power source in many areas but, with a maximum output of only about 0.075 kW (0.1 hp) per person, the area of cropping which can be handled with hand tools only is usually restricted to a maximum of 2–3 ha per family. There is a wide variety of tasks which involve considerable deployment of human energy, even when animal or tractor power contribute to total power use. Weeding operations in many crops are particularly suited to hand labour due to the degree of selection which is possible.

Hand tools for primary and secondary cultivation and weeding (Metianu, 1992) were formerly developed to suit the local availability of materials and the local soil types, but nowadays by far the commonest cultivating tool is the mattock-type hoe with an iron blade angled inwards at about 80° to a strong wooden handle if it is to be used for digging, or at a more acute angle if it is to be used for shallow scraping of the soil as is usually done in weeding. The length of the handle varies, often being quite short for weeding but longer than usual for forming ridges. Hand-held planting tools are primarily designed to open a hole into which seed is dropped by hand. The ability of such tools to operate in the presence of considerable amounts of plant residues on the surface is especially important with mulched crops.

Animal draught power

Draught animals are used on no more than 10% of farmlands in Sub-Saharan Africa and are mostly employed in savanna areas where they practise rather inefficient ploughing, leaving weeding, harvesting and most transport to be done by human labour. By contrast, in many parts of Asia, farmers have for centuries prepared land with animal-drawn ploughs and harrows and there is widespread use of carts drawn by oxen or buffaloes. In the rice fields, apart from ploughs, diverse animal-drawn harrows, wooden rollers, serrated boards and rakes are used for puddling, levelling and removing or burying weeds.

Types of animals

It has been estimated that among the world's draught animals are 300 million bovines, 80 million equines and 20 million other animals (yaks, camels, llamas and elephants). Larger animals (oxen, horses, mules) can provide about 0.5 kW (0.7 hp) for a 5–6 hour day whereas smaller animals (donkeys) provide 0.1–0.2 kW (0.13–0.27 hp) and are thus suited only to low draught and light transport tasks. While local traditions have a strong influence on the choice of animals for draught work, improvements are possible by adopting new species or breeds and by improving animal health and nutrition. The introduction of mules is seen in some areas as an opportunity to increase the rate of work. The use of cows, rather than oxen, would enable fewer animals to be maintained and thus could improve the nutritional level of individual animals.

The selection of animals for improved draught qualities has tended to be based on body weight. Optimum pull for bovines (oxen, cows, buffaloes) is about 10–12% of body weight, whereas for equines (horses, donkeys, mules) and camels the corresponding value is 12–15% of body weight (Inns, 1992). In Ethiopia the International Livestock Centre for Africa found that two 500 kg Boran × Friesian cross-bred oxen could pull a tool carrier whereas two 300 kg zebu cattle could not. In Brazil draught oxen may weigh as much as 750 kg. There are opportunities for breeding programmes which take account of pulling power, stamina, efficiency of food use, temperament, body heat production and capacity for heat loss and disease tolerance. Improvements in draught animal performance could permit a single animal being adequate for many operations which are currently undertaken by two animals, with associated advantages on smaller farms.

Problems with animal power

For farmers who have not previously used animals, there is a considerable technological barrier to be overcome, particularly in relation to the skills needed for the selection and management of draught animals (Munzinger, 1982). Nutrition exerts a profound influence on the draught capability of animals. At the end of the dry season, animals may be so weak as to be unable to work just when the demand is greatest at the start of the growing season, thus delaying field operations until they have benefited from the growth of herbage after the first rains and seriously delaying land preparation and planting. To overcome this problem opportunities should be taken to complete as much primary cultivation as possible at the end of the rains when animals are in good condition.

Diseases markedly influence the availability and strength of animals and a lack of familiarity with the management and handling of animals will reduce the success of their use in field operations. Extending the working life of trained draught animals must be encouraged. The training of animals for draught operations reaches a high level in Asia or Southern Africa where a team of two or more oxen may be controlled by a single driver, whereas it is not uncommon in Central Africa for a team of several attendants to be involved in field operations, thus failing to achieve the expected increases of productivity per worker. Training is particularly important for animals to be used in row crops where erratic cultivation can result in substantial crop losses. Well-trained animals, however, can undertake weeding operations in row crops at a later stage of crop development than would be possible by mechanised equipment.

The productivity of draught animals may be very low on an annual basis since they will usually spend only a small amount of their time on field work, depending on the seasonal nature of cropping. In Rajasthan, India, Sharma (1991) reported that farm animals, which provide the main source of power, work for 65–75 days per year while in Africa the annual work load may occupy only 40–50 days. In Africa, the limited adoption of animals for draught purposes is often attributable to the lack of suitable animals among arable farmers, while cattle-owning pastoralists, even after abandoning a nomadic existence, lack experience of arable farming.

Opportunities for animal power

There has recently been a resurgence of interest in advancing the uptake and efficiency of animal power (Starkey and Faye, 1990; Panin and Ellis-Jones, 1994), particularly in those areas where there is no long-standing tradition for their use. Draught animal power is seen as providing an opportunity to avoid much of the drudgery and low productivity of hand labour while at the same time offering an achievable and appropriate level of technology without the problems of tractor use. Apart from land preparation, there are numerous opportunities for animals to be employed in other work, such as transport, thus considerably reducing the costs attributable to their use in crop production. The possible role of draught cattle as a source of meat will also influence their value.

Generalised data for work rates and daily output of ploughing using different power sources (Table 5.3) indicate that the adoption of draught animal power in place of hand labour can provide opportunities for raising yields through improved timeliness, and 'new' operations such as ridging and better weed control. It can, at the same time, encourage extensification by permitting a farmer to cultivate a larger area.

Table 5.3 Approximate indicative work rates and daily output of ploughing using different power sources. (Source: Morris, 1983.)

	Tractor 50 kW	Animal 2 oxen	Human 1 man
Work rate (hours/ha)	2–3	25	100
Work day length (hours)	8–16	5	5
Daily output (ha/day)	3–7	0.2	0.05

There is a need to improve the efficiency of draught animals (Smith, 1984; Starkey and Faye, 1990). The nutrition of draught animals has an important influence on their work output, and will also greatly affect their reproductive efficiency bearing in mind that both sexes are usually employed equally for draught purposes. Reductions in an implement's draught can be achieved by reducing its weight, adjusting the harness and linkage to increase the angle of pull and counteracting a greater proportion of the implement's weight by increasing the support forces acting on it (Inns, 1992).

Harnesses for draught animals

Many types of harnesses, including yokes (head, neck, shoulder), collars (full, split) and breastbands, are traditionally employed depending on the type of animal involved (Inns, 1992). Simple shoulder yokes are practical for humped zebu cattle in Africa and Asia. However, in Latin America where animals do not generally have humps, yokes are attached to the horns. Split collar harnesses are used in some localities for oxen, buffaloes and donkeys, in contrast to full collars which have been widely used for both horses and cattle throughout Europe and North America. The most appropriate harness will depend upon whether a draught implement is either beam- or pole-pulled or, alternatively, chain-pulled. There are opportunities to improve harnesses of animals in the tropics and tests are in progress on a number of types of collars.

Implement systems for draught animals

The successful application of animal draught power for cultivation and weeding operations is dependent on the availability of appropriate, efficient implements which are economically acceptable to farmers. Such implements are generally simple enough for their successful local manufacture and repair, especially in Asia where local free-trade conditions may provide competitive opportunities for improvements in design and after-sales service (Rodriguez, 1992). In view of the perceived limitations to the available animal-drawn implements in many parts of the tropics, there have been vigorous attempts to introduce new designs, particularly for systems providing interchangeability of implements on either toolbars or wheeled tool carriers of varying complexity. Although these developments have attracted much attention, single-purpose implements remain the predominant choice throughout most of the tropics. Toolbars have evolved from single-purpose cultivator frames and are of simple construction, similar to a plough frame, on which a number of implements (plough body, ridger, discs, tines and crop lifters) may be attached. They are simple to manoeuvre and suited to single interrow work with smaller animals, perhaps even a single donkey. They usually lack main wheels but may sometimes have one or two depth-control wheels (Fig. 5.7). Improvements in design have resulted in widely acceptable implements (Nolle, 1986) and some types have achieved sales of 350 000 units worldwide. Larger toolframes, having a rectangular frame supported on two wheels, are suited to multi-row seeding or weeding. Like toolbars, they usually rely on a chain for traction.

Wheeled toolcarriers (termed 'polyculteurs' in French) consist of a transverse rectangular frame mounted on two large, usually pneumatic, wheels with a long traction beam to a pair of animals. They provide a wide interchangeability of implements for multi-row operation and conversion to a transport cart. The development of wheeled toolcarriers over a period of some 35 years, particularly in Africa and India, has been described in detail by Starkey (1988). Nolle (1986) was responsible for much of the pioneering work in Africa. He promoted the three prin-

Fig. 5.7 Interrow cultivation using a single ox-drawn wheel-mounted toolbar in Burkina Faso. (Source: FAO, Rome.)

ciples of simplicity, multi-purpose use and standardisation of components which were a feature of his first design, developed in Senegal in 1955. In 1958, NIAE (now the Silsoe Research Institute, UK) started development work on wheeled toolcarrier systems which eventually involved studies in 25 tropical countries (Willcocks, 1969). Subsequent co-operative research and development projects at the International Crops Research Institute for the Semi-Arid Tropics (ICRISAT) in India led to a number of further designs specifically to promote the growth of crops on ridges and beds. Independent development work was also undertaken in Botswana ('Makgonatsotlhe'). Development and testing work has continued in Mexico (Sims *et al.*, 1990) where a simplified wheeled toolcarrier ('Yunticultor') has been subjected to extensive trials in cooperation with farmers.

Starkey (1988) has analysed and summarised all such development programmes throughout the tropics. He listed 45 different designs promoted during the period 1955–86, of which a total of some 10000 units were built. Although the engineering aspects of the designs have usually been sound, wheeled toolcarriers have not generally been accepted by farmers, largely because of their complexity, cost and poor performance when used for mouldboard ploughing. The pneumatic tyres punctured in many areas while the fixed wheel track gave rise to offset problems when operated with a single mouldboard plough. Interrow cultivation of crops at different row spacings was restricted. Certain designs (e.g. the 'Makgonatsotlhe' in Botswana), however, allowed adjustable wheel spacing. The concept of multiple use as both a cultivation and a transport cart failed because the necessary conversion and adjustments proved to be too complex and time-consuming. In addition, the floor of the cart had to be unacceptably high in order to clear the high level of the toolcarrier chassis required for 'late' weeding operations. Wheeled toolcarriers, without the problems of interchangeability to a cart, appear likely to find application on larger farms, especially for row crop-work and where technical training in their use can be provided (Sims *et al.*, 1990).

Tractor power

The adoption of tractors in the tropics is extremely variable (Table 5.2) and depends on socio-economic circumstances, the type of cropping undertaken and the extent of governmental economic support. In some countries in Southeast Asia, 11–15 kW (15–20 hp) tractors are widely used. In India the availability of tractor power has reached 0.6 kW/ha (0.8 hp/ha) compared to 0.3 kW/ha (0.4 hp/ha) in Indonesia (Yadav and Suryanto, 1990). In Latin America, where 22% of total power use is provided by tractors, their use has been claimed to be responsible for considerable increases in productivity per unit area and per labour unit.

The governments of many tropical countries have provided generous but often poorly targeted subsidies of various types to support the utilisation of tractor power, an aim which was also encouraged by many international agencies and aid organisations which frequently gave away large numbers of tractors. Large-scale tractor utilisation schemes were initiated in many countries during the 1960s but the failure of most of these schemes has had an important effect on subsequent mechanisation policies. During the 1980s the upward rise in tractor numbers ceased in many countries owing to foreign currency problems and the perceived low efficiency of tractor programmes attributable to a shortage of technical skills, poor maintenance and inadequate availability of spares (Fig. 5.8). Ideally, suppliers of imported tractors should be satisfied that local dealers have the skills and resources to handle after-sales services efficiently and that adequate facilities are available for training operators in the use and maintenance of their products (Rodriguez, 1992).

Types and utilisation of tractors

Two-wheel walking tractors (Fig. 5.9), are often considered to be the most appropriate machines for small farms (Krause *et al.*, 1984) and are widely used for rice production in Asia and for horticultural crops in some other areas. Because of their low weight, traction is limited and

Fig. 5.8 Premature breakdown of tractors is all too common under the conditions in the tropics where the availability of spares and service facilities may be inadequate. (Source: Overseas Division, Silsoe Research Institute.)

Fig. 5.9 'Walking' or single-axle tractors are widely used in some areas of Southeast Asia. (Source: FAO, Rome.)

two-wheel tractors are usually used with direct-coupled rotary cultivators. Four-wheel tractors of several types of design and power are available. When the engine power is less than about 25 kW (33 hp) they may have conventional design or be of simplified design intended for indigenous manufacture (Holtkamp, 1990). For primary cultivation operations, a sustainable pull of at least 5 kN (500 kgf) is desirable, for which a tractor having an engine power of at least 20 kW (27 hp) and a total mass of 1200 kg will generally be necessary (Inns, 1992). Attempts are often made to introduce imported standard 48–75 kW (65–100 hp) tractors rather than simplified lower power models, specifically designed to meet tropical requirements, which, in many cases, exceed the cost and lack the reliability of larger standard models. There is a considerable trade in little used second-hand tractors and the assembly of re-built machines using local labour. There are obvious needs to standardise tractor makes and models within regions to make it possible to maintain stocks of spares locally and to simplify repair operations. With the rising agricultural productivity in several parts of Southeast Asia, there has been a strong development of the local manufacture of tractors. This has been particularly pronounced in India in response to the ban on the import of tractors imposed in 1977–78, and tractors manufactured in India are likely to be suitable elsewhere in the tropics (Fig. 5.10).

The achieved efficiency of the use of tractors in the tropics is often very low. This may be attributable to excessive delays in obtaining spares and repairing punctures due to tyres not being sufficiently durable to resist damage from stumps, stones, etc. Power losses attributable to altitude (1% loss per 100 m above sea level) are comparatively less important than the reliability of the vehicle as a whole. Due to lack of training, motivation and maintenance facilities, drivers may lack the incentive to ensure that modern tractors achieve the high efficiency of which they are capable. High operating costs are likely to outweigh the potential advantage of high output. Lack of reliability in the supply of reasonably priced tractor fuel due to problems with communications, transport and foreign currency may restrict the operational efficiency of tractors. In many parts of the tropics the disproportionately high cost of fuel may restrict the use of tractors even though the supply of tractors *per se* may not be a major limiting factor.

Fig. 5.10 A tractor manufactured in India undertaking cultivation work for seedbed preparation. (Source: Messrs Escorts Ltd., Faridabad, India.)

On small farms individual ownership of a tractor is not economically possible; about 40 ha of arable land being considered necessary to support the costs of one 20 kW (27 hp) tractor (Inns, 1992). Multi-farm use arrangements, such as co-operative ownership, contract or hire services are therefore necessary. However, on larger farms, private ownership by farmers possessing the necessary managerial skills is the most likely route to successful utilisation. Government ownership with utilisation operated through tractor hire services has frequently led to low efficiency, high costs and discouragement of private hire schemes. However, government-sponsored schemes may have a pioneering role, especially in the absence of alternative ways of providing tractors. Where there is a centralised organisation responsible for an intensive crop production scheme, such as the 880 000 ha of the Gezira scheme in the Sudan, the level of tractor utilisation efficiency for land preparation and other operations can in theory be greatly increased. The use of mechanisation in the tropics is likely to be most effective where its use is restricted initially to selected power-intensive operations, retaining animal power for other operations. Only later, when tractor expense is justified, should their use be extended to other uses.

WEED CONTROL TECHNIQUES

Weeding accounts for the major part of the hoe farmers' labour in crop production and may amount to as much as 80% of the total (Table 5.4; Miller, 1982) (Fig. 5.11). This high labour requirement limits both the area of land that a family can crop and the time for which it can be cropped before it must be allowed to revert to bush fallow. After starting with a relatively clean plot following clearing and burning, weed infestation tends to increase year by year under cultivation and is often the main reason for abandoning the plot and clearing a new one.

Poor weed control is one of the major causes of low yields of both annual and perennial crops on smallholdings. Failure to keep plantations of perennials (such as rubber, oil palm and cocoa) sufficiently clean during their immature phase in the first few years after planting commonly results in poor growth, delayed cropping and consequently reduced yields. With annual crops it is particularly necessary to weed early. There is experimental evidence that many crops, especially maize, cotton, groundnuts, beans and sweet potatoes, are severely checked in their early growth stage by even a moderate cover of weeds and that this check reduces yields. No amount of subsequent weed control will offset this check. Conversely, if the crop is clean weeded in its early stages, considerable weed infestation later may not affect yields adversely. Weeding is an important component of land clearing, primary and secondary cultivation operations, as well as post-planting interrow cultivations. However, successful weed control also relies on the additional benefits which accrue from cultural practices such as mixed cropping, crop rotations, minimising weed infestation and the use of false seedbeds, which encourage weed seeds to germinate and allow the seedlings to be destroyed before crop plants are planted (Radley, 1992). Unconventional methods of weed control may be adopted to suit local conditions (Reijntjes et al., 1992). After crops are harvested, grazing animals may be used to eliminate residual weeds, goats being particularly effective in eliminating woody species. Cover crops, such as *Mucuna utilis*, can suppress weeds as aggressive as *Imperata cylindrica*. Intercropping will often result in a lower weed population than when crops are grown singly. Certain crops, such as sunn hemp (*Crotalaria*

Table 5.4 Average labour requirement for various farm operations on savanna and forest land in Nigeria. (Source: Miller, 1982.)

Operation	Savanna		Forest	
	days/ha	%	days/ha	%
Clearing	7	2	13	4
Planting	33	12	13	4
Weeding	189	66	256	79
Harvesting	57	20	44	13
Total	286	100	326	100

Fig. 5.11 Shallow cultivation and weeding in a young banana plantation using short-handled hoes in Somalia. (Source: FAO, Rome.)

ochroleuca), have a specific suppressant action on weeds when grown either in rotation or as an intercrop. Dead weeds (providing they are not at the seeding stage) serve as a useful mulch, while weeds growing prior to a crop can be treated as a green manure.

Small farmers commonly do not start weeding their annual crops early enough. Having waited for the first rains to soften the land before beginning preparatory cultivation they are late with planting and, by the time this has been completed, weeds are growing vigorously on the earliest planted part of the holding. The heavy weed growth which develops often means that an insurmountable peak labour demand builds up. More work is required to do inefficient late weeding than would be needed for effective control of earlier, lighter weed growth. Baker (1975) observed that the time taken to undertake a single late weeding with hand hoes was appreciably longer than the total time required for two timely light weedings. If weeds could be tackled at an earlier stage they could probably be controlled more easily and speedily with a light Dutch hoe than with the heavier digging hoe customarily used by small farmers. By reducing yields, delayed weeding is liable to make other inputs uneconomic, especially fertilisers, which increase weed growth thus adding to the work load and decreasing the proportion of the applied nutrients absorbed by the crop.

Hoes designed for digging, chopping, pushing and pulling are the primary tools for manual weeding, the preferred type being adapted to suit local conditions of soil, topography and crop (Radley, 1992). Apart from hoeing, the commonest means of weed control is hand slashing with a machete or similar tool, but this is mainly used in tree and bush crops. It can control the vigour and competitive power of some species (but not of many grasses) to acceptable levels without exposing the soil to erosion. Hand pulling may also be used, but is very laborious and mainly used for the species of *Orobanche* and *Striga* which parasitise some crops. Interrow weeding can be undertaken with hand-pushed wheeled hoes, or animal- and tractor-powered cultivators. For maximum efficiency the width of such operations should be adjustable to suit soil and crop conditions. Small engine-powered rotary cultivators, with or without wheels, are designed specifically for interrow cultivation (Radley, 1992).

Herbicide techniques

There is some opposition to the use of herbicides by small- and medium-scale farms in the tropics on the grounds of danger to human health resulting from the use of spraying equipment without proper safety precautions, the unsafe disposal of used containers and the possible presence of harmful herbicide residues in edible crop products. It is also thought by some that the persistence of some herbicides may have harmful effects on the soil flora and fauna. However, as modern herbidices are usually very effective in tropical crops and can save much of the enormous amount of labour involved in hand weeding, it is likely that their use will expand, especially as greater efforts are being made to improve herbicidal action, increase product safety and to reduce the amount of water required for their application (Akobundu, 1987). Applying herbicides to supplement hand weeding may also be useful and economic since trials have shown yield increases of 10–30% when herbicides were used in addition to hand weeding, suggesting the latter fell short of an adequate level of control.

By 1983, 11.5% of the world's use of herbicides lay outside North America, Europe and Japan. Significant areas of small-scale maize grown in East and West Africa are treated with herbicides and in West Africa herbicides are also widely used on small-scale cotton and groundnuts (Parker, 1984). In India herbicides are extensively used in the Punjab for the control of *Phalaris minor* in wheat while in Central and South America herbicides are used to a considerable extent on maize, barley and bananas.

Herbicides have been found to be particularly useful in high rainfall areas and for the control of parasitic weeds (e.g. *Striga, Alectra, Orobanche* spp.) which are very difficult to control by hand. The application of herbicides will not risk exposure of the surface soil to erosion risks, a problem with hand weeding especially on steeply sloping land.

Need for information and training

For most farmers in the tropics, the use of herbicides represents a very large advance in technology and, to be successful and safe, requires a level of training and experience which is still largely lacking. There are, as yet, few nationally recognised recommendations or codes of practice concerning the use of herbicides in tropical countries. Guides for regions, such as that prepared by Terry (1984) for East Africa, fill a valuable role but, in view of the continual advances in technology, users and advisers need to maintain close links with sources of information supplied by the chemical companies and be able to understand written instructions on product labels and accompanying literature. The very wide and constantly extending range of herbicides now available, and the diversity of crops grown in the tropics, makes the selection of appropriate herbicides a matter of some difficulty. Terry (1984) listed some 99 herbicides for possible use in East Africa and their complex names are, understandably, often the source of confusion among users. In spite of the apparent proliferation of herbicide chemicals, there are still some tropical crops and weeds for which no suitable herbicide is yet available. Because of the relatively small market, chemical manufacturers are unwilling to develop herbicides for a limited application on uncommon tropical crops.

A problem which is still comparatively new, but which should be carefully monitored, is the shift in weed flora as resistance to a much-used herbicide is acquired by certain species. For example, following the use of atrazine or alachlor on maize, a build-up of *Rottboelia exaltata* has sometimes occurred, with serious results. Apart from the problems of understanding the selectivity and application techniques of herbicides, there are often important factors related to the age and physiological condition of both weed and crop which will influence whether the expected control is achieved. These factors will not be appreciated by farmers unless suitable training and experience can be gained.

Economic and logistic constraints

Apart from the need for more information and training, other difficulties in the way of greater use of herbicides by small-scale farmers include the need for substantial quantities of water which may have to be carried some distance, remoteness from sources of supply of the chemicals, the fact that available formulation containers may be excessively large for use with small sprayers on small plots and a lack of credit facilities for purchasing equipment and chemicals. There is commonly the further economic constraint that much of the labour used by small farmers for weeding is unpaid family labour. Parker (1984) considered that the greater use of herbicides on small farms in the tropics may sometimes 'be quite unnecessary or even socially undesirable'. At any rate, in view of the undoubted problems associated with herbicide use on small farms, their use cannot be recommended unless the farmer can be shown to benefit from increased yield, reduced cost of weeding, increased area of cropping or the release of labour for other work.

On larger arable farms and plantations of perennial crops, economic and educational constraints are usually less pervasive and where crops are grown for export (e.g. groundnuts, sugar cane, bananas, cotton, coffee, tea and tree crops) the use of herbicides is likely to show economic advantages over other methods of weed control. Similarly, on large-scale settlement or irrigation schemes where there is centralised management, herbicide use can be much more effectively organised than on scattered holdings. On the Gezira scheme in the Sudan, the use of herbicides advanced from 8% of the area in 1977–79 to 90% of the area in 1986.

Types of herbicides and mode of action

Herbicides are divided into:

(1) Total or non-selective herbicides which are effective against all or most vegetation.
(2) Selective herbicides which suppress or kill weeds but do not affect specified crop plants.

Herbicides may be foliar-applied or soil-applied (Fig. 5.12). Foliar-applied herbicides may have a contact action or be translocated within the weed. Soil-applied herbicides can persist in the soil for short or long periods and act on the underground parts (either roots or pre-emergence seedlings). Herbicides are developed for application at specific stages of crop growth such as pre-sowing, pre-emergence or post-emergence. The mode of application may involve 100% coverage or may be directed only towards the main weed concentration and away from the crop or, alternatively, may be directed only to the vicinity of the crop.

The persistence of different herbicides varies widely and is influenced by weather and other factors such as microorganisms, chemical decomposition, absorption on clay or organic colloids, leaching, photo-decomposition and volatilisation. In view of the indefinite occurrence of these processes, the assessment of persistence of herbicides is necessarily inexact. Persistence longer than the life of the crop would run the risk of injury to the growth of subsequent susceptible crops. Very short persistence, however, may be of little lasting benefit for the control of weeds which emerge later and compete with the crop for a long time.

Application techniques

The successful and safe use of herbicides depends on the correct use of appropriate application equipment. Many types of such equipment are available, varying in scale from very small hand-held devices up to self-propelled spraying vehicles (Matthews, 1984, 1992). Attention will be given here to equipment suited to smaller farmers rather than estates and plantations.

The choice of application equipment is likely to depend as much on cost, local availability and servicing arrangements as on the technical suitability of the equipment. The availability of clean water, particularly at the start of the cropping season, may well have an overriding influence on the whole process of herbicide application. In these circumstances it is necessary to consider carefully the implications of the spray application rates which can vary from over 700 litres/ha

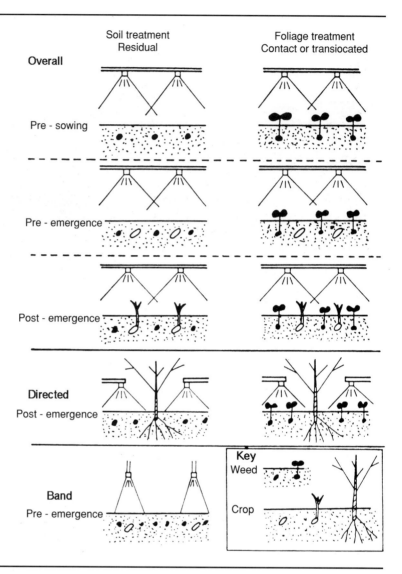

Fig. 5.12 Situations in which herbicides may be used for weed control in crops. (Reproduced from Fryer, J. D. and Makepeace, R.J. (eds) 1977 *Weed Control Handbook, Vol. 1: Principles including Growth Regulators*, 6th edn, by permission of Blackwell Science, Oxford.)

for high volume systems to less than 10 litres/ha for ultra-low volume systems. However, where very low volumes are applied the chemical concentration in the applied spray will be much more concentrated and this may represent an additional safety problem.

Certain herbicides are now available in granular or microgranular form and can be spread by hand or simple hand- or machine-operated blowers. Such materials find application for the control of aquatic weeds and weeds in paddy rice.

However, most herbicides are formulated as emulsifiable concentrate or wettable powder designed for application in water either to the soil or to the weed foliage.

Types and selection of sprayers

Sprayers consist essentially of a tank holding the diluted herbicide and a propulsion system for distributing the spray at the correct application rate to the required position. For use in the trop-

Fig. 5.13 Spraying with a pressure-type knapsack sprayer in Pakistan. (Source: FAO, Rome.)

ics, the most relevant types of propulsion systems are:

(1) hydraulic nozzles (knapsack or boom-mounted)
(2) spinning discs
(3) electrostatic systems

Each type of propulsion system has fundamentally different characteristics which control the performance achieved.

Hydraulic nozzles These are designed to operate with the spray material within a specified range of pressure. The orifice of the nozzle must be kept clean, unblocked and unworn if the design application rate is to be obtained. For this reason, water should be filtered before filling the sprayer tank and, in addition, a filter should be fitted in the supply line or within the nozzle. A wide variety of nozzle types is available, giving different distribution angles, and the spray pattern and droplet size can be modified to suit requirements. The type of nozzle used must be related to the available pressure, the spacing of nozzles on the boom and the height of the boom above the surface of the soil or crop. The throughput of nozzles and the distribution of spray should be checked regularly (Radley, 1992).

Knapsack and pressure cylinder sprayers Knapsack and pressure cylinder sprayers are the standard method of applying herbicides on small farms in many parts of the tropics (Fig. 5.13), with pressure maintained either by continuous pumping during spraying (knapsack) or by intermittent pumping between spraying (pressure cylinder) (Radley, 1992). Maximum tank capacity is necessarily limited to about 25 litre for carrying purposes. However, the advantages of simplicity of design are offset by the rather high application rate (typically around 200–400 litres/ha) and the considerable physical effort required for manual pumping. Pressure regulation valves and gauges are often either not fitted or are defective, resulting in fluctuating application rates. While, in theory one operator can cover 1 ha in 3 hours if walking along rows 1 m wide at 1 m/s, in practice the problems associated with mixing spray material, obtaining the required amount of water and clearing blockages may result in up to 4 man-days being required to cover 1 ha. The application rate, however, may vary widely as it is dependent on both the rate of pumping and the length of the stroke, both of which are difficult to maintain constant in arduous operating conditions.

Spinning disc sprayers   This type of sprayer has important features, particularly its ability to apply as little as 10–20 litres/ha. A DC motor powered by dry batteries drives a toothed disc at a speed of about 2000 rev/min with gravity feed of spray liquid on to the disc. Droplet size can be selected to suit different requirements but is dependent on the peripheral edge speed of the discs thus providing controlled droplet application (CDA) for any given speed. A single 750 ml 'Bozzle' container can treat up to 1.5 ha of crop. Where the supply of replacement batteries is difficult there are alternatives, such as the use of solar energy (with or without batteries), a ground-wheel generator or use of an air-driven disc with a knapsack-type lever-operated air pump. Apart from the cost and often poor availability of batteries, spinning disc applicators have poor directional control and the electric motors may be unreliable.

Electrostatic sprayers   A variety of different types is available and droplet size can be varied to suit requirements. The 'Electrodyn' sprayer was developed for small farmers and has the advantage of having no moving parts and no calibration is needed. Pre-mix formulation reduces contamination risks and avoids the problems of measuring, mixing and water supply problems and the application rate may be as low as 1 litre/ha with a power requirement of only 0.1 W compared to 2.7 W for a small spinning disc sprayer. The electrically charged drops are mutually repellent but are attracted to the 'earthed' surface of the foliage. As upward movement can occur application to the under surfaces of leaves is improved. Because of the very small droplet size it is not as easy to see the spray as with hydraulic nozzles and swath matching is therefore more difficult.

Weed wipers   Wipers of different designs are impregnated with herbicides for direct contact application to the visible and accessible parts of weeds. They are cheap and especially suited to the application of non-selective systemic herbicides, e.g. glyphosate, to a limited infestation of deep-rooted weeds. Although using concentrated chemicals, wipers are safe if used correctly.

Herbicide safety

Simple codes of practice for herbicide users, such as that given by Terry (1984), should ideally be studied before purchase but in practice this is rarely done. Herbicides, like all chemicals, have to be handled with care, especially in concentrated form. The problems of ensuring the safe application of herbicides in the tropics are appreciably more acute than in temperate areas, where attendance on training courses and the distribution and understanding of manufacturers' instructions are more readily achieved. Farmers who are illiterate or innumerate may experience great difficulty in handling, diluting and dispensing spray material to obtain the specified dilution without risk of exposure to toxic chemicals. There are opportunities for spray concentrate to be marketed in single-mix quantities to minimise these risks.

The use of hand-held or knapsack sprayers gives rise to considerably greater risks of exposure to the spray material than is the case for drivers of tractor-mounted sprayers. Legs and feet are liable to be in direct contact with the wet foliage since the nozzles are held in front and operators will usually be barefoot and without any protective clothing. There are several problems associated with the drift of spray droplets which are rarely appreciated by operators. In very low humidities rapid evaporation of water from spray droplets will give rise to small, easily drifted and highly concentrated droplets which may travel up to 500 m, even from a knapsack sprayer.

The correct use of protective clothing is an important component in ensuring safety in the use of herbicides (Radley, 1992). In some countries, the use of protective clothing is specified in legislation while in others it is left to the farmer or operator to follow such advice as may be available. Full sets of high quality protective clothing, including face masks and gloves as used in temperate countries, may be prohibitively expensive and seriously uncomfortable when used in the tropics, although their use during cooler hours may be acceptable. Sets of protective clothing of a suitable quality for use in hot climates are available from some of the pesticide manufacturers. Particular care must be taken when handling concentrated formulations, when the use of undamaged gloves and footware of suitable quality is essential. Protective clothing should be washed and stored carefully after use, with regular inspection and replacement when worn or damaged.

The problem of safety in the use of pesticides is increasingly seen as an international one, demanding solutions on an international level (Adam, 1986). There is also a recognition that the hazards in the handling and application of pesticides in the tropics tend to be different from those in temperate areas. Of particular importance is the lack of a long-established tradition in safety precautions with agricultural chemicals, the difficulties arising from failures to issue instructions in a form readily understood by farmers with limited training or technical experience, and also the very practical difficulties which arise in relation to the wearing of protective clothing in high temperatures. The distribution and marketing of pesticides in the tropics may not be subject to the stringent standards specified in temperate areas while the facilities for storage of chemicals and the safe disposal of empty containers may, in some cases, be non-existent.

Labels on herbicide containers may be of little value to operators in the field owing to illiteracy, the use of a foreign language or unfamiliar technical terms, or because of them being illegible or missing. The lack of on-farm first-aid facilities or nearby medical practitioners familiar with the treatment of pesticide poisoning accentuates the hazards (Radley, 1992).

Following extensive consultation over several years, the Food and Agriculture Organization of the United Nations (FAO) adopted a voluntary International Code of Conduct for Pesticides in 1985. The code is supported by the Groupement International des Associations Nationales de Fabricants de Products Agrochimiques (GIFAP), the pesticide manufacturers' association, and backed-up with technical guidelines which are available in different languages. The code comprises 12 articles dealing with all aspects of safety relevant to the distribution, labelling, packaging, use and disposal of herbicides (Adam, 1986). The code is seen as a dynamic text, with a recognised need for updating as technology and practices change. The document is primarily designed to help countries which have not yet adopted national pesticide registration and control schemes and it should be consulted by all those concerned with the distribution and use of pesticides.

6
Rain-fed Arable Farming Systems and their Improvement

J.K. Coulter

Farming systems are determined partly by the environmental factors of climate, soil, natural vegetation and topography and partly by socio-economic factors such as the customs of the people and the level of technology that they have attained, population density, the financial resources available for capital and recurrent inputs, and the family or market demands for food and cash crops. The wide variation in these factors in different parts of the tropics is paralleled by a great diversity of farming systems. However, the majority of the more widespread systems can be regarded as falling into four broad categories:

(1) Rain-fed arable farming systems other than those involving wetland rice.
(2) Rain-fed and irrigated systems based on wetland rice.
(3) Monoculture of perennial crops.
(4) Predominantly livestock systems, e.g. nomadic pastoralism and ranching.

This chapter deals with rain-fed arable systems excluding those based on wetland rice which, because of their distinctive features and frequent association with irrigation, are discussed in Chapter 7. Monoculture of perennial crops is covered in Chapter 8 and pastoral systems in Chapter 11. Alternate husbandry, involving the rotation of arable crops with sown grass leys grazed by animals, is of very limited importance in the tropics but is referred to in Chapter 11.

SHIFTING CULTIVATION

Many annual rain-fed cropping systems consist of one or more years of cropping followed by a resting period in which the land is allowed to revert to a natural fallow, consisting of regenerating bush or grass or a mixture of the two. Thus shifting cultivation is still by far the most widespread system of rain-fed arable farming in the tropics. It is practised under a wide range of soils and climatic conditions and by people with many different customs and cultural traditions; thus there is naturally a great variation in the kinds of crops that are grown and in their cultural techniques. However, certain essential features are found wherever the traditional methods exist. Land is cleared from forest or savanna, the debris is burned, crops are planted on the ash-enriched soil and cropping is continued for a period which may vary from 1 to 10 years but is usually from 2 to 4 years. After this the land is abandoned to lie fallow under the regenerating natural vegetation for some years and a new area is cleared for cultivation. Almost universal reliance is placed on the fallow for the maintenance of fertility; no fertilisers or manure are used and, even if livestock are kept, it is only in specialised farming that use is made of the dung. A major feature of shifting cultivation in many regions is the almost complete dependence on human energy for all of the farming operations.

A broad distinction can be made between two types of shifting cultivation which arise from the settlement patterns of the people concerned. The first consists of the usually fairly small and scattered groups of tribal peoples and ethnic minorities such as the Iban in Sarawak described by Chin (1985). There are many other descriptions of these systems, including those of Amazonian Indians (Meganck and Goebel, 1979), Papua New Guinea (Clarke, 1971) and northern India (Murthy and Pandey, 1981). These groups fell and burn relatively small areas of forest and are involved in hunting and gathering as well as cropping. Periodically the whole community will move to another area and start a fresh cycle of felling the forest. Wills (1962) defined this as 'shifting cultivation' *sensu stricto*. This system is coming under increasing pressure, not only from the growth of tribal populations but also from landless migrants, from the dominant ethnic groups, moving into the area. One consequence of this is increasing conflicts over land rights.

The second and very much more important kind of shifting cultivation is that of settled farmers who practise land rotation (Tiffen, 1982). In this system the people live in permanent villages and towns and their cultivated and fallow lands cover a fairly large area round about. The prolonged use of a relatively limited area naturally results in a more rapid rotation of the cultivated clearings, and fallow periods tend to become gradually shorter, especially close to the village or town. As the productivity of land in the immediate vicinity of the village declines, the distance from the village to the main cultivated area may be up to 10 km and frequently temporary huts are built on the 'farm' and occupied for periods of days or weeks at a time during the growing season. This type of farming is the dominant form of rain-fed arable farming over very large areas, including much of Sub-Saharan Africa.

Farmers involved in shifting cultivation operate under many different land tenure systems. Usually, but not invariably, the land is owned communally by the clan or tribe, with control powers vested in a chief who exercises them on the advice of a council of elders. Individuals have only user-rights, but these are permanent as long as they continue to use the land and often, but not always, extend to fallow land. Traditionally, some degree of beneficial control of the use of the land was exercised by the chiefs and elders.

As part of the control of land use an individual may have to obtain permission before clearing virgin land and felling might be prohibited in certain areas that are used for hunting or as a source of building poles, fuel and wild food or medicinal plants. Cultivation might be forbidden near salt licks, watering places or along river banks. In some areas the time and extent of burning for hunting used to be controlled. The extent of these controls has declined markedly in modern times; population pressure, changes in market opportunity, especially the sale of food crops in the towns and cities, and migration have all played a part. Russell (1993) detailed the changes that have taken place in some villages in Cameroon, including greater competition for land from young men who defy the traditional system for allocating land. In general, however, the traditional system still exerts an influence and, as a rule, care continues to be taken in the allocation of land by the tribal authorities, or head of the extended family, to ensure each holding has a fair share of the different qualities of land.

In the past, shifting cultivation was operated successfully for centuries; Bose (1984) suggested that it had operated for many millenia in some parts of India. With sparse populations, using it for subsistence, the system was not labour intensive and fallows were long enough in the closed forest to restore soil fertility. Even today, in parts of Zaire where land is still plentiful and where there is mainly subsistence agriculture, men and women farmers spend between 20 and 25% of their time on income-generating activities; of this only about 8% goes on farming, while about 18% of the time is spent on hunting, fishing and gathering edible fruits, insects and honey (Tshibaka, 1992). In the drier woodlands and savannas the system was less effective and there was some overall decline in soil fertility. Fire has contributed to the extension of derived savanna areas, through the destruction of tree seedlings, and the soils of many of these, particularly in northern South America, are extremely low in plant nutrients.

It is impossible to estimate accurately the numbers of people completely dependent on shifting cultivation for farming. Chapman (1975) estimated that the system was practised by 30 million people in Southeast Asia but many of these had sedentary systems on part of their holdings. Altogether it was estimated that 300–500 million people were farming in forest land (Myers, 1989). To this must be added those farming in savanna areas who also practise various forms of shifting cultivation. The area of land that is cleared and cropped annually by a family may vary; Wiersum et al. (1985) reported that, in one district in Sierra Leone, 1.3 ha was cleared and planted. In Papua New Guinea, Wood (1979) reported that a family would also use about 1.3 ha/yr while in Côte d'Ivoire an area of 2 ha was quoted as being needed for upland rice. Using an average of 1.3 ha per family per annum would suggest that about 800–1000 Mha of land are farmed under shifting cultivation each year. With population growth rates of 2.5–3% per annum the yearly demand for additional land may thus be of the order of 30 Mha. This must come from intrusion into forest land and from shortening the fallow periods.

Increased population pressure and decline of yields of over-exploited land is taking place throughout the rain-fed arable farming areas of the tropics. Thus the replacement of the land-extensive system of shifting cultivation with a system that is more productive in terms of land and labour is probably the most important problem facing tropical agriculture. This is reflected in the search for 'sustainable' systems, systems that will preserve the land for future generations and yet give the present generation of farmers an improved livelihood. Thus it is important to understand the traditional systems in relation to the environmental and socio-cultural limitations before attempting to consider possible improvements.

The fallow period

The extent to which fertility is restored during the fallow depends on the soil, climate, duration of the fallow period and on the nature of the vegetation succession that develops. The last may be influenced by the method used for the initial clearing and by the duration of the cropping period. Mechanical clearing of forest or woodland may favour the development of a grass succession and prolonged or frequent cropping leads to dominance of *Imperata* and *Chromolaena*. Eden (1985) observed that in some forest areas more than 20% of the fields, abandoned after cultivation, were colonised by grasses while in others all of the fields reverted to secondary forest. In studies in Asia and the South Pacific the succession of grass and herbaceous weeds (mainly *Imperata* and *Chromolaena*) and methods for their control have received considerable attention (Robison and McKean, 1992).

Nutrients are taken up by the fallow vegetation from a varying depth of soil; some of the nutrients are stored in the vegetation (which may also accumulate a certain amount of nitrogen by symbiotic fixation), and some are returned to the surface soil by rainwash from leaves and twigs, by litter and timber fall and in the form of dead roots and root exudates. The soil humus is increased during the fallow period, chiefly from the litter, which falls continuously under regenerating forest vegetation, though there may be seasonal peaks. Swift et al. (1981), working in secondary forest in Nigeria, found a high rate of litter fall during the dry season but very little decomposition and an accumulation of nitrogen, phophorus and bases. The onset of the rainy season caused a rapid decomposition of the litter and the release of much of the nutrients. Under continuously moist conditions there will be no large accumulation on the soil surface because it decomposes quickly. In the savanna the much smaller amount of litter deposited by the predominantly grass vegetation is mostly destroyed by the annual fires.

In the process of decomposition the greater part of the organic matter in the litter is lost through oxidation, but a fraction is broken down to humus and incorporated in the soil. As the humus itself also undergoes decomposition, the amount in the soil does not build up infinitely during the fallow but tends to rise to an equilibrium level (dependent on the soil, vegetation and climate) at which its rate of decomposition is roughly equivalent to the rate of addition of humified organic matter to the soil from the litter. This build-up affects both the physical and the chemical characteristics of the soil; the percentage of water-stable aggregates is increased and aggregate stability of the cultivated soils may range from one-fifth to one-third of those in fallow soils; in addition the bulk density of the cultivated soils is increased (Aina, 1979). The total nitrogen content of the surface soil rises with the increase in humus and may also be enhanced by non-symbiotic fixation. During the fallow period there will also be some losses from the soil–vegetation system as a result of runoff and erosion, leaching and denitrification.

The amount of nutrients stored in the vegetation during the fallow, the quantities returned to the soil, the losses from the soil–vegetation system and the net amount stored in the surface soil, vary with the soil and the nature and age of the fallow vegetation. In particular, there are marked differences between fallows in the forest and savanna zones.

The forest zones

In the forest zones the short cropping period, the practice of leaving the stumps and roots of trees undisturbed after felling and the prevailing high rainfall and temperature, lead to rapid regeneration after cultivation is abandoned, and large amounts of nutrients are quickly accumulated in the secondary forest growth.

The relative importance of litter fall, timber fall and rainwash in returning nutrients from the fallow vegetation to the soil is indicated by figures obtained by Nye and Greenland (1960) for high forest, 40 years' old, at Kade, Ghana (Table 6.1). The major contribution of organic matter, nitrogen, calcium and magnesium to the surface soil comes from the leaf and twig litterfall, which is estimated at between 10 and 11 t/ha/yr in moist evergreen and deciduous forest, this high rate of litter production being attained early in the fallow period. In older forest fallows there is considerable annual fall of branches and stems, estimated at about 11 t/ha/yr of dry wood at Kade, but compared with litter this contains only small quantities of nutrients except for calcium and, perhaps, phosphorus. In addition, Nye and Greenland estimated that, in mature forest fallows, dead roots, root slough and root exudates also added to the soil about 5.6 t/ha/yr of dry matter. There is very little information on the amounts of nutrients washed from leaves by rain in tropical fallows, but the Kade figures indicate that rainwash contributes remarkably large quantities of potassium and significant amounts of phosphorus.

These figures are very close to those reported by Ewel (1976) for a tropical forest succession in Guatamala where litter falls increased with increasing age of the young stands to a maximum of 10 t/ha/yr at 14 years. Brubacher *et al.* (1989) did similar measurements in Belize after *milpa* farming; nitrogen, phosphorus and potassium rapidly reached a plateau in the leaves in 1-year-old fallows and there was little increase in 2- and 3-year-old sites. Potassium also accumulated rapidly in the stems while nitrogen and phosphorus showed steady rates of accumulation over the 3-year period. The total biomass in a 3-year

Table 6.1 Nutrients returned to soil from vegetation in high forest fallow at Kade, Ghana. (Source: Nye and Greenland, 1960.)

	Oven-dry material (kg)	Nutrient elements (kg/ha/yr)				
		N	P	K	Ca	Mg
Rainwash from leaves		12	3.7	210	29	18
Timber fall	11 200	34	2.9	6	82	8
Litter fall	10 500	200	7.3	67	206	45
Total addition to soil surface		246	13.9	283	317	71

fallow was 20.7 t/ha, with 103 kg nitrogen, 73 kg phosphorus and 132 kg potassium per ha. Woody species accounted for 80% of the total biomass after 1 year and these had eliminated the herbaceous species after 2 years.

Although a high proportion of the total root weight of most forest trees occurs at shallow depth, soil analyses made at different stages in the crop–fallow cycle have shown that subsoil feeding by the fallow vegetation has an important effect in transferring nutrients from the subsoil, via the vegetation and the litter fall (and in the ash, after burning) to the surface horizons. This counteracts the losses suffered by the surface soil from crop removal and leaching during the cropping period. The transfer of nutrients from subsoil to topsoil probably does not begin until after the first year or two of fallow. During this initial period the topsoil is further depleted by leaching and by the uptake of nutrients by the regenerating vegetation, which has few active roots in the subsoil at this stage. Figures obtained by Popenoe (1959) for the nutrient content of various soil horizons under regenerating forest fallow in Guatemala showed an initial depletion of the surface soil, after which the content of potassium, calcium and magnesium rose in the top 30 cm and fell in the horizons below, as the vegetation developed.

Once a good growth of trees has developed in the fallow it is probable that leaching of nutrients through the soil is fairly well balanced by the uptake of nutrients by plants and their subsequent storage in the vegetation, or return to the soil in rainwash, litter and timber fall, so that a closed cycle of nutrients exists. Nye and Greenland (1960) pointed out that losses to the soil–vegetation system by leaching will occur only if water percolates through the profile to beyond the root range, provided that cations are held in the soil solution by balancing anions, the nitrate ion being quantitatively the most important anion in the slightly acid to acid forest soils. Under fallows in the deciduous and semi-deciduous forest zones (rainfall 950–2300 mm), although there will be relatively large amounts of nitrate in the soil, leaching will be restricted both by the uptake of anions by the vegetation, and because the transpiration of a large proportion of the rainfall will reduce through percolation.

However, some through-percolation is likely to occur where the rainfall exceeds 1500 mm. In the wetter evergreen forest zone where precipitation (over 2000 mm) considerably exceeds transpiration, through-percolation is bound to occur, but leaching losses will be restricted, not only by the absorption of anions by the vegetation but also by low nitrate levels in the soil, since nitrification will be limited by high acidity. Jordan et al. (1983) measured the nitrogen fluxes over a 4-year period in an experimental slash-and-burn site in the Amazon Basin compared with a control site. During this period there was a decrease of about 15% in the nitrogen levels in the system as a whole due to leaching, harvesting and denitrification, but the soil nitrogen level did not change as it was replenished by the decomposing slashed vegetation, suggesting that loss of nitrogen does not affect crop yields. Complete burning will remove this source. However, once a forest fallow has developed, leaching losses are probably not very great in most places, except in the early stages of the fallow. Similarly, although there may be some loss of nutrients by erosion in the earliest stages of the fallow, once the regenerating vegetation has developed a full canopy and a litter layer to protect the soil surface, such losses will be slight (except, perhaps, on very steep land).

Where cultivation is too prolonged and grass (for example, *Imperata cylindrica*) takes over, this is usually accompanied by frequent fires so that there is no accumulation of litter and little protection of the soil. This is common on the steep lands of Southeast Asia and the Pacific Islands and leads to serious soil deterioration.

The savanna zones

There has been much less research on nutrient cycling in the savanna zones compared with the forest areas. In the former the amounts of nutrients stored in the fallow vegetation, and the organic matter and nitrogen content of the soil after fallow, are much less than in the forest zone. Except in the tall grass savannas on the margins of the closed forests, grasses and herbs contain only relatively small amounts of nutrients and, as a rule, the amount of nutrients stored in the fallow vegetation depends largely on the extent to which trees and shrubs are present. The amount of woody vegetation varies greatly, depending on rainfall, soils and farming systems. In more recent times, however, savanna woodland has been seriously depleted, especially in areas near to roads and towns because of charcoal production. Futhermore, the growth of trees is much slower, being limited by the low rainfall and being annually checked by fire. Since the leaves and stems of grasses are destroyed annually their nitrogen and sulphur are lost and sulphur deficiency is thus very common, especially on the derived grasslands of Southeast Asia and the South Pacific. However, the roots are not destroyed and Nye and Greenland (1960) found that in the moist high-grass savanna of Ghana there was an annual contribution to the soil of about 3 t/ha of dry matter from the roots. In the drier savanna the amounts would be less.

The extent to which fallows in the savanna transfer nutrients from the subsoil to the surface soil is probably rather variable. Nye and Bertheux (1957) have shown in Ghana that the accumulation of phosphorus in the surface soil is less marked in the savanna than in the forest, and it may be that in areas with only a comparatively short dry season the savanna grasses are predominantly surface rooting. On the other hand, in parts of East Africa with a pronounced dry season, many of the grasses may root to 2 m or more and in these conditions they will transfer nutrients from the subsoil to the surface.

Leaching losses are probably comparatively small in most of the savannas, partly on account of the lower rainfall and high transpiration levels and partly from the absence of nitrate ions. Since nutrient levels are low after the grassy fallow, yields are quite small and the decrease with time after clearing is less marked than after forest clearing. Pieri (1992) stated that soil deterioration was due more to a decline in physical conditions than in nutrient losses. The former are more difficult to measure, though farmers identify them as capping and increased runoff.

Clearing and burning

Clearing

Under traditional practices in the forest zone the length of the fallow period is usually between 10 and 20 years, but may be longer where the population is sparse. Farmers normally prefer to clear good secondary growth rather than virgin forest because the latter requires more labour for felling and burning, takes longer to dry and is more difficult to burn thoroughly. The clearing and burning involve little soil disturbance and hence leave the soil more or less as it was under forest.

The traditional duration of the fallow in the woodlands and savannas varies. When the vegetation is mainly grass, the land is usually fallowed for less than 10 years but in woodland areas it may be between 10 and 20 years. The trees are felled and burnt with the grass during the dry season. The stumps and roots are left and larger trees may be left standing, though most are felled. Although burning is usual there are some places in predominantly grass savanna where the grass is not burned but is slashed or hoed and buried in mounds; in some places the mounds are burned.

Burning

The effects of burning have been described in detail by Peters and Neuenschwander (1988); not only are there changes in chemical, physical and biological conditions of the soil but the impact of repeated burning is to replace tropical forest with grass savanna. In Southeast Asia the fire climax is often *Imperata cylindrica* and fire exerts a strong selection pressure in favour of fire-resistant species. Fire destroys most tree seeds in the soil. Brinkmann and Vieira (1971) showed that when seeds of Amazon trees were buried in the soil at 2–20 cm depth, 100% of those at 2 cm and 80–100% of those at 5 cm were destroyed by fire. They concluded that forest successions, after slash-and-burn, will come almost exclusively from air-borne seeds or tree suckers. Thus clearing large areas deprives them of this source of regeneration.

Nevertheless, burning after clearing is the general rule, and is usually essential, if only because it is the only practical method of clearing the land of debris before cropping. Burning temporarily enriches the soil in phosphates and cations but in the longer run leads to a run-down in soil fertility due to leaching and losses of nitrogen and sulphur.

Physical conditions of the soil after clearing and burning the fallow

In the forest zone, after a long fallow with a strong regeneration of shrubs and trees, the physical condition of the surface soil, and to a lesser extent that of the subsoil, will be good immediately after clearing and burning. The fresh organic matter of the forest litter is largely destroyed by the burn, but some of the partially decomposed organic matter immediately below remains, and the humus in the top 5–8 cm of the soil is little affected. The condition of the deeper, less permeable horizons will have been improved by the channels of the tree roots, left by their death and decay.

The physical condition of the soil is also improved after a predominantly grass fallow in the savanna zone, but not as much as under the forest. Owing to the nature of the vegetation and the occurrence of annual fires, litter deposition will have been much less, and the humus content of the surface soil correspondingly lower, than in the forest. In addition, the improvement of the structure will have been annually checked by the exposure of the soil to the early rains after the burn during the dry season.

The cropping period

Crops and cultural methods

As might be expected, the crops grown and the cultural methods adopted vary widely, being influenced by a variety of factors, among which the most important are climate – especially the amount and distribution of the annual rainfall – soil, and the customs and dietary habits of the people.

One almost universal feature is mixed cropping, several species, and often several cultivars of a particular species, being grown simultaneously on the same plot, although they may be planted and harvested at different times. Where the rainfall regime permits, an overlapping sequence of crops is commonly planted which maintains a more or less complete soil cover throughout the cropping period. The main reason for the practice of mixed cropping is probably that it maximises returns for a given effort. The more complete cover provided by a mixture of crops shades weeds and checks their growth, thus reducing the labour of weeding. As noted later, growing a mixture of several crops, which have different growth habits and root systems and make different demands on the soil, enables the best use to be made of soil, light and rainfall. Solar energy during the growing season may be only 300–350 cal/cm^2/day so periods of sun and rain can be maximised by combining as many crops as possible, thus providing the maximum return from the land cultivated. Mixed cropping also provides an 'insurance policy' as some crops are likely to give a fair return even if bad weather, pests or diseases cause the partial or total failure of others.

Owing to the excellent physical condition of the soil after clearing and burning a forest fallow, little preparatory cultivation is needed. In some places none is attempted, the seeds merely being scattered on the surface of the friable, ash-enriched soil. More commonly seeds are dibbled into holes made with a digging-stick or a hoe, without any cultivation of the soil between the planting holes. Less frequently a superficial scratching may be done with hoes and, in some areas where roots are planted as the first crop, ridges or mounds may be made.

The commonest practice in the forest zone is to plant a cereal as the first main crop, usually with some admixture of other crops, such as pulses and vegetables; then to interplant with annual or semi-perennial root crops, and subsequently to interplant again with perennials such as bananas. The cereals and other minor annual crops are harvested in the first year and the roots mainly in the second year, although some, such as eddoes and short-term varieties of cassava, may be dug in the first year and others, such as longer-term cassava varieties, left until the third year or even later. Harvesting of perennials starts in the second year and continues in the third, after which cultivation usually ceases, but perennials and semi-perennial root crops may continue to be harvested for several years after the clearing has been abandoned to natural regeneration.

The above sequence was followed in the forest zone of Ghana as described by Nye and Greenland (1960). Maize was sown as soon as the rains broke and was almost immediately interplanted with relatively small amounts of other crops, such as pepper, okra, spinach and other vegetables. Later, either during the growth of the maize or shortly after it had been harvested, the farmer planted root crops such as cassava (*Manihot utilissima*), yams (*Dioscorea* spp.), coco-yams, eddoes, or dasheen (*Colocasia antiquorum*), tannia (*Xanthosoma sagittifolium*), and bananas. During the second year some of the bananas and roots were harvested, while the remainder were left to be harvested later as needed, or, if not required, abandoned in the regenerating forest. More recent descriptions of the system suggest that the cropping has not changed very much though the fallow period is now down to 3–5 years in some areas (Brookman-Amissah, 1985). In high rainfall areas, such as those in Sierra Leone and Liberia, upland rice is the first crop which may be followed by cassava. Similar patterns were followed in traditional shifting cultivation in Southeast Asia (Conklin, 1957; Terra, 1958; Barrau, 1959).

In the savannas and woodlands there is usually more preparatory cultivation than in the forest zone because the soil is not left in as good a physical condition after clearing and burning the fallow, and on account of the need to get rid of the grass rootstocks. Land may be ploughed where draught animals are available or it may be dug with a hoe to a depth of about 10 cm before planting the first crop and this procedure is repeated each season before replanting at the beginning of the rains. Mounds or ridges are often made if a root crop is to be grown, or for the purpose of burying grass if burning has not been done after clearing. Further hoeing is usually carried out during the growth of each crop to control weeds.

In the Sahelian and Sudan zones, millets and sorghum are grown but in the higher rainfall zones, cereals are not invariably the first crop. Where the dominant vegetation is grass, very little nitrate occurs in the soil during the fallow and, after clearing, a low level of available nitrogen persists for a year or more until the rate of mineralisation of nitrogen improves. Quick-growing cereals are sensitive to nitrogen deficiency since they need to take up relatively large amounts of this element during a short period of rapid growth. Consequently, where grass has been the main component of the fallow vegetation, and especially where the grass is buried instead of being burnt after clearing, legumes or roots are usually the first crop to be planted. On the other hand, a cereal is commonly the first crop where predominantly woody vegetation has been felled and burned.

Cropping continues for a period which may be anything from 2 to 10 years, but is usually from 3 or 4 years, during which a sequence of cereals, legumes and roots is grown, with lesser amounts of interplanted vegetables and other minor crops. Very commonly pigeon pea or cassava is planted at the end of the cropping period, weeding being discontinued after their establishment, and the crop subsequently harvested during the first year or two of fallow regeneration. The forest zone practice of growing perennial crops and harvesting them after abandoning the land is ruled out by the prevalence of annual fires. Crops and cultural methods vary greatly and only a few illustrative examples can be given here.

In the Sahelian and Sudan–Sahelian zones some of the villages have three crop cultivation areas. The fields close to the villages are permanently cultivated and manured, millet or sorghum intercropped with cowpeas being the main crops, but with little rotation; the bush area, forming the middle zone, is sown with intercropped millet and cowpeas, fornio (*Digitaria exilis*) and groundnuts; the fields are rarely supplied with manure and are left fallow whenever possible. The third and outer zone between adjacent villages is used for pasture and may comprise fallow fields and bush.

In the wetter savanna zones of southern Nigeria and Ghana, yams are usually planted first on large, well-prepared ridges or mounds at the beginning of the rains, and a variety of other crops, such as maize, beans and various vegetables, may be sown shortly afterwards on the sides of the ridges. In the second year maize and sorghum are planted on the remains of the yam ridges and interplanted with other minor crops. In the third year the main crops are commonly groundnuts and millet and thereafter the land may be abandoned, or cassava may be planted in a fourth year.

It will be seen that shifting cultivators, both in the forest and savanna, follow fairly regular crop sequences that are adapted to local conditions. These rotations and the cultural methods accompanying them are primarily determined by the climate, the type of natural vegetation and the dietary habits of the people, but they are also influenced by other considerations. First, the inherent fertility of the soil will affect the crops grown and the duration of cultivation. For example, on poor, sandy soils cassava will be grown rather than yams, millet may be more satisfactory than maize or sorghum and the land will be cultivated for only 1–3 years, whereas more fertile soils will grow a wide variety of crops and can be cropped longer. Crop sequences are also related to the declining fertility of the soil during cultivation. Certain crops, such as finger millet, being found to benefit especially from first-year land while others, such as cassava and pigeon pea, will give a fair crop on relatively impoverished soil and are consequently taken as the last crop.

Second, the characteristics of certain crops will affect their place in the sequence and their association with other crops in mixtures. Yams, being a long-term crop and demanding good preparatory cultivation, are often a first crop and succeeding crops are planted on the yam ridges with little or no further cultivation. Maize, which responds well to fertile soil, is usually grown early in the rotation; sorghum may be planted at any time but is commonly a second crop. Two consecutive crops of groundnut are rarely taken, since the second crop is almost invariably poor.

Third, crop sequences and mixtures have undoubtedly been chosen in order to facilitate weed control. In the forest, provided cropping starts with a clean seedbed as a result of a good burn, the farmer has little difficulty in controlling herbaceous weeds in the first cereal crop. Subsequently the mixture of roots and semi-perennials reduces weed growth by shading and, as these crops are also less affected by weed competition than cereals, the farmer is able to maintain adequate weed control for a further period of 2 years simply by slashing. As already mentioned, if the farmer departs from the traditional system to grow a second cereal crop, weed control becomes more difficult. In the savanna, crop sequences and mixtures may also be chosen with a view to minimising the work of weed control, but here weeds are more troublesome and the farmer often fails to deal with them early enough to prevent adverse affects on his crops. Some rhizomatous grasses such as *Imperata cylindrica* or *Digitaria scalarum*, are very difficult to eradicate with the hoe and become increasingly troublesome as cropping is prolonged, especially in the wetter savannas. In the drier savannas weeds are generally less troublesome, although in some places *Striga* spp., which are parasitic on some cereals, can be serious.

Fourth, the customary crop sequences and combinations were chosen with a view to spreading labour requirements evenly through the cropping season, as well as minimising the total effort required. For example, the preparation of subsidiary mounds during the previous rains reduces the work of preparatory cultivation during the busy time at the onset of the wet season; early maturing beans are planted after, and harvested before, the maize; new mounds are prepared in conjunction with weeding, and the late-sown legumes on them are harvested after the maize.

While shifting cultivators normally tend to adopt some locally adapted pattern of crop sequence, this does not usually constitute a well-organised rotation that is rigidly followed. Deviations are inevitable because crop sequences, and the areas under different crops, are affected by a number of factors over which the farmer has no control. These include the weather, the incidence of pests and diseases and the family labour supply, which may be reduced by illness, poor food supply, social and ceremonial commitments, or opportunities for off-farm employment.

The decline in yields and fertility

Generally yields fall in successive seasons during the cropping period, but the rate of this decline varies greatly, depending on the soil, climate, topography, crops grown and cultural methods. In the moist evergreen and semi-evergreen forest zones, where rainfall is heavy and most soils are highly leached, yields fall rapidly. The decline is slower on the better soils, and under the lower rainfall of the semi-deciduous and deciduous forest zones. In the savannas the rate of decline is certainly slower than in the moist evergreen forests but is very variable, depending on soil and rainfall; the latter severely limits the crop growth in many areas, and thus the removal of soil nutrients in the crops.

Under traditional non-intensive methods of shifting cultivation it is unlikely that a build-up of pests and diseases is an important cause of the decline in yields. Where population pressure has led to reduction in the length of the fallow or to the cultivation of the same crop on the same land in successive seasons, and especially where large areas of cash crops, such as maize, cotton or tobacco, are grown in pure stands, a build-up of pests or diseases may certainly occur. This increase in soil-borne pests and diseases, such as nematodes, is an increasingly serious problem, particularly in crops such as bananas. Weeds become more important as the cropping period extends beyond 1 or 2 years and Moody (1974) asserted that the greatest limitation to continuous cultivation by the African farmer was the inability to control weeds; Eden and Andrade (1987) reported that, in Colombia, weed competition rather than loss of soil fertility caused the farmers to abandon cassava cultivation after 2 or 3 years. Thus the longer the cultivation period the more important weeds become, particularly the invasion by grasses such as *Imperata* and herbs such as *Chromolaena odorata*, which are very difficult to control with hand tools.

Though weeds, pests and diseases are important factors in lowering productivity under rain-fed arable cropping, a decline in soil fertility usually plays a major role. Plant nutrients and soil organic matter are reduced and there is usually a fall in pH but the situation is far from simple because the level of plant nutrients may be considerably higher in one area, recently abandoned, than in another newly opened area.

The physical condition of the soil In the customary farming practices in the forest zone there is little preparatory tillage that might break down the good structure of the surface soil, after clearing and burning the fallow vegetation. Soil organic matter oxidises rapidly and some damage will be caused by heavy raindrops until the soil is covered by the crop mixture. The latter will protect the soil as there is relatively little need for weed control initially. A succession of annual crops, with the accompanying hoeing for weed control, will cause a more rapid decline in soil structure.

Several factors contribute to the rapid decline of soil structure in the savanna areas. The soils, often Alfisols, have less structural stability than the Oxisols and Ultisols of the wetter tropics and often have a large percentage of fine sand and coarse silt fractions, which cause surface capping. Little organic matter is added under grass fallows and this oxidises rapidly under cultivation. The surface capping causes rapid run-off and difficulties in seed germination. Water permeability may fall to less than 20% of that under woodland vegetation (Charreau and Nicou, 1971). Pieri (1992) stated that although farmers recognised the deterioration in soil structure there was no convincing evidence that farm yields were deteriorating. This appears to be confirmed by Speirs and Olsen (1992) who showed that sorghum and millet yields in several countries in West Africa have remained more or less constant over the past 15 years. However, long-term experiments in the same region show a steady decline, an increase in year-to-year variation and a declining response to fertilisers.

Erosion Immediately after clearing and burning forest on sloping land there is bound to be some runoff and erosion but this is minimised by minimal tillage, the presence of tree stumps and roots and by the cover soon provided by the farmers' crop mixtures. Furthermore, with a sparse population the widely scattered plots in the forest also protect catchments from severe erosion. Hence, under traditional practice on moderate slopes little erosion is likely during the cropping period but, as mentioned by Lal and Okigbo (1990), there is little quantitative information about losses on actual farms under different systems. However, where cultivation of long stretches of steeper slopes is now usual, there is plenty of evidence of severe erosion, even where conservation measures such as grass strips are used.

Erosion in savanna areas is markedly influenced by soil type as well as by slope and cropping systems. For example, runoff and soil losses are much higher in West African savanna areas where the soils have a permeability of 15–25 mm/h than in the savannas of northeast Brazil where soils have permeabilities of 50–400 mm/h. In general, erosion under shifting cultivation is more serious in the savannas than in the forest zone because of the cropping systems and the cultivation in preparation for planting and for weed control which tends to destroy the normally rather weak soil aggregates and thus enhance runoff.

Changes in the nutrient status of the soil

Humus
Humus declines after clearing but the loss may not be as rapid as is sometimes suggested. Measurements made by Tulaphitak *et al.* (1985) in a forest area of Thailand showed a steady-state rate of decomposition of organic matter of 6.5 t/ha/yr of carbon. A field under shifting cultivation had twice as rapid a rate and nearly 30% of the organic matter was lost in the first two years of cultivation. With increasing time of cultivation, the humus level falls to a new equilibrium level but a series of short cropping periods, alternating with long fallows, is unlikely to cause appreciable depletion of humus.

The rate of decomposition in savanna areas is quite high and this from a low initial level. Pieri (1992) gave values of 4.7% per annum for very sandy soils in West Africa and 2% in loamy sands. This means that 15 years of cropping in the very sandy soils would reduce the humus by 50%, while it would require 35 years to halve the levels in the loamy sands.

Nitrogen
There are large quantities of nitrogen in the fallow system which are mineralised on clearing and burning. Sanchez (1982) reported that 20–25% of the nitrogen in a forest ecosystem in Latin America was lost through clearing and burning. Mueller-Harvey *et al.* (1985) measured the changes in carbon, nitrogen, phosphorus and sulphur in a soil cleared from forest in southern Nigeria; they calculated that the half-lives of these four elements were 3.5, 3.3, 4.7 and 2.3 years repectively. There was more than sufficient nitrogen and phosphorus mineralised to supply the uptake of one crop of soya beans and three crops of maize. This supports the evidence from experiments elsewhere that nitrogen responses are rare in former forested soils during the normal cropping period.

In the savannas very little nitrate nitrogen occurs in the soil under a grass fallow. After clearing and burning the total nitrogen content of the soil is less than in the forest and the nitrate content is very low, so that deficiency occurs in varying degrees depending, in part, on the age and luxuriance of the grass. As a rule there is an improvement in the rate of mineralisation in the first one or two years of cultivation, after which there is little change for some years. Compared with the forest zones, leaching losses are limited by the low average nitrate levels and the lower percolation resulting from lower rainfall but, on the other hand, they are enhanced at the beginning of the rains owing to the flush of nitrates and the absence of a crop to absorb the nitrate. Responses to nitrogen fertilisers are very common in savanna soils and lack of nitrogen contributes to the low yields on such soils.

Phosphorus and sulphur
The amount of phosphorus stored in the fallow vegetation and returned to the soil on felling and burning varies considerably, depending on the soil type and the nature and age of the vegetation. Nye and Greenland (1960) gave a value of 27 kg/ha of phosphorus in a 10-year-old fallow in Ghana; Oya and Tokashiki (1984) reported 43 kg/ha of phosphorus in a broad-leaved evergreen forest in Okinawa and Rhodes (1988) gave an average value of 4–6 kg/ha for bush fallow in Africa. In addition there is a considerable amount supplied by mineralisation of the organic matter in the soil. Under intensive cultivation in Nigeria, Adepetu and Corey (1977) found that 25% of the organic phosphorus was mineralised in the first two cropping periods; the mineralisation took place during the wet season and was about three times as much as that taken up by the crop. In savanna soils the amounts will be less and the value may be of the order of 3 kg/ha/yr. The decline in phosphorus status of the soil during the cropping period is due to phosphorus removal in the crop and, in some soils, strong adsorption on the colloidal complex. Leaching is not a problem but erosion of the surface soil, where much of the phosphorus may be concentrated, can account for substantial losses.

Sulphur is lost when vegetation is burnt and, as there is very little sulphur in tropical rainfall, deficiencies are quite common in areas where grassland has taken over from forest and also in sensitive crops in the savanna areas, especially legumes.

Nutrient cations
Although part of the ash may be washed away by the first rains, the burning of a forest fallow will add considerable quantities of nutrient cations to the exchange complex of the soil and these can vary between 20 and more than 200 kg/ha. The extent to which cations are lost by leaching during the cropping period will depend on the crop, the soil and the rainfall; these losses could be considerable taking into account the substantial amounts of nitrates which are in excess of crop uptake. Root crops absorb large amounts of potassium and Rhodes (1988) estimated that a cassava crop would require 60 kg/ha, even on newly cleared land, in Sierra Leone. Some of the potassium may be restored by weathering and Arcoll et al. (1985) suggested that traces of 2:1 layer silicate minerals in Oxisols in Brazil may be important in this respect. However, it is highly likely that potassium fertilisers will become increasingly needed as agriculture is intensified since many of the highly weathered and highly leached tropical soils are very low in this nutrient. The loss of calcium is also important as this will lead eventually to a drop in pH and consequently an increase in exchangeable aluminium. Excess quantities of this can only be counteracted by liming or by high organic matter levels.

In savannas, burning the fallow vegetation adds to the soil appreciably smaller amounts of nutrient cations than in the forests. The potassium balance is dependent on the crop uptake; sorghum, millet and upland rice can take up as much as 100 kg/ha K_2O and, if there are large amounts of potassium available, luxury uptake will take place; in these conditions the potassium concentration in the straw may reach 2.45% (Pieri, 1992). If the straw is left on the land, the cations will be returned, but in most savanna areas straw is a valuable commodity for fodder, fuel and fencing and thus most straw leaves the field and the nutrients in it are not returned.

Conclusions on the traditional practice of shifting cultivation

Shifting cultivation did provide a stable farming system for milennia, not only in tropical lands but formerly in temperate countries as well. One of its most important features is its ubiquity over a very wide range of soil and climatic conditions. It is a system which has generated an enormous amount of interest among both biological and social scientists and Robison and McKean (1992) noted more than 5500 abstracts of publications, between 1972 and 1990, which referred to this topic.

In the tropical forest zones, traditional cultural methods and cropping practices during the short periods of cultivation restrict losses by leaching and erosion; they control weeds, pests and diseases, maintain the ecosystem and may sustain biological diversity. Where land is plentiful and labour scarce, and where machines and animal traction are not available, it gives a better return per labour-day than many other systems of farming.

The long fallow leaves the soil in excellent physical condition and restores humus and nitrogen contents to close to their original levels under high forest. However, there are virtually no closed nutrient systems in any of these farming practices for there is always some loss of plant nutrients from the field, the village and the region. Nitrogen can be replaced in the system by biological means but phosphorus and sulphur and the cations can be replaced only by weathering of soil minerals or from the atmosphere. On medium to rich soils, such as those that occur on young alluvial or volcanic deposits, there are ample weatherable minerals to supply these nutrients. However, very large areas of tropical soils are formed on very old landscapes, which have been through several weathering cycles without rejuvenation. These have very low reserves of weatherable minerals and thus can only replace lost nutrients extremely slowly through weathering. Experiences with shifting cultivation suggest that a ratio of about 10 ha of fallow land to 1 ha of cultivation will maintain a reasonably stable system. Even if this ratio could be maintained there is still the important question whether such a system would be sustainable on poor soils or whether there would be a slow decline in soil fertility over the very long term.

In the savannas, traditional shifting cultivation is less satisfactory for the maintenance of fertility. The dominant grassland vegetation and the annual fires mean that the fallow is much less effective than a forest fallow in regenerating the fertility of the soil by improving the physical condition, building up the humus and nitrogen, and transferring nutrients from the subsoil to the surface soil. During the cropping period, deficiencies of available nitrogen and phosphorus will limit yield, while exposure of the soil and tillage for weed control adversely affect soil structure and increase the risk of erosion. Pieri (1992) reported that measurements of yield declined in Senegal and Burkina Faso suggesting that the annual yield loss over a 25-year period, due to a decline in soil fertility, was of the order of 3–5% per annum. He commented that such a loss was small in comparison with the annual yield fluctuations, which may vary by a factor of 10 due to the vagaries of the rainfall; consequently the farmer may not be aware of this yield decline.

Traditional shifting cultivation has worked well when practised by a relatively sparse population, solely for the production of subsistence food crops. In the forest it was the best system that could be devised and resulted in, at worst, a slow decline in soil fertility. In the savannas it was less satisfactory and somewhat wasteful because of the annual burning but, as the population was usually thinly scattered, it served to support the people at a rather low level of subsistence without a disastrous decline in fertility. Even in the forest, however, the system is one which precludes any long-term improvement in soil fertility and, if it is intensified as a result of the population exceeding a certain density, the land deteriorates and yields fall.

Shifting cultivation in present-day circumstances

During the twentieth century the practice of shifting cultivation has gradually been intensified, and in most areas it is no longer operated by sparse populations merely for the production of subsistence crops. Indeed throughout Sub-Saharan Africa it has long been recognised that shifting cultivation must eventually become untenable as population increases. The rapid population growth since the 1960s, when the population of Sub-Saharan Africa has more than doubled, combined with the cultivation of large areas of cash crops in addition to food crops, has increased the pressure on the land, with the result that, in many places, fallows have been reduced in duration to such an extent that they are quite inadequate to maintain fertility. In villages in southeastern Nigeria the International Institute of Tropical Agriculture (IITA, 1990) reported a close correlation between the fallow length and population; with a density of about 100/km^2 the fallow length was between 6 and 7 years; when the density reached 1100 the fallow period virtually disappeared. In the same area food crop yields have fallen and Carr (1989) reported that a comparison of three villages, selected for uniformity of soil type, showed that after fallow periods of 7, 4 and 3 years, cassava yields were 10.8, 3.8 and 2.0 t/ha respectively. Other examples of the impact of falling fertility in the same region are given by Lal and Okigbo (1990); the complexity of the species of the secondary forest was reduced, the area planted to cassava increased and that to yams decreased and there was increased colonisation by noxious weeds such as *Imperata* and *Chromolaena*. These lead to the land being abandoned, and in a survey in Oyo State such abandoned fields were found in 31 out of 39 villages (IITA, 1990).

The production of cash crops not only increases the pressure on land but, in addition, farmers growing such crops have been encouraged to clear their land of stumps and roots, to plant the cash crops in pure stands, often in widely spaced rows, and to weed them more thoroughly than they did their food crops. Compared with the customary methods, these practices, although they have many advantages in other respects, tend to increase erosion in the absence of proper conservation measures. Pieri (1992) stated that in the north of the Côte d'Ivoire cotton had been an indirect cause of soil degradation, even though there were only 11 people/km^2, because of the complete clearing of bush before its introduction. In Togo, on the other hand, where the cotton has been integrated into the existing farming system, the soils appear to have improved, even though the population density is about four times as great. Not infrequently, introduced cash crops have not been well suited to the customary, locally adapted sequence of crops and, in particular, there has been conflict between the labour demands of cash and food crops. In such circumstances the primary importance of the food crop is likely to lead to late sowing of the cash crop, late weeding, and delayed harvest, which will not only result in low yields but will tend to encourage erosion where planting is delayed into the rains.

When the plough has been introduced into areas where cultivation was formerly done entirely with hand tools, its use has often facilitated the opening up and exploitation of larger areas of land without the adoption of any proper measures for controlling erosion and maintaining soil fertility. More thorough removal of stumps and roots must be done if the plough is to be used; this means that the land is more liable to erosion during the cropping period and there is less opportunity for the regeneration of woody species during the fallow. Ploughing itself effects greater disturbance of the soil than hoeing, thus resulting in more rapid decline in physical condition and accelerating the decomposition of organic matter. This is liable to reduce the infiltration of rain and increase the risk of erosion.

It will be seen that, due to higher population densities and the cultivation of cash crops, cropping becomes more intensive and prolonged, fallows are shortened and traditional cultural methods and crop sequences are modified, resulting in deterioration in the physical condition and nutrient status of the soil and increased erosion. As shifting cultivation is essentially suited to the maintenance of the subsistence economy of a sparse population, it is certainly incapable of providing for the higher standard of living expected by the increasing population.

IMPROVEMENT OF ARABLE RAIN-FED FARMING

Jacks (1956) described three stages in the evolutionary processes of people and soils. The first of these is shifting cultivation *sensu stricto*, in which man's economic activities do not upset the ecological balance so that the system can be maintained in perpetuity. Such systems would have a small element of cropping but extractive activities, like rubber tapping in the Amazon basin, will contribute to the family income. Such systems may require 300–500 ha to sustain a family and clearly are only feasible in very sparsely populated regions. The much more important system, involving land rotation, described in the previous section, requires about 10–12 ha per family; it may be sustainable for several generations, but probably not in perpetuity, especially on the poorer soils. It is thus a step towards the second stage, described by Jacks, as that of soil exploitation whereby man, operating as a settled agriculturist, damages the environment by 'mining' the soil and not replacing the plant nutrients. This is what now happens in much of the rain-fed arable farming in the tropics.

The third stage is the soil conserving and improvement of fertility stage whereby as much or more is returned to the land as is taken from it. The return of plant nutrients, usually in the form of inorganic fertilisers, and conservation of the soil represents a stage of good husbandry whereby the land can be farmed indefinitely. In the industrialised countries that stage is often helped by the transfer of resources from the industrial to the rural sector.

The challenge facing rain-fed arable agriculture in the tropics is therefore to make the transition from the second stage of soil mining to the resource-conserving stage. It has to do this in the context of an unprecedented rate of population growth, often exceeding 3% per annum, and in situations where there can be no transfer of resources from other sectors. The problem is particularly acute in Sub-Saharan Africa where agricultural productivity needs to grow at 4% per annum (World Bank, 1989). Agricultural intensification has been an evolutionary process in the past but it will have to make much more rapid changes in the future and slow processes of change may not be able to cope with these challenges.

Development of permanent systems

Farmers have been modifying the system of shifting cultivation for years and most now practice some form of land rotation to varying degrees of intensity, usually while maintaining their homesteads at permanent sites. Even many ethnic minorities which once moved their settlements regularly are now permanently settled, often with the encouragement of, or sometimes even enforcement by, the government concerned. These modifications demonstrate the ability of farmers to adapt to changing circumstances. According to Boserup (1965, 1981) technological change can be brought about by population growth. As agricultural land grows scarcer farmers adopt technologies, which they have ignored or used only on a small scale, on a much wider scale in order to increase yields per unit area. However, there must always be a question as to whether farmers can adapt rapidly enough, taking into account the rapid population growth, to develop stable, permanent farming systems.

Permanent systems are common in the densely populated, drier monsoon areas of India and Pakistan, where several centuries of permanent cropping on the mostly small and often fragmented, farms resulted in impoverishment of the soil. Now that fertilisers are more widely available in these countries, they are being used in dryland farming to remedy this situation. Jodha (1979) described the traditional system in a number of villages in semi-arid areas of India; he stated that intercropping was more important in small rain-fed farms than in large irrigated farms. Farmers, with their need to satisfy both profit and subsistence, grew high-value cash crops like cotton and groundnuts and subsistence crops like sorghum and pigeon pea. Drought resistance, maintenance of soil fertility and fodder requirements for animals all affect the decisions on which crops to grow.

In West Africa the pattern for intensification is demonstrated by the 'Kano' system, existing not only in northern Nigeria but in several other parts of West Africa (for example in northeast Ghana) in which the land around the village or homestead is under permanent cultivation with annual crops, and is manured regularly, animal manures being often transported over considerable distances.

Tiffen (1993) has made a detailed historical study of agricultural changes in the Machakos area of Kenya for the period, 1930 to 1990. During that time the population increased five-fold and now stands at more than 1.4 million; the land situation changed from one of abundance to one of scarcity. From the 1930s onwards great concern was being expressed about soil degradation and overstocking but this acute degradation was reversed by the farmers themselves investing labour and capital in improving the land. Crops and cropping have changed so that there is now little fallow land. The large population in nearby Nairobi has created a substantial market for fruit and vegetables and opportunities for off-farm employment have increased greatly. Land titling has promoted the demarcation, registration and enclosure of grazing land, resulting in decreased pressure on the communal grazing land and the farmers have planted large numbers of trees in their field boundaries. As the population grew the percentage of land cultivated increased and there is now 100% terracing on the long-cultivated lands. The study showed an interesting relationship between population growth and agricultural change. The land degradation in the earlier years was taking place with relatively low population densities but at a critical stage in the people-to-land ratio and, with a profitable market for their produce nearby, the farmers changed their farming systems so that output per ha kept pace with population growth. Thus, to some extent, population increase drove agricultural innovation. However, Tiffen expressed the view that the people-to-land ratio may have reached a second critical stage when output per ha was again falling behind population growth.

Home gardens

These are the compound gardens, food forests or 'Kandy' gardens which are a common feature of densely populated areas in the humid tropics. A variety of annuals may be grown, including vegetables, condiments, spices, tobacco, medicinal plants, cereals such as upland rice and maize and some root crops. Fernandes and Nair (1986) stated that the average size of these gardens was less than 0.5 ha; although such gardens often gave the appearance of being of an unsystematic nature they found that they were carefully structured to form a vertical canopy strata with each component having a specific place and function. The woody components produced fruits and other types of food in addition to fuel, building poles and materials for making mats, roofing and handicrafts. The species diversity ensured production throughout the year. Soil fertility was maintained by adding refuse, mulches and animal manure, where available, thus transferring fertility from the more distant parts of the farm.

Similar systems have long been established where the banana is a staple food. Examples are several parts of East Africa, notably in the Buganda province of Uganda, where the natural vegetation is closed forest and the initial secondary generation is Elephant grass. Tothill (1940) and Masefield (1944) have reported that a Baganda family lived permanently on its own land and devoted much care to the banana garden which was manured with household refuse and mulched with old leaves and pseudostems and sometimes also with Elephant grass cut for this purpose. These gardens averaged about 0.8 ha/holding and were maintained for long periods, some being as much as 60 years old. During that period population growth was quite slow, probably not exceeding 1.5% per annum, and farmers were able to supplement their needs by practising shifting cultivation on other land. The introduction of cash crops such as cotton and coffee and the rapid population increase in more recent years has, however, put increasing pressure on the system.

These examples from several tropical areas illustrate how farmers adapt to increasing land pressures by adopting more capital and labour intensive systems; but they also suggest that there comes a certain critical stage, no doubt depending on the soils, the rainfall, the market for farm produce and the off-farm employment opportunities, when these systems too may break down.

Research to develop improved rain-fed arable farming

While farmers in some areas have been evolving 'permanent' systems, with higher inputs of labour and capital, and higher outputs per unit of land, agricultural scientists have also been attempting to design more productive systems. Some of these are aimed at making more radical changes, while others have been of a more evolutionary nature, designed to build on the most useful features of the farmers' systems. Padwick (1983), in his review of experiments on maintaining soil fertility in Sub-Saharan Africa, recorded the large number of experiments, several of a long-term nature, which have been done on this topic, over the past 50 years.

Much of the research has been designed to find ways by which small farmers can maintain or improve the fertility of their land. Clearly farmers need to conserve their soils from erosion and many attempts have been made to force them to do this, often unsuccessfully, but as the Machakos study of Tiffen (1993) showed, farmers may adopt intensive soil conservation measures when these become necessary for survival.

One of the obvious ways of restoring and upgrading soil fertility would be the use of fertilisers but there have been strong efforts by many development agencies to avoid this route. Some of the objections are based on the perceptions about environmental degradation, reportedly caused by the high fertiliser usage in developed countries, leading to atmospheric and water pollution and a 'lower' quality of food. The more serious difficulties are, however, the economic problems of high fertiliser prices in relation to the value of agricultural products, the lack of credit to purchase fertilisers and the foreign exchange costs for poor countries. Added to this is the often poor infrastructure, so that inputs are delivered too late for the planting season; this increases the inherent risks in rain-fed agriculture. The consequence of these difficulties is that in certain areas, especially in Africa, fertilisers, at present, are not a solution to the problem of maintaining soil fertility.

Improvement to the fallow in shifting cultivation

In the savanna zones, and to a lesser extent in those of deciduous forests, a good deal of effort has been devoted to investigating the value of planted grass fallows as an alternative to natural regeneration. Where the rainfall is adequate for the establishment and maintenance of a good grass cover, and provided that the soil is not initially infertile nor exhausted by intensive cropping, it has usually been found that a rotation of about 3 years' grass and 3 years' cropping has maintained fertility at a moderate level over a long period. But it has also been generally observed that a planted grass fallow has been no more effective than natural regeneration, provided that the latter was not burnt annually. Furthermore, in most experiments continuous cropping has given the highest total yields; although yields increased after a grass rest, these increases were insufficient to make up for the loss in total yield resulting from absence of cropping during the rest period. These results were obtained in trials that ran for 5–15 years, and it is probable that yields would have declined further on land continuously cropped for longer periods. However, it seems unlikely that the labour involved in establishing grass, instead of merely allowing natural regeneration to occur, would be worthwhile unless a good return could be obtained from animal production on the grass. It is also improbable that fertility can be maintained at a satisfactory level unless legumes are planted along with the grass, or unless fertilisers and imported feeding stuffs are used. For these reasons, further discussion of grass rests and leys is deferred until Chapter 11, after the subject of grassland has been covered.

A number of experiments have been made with 'fallows', of 3 or 4 years' duration, in which the land has been planted with various perennial legumes, such as *Tephrosia candida*, *T. vogelii*, *Cajanus cajan* (pigeon pea), *Centrosema pubescens*, *Glycine javanica*, etc. The general experience has been that the effect of several years under these legumes on soil fertility, as measured by following crop yields, has been very similar to that of an equal period of natural regeneration or of planted grass.

Pigeon pea has been more frequently employed as a restorative crop in experiments of this kind than any other legume. For small farmers, who are averse to planting restorative crops that yield no direct return, it has the great advantage that it will provide human or stock food for at least 2 years of a 3-year 'fallow'. Many varieties will persist for 3 or 4 years, providing a good cover and abundant leaf fall, and requiring no weeding after the establishment phase in the first year. The crop is tolerant of a wide range of soils and climates and, being deep rooted, it has some effect in enriching the surface soil with nutrients brought up from deeper layers. A number of experiments have shown that 3 or 4 years under this crop is equal to, or even slightly better than, a similar period under natural regeneration or planted grass fallows in its effect in improving soil fertility. The yields given in Table 6.2 for crops following 3-year fallows of pigeon pea, Gamba grass (*Andropogon gayanus*) and natural regeneration in three trials at Yandev in the southern Guinea savanna zone of northern Nigeria, showed no significant differences between fallow treatments (Dennison, 1959).

Legumes and nitrogen fixation

Legumes, which have the ability to fix atmospheric nitrogen in symbiosis with the *Rhizobium* bacteria in their root nodules, are widely grown by small farmers for their seeds, green pods and leaves which are an important source of dietary protein. They may also be used in the restorative 'fallows', mentioned above, as green manures, cover plants in tree plantations, fodder crops and as components of pastures. This section considers their potential as aids in sustaining soil fertility in rain-fed arable farming by adding organic matter and nitrogen to the soil.

Reliable information on the amounts of nitrogen fixed by legumes under field conditions in the tropics is limited owing to the difficulty in distinguishing between the nitrogen obtained from the atmosphere and that taken up from, and subsequently returned to, the soil. As mentioned earlier there are limitations to the methods used for estimating fixed nitrogen.

Table 6.2 Yields (kg/ha) of crops following different three-year fallows at Yandev, northern Nigeria. (Source: Dennison, 1959.)

	Type of fallow		
	Pigeon pea	Gamba grass	Natural regeneration
Trial No. 1			
First year, yams	9141	9546	8777
Second year, millet and sorghum grain	934	948	887
Third year, sesame	237	267	235
Trial No. 2			
First year, yams	9469	8531	8873
Second year, cowpea	139	135	117
Second year, millet and sorghum grain	882	866	909
Third year, sesame	515	461	453
Trial No. 3			
First year, yams	3035	2869	3037
Second year, millet and sorghum grain	1435	1288	1536
Third year, sesame (3-year fallow)	303	266	267
Seventh year, yams	14191	13232	12899
Eighth year, millet and sorghum grain	775	895	954
Ninth year, sesame	392	400	385

There is also wide variation in the amount fixed by individual crops of the same species owing to differences in growing conditions.

Estimates of the amount of nitrogen fixed by groundnuts and grain legumes range from 25 to 200 kg/ha during growing seasons of 60–120 days (Table 6.3). Bouldin *et al.* (1979), reviewing data on nitrogen fixation (defined as the net accretion of nitrogen in the soil–plant system), gave figures as high as 535 kg/ha fixed by *Crotalaria* sp. in 6 months and values of 400 kg/ha have been recorded for other pure legume green manure crops. Moore (1962) found that with unfertilised Star grass – *Centrosema pubescens* – the legume fixed nitrogen at the rate of about 280 kg/ha/yr over 2.5 years but, in general, it seems that grass–legume pastures add nitrogen to the ecosystem at the rate of 100–200 kg/ha/yr.

As the symbiotic activity is dependent on the supply of energy in the form of photosynthate, it can be expected that environmental conditions conducive to vigorous growth of the host plant would generally favour nitrogen fixation. In addition, the production and efficiency of nodules are directly affected by several factors which have been summarised by Giller and Wilson (1991). High soil temperatures can prevent nodulation or, if nodulation occurs, can inhibit nitrogen fixation. There appear to be optimum air and soil temperature ranges for nodulation and nitrogen fixation but these vary considerably between host species. Drought can markedly reduce the rate of nitrogen fixation which is more sensitive to lack of moisture than photosynthesis. Deep rooted species, such as cowpea and pigeon pea, which can exploit water in deeper soil horizons, can thus function more efficiently in dry conditions than shallower rooting crops such as soya bean. Light intensity and interception are important and shading has been found to reduce nodulation in relation to plant weight in a number of species.

Shortage of phosphate, in which many tropical soils are deficient, can severely limit nodulation and nitrogen fixation. Application of phosphate fertiliser is often beneficial despite the fact that most legumes are very efficient in their uptake of this element because of their associated mycorrhiza. As effectively nodulating legumes are able to meet their nitrogen requirements by fixation, the application of nitrogen fertiliser usually has no effect or raises soil nitrate to a level which depresses nodulation and nitrogen fixation. There are no reports of direct effects of other major nutrient elements on nitrogen fixation although adequate supplies of all of them are needed for good legume growth. Of the trace elements, molybdenum is particularly important for nitrogen fixation as it is a constituent of the enzyme nitrogenase and nitrate reductase,

Table 6.3 Estimates of nitrogen fixation by grain legumes. Small amounts of nitrogen fertiliser and adequate amounts of phosphate fertiliser were applied. (Adapted from Giller and Wilson, 1991.)

Grain Legume	Nitrogen fixed (kg/ha)	Time (days)	Country
Arachis hypogaea	68–116	110	Brazil
	101	—	Ghana
	100–152	89	India
	139–206	120	Australia
Cajanus cajan	68–88	—	India
Cicer arietinum	60–84	160	Australia
Glycine max	26–57	64–73	Thailand
	114–188	66	Nigeria
	85–154	110	Brazil
Phaseolus vulgaris	25–65	60–90	Brazil

and responses to its application have been quite common.

Observations on host *Rhizobium* specificity led to the concept of 'cross-inoculation groups' whereby the species within each group will nodulate effectively with each others' strains of *Rhizobium* but not with the bacteria of other groups. There are many anomalies to this concept but some legumes are quite specific in their *Rhizobium* requirements. When they are grown on land lacking the strain they require it is necessary to inoculate the seed with an appropriate 'effective' strain which will nodulate the plants in competition with 'ineffective' strains, native to the soil, which form nodules that fix little or no nitrogen. However, many of the legumes grown in the tropics, with the notable exception of soya bean, are nodulated and fix nitrogen in symbiosis with a range of promiscuous *Rhizobium* strains, known as 'the cowpea miscellany', which are commonly present in the soil.

It may well be that the performance of some of these legumes could be improved by inoculation of their seeds with selected *Rhizobium* strains more effective than those native to the soil. Research continues on the testing and selection of strains but a major difficulty is that the indigenous soil strains, with which a selected inoculant must compete, vary between regions and soil types. Experimentally, a number of legume species have been inoculated with selected strains and there have been some reports of improved performance. However, for most species, except soya bean, there does not appear to be any good evidence of the efficacy of inoculation under field conditions and small farmers make little or no use of inoculation.

There is also the likelihood of improving nitrogen fixation by exploiting the variability existing in some host species in such characters as early or increased nodulation or symbiotic efficiency. Since the photosynthesis of the host plant is the major factor in the symbiosis, it is likely that legume breeding for improved growth and yield will also result in increased nitrogen fixation.

Grain legumes An increase in nitrogen fixation from legume breeding or *Rhizobium* selection should result in greater enrichment of the the soil with nitrogen by legumes grown in recuperative 'fallows' or as green manures. With the much more important grain legumes the effect will depend on how much nitrogen is fixed and how much is removed in the crop. Usually, both seeds (which receive a major part of the fixed nitrogen) and straw are harvested, the latter being used as fodder or fuel. However, there is good evidence that even when only the roots and, perhaps, some leaf fall, are left in the field, grain legumes commonly, although not invariably, increase the soil nitrogen content and can thereby improve the productivity of a following non-legume crop. This has been observed in many places but may be illustrated by some experiments in the drier parts of India where important objectives are minimising the use of costly nitrogen fertilisers and producing cheap dietary protein. Thus, Gire and De (1979) found that yields of pearl millet following groundnut, cowpea or pigeon pea were significantly greater than after a previous pearl millet crop. Ahlawal *et al.* (1981) reported that yields of a summer maize crop were increased by 25–44% when following winter crops of chickpea, lentil, or *Lathyrus sativa* compared with yields after wheat or bare fallow. Kumar Rao *et al.* (1983) obtained maize yields after pigeon pea which were very significantly greater than those after sorghum or fallow. In most of these experiments an increase in soil nitrogen was found after the legume.

Green manuring Growing an annual legume and incorporating the whole of it in the soil usually enhances the yield of one or more following crops. This was demonstrated by several long-term experiments carried out in Africa many years ago. In Zimbabwe, the alternation in successive years of maize with a green manure crop, such as velvet bean (*Mucuna utilis*) or sunn hemp (*Crotalaria juncea*), accompanied by applications of phosphate either to the legume or to the maize, maintained good maize yields (for the varieties then grown) for over 20 years. The maize yields thus obtained were more than double those from continuous maize growing (Rattray and Ellis, 1953). In the 1950s, green manuring was recommended to supply the nitrogen requirement of maize but nowadays the application of 150 kg nitrogen/ha is recommended instead (Tattersfield, 1982).

At Ibadan, Nigeria, where the rainfall regime permits two cropping seasons a year, a large-scale experiment which ran for over 30 years showed that a rotation with three or four green manure crops of velvet beans in 4 years (eight cropping seasons) maintained yields of various crops for many years. However, eventually it appeared unlikely that green manuring could continue to maintain yields without fertilisers (Webster, 1938; Vine, 1953). Bouldin et al. (1989) grew four green manures (*Mucuna aterrima*, *Crotalaria paulina*, *Canavalia ensiformis* (Jack bean) and *Zornia latifolia*) and measured the yield and nitrogen uptake of maize planted after their incorporation compared with those of maize receiving nitrogen fertiliser (Table 6.4). Some of the plots were left fallow and the mineral nitrogen contents of these plots measured 100 days after incorporating the green manures were well correlated with the maize yields and the nitrogen contents of their above ground dry matter.

All these experiments demonstrated that the main benefit of a green manure is an increase in the available nitrogen content of the soil resulting from the rapid decomposition of buried plant material with a relatively low carbon:nitrogen ratio. As might be expected from this, green manures produce no lasting benefit in soil humus or total soil nitrogen and their beneficial effect rarely extends beyond the first following crop. It is desirable to dig in the green manure plants before they seed as their effect is lessened if seed is harvested and senescent material dug in. As adequate moisture is needed both for the growth of the plants and for their decomposition in time to benefit the following crop, green manuring is impracticable in drier areas.

Green manures may find limited use in some places where they can be grown as short-term crops occupying the land for only part of the rainy season, either before or after a main crop. For the most part, farmers will make little or no use of green manuring because they are naturally averse to setting aside land during the whole, or the greater part, of the rainy season for a crop which offers no direct return and which demands labour for land preparation, planting, weeding and digging in at times when there may be conflict with the labour demands of the all-important food crops.

Rotations and mixed cropping

Rotation of crops alone cannot maintain fertility but a good rotation will usually give better average yields than continued cultivation of the same type of crop, e.g. cereals. Part of the value of a rotation is in checking the build-up of pests and diseases but it may also reduce weed infestation and frequency of cultivation and lessen the risk of erosion. More important is the contribution which the inclusion of a legume in a rotation can make to increasing the soil nitrogen level and enhancing the yield of a following crop, as already mentioned.

Most small farmers in the tropics do not employ rotations of crops in pure stand but follow a flexible, locally adapted sequence of mixtures of crops. Crop mixtures have advantages in checking weed growth, reducing and spreading labour demand and in providing greater assurance of a food supply, since some crops are likely to give a fair yield even if bad weather, pests or

Table 6.4 Grain yield and above-ground nitrogen uptake of maize with green manures and fertilisers. (Source: Bouldin et al., 1989.)

Treatment	Maize yield (t/ha)	Nitrogen content in above ground material (kg/ha)
Nitrogen fertilizer		
0 kg/ha	3.74	69
50 kg/ha	4.60	96
100 kg/ha	5.57	123
200 kg/ha	6.72	172
Legume		
Zornia	4.72	116
Crotalaria	5.81	149
Jack beans	6.14	169
Mucuna	6.43	159

diseases cause partial or total failure of others. The importance of yield stability in dry areas led Rao and Willey (1980) to examine this feature in a series of sorghum–pigeon pea intercrop systems. They estimated that sole pigeon pea would fail, on average, one year in five, sole sorghum one year in eight but the intercrop system only one year in 36.

Quite apart from the effect on yield stability, many trials have demonstrated that mixed cropping usually gives greater productivity per unit area of land than pure stands not only on low input–output farms but also in more intensive, well-managed farming. The earlier experiments used mixtures obtained by adding together the plant populations considered appropriate for pure stands. The total population in the mixtures was therefore greater than that in either of the pure stands so that the advantage of the mixtures might have been simply due to the fact that they were the only treatments with adequate population pressure. This possible source of confusion was eliminated in later trials. For example, in Uganda, Willey and Osiru (1972) used a replacement technique in which a mixture of two crops was formed by replacing a certain proportion of one species by the other, thus keeping the population pressure constant. They first compared a 'replacement series' of pure maize, two-thirds maize and one-third beans (*Phaseolus vulgaris*), one-third maize and two-thirds beans, and pure beans, at four populations per hectare. Later they studied a similar series using dwarf sorghum and beans (Osiru and Willey, 1972). Yields of mixtures of maize and beans were up to 35%, and those of sorghum and beans up to 55%, higher than could be achieved by growing the crops separately. In both cases the largest increases from interplanting were obtained at the higher plant populations, indicating that the mixtures had higher optimum populations than the pure stands.

Even in drier regions, larger yields have been reported from increased plant populations. Shetty (1989) showed that doubling the population of pearl millet, from 15 000 to 30 000 plants/ha, in a maize/pearl millet mixed crop increased the millet yield from 650 to 1010 kg/ha but reduced the yield of maize only from 3240 to 3110 kg/ha. He also reported that doubling groundnut density improved groundnut yields without significantly lowering sorghum yields.

Yield advantages for mixtures of two crops over sole cropping have now been demonstrated with a variety of species under a wide range of conditions by numerous experiments, including those conducted by the international agricultural research centres in Nigeria (IITA, 1990), Kenya (ICIPE, 1990) and West Africa (ICRISAT, 1988).

The main reason for the increased productivity is that appropriate mixtures of crops are able to make more efficient use of the environmental resources of light, water and nutrients than the same crops in pure stand or, to express it another way, that inter-species competition is less than intra-species competition. This comes about because of differences between the companion crops in stature, growth habit and root development and depth, or on account of differences in their growth cycles which result in their maximum demands on the environment, especially for nutrients and water, occurring at different times.

More efficient light interception by interplanted crops of different height and canopy structure is often a major factor. Measurements have shown that for most of the growing season less light is transmitted to the ground through mixtures of maize with groundnut, mung bean or sweet potato than through equivalent populations of these crops in pure stand. Willey and Osiru (1972) considered that the yield advantage of their maize–bean mixture was due to more efficient light utilisation by the combination of tall maize and short beans. On the other hand, they thought that the larger yield increase from the dwarf sorghum–bean mixture, the structure of which was less advantageous for light interception, was due to differences in rooting habit and growth cycles between the two crops.

Differences in growth cycles, enabling more effective use to be made of water and nutrients, can be important. In northern Nigeria, interplanting early maturing millet with a longer-term sorghum made better use of the 6-month rainy season than the latter alone. Grain development of the sorghum at the end of the rainy season depends on soil moisture reserves which are often marginal for the high plant populations required to obtain good yields from this crop in pure stand, so that production frequently suffers from drought. With the mixture, the early millet completed its growth cycle in the first 80–90 days of the rainy season, when the moisture supply is usually ample, and the lower population of the deeper-rooted sorghum, which flowered after the millet had been harvested, was at less risk from moisture stress (Andrews, 1972).

Although legumes are usually included in traditional crop mixtures, the amount of nitrogen that they transfer to companion crops is probably small. On the other hand, as might be expected, it has sometimes been shown that incorporating the residue of the legume component of a mixture in the soil has benefited a following crop.

The productivity of mixed cropping depends primarily on the selection of crop combinations capable of making the most efficient use of the environmental resources and on planting the crops in proportions, and at densities, which will maximise the total yields while achieving the desired balance of production between the species. The selection of the most suitable varieties of the crops to be planted can be important. Thus, Rao and Willey (1983) compared several genotypes of sorghum and pigeon pea, which varied in height, habit and maturity time, to find combinations which would raise pigeon pea yields above the poor level achieved in traditional mixtures in semi-arid tropical India. They found that a combination of a short, early sorghum with a pigeon pea that matured as late as possible without incurring undue risk of moisture stress, gave good pigeon pea yields and sorghum yields of 80% or more of those from pure stands of the cereal.

The spatial arrangement of the crops may have a smaller, but significant, effect on productivity. With mixtures of two crops planted in separate rows, it has generally been found best to sow alternate rows of the two species rather than to have two or more adjacent rows of the same species. Sowing both crops in the same row has usually given no better, and sometimes poorer, yields than planting them in alternate rows.

The time of planting the components of a mixture may influence productivity by affecting the degree of competition between the species. Thus, when interplanting maize and mung bean, it was found best to plant both crops at the same time since later sowing of the legume resulted in shade from the cereal reducing its growth and yield. On the other hand, the slower growing soya bean benefited by being sown about 20 days before the maize (IRRI, 1974). However, most subsistence farmers are unlikely to change their established custom of planting the staple cereal as early as possible (since it is known that cereal yields markedly decline with delay in planting) and then interplanting legumes later at the time of the first weeding.

There is ample experimental evidence that mixed cropping improves productivity per unit area, more than 1 ha of crops in pure stand being required to produce the yield of 1 ha of intercropping. Since it also improves production per man-day, it is likely that improved mixed cropping can be a useful component in the development of more intensive systems of rain-fed arable farming.

Agroforestry

Agroforestry is another potential source of improvement of nitrogen and organic matter in soils. Robison and McKean (1992), in their review of shifting cultivation, found that interest in agroforestry (trees with crops) and agrisilviculture (trees with pastures and livestock) represented the largest change in research on shifting cultivation in recent years. In the more humid regions of the tropics, agroforestry is regarded as a potential alternative to the natural fallows in shifting cultivation with trees as part of a planted fallow. Trees can be included in a 'taungya' system, in alley cropping, in wind breaks and in shelter belts or, where land is more plentiful, as blocks. Much of the attention has been given to nitrogen-fixing leguminous trees, of which the most important genera are *Acacia, Calliandra, Gliricidia* and *Leucaena*. Giller and Wilson

(1991) stated that trees were the most problematical for measurements of the amounts of nitrogen fixed since it was difficult to distinguish between the nitrogen fixed from the atmosphere and that taken up from the soil. Measurements of the nitrogen content of the above-ground parts of a densely planted stand of *Leucaena* gave values of 500–600 kg nitrogen/ha/yr; this may be an under-estimate, however, since pot experiments have shown that 60% of the total nitrogen can be in the roots of young *Leucaena*. Apart from adding nitrogen to the system, trees also recycle other nutrients, taking these up from both the surface soil and from deeper horizons. The amounts in the prunings of three legume trees, from an alley cropping experiment in Nigeria, are given in Table 6.5. The alleys were 4 m wide and the intrarow spacing 0.5 m; the first prunings were made two years after planting and further prunings at 6, 22 and 30 weeks later. Although *Senna* does not nodulate or fix nitrogen, its nitrogen contribution was the largest initially, but *Gliricidia* provided larger quantities in subsequent prunings although it is not known how much of this nitrogen was fixed and how much was merely extracted from the soil and returned in the prunings.

Table 6.6 shows the results of an experiment, established at Chalimbana Research Station, Zambia, in 1987 in which the effects of *Leucaena leucocephala*, *Flemingia congesta* and *Sesbania sesban* on interplanted maize yields were measured (ICRAF, 1989a). The hedgerows were cut to 0.5 or 1 m one year after planting and subsequently at monthly intervals. It will be seen that yields from the fertiliser plus prunings treatments are larger than from fertiliser alone but the yields from the prunings treatment were consistently lower than those from fertiliser treatment. Kang *et al.* (1981) in Nigeria reported that 5–6 prunings per year from *Leucaena* could sustain annual yields of maize of nearly 4 t/ha. In Cameroon, ICRAF (1989a) reported maize yields of 4.11 t/ha from *Leucaena* prunings compared with 2.18 t/ha from the control without trees. Many of the experimenters reported that they obtained higher yields when fertilisers were used in addition to the leaf prunings, suggesting that even though there were substantial quantities of nitrogen in the prunings, this was insufficient to give a maximum crop. The yields from tree prunings alone will depend on the phosphate, potassium and sulphur reserves in the soil as well as the pH; the adequacy or otherwise of these nutrients will affect the growth of the trees in addition to that of the food crop.

Wide adoption of alley cropping by farmers has not been reported. Yamoah *et al.* (1986) surveyed some farming areas in Nigeria and reported that *Gliricidia*, widely used in some areas for staking and fencing, was not proving popular for alley cropping; rapid shoot growth

Table 6.5 Nutrient additions from prunings in alley cropping in Nigeria. (Source: Giller and Wilson, 1991.)

Hedgerow tree	Nutrient	Nutrient content in prunings (kg/ha)				
		First		Second	Third	Fourth (pruning)
		Prunings (Litter)				
Flemingia macrophylla	N	62	(44)	46	25	23
	P	6	(4)	5	3	3
	K	40	(7)	35	16	23
Gliricidia sepium	N	126	(23)	119	144	88
	P	8	(2)	7	9	5
	K	86	(7)	75	12	52
Senna siamea	N	274	(102)	16	78	5
	P	27	(13)	2	10	1
	K	123	(40)	12	59	5

Table 6.6 Yields of alley cropped maize, with and without fertilisers (figures in brackets are standard deviations). (Source: ICRAF, 1989a.)

Species	Cutting height (cm)	Fertiliser application (kg/ha)	Grain yield (kg/ha)	Increase/decrease over control (%)
Flemingia congesta	50	0	2819 (441)	−49
	50	150	6912 (1216)	26
	100	0	3699 (1021)	−33
	100	150	6133 (876)	12
Leucaena leucocephala	50	0	1859 (613)	−66
	50	150	7204 (337)	31
	100	0	1947 (337)	−64
	100	150	7301 (773)	33
Sesbania sesban	50	0	2823 (1608)	−49
	50	150	5841 (1012)	6
	100	0	2725 (337)	−50
	100	150	5739 (2490)	4
Control (maize only)		150	5497 (1769)	—

was seen as a problem. Nair (1990), in a comprehensive review of agroforestry, was of the view that while alley cropping had some advantages, the relatively high labour requirement was a constraint. Other constraints are the lack of aluminium toxicity tolerance by several of the useful species and competition for water in the drier areas. Several studies of agroforestry systems in the drier regions of India have shown unexpectedly high losses in crop yield and these studies have emphasised the strongly adverse effect of the tree root competition (Ong et al., 1990). Alley cropping is only one possible configuration for trees grown on the farm and it would appear that farmers are reluctant to grow trees for mulch or fertility improvement alone though they place a high value on them for use as fuel, stakes or animal fodder.

Customary land tenure systems will be important and, where trees are regarded as conferring some additional rights to land, this may prove an obstacle to the development of agroforestry systems. In conclusion, it would appear that, while agroforestry systems may constitute a useful approach to finding replacements for shifting cultivation, there are many technical, social and labour problems to overcome.

Table 6.7 Variability of manures (% on dry matter basis) from 24 farms in central Senegal. (Source: Pieri, 1992.)

Nutrient	Mean	Standard error	Range
N	1.41	0.57	0.49–2.65
P_2O_5	0.69	0.28	0.29–1.55
K_2O	1.47	0.76	0.24–3.52
CaO	1.60	0.68	0.51–2.89
MgO	0.81	0.33	0.20–1.33

Farmyard manure

Farmyard manure, or pen manure, consists of the excreta of animals and the straw provided for their bedding. Even when protected from the weather, its composition is very variable. Table 6.7 gives a range of values from farms in Senegal (Pieri, 1992). Analyses given by Phillips (1956) of pen manure made under cover at various places in Nigeria showed similar ranges, except for a higher extreme value for potassium. The moisture content varied from 24.4 to 70.8% and the percentage of nutrients in the dry matter ranged from 1.03 to 2.12% for N, 0.49 to 1.30% for P_2O_5 and 2.33 to 5.54% for K_2O. Clearly, the quality will depend on the method of storage as well as on the quality of the feed available to the animals.

On many African farms the 'boma' or 'kraal' manure is of poor quality since it consists of the weathered excreta of animals kept in uncovered pens at night, usually without bedding, and it often contains much soil. Stall feeding, common in Asia, is becoming more widespread in Africa, especially in the more densely populated areas. Even in good storage conditions about 50% of the nitrogen will be lost before the manure is applied on the field.

Numerous experiments, in many parts of the tropics, have shown that most annual crops respond to applications of farmyard manure. There is no doubt that, on most soils, adequate dressings of this material can maintain crop yields under continuous cultivation. Obviously, the rates and frequencies of application of the manure required to maintain yields must vary with climate, soil, crops grown and quality of the manure.

Russell (1958) suggested that in East Africa dressings of the order of 2.5 t/ha/yr, or 7.5 t/ha every third year, or 12.5 t/ha every fifth year, would probably maintain yields in the absence of fertilisers. This is probably a fair generalisation for many East African soils, but on some soils heavier dressings are likely to be needed in order to maintain satisfactory yields. On poor, sandy soils in the Kenya Coast Province, 7.5 t/ha/yr was needed to maintain yields at a modest level under continuous cultivation (Grimes and Clarke, 1962) while in the Machakos area of Kenya 8 t/ha/yr seems to be needed to give consistently good yields in all seasons (Ikombo, 1984). Experiments have shown little difference in overall yields between applying manure annually and applying an equivalent amount in larger dressings at intervals of several years.

In some experiments fertiliser dressings of equivalent nitrogen, phosphate and potassium (NPK) content have proved just as effective as farmyard manure. However, in many other experiments, especially long-term ones, farmyard manure has given better results than equivalent amounts of NPK in inorganic fertilisers. This was so, for example, in long-term, continuously cropped experiments in northern Nigeria (Dennison, 1961), in a series of trials with a variety of crops in farmers' fields in the Lake Victoria area (Doughty, 1953) and in a number of factorial trials in Ghana (Djokoto and Stephens, 1961).

On many soils it is probable that part of the effect of farmyard manure is due to a beneficial effect of its organic matter on the physical condition of the soil and perhaps also on microbiological activity. Pieri (1992) stated that, in West Africa, regular application of farmyard manure to sandy soils had greatly improved the water supply to crops and that this was due entirely to the stimulation of root growth, not to any improvement in the water retention capacity of the soil. In addition, the relatively slow decomposition of the organic matter should result in a steady supply of balanced available nutrients, whereas the application of fertilisers may provide a somwhat unbalanced supply of nutrients, especially in soils of low nutrient status and low buffering capacity.

Farmyard manure is unquestionably a most useful aid in maintaining fertility and should be used wherever it is practicable and economic. There are, however, several limitations to its use by small farmers, even though most of them are well aware of its value. First, there are large areas, especially those in Africa infested with tsetse fly, where no cattle are kept. Second, in some places, cattle keeping has not been integrated with crop husbandry and no use is made of available manure, although this is rarely the case in Asia. Third, where there is no wheeled transport a great deal of labour is involved in carrying crop residues, etc., to the cattle sheds and in the transport and application of the manure to the fields.

Finally, the most important limitation is that, except for the rather limited number of intensive pig and poultry units, there is very little use of animal feeds imported onto the farm in the tropics. Hence, the use of farmyard manure to maintain crop yields generally involves the transfer of nutrients from a fairly extensive grazing area to a smaller cropped area. Obviously, such a system must become increasingly difficult to operate as population pressure on the land increases. Indeed, in many places the amount of farmyard manure currently available from the animals that can be kept on the limited areas of poor grazing is quite inadequate to maintain the fertility of the cropped land.

Compost

Compost consists of a partially decomposed mixture of household refuse, crop residues, weeds and other vegetable material, either with or without the addition of some animal or human excreta. In some places where stock are kept, the preparation of composts containing a small proportion of animal excreta is advocated as a means of making available larger amounts of organic matter than could be obtained by the ordinary method of making farmyard manure. The nutrient value of composts varies with the nature of the materials used in their preparation, but even those made without any animal excreta may contain useful amounts of the main plant nutrients. For example, analyses of five composts made from grasses, weeds and household refuse in Nigeria showed that moisture content varied from 21.9 to 50.4% and the percentage nutrients in the dry matter from 0.43 to 0.91% for N, 0.16 to 0.56% for P_2O_5 and 0.42 to 2.7 for K_2O (Phillips, 1956). It is worth noting that 5t of compost, with 0.5% of nitrogen, would contain about the same amount of this nutrient as a 50 kg bag of urea.

The advantage of composting as opposed to burying raw crop residues and weeds, is that it enables partially decomposed organic matter to be applied to the land at the best time so that available plant nutrients are rapidly released for the growth of a crop. Furthermore the temperatures of 45–75°C reached in the compost heap, destroy pathogens and weed seeds. Burying raw vegetable wastes of wide carbon:nitrogen ratio may easily result in a temporary shortage of available nitrogen for the following crop. On the other hand, burying fresh vegetable material of narrow carbon:nitrogen ratio is liable to result in decomposition reaching an advanced stage and some nitrogen being lost by leaching before a crop can utilise it.

Crops usually respond to dressings of good compost in much the same way as they do to farmyard manure, although larger dressings of compost are required to produce equivalent results. The use of compost is, therefore, to be encouraged wherever it is practical, but there are serious limitations which make it unlikely that its use will make a major contribution to maintaining soil fertility. The first is the small quantity of materials available for composting in relation to the very large areas of cultivated land. The second is the high labour requirement for carrying materials and water to the pits or heaps, turning the latter and for transport and application of the final product in the field. However, in some areas, low soil fertility, which is the main constraint on productivity, pressure on land and the smallness and continued subdivision of holdings means that farmers have to resort to this labour-intensive system. Composting is yet another means of concentrating plant nutrients from a wide area into a small area, but again it does not introduce new sources of nutrients into the system. While it may maintain or increase the output per unit area, it has a cost in terms of a smaller return per labour-day.

Other organic manures

Except for nightsoil other organic manures are little used on annual crops. Residual cakes and meals from oil seeds are not generally available and are expensive to buy and transport. Moreover such materials have generally given similar, or smaller, yield increases than an equivalent quantity of NPK applied more cheaply as fertiliser. Nightsoil, which is solid and liquid human excrement, has been used for many centuries in China, Japan and Korea and has played a very important role in the maintenance of soil fertility, even though the soils in these countries are generally richer, many being derived from young alluvium or volcanic deposits. King (1911) recorded that in the early part of the twentieth century, about 24 Mt of nightsoil were applied annually to the fields of Japan, equivalent to more than 4 t/ha of farmed land. Before being used for direct application, nightsoil is normally diluted and stored for a variable period, partly because this is necessary to avoid damage to the plants, and partly because the main need for nightsoil in the fields is seasonal. During storage, which may be in pits, concrete tanks, wooden barrels or stone jars, appreciable losses of organic matter and nitrogen occur. Although the use of nightsoil was a major factor in the maintenance of fertility in these countries, it has undoubtedly been the cause of large amounts of human disease and ill-health.

Mulching

The effects of mulches on soil and crop yields have been described in Chapter 5. Hitherto they have been little used for arable crops except for relatively small areas of vegetables and for yams. They have been shown to increase yields of a range of crops in drier areas but the use of crop residues and grass for fodder, fuel, fencing and construction usually means that insufficient material is available for an effective mulch.

Fertilisers

Thousands of experiments in many countries have shown that, on most tropical soils that have been cropped for 2 or 3 years or more, the food and cash crops commonly grown by small farmers respond to nitrogen and phosphate and less frequently to potash fertilisers. A large proportion of these experiments have been 1- or 2-year trials, many designed as demonstrations to measure yields under farm conditions. A more recent example is the work of the SG2000 programme in Ghana with several thousand 0.4 ha maize plots on farmers' fields (Yudelman *et al.*, 1992); the yields on the fertilised plots averaged between 3.0 and 4.3 t/ha, whereas the farmers' plots, without fertiliser, yielded between 1.2 and 2.5 t/ha. Zake (1987) reported that, in Uganda, over 6000 fertiliser trials have been done on farmers' fields since 1963 to test the effectiveness of nitrogen and phosphorus fertilisers on both cash and food crops. Based on the results, fertiliser recommendations were formulated for cotton, maize, millet, beans and groundnut but these were generally at two levels only, 125 kg/ha single superphosphate and 125 or 250 kg/ha ammonium sulphate.

Long-term experiments, which provide information on the management of soil fertility and the impact on yields, have been less numerous although the information they can provide is essential for planning soil fertility management under different farming systems. Earlier examples of large long-term programmes of such experiments are those reported from Nigeria by Goldsworthy and Heathcote (1963), from Ghana by Nye and Stephens (1962), from Tanzania by Scaife (1968), from Kenya and Uganda by Doughty (1953) and from India by Panse and Khanna (1964).

Fertilisers and farmyard manure, alone and in combination, have been tried in many experiments. Thus Butai (1987), reporting on a 5-year experiment in Zimbabwe on the effect of nitrogen at 90, 135 or 180 kg/ha, ploughing in or removal of stover and applying 4.5 or 9 t/ha of manure, gave the results shown in Table 6.8.

This experiment supported the results of many others which showed that, in cereal crops, nitrogen almost always showed the most severe shortage, especially where the land had been cropped for any length of time. Responses to farmyard manure or composts were variable but, in general, unless used in large quantities, they did not contain adequate amounts of nitrogen to give maximum yields. Some of the earlier

Table 6.8 Effects of nitrogen fertiliser, cattle manure and stover management on maize yield (t/ha). (Source: Butai, 1987.)

Stover management	Nitrogen levels (kg/ha)			
	0	90	135	180
Stover removed	0.45	2.81	3.99	3.98
Stover ploughed in	0.52	2.39	3.50	4.88
Stover removed plus 4.5 t/ha manure	0.97	3.01	3.92	4.70
Stover removed plus 9.0 t/ha manure	1.59	3.76	4.22	4.90

experiments, comparing fertilisers with manure used only quite small amounts of nitrogen; thus the trials by Bache and Heathcote (1969) in Nigeria used levels of 56 and 112 kg/ha sulphate of ammonia, which would supply only 12 and 24 kg/ha of nitrogen. Even at these low levels, however, the authors noted the strong effect of sulphate of ammonia in reducing the pH over the 15-year period.

Long-term experiments to determine the ability of fertilisers to maintain productivity in tropical soils have been reported from a number of countries. In a continuous cropping experiment in the Amazon area of Peru, Alegre and Sanchez (1989) reported an upland rice–maize–soya bean rotation which, over a 13-year period gave stable yields, with NPK fertilisers together with small quantities of lime and trace elements; yields for the thirty-second crop was 1.5 t/ha soya bean and for the thirty-third crop was 3.5 t/ha maize. A report on continuous cropping with maize in Nigeria gave a different picture; the fertilised crop yield dropped from 3 to 1 t/ha over a seven-year period; the drop was attributed to loss of organic matter (IITA, 1990).

In addition to the experimental evidence on fertiliser responses, there are reliable statistics from commercial farming. In Zimbabwe, Tattersfield (1982) gave the statistics for average yields on commercial farms over the period 1951–55 to 1976–80; maize yields rose from 1421 to 4726 kg/ha, sorghum from 568 to 2499 kg/ha, groundnut from 560 to 2335 kg/ha, cotton from 256 to 1709 kg/ha and soya beans from 470 to 2011 kg/ha respectively. The most important factor in increasing maize yields, for example, had been the use of nitrogen fertilisers, as shown by Table 6.9.

Although this information was collected from large, well-capitalised commercial farms there is also evidence that considerable numbers of small farmers in Africa use fertilisers on some crops. Byerlee and Heisey (1993) reported the results of a survey of maize farmers in Malawi and eastern Zambia in 1990 and 1991 which showed that in Malawi between 33 and 48% of the farmers with less than 1 ha used fertilisers, compared with 86% of those with 2–3 ha. In Zambia between 40 and 50% of the crop area was fertilised on the farms with less than 5 ha; on farms with more than 5 ha, the percentage was over 70%. Pieri (1992) recorded very large increases in yields of rain-fed cotton grown by small farmers in Mali and Burkina Faso; between 1960 and 1980 yields increased from about 200 kg/ha to 1100 kg/ha in Mali and from about 150 kg/ha to 800 kg/ha in Burkina Faso. Use of fertilisers played an important part in these increases, although plant protection, improved seeds and better agronomy were also important. However, as will be discussed later, fertiliser use depends very much on fertiliser–crop price ratios and in Senegal, for example, the use of fertiliser on groundnut has virtually ceased with the withdrawal of subsidised credit for purchase of seed, fertiliser and equipment (Speirs and Olsen, 1992).

These examples of fertiliser use in Sub-Saharan Africa are, however, the exception rather than the rule. While there was a 12-fold increase in the use of NPK fertilisers in this region between 1960 and 1987, this was from a very low base of 100 000 t in 1960 (Bumb, 1989). The contrast with South Asia and East Asia is quite revealing; in East Asia over the same period consumption rose from 2.8 Mt to 24.9 Mt but in South Asia the rise was even greater, from 430 000 t to 12.5 Mt. The values for Sub-Saharan Africa translate into very low per ha consumption figures; in 1985 many of the countries consumed less than 10 kg of NPK nutrient per

Table 6.9 Estimated percentage improvement in maize yields from improved technology since 1950. (Source: Tattersfield, 1982.)

Factor	Percentage increase in yield
Nitrogenous fertiliser	200
Hybrids	45
Increased plant population	20
Early planting	15
Weed control	30
Pest control	10
Early reaping	5
Total	325

cropped ha with Kenya the highest at 42 kg. Fertiliser is used mainly on cash crops and on selected food crops such as maize. By contrast, the figures for Bangladesh, India, Indonesia and Malaysia were 59, 50, 95 and 117 kg/ha of NPK nutrient respectively (Baanante et al., 1989). Although it is often suggested that the relatively high consumption in Asia is due mainly to the seed–fertiliser technology for irrigated crops, Byerlee and Heisey (1993) pointed out that, while the rate of progress has been less spectacular in rain-fed areas, the uptake of this technology has been very widespread in these areas as well.

There are many factors which determine the amounts and kinds of fertilisers that farmers use; if land is plentiful then the farmer will often prefer to cultivate a larger area, especially if animal traction is available. Purchase of fertilisers requires access to credit, storage, marketing and delivery systems and thus an infrastructure. Perhaps most important, the ratio of the price of farmers' produce to the price of fertiliser must be favourable. Byerlee and Heisey (1993) gave an example of the ratios which were current in 1989 (Table 6.10).

Obviously these ratio can be readily changed by government policies on pricing and subsidies but they serve to illustrate the wide divergences in the cost of inputs and hence the incentives, or lack of them, for farmers to use fertilisers. To make fertiliser use financially attractive to small farmers, the value of the increased production needs to be between two and three times the cost of the fertiliser; in many countries in Africa this would require yield increments of 15–20 kg of maize per kg of nitrogen, whereas in much of Asia it would require only 5–8 kg.

These factors emphasise the need to use the right kind of fertiliser in the proper way and at the correct time to maximise the benefits. The most concentrated fertilisers, and therefore the cheapest to transport per unit of nutrient, are urea (45% N) and triple superphosphate (43–52% P_2O_5). Urea has the disadvantage that, if applied to the soil surface, the ammonia volatilises unless cultivated or washed in by rain. Since nitrates leach readily, timing of application is important and split dressings are often advocated; however, small farmers will often wait until they are convinced that the rains will be adequate and thus are often late in applying the fertiliser.

Many tropical soils are very deficient in phosphate and several crops respond well to relatively small dressings, which, as noted earlier, often have quite substantial residual effects. Pieri (1992) stated that the sandy soils of the West African savannas responded very favourably to small dressings of the order of 10–40 kg/ha P_2O_5, applied in soluble form. Superphosphates, containing water-soluble monocalcium phosphate, are the commonest phosphatic fertilisers used for annual crops. Single superphosphate (18–21% P_2O_5) also contains calcium sulphate and is therefore beneficial on sulphur-deficient soils, but is more expensive to transport, per unit of nutrient, than the sulphur-free triple superphosphate (43–52% P_2O_5). Banding and other methods of placement of soluble phosphates have proved beneficial in some soils and for some crops, especially in soils with a high phosphate retention capacity.

Cheaper sources of phosphate would be desirable and rock phosphate, usually in a finely ground form, has been used extensively as a

Table 6.10 Prices of nitrogen fertiliser in relation to those for maize grain. (Source: Byerlee and Heisey, 1993.)

	Nitrogen-to-grain price ratio
Africa	
Cameroon	7.3
Ghana	8.0
Kenya	5.0
Malawi	11.1
Tanzania	6.0
Zambia	2.8
Zimbabwe	7.2
Other regions	
Brazil	6.0
India	2.1
Mexico	1.6
Pakistan	2.6
Philippines	2.9
Thailand	7.9

phosphate fertiliser for perennial crops, especially rubber and tea. Large areas of tropical soils have characteristics that make them suitable for the use of reactive, finely ground phosphate rock; low pH, low levels of phosphate in the soil solution, low exchangeable calcium and a long and warm growing season in the humid tropics are all conducive to its use. Sub-Saharan Africa, with very large areas of phosphate-deficient soils, also has large deposits of rock phosphate in several countries (McClellan and Northolt, 1986). If these could be used they would provide an attractive alternative to imported phosphate fertilisers. However, few of them are suitable for direct application and one of the improvement techniques being tried is partial acidulation. This process involves using a percentage of the sulphuric or phosphoric acid that would be used to make single or triple superphosphate, respectively, for a particular rock phosphate (Hellums, 1992). A rock phosphate from Niger was ineffective when used without acidulation on millet, but partial acidulation with 40% of the sulphuric acid required to make single superphosphate raised its agronomic effectiveness to 75% of that of the single superphosphate. However, partial acidulation does not improve the performance of some rocks, particularly those with high levels of iron and aluminium oxides (Hellums, 1992).

Response to potash is less common than to nitrogen and phosphate. Butai (1987) reported that in 300 long-term trials with a wide range of crops in Zimbabwe, only eight showed significant response to potassium; treatment levels ranged from 0 to 320 kg/ha K_2O. Responses have been infrequent also in the savanna zone in Africa, even on land that is intensively cropped, but have been obtained more frequently in forest regions after short fallows or on land that has been cropped for several years. They usually occur on the lighter, sandier soils and tobacco, yams, sweet potatoes and cotton are liable to suffer from potassium deficiency on these soils. Straw and the tubers of root crops contain substantial amounts of potassium and their removal from the farmers' fields depletes the potassium reserves of the soils; this imbalance will become more pronounced as and when nitrogen and phosphate fertilisers are used in larger quantities, thus leading to bigger crops and more loss of potassium, unless all the crop residues are returned to the fields.

Soil acidity and the need for liming has been discussed in Chapter 3 and there is a considerable amount of information which shows that, on many soils, yield responses to fertilisers decrease with their continuing use, which can be attributed to increasing soil acidity. Pieri (1992) reported that, in the savanna areas of West Africa, fertiliser efficiency declined to one-half or one-third of its initial response after 10 years or less; the initial efficiency could be restored in 1 or 2 years by liming. In a nutrient balance study with a maize–cotton–sorghum–groundnut rotation in Senegal, he calculated that 2.3 kg CaO is lost for each 1 kg of nitrogen applied; a similar effect could come from the large amount of nitrate released, as noted earlier, after the bush fallow. There appear to be few, if any, long-term records of changes in pH under the natural fallow system but length of fallow, length of cropping, crop and fallow species, soil type and rainfall intensity will all influence the balance between bases and nitrates and hence the gain–loss balance and the effect on pH.

Soil acidity can be corrected by liming but the lack of calcium carbonate deposits in some countries, the cost of transporting even minimal dressings to distribution points and from there to the fields, especially where there is no wheeled transport, and the problems of application with only hand tools, make the use of lime a formidable problem in many farming systems.

Since the major toxicity in acid soils is aluminium in the soil solution, this can be counteracted to some degree by breeding acid-tolerant cultivars and by growing acid-tolerant crops, e.g. cassava and cowpeas. Soil organic matter also counteracts aluminium toxicity and experiments in alley cropping in Nigeria by IITA (1990) showed that, over an 8-year period, maize yields could be maintained at about 3 t/ha with a combination of fertilisers and organic matter from the tree loppings; this compared with a decrease from 3 t/ha to 1 t/ha over the same period with fertilisers alone.

There is no doubt that tropical farming, including rain-fed arable cropping, is relying more and more on inorganic fertilisers to meet the rapidly growing demand for food and cash crops. Even China, which has been the example *par excellence* of intense recycling of crop residues as well as animal and human waste, has now become a very large consumer of fertilisers and, by 1985, was reported to be using 167 kg/ha of NPK nutrients (Baanante *et al.*, 1989). It is impossible to envisage any widely applicable farming system, which relies only on fallows and limited supplies of locally available organic manures, yet which will be capable of improving soil fertility sufficiently to sustain or enhance the productivity of the land.

Good husbandry

The need for increased productivity per unit of land and per unit of labour has been stressed throughout this chapter. The improvements needed for these increases may include additional inputs, such as fertilisers, pesticides, better yielding crop varieties, more labour for making and applying farmyard manure or compost or for constructing soil conservation works. The full benefits of these inputs will not be realised unless they are accompanied by improved husbandry which includes timely planting, good seed, appropriate plant population, weed, pest and disease control, and improved soil fertility. The greatest improvement in yield and profit is normally obtained by a combination of inputs in a package of improved technology. However, small farmers will seldom apply the whole of an improved package simultaneously but will choose some parts of it, depending on their resources of labour and money. Timely planting, proper population density and good weed control are crucial aspects of good husbandry which can usually be undertaken without cash outlays and they have long been a major theme of extension advisers. There is ample evidence that these practices will improve yields, an example being a series of factorial experiments in Kenya (Harrison 1970), which showed that for six practices, the order of importance for improving yields were:

(1) early planting
(2) use of a hybrid variety
(3) adequate weed control
(4) dense planting
(5) nitrogen fertiliser
(6) phosphate fertiliser.

With poor husbandry and no fertiliser the local maize yielded 2.1 t/ha; the hybrid variety with fertiliser but poor husbandry gave 3.5 t/ha but the hybrid with both fertiliser and good husbandry produced 8.7 t/ha. These results appear somewhat at variance with the relative importance of the various management factors reported for commercial farms in Zimbabwe, where nitrogen fertiliser made the largest contribution to yield increases (Table 6.9). However, the soils in this region of Kenya are much better than those in many of the maize-growing areas of Zimbabwe and, without improving the fertility of the soils of the latter, the impact of the other practices would be small.

Many, if not most, farmers are well aware of the benefits of these practices but will often ignore some or all of them. One of the major reasons for this is that many of them require additional labour and, where virtually all of the energy going into agricultural production is human energy, bottlenecks are bound to occur. It was noted earlier that farmers involved in a purely subsistence agriculture had ample spare time during most of the year, but this does not apply where the farmer is working in the cash economy. Pieri (1992) reported that, in the Côte d'Ivoire, a farm family devoted about 360 labour-days to the production of food crops, cotton and rice; about 20% of this was for food crop production, 35% for cotton and 45% for rice production. Baker (1975) found that the use of an improved variety, early planting, more thorough weeding, denser planting and fertiliser could markedly increase maize yields, but the application of this 'package' would increase the labour requirements at some periods so much that the average family farm would either have to employ labour, which would be difficult in an agricultural community, or reduce the area under cultivation. In the latter event, a 60% reduction in area would be needed, and a 150% response required to produce the same gross yield as would have resulted from the larger area without the proposed inputs. This again illustrates how intensification may greatly increase labour demands and explains why farmers do not always respond to new technology as expected.

CONCLUSION

It seems that no proven and practicable solution has yet been found to the problem of replacing shifting cultivation by a widely applicable, sustainable system of rain-fed arable farming sufficiently productive to meet the needs of rapidly increasing populations.

To maintain the productivity of a soil indefinitely will sooner or later require the replacement of the nutrients removed in the crops. The need for this may be delayed for some considerable time by the release of slowly available soil reserves and/or by the transfer of nutrients from subsoil to topsoil by growing deep-rooted plants, but on the majority of tropical soils inorganic fertilisers will be needed to sustain a moderately intensive cropping system.

It will also be necessary to maintain a level of organic matter in the soil sufficient to keep it in a physical condition that facilitates water infiltration, aeration and root growth and which also permits microbiological activity and counteracts the increasing amounts of exchangeable aluminium that result from increasing acidity. Under cultivation, the organic matter content of tropical soils soon declines, owing to the rapid decomposition of this material, and means must be found for regularly augmenting it. The use of green manures, farmyard manure or composts and the return of crop residues to the soil can all contribute some organic matter but the amounts of these materials available to, or practicable for, small farmers are usually insufficient to maintain indefinitely a level adequate for productive cropping. Furthermore, the use of farmyard manure involves transferring fertility from an extensive grazing area to a smaller cropped area and this becomes increasingly difficult as population pressure on the land increases.

There is evidence that, in regions of good rainfall and on soils not initially infertile or exhausted by cropping, a rotation of about 3 years of grass or legume and 3 years of cropping, together with the application of fertilisers, could maintain crop yields for many years. Such a system might be practicable in some areas, although yields of food and cash crops averaged over the rotation could only be modest unless the restorative break could be a profitable perennial, such as sugar cane. However, in many places population pressure precludes setting aside land for an unproductive restorative break.

There remains the possibility of intercropping food and cash crops with fertility-restoring perennials in some kind of agroforestry. Alley cropping, which might be combined with improvements that have been developed in mixed cropping, is claimed to have given promising results. The 'hedges' of trees or shrubs in this system must obviously compete for light, water and nutrients with the crops grown between them but it is claimed that this competition is offset by the organic matter and nutrients contributed by the loppings and leaf fall of the perennials. However, alley cropping has so far been little adopted by farmers and, as yet, there does not appear to be sufficient evidence of its ability to maintain crop yields at a satisfactory level.

Whatever the system, inorganic fertilisers will be needed on the majority of tropical soils. Their high cost in relation to the value of most farm produce will make it desirable to minimise the amounts required by using other aids, such as farmyard manure and legumes, where this is practicable. It will also be important to maximise the response to fertilisers by providing other inputs to ensure good growth and yields, such as improved varieties, timely planting, optimum plant density and good weed, pest and disease control. All these inputs need to be put together, according to local requirements, in farming systems which will need to be tested, and probably modified, for some years before it can be established that they are sustainable and practicable. Unfortunately, there now appear to be few long-term experiments in this field.

FURTHER READING

Hunt, D. (1984) *The Labour Aspects of Shifting Cultivation in African Agriculture*. FAO: Rome.

Lal, R., Sanchez, P.A. and Cummings, R.W. Jr. (1986) *Land Clearing and Development in the Tropics*. A.A. Balkema: Rotterdam.

Ruttenberg, H. (1980) *Farming Systems in the Tropics*. Oxford University Press: Oxford.

7
Rice and Rice-based Farming Systems
D.J. Greenland

Rice is important as the staple cereal for more people than any other crop. Worldwide rice is grown on a total area of around 150 Mha and produces about 500 Mt annually. Most rice is produced in the tropics, and most is consumed near where it is produced. International trade in rice is only 12–15 Mt each year, and so is almost negligible compared with production. However, international trade is significant in its influence on rice prices. Asia produces and consumes more than 90% of all rice, although rice production and consumption in Africa, the Middle East and Latin America is increasing rapidly.

Rice is unique amongst the major food crops in that it grows better in standing water than in dryland conditions. It can, however, be produced with or without standing water, although yields from the crop grown with controlled flooding are almost always better than when it is grown as a dryland crop. Thus, the crop is most widely grown in areas where water is readily available and the supply of water to the fields can be easily controlled.

Such areas include the deltas of the great river systems of South and Southeast Asia where rivers and rainfall give an abundant water supply, and where the sedimentary soils are easily worked to enable fields to be levelled and bunds to be built to control water movement. Almost half of the total rice area in the vicinity of the great rivers has controlled water delivery systems, where the natural water channels are supplemented by appropriately constructed canal systems. In most of the remaining areas close to the major rivers, rice is produced where the soil is inundated every year by natural flood waters, usually controlled, to a greater or lesser extent, by simple earthworks. In these areas water depth is difficult to control and flood water may reach depths of 1 m and occasionally 5 or 6 m. There are rice varieties adapted to survive in such areas – the so-called deep-water and floating rice varieties.

Away from the deltas and flood plains, smaller valleys may be terraced. As the streams and rivers may not always carry sufficient water for the terraces to receive an uninterrupted supply, rice grown on them is likely to suffer from periodic water shortages. Above the terraces rice may also be grown, in much the same way as any other cereal crop. The rice varieties adapted to such dryland conditions are known as upland or dryland rice. Their tolerance to limited water availability is mostly low.

Except in very favourable soil conditions, upland rice seldom yields more than 1 t/ha. In contrast, yields of irrigated rice are normally high, quite often exceeding 10 t/ha and yields as high as 16 t/ha have been recorded. The national average yield of rice in China has exceeded 5 t/ha in recent years and the average for irrigated rice in Indonesia is 6 t/ha. Yields of rain-fed lowland rice vary considerably. The uncertainties related to flooding or drought tend to inhibit farmers from using inputs so that yields are usually rather low.

THE ECOSYSTEMS IN WHICH RICE IS GROWN

Outside the tropics almost all rice is grown under controlled irrigation, and as continuous sole crop rice. In the tropics, systems which are based on retention of water from natural flooding are rather more common than those which have a fully controlled water supply. In these rain-fed systems the rice crop may be followed by an upland crop, commonly a short duration legume such as mung beans, largely dependent on the residual moisture stored in the soil profile after harvest of the rice crop.

The relationship between rice production systems and landform has been well described by Moormann and van Breemen (1978). Amongst the landforms of the lowland rice-growing areas they recognise the alluvial fans and terraces of the major river systems of monsoonal Asia. The soils of the upper part of an alluvial fan may be lighter than those of the lower part but, provided that textures are loamy or heavier, rice is normally grown throughout the area occupied by the fan, and often with well-managed irrigation. This is also true of most of the alluvial terraces. In addition to the alluvial fans, the meander flood plains in the middle courses of rivers, lacustrine flood plains, and marine flood plains formed behind coastal sediments provide suitable conditions for flooded rice production, although the depth and the persistence of the flood can vary considerably in place and time. Much rice is also grown along the lower slopes of inland valleys of modest to very small size. Usually, small terraces are constructed and water diverted from the stream or river occupying the valley to irrigate the paddy fields.

RICE FARMING SYSTEMS

Rice farming systems are commonly divided into upland, rain-fed lowland and irrigated. The different rice ecosystems are illustrated in Fig. 7.1.

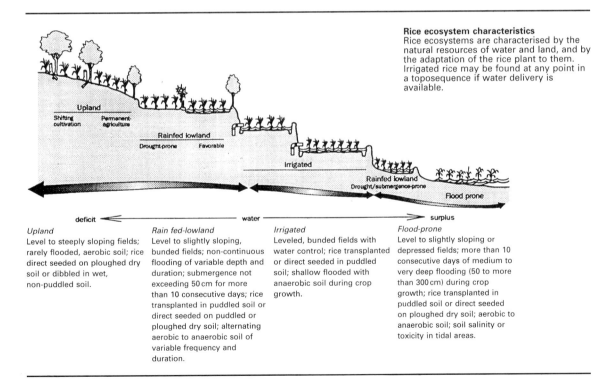

Fig. 7.1 Rice ecosystem characteristics. (Source: IRRI, 1993.)

Fig. 7.2 A typical valley developed for rice farming, with irrigated rice on the valley floor and lower terraces and rain-fed rice on the higher terraces. Upland rice might be grown on the land above the higher terraces. (Photo: M. Raunet.)

A typical landscape where different rice farming systems are practised is illustrated in Fig. 7.2.

Irrigated systems

The term *irrigated* may be restricted to those systems where water is stored, or where water is available throughout the year from a perennial river system. The term *partially irrigated* is used for the systems, usually in the flood plains, where, after the flood, water may be diverted to the fields from the river. Many villages in Asia supplement water supplies to the rice paddies by stream and river diversions, and such village schemes are sometimes referred to as 'non-technical', in contrast to the 'technical' schemes based on major engineering works.

Irrigated rice is found wherever water is available and can be appropriately controlled. Most irrigated rice is found in lowland areas but the construction of terrace systems enables it to be grown on many hill slopes, and even in mountainous areas.

Lowland rain-fed rice

Lowland rain-fed systems include many which are partially irrigated, where water is diverted from a seasonal stream or river to supplement the natural flooding, as well as many where only natural inundation occurs. All the lowland rain-fed areas are characterised by limited water control, and the fields are flooded to different depths, depending on landscape position and rainfall in the catchment area of the river system.

On the alluvial terraces the major problem may be occurrence of drought or on the lower terraces and in the flood plains submergence, and these terms are often used as modifiers to describe rain-fed lowland rice. The length of time for which the floods persist, and the depth of flooding, also determine the system of rice production that can be practised. The lowland rain-fed rice areas may accordingly be further subdivided by normal depth of flooding, traditionally ankle deep, knee deep and thigh deep, or shallow rain-fed (water depth not normally exceeding 15–30 cm), medium deep (up to 50 cm) and deep (up to 1 m). There are many variations on this terminology, reflecting local importance of different flooding patterns. In flood plains which are flooded to depths of 1 m or more the specialised rice varieties commonly known as deep-water and floating rice are grown. These varieties have tolerance to submergence for several days and elongation ability sufficient in

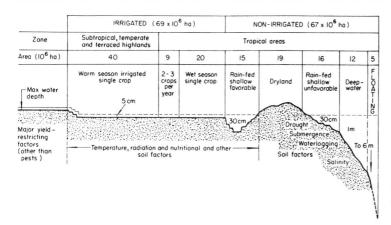

Fig. 7.3 The relative extent of different rice production systems. (Source: Greenland, 1984.)

some varieties to enable them to survive and yield adequately in depths up to 3 m, and occasionally as deep as 5 or 6 m.

In the flood plains excess water, and the problems of excessive soil reduction associated with waterlogging when no water movement occurs at all in the soil, are a major limitation. These areas are usefully subdivided into submergence prone and waterlogged areas.

The areas of deep-water rice are largely limited to the deltas of the Ganges, Brahmaputra, Irrawaddy, Mekong and Chao Phraya rivers, and even in these areas the extent is tending to decrease (IRRI, 1987b). In 1987 there were believed to be about 5 Mha of deep-water rice in Asia, of which about 3 Mha were in Bangladesh (Huke and Huke, 1983). In the flood plains of the upper reaches of the Niger River in Africa a separate rice species, *Oryza glaberrima*, is also grown under deep-water conditions but production is relatively small (Vallee and Vuong, 1978).

Upland rice

The rain-fed upland areas where rice is produced with no standing water on the soil at any time are quite distinct from the others. They are probably most easily distinguished from the rain-fed lowland areas by the absence of any form of bunding to retain water in the fields. The term dryland rather than upland is often used to describe these areas. Neither upland nor dryland is an entirely satisfactory term as much upland rice is grown at low elevations on soils with favourable moisture characteristics and under rainfall such that the use of the term *dryland* is equally inappropriate. The term upland rice will be used here.

Upland rice accounts for about 15% of the total rice area, but less than 5% of production. Although most upland rice is produced in Asia, it is relatively more important in Latin America and Africa where it is the major rice production system. In both Latin America and Africa the area of rice grown is increasing rapidly.

The relative areas devoted to irrigated and rain-fed rice, and the subdivisions of rain-fed rice, are illustrated in Fig. 7.3. The distribution of the different rice production systems in Asia is illustrated in Fig. 7.4.

CROP MANAGEMENT IN RICE FARMING SYSTEMS

The principal characteristics of the major rice farming systems are given in Table 7.1. The principles and practice of rice production in each of these systems are fully described by De Datta (1981).

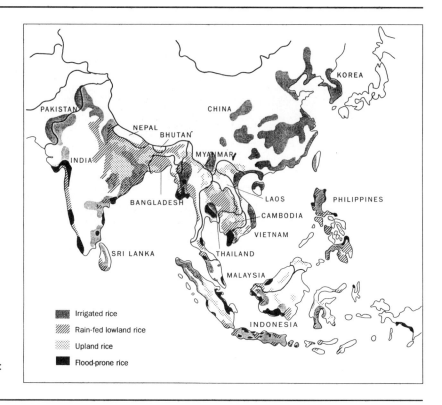

Fig. 7.4 Land use for rice production in Asia. (Source: IRRI, 1993.)

Irrigated rice

Land preparation

When rice is produced under flooded conditions the normal land preparation method is to flood the soil and allow it to remain under water for some time, a practice known as 'land soaking'. After several weeks the soil is cultivated when wet. Primary tillage usually involves an initial ploughing and/or rotary harrowing, and secondary tillage harrowing and puddling. Puddling is normally done by dragging a simple multiple set of tines through the soil. These wet cultivation treatments tend to disperse the soil and create a clay pan at the base of the ploughed layer which minimises water losses by seepage. The cultivations also soften the soil and destroy weeds.

The initial period of 'land soaking' appears to be wasteful of time and water, but it may be important in some instances to ensure that the toxic materials which are produced during the anaerobic decomposition of organic matter in the soil have time to dissipate prior to transplanting the rice crop (IRRI, 1984b).

Most loamy and heavier soils used for rice (Entisols, Inceptisols and Vertisols) are rather easily dispersed. Where rice is grown on soils with more resistant structural characteristics (Oxisols and Ultisols) more rigorous treatment may be needed to induce dispersion. The use of large teams of water buffaloes repeatedly dragging the puddling equipment around a small rice paddy is one method used to accomplish this.

Crop establishment

After land soaking, the traditional and most widely practised method of establishing the crop is to germinate the seeds in a seedbed and trans-

Fig. 7.5 Transplanting rice.

Table 7.1 Characteristics of rice farming systems in the tropics.

A. Irrigated rice, with assured year-round water:
Continuous rice, with two or three crops per year.

B. Partially irrigated and rain-fed lowland rice.
   (i) Where monsoon season is 6 months or more:
   Dry seeded (aus) rice, followed by irrigated rice, followed by upland crop.
   (ii) Where monsoon less than 6 months:
   Irrigated, transplanted rice, followed by upland crop.

C. In rain-fed, drought prone areas (mostly on alluvial terraces):
One dry or wet seeded rice crop. An upland crop may follow in good seasons or in favourable sites.

D. In flood-prone (deepwater and floating rice) areas:
One transplanted or wet seeded rice crop. (In some deepwater areas double transplanting is practised.)

E. In upland rice areas.
   (i) Under shifting cultivation:
   Dry seeded rice, often interplanted, e.g. with maize, and followed by another upland crop, e.g. cassava.
   (ii) Under mechanised cultivation:
   Sole crop dry seeded rice, grown annually for several years, after which a grass pasture may be established.

plant the seedlings when they are about 20 days old. The seedlings are usually transplanted by hand into shallow standing water (Fig. 7.5). Fertilisers and organic manures should normally be applied before transplanting, and preferably incorporated in the soil to minimise nitrogen losses by volatilisation. Herbicides, if used, are also applied at this time. A further application of nitrogen and of insecticides may be needed after transplanting, as well as one or two weedings. However, the crop is often given only limited attention until a few weeks before harvest, when water is drained from the field to assist the grain to ripen.

Transplanting by hand is not only hard work, it is also very costly except where labour is cheap or provided only by the farm family. Wherever labour costs are high it is tending to be replaced by mechanical transplanting, as in Japan, or by direct seeding with pregerminated seed broadcast on the saturated, but not flooded, soil, as in the United States and Australia, where the pregerminated seed may be distributed by low-flying aircraft.

Water management

Water management during the growing period usually consists of maintenance of 5–10 cm of water on the paddy until about 10 days before harvest. In China and Japan it is widely held that a seepage rate of 10–15 mm/day is essential for high rice yields, and in southern China mid-season drainage (withholding water from the paddy for a few days to allow the soil to drain) is widely practised. Experiments in the tropics have mostly failed to show a yield advantage from these practices (Sharma et al., 1989) except for some acidic soils and soils high in organic matter. It is likely that the advantage gained from draining and partial drying of the soil is due to removal of plant toxins formed under extreme anaerobic conditions.

Organic manuring is also common in the high-yielding rice areas of East Asia. The organic materials are beneficial but also give rise to toxic materials which reduce the redox potential so that concentrations of iron and manganese may reach toxic levels. Whenever organic manures are used it is probably good practice to use mid-season drainage, although the faster decomposition rate in the tropics may make aeration of the soil by drainage less important than in cooler conditions.

Mechanisation

Most rice farms in Asia are smaller than 5 ha and many are less than 1 ha. Hence farm machinery consists mostly of simple manual or animal-drawn implements. Small power tillers are gradually becoming more widely used and rotary threshers have an established niche in Thailand and the Philippines (IRRI, 1986c).

In Japan farms are also small but the farmer receives a heavily subsidised price for his rice; hence the use of small but highly sophisticated machinery becomes economic. Power tillers, transplanters and combine harvesters adapted to fields smaller than a hectare are widely used.

As mentioned earlier, there are a very few rice production schemes in the tropics where seed is sown from aircraft into large paddies which have been levelled using large laser-controlled tractors to ensure that water is evenly distributed. Harvesting may be done by large combines. These methods, developed in the United States and Australia, are not suited to small farm systems.

Multiple cropping of irrigated rice

Where water availability is sufficient it is common in the tropics, where temperatures remain high throughout the year, for a second, and occasionally a third, rice crop to be grown. This requires that the soil be flooded soon after harvest, and the land again puddled for the succeeding crop. Most rice is grown during the monsoon season when water is plentiful. The great advantage of having water available all year round is that a second crop can be produced in the dry season when solar radiation is most intense, and pest attacks least. There are probably about 10 Mha in the tropics where sufficient water is available to enable both a wet season and a dry season crop of rice to be harvested each year. In a few places a third rice crop may be grown. The Chinese talk of 'three colours in a day', meaning the gold of the ripe rice, the grey-brown of the puddled paddy field, and the green of the newly transplanted rice. In favourable circumstances this is possible and annual production of more than 20 t/ha has been obtained from such triple or continuous rice production systems. However, the labour demands are considerable and seldom is the system economic. Even in southern China, where triple cropping has been most widely practised, it is tending to become less common.

Rain-fed lowland rice

The problems of wetland rice production in systems where water supply is difficult to control are considerable. Inadequate definition of the systems has tended to restrict understanding of the special features of the many systems currently used. Much of the current knowledge of rain-fed lowland rice is encompassed in two conference proceedings of the International Rice Research Institute (IRRI, 1979b, 1986a).

In favourable conditions rain-fed lowland rice can be managed exactly like irrigated rice. However, without controlled irrigation, these areas are more likely to be affected by water shortages or by submergence of the crop under excess water. Drought is more commonly a problem on the alluvial terraces and submergence on the flood plains. The occurrence of both is also related to the weather pattern.

Drought-prone areas

Land preparation in these areas is often difficult. Puddling is important in order to retain water once the rains have developed but it can only be done in a saturated soil. On light textured soils, an initial ploughing may be possible before the rains are established but, with limited power available, this is not always possible. When ploughing is done early, weed control may subsequently pose more serious problems. If puddling is inadequate, and the soil drains too rapidly, the crop may be affected by drought if later rainfall is not sustained. The tillage treatments should also open the soil to a sufficient depth to provide for deeper rooting than under irrigated conditions in order to help the crop tolerate periods of inadequate rainfall.

The varieties grown in drought-prone rain-fed areas need to embody certain characteristics, such as the features which enable a variety to be tolerant of drought or to escape drought. These include appropriate rooting characteristics, leaf characteristics such as early leaf rolling and stomatal closure, and maturity adapted to the climate – the crop needs to mature as the rain ceases or earlier, as drought at the time of flowering and grain filling has a severe effect on yield (IRRI, 1982a).

Dry seeding of rice in rain-fed drought-prone areas is practised in several parts of Asia. The practice is known as 'aus' cropping in northeastern India and Bangladesh and as 'gogorancah' in Indonesia. It is normally associated with multiple cropping, the dry-seeded rice normally being an early maturing variety, and its early establishment and short duration enabling a second rice crop to be transplanted following harvest, or an upland (non-rice) crop to be grown as the rains cease.

As well as rice–rice and rice–upland crop systems an upland crop–rice system is sometimes practised, with an upland crop established by conventional means and the soil puddled for transplanted rice after the upland crop has been harvested and the rains intensify. Such systems are adapted to areas where there is a slow start to the onset of the monsoon.

Submergence-prone and deep-water areas

These are the flood plain areas where water may accumulate to depths of 30–50 cm (medium deep-water), or 50 cm to 1 m (deep-water). Where floods normally exceed 1 m special considerations apply and these areas are considered separately. Establishment in the flood-prone areas is normally by direct seeding prior to the onset of the monsoon. Again depending on the initiation of the monsoon rains, it is usually necessary that the land be dry ploughed prior to the start of the rains. In Bangladesh deep-water rice may be transplanted following an aus crop.

In the flood-prone areas little can be done to control the rate of rise of the flood water and it is therefore important that the rice varieties grown have some tolerance to submergence. It is also important that submergence does not occur at or after flowering, so that it is normally necessary that photoperiod-sensitive varieties are used to ensure flowering after the floods start to recede. Submergence patterns are difficult to predict. They may arise as a result of extra heavy rainfall in a local area (climatic submergence), or as a result of overflow of river water transported from a more remote area (fluvial submergence). The two causes interact. Given the uncertainties introduced by the problems of unpredictable flooding it is not surprising that deep-water rice farmers tend to use very limited inputs other than seed.

As well as the flood problem the deep-water areas suffer from the disadvantages associated with stagnant water; the soils are highly reduced and various toxicity problems arise. Not surprisingly, the areas of eastern India and Bangladesh where these conditions are common (Fig. 7.4) are amongst the poorest anywhere.

As the flood waters recede, and after the rice crop has been harvested, it may be possible to establish an upland crop. Such crops may be able to use water from the falling water table but in the later stages of growth will be dependent on moisture stored in the soil profile, only occasionally supplemented by rain falling after the end of the monsoon. An important example of this type of system is the annual rice–wheat system of northern India, Pakistan, Bangladesh and central China.

Very deep water and floating rice areas

In general, management of the rice crop is similar to that of rice in the deep-water areas, but because of the depths attained by the flood waters the fields are not normally bunded. As well as submergence-tolerance and appropriate photoperiod sensitivity, the varieties grown must have rapid elongation ability. Even with tolerance to submergence for 10 days an elongation ability of up to 20 cm/day is necessary to cope with rates of water rise in these areas. The floods often persist for several months in the floating rice areas and the individual plants may attain lengths of several metres. Both deep-water and floating rice varieties show secondary and tertiary tillering from nodes well above the base of the plant. Adventitious roots also appear from the secondary and tertiary nodes, presumably collecting some nutrients from the flood water. Harvesting has normally to be done when the floods recede and leave a tangled mass of mature rice plants on the mud. Yields of 2–3 t/ha are normally recorded (Catling *et al.*, 1983; IRRI, 1987b; Catling, 1993).

Upland rice

Upland (or dryland) rice is produced in two very different systems: shifting cultivation which is most common in Africa and large-scale mechanised cultivation which is the more important method in Latin America. Although accounting for less than 5% of total rice production upland rice has considerable importance as a subsistence crop for many of the world's poorest. It has been the subject of important conferences (IRRI, 1984a, 1986b), a monograph (Gupta and O'Toole, 1986) and much information was also given by Buddenhagen and Persley (1978).

Upland rice is grown in a wide range of environments in the tropics, at both high and low altitudes, on fertile and on poor soils, although rarely on heavy clay soils. Although upland rice varieties are often described as 'drought tolerant' the term is relative to other rice varieties; rice has limited tolerance to water shortage when compared to crops such as sorghum and millet (IRRI, 1982a). Hence upland rice is seldom grown where there is not a wet season of four months or more.

Upland rice in shifting cultivation systems

Upland rice is produced as a subsistence crop in shifting cultivation systems in many parts of the world. It is often intercropped with maize, commonly as an opening crop after clearing forest regrowth, particularly in Asia where rainfall is higher and more consistent in the areas of shifting cultivation than in Africa. Rice–maize may be followed by cassava and the land then allowed to revert to a natural fallow, or rice may be followed by maize, or rice may be grown continuously for two or three crops, by which time yields are usually very low indeed (Okigbo and Greenland, 1976; Nakano, 1978, 1980; Watabe, 1981; Greenland and Okigbo, 1983). Seeds are usually hand sown or dibbled into lightly prepared land. Fertilisers are seldom used and weeding is by hand, at least two weedings normally being required.

In many parts of Asia shifting cultivators also grow some rain-fed lowland rice if suitable wetland areas are accessible to them and rainfall is sufficient to enable some water to be trapped by bunding the land. Although bunding is rare in Africa wet spots are often identified and used for rice and vegetables while upland areas are used for other cereals. In India it is quite common practice for all rain-fed land to be bunded if rice is to be grown, even when the drainage characteristics of the soil and the rainfall are unlikely to lead to water being ponded on the soil. The bunds prevent run-off and erosion and ensure that whatever rain does fall enters the soil.

Upland rice under mechanised production systems

In Latin America the area used for upland rice is about three-quarters of the total rice area (Fig. 7.6) and yields almost half of the rice produced (CIAT, 1984). A large part of the production comes from large farms established following forest clearing (EMBRAPA, 1984). In Brazil, rice may be grown for several years but, as yields tend to decline, the rice is replaced by pastures or sometimes, in more favourable situations, by maize or soya beans. Large machinery is used to prepare the land, and cultivation, sowing and weeding are all done with machines, although herbicides may be used to supplement

Fig. 7.6 The irrigated and upland rice regions of Latin America. (Source: CIAT, 1984.)

mechanical weeding. Harvesting is done using large combine harvesters.

The rainfall in the areas where this cultivation system is practised is high, almost always more than 1000 mm in a rainy season of 3 to 6 months. The soils are of rather low or very low inherent fertility. Nevertheless, only low rates of fertiliser are normally used and yields are correspondingly modest. Low planting densities are also used to mitigate the effects of dry spells, locally known as *veranicos*. In the climatically more favoured areas, less likely to suffer from *veranicos*, rather higher inputs of fertilisers and pesticides may be used. The system is economic because of the low cost of land and the relatively high value of rice.

SOILS AND SOIL MANAGEMENT FOR WETLAND RICE PRODUCTION

The chemistry, physics and biology of wetland soils differ considerably from those of the dryland soils discussed in Chapter 3. Their management also differs. Some basic properties of wetland soils have been fundamental to the sustainability of rice farming systems in Asia. These include the fact that they do not become acid after continuous cultivation for reasons associated with their physical chemistry; because of their position in the landscape nutrients tend to be leached into the soil rather than out of it; phosphorus is usually more readily available because iron is present in the ferrous rather than ferric state; nutrients are replenished from time to time by silt deposited from flood waters; erosion is unlikely to occur because the rice paddies are surrounded by bunds and covered by water; and finally the active population of nitrogen-fixing organisms in rice paddies helps to maintain a level of organic nitrogen in the soil sufficient to support a modest level of rice production for many years.

There are also negative factors associated with anaerobism induced by flooding such as the toxic levels of some elements that may occur and the formation of phytotoxic compounds during the decomposition of fresh organic material, as well as the tendency for salts to accumulate if there is no drainage from the soil. The puddling process used in rice cultivation is important to reduce water loss by downward percolation. However, it normally leads to the formation of dense clods when the soil dries. This, together with the effect of the clay pans formed at the base of the plough layer, makes it difficult to grow upland crops before or after rice.

The characterisation of wetland soils

Flooding a soil excludes oxygen with the consequence that anaerobic organisms control soil processes. Provided that some organic material is present to provide the energy for the microorganisms, they will reduce most oxidised soil constituents to the reduced state (Rowell, 1981). Most noticeably, red and brown compounds of ferric iron are reduced to blue-grey compounds of ferrous iron. Nitrates are reduced to nitrogen and sulphates to sulphide. Carbon metabolism leads to the production of methane rather than carbon dioxide.

In upland and alluvial soils where the water table is normally well below the soil surface, puddling and flooding the soil for rice leads to reduction of the surface soil to the depth of puddling and a little below, but most of the subsoil is likely to remain oxidised. A 'perched water table' is formed in the puddled layer. The reduction process in soils is often referred to as 'gleying' and these soils are sometimes referred to as 'surface water gleys'. Moving from upland and alluvial areas towards the river flood plains the natural water table moves closer to the surface and, eventually, all the soil profile will remain waterlogged and reduced for most or all of the year (Fig. 7.7). These soils are 'ground water' or 'stagnogleys'. The soils on the alluvial terraces, in the higher positions, usually show some internal drainage whereas in the groundwater gleys there is no internal water movement. Movement of water into the soil will bring some oxygen with the water and help to prevent any accumulation of toxic compounds which may reduce rice yields. Hence the soils in

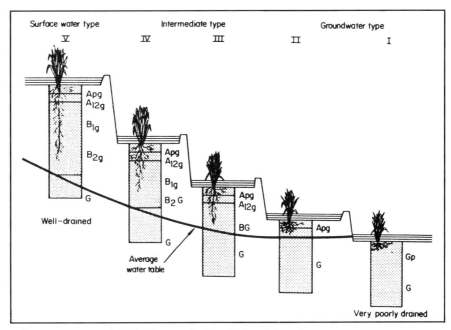

Fig. 7.7 Characteristics of rice soils in different landscape positions. Ap is the surface horizon, coincident with the plough layer; A_{12} is the transitional layer between the A and B horizons; g indicates minor gleying, usually revealed as a mottling of reddish and blue-grey colours; G indicates a completely gleyed horizon of grey-blue colour. (Source: Moormann and van Breemen, 1978.)

the higher landscape positions are those usually associated with higher rice productivity. Bringing in fresh oxygenated water is important. In practical terms it means that it is essential to see that surface drains are kept open, to allow water to drain through and across rice paddies.

Landscape position and drainage characteristics are of course superimposed on other soil characteristics which determine texture, nutritional status and other soil properties. The Chinese use the terms 'surface water paddy soils' and 'well-drained paddy soils' for the surface water gleys in the higher and lower parts of the upper landscape positions, and the terms 'groundwater paddy soils' and 'bog soils' for those in the lower positions, separating the groundwater and bog soils on the basis of whether the soil remains completely waterlogged throughout the year or not (Gong Zi-Tong, 1985). More information about the characterisation and classification of wetland soils is given in IRRI (1978, 1985) and Institute of Soil Science Academia Sinica (1981).

Chemistry of wetland soils

When soils are flooded the pH of the soil is determined by the concentration of carbon dioxide in the water. If the water is in equilibrium with the atmosphere this will mean that it is close to 6.0, or nearly neutral. At this pH most plant nutrients are able to maintain solution concentrations well suited to plant growth. Soils which are initially acid when flooded tend to become less so in a

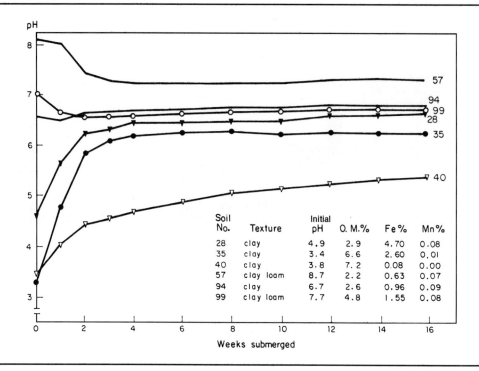

Fig. 7.8 Changes in the pH values of soils following submergence. (Source: Ponnamperuma, 1972.)

matter of days or weeks, and soils which are initially alkaline move towards neutrality, usually a little more slowly (Fig. 7.8). The controlling processes are the reduction of iron, which releases hydroxyls, and the higher partial pressure of carbon dioxide in the soil solution buffering the carbonate in otherwise alkaline soils (Ponnamperuma, 1972).

Around pH 6 the concentration of phosphate in the soil solution is largely dependent on the concentration of iron, with which it forms sparingly soluble iron phosphates. As ferrous phosphates are slightly less insoluble than ferric, reduction usually implies a release of phosphate to the soil solution so that its diffusion to the root is more rapid. Although the level of phosphorus in many soils used for flooded rice production tends to be higher than in upland soils, responses to phosphorus fertilisers are common. Even in soils with initially high levels, responses are likely to be obtained after several years of intensive cropping. An example is given by a long-term experiment in the Philippines (Fig. 7.9).

The commonest nutrient deficiency in paddy soils is of nitrogen. In the tropics, mineralisation of organic matter proceeds almost as rapidly under saturated conditions as it does in aerated soils. Hence even in rice paddies which are almost always flooded, organic matter levels, and so organic nitrogen levels, are low. The mineralisation of organic nitrogen proceeds to ammonium but, unless the soil is allowed to dry, does not proceed to nitrate. If the soil does become dry and is then again flooded a rapid loss of nitrogen occurs as the denitrifying bacteria use the nitrogen released and convert it to nitrogen gas (Fig. 7.10).

In the paddy water above the soil other important changes occur. Photosynthetic algae are almost ubiquitous in paddy water. During the

Rice and Rice-based Farming Systems

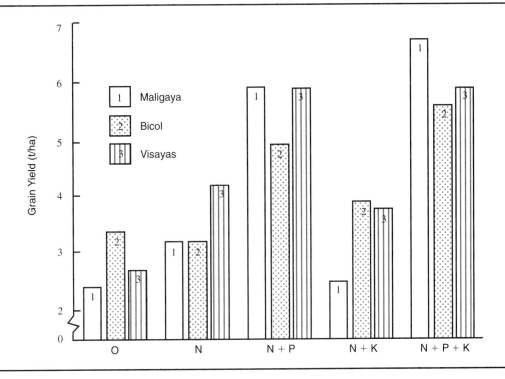

Fig. 7.9 Effects of N, N + P, N + K, and N + P + K on the yield of rice at the experimental stations of the Philippine Bureau of Plant Industry, dry season 1984. The results are means for two varieties grown each year at three sites, in a continuous cropping experiment with two crops each year, and in the seventh year of the experiment. (Source: De Datta *et al.*, 1988.)

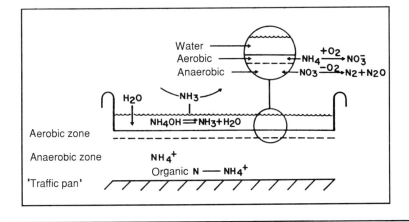

Fig. 7.10 The sites of nitrogen transformation in an idealised paddy field. (Source: Bouldin, 1986.)

day when they are photosynthesising they take carbon dioxide from the water, with the consequence that the pH of the water rises, usually sufficiently for the water to become quite alkaline. Consequently any ammonium in the water becomes ammonia and significant amounts are lost to the atmosphere. At night the pH falls as the algae are no longer photosynthesising. Thus while rice commonly responds to the addition of nitrogen fertilisers, it is best that they are added to the soil and incorporated before transplanting rather than added to the paddy water

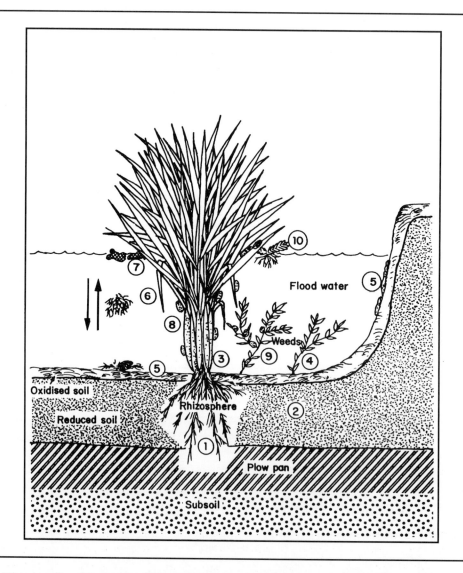

Fig. 7.11 Diagram of environment and nitrogen-fixing components in a rice field ecosystem. *Nitrogen fixing bacteria*: (1) associated with the roots; (2) in the soil; (3) epiphytic on rice; (4) epiphytic on weeds. *Blue-green algae*: (5) at soil–water interface; (6) free floating; (7) at air–water interface; (8) epiphytic on rice; (9) epiphytic on weeds. *Azolla*: (10). (Source: Roger and Watanabe, 1986.)

(Fillery *et al.*, 1984; Bouldin, 1986). A late top dressing of nitrogen is usually advantageous after the crop is well established, as ammonia is then less rapidly removed from the water because the air at the water surface is static. The various processes affecting nitrogen in wetland soils are well described by De Datta and Patrick (1986).

Great importance has sometimes been given to the fixation of nitrogen by blue-green algae in the paddy fields. They do undoubtedly make a contribution but there are several other sites in the paddy ecosystem where nitrogen fixers flourish (Fig. 7.11). Although claims have often been made for the advantage to be obtained by inoculating soils with blue-green algae, the evidence is equivocal. Normally the adapted species of nitrogen-fixing algae will already be present in the paddy soil and their activity controlled by factors such as predation by snails. Inoculation is unlikely to effect more than a very temporary increase in numbers (Roger and Kulasooriya, 1980).

Many years ago the Chinese discovered that the small water fern *Azolla* was a good manure for rice. More recently the reason has been shown to be that the underside of the leaf provides an ideal environment for blue-green algae where they use the filtered sunlight to fix nitrogen very efficiently in an environment where they are free from predators. However, they still require an adequate supply of phosphorus to flourish and insect predators can attack the *Azolla*. Nevertheless, in favourable circumstances *Azolla* can provide an excellent source of nitrogen for rice and can also support fish production in the rice paddies (IRRI, 1987a).

Many leguminous plants have been used as green manures for rice. They include woody perennials, annuals from which a harvest of beans may be taken prior to incorporation in the soil, and a number which are grown solely and specifically to provide nitrogen for the rice crop. There are ample data to show that the amount of nitrogen provided is most commonly equivalent to the amount supplied by 50–100 kg nitrogen/ha of inorganic fertiliser (IRRI, 1988a). Sunn hemp (*Crotalaria juncea*), dhaincha (*Sesbania aculeata*), pillipesera (*Phaseolus trilobus*), berseem (*Trifolium alexandrinum*), mung bean (*Phaseolus aureus*) and cowpea (*Vigna unguiculata*) are among the species which have been most commonly used as green manures. Recent interest has been focused on the stem nodulated legumes, *Sesbania rostrata* and *Aeschynomine afraspera*, (Fig. 7.12). They are fast growing and active nitrogen fixers even when growing in standing water (Rinaudo *et al.*, 1988). Thus, unlike other green manures used for rice, they can be grown on the wetland sites where the rice is to be planted, and ploughed in prior to planting the rice. Other green manures may need to be grown on the bunds, or before planting the rice crop, and preferably before the soil is flooded. Although green manures are still widely used, especially in India and China, the economics of their use is becoming less attractive as labour costs increase and the extent of the practice is decreasing.

The case for green manuring and the wider use of organic manures is often made on the grounds that improving soil conditions by addition of organic matter is essential to maintain soil fertility and to decrease the amount of carbon dioxide contributing to the greenhouse effect. In fact, as it is important to be able to disperse paddy soils by puddling, as far as physical condition is concerned low rather than high organic matter levels are advantageous in most rice soils. As far as the greenhouse effect is concerned, adding organic matter to flooded soils is highly disadvantageous because decomposition of the organic material produces methane rather than carbon dioxide, and the greenhouse effect of the methane is several times greater than that of carbon dioxide (Neue and Scharpenseel, 1984). However, organic matter can be critically important to the supply of nitrogen and sulphur to the rice plant.

As noted earlier, rice is most commonly grown on colluvial and alluvial soils. Such soils tend to have relatively high levels of cationic nutrients (cations tend to be washed in rather than washed out). Thus calcium and magnesium deficiencies are rare, and potassium is quite often adequate, although it will usually become deficient under

Fig. 7.12 Stem nodules containing nitrogen-fixing rhizobia on *Sesbania rostrata*. (Photo: P. Roger.)

continuous intensive cropping (Fig. 7.9 and De Datta *et al.*, 1988).

Sulphur deficiency has been reported for rice in several areas, notably in Bangladesh and Indonesia (Jones *et al.*, 1982). Sulphide is often thought to cause toxicity problems in rice. Although sulphide may be formed if the redox potential falls sufficiently low, there is normally sufficient ferrous iron in the soil solution for all the sulphide to be precipitated (Yoshida, 1981). Although it is possible that the concentration of sulphide may persist long enough before precipitation to have important phytotoxic effects, it is doubtful whether this is a general problem.

Of the micronutrients, zinc has most commonly been reported to be deficient (Randhawa *et al.*, 1978). Ferrous iron and manganous manganese can both attain toxic levels in flooded soils if the soils do not have some residual drainage. Most rice varieties can, however, tolerate high levels of manganese (up to 7000 ppm in leaf tissue) and many varieties are known with good tolerance of high iron concentrations (Ikehashi and Ponnamperuma, 1978; Yoshida, 1981). Silicon has sometimes been claimed to be an essential element for rice. Responses to silicate applications have been quite frequently recorded and the rice plant commonly accumulates 10–12% in its tissues. Most of the silicon in the plant enters in the transpiration stream, as paddy water is normally saturated with silicate and of course water is freely available for transpiration.

Levels of silicon in the plant have been related to insect and disease resistance, but it has not been possible to demonstrate that silicon is an essential nutrient (Yoshida, 1981). It is possible that silicate helps the rice plant to absorb phosphate, by blocking phosphate absorption sites on the iron oxides which tend to precipitate around the roots of the plant.

Salinity is also a problem in many areas where rice is grown, such as tidal wetlands and any areas where no internal drainage occurs so that any salts in the irrigation water gradually accumulate in the soil. The best remedy is to ensure that some internal drainage, however slight, occurs. This will require that drains and ditches are well maintained. The rice plant is moderately tolerant of salinity and some varieties have considerable tolerance. The advent of tissue culture techniques for rice is enabling selection for salinity tolerance to be more readily achieved (IRRI, 1982b; Dykes and Nabors, 1986).

The problems of acid sulphate soils have been described in Chapter 3. If such soils are used for rice production they must be kept flooded at all times to avoid the oxidation of sulphides and the release of sulphuric acid.

MANAGING RICE PESTS

Weeds

As mentioned earlier, wet cultivation is an excellent method for controlling weeds. Nevertheless, aquatic weeds still cause significant yield losses. In irrigated, transplanted rice yield losses, if no weeding is done, may be as high as 30%; in dry seeded rice or upland rice losses may reach 80% (IRRI, 1983). A study in Indonesia showed average yield losses due to weeds of 50%. In this instance the losses were reported to be independent of production system or season (Zoschke, 1990).

In transplanted rice the C4 plants *Echinochloa crus-galli* and *Cyperus rotundus* are the most serious weeds, whereas in dry seeded rice these species as well as other members of the Poaceae and Cyperaceae are joined by many C3 plants (IRRI, 1983). The most important weeds of tropical rice are listed in Table 7.2.

Hand weeding is still by far the most commonly used method of weed control in the tropics, although use of small hand-pushed rotary weeders is gaining in popularity wherever row planting is used. Herbicides are used only where labour costs are relatively high. Both a pre-emergence treatment, e.g. with butachlor, and a post-emergence treatment, e.g. with propanil, are needed for good control. The use of 2,4-D has also proved surprisingly effective for the control of both sedges and grassy weeds of rice (De Datta, 1981), although its use is now restricted.

Insect pests

There are a large number of insect pests of rice (see Table 7.2). Those most commonly considered to be of major importance include the stem borers (*Chilo* spp., *Tryporyza* spp., *Sesamia* spp.), leafhoppers (*Nephotettix* spp.), planthoppers (in Asia the brown planthopper *Nilaparvata lugens*; in Latin America *Sogatodes* spp.), the gall midge (*Orseolia oryzae*), the whorl maggot (*Hydrellia* sp.), the leaf folder (*Cnaphalocrocis medinalis*) and rice bugs (*Leptocorisa* spp.).

Of these pests, stem borers probably cause the most widespread damage whereas brown planthoppers cause the most spectacular damage, 'hopper burn' destroying large areas of rice. When the semi-dwarf, high-yielding rice varieties were initially widely grown in Asia in the 1960s, they had little resistance to planthoppers, and a conference was held at the International Rice Research Institute (IRRI) in 1978 under the title '*Brown Planthopper: Threat to Rice Production in Asia*'. The use of chemical sprays for control of hoppers and other pests was widely advocated. During the 1970s a major effort was made to breed resistance to brown planthoppers and other pests into rice, with considerable success. Although new populations ('biotypes') of brown planthopper appeared able to overcome some of the sources of resistance, new sources of resistance were found (Khush, 1980). It was also shown that the use of broad-

Table 7.2 Economically important pests of tropical rice. (Source: Grayson *et al.*, 1990.)

Insect pests	Rice diseases
Vegetative stage	Bacterial blight
Army worms and cutworms	Bacterial leaf streak
Grasshoppers, katydids, and field crickets	Bakanae
Mealybug	Brown spot
Rice black bug	False smut
Rice caseworm	Grassy stunt virus
Rice gall midge	Narrow brown leaf spot
Rice green hairy caterpillar	Rice blast
Rice green semilooper	Rice ragged stunt
Rice hispa	Sheath blight
Rice leaffolder	Sheath rot
Rice stem borers	Stem nematode
Dark-headed stem borer	Stem rot
Pink stem borer	Tungro virus
Striped Stem borer	White tip
White stem borer	Yellow dwarf disease
Yellow stem borer	**Weed pests**
Rice thrips	*Commelina benghalensis*
Rice whorl maggots	*Cyperus difformis*
Seedling maggots	*Cyprus iria*
Reproductive stage	*Cyperus rotundus*
Rice brown planthopper	*Dactyloctenium aegyptium*
Rice green leafhopper	*Digitaria ciliaris*
Rice greenhorned caterpillar	*Echinochloa colona*
Rice skippers	*Echinochloa crus-galli*
Rice white leafhopper	*Eleusine indica*
Rice whitebacked planthopper	*Fimbristylis miliacea*
Rice zigzag leafhopper	*Monochoria vaginilis*
Smaller brown planthopper	*Paspalum distichum*
Ripening stage	*Portulaca oleracea*
Rice panicle mite	*Scirpus martimus*
Rice seed bug	*Sphenoclea zeylanica*

spectrum insecticides could be highly disadvantageous. Such insecticides kill both the pest and its natural enemies so that there is a resurgence of the more rapidly multiplying pest, and more damage is caused than if the insecticide were not used. The more recent introduction of selective insect growth regulators such as buprofezin (Konno, 1990) avoids the resurgence problem. Undoubtedly the best solution to pest problems is to use integrated pest management techniques (Shepard, 1990) involving varietal resistance, crop management strategies to minimise opportunities for pest multiplication, and use of pesticides only when essential and in such a way as to ensure that natural enemies of the pest are preserved.

Rice diseases

There are many diseases of rice. They have been well described by Ou (1973, 1985). The more important are listed in Table 7.2. Of the fungal diseases, blast caused by *Pyricularia oryzae* has probably received most attention (IRRI, 1963, 1979a). It is common on upland rice and wherever low night temperatures lead to condensation on leaf surfaces and a favourable environment for the spread of the pathogen. It can be controlled chemically and by varietal resistance. However, breeding for resistance is complicated by the many races of the organism which exist. Crill (1982) has been a powerful advocate of monogene resistance and varietal rotation to maintain resistance although, more generally, polygene resistance has been sought in a single variety. Bacterial leaf blight, or kresek disease, due to *Xanthomonas oryzae* is also rather common and also occurs as several races. It is difficult to control except by varietal resistance. Of the virus diseases tungro is most widespread and causes serious yield losses. It is spread by the green leafhopper. Some sources of resistance to the virus are known, but mostly resistance is associated with resistance to the vector.

Nematodes are not generally thought to be a serious problem for wetland rice, but 'ufra' disease due to the stem nematode *Ditylenchus angustus* is a problem in deep-water rice, and several nematode pests of upland rice are known to cause significant yield losses.

RICE VARIETAL CHARACTERISTICS

It was at one time thought that there were three subspecies of *Oryza sativa*, *indica*, *javanica*, and *sinica* or *japonica*, originating from India, Indonesia and China respectively. It was later accepted that these were ecogeographic races. The *indica* and *javanica* rices are adapted to tropical regions. Both are tall, but the *indicas* form many tillers, whereas the *javanicas* are low tillering. The characteristics of the grain are also different, the *indicas* normally having slender, rather flat grains which are not sticky when cooked. The *javanicas* tend to have large, bold grains borne on long panicles. The *sinicas* or *japonicas* are mostly shorter, adapted to cooler conditions, only moderately tillering, and with shorter, rounded grains which cook to a stickier texture. In recent years many crosses have been made between the *indicas* and *sinicas*, and to a lesser extent with the *javanicas*, so that the distinctions have become less clear.

There was for many years a widely held belief that rice grown in the tropics did not respond to nitrogen fertiliser. Various reasons were put forward but the most common problem was the inability of tall *indicas* and *javanicas* to utilise nitrogen to produce grain rather than green matter. Addition of nitrogen made the plants lodge more easily and the grain rotted if it fell into the water. The advent of the semi-dwarf rices from China and Japan, and their crossing with more vigorous tropical varieties, changed the responsiveness of rice to nitrogen dramatically. The now classic response curves (Fig. 7.13) showed the yield gains to be obtained when a cross of the tall *indica* Peta was made with the short *sinica* Dee gee woo gen to produce IR8.

The use of the new semi-dwarf rice varieties paved the way for the 'green revolution'. The

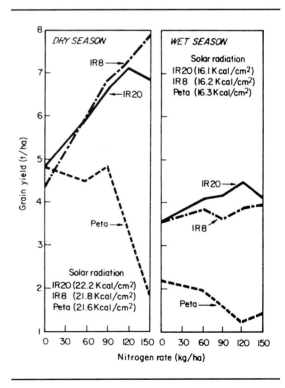

Fig. 7.13 Response of the semi-dwarf rices IR8 and IR20 to nitrogen fertiliser, compared with that of a traditional *indica* type rice, Peta. Results are the average for the five years, 1968–73. (Source: De Datta and Malabuyoc, 1976.)

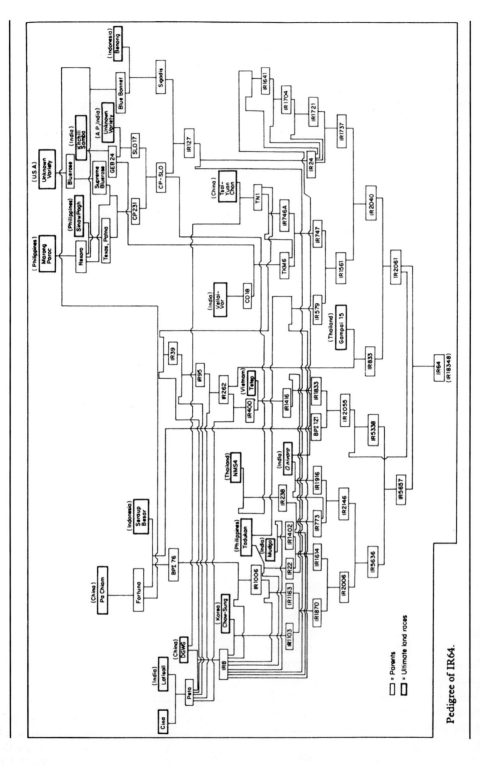

Fig. 7.14 Pedigree of IR64. (Source: Swaminathan, 1985.)

short stiff-strawed varieties had a much higher harvest index than the traditional tropical rices, putting a much higher proportion of photosynthate into the grain than into leaf and stem production. Thus even when grown under average fertility conditions, they yielded more than traditional rices and, if given additional nitrogen, they yielded spectacularly more. Hence IR8 was dubbed 'miracle rice' by some journalists. It had, however, several deficient features, notably poor pest resistance. This was remedied in varieties released subsequently from the IRRI in the Philippines, of which IR20 was the first to be widely grown. Grain quality was improved in later varieties such as IR36 which became the most widely grown variety of any species. The wide popularity of the IRRI rices raised concern about a narrowing of the genetic base. The conscious effort made to ensure that there is a wide genetic base is illustrated by the pedigree of IR64, first released in 1985 (Fig. 7.14). With more than 80 000 accessions available in the world rice germplasm collection at the IRRI, there is ample material available to assist plant breeders to overcome most threats from the wide spectrum of rice pests.

The main challenge for the future of rice farming is to raise the yield potential of the crop, given that the population of Asia and the demand for rice is likely to be twice what it was in the late 1980s by the year 2020 (IRRI, 1989). The extra production has to be achieved with little new land or water reserves available. One possibility for doing so arises from the discovery in China of male sterile lines of rice (Shen, 1980). This discovery opens the way to the development of hybrids for use in the tropics, and progress is being made towards wider use of hybrid seed (IRRI, 1988b). However, there is still much to be gained from better soil, water and pest management, and continued development of better plant types for the many environments in which rice is grown.

FURTHER READING

Catling, D. (1993) *Rice in Deep water*. Macmillan: London.

De Datta, S.K. (1981) *Principles and Practices of Rice Production*. John Wiley & Sons: New York, NY.

Greenland, D.J. (1997) *The Sustainability of Rice Farming*. CAB International: Wallingford.

Grist, D.H. (1981) *Rice*, 6th edn. Longmans: London.

Gupta, P.C. and O'Toole, J.C. (1986) *Upland Rice: a Global Parspective*. International Rice Research Institute: Los Banos, Philippines.

Khush, G.S. and Toenniessen, G.H. (1991) *Rice Biotechnology*. CAB International: Wallingford.

Ou, J.H. (1985) *Rice Diseases*, 2nd edn. CAB International: Wallingford.

Yoshida, S. (1981) *Fundamentals of Rice Crop Science*. International Rice Research Institute: Los Banos, Philippines.

8
Plantation Crops

W. Stephens, A.P. Hamilton and M.K.V. Carr

OVERVIEW OF PLANTATION AGRICULTURE

Introduction

Since the 1960s there have been fundamental demographic, social, economic, political and technological changes in the tropics, all of which impinge on existing agricultural systems. The plantation sector, like other agricultural industries, has had to change and to develop in order to remain viable. Included in these transformations are many technical developments introduced to improve productivity whilst maintaining product quality. However, they would not have been developed or implemented without concurrent changes in management capabilities and methods. The aim of this chapter is, therefore, not to provide a detailed description of how to grow and process plantation crops in various parts of the world, but to discuss recent trends in plantation agriculture, highlighting the key technical and management factors affecting the performance of the sector.

The chapter first defines plantations and presents an overview of their importance in world agriculture. There is then a section providing a framework for considering the potential yield of plantation crops. Two case studies follow on contrasting crops: tea and oil palm. Finally, some general conclusions are drawn as to the future prospects for plantation agriculture.

Definitions of plantations

There are many definitions of plantations some of which have outdated connotations of oppressive regimes involving slavery and indentured labour. Although the early plantations in America and elsewhere were often of this type, modern plantations are now commercial enterprises that compete for labour and capital with other industries on the open market.

Plantations are normally defined as large farms that specialise in monocropping perennial crops. Ruthenburg (1980) goes further and distinguishes three main cropping systems on the basis of the length of the cultivation cycle and the amount of cultivation involved. These are:

- *Perennial field crops* e.g. sisal, sugar cane, bananas and pineapples.
- *Shrub crops* e.g. tea and coffee.
- *Tree crops* e.g. rubber, cocoa, coconut and oil palm.

Tiffen and Mortimore (1990) characterised plantations in more detail as areas that are typically monocropped with perennials, producing tropical or subtropical products that commonly require prompt initial processing and for which there is an export market. Plantation production systems generally require large amounts of fixed capital investment on planting material, processing/packaging equipment and infrastructure such as roads and housing. They normally generate some activity during most of the year and therefore tend to have a large permanent labour force.

These characteristics mean that there is limited opportunity for rapid changes in either product or process and plantation organisations are therefore vulnerable to fluctuations in the cost of inputs and the price of the commodity.

The importance of plantation crops in world agriculture

Plantation crops play an extremely important role in world agriculture for a number of reasons. The nature of the crops and the intensive management required for maintenance and harvesting mean that plantations are important providers of employment in the tropics. As perennial crops, the ground cover they provide controls soil erosion in the high rainfall areas in which many are grown. The produce from plantations generates valuable foreign exchange and helps to stimulate the local economy as well as allowing developing countries to participate in international trade by exploiting their competitive advantages of a beneficial climate and cheap labour.

In the early 1950s, coffee was the most important plantation crop in financial terms with the annual worldwide production of dry beans averaging about 2.5 Mt compared with just under 2 Mt of natural rubber and less than 1 Mt each of palm oil, black tea and cocoa. Over the following ten years, the increases in production of coffee and cocoa were driven by the development of new plantations. The area planted to coffee increased from 6 to over 10 Mha; yet the area under cocoa rose by only 0.8 Mha to 4.3 Mha. By contrast, during this period, the area of tea remained virtually static at around 1.3 Mha and the increases in production came mainly from greater yields.

The 1960s were characterised by wide fluctuations in coffee production, mirrored to a lesser extent by cocoa. From 1965 onwards, the area of oil palm began to expand rapidly and, since the early 1980s, the production of palm oil has grown at a rate of over 0.8 Mt per annum, to 17 Mt in 1996, almost three times that (in terms of weight) of any other plantation crop (Fig. 8.1).

Production of cocoa and tea have increased at average rates of 38 000 and 47 000 t per annum, compared with 85 000 and 88 000 t per annum for coffee and rubber respectively. By the 1990s, the annual production of rubber, cocoa, coffee and tea were each three to four times those achieved in the 1950s.

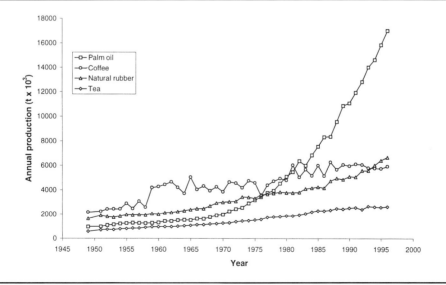

Fig. 8.1 World production trends of mayor plantation crops: 1949–96. (Source: FAO, 1997.)

By comparison with major grain crops such as wheat, maize and rice, the weight of plantation crops harvested is small. However, since the value of the produce per unit weight is up to ten times that of an arable crop the monetary values are more comparable. For example, in terms of gross value, coffee is second only to crude oil in internationally traded commodities.

The extent to which these increases in production can continue depends on the rate of increase in planted area and the development of yield. With available land resources restricted, the emphasis will inevitably turn to improving the yield per unit area of plantation crops, and reducing the unit cost of production, whilst maintaining quality.

Yield potential of plantation crops

Introduction

In order to understand the extent to which yields can be increased above present levels, or to identify current constraints to production, it is helpful to estimate the potential yield of a crop for any given environment. The capacity for further improvements in yield can then be determined and limiting factors identified. This section outlines a rational method for determining potential yield and presents examples for some of the major plantation crops.

The yield of any crop (Y) can be considered as follows:

$$Y = S \times f \times e \times HI$$

where S is the total solar energy received at the surface of the crop; f is the fraction of the energy intercepted by the leaf canopy; e is the conversion ratio of solar radiation to dry matter (sometimes called the conversion efficiency); and HI is the ratio of energy in the economic product to the total energy fixed by the crop (in non oil-bearing crops the dry matter ratio is often used instead).

Incident solar radiation (S)

The potential total solar radiation varies with latitude and is greatest at the equator. However, cloud cover can restrict the incoming radiation so areas with very high rainfall generally have lower annual incident radiation totals. Typical values range from around 55 TJ/ha in the high rainfall tea growing areas of Assam and Bangladesh, to 63 TJ/ha for oil palm growing areas of Malaysia and up to 70 TJ/ha for some areas of eastern Africa with clear skies during the long dry seasons.

Fractional interception (f)

The potential production of most perennial plantation crops is greater than for annual cropping systems because the growing season can last for 12 months a year and a complete leaf canopy is usually maintained throughout the year. This enables greater interception of solar radiation, and hence production of dry matter, than annual crops.

The leaf area required to intercept a given proportion of incoming solar radiation depends largely on the canopy geometry of the crop. Those with erect leaves held in clumps require a larger leaf area index (LAI), defined as the leaf area per unit of ground area, to intercept the same amount of radiation than crops with horizontal, uniformly spaced leaves. Thus palms, such as oil palm or coconut, require a LAI of almost 7 to achieve 95% interception of solar radiation whereas crops such as cocoa and tea will intercept the same proportion with a LAI of 4.

The largest overall yields tend to be achieved by planting systems that maximise the interception over the whole life of the plantation including an immature period which is normally between 1.5 and 5 years from planting, depending on the crop. In oil palm, the maximum leaf area is achieved when the palms are 8–10 years old. For a plantation intercepting 96% of incoming sunshine at this stage, the average proportion of radiation intercepted over 25 years (until the palms are too tall to harvest and are felled) would be less than 88%. By contrast, well-managed tea plantations in East Africa can achieve 95% light interception within 3 years of planting but are artificially defoliated every 3 or 4 years for a period of about 3 months when the bushes are pruned. The average light interception over a similar period is thus very similar to that of oil palm.

Where crops such as tea and cocoa are grown under shade, the proportion of light intercepted by the shade cover must also be taken into account. For instance, in cocoa grown under coconuts at 120 palms/ha, the cocoa may only receive 50% of the total incident radiation.

Solar radiation conversion ratio (e)

Although the interception of radiation is greater in perennials than in annuals, the solar radiation conversion ratio, e, tends to be considerably lower. For example, many temperate annual crops have conversion ratios close to 1.4 g/MJ compared to 0.8 g/MJ for oil palm, 0.6 g/MJ for coconut and as low as 0.2–0.4 g/MJ for tea.

One factor reducing e is the photosynthetic response of plantation crops to increasing solar radiation. Typically these crops become light saturated at only about 30% of full sunlight. By comparison, tropical C4 crops such as sugar cane and maize continue to increase photosynthetic rates up to full sunlight. However, when crops are grown under shade they utilise the available light more efficiently. For example, the conversion ratio of shaded cocoa has been reported to be as high as 3.0 g/MJ.

The radiation conversion ratio is also dependent on a number of other environmental factors including temperature, water stress and the nutrient status of the soil. The optimum temperature for photosynthesis in oil palm, cocoa, tea and coffee is between 30 and 35°C. In the lowland tropics, leaves exposed at the top of the canopy can experience temperatures in excess of 40°C which greatly reduces the efficiency of photosynthesis. In addition, stomatal closure, caused by low atmospheric humidity or soil water stress, reduces the cooling effect of transpiration and also leads to increased leaf temperatures.

The amount of dry matter produced per unit of intercepted radiation depends on the balance between photosynthesis and respiration. In perennials such as oil palm, as much as 75% of the gross photosynthetic production may be respired to maintain the large standing biomass of the crop, another reason for the apparently low solar radiation conversion ratios achieved.

Harvest index (HI)

In annual cereals such as wheat and rice, the proportion of energy partitioned to the grain can be up to 50% in well-managed crops. The range of harvest indices for tropical plantation crops is very wide, from about 20% in cocoa up to 57% of total energy fixed in oil palm, and depends on the harvested product (Table 8.1). In tea, the maximum values for HI reported in the scientific literature are less than 25%. However, the

Table 8.1 Theoretical potential dry matter production*, harvest index and yield, and record yields for some tropical perennial plantation crops. (Source [except tea]: Corley, 1983.)

Crop	Botanical product	Total dry matter production (t/ha/yr)	Potential harvest index		Potential yield of economic product (t/ha/yr)	Record yield (t/ha/yr)
			% of total dry matter	% of total energy fixed		
Cocoa	Seed	56	20	30	11	6
Coconut	Kernel	51	26	38	13	6
Rubber	Latex	46	34	52	15	5
Oil palm	Oil	44	40	57	17	10
Tea	Leaves	30†	40	40	12	11

* Corley's calculations assume 63 TJ/ha incoming solar radiation which is an average value for Peninsular Malaysia.
† Estimated using a solar radiation conversion ratio of 0.5 g/MJ.

recorded yields in commercial fields suggest that, for the best clones grown in an ideal environment, the *HI* should be closer to 40%.

Potential yield

With the exception of tea, the theoretical potential dry matter production of the major perennial plantation crops (calculated using potential solar radiation receipts and conversion ratios) is in the range 44–56 t/ha/yr (Table 8.1). The potential economic yields, calculated using the maximum recorded *HI*, are in a narrower range of 11–17 t/ha/yr. Oil palm and rubber have the greatest yields especially considering the high energy value of the harvested produce. By contrast, the potential yields of coconut, cocoa and tea are achieved through different mechanisms since coconut and cocoa both have higher conversion efficiencies but lower harvest indices than tea.

Actual yield

Corley's (1983) estimates of potential yields are still well above the record actual yields achieved on a field basis. Comparisons between crops are, however, difficult because of the wide range of economic products.

The average yields achieved for the various plantation crops are much lower than the potential and record yields (Table 8.2) and are largely limited by the intensity of management.

Smallholder yields tend to be less than 75% and as little as 16% of estate yields in the same country. There are exceptions, however, where smallholders have access to modern clones, technologies and finance. For example, smallholders with the Federal Land Development Authority (FELDA) in Malaysia have achieved rubber yields of 2.4 t/ha/yr by using the best available clones and modern latex stimulation techniques.

Tiffen and Mortimore (1990) explained the observed differences in yields by examining the ratio of capital to land in the enterprise. Where there was sufficient capital to ensure good management, appropriate research and development, and the application of the latest technology, then yields could be expected to be higher than on estates or smallholdings with a lower capital:land ratio. The organisational structure of plantations is, therefore, very important in determining whether or not they are successful.

Table 8.2 Typical yields on estates and smallholdings. (Source: Tiffen and Mortimore, 1990.)

Crop	Economic product	Country	Year	Estates (t/ha/yr)	Smallholdings (t/ha/yr)
Cocoa	Dried seed	Papua New Guinea	1984	0.4	0.3
Coconut	Dried kernel	Papua New Guinea	1984	0.9	0.5
Coffee	Dried seed	Papua New Guinea	1984	2.0	0.7
		Kenya	1984	1.4	0.5
Palm fruit	Fresh fruit	Papua New Guinea	1984	21.5	11.9
		Côte d'Ivoire	1980s	10.2	5.8
		Malaysia	1980s	24.1	15.7
Rubber	Dried latex	Sri Lanka	c.1984	1.0	0.6
		Papua New Guinea	1984	0.6	0.4
		Peninsular Malaysia	1980	1.3	0.9
		Indonesia	1978	0.6	0.4
Tea	Dried leaves	Sri Lanka	c.1984	0.9	0.7
		Kenya	1990	3.5	1.4
		Tanzania	1990	3.0	0.5

Plantation organisation

Organisational structure

Just as a consistent approach can be adopted to compare the potential yields of widely different plantation crops, so plantation organisations can also be compared regardless of the crop being grown. The nature of plantation crops has, in many ways, determined the management method (Goldthorpe, 1994).

Large areas planted to a single crop and maintained and harvested by a large workforce tend to foster a very bureaucratic management style with intense supervision to ensure that the same operations are carried out uniformly throughout the estate. In the past, decision-making power was vested in the hands of a few senior managers who were often not even resident on the estate. In the extreme case, orders would be passed on from headquarters in Europe for implementation by the field staff.

The earliest plantations often used coercive power on the labour force through the use of indentured, contract or even slave labour. Modern estates exert a more benevolent authority with increasing consultation of the workforce and greater attempts to involve more junior management in the decision-making process. This evolution has been facilitated by the development of plantation workers' unions, the increasing levels of education amongst the workforce, the employment of qualified managers and the gradual adoption of modern business practices.

Tiffen and Mortimore (1990) contended that the advantages of estates as a form of production were not in the economies of scale for agricultural operations, but were due to managerial efficiency combined with adequate capitalisation. This can result in a suitable level of research funding, the rapid adoption of new technology, responsiveness to changes in market conditions, attention to product quality, proper maintenance of replanting programmes and the maximum use of the land resource. However, where management is inefficient the estate loses much of its economic advantage over smallholdings.

Unsurprisingly, the importance of smallholder production has increased over the last 30 years in most countries. However, the nature of plantation agriculture, with the requirement for rapid processing of produce using capital intensive equipment means that, with the exception of some crops such as coffee, smallholders are unable to process their own produce. As a result therefore, a commercial or parastatal organisation is often involved in the manufacture and marketing of this produce. For commercial companies, the attraction is a reduction in the capital investment in land although small 'nucleus' estates are often planted immediately around the factory. These organisations operate with varying degrees of smallholder participation in operational and strategic decision making.

Where these organisations are themselves efficient, smallholders can have confidence in receiving good support services in terms of extension advice, extended credit for agronomic inputs, transport for produce and, perhaps most importantly, timely payment. However, smallholders are naturally risk averse and will rapidly lose confidence in the plantation crop if the support services falter, concentrating instead on other aspects of their agricultural enterprise. In these circumstances, the profitability of the parastatal organisation can quickly be jeopardised as supplies of the raw material decline and production fails.

In summary, the organisational structure of the plantation industry covers the whole range from autocratic bureaucracy on some large estates through to loose democratic alliances of smallholders. The success of any of the these organisational structures depends ultimately on the acumen of management and its ability to respond to changes in the external environment.

External factors affecting plantation agriculture

A large number of external factors have an important bearing on the success or failure of plantation companies. Many of these arise from the increasing integration of plantation agriculture into the local economy. In the early days, plantations were outposts, often surrounded by jungle, that had to survive autonomously. In order to maintain a steady supply of labour, outsiders were often brought in who had to be housed and fed. The children needed schools and the workforce as a whole required medical care. In addition, the infrastructure of the estate was owned and maintained by the company so that the estate manager or proprietor was responsible for the well-being of the whole community. Once harvested and processed the produce from the estate was sent to auction, usually in Europe. Thus the main external factor was the price received for the product.

In most countries, with the exception perhaps of some areas of Papua New Guinea, Kalimantan and the Democratic Republic of the Congo (formerly Zaire), the estates are now part of a local community. The workers often have a choice of working in local industry and thus comparability of wages becomes an issue. As educational standards improve, workers wish to be more closely involved with the companies and managers must then develop negotiating, team management and leadership skills.

With the advent of satellite technology, communications now allow rapid transfer of information and decision making. This means that managers are increasingly aware of market demands and also of their accountability to shareholders. They therefore need improved financial management capabilities in order to meet these wide responsibilities. In short, the external environment is undergoing dramatic and irreversible change to which the manager must be ready and able to respond.

CASE STUDIES

Tea (*Camellia sinensis* L.)

Introduction

History Tea is thought to have originated from a region bordering Assam, north Burma, southwest China and Tibet and was brought into cultivation by the Chinese in about 2700 BC. The earliest remaining authentic account of tea, the Cha Ching or tea book, was written in 780 AD by Lo Yu and described the manufacture of tea (Weatherstone, 1992).

The first tea to reach Europe was brought by Dutch traders in the early part of the seventeenth century and from there it reached England and was first available to the public in 1657 in the coffee houses of London. For the next 200 years China dominated the export trade. In order to develop other sources of supply, the British attempted to establish plantations of tea in north India using seed imported illicitly from China. However, at about the same time wild tea plants were discovered in the Assamese jungles and these were soon adopted as they grew better than the imported Chinese plants. The original plants planted by Bruce in 1838 are still growing today, more than 150 years later.

The first tea estates in Sri Lanka (then Ceylon) were established in 1825, nine years before the start of the Indian tea industry. Tea cultivation spread rapidly as a replacement for coffee that had been devastated by coffee leaf rust (*Hemileia vastatrix*) and by 1900 there were over 150 000 ha of tea. The tea industry also had an early start in Java but did not begin to develop rapidly there until the 1890s. By the beginning of the twentieth century, exports from Java were only 10% of those from India, China or Sri Lanka.

Tea rapidly became the British national drink leading Gladstone, the Prime Minister in 1865, to remark:

> 'If you are cold, tea will warm you; if you are too heated, it will cool you; if you are depressed, it will cheer you; if you are excited, it will calm you.'

By 1938, Britain was importing 50% of the world exports in tea but by 1988 the proportion had declined to only 17% (van de Meerburg, 1992). The international significance of tea as a plantation crop is now surpassed by coffee and cocoa but it is still a crop of major economic importance in many countries.

Production methods The following section gives only a very brief introduction to production methods used in tea. For more detailed information the reader is referred to Willson and Clifford (1992).

Tea is an evergreen tree or shrub that is generally cultivated to maintain an even flat canopy or 'plucking table'. Nowadays, it is almost entirely propagated vegetatively and planted out at densities of 10 000–18 000 plants/ha after 12 to 18 months in a nursery. Under good management, a complete crop ground cover should have developed within 3 years in all but the coolest areas after which there is little further risk of soil erosion (Fig. 8.2).

Tea is harvested by removing young shoots with two or three leaves and an apical bud that protrude above the canopy. The harvest interval varies with location and season, from once a week during the warm season in low-lying areas to only two or three times a year in the seasonal climates of Japan and Turkey, and is primarily dependent on temperature. The harvested shoots are then transported to factories to produce black (fermented) or green (unfermented) tea.

The tea canopy increases in height over time so that it eventually becomes increasingly difficult to harvest by hand. At this stage the tea is pruned to a more manageable height and the process begins again. In eastern Africa, tea is pruned every 3–5 years but in some areas of Asia, such as north India and Bangladesh, there is a more complicated system of bush management which involves the annual removal of a proportion of the canopy in a process known as 'skiffing'.

The quality of the tea, assessed subjectively in terms of appearance and taste, depends on the environment in which it is grown, the number of leaves harvested on each shoot, and the manufacturing process. The value of most tea on the market is largely determined by its area of origin with all prices falling in a relatively narrow price band. However, the price obtained for top quality teas in areas such as Assam and Darjeerling, in India and Uva in Sri Lanka can be more than ten times the regional average.

Fig. 8.2 View of a typical tea estate in Sri Lanka.

Areas of production

Tea has the widest geographical distribution of any plantation crop with production from 33°S in Natal, South Africa to 49°N in Georgia. It is also cultivated at a wide range of altitudes from sea level in Bangladesh to as high as 2600 m above sea level in Kenya.

The production of tea worldwide has steadily increased from just under 0.6 Mt in 1949–50 to almost 2.6 Mt in 1993 (Fig. 8.3). This has been predominantly as a result of the increase in planted area from 1 to 2.6 Mha. The two main periods of expansion were up until 1960 and then an accelerating trend from 1965 to 1980. However, since 1980 the area of tea in production has been fairly static despite large fluctuations in the reported area, due in part to civil or economic strife in some countries.

By contrast, yields increased steadily until the mid-1960s but then stagnated or even declined until the early 1980s. Since then the upward trend has started again and yields in 1993 were 17% greater than in 1983, an average increase of 20 kg/ha/yr.

In 1996, 22 countries produced more than 6000 t/yr but over half the total annual production was in India and China with a further 30% coming from Sri Lanka, Kenya, Indonesia and Turkey (Table 8.3). The most dramatic production increases in recent years have come from Kenya and Turkey, largely through the development of large smallholder sectors. Whilst Kenya is considered to have a favourable climate for tea (see later) and tea is harvested all year round, the same cannot be said for Turkey which is towards the northern extremity of the tea growing areas with a distinctly temperate climate and two or three harvests only during May to September.

Climatic requirements

The wide variety of locations where tea is grown indicates its tolerance to a broad range of climates. Laycock (1964) stated that a 'good growing season' for tea has 'warm days, long sunshine hours, high humidity and adequate rainfall, preferably in overnight showers'. More recent research on the environmental physiology of tea has allowed more precise definitions of the limits to growth, which are reviewed in detail by Carr and Stephens (1992).

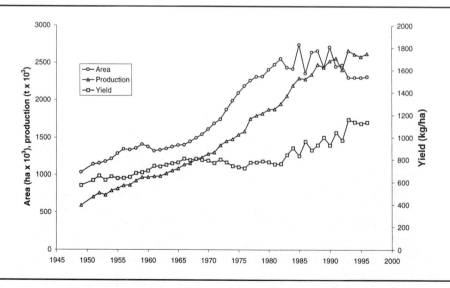

Fig. 8.3 World trends in area, production and yield of tea: 1949–96. (Source: FAO, 1997.)

Table 8.3 Estimated world production of tea by country in 1996. (Source: FAO, 1997.)

Country	Estimated production (t × 10³)	Proportion of world production (%)
India	715	27.2
China	609	23.2
Kenya	255	9.7
Sri Lanka	246	9.4
Indonesia	169	6.4
Turkey	124	4.7
Japan	90	3.4
Iran	56	2.1
Bangladesh	48	1.8
Vietnam	40	1.5
Argentina	40	1.5
Malawi	37	1.4
Georgia	34	1.3
Tanzania	23	0.9
Uganda	17	0.6
Myanmar	16	0.6
Zimbabwe	15	0.6
South Africa	12	0.5
Brazil	10	0.4
Papua New Guinea	9	0.3
Malaysia	6	0.2
Rwanda	6	0.2
Burundi	5	0.2
Thailand	5	0.2
Cameroon	4	0.2
Azerbaijan	4	0.2
Other African countries	11	0.4
Other South American countries	8	0.3
Other Asian countries	7	0.3
Total	2624	100.0

Temperature The minimum (base) air temperature for shoot extension and development varies from clone to clone in the range 8–15°C. Above the base temperature, the rates of extension and development are linear up to a maximum of about 30–35°C. The time taken for a new shoot to grow to a harvestable size of three leaves and a bud during the warmest part of the year varies from about 120 days at 2200 m above sea level in Kenya (15°C) to 70 days in the Southern Highlands of Tanzania (18°C), 42 days in Malawi (23°C) and as little as 30 days in Assam and Bangladesh (26–29°C). Through this effect on shoot growth and development, temperature has a direct effect on the seasonal yield of tea throughout the world.

Soil temperatures (measured beneath grass) below 20°C or above 25°C can also be important in controlling the rate of shoot growth and hence the yield of tea (Carr and Stephens, 1992).

Saturation deficit The effects of saturation deficits (a measure of the dryness of the air) are closely linked to the adverse effects of high temperatures. Tea is generally produced in the humid areas of the world but, in areas with distinct warm dry seasons such as Malawi, the saturation deficit can be large enough (>2.3 kPa) to cause stomatal closure and reduced shoot extension during the hottest part of the day. In these areas, responses to irrigation can be less than expected since the principal factor limiting yields is not the amount of water in the soil but the dryness of the air during the middle of the day.

Solar radiation Healthy tea, with a relatively flat canopy completely shading the ground, has a leaf area index of about 6 and intercepts 95% of the incident solar radiation in the top 0.3 m. The incident flux of solar radiation at high altitudes near the equator can exceed 1000 W/m². However, photosynthesis of single tea leaves and the tea canopy as a whole become light saturated at about 350 and 750 W/m² respectively.

Less than 3% of the available energy is actually used for photosynthesis and the rest is dissipated through latent heat (evaporation) and as sensible heat (heating the tea bush and the surrounding air). In the hot production areas leaf temperatures can exceed air temperatures by about 10°C leading to supra-optimal temperatures for photosynthesis and growth. High leaf temperatures also result in greater leaf-to-air saturation deficits leading to further reductions in shoot extension.

In order to counter these effects, shade is widely used in Assam and Bangladesh to reduce the level of incident radiation and hence the heat load on leaves. By contrast, in most of Africa, where daily maximum temperatures rarely exceed 30°C, shade is unnecessary and usually leads to a reduction in the yield of tea.

In tropical tea growing areas, the annual variation in day length varies from a few minutes at the equator up to about 3 h 20 min at 26°N in Assam. Experiments in Assam have shown that increasing the day length resulted in greater crop yields. However, shoot extension rates are also restricted by low temperatures at the same time (see earlier) which confounds the effects of day length and can explain the majority of the seasonal yield differences. There may, however, be some effect of day length in synchronising shoot growth during short days leading to a larger peak when temperatures rise again at the end of the cool season (Matthews and Stephens, 1998a).

Water availability Tea is grown in areas with annual rainfall totals from less than 700 mm in Zimbabwe (where irrigation is essential) up to more than 5000 mm in parts of Sri Lanka (Carr and Stephens, 1992). Whether the rainfall in an area is adequate for high yields of tea depends on the seasonal distribution of rainfall and the potential evaporation rate. The rainfall pattern is dependent on latitude but is modified by altitude and topography. Close to the equator there are normally two rainy seasons. At higher latitudes there is one rainy or monsoon season and one longer dry season that can last for 6 months in Tanzania and Malawi. The potential evapotranspiration rate of tea is between 90 and 180 mm per month in most areas of the tropics, depending on location and season. The ideal rainfall distribution would therefore be sufficient to meet the evapotranspiration demand, preferably falling at night to allow maximum solar radiation interception during the clear days.

Where the rainfall is less than the evapotranspiration demand, the balance of the water must be met from storage in the soil if the tea is not to suffer from water stress and hence reduced growth and yield. The amount of available water stored in the soil depends on the texture and structure of the soil and also on the rooting depth of the plants. For a 1 m depth of soil, the water held can vary between about 80 mm in sandy soils and up to 220 mm in silty clays and clays.

The depth achieved by roots also depends strongly on the soil profile and on structural restrictions such as compacted layers and rocks. In addition, seasonally high water tables, such as those experienced in the low flats in Bangladesh and Assam, kill roots and restrict the root depth to less than 1 m in many tea gardens. Here the amount of water available to the plant from the root zone is about 150 mm. At the opposite end of the range, tea roots have been recorded at depths of 6 m in Tanzania and Kenya with available water capacities of 420 and 1200 mm respectively.

The capacity of the plant to withstand dry periods is therefore very variable depending on the amount of stored water and the evaporative demand from the atmosphere. However, when as little as 10% of the available water has been used, shoot extension rates can be restricted and yields reduced. The yield loss reported from irrigation experiments in Tanzania ranged from 1 to 4 kg/ha of made tea (i.e. manufactured at 3% moisture content) for each 1 mm reduction in water used by the plant (Stephens and Carr, 1991). Since the potential soil water deficit at the end of the dry season can be as large as 800 mm, yield losses can exceed 3 t/ha. Even in areas such as Kericho, Kenya, where the dry season is rarely longer than 2 or 3 months, the average annual yield loss due to drought is 20% (Othieno et al., 1992).

Site requirements

It is difficult to be specific about the exact site requirements of tea since the wide geographical distribution of tea growing areas also encompasses great diversity in topography and soils. In most regions, the climatic requirement for high rainfall means that tea is grown in hilly areas on slopes, in some cases, up to 30%. The soils in these regions are also very variable ranging from very deep volcanic clays in Kenya to shallow, stony soils in mountainous areas such as Darjeeling. Tea will grow in acid soils with pH values as low as 4 though values in the range pH 5–5.6 are considered ideal. Soils with pH values above 6.5 are normally regarded as unsuitable for tea.

At a smaller scale, the aspect and slope of a field modifies the microclimate and thus the growth and yield of tea. These factors also affect the ease of harvesting and transport of leaf and therefore the costs of production. Similarly, the soil texture, structure and depth determines the water-holding capacity and therefore the capacity of the tea plants to withstand long dry periods.

This diversity has been possible because of the manual nature of tea production yet, as pressure to intensify production increases, producers on flatter land with better soils will have a distinct advantage over those operating on more difficult terrain.

Yield and price trends

From the time when tea was harvested from semi-wild 'tracts' in the Assamese jungle to the modern day intensive tea plantation there have been considerable improvements in yield in many regions. Since the mid-1950s the yield increases achieved from the same bushes in some East African countries are particularly impressive. The best estates in Tanzania have increased mean yields over a four-year pruning cycle from just over 0.5 t/ha/yr in 1955 to around 3.7 t/ha/yr in 1995; experimental yields in the same area suggest that 6–7 t/ha is a realistic commercial target. Similarly, estate yields in Kenya have risen from 1 to over 4 t/ha/yr over the same period and are now more than four times the global average yield. Kenya also lays claim to the largest recorded commercial yield from a single field of 11 t/ha, very close to the theoretical potential yield (Table 8.1). By contrast, in other estates and, more generally in countries such as Sri Lanka, yields have changed little or in some cases declined since the 1970s in line with declining investment and poor management.

In the smallholder sector, there have also been yield increases in some countries, particularly in Kenya, but these have not been as large as in the estates. These increases, or lack of them, are not due to climate change or other outside influences but to the intensity of management effort.

Changes in technology and management

The following section outlines some of the major developments that have occurred in the technology and management of tea. The changes highlighted have by no means happened uniformly across the tea industry and pockets of inertia or, conversely, innovative excellence occur in all countries.

Technology Intensification of field management has led to a number of fundamental developments, some of which have only been made possible by transformation outside the industry. Tea fields are now generally planted at greater densities than before, with clonal or occasionally composite plants. (Composite plants have high quality or large yielding scions grafted onto drought- or disease-resistant rootstocks in the nursery.)

In many areas, manual weeding was replaced by herbicides in the early 1960s allowing improved root development and nutrient extraction from the upper, organically richer, soil horizons. Where the pruned wood has been allowed to remain in the fields, the soil structure and nutrient status have tended to improve over time as a result of the organic matter input. However, in many countries, especially in Asia, the prunings are traditionally taken by the workers for firewood.

In most areas, fertiliser applications have increased since the mid-1950s and are, in many cases, now based on rational plans to replace the nutrients stored in the standing biomass or removed in the harvested crop. In India, Bangladesh and Sri Lanka there are moves away from blanket spraying of insecticides towards integrated pest management using more specific agrochemicals and, where possible, biological control methods. This trend is likely to continue as international requirements on chemical residues in food products are tightened. In Africa, pests and diseases are far less of a problem than in Asia at present, but this may change with time. However, agrochemicals are a major production cost in all tea growing countries, which has led to some companies producing at least some 'organic' tea, with no agrochemical inputs, in an attempt to improve profitability.

The other major input cost in areas with a protracted dry season is irrigation, which has gained in importance in countries such as Tanzania where initial research proved its worth in the late 1960s. Research in the late 1990s in these areas has concentrated on making the most effective use of the irrigation water to reduce pumping and storage costs. The arguments for shade and shelter have already been presented but management opinion on this subject tends to be cyclical.

The vast majority of tea is still harvested by hand but the labour requirements (0.2–0.8 ha per worker) are greater than for any of the other major plantation crops (see earlier) and there is continual pressure to reduce production costs by improving the productivity of the workforce. In addition, as education standards rise and opportunities for work in other industries develop, it is increasingly difficult to employ sufficient workers to harvest at the correct intervals.

One response has been to mechanise harvesting, though the extent depends on a number of factors such as topography, local labour availability and cost, and also union representation and government policies. Large tracked or wheeled machines capable of harvesting up to 10 ha/d have been designed and used in areas such as Uganda, where civil war in the 1970s resulted in the largely Rwandan labour force returning home, and in Argentina, Papua New Guinea and Australia, where wage rates are too high for hand 'plucking' to be economically viable. These machines require high levels of mechanical support and, unless designed and operated correctly, can cause additional problems associated with soil compaction and poor harvested leaf quality.

An increasing number of single- or double-handed, engine-powered harvesters that cover one or two rows of tea are used in Japan though their durability with more frequent use and under harsher operating conditions elsewhere in the world is not yet fully known. Finally, there is a range of semi-mechanical harvesters such as shears and knives that have been designed to increase plucker productivity (Fig. 8.4). These are used routinely in some areas of south India and Africa for example but, without training and adequate supervision, the yield and quality of harvested leaf can be adversely affected.

One key management task is to ensure that the fields are harvested when the shoots are at the correct stage to ensure the optimum quantity of the required quality leaves. If the harvest is delayed then the yield increases, but at the expense of quality and hence price. In the past this decision has relied on the skill and experience of the manager. However, one recent development has been to relate the harvest interval to temperature to ensure that the harvested leaves are at the same stage of development despite fluctuating temperatures. There is considerable potential in this approach which has now been incorporated into a physiologically based model of shoot growth and development that will predict yield and its distribution through the year (Matthews and Stephens, 1998b).

Other ways of increasing the productivity of field workers have included replacing the tradition 'plucking basket', used to carry up to 15 kg of harvested leaf, with bags that are dragged across the top of the bushes and the provision of more frequent weighing points to reduce the amount of time spent walking to and from the field. In addition, bulk handling of leaf is increasing coupled with measures aimed at minimising the time taken for harvested shoots to get to the factory for processing.

In the factory, the main technological developments have been aimed at reducing labour and energy requirements, whilst increasing factory throughput and the degree of control over the manufacturing process. In addition, some factory managers frequently modify their manufacturing process to produce tea meeting the tight specifications of niche markets.

Fig. 8.4 A worker using shears to harvest tea in Tanzania.

Management The pressures on management in the tea industry are similar to those in other plantation crops with falling prices and rising costs. The effects on organisational structure and management approach have also been similar and are outlined below.

A proportion of the world tea business is still controlled by multi-national companies but this is much less than previously. As in other plantation industries, the trend has been towards private national companies, after periods of nationalisation in countries such as Sri Lanka and Tanzania. There is now also a large smallholder sector in many countries, notably Kenya.

One major development within some companies has been the move from training on the job towards specifically designed technical and management development programmes. These are aimed at producing a competent cadre of indigenous managers to replace expensive expatriate staff. This tends to be accompanied by a reduction in the number of management staff in an effort to reduce fixed costs. However, this trend has not been universal and the tea industry remains one of the most conservative within the plantation sector.

Future prospects

The future of the tea industry globally will largely be determined by the level of consumption in the producing countries. At a national level, government policy is also crucial in sustaining or constraining tea production. An extreme example is that one extra cup of tea drunk per person per day in India could use the entire Indian annual production of tea. Conversely, in the south Indian tea growing region, the viability of the whole industry is dependent on government import restrictions that prevent cheaper tea being imported from Africa and Indonesia to supply the strong local market.

Tea sales in Western Europe are under ever increasing competition from soft drinks and mineral water, despite innovative new products such as canned iced tea. It is, therefore, only by developing local markets that the industry as a whole will avoid the current chronic condition of oversupply. This position could become critical if countries such as China and Vietnam start to manufacture more black tea and sell it on the world market.

Faced with continually falling prices and increasing costs in real terms, the industry can either respond by intensifying production to increase yields still further or by moving to a more extensive approach. The latter is unlikely to be successful in the long term because of the competition for land in the high potential areas where tea is grown. Tea can though be abandoned for many years, most recently for example in Uganda and Mozambique, and then reclaimed when political and other conditions ameliorate.

There will be an ever greater emphasis on breeding for enhanced quality and yield, maximising the use of existing land holding and replanting with improved clones. To control production costs, the mechanisation of harvesting will inevitably increase and the application rates of agrochemicals will be more closely controlled to reduce wastage, and to avoid the risk of residues in the manufactured tea.

Oil palm (*Elaeis guineensis*)

Introduction

History The centre of biological diversity of the oil palm lies in West Africa where it has been cultivated since at least 3000 BC. By the fifteenth century, Portuguese explorers found the use of palm oil food products to be widespread along the Gulf of Guinea (Hartley, 1989).

Production of oil by smallholders in West Africa was encouraged by the British in the mid-nineteenth century to substitute for the loss of revenue from the slave trade which had recently been abolished. The resultant increased production also helped to satisfy the demand for cooking oil, candles, lubricants and soap arising from the industrial revolution in Europe. Estate development was started early in the twentieth century in the former Belgian Congo by British interests when concessions were granted to Sir William Lever. At the same time the Dutch had transferred oil palm to the Dutch East Indies (now Indonesia) and established plantations in Sumatra.

By 1925, the area of oil palm in Sumatra had increased to over 30000 ha and the rapid expansion continued reaching 90000 ha in 1938. By this time 30000 ha had also been planted in Malaysia, far more than the 14000 ha in West and Central Africa. World War II and subsequent internal unrest and economic difficulties in Indonesia disrupted expansion efforts until the early 1960s but after World War II the recovery was much faster in Malaysia where, by the mid-1950s, production of palm oil had exceeded the export peak in the late 1930s. There then followed an extremely rapid expansion in production with a 30-fold increase in Asia from 0.3 Mt in 1963 to just under 13 Mt in 1995. By contrast, production in Africa only increased by 65% over the same time period (Fig. 8.5).

Originally, the main thrust of the Asian expansion was in Malaysia but, although it is still the world's largest producer of palm oil with over 50% of the world production (Table 8.4), the major expansion is now taking place in Indonesia where new estates are being developed in areas such as Sumatra and Kalimantan. Indeed, several famous estates in west Malaysia are being redeveloped for industrial, recreational or residential use. Since 1980 producers in South America (particularly Colombia and Ecuador) and the South Pacific (Papua New Guinea and the Solomon Islands) have also increased production, from 0.2 Mt in 1980 to just over 1.2 Mt in 1996. However, it is unlikely that the domination of palm oil production by Malaysia and Indonesia will be challenged for many years.

A variety of reasons contributed to the shift in export production from Africa to Asia. These include:

- The proximity of Southeast Asian producers to emerging markets in Indonesia, Japan, China, India and Pakistan.
- Substitution of home-grown oil crops such as soya, oil seed rape and olive oil in Europe and the United States which were traditional markets for West African palm oil.

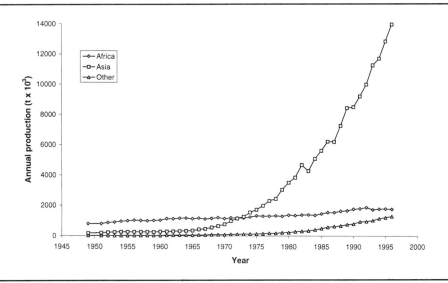

Fig. 8.5 Annual production of palm oil in Asia, Africa and other countries (principally South America): 1949–96. (Source: FAO, 1997.)

- A more conducive political, business and institutional environment for large-scale export oriented investment and diversification in Southeast Asia than in West Africa allowed the efficient exploitation of good prices after World War II, fuelled by the burgeoning demand for oils and fats in overseas, regional and domestic markets.
- Established rubber plantations in Malaysia were easily able to diversify into oil palm.
- Willingness of Southeast Asian countries to invest both public and private funds in large-scale agricultural development.
- Larger yields and greater manpower productivity in Southeast Asia than in Africa.
- Disruption of production in West and Central Africa by internal unrest in the 1960s and 1970s coupled with a rapid population increase in Nigeria leading to an increase in internal consumption and competition for land for food production.

In West Africa the crop remains important to small-scale producers and subsistence farmers who rely on semi-cultivated 'groves' from which they obtain roofing (fronds) and fencing materials (rachis) as well as food (the annual per capita consumption of palm oil is as high as 45 kg compared with less than 1.5 kg in Western Europe) and drink (palm wine).

Production methods Oil palm is normally grown at a density of 120–140 palms/ha, in a triangular arrangement to maximise light interception. However, some recent plantings have been at up to 220 palms/ha in order to increase yields in the early years of the planting cycle, to compensate for slower canopy development on poor soils and, in Africa, to allow for inevitable losses through plant vascular wilt disease. Plantations are generally established with a leguminous cover crop (commonly *Pueraria* spp.) to reduce soil erosion and to provide inputs of nitrogen into the system. This then dies out as the canopy closes. The area round the base of the palms is generally clean weeded in order to facilitate harvest, especially of the loose fruit that detach from the bunch as it ripens and is cut down.

A regular harvest interval of 7–14 days is common with bunch ripeness being judged by the number of loose ripe fruit on a bunch and on

Table 8.4 Estimated world production of palm oil by country in 1996. (Source: FAO, 1997.)

Country	Estimated production (t × 10³)	Proportion of world production (%)
Malaysia	8 385	49.2
Indonesia	4 998	29.3
Nigeria	776	4.6
Colombia	446	2.6
Thailand	400	2.3
Côte d'Ivoire	267	1.6
Papua New Guinea	250	1.5
Ecuador	234	1.4
Congo, Democratic Republic	181	1.1
Cameroon	161	0.9
China	150	0.9
Ghana	100	0.6
Costa Rica	97	0.6
Honduras	76	0.4
Brazil	76	0.4
Philippines	65	0.4
Guinea	55	0.3
Angola	52	0.3
Sierra Leone	45	0.3
Liberia	38	0.2
Peru	28	0.2
Solomon Islands	26	0.2
Other African countries	82	0.5
Other South American countries	60	0.3
Total	17 046	100.0

Fig. 8.6 A bunch of ripening fruit on a young oil palm.

the ground (Fig. 8.6). As the palms grow taller, a chisel and later a sickle on the end of a long bamboo or extendible aluminium pole are used to cut the bunches which are then transported (by hand, wheelbarrow, or mini-tractor) to the access roads for collection. The effective life of the plantation is determined by the height and accessibility of the bunches. After about 25 years, the bunches can no longer be reached for harvest and replanting is then required.

The oil content and composition of the harvested fresh fruit bunches (FFB) deteriorate rapidly once harvested, necessitating an efficient and rapid transport network to deliver the FFB to the mill. In particular a free fatty acid content of less than 5% is required for acceptable quality crude palm oil to be produced. Clean bunches (c. 0.005–0.13% dirt) and low moisture contents (c. 0.1%) are also required by the mill to ensure the production of good quality oil.

Workers in oil palm plantations can look after larger areas than in other perennial plantation crops. Figures reported in the literature range from 3.6 ha per worker in Malaysia up to 7.5 ha per worker in Africa.

Climatic requirements

The oil palm will survive in a range of tropical climates from wet savanna to the humid tropics.

However, the complex physiological process of bunch development, starting from floral initiation some 36 months before harvest, means that climatic stress tends to have a long-term effect on yield. This can be through a reduced ratio of female to male inflorescences (and therefore fewer fruit bunches), fewer fruit on the bunch, and smaller size and oil content of fruit. In addition, the fruit bunches have the lowest priority for assimilate partitioning so that any reduction in net photosynthesis leads to disproportionately smaller yields.

Temperature The optimum temperature for photosynthesis in oil palms is between 30 and 35°C and growth ceases below about 15°C. The best production is achieved in warm (25–35°C), humid conditions with a small diurnal temperature range and minimum temperatures above 22°C. The crop is therefore restricted to the lowland tropics below 500 m in altitude, and generally within about 10° of the equator.

Saturation deficit Transpiration in oil palm is linearly related to the saturation deficit of the air in the range 1–4.5 kPa. By contrast, photosynthesis appears to be unaffected until the saturation deficit is greater than about 2 kPa (Dufrene and Saugier, 1993). Other work has suggested that there is a linear decrease in *canopy* photosynthesis at saturation deficits greater than 0.8 kPa (81% relative humidity at 30°C) indicating that, in dry weather, the production of dry matter may be reduced even in a wet soil.

Solar radiation The yield of oil palm is directly dependent on solar radiation income (see earlier) so, in cloudy high rainfall areas that have less incident radiation, potential yields are less. For the same reason, increases in atmospheric haze due to forest fires or volcanic eruptions also reduce yield.

The planting pattern used for oil palm is therefore designed to minimise the competition for light between palms, whilst maintaining the maximum possible solar radiation interception by the whole crop, especially in the first few years of growth. A high planting density will give early returns at the expense of smaller yields when the canopy closes and interception by individual palms is reduced.

Water availability In oil palm, the effects of water stress are more complicated than in tea since the reproductive system is involved (see above) as well as the direct effects on vegetative development and photosynthesis. Traditionally, an annual rainfall of between 2000 and 4000 mm, evenly distributed through the year, is considered desirable; yet oil palm is grown in Thailand, West Africa and other regions where distinct dry seasons occur and potential evapotranspiration exceeds rainfall for up to 4 months.

The root system of the oil palm is extensive, and may extend horizontally for over 30 m, but the roots rarely penetrate much deeper than 1 m. The total rooted volume of soil under oil palm may therefore be 20 times greater than even deep-rooted tea but the available soil water per unit area tends to be less than for tea. Although good data are sparse, oil palm evapotranspiration rates are probably between 90 and 150 mm per month (similar to tea). Large soil potential water deficits can therefore develop during even short dry seasons.

In these areas, irrigation is required to maintain the continuity of production and, in some areas of West Africa, has been shown to result in large yield responses. Elsewhere, however, it has been difficult to prove any benefits from irrigation because of the short dry seasons and the long delay (up to 3 years) before any yield response occurs. The root system of oil palm also means that irrigation experiments must be carefully designed to ensure the separation of treatments.

Site requirements

Although oil palm is grown on sloping land on contour terraces in many places, it is best suited to flatter areas that allow easy extraction of the fruit bunches after harvest and lower transport costs. As with most crops, the best soils are well structured with a high organic matter

Fig. 8.7 Oil palm nursery in Malaysia.

content, large water-holding capacity and free drainage.

In estates on deep peat soils in Malaysia and Kalimantan, severe problems can occur since the soil does not have sufficient structural strength to support the palms which can fall over, making access for cultural operations and harvesting very difficult. In these areas, high water tables can also cause waterlogging problems and shrinkage of the peat may lead to lower land levels and hence inundation by the sea.

Changes in technology and management

Technology The oil palm industry has generally been characterised by an open-minded attitude to research collaboration and, led by the Palm Oil Research Institute of Malaysia, there has been an accelerating development of specialised technical knowledge. Over the period from 1960 to 1990 this resulted in a steady 3% annual increase in yields in Malaysia.

Gray and Siggs (1994) have listed, in a conference paper, the major technical developments in the oil palm industry since the mid-1950s, the most important of which are detailed below. The impact of these innovations will obviously depend on location specific constraints.

The breeding and use of improved seed including *Tenera* material obtained by crossing the *Pisifera* and *Dura* forms of the palm have played a major role in increasing yields. The *Tenera* form has a larger fruit oil content due to greater oil concentrations and reduced shell thickness and kernel size (Fig. 8.8). Tissue culture techniques are increasingly important in developing uniform clonal planting material and in reducing the time taken to incorporate improved genetic material into the breeding system.

Efforts to improve the economics of oil palm plantations by reducing the time between replanting and the start of bunch production have included underplanting existing plantations with replacement palms before the old palms are felled, and the use of older and therefore larger plants for replanting. These techniques can reduce the yield gap from 2.5 to 1.5 years.

One of the most outstanding innovations in the oil palm industry was the introduction of

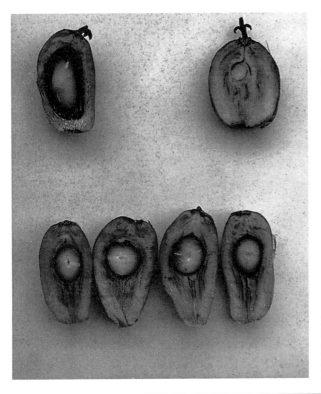

Dura × Pisifera

=

Tenera

Fig. 8.8 Cross-sections of fruit from *Pisifera*, (top left) *Dura* (top right) and *Tenera* (bottom row) palms showing characteristic differences in mesocarp and kernel size. (Courtesy of Unilever Plantations and Plant Sciences Group.)

the Cameroonian oil palm pollinating weevil *Elaedobius kamerunicus* to Malaysia in 1981 (and from there to Southeast Asia and the Pacific). This has eliminated the laborious and inefficient hand pollination of female flowers and has led to increased bunch weights especially in smallholder fields where hand pollination was rarely effective.

Developments in integrated pest management, the use of specific herbicides (applied with ultra-low volume sprayers) and the accurate determination of fertiliser requirements according to soil type, yield profile and leaf analysis have also helped to reduce both costs and environmental problems. After processing, the empty fruit bunches are now more commonly returned to the field as a mulch around newly planted palms and in mature areas. The bunches are a valuable way of recycling nutrients, increasing soil organic matter contents and conserving soil water.

Mechanisation of the plantation industry is progressing as labour becomes difficult to recruit and retain. A major constraint to reducing the cost of production remains the intensive physical and manual task of harvesting the fruit bunches and collecting the loose fruit. Recent management innovations include the separation of the harvesting tasks into cutting, loose fruit collection and in-field transport. Mechanised harvesting is not yet a reality but fertiliser applications, herbicide spraying, drain maintenance and the extraction of harvested bunches from the field are mechanised in many estates. The progression from hand carrying bunches through the use of wheelbarrows and water buffalo to mini-tractors and trailers demonstrates the pace of change in this area (Fig. 8.9).

Most modern estates are also paying great attention to the efficient transport of fruit bunches from field to factory. This generally involves reduced manual handling and improved road layouts to allow large lorries or tractors to collect rapidly the bunches from bulked collection points. Improved loading methods include the use of ramps, nets and, in some cases, high-lift tipping trailers.

The processing of oil palm is often regarded as an entirely separate operation to field production but there have also been important developments in this area. Some examples include the replacement of hydraulic presses with screw presses, improved bulk handling of bunches through the direct loading of steriliser cages in the field, better oil recovery systems and enhanced environmental safeguards such as waste water treatment plants. In the most progressive companies, downstream processing is now being incorporated to add value and diversify the market for the palm oil.

Management The movement of management control from trans-national to national corporations has been particularly marked in Asia with the decline in relative importance of well-known companies such as Harrisons & Crossfield, Socfin, HVA and Uniroyal. There has also been an increasing number of 'nucleus estates' developed, where most of the production is by smallholders supplying a mill owned by a company that may also organise transport, provide planting material and provide extension advice.

Where companies retain estates, the number of management tiers are being reduced and decision-making powers devolved down the management hierarchy. As with any industry, there will always be a few companies that achieve considerable success through the personal drive and charisma of autocratic leaders. In addition, most companies will look to compare their technical and management performance with that of other estates. This 'benchmarking' is already common in many industries and allows the rapid uptake of the best current management practices and comparisons of relative performance.

Future prospects

Oil palm is arguably the most efficient source of vegetable oil on a unit area basis with annual production levels on the best estates of around

Fig. 8.9 A mini-tractor with loading grab for transporting harvested oil palm bunches out of the field.

8 t/ha. However, the investment costs in mill capacity, transport infrastructure and management are high whilst the market is characterised by a long-term downward trend in prices alleviated only by occasional peaks created by shortages or political instability. Since 1970, prices have fallen from over US$1000/t to less than US$500 in 1990. World Bank projections (at constant 1990 US$ value) of US$270/t by 2005 are probably a fair estimate of the likely continued pressure on the industry.

Against this backdrop, costs of production per unit area have not fallen as rapidly and, in 1992, ranged from US$119 in Indonesia, $202 in Malaysia, $294 in Colombia and up to $400 in the Côte d'Ivoire. The production costs of the lowest cost producers in Indonesia and Malaysia were about 30% less than the typical values.

In order to remain in business the oil palm industry of the future needs to continue to invest in research and development to increase yields of improved quality oil. This will allow larger volumes to be processed thus exploiting economies of scale. New and existing knowledge on improved production methods must then be applied through better management to reduce the cost of production. A large part of this reduction will also be achieved by reducing overheads and, in particular, the cost of management.

CONCLUSIONS

The continuing success of plantations as an agricultural system in the tropics will inevitably require large changes in the technology and management of a traditionally conservative industry. Despite population increases in most of the tropical countries where plantations are grown, improved education coupled to competition for labour with other industries will lead to progressive intensification of all aspects of production. In all plantation crops, there is still the potential to increase average yields considerably but, in the long term, only those sections of the industry that manage to combine advanced technology, enlightened management and a high level of capitalisation will survive.

FURTHER READING

Hartley, C.W.S. (1989) *The Oil Palm, Elaeis Guineensis*, 3rd edn. Longman: London.
Tiffen, M. and Mortimore, M. (1990) *Theory and Practice in Plantation Agriculture*. Overseas Development Institute: London.
Webster, C.C. and Baulkwill, W.J. (1989) *Rubber*. Longman: Harlow.
Wilson, K.C. and Clifford, M.C. (eds) (1992) *Tea: Cultivation to Consumption*. Chapman & Hall: London.
Wood, G.A.R. and Lass, R.A. (1985) *Cocoa*. Longman: Harlow.
Wrigley, G. (1988) *Coffee*. Longman: Harlow.

9
Agroforestry

P.A. Huxley

AN INTRODUCTION

What is agroforestry?

Agroforestry is a form of multiple cropping (Francis, 1986); that is where we purposefully grow more than one kind of plant on the same piece of land. This can be done either where crops grow on the land *simultaneously* (intercropping), or where they follow one another *sequentially* (rotations). Relay cropping is a mixture of the two. Where woody plants are present the form of the cropping pattern can raise important issues, as we shall see below.

Multiple cropping forms the basis for much of tropical agriculture – particularly subsistence farming (Francis, 1986; and see Chapter 6). Also plantation crops such as oil palm and rubber are commonly grown with leguminous cover crops during the early part of their life cycle. Others such as tea and coffee may be grown among shade trees and some, such as coconuts and other palms, are often underplanted with food crops (see Chapter 8). Tropical pastures also have mixtures of grasses and herbaceous plants, including legumes, growing simultaneously and in rangelands these are mixed with trees and shrubs (Chapter 12). Thus there is a wealth of experience regarding multiple cropping with tropical woody plants. Such complex associations of plants can afford many opportunities to manipulate the proportions of species and/or to arrange them in space and time. This, again, is a key issue in agroforestry.

The land itself can be intermittently or continuously occupied by useful plants. Intermittent cropping takes advantage of just the favourable climatic periods (the 'crop seasons') to grow arable crop species selected to do this productively. Occupancy can be shortened on purpose. A 'bare fallow' is sometimes practised in dry regions (e.g. North Africa) omitting one crop so as to store water in the soil for the next. However, bare ground can often be subjected to erosion and leaching in seasonally arid regions. Even so, as in temperate climates, many forms of high-input monocultures can be very efficient in the tropics, although concerns are growing about maintaining the levels of inputs required (fertilisers, irrigation, pesticides) in order to satisfy the output needs required by a growing world population, and about doing this sustainably. The equitable distribution of food from surplus to deficit regions will remain a problem so that improvements in *local* production systems demands continuing attention. Agroforestry practices occupy land continuously and the potential advantages of this are discussed later in this chapter.

In many temperate regions increased agricultural productivity has resulted in surpluses, so that temperate agroforestry is often one of the alternatives being explored as a beneficial way to *reduce* crop production whilst benefiting the land in the long term, e.g. in 'set-aside' lands. Tropical agroforestry, however, is about *occupying land* so as to obtain a whole range of benefits. It is *not* just about productivity, nor solely about sustainable production.

Agroforestry has been described by Lundgren (1982) as

'the deliberate inter- or sequential cropping of woody and non-woody plant components (sometimes with animals) in order to generate multiple products and "services". There are both ecological and economic interactions between the plant components.'

Table 9.1 provides a list of agroforestry practices. It may seem strange that agroforestry has only relatively recently emerged as a new field of scientific study when we have such extensive experience available from agricultural multiple cropping. Surprisingly, scientists failed to appreciate either the prevalence of this form of land use, or its potentials, until the late 1970s (Bene et al., 1977; King, 1987), when it suddenly became apparent that, although we had applied scientific principles to the cultivation of woody plantation crops (see Chapter 8), and commercial forestry species (Evans, 1992), we were generally much less aware of how to set about using trees and other woody plants in mixtures with agricultural crops. Only a little had been done to promote the integration of cropping with afforestation, or to encourage the off-take of 'secondary' products from forested lands.

Agroforestry, as it is now called, has been around for a long time. Early forms of land use practised by mankind at the onset of settled agriculture 10000 years or so ago would certainly have included woody plants, perhaps opportunistically at first and then by deliberate nurturing or planting and by managing selected, valued trees, shrubs and vines (Plucknett and Smith, 1986). The development of monocultural systems over the last few centuries has been a response to the search for yield improvement obtainable through the more precise management that handling just a single crop at a time can offer. The use of a range of inputs such as fertilisers, irrigation and chemicals for the control of pests and diseases are all facilitated in monocultures, which can be highly successful (Ewell, 1991), and plant breeders have created cultivars that can make the best use of such management conditions (see Chapter 10).

Agroforestry is *not* simply a combination of agriculture and forestry (e.g. Steppler and Nair,

Table 9.1 Agroforestry practices.

Trees with crops (agrosylviculture)

- Rotated in time (*sequential practices*):
    - Shifting cultivation (sometimes with enrichment of the woody components).
    - Improved tree fallow.
    - Taungya (i.e. cropping during the establishment phase of commercial forest tree plantations).
- Spatially mixed (*simultaneous practices*):
    - Trees on cropped land.
    - Multiple use of trees in crop plantations.
    - Mixed multistorey tree and crop arrangements (e.g. tropical home gardens).
- Spatially zoned (*simultaneous practices*):
    - Hedgerow intercropping ('alley cropping'), contour hedging or barrier planting and other types of linear tree plantings.
    - Boundary planting for various products, or live fencing.
    - Strip planting in forests or timber plantations ('corridor farming').
    - Windbreaks, tree-enriched windstrips, shelterbelts and wild animal habitat plantings.

Trees with grass and animals (sylvopastoral) – simultaneous systems

- Spatially mixed:
    - Trees planted on rangelands, in permanent pastures and grass and grass–legume leys for leaf and/or pod fodder (sometimes for edible flowers, additionally).
    - Tree–crop plantations with pastures.
- Spatially zoned:
    - Alley farming (sometimes with 'energy' gardens of shrubs and grass mixtures).
    - Live fences.
    - Boundary plantings, mainly for fodder.

Managed tree plots

- Fodder banks using woody species.
- Fuelwood lots.
- Mixed orchards (especially if for several products, e.g. fruits and honey).

1987; Nair, 1989), but some of its forms, e.g. 'village woodlots', where trees are grown separately, can really be considered as 'microforestry'. In these cases the tree species used may even be those found in commercial forestry practice (e.g. *Cupressus, Pinus, Eucalyptus* spp., etc.).

Other practices involve woody crop plants such as cocoa and coffee with a mixture of shade trees that may be used for timber so as to provide more than one product from the land. Coconut with mixed under-storey food crops, or scattered fruit trees in pasture, are commonly considered to be 'agriculture'; thus the boundaries of agroforestry are ill-defined. However, the key issue in any such land use *practice* is how best to design and manage mixtures of plants, often mixtures of woody and non-woody species. Farmers will all have their individual ways of utilising such mixtures, emphasising their own resource limitations and needs, so there will be an infinity of different agroforestry *systems*. Furthermore, in agroforestry, farmers do not utilise trees just for the products they provide but also for 'services'; that is for shelter or shade, for amenity purposes, to help restore soil fertility, to mark boundaries and to aid in the control of soil and water conservation. Animals may often play an important role in many forms of agroforestry (e.g. Torres, 1983; Baumer, 1991), and the use of tree fodder is currently receiving renewed attention (e.g. Le Houérou, 1980; Ivory, 1990; Leng, in press).

Readers can find publications giving general accounts of agroforestry (see 'Further reading' at the end of this chapter) but, in this chapter, rather than elaborate on agroforestry practices, or describe details of individual tree components, various issues are considered concerning the 'nature' of agroforestry and the ways agroforestry systems function, together with some important features associated with their management.

Figures 9.1–9.6 show some examples of agroforestry that serve to illustrate some of the points made later but it is useful, first, to consider briefly some underlying features common to agroforestry practices in general.

Simultaneous cropping and 'the crop season'

Successful agricultural cropping depends on timeliness and in using 'the crop season' wisely.

Fig. 9.1 Barrier planting with *Leucaena leucocephala* on a slope at Machakos, Kenya. This is a form of hedgerow intercropping where the hedges are heavily pruned and, although prunings are distributed as mulch to improve fertility in the cowpea area, the main function of this practice is to prevent soil and water erosion (Kiepe and Rao, 1994). (Photo: Anthony Njenga, ICRAF.)

Fig. 9.2 A living fence of *Erythrina fusca*, Costa Rica. A low-cost form of protective boundary planting. The woody species has to be capable of rooting easily from hardwood cuttings and/or from seed and to respond to pollarding. If some wire is used the long shoots can also be utilised as 'risers' to help form a secure barrier against animals. (Photo: Gerardo Budowski.)

Fig. 9.3 *Gliricidia sepium* grown under coconuts together with a cover crop (*Calopogonium mucunoides*) at the Coconut Research Institute, Luniwila, Sri Lanka. The *G. sepium* is cut and the prunings are placed around the base of the palms to provide nitrogen (Gunathilake and Liyanage, 1996). (Photo: Peter Huxley.)

Fig. 9.4 A 'home garden' in Kalimantan, Indonesia. These multistrata practices have been developed over time by farmers themselves and have remained sustainable. (Photo: Andri Wahyono.)

Fig. 9.5 Mature trees of *Faidherbia albida* scattered in millet lands in southern Sudan. The trees are in leaf in the dry season prior to the onset of the rains when the millet will be sown. (Photo: Peter Huxley.)

Fig. 9.6 An example of 'slash-and-burn' agroforestry in northern Zambia in miombo woodland (a *'Chitomene'* system). This is a sequential practice where the cropping phase alternates with a 'bush' fallow. The natural fallow vegetation is cut, heaped in a circle and burned and eleusine millet is often grown in the first cropping season, followed by field beans and then cassava and bambara groundnuts before the site reverts to fallow again. (Photo: Ken Giller.)

Especially with monocultures, plant breeders have 'tailored' cultivars to grow and make the best use of available environmental resources depending on the expected length and nature of a crop season. However, when trees and/or shrubs are part of a system, cropping patterns will invariably be more complex. This is because woody plants in the tropics show an incredible diversity of behaviour with regard to the periods over which they grow, flower and fruit, that can vary even between individual trees of the same species that are growing at the same site. This diversity has come about as a result of evolutionary changes to enhance survival and/or establish competitive advantages. Woody plants have become not so much 'opportunists' as 'wily strategists'! In agroforestry, therefore, it is best to consider the progression of active vegetative growth, flowering and fruit development of a tree as a series of reiterated but sometimes intermittent, rather than regular, episodes occurring over their whole life span; and these episodes may or may not coincide with 'crop seasons'.

In those agroforestry practices where trees and crops occupy the same piece of land (simultaneous practices) this is important because it defines the extent of the woody and non-woody plant component interactions in time. Competitive processes above and/or below ground will occur when the plant components are *actively growing* at the same time, but leaf-retaining trees continuously shade lower plant components. *Pre-emptive* competition can also sometimes take place. This is when one plant component occupies space before another can get to it, e.g. where the roots of a woody perennial can occupy part of the soil profile at the expense of the crop plant associated with it. Foresters call this 'site capture'.

Sequential cropping

Agricultural crops can occupy land sequentially (be 'rotated') with trees. In such forms of agroforestry the trees/shrubs will be purposefully planted for some products (timber, fuelwood, building materials such as poles, fruits, leafy or pod fodder, etc.). For example, woodlots or fodder banks may be established and ultimately alternated with food crops. Whether considered as 'agriculture' or a form of 'agroforestry', shifting cultivation practices ('slash-and-burn' or 'swidden' agriculture; see Chapters 3 and 6; Lanly, 1984; Warner, 1991), have a soil restorative fallow phase of natural vegetation immediately following the cropping sequence (Chapter 6); and shifting cultivation can evolve into more advanced agroforestry practices (Raintree and Warner, 1986). A fallow of natural vegetation can be enriched with useful woody species established in the last part of the cropping phase, for example, rattan palms, or useful timber or fruit species. This is how Sudanese gum arabic 'gardens' are continually re-established (Seif el Din, 1981). Although in simultaneous agroforestry plant components may still avoid interacting, at least at certain times, in sequential systems where trees and crops or grass pastures occupy the land separately in time, they clearly do not interact at all in the short term; although the legacy of planting one or the other may spill over. For example, there may be detrimental allelopathic effects (Horsley, 1991) where it is difficult to plant crops or even trees themselves to follow after the first rotation of trees has been cleared (the 'replant problem').

One of the apparent paradoxes arising during the early adoption of agroforestry in its new forms (e.g. 'alley cropping') was the proposal that just a fractional cover of trees or bushes growing continuously among agricultural crops can accomplish soil fertility maintenance or restoration that will, over time, equate or better that achieved by a long-term rotational bush fallow. The likelihood or otherwise of such a 'fertility effect' in simultaneous cropping of woody and non-woody plants is now becoming better understood, and its magnitude under different circumstances has become the object of a good deal of research (see below).

Although sequential practices are biologically simpler than simultaneous ones there are considerable social and economic implications in separating the phases in this way. Not least are the labour inputs required for clearing forest/trees, or the costs of planting up and protecting trees in a temporary plot, as well as the eventual tree harvesting and clearing costs. The costs of *not* committing resources to such intermittent activities may be high. Furthermore, if land occupancy is not absolutely continuous then erosion may often be a problem (Chapter 4). The concepts of tree fallows (sequential) and hedgerow intercropping (simultaneous) can be combined in 'rotational hedgerow intercropping' so as to provide a flexibly managed practice for settled farmers with enough land (Fig. 9.7).

'Patchiness'

Farmers and, indeed, pastoralists may combine their knowledge of disorderliness in time with an appreciation of spatial patchiness, so as to exploit their environment and put the lands they occupy to the best possible use (Fig. 9.8); unlike researchers who often try to measure spatial variability in order to remove its influence! If farmers utilise a broader range of plant types, including woody ones, then there are clearly more opportunities to exploit space and time effectively. However, achieving a better utilisation of environmental resources for the system as a whole throughout the year is not achieved by just mixing up trees and crops in any kind of combination. This is where the application of science to agroforestry is essential in order to complement a thorough understanding of the socio-economic attributes of trees and tree-planting practices, as discussed in Raintree (1991).

Dependence on skill and judgement

Successful management of woody species nearly always depends on relatively high levels of skill, e.g. for propagation, planting out and protecting young trees, training and pruning and, often,

Fig. 9.7 Demonstration plot of rotational hedgerow intercropping with *Senna siamea*, Machakos, Kenya (midfield). Hedgerow intercropping with maize to the left, the 'rotational' plot of trees alone to the right. When crop yields have declined the hedgerows can be allowed to grow out and those in the rotational plot can be pruned back to establish a hedgerow intercropping sequence. (Photo: Peter Huxley.)

Fig. 9.8 This aerial photograph of small farms in Embu, Kenya shows how 'patchy' such land is. Natural variation caused by soil and micro-relief is enhanced by innumerable 'tree–crop interfaces'. (Photo: Peter Dewees.)

harvesting in an appropriate way (for example as with trees and shrubs which are to provide leafy fodder). Judgement is needed with intercropping systems so as to get the 'best' from the woody plants whilst limiting their competitiveness.

There is often a rich source of local knowledge about handling indigenous plants onto which scientific inputs can be grafted (Warner, 1991; Barrow, 1996). Much of the intercropping found in the tropics has arisen as a consequence of a perceived need for, and a realisation of, the benefits to be derived from diversity despite the difficult decisions that it imposes on farmers. Then again, farmers will not normally wish to change the agricultural crops that provide them with their subsistence or cash income. This is not to say that there may not be a need to provide them with crop cultivars or pasture grasses that are better suited to intercropping with woody species, but this is likely to be the result of long-term attention to specifically identified needs. At this time farmers are increasingly aware of the different kinds of agroforestry practices that may be suited to their situations. Since the 1970s there has been an explosion of interest at the grass roots level in agroforestry and other forms of tree planting, fostered and supported by extension services, the media and by international, national and non-governmental organisations (NGOs). But there are many unanswered questions. For example, what if farmers come now to ask us questions about any particular agroforestry practice such as 'Are we using the best tree species?', 'How many trees do we need?', or 'How best do we manage them?' Answers to such questions clearly require an understanding not only of their system, their objectives and the resources available to them but also the application of scientific principles. So biophysical scientists have to combine with those from social and economic disciplines for any success to be achieved. It is this need for a multidisciplinary approach that makes agroforestry research so stimulating.

The rest of this chapter describes the understanding needed in order to respond to such questions sensibly, the emphasis being on biophysical aspects.

ARE WE USING THE BEST SPECIES?

Appraising farmers' needs and using their indigenous knowledge

Questions about the best species to use will nearly always refer to the woody plants. Currently, the world is an exciting place for agroforestry practitioners and tree breeders. At no time in its past history has so much germplasm been distributed around the tropics for potential use in agroforestry, whether for 'high' or 'low' input practices. Testing and selection of so-called 'multipurpose trees' (MPTs) is being energetically carried out by national, international and non-government organisations. This is not without concern for some (e.g. Hughes, 1994) because of the dangers in introducing potentially invasive, weedy species (e.g. *Prosopis juliflora* in India and the Sahel, or *Acacia nilotica* in Queensland, Australia). Farmers, themselves, may not be aware of the potential benefits of introduced ('exotic') MPT species and sometimes they prefer to choose known indigenous trees and shrub species, collecting their own seed. There are arguments both ways but, certainly, despite any apparent benefits (e.g. fast growth, nitrogen fixation potential), exotic species should not be distributed without a careful examination of any potential dangers. Above all the specific requirements of the farmers need to be identified. This process can be done through a rapid rural appraisal activity such as the International Centre for Research in Agroforestry's (ICRAF) 'diagnosis and design' methodology (Raintree, 1987a, 1987b).

Rural communities often have a very extensive knowledge of the indigenous species that can be useful to them, including the woody ones (Walker *et al.*, 1994) and it is imperative to access such knowledge and experience. Any potential technical benefits from agroforestry certainly have to be evaluated alongside economic considerations, and the social context can often be all-important. For example, land tenure can be a critical issue (Fortmann and Riddell, 1985), and also the ownership of rights to harvest individual trees, or small groups of trees. Tree planting may help establish a right to crop land (as in parts of Uganda), or lack of land tenure may be the over-riding reason for not planting trees (Burley, 1984). The rewards of tree planting may need to be shared throughout a community and this may not be easily achieved where there is a mixture of large and small holdings and possibly even landless poor. At the household level the work required to establish and maintain trees in the system will most readily be obtained if the benefits are shared by all the components of the family.

Land users may well have very definite ideas, based on their own experience and that of the communities in which they live, not only about what species are useful but on how they may best be utilised. A good example is the wealth of knowledge that pastoralists have acquired over many generations about the vegetation that supports them and their animals. With such long-established communities, merely supplying a range of supposedly 'useful' tree seedlings can be a fruitless venture, and field development projects need to be genuinely participatory if they are to be successful (Fig. 9.9) (Barrow, 1996).

Perhaps less obviously, but clearly along similar lines, will be the situation of settled farmers offered planting materials and/or instructions on how to use them (e.g. in alley cropping), also shifting cultivators offered species to 'enrich' their fallows, or those cultivating home gardens when given 'new' fruit species. The use of indigenous knowledge in all such cases is essential, especially so where practices that have evolved over a long time are to be 'improved'. For example, the current need for research on sophisticated home gardens in Sri Lanka may well be for compatible, and probably indigenous *timber* species in view of the fact that by the year 2005 estimates suggest that some 80% of Sri Lanka's timber needs will be met from this source (Weerakoon, 1996).

Fig. 9.9 Micro-water catchments help to establish various species of multipurpose tree in Turkana, northern Kenya. Rainfall is erratic and averages <400mm per year. Constructing simple earth ridges to lead rainfall to a small pit can greatly increase the water supply to the young tree seedlings which are planted on the ridge above the pit. (Photo: Peter Huxley.)

'Multipurposeness'

A feature of agroforestry is that the woody components fulfil a wide range of tasks, some of which have already been mentioned, but there are many more: to provide fibres, tool and craft wood, medicinal products, honey, gums and extractants, fish fodder, to name some. But why '*multipurpose*' tree? These have been defined earlier as 'Woody perennials (trees, shrubs, woody vines, bamboos) that are purposefully grown to provide more than one significant contribution to the production and/or service functions of the land-use system they occupy' (Huxley, 1985). But does this imply that farmers require a number of these products together with one or more services from an individual tree? Are such trees expected to provide them simultaneously? The definition of MPTs given above is not too clear on these points!

Indeed, *all* tree species can be multipurpose but, clearly, some woody species are innately so (e.g. palms, where nearly every part is utilised in some way). Others can readily be managed to supply more than one product/service at the same time (e.g. *Leucaena leucocephala*, Fig. 9.10). The former we might term 'obligate', and the latter 'facultative' MPTs (Huxley, in press). There are some biological rules that have to be obeyed, however! Quite simply, the various parts of a plant compete, internally, for resources (e.g. carbohydrates, nutrients) and if a tree is managed at any one time in such a way as to maximise the production of one part it is unlikely that the other parts will also respond as favourably. Thus if a particular product is to be encouraged (through management or selection) it may not be feasible to maximise another at the same time. For example, it will not be possible to obtain a maximum amount of both fruits *and* leafy fodder (or fuelwood, timber poles or other woody parts). Similarly, if a shrub is managed to

Fig. 9.10 *Leucaena leucocephala* cv. K8 at Machakos, Kenya. Grown unpruned this is badly adapted in this region, as can be seen by this plant which is heavily overbearing. *L. leucocephala* flowers terminally on new wood and delayed vegetative flushing (due to overbearing stress from the previous season) recurringly commits the plant to filling out pods as best it can well into each dry season. (Photo: Peter Huxley.)

provide the maximum amount of leafy fodder then it will not produce as much fuelwood as it might. Of course, it is possible just to select very vigorous woody species (e.g. fast-growing nitrogen-fixing trees) so that there is plenty of everything. However, as discussed below, it is not always wise to grow vigorous woody components in an agroforestry system.

The bottom line is that woody plants can be managed to provide several outputs over the same short period of time, but this cannot usually be done as efficiently as it may be for a single product. In addition, MPTs will provide particular services well only if they are grown in the right way and this may interfere with the amount of products that they can supply. For example, trees and shrubs used in windbreaks need to be an appropriate shape and have an appropriate canopy density so as to provide a semi-permeable barrier and the right kind of windflow profile. On steep slopes hedgerows aligned for soil and water conservation and, perhaps, the provision of 'self-forming' terraces (Fig. 9.1) may need to be pruned back severely at the start of the crop growing season in order to prevent competition, but this will offset their ability to play a soil fertility-raising role.

Perhaps the chief benefit of 'multipurposeness' for 'facultative' species is the possession of biological characteristics that confer *flexibility of management* that will then enable a farmer to convert plants for the production of alternative outputs as and when they are required in the future. Planting-up and protecting young trees and then waiting for them to yield often represents a major investment of a farmer's time and resources. There is an analogy with woody plantation crops where replanting is a major economic event (Chapter 8). Thus, in agroforestry, a farmer who is using trees to produce fuelwood, but who may then need leafy fodder because animals have been acquired, is able to supply this from *existing* MPTs as long as they have the potential to produce either product. This has important implications for the selection and breeding of MPTs because, clearly, this can only be done effectively if breeding objectives are set out explicitly.

Some other important characteristics

A major distinctive feature of woody perennials that needs to be considered in tropical agroforestry is whether one uses species which are evergreen (leaf-retaining) or deciduous (leaf-exchangers) (Schoettle and Fahey, 1994), especially because this can affect both the time and duration over which they use water and/or capture sunlight and cast shade. In fact, these terms may often be somewhat misleading because an individual species may be largely deciduous in one environment but remain evergreen in another. Whichever they are, tropical plants show an extraordinary range of diversity in their patterns of growth (Huxley and Van Eck, 1974; Huxley, 1983b; Akunda and Huxley, 1990). So far we can make this statement with assurance about tree canopies, but very little has been done to explore patterns of growth of root systems, and relate these to the ones we see above ground. Clearly this is an extremely important issue in understanding and managing agroforestry systems; it will, for example, affect the potential for dry matter production and water use. This lack of information is going to take some time to remedy.

Most MPT species are outcrossing and, therefore, genetically highly heterozygous. Many of the 2000 or so species that have been recorded as being used somewhere, and potentially of value in agroforestry practices (Burley, 1984), are as yet unselected. Some genera, e.g. *Leucaena*, *Gliricidia*, *Calliandra*, *Sesbania* and others, have been the subject of range-wide provenance collections and have been (e.g. *Leucaena* spp., see Brewbaker, 1987), or are now, undergoing intensive selection. Collectively, the potential for genetic gain is enormous. Studies comparing growth rates and pod production of individual genotypes of various *Prosopis* spp. have shown ten-fold differences, for example (Felker *et al.*, 1983, 1984), and with some species of MPT we are virtually where our major agricultural crop plants started (see Chapter 10). However, we need to realise that this means that MPT species are likely still to possess some plant attributes that we may not be seeking, i.e. those that may confer survivability such as seed dormancies, thorniness, unpalatability, very slow growth, etc.

Myths, hypotheses and assumptions

'Trees are good'

Enthusiastic agroforesters often assume that it is *always* 'good' to use woody plants, but 'woodiness' confers some benefits and some disadvantages. There are 'fast'- and 'slow'-growing trees but, as a generality, woody plants can certainly be as efficient in terms of carbon assimilation as C3 herbaceous crop species (Cannell, 1989b). When agroforestry first became of scientific interest it was largely promoted as a potential way of reclaiming 'wasted lands' (King and Chandler, 1978), and trees have also been shown to be useful in the reclamation of saline and/or alkaline lands (e.g. Grewal and Abrol, 1986; Ahmed, 1991), and to lower the water table, a rise in which can sometimes cause secondary salinity (e.g. in Australia; Bell, 1988).

Some positive general characteristics of woody perennials can be summarised as:

- Improved 'survivability' – few tropical trees can survive frost, but other adverse environmental episodes may not damage them too badly; they can often endure grazing by herbivores and severe removal of parts (coppicing and pollarding).
- 'Shapeability' – a distinct 'plus' where manipulation of the canopy is required for various reasons.
- Longevity – a useful benefit for long-term planning.
- Perennial root systems – these fill a space in the soil from which further root growth and exploration can continue. Tree root systems can have characteristics that differ from those generally found in herbaceous plants and grasses (Huxley, 1996).
- Having a range of different patterns of growth and development – these can occur both between- and within-species between seasons.
- Having wide-ranging levels of constituent nutrient elements and ancillary compounds (e.g polyphenols) both between- and within-species – a feature that may be important in providing supplementary animal fodder and potential soil-improving residues, and one that can be modified by environment and management.
- Having a potential to improve their microsites – this can occur in a number of ways depending on how they change the above- and below-ground environment (as discussed in more detail, later).

It must be remembered, however, that because woody perennials are there from season to season there is a higher chance of woody plants hosting unwanted pests that can spread to adjacent crops. Furthermore, serious pests are encouraged by tree canopies such as birds that attack crops (e.g. *Quelea* species), and tsetse flies (*Glossina morsitans*), which cause disease in man and cattle as they lay their eggs in the moist litter below bushes. Above all, woody plants can *compete* with crops and pasture grasses which, in an agroforestry system, may have the farmer's highest priority.

From the farmer's point of view the long-term nature of woody perennials means a commitment to an investment in land and labour which is ongoing as long as that land-use practice continues. This means that planning the layout needs to be done carefully. Furthermore, woody plants need to be 'raised', i.e. they often require nurseries (Fig. 9.11). Land preparation for planting woody species is, or should be, more thorough than for annual crops and all perennial weeds, especially grass weeds, need to be removed *before* any trees are planted out. Most often, young trees are planted out in prepared holes so that the cost of establishment can sometimes be high; especially as they then require some protection. Woody species used in hedgerow intercropping are chosen because they can be easily propagated from seed or cuttings. Young trees tend to be 'dominated' in their early growth phase by faster-growing crop plants or C4 grasses and herbaceous weeds. Shaping the

Fig. 9.11 Two different forms of tree nursery. (a) This Kenyan farmer is an entrepreneur using all kinds of waste, locally found materials, to raise seedlings for sale in his village in Machakos; (b) a more formal nursery (at Makoka, Malawi) raising tree seedlings in polythene sleeves. Advanced Western nurseries use many sophisticated techniques for raising plants that yet remain to be adapted to low-cost local conditions. (Photos: Peter Huxley.)

woody structure and maintaining it requires time and skill. One other feature of woody plants, in general, is that they often exhibit erratic flowering patterns and this has to be taken into account in their management.

Expected performance

In general terms agroforestry systems are expected to confer increased productivity in a sustainable way, to diminish risk and to enhance their environment in various ways. These statements, though, are more in the nature of intuitive intentions that need to be given substance through rationalised analysis of individual situations. There are a number of assumptions ('hypotheses') about agroforestry, not all yet proven (Sanchez, 1995) that are based largely on the presumed benefits of one kind or another that trees may, or may not, confer that are inferred from generalised assumptions about their biological attributes and functional capacities.

- *Trees are deep-rooting.*
  Woody plants are supposed to enrich the upper soil layers, in part because they are deep-rooting. In fact, they exhibit a range of genetically controlled *potentials* for different rooting characteristics including 'generally deep', or 'generally shallow and extensive', etc. Palms invariably have relatively shallow, dense root systems and these can be widespreading (as in oil palm). Many species from arid or semi-arid regions may have some deep roots, but these usually also extend laterally for long distances (e.g. some species of *Acacia* and *Prosopis*); *Senna siamea* also behaves in this way. However, in practice, environmental conditions, especially those relating to soil type, can be expected to modify root systems extensively and plants in general show considerable phenotypic 'plasticity' in this respect. Where subsoils have a low pH, impeded layers, or are waterlogged, then tree roots will almost certainly *not* reach deeper levels. Trees in the Amazon were often thought to be shallow-rooted but recent research has suggested that half of the closed forest of the Brazilian Amazonia depends on deep roots to maintain green canopies during the dry season (Nepstad *et al.*, 1994). Some tree species do require deep, free-draining soils if they are to flourish, e.g. *Citrus* spp. In general, a high proportion of the fine roots of trees and shrubs are found nearer the soil surface with far fewer in deeper soil horizons, and this has implications when considering nutrient recycling. Trees and crops (or grasses) may often be occupying the same part of the soil profile. To what extent they share the water and nutrients found there is a key issue in agroforestry.

- *Trees can control soil erosion and runoff.*
  This is certainly a possibility but, like the other assumptions, generalisation is to be avoided. Trees and bushes have to be appropriately planted and managed otherwise they can actually *increase* erosion (Fig. 9.12). Well-designed vegetation strips are permeable barriers that can slow down and disperse runoff. The plant residues so provided can improve surface soil infiltration rates, and force runoff to drop its sediment load on the field itself, resulting in terrace formation (Kiepe and Young, 1992; Kiepe and Rao, 1994).

- *Trees are efficient at nutrient recycling.*
  This is, perhaps, one of the most important assumptions relating to the use of trees in agroforestry land use. A high proportion of nutrients may be recycled within a system under woody plants, and the production of soil organic matter (SOM) from their residues is an important feature (Fig. 9.13). But nutrient recycling happens under *any* kind of vegetative cover, and especially in grass swards. Closed nutrient recycling is thought to be typical of natural mixed vegetation in wetter regions. In drier areas, where plant cover is more widely spaced, a mixture of species could also favour a greater total root mass but the system is unlikely to capture and recycle all nutrients, some will be lost by leaching when it does rain (i.e. the system will be 'open' in this respect). Unfortunately, there are only a few detailed studies of nutrient recycling of managed, multi-layered, woody plant systems (e.g. that of *Erythrina poeppigiana*, *Cordia alliodora* and *Coffea arabica* in Costa Rica – Glover and Beer, 1986; Fassbender *et al.*, 1988; Heuveldop *et al.*, 1988; Fassbender *et al.*, 1991), although some older work can be sought out, e.g. on nutrient recycling under shade trees in tea in Assam (Wight, 1958: pp. 75–150). Studies of this kind suggest that there are high levels of nutrient recycling in such systems. The work on litter deposition in forests also supports a high rate of SOM build-up under tree cover

Fig. 9.12 Gulleying in Baringo District, Kenya. These indigenous acacias are adding to the problem started by grazing out the grass cover. Water is now being deflected to flow at increased speed as it is being channelled between the trees. (Photo: Peter Huxley.)

Fig. 9.13 *Acacia cyanophylla* being used for sand dune fixation in the Coastal Plain, Libya. Dried grass has first been inserted into the sand to form a stabilising 'checkerboard'. Note how the litter from these 3-year-old trees is already changing the 'soil'. (Photo: Peter Huxley.)

(e.g. Malaisse *et al.*, 1975) but this, in itself, does not confirm that the system is closed. As yet there are no studies of *all* the nutrient flows and pools in forested systems and some forests definitely are 'leaky' (Whitmore, 1989).

Mixtures of woody and non-woody plants are supposed to be more efficient than many other forms of plant cover at nutrient recycling because they are thought to have a combined root system that can explore more of the soil volume more thoroughly (the 'safety net' hypothesis, Fig. 9.14). Quite clearly there is a priority need to obtain more information about the rooting behaviour of the plant components in particular agroforestry practices and some early investigations of agroforestry systems show that these assumptions are probably correct in some circumstances (van Noordwijk *et al.*, 1996), but the key issues in establishing how effective woody/non-woody plant mixtures can be at capturing and recycling plant nutrients under different circumstances are:

○ How effective are particular species mixtures and arrangements in preventing leaching of plant nutrients to deep drainage?
○ At what rate, and over what time, are different nutrients moved from one part of the system (e.g. the lower soil layers) to another (the top part of the soil profile)?

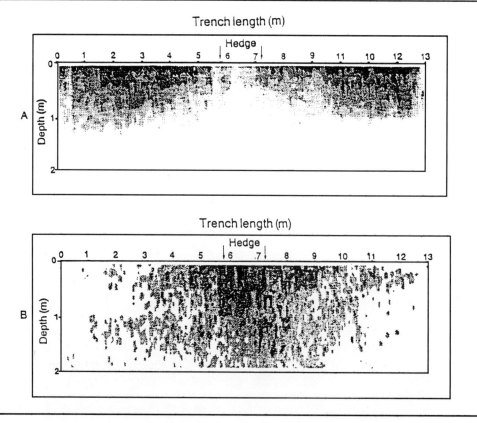

Fig. 9.14 Fine root profiles (root counts) of (A) maize sown adjacent to (B) a 4.5-year-old *robusta* hedge at Machakos, Kenya. In this example the *G. robusta* rooted more deeply than the maize and they shared the top 1.3 m, with some maize roots found even beneath the hedge (details in Huxley *et al.*, 1994).

○ And, particularly important, how does this affect both the total SOM and the particular fractions of SOM that are functionally important for facilitating transfer of plant nutrients to roots? Bearing in mind that the total amount of SOM in soils is not, in itself, a satisfactory indicator of soil fertility (Szott et al., 1991; Blair et al., 1994).

To summarise, tree root systems create macropores in the soil and fine roots die and, together with root exudates, supply carbon compounds to create SOM. Leaf litter, and other forms of residue mulch such as bark, twigs, fruits and seeds, can improve soil surface characteristics and so enhance rainfall infiltration rates, although if above-ground residues are compacted on the soil surface this beneficial effect will not necessarily occur (Huxley et al., 1994). That trees can make systems less 'leaky' for plant nutrients than those consisting solely of single species (e.g. Ewell et al., 1991), and sometimes improve topsoil fertility, is known from examining soil fertility changes over time in forested lands (including chemical, physical and biological states of the soil, Sanchez et al., 1985), and also from plantation crop studies and from a few investigations on single trees (e.g. Belsky et al., 1989). The consequences of *removing* woody plant cover can also be instructive. Examples suggest that nutrient recycling is diminished, leaching of plant nutrients is enhanced and, in some circumstances (as in parts of Australia) evapotranspiration from the plant cover is reduced with a consequent rise in the (saline) water table. However, quantitative evidence about recycling is needed for particular agroforestry practices with potentially different arrangements of tree and crop root systems. There is also a need to demonstrate whether or not more of the soil profile is being exploited and/or that a greater total root length density is present (i.e. more roots per unit volume of soil). Existing evidence from similar land-use systems is inadequate for predictive purposes in agroforestry. The question of how 'closed' or 'open' agroforestry systems may be is one that is addressed again below with regard to sustainability.

- *Nitrogen-fixing trees may add substantial nitrogen inputs to soils.*
  This statement is potentially true. For example, leguminous trees have bacterial associations with rhizobia, and actinorhizal species such as *Casuarina* have associations with *Frankia* spp. Many have been shown to be active in fixing atmospheric nitrogen (Sprent and Sutherland, 1990), although some members of the Caesalpiniaceae used in agroforestry, such as *Senna* (syn. *Cassia*) *siamea* and *S. spectabilis* do not nodulate. The question is to what extent is nitrogen being fixed in any particular situation? Some estimates of dinitrogen fixation exceeding 100 kg/ha/yr of nitrogen for well-nodulated species such as *Leucaena leucocephala* and *Sesbania sesban* have been made (e.g. Dommergues, 1987). And, if the soil is reasonably well provided with nitrogen then total nitrogen accumulation in the above-ground biomass several times this has been recorded for actively growing, close-planted tree legume species (Giller and Wilson, 1991).

  It has to be remembered, however, that nodulation only occurs when the host plant is flourishing, and active fixation takes place only if effective *Rhizobium* strains are present and the plant is not under stress. In addition there has to be an adequate supply of major plant nutrients available as well as the minor elements such as molybdenum (for nitrogen fixing activity) and boron (for nodule growth). The potential for nitrogen fixation can certainly be an important 'plus' when selecting MPT species – but it is by no means the only attribute that has to be considered. Indeed as dinitrogen fixation occurs in crop species (peas and beans), cover crops and pasture legumes, these may equally well be used to add nitrogen to the system. Finally, any benefit to the total nitrogen balance of the system, and thus any residual effects, will depend not just on how much nitrogen is fixed but also on how much is exported in crop yields and tree biomass (e.g. Giller *et al.*, 1994).

- *Trees can improve soil pH and reduce aluminium toxicity.*
  In this respect the leaf, branch and bark litter from trees can contain varying percentages of plant nutrients depending on the species. *Gmelina arborea* is a calcium accumulator, for example. However, the rotting litter from pines and other conifers tends to *reduce* soil pH (Wild, 1988), and the accumulating leaves of resinous species such as eucalyptus may inhibit the growth of some ground cover plants.

- *Trees enhance their microsites.*
  The topsoil around trees growing apart may often appear more fertile than that of the surrounding soil. Some of this is due to the factors already discussed above with regard to the cycling of nutrients, but there are other considerations also. Trees tend to be 'accumulators'. In the dry season they collect wind-blown dust which is then washed off when the first rains come, mainly as stemflow. Similarly, trees will intercept rainfall in which small traces of nutrients are consistently found. Birds perch and insects visit and/or make homes in tree canopies; in some cases nitrogen-fixing bacteria live in the gut of these insects. Thus residues containing plant nutrients fall to the ground. Trees also attract other small animals which live in and around them, and people and domesticated animals tend to rest in tree sites. All of this may bring and deposit nutrients which will build up in the topsoil over time. Such accumulations may seem minor but they can build up over time and, certainly in dry regions, they contribute a significant part of 'microsite enrichment' as seen in the field (Belsky *et al.*, 1989). Here again, these nutrient flows need to be quantified and some idea of the time-scale over which accretion occurs obtained. Clearly, measurable changes in the nutrient status of the topsoil may, depending on the circumstances, take some time to occur, especially if capture from nutrient-poor subsoil layers with few roots is also low.

The new equilibrium of SOM in these topsoil layers that will be achieved once a tree has become established on a new site will thus be occurring over what might be termed 'ecological' time. In agroforestry, however, we have to deal with a farmer's time horizon which is likely to demand a perceived change for the better in just a few years – i.e. in 'agricultural' time. Thus, although we know the *processes* that are occurring with nutrient accumulation, nutrient cycling and the degradation of plant residues, either on or in the soil, there is clearly much to be learnt with regard to how and to what extent processes may benefit any particular agroforestry system. Because there are so many factors to consider quantitative proof about the magnitude of the processes that trees bring about in particular agroforestry situations is only slowly being acquired. Meanwhile it can be unwise to make too many generalisations.

There is a great deal of circumstantial evidence that, on some soil types, quite small amounts of SOM can have large effects on crop productivity (Jenkinson, 1988; Woomer and Swift, 1994; Young, 1997). The changes in different SOM fractions, and in the chemical and physical improvements in soils brought about by trees, is partly due to the proliferation of soil biota which occurs after the deposition of tree litter and the amelioration of the surface soil environment that occurs in these circumstances (lower soil surface temperatures, an improvement in topsoil water status). The consequences of these changes on soil biota are often profound but, because the numbers of different groups of soil organisms are very volatile, it is not at all easy to relate SOM through soil biota to a fertility response. Both total numbers and kinds of soil inhabiting organisms can fluctuate greatly over short periods and these changes can be difficult to measure (Swift *et al.*, 1979; Anderson and Ingram, 1993), and to interpret in terms of their effects on soil fertility (Woomer and Swift, 1995).

There are, indeed, many factors that can contribute to improved soil fertility, and so to increased productivity and sustainability of systems in which trees are growing. Certainly, sustained production from low-input agroforestry systems in the tropics depends largely on an adequate supply of plant residues, i.e. on a suitably high level of total biomass production. Our task is to be able to predict, for any agroforestry system, what biomass is to be expected, how much plant residue material will be formed, what will happen to it and what its impact will be.

Physical changes take place in the soil under trees. More particularly rainfall infiltration at the soil surface is increased as a consequence of the deposition of above-ground litter, and tree roots and associated soil biota form a network of macropores throughout the profile they occupy. These combine to enhance water flow into and through the soil helping to fill the lower layers of the soil profile but, under some circumstances, encouraging leaching of nutrients to deep drainage. Obviously, the likelihood of this occurring will depend on the amount and character of the various rainfall episodes, the nature of the soil and the characteristics of the tree root system. Fortunately, water tends to flow relatively rapidly down macropores and movement through the soil volumes between is much slower. In fact, soluble nutrients such as nitrogen, potassium, and magnesium may be 'channelled' along root surfaces in this way.

The need for scientific knowledge

Are our information sources adequate for agroforestry? MPTs are not a particular 'breed' of woody plant and a great deal of information is available from a range of literature sources on how trees and other woody plants behave, and exactly what the possession of 'a woody habit' implies. As an example, tea and coffee crops have been studied in detail (see Chapters 8 and 10) and the reports of detailed scientific investigations on these and other plantation and forest tree species provide extremely rich sources of information that can underpin *any* further work with MPT species. There are also many excellent general publications on trees and woody vegetation that can be useful in agroforestry (e.g. Le Houérou, 1980; von Maydell, 1986; McKell, 1989; and the publications of the Nitrogen Fixing Tree Association, the Commonwealth Scientific and Industrial Research Organization (CSIRO) (Australia), the Oxford Forestry Institute, the Food and Agriculture Organization of the United Nations (FAO) and ICRAF), including information about their propagation (e.g. Leakey *et al.*, 1992).

If such a wealth of information is available why do we need more? It is worth pausing to consider the reasons. First of all, although the *processes* involved in plant and plant–environment interactions when growing woody and non-woody plants together are well known, the actual *extent* to which they occur in any particular agroforestry situation still needs to be established. Agroforestry mixtures involve many tree/crop interfaces (TCIs) the outcome of which will be highly site-dependent, and so require testing in the field. Finally, although many of the sources of information mentioned above provide a sound scientific foundation on which to base progress in agroforestry, the actual questions addressed in ecology, forestry, horticulture, etc., have not gone far enough in some instances. Surprisingly, perhaps, considering the range of detailed studies undertaken over the last 100 years or so on woody/non-woody plant associations, ecologists do not have all the answers we require. For example, although competition has been studied in considerable depth, ecological studies are usually concerned with which species ultimately 'win' and which 'loose'. In agroforestry winners or losers are not required. It is desirable that all the plant components not only survive but flourish! Thus, although ecological research is often relevant, and points in the right direction, the results to hand may not be immediately useful.

Agricultural intercropping research is another well-established source of information and relevant ideas. The concepts of environmental resource capture and resource utilisation that have arisen from studies on crop mixtures, as they have been extended from original work on monocultures, are basic also to agroforestry (Ong et al., 1996). However, in dealing with the particular mixtures of plants of diverse stature and of varying lengths of life cycle found in agroforestry, intercropping concepts derived from agriculture have to be modified and extended (Ong and Huxley, 1996).

In some cases information needed for agroforestry *is* still largely lacking. Investigations on roots and rooting in woody/non-woody plant mixtures is just inadequate (Huxley, 1996; van Noordwijk et al., 1996). Here the task is to define more clearly what questions need to be answered, and then to mount sufficient effort to obtain the necessary information; such initiatives are now being made (e.g. ICRAF, 1996).

All the factors mentioned so far have to be considered alongside the economic and social consequences of choosing trees for inclusion in a particular land-use system. The possible benefits of biological diversity may or may not be accompanied by greater security and economic stability. The opportunities to create a greater amount of biomass by improving the efficiency of the system in technical ways has to be tempered by a farmer's ability to provide the skills for producing it and the labour needed. Improvements in nutrient capture, nitrogen fixation and nutrient recycling will only have a beneficial economic consequence if a farmer takes advantage of the improvement in topsoil fertility that may occur by growing appropriate crop plant species.

Many of the other characteristics of MPTs that need to be evaluated relate to how they may (a) contribute to the overall performance of the land-use system as a whole, and (b) to what extent they are 'compatible' with associated plants. Obviously (a) depends to some extent on (b). These factors are considered in more detail below.

Agroforestry systems contain both MPTs and crops (and/or grasses), but some crop species will be more susceptible to competition from MPTs than others. However, the choice of crops is usually dependent on the needs of the farmer, who will choose from what is familiar and what is wanted. There have been some attempts to breed certain crop species to fit agroforestry practices. For example, the International Potato Institute (CIP) in Peru has screened Irish potato varieties for shade tolerance; a useful characteristic in an under-storey crop. The Centro International de Agricultura Tropical (CIAT) in Colombia have, similarly, screened grasses for shade tolerance for use in tree–grass mixtures. The International Crop Research Institute for the Semi-Arid Tropics (ICRISAT) in India has selected pigeon pea (*Cajanus cajan*) so as to reinstate the older more-branched, perennial types (Daniel and Ong, 1990), which farmers tend to prefer in agroforestry because they can be ratooned, and they can provide fuelwood as well as yielding a crop of seeds. Pigeon pea is, in any case, a short-term woody perennial, as is cassava (*Manihot utilissima*); both of which are useful in agroforestry systems because they can extend the cropping season.

The choice of crop will, therefore, be farmer-preferenced, and the variety will be that best adapted to local conditions and providing the kind of product required, i.e. it will fit the growing season, and it will be managed in a way that suits normal agronomic practice. Thus, after selecting an appropriate MPT that will best suit the technology and the site, the proportions of tree and crop need to be assessed and the spatial arrangements decided upon.

HOW MANY TREES? AND HOW ARE THEY BEST ARRANGED?

The answers to these questions are closely linked with the topics discussed so far. Clearly, the choice of species is important, but then so are other factors such as planting density and the way in which the different plant components can be mixed together. Furthermore, as discussed later, plant management can also be used to regulate tree growth, particularly training and pruning. All these affect how 'compatible' or otherwise the woody plants are when intercropped with non-woody crop plants or pastures. In other words, to what extent they will *compete*.

Planting density and yield

A great deal is known about the relationships between planting density and yield. These are fundamental issues that apply to both herbaceous and woody plants. Basically, this relationship can be described by a curve that rises to an optimum yield where a plant part such as the grain is to be harvested, and by one that reaches a plateau where total biomass is the eventual output. Actual responses are obtained for individual crops by field experimentation (see Cannell, 1983). However, farmers rarely use optimum planting densities to maximise production because they are concerned with conserving seed in case an early crop failure requires them to resow. They are cautious and usually expect and prepare for adverse weather conditions such as poor rains, in which case optimum production occurs at a lower planting density anyway (Huxley, 1983a).

'Intimacy' and the 'tree/crop interface'

The yield of each of the components in an agroforestry system would be dependent upon its yield–planting density relationship if it were being grown as a sole crop (i.e. if each component was being grown in separate but contiguous zones). However, it is necessary in most agroforestry practices to consider another factor – the *intimacy* of the components. Plants only compete if they are close enough to do so, so that the opportunity for competition between the woody and non-woody plants to occur will, in part, depend upon the extent of the 'tree/crop interface' that exists (Huxley, 1986; and see Fig. 9.15). The *degree* of competition will depend on

Fig. 9.15 A 'tree/crop' interface with *Senna* (syn. *Cassia*) *siamea* and maize at Machakos, Kenya. This aboveground interface profile is typical of those often found between woody and non-woody plants on flat land. However, in such cases the reduction in crop growth close to the hedge is probably mainly due to shading and *all* the crop will suffer a yield reduction because the tree roots extend right across the plot. Sometimes a positive effect on the crop can be found (e.g. see Huxley *et al.*, 1989), and on slopes 'scouring' of topsoil and other artifacts can bring about asymmetrical crop yield profiles (see Huxley *et al.*, 1989; Garrity, 1996). (Photo: Peter Huxley.)

the various species and their growth attributes, the characteristics of the site (i.e. the level of availability of environmental resources light, water and nutrients), and the way the plants are managed. From what has already been said about the nature of woody perennials it is also going to be affected in simultaneous cropping practices by the *time* over which the various components are actually growing when together.

Environmental resource capture and utilisation, competition and complementarity

It is possible to measure the efficiency with which any crop grows by the amount of environmental resources it can capture and how these captured resources are then utilised. These are complex issues which are beyond the scope of this chapter, but which are discussed in greater depth elsewhere (e.g. Squire, 1990; Anderson and Sinclair, 1993; Ong *et al.*, 1996). Agroforestry can be more successful than conventional agriculture in biological terms only if the combination of woody and non-woody plants, together, capture more environmental resources and the trees do *not* compete unduly with the crops. In other words, will a mixture of trees and crops capture more light, water and nutrients than when the component species have been grown separately? The answers are going to depend on how the individual components develop and maintain active leafy canopies, and proliferate root systems, either more intensively or throughout a larger part of the soil profile. If, together, a mixture of plant components *do* capture more resources then, potentially, more dry matter will be produced in the system as a whole as long as utilisation efficiencies remain the same or are, perhaps, enhanced. However, a key issue in agroforestry is that this has to be done so that the environmental resources are *shared* appropriately; the system will not be successful in the farmer's view if one component succeeds to the detriment of others (Fig. 9.16).

Fig. 9.16 An experimental plot with *Sesbania sesban* and maize at Makoka, Malawi. Clearly, there is not enough water in this system to sustain growth of both the tree and the crop (see Cannell *et al.*, 1998). Simultaneous agroforestry systems will be successful where the woody component uses resources that the crop cannot. (Photo: Peter Huxley.)

When proximal plants grow together in mixtures there will always be a degree of competition for some of the environmental resources over some of the time. If competitive interactions are severe the mixture will yield less well than if the plants were grown apart ('underyielding'). However, there may also be an element of 'facilitation' taking place (Vandermeer, 1989). This is when one plant component can alter the environment so as to aid another to change in form, modify its life cycle and/or behave physiologically so as to capture and/or utilise resources more efficiently. One example would be if a taller plant was casting light shade that benefited the under-storey crop (Ong and Black, 1994, for groundnuts under pigeon pea) although they are both still competing for water and nutrients. These situations are obviously very complex both above- and belowground but, if the outcome of competition and facilitation is positive, that is the plant mixture yields more than it would have if grown as separate sole crops, then they are said to 'complement' one another ('overyielding'). Where plant components are growing together at the same time this is 'spatial complementarity', although even when planted together they may actually grow at different times, at least to some extent. This can often happen with tropical woody species that will grow at times other than in the crop season and thus use more environmental resources (e.g. capture more sunlight) than they otherwise would. Such extension of resource capture time is termed 'temporal complementarity'. We can express competition/complementarity diagrammatically in a diagram that relates the output of a system containing two plant components grown together (see Fig. 9.17).

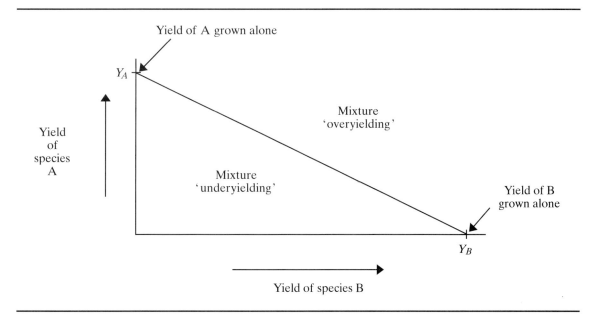

Fig. 9.17 Representation of the relationship of potential yield (or biomass production) with changes in population density in a two-plant component mixture. Y_A and Y_B represent the yields/biomass of the monocultures (each at its optimum planting density). If the combined yield/biomass of the mixture A plus B is *greater* than that expected according to the ratio in which they are present (i.e. if it is above the line $Y_A - Y_B$) then 'facilitation' interactions have exceeded 'competitive' ones and/or environmental resource capture has been enhanced. If the mixture yield/biomass is *less* than expected (i.e. it falls below the line $Y_A - Y_B$) then 'competitive' processes are the greater. (Based on Vandermeer, 1989.)

Clearly, in agroforestry we are attempting to utilise a greater proportion of available environmental resources *and* to share and utilise them in such a way that complementary relationships outweigh competitive ones (see Cannell *et al.*, 1997).

To return to the questions about the number of trees and how to arrange them, it can be seen that before these can be answered it is necessary to know what the effects of the TCIs will be (Huxley *et al.*, 1989); and to remember that competition above-ground will be for light (i.e. one-sided competition, where taller-statured plants override their lowlier companions), and below-ground where competition for water and nutrients will be two-sided (i.e. the trees may be adversely affected by competition from lower crop plants or grasses).

If there is an adverse affect at the TCI it is often possible to manipulate the design of a practice so as to minimise the extent of the TCIs (e.g. by growing the various plant components in strips or 'zones' rather than a scattered one among the others). If the adverse effects are small (i.e. trees and crops interact favourably one with the other), then issues other than rearranging the TCIs can take priority. Clearly TCIs can be minimised if systems are arranged in blocks or strips ('zonal' practices), and they will be maximised where trees are scattered in amongst crops or grassland ('mixed' practices). The intimacy of the different plant components at the TCI can also be affected by arranging for crops and trees to be nearer or further apart from one another, or a strip of another compatible crop component could be introduced next to the trees in a multicrop system.

The number of MPTs in a particular system will, therefore, depend on the farmer's preference for the proportion of woody outputs and effective services, plus the need to arrange them best so as to minimise competitive interactions and emphasise facilitatory ones. This highlights the need to understand the relevant plant characteristics, both in terms of form and function, so as to be able to evaluate and adjust interactions in these woody/non-woody plant mixtures. This is another important area where practical people and scientists need to get together. The alternative to this is to try all the possible combinations and management practices and see which turns out to be best! It is for this reason that computer modelling of agroforestry systems has a very important part to play in predicting what will happen when either we change the components, or we move the same components into a different environment, or manage them in a different way (Lawson, 1995).

MANAGING AGROFORESTRY SYSTEMS

Management of crops and pastures

This will usually be the 'standard' practice for any particular crop species or grass sward, if these are the farmer's priority enterprises. However, where emphasis is being given to the tree products there may be a case for picking a shorter-season crop cultivar and/or for maintaining a pasture so that it is less competitive. Particularly in drier regions, grasses can severely restrict the growth of woody plants, mainly by below-ground competition for water and nutrients. Grass fires can also be a severe hazard to young trees. For crops, sowing time may also be manipulated and, in establishing any practice, there may be a case for planting the woody species before the grasses are established.

The crop components will determine the type of soil management to be practised. Deeper-rooting tuber crops such as yams (*Dioscorea* spp.) and cassava (*Manihot utilissima*) will create some soil disturbance at harvest which will interfere with the root systems of the woody components, but this may be acceptable. Where there are sufficient plant residues to afford a ground cover of mulch, some form of minimum or zero tillage might be the most appropriate for the system as a whole. However, this is likely to encourage tree roots to proliferate extensively in the soil surface region and deep cultivation near to, say, hedgerows might be done in order to limit the below-ground competition from the trees.

Agroforestry will obviously not automatically

guarantee high outputs and, depending on crop, soil and climate some levels of inputs will often be needed if this is the aim. If fertilisers are to be used they will normally be applied to the crops, but there is still much to be learnt about the best ways to apply fertilisers in agroforestry practices; the opportunities for various forms of placement are greater than they are in conventional agriculture. Similarly, irrigation and pesticides will need to be used with an awareness of the greater complexity and, perhaps, wider range of opportunities that agroforestry systems offer.

Managing the woody plants

Practical concerns about the choice of tree characteristics that arise when considering agroforestry applications from biological, managerial and socio-economic points of view are discussed in Huxley (1983c) and listed in Huxley (1983d).

MPTs: what size and shape?

One of the major decisions that will have had to have been made when helping to improve an existing agroforestry system, or to design a new one, will be to what extent it is desirable to have large or small trees/bushes/hedgerows. This will initially be established by the choice of species but, with woody perennials, there is a considerable amount that can be done to either exploit or contain growth potential.

Early training will establish the shape and size of the woody framework within the limits set by the species. An elevated tree canopy will allow for better light distribution (Wallace, 1995), and each tree may be single or multi-stemmed. Or, to avoid competition with a crop, it may be useful to reduce the size of the trees/shrubs and keep them that way, as in hedges. Restricting the seasonal growth and overall size of a woody plant can be labour-consuming, so that it is best to choose species/provenances that inherently meet our current and future needs. As indicated below, one consequence of restricting the canopy will also be to restrict the size of the root system, and possibly also its form.

So do we want vigorous multipurpose trees?

Even in these circumstances, if a vigorous genotype is chosen there may be an unwanted level of competition exerted by the tree/shrub. The question of 'Do we want vigorous MPTs?' is an important one. Vigorous trees will produce more biomass, but an increased competitiveness is to be expected that may well offset the potential benefits that can arise from improved soil fertility through the provision of woody plant residues, shelter, and so on. These two opposing factors in, say, a hedgerow intercropping system can be represented by a simple equation (Ong, 1996):

$$I = F + C$$

where I is the overall tree/crop interaction, as a percentage of sole-crop yields; F is a 'fertility' effect, and this would strictly include any other benefits such as improved soil surface temperatures, shelter, etc.; and C is the competition effect, i.e. the decrease in crop yields caused by competition from the tree for light water and nutrients. (There are other factors that can be added, such as the long-term effect on site soil fertility, offsetting soil and water erosion, see Ong (1996) for a full account).

If the appropriate control plots exist in a field experiment then I, F and C can be obtained. Figure 9.18 shows a comparison of F and C values obtained from an experiment involving hedgerow intercropping with maize and both *Senna siamea* and *Leucaena leucocephala* at Machakos in Kenya. Clearly, there can be considerable differences in both factors depending on environmental conditions and management as well as on species. Analysed in this way a number of hedgerow intercropping field experiments have been shown to provide only limited positive tree/crop interactions as far as crop production is concerned (Ong, 1994; Sanchez, 1995).

Depending on farmer requirements the products and/or services from the woody plants might be more important than the crop yields in some circumstances. So in order to evaluate an agroforestry system thoroughly the above

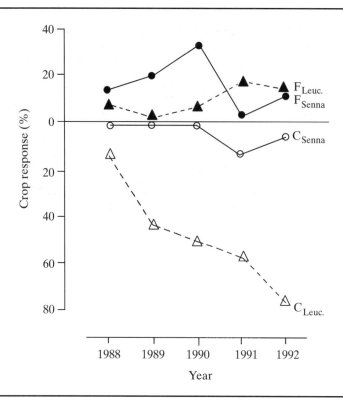

Fig. 9.18 Effects on maize crop yields (as ±%) of the 'fertility' (*F*) responses (due to hedgerow mulch effects) and the competitive effects (*C*) of the hedgerows of *Senna* (syn. *Cassia*) *siamea* and *Leucaena leucocephala*. The competitive effects of *Leucaena* far exceeded its fertility benefits in this case especially as the system ages and *Leucaena* roots occupy more of the soil. (Redrawn from Ong, 1996, from data provided by Paul Kiepe.)

analysis should be carried out so as to indicate also how *F*s and *C*s affect *tree* production. For example, one estimate in a *Grevillea robusta* and maize experiment at Machakos in Kenya of the affect of the crop on a hedge over 4.5 years showed a 30% decrease in hedgerow growth next to the maize (Huxley *et al.*, 1994). Grasses are very competitive and can severely restrict the growth of adjacent trees (e.g. Howard, 1924). Unfortunately, very few data are currently available in agroforestry situations, largely because most agroforestry field experiments have failed to establish all the necessary control plots (crops ± mulch and ± hedgerow, and hedge-alone ± mulch).

Although in many drier regions, where competition for water at some stage can be a predominant factor, hedgerow intercropping (alley cropping) has not proved as successful as it was first thought it might be, it does have a very useful function in providing an effective method of soil and water conservation in such regions (Kiepe and Rao, 1994). If established mainly for this purpose the hedgerows ('barrier plantings') can be kept pruned back in order to restrict competition with the crop (Fig. 9.1).

The beneficial effects of windbreaks are well documented (Brenner, 1996), but there is some evidence that even small barriers, such as hedgerows, can protect crops on the downwind side in

Fig. 9.19 The effects of shelter from a 1 m high inert barrier (a papyrus fence) on small plants of maize at Machakos, Kenya. Exposure to prevailing winds occurred from the bottom right where the maize can be seen to be growing less well than behind the fence to the left. (Photo: Peter Huxley.)

some circumstances (Fig. 9.19). Because this is a situation that can occur frequently in many different agroforestry practices it needs to be investigated further.

Improving the efficiency of the system as a whole

Maximisation of biomass can only be achieved by designing and managing the system so as to maximise the capture of light (PAR – photosynthetically active radiation). This is because, other factors not being limiting, there is usually a linear relationship between accumulated light interception and biomass production (or yield if this is a consistent part of biomass). Other things often are limiting; most usually available water. Water may be lost from a system through canopy interception, surface soil runoff and leaching to deeper regions. In addition, the water evaporated from the soil surface is not used by the plants and therefore does not contribute to biomass production. It may constitute a significant fraction of 'wasted' water in systems depending on how widely spaced the plants are and the type of rainfall, i.e. whether there are just a few heavy showers or many light ones. Agricultural crops often use relatively little of the total rainfall (e.g. at Hyderabad in India the millet crop may use as little as only 40% (Ong, pers. comm.). Interestingly, there is also often a linear relationship between the dry matter produced and the water used by plants in a cropping system (the water use ratio or transpiration:dry matter production ratio, Squire, 1990).

Within the genetic limits on the efficiency with which light and water can be used by the component plant species these two relationships, well established with agricultural crops, suggest how to maximise the biomass from agroforestry systems. However, this ignores competitive interactions, which can be overriding in dry regions (Cannell *et al.*, 1998). The first need is to establish how much water is available in the system, how it is likely to be partitioned to the various tree and crop components, and how its use can be maximised over the year. In regions where water is freely available it is essential to maximise the capture of light by whichever components are able to grow, or be grown at those times (i.e. to maintain an active canopy). Again, over-storey and under-storey plans have to share incoming PAR and this may have to be regulated by a combination of design (type of species, proportions of components, planting densities and plant arrangement), and management (time of sowing the crop, time and amount of pruning the woody perennial). By these means biomass will be maximised and the relative yields of the different components maintained. The system will operate efficiently as long as sufficient plant nutrients are available. The greater the production of biomass the more nutrients are required, but they are also needed at the right times and all essential elements must be available in the right amounts. In this respect, SOM will encourage root–soil interfaces in the nutrient-rich topsoil, and the labile fractions will boost availability, especially of nutrients such as phosphorus that are commonly fixed in many tropical soils (Chapter 3).

Obviously, it is no easy task to balance all these factors, but it is now known how they are interrelated and how to identify what factors may be limiting so as to avoid or overcome them through appropriate interventions. If agroforestry practices are to be technically successful it is necessary to apply the knowledge gained from agricultural research and apply it to these more complex systems. These aspects are dealt with more thoroughly in Ong *et al.* (1996). The possibilities for exploiting the environment in a more thorough way by using tree/crop mixtures is one of the promising aspects of agroforestry. However, it is clear that this can be achieved only by understanding how the plants are growing together and by designing and managing the system accordingly. Just hopefully throwing together some trees and crops may achieve very little!

Replacing the system

Agroforestry systems will 'age'; as they grow older the woody plants extend their canopies above the ground and their root system below it and it is necessary to understand these processes in order to manage such systems successfully on a continuing basis. Eventually, when the woody plants have become senescent they will have to be removed and replaced. It is usual to find that little thought has been given to the best way to renew an existing agroforestry system. For example, if the protection from a windbreak is not to be lost there has to be some plan for replacing parts of it (e.g. perhaps individual rows), or replacement strips have to be planted between existing ones in good time. Planting up young trees in gaps among older ones is seldom successful because the roots of the latter are well established and the larger trees shade out the younger ones. Replacing hedgerows also needs careful planning and renewing old trees scattered among crops or in pasture will also present problems.

Dry matter distribution, rooting and fruiting

Sources and sinks

In agroforestry it is important to maximise the yield of the required plant parts, as in agriculture. To this end it is necessary to understand how plants distribute their dry matter and what precisely can affect this process. Depending on the species and plant genotype, dry matter is distributed in plants according to a strict set of 'rules' which are well understood (Ledig, 1983; Cannell, 1985). Within a particular environment a plant will allocate the carbon assimilates formed by its photosynthetic 'sources' (leaves and other green parts) to various 'sinks', or sites, where they are used, such as stem cambia (which include trunks, main branches and twigs), or roots, storage organs such as tubers, and flowers, developing fruits and seeds. But within the plant these parts *compete* for resources, including carbon assimilates, and it is for this reason that one cannot expect to maximise the production of any two growing parts if they are drawing substantially on the same resources. For example, fruits and seeds are powerful sinks for carbon assimilates, either newly formed or stored, even at the expense of the growth of other parts such as growing shoots and roots. This 'demand and supply' situation will depend on the stage of development reached by the plant, and especially whether the plant is flowering or fruiting. In any growth phase there will be a constant relationship ('allometry') between parts, including the overall shoot mass and the overall root mass. If anything occurs to disturb this relationship (i.e. the farmer prunes part of the canopy), then subsequent growth of the plant will occur so as to restore the balance (root growth will be restricted compared with what it was before in order to allow for more shoot growth).

This needs to be appreciated when managing agroforestry systems as, together with ideas of resource capture and resource utilisation, it is one of the main functional aspects of plant growth that can be manipulated. If trees and shrubs have aged, and the canopy is getting old and inefficient, then some form of 'renewal pruning' will help restore its photosynthetic efficiency, only temporarily reducing light capture. Roots can also be pruned with a parallel effect that subsequent growth will tend to restore the balance of roots.

Fruiting has an important influence on dry matter distribution. Not only does it alter the proportion of carbon assimilates that are allocated throughout the plant (in some cases as much as 60% or more of the annual dry matter increment can move into developing fruits, e.g. as in coffee), but fruiting also enhances the overall growth of the plant (Cannell, 1971). The problem with woody perennials is to prevent both over- and under-fruiting. The number of fruiting sites that develop in any one year depends not only on the current season but often on the previous history – in fact flower initiation might have occurred in the previous season, or even on older wood and/or short shoots ('spurs'). If a plant is growing in a stressed situation then irrigation and fertiliser can enhance fruiting potential, but excessive amounts of either will tend to promote vegetative growth and inhibit flower initiation, as will shade. When pruning, initiation sites can be left or removed and the way the tree is pruned will either promote or not promote new structures that will become future fruiting sites. This is where some of the skill in managing woody plants is needed.

It may be of interest to *prevent* flowering and fruiting, as with *Leucaena leucocephala* grown for leafy fodder. This species flowers terminally on new wood, thus pruning or grazing young shoots will remove flowering sites and promote the growth of new vegetative laterals, which it does readily; it is this that makes *L. leucocephala* a productive hedgerow species. Pod production, with the consequent fall of seeds, often encourages a heavy growth of young *Leucaena* seedlings which have to be weeded or grazed in order to control them. New strains are now being bred to minimalise fruiting potential (Brewbaker, 1987).

Fine root turnover

One other aspect of dry matter distribution is of considerable importance in agroforestry. As mentioned above, there is an equilibrium established depending on species, site and management, between the ratio of above-ground to below-ground dry matter. Only some 30–40% of the annually produced below-ground biomass may be recovered when trees and shrubs are dug up. Trying to recover all the currently existing fine roots is notoriously very difficult. However, some critical studies of woody plants (mainly north temperate coniferous species) have shown that considerable amounts of photosynthates are transferred below-ground during the year, much of which is lost as 'fine root turnover' (Schoeetle and Fahey, 1994). The amounts involved for even the most important of MPT species is yet to be determined. It is to be expected, however, that even if tropical woody species do not show as much variability in their patterns of growth below-ground as they do above, there will still be large differences to be observed in fine root turnover. Nevertheless, as the potential for depositing large amounts of root litter below-ground may be present this is a factor that needs to be taken into account when considering the contribution of trees to the plant residues in an agroforestry system.

From what has been said about the allocation

of plant dry matter it might be expected that there is a likelihood that fine roots will be shed if the aerial parts are pruned hard. Again, species will probably differ in this respect and this is another area in which more data are needed. In addition, root residues from the crop plants and/or grasses also need to be considered (Bowen, 1985).

Tree form, buds and pruning

Manipulation of the tree canopy for one purpose or another in agroforestry is clearly important. It is achieved by training and pruning. Woody perennials conform to distinctive architectural patterns (Hallé et al., 1978; Barthélémy et al., 1989a, 1989b) which can indicate what the overall shape will be, but of immediate practical importance in pruning is to know what buds exist in the axils of leaves and how these will respond when a stem apex is removed. Pruning consists of two basic operations, 'heading-back' and 'thinning-out'; terms which are self explanatory (Huxley, 1985). Thus a woody plant can be encouraged to fill out the canopy volume or to attenuate it, while bearing in mind the number and positions of fruiting points that are needed. Considerable skill is required to do this, and there are few, if any, detailed studies for MPT species, where pruning usually means just cutting a hedgerow back to a stated height, or lopping off a few branches from a tree.

Pest management

There is some evidence that plant mixtures can, in some circumstances, limit the incidence of insect pests and plant pathogens (Altieri and Liebman, 1986; Altieri et al., 1987; Bird et al., 1990). One species may act as a mechanical barrier for the transfer of pests of another; there may be changes in microenvironments that inhibit pest development, the greater the variety of plants in a mixture the more likely it is to be a favourable habitat for pest parasites and predators, and so on. So far there are few studies in agroforestry situations which can help us determine where such cases are important (Huxley and Greenland, 1989). The converse (that perennial woody plants can act as perpetuating sites for pests which can then spread to crops) is certainly known to be the case. Woody plants will also encourage termites. This may be helpful as part of the process to break down dead plant residues in the process of SOM formation, or detrimental where plant-attacking termite species are encouraged (Roy-Noël, 1979). One beneficial effect of an elevated canopy is that, combined with a lower level crop canopy, there may be a considerable amount of shading imposed on any weeds present, thus retarding their growth. This is still considered one of the benefits of shade in coffee growing.

Even if a generally higher level of biological pest control is achieved in agroforestry systems (especially in such practices as multistrata home gardens), there may still be a case for chemical pest control. However, the application of sprays and dusts in an appropriately precise but economical manner, where canopy shape and the disposition of species may vary considerably, is likely to be difficult; as will determining an appropriate time to spray for particular pest outbreaks. Hedgerow intercropping practices present a more formalised arrangement than random species mixtures and they may therefore lend themselves more to such treatments.

There is much to be learnt about the effects of various agroforestry practices on the incidence, etiology and epidemiology of various pests and the best ways to tackle pest management. Despite attempts to draw attention to the importance of this subject, which can result in the success or failure of individual farmer's agroforestry enterprises, there has been remarkably little attention paid to it other than for a major pest of *Leuceana leucocephala* (the psyllid, *Heteropsylla cubana*).

AGROFORESTRY AND SUSTAINABLE PRODUCTION

Similarities and differences

Agroforestry in its various forms is often promoted as a sustainable way of using land, but more hard evidence is needed to support this claim. In its early days agroforestry was also promoted as a means of repairing damage done to fragile lands (King and Chandler, 1978) although, from what has been said so far, there is nothing to prevent agroforestry being practised on highly fertile land and/or with high inputs. The reason why this is not usually found in the tropics is because such lands are allocated to monocropping, often for the production of cash crops. The benefits that diversity provides, and that trees may confer on soils under the right circumstances, influences farmers to invest in agroforestry practices in less fertile areas. But generalisations are again, unhelpful. There are, certainly, some agroforestry practices that are likely to be highly sustainable in the short to medium term, for example multistrata systems such as home gardens (Torquebiau, 1992), and their replacement by agricultural cropping alone is likely to lead swiftly to land degradation; as is found when forest cover is removed.

It is hard to imagine anything that happens to land that does not affect its sustainability, so that again it is important to focus on key issues that relate to individual forms of land use appropriately. In this respect, agroforestry practices are no different from any other ways of using land. For example, at any level of productivity the sustainability of a system will depend on how much is being removed from it; it does not matter whether this is in the form of crop yields or products from trees. Thus having woody perennials in a land-use system does not automatically make it sustainable, but some of the features already discussed in this chapter suggest that, managed wisely, it may be made to be so in the short to medium term (ICRAF, 1989b; Young, 1990).

Sustainable land use has been defined as 'that which achieves production sufficient to meet the needs of present and future populations, while conserving and enhancing the land resources on which that production depends' (FAO, 1989b, 1992). But herein lies a dilemma! In terms of small areas it may not be very easy to predict what the future populations will be, although on a national or world scale the position may be much clearer (McCalla, 1994). Thus sustainable land use needs to be thought of in terms of appropriate scales (Fresco and Kroonenberg, 1992), both in space and in time over the short (less than 10 years or so), medium (up to 100 years) and long term (100s to 1000s of years).

The longer-term view is seldom addressed but information on this time scale is available (e.g. about pedogenesis) and it highlights the need to define the whole extent of the agroforestry system being considered and not, for example, just the fertility changes in the upper soil horizons. Different types of forests can exist on both deep, fertile and shallow, potentially highly degradable soils, and there is a lesson for agroforestry in considering what is known about the nutrient dynamics in forest systems (Proctor, 1989) although, as Whitmore (1989) points out, there have been no studies of *all* the nutrient pools and flows. In 'open' forest systems the slow addition of nutrients from weathered rock (saprolite) and from atmospheric sources balance those lost through leaching. In so-called 'closed' systems the weathered rock lies too deep to be a source of nutrients to the forest system (Burnham, 1989) and efficient recycling maintains the level of nutrients historically accumulated when this was not so, and from atmospheric sources, so as to maintain the forest biomass. Although it was thought that a high proportion of the total nutrients in the system were to be found in the aboveground biomass in forests, this is now known not to be so in many cases; they are conserved in the upper soil horizons (Whitmore, 1989). The important issue is that these processes are all occurring very slowly, time scales may be measured in hundreds of years.

In managed forests and plantations there is an export of nutrients in the timber removed which reduces nutrients in the system at a faster rate depending on how much is taken off. This, together with enhanced leaching during and after logging operations (when the soil is bare), often brings about a decline in fertility (e.g. Sanchez *et al*. 1985). If in tree-based systems natural replacement of nutrients is occurring over a long time scale, and nutrient decline through export (and any 'leakiness') over a much shorter one, the system is running down. That is it is ultimately 'non-sustainable', even if there are plenty of stored nutrients at present and the time when it will cease to function is well into the future. Agroforestry practices are subject to the same kind of processes, although inputs in the form of fertilisers and/or mulch derived from other systems may be provided from time to time. Because current concerns relate specifically to what will happen in the next few decades, such long-term assessments of sustainability may seem remote. Nevertheless, the human race will have to address them at some time!

Some of the factors that can increase productivity in tree–crop or tree–grass mixtures will also promote short- to medium-term sustainability. For example, the establishment and maintenance of denser plant canopies with consequent production of more plant residues, and a better level of recycling of nutrients and the establishment of higher levels of SOM. Some may not, however. Thus tillage may increase crop yields over a season but also exacerbate soil erosion (Chapters 4 and 5). The application of pesticides can, similarly, increase production in the short term but, over a longer period, it may destabilise the ecological system and act detrimentally against the parasites and predators of the crop pests. Using fast-growing MPT species implies that more water and nutrients will be utilised within the system and this may adversely affect the water table and, if yields (i.e. exported products) are high then it can ultimately lead to nutrient depletion in the system. There can, therefore, be a conflict between 'productivity' and 'sustainability' which has to be resolved depending on particular circumstances. This applies as much to agroforestry as it does to any other form of agriculture.

Assessing sustainability in agroforestry systems

Because of variability in weather, and also in management practice, crop yields can vary considerably from season to season. The situation is more difficult to assess with trees because harvests may be seasonal, continuous, intermittent or terminal at the end of the life of the tree, or some or all of these. For straightforward timber production foresters assess, non-destructively, the yearly *increment* of wood in forest trees ('current annual increment' and 'mean annual increment') so as to obtain 'yield classes' (Hamilton and Christie, 1971). In agroforestry, where the tree products can be fuelwood or pod or leaf fodder, etc., these outputs can be measured and some sort of yearly average taken. However, in agroforestry systems the long-term trend of yearly outputs will not be so easily assessed as in many agricultural practices and the evaluation of services may be equally difficult to measure over time.

The sustainability of land-use systems is often judged in terms of production, efficiency, stability and resilience. *Production* (the output per unit area of land over time) is important because this is what is usually exported, but it needs to be related to total biomass production in agroforestry so as to take into account the positive inputs of plant residues. *Efficiency* or *productivity* will clearly be diminished if a system is becoming non-sustainable, but to evaluate the system it is necessary to express production in terms of some measure of efficiency that is considered to be particularly important. For example, output per unit input of solar radiation (PAR), or of an economic unit such as labour. These can be measured and compared over time. *Stability* refers to the actual fluctuations of production season by season compared with those that might be expected. If there is an increasing departure from the 'normal' then the system may be becoming unstable. *Resilience* refers to the ability of a system to return to its normal production level after some form of perturbation; if it does not return then it has become unsustainable. It may be useful to look for key factors that might create unsustainability such as the loss of more than a certain amount of soil, or a lowering of pH. Farmers judge risk in terms of the probability of achieving certain goals together with the penalties for failing to achieve them and, in this respect, understanding the resilience of their system is vital. In agroforestry there might be a case for expecting increased production and, certainly more diverse production if the efficiency of a system is improved by using a tree–crop or tree-grass mixture. Greater overall stability is to be expected because of diversity, and the nature of woodiness suggests that such systems may have greater resilience.

Clearly it will be hard to assess or monitor sustainability in agroforestry practices. It is first necessary to decide on the kinds of attributes to be considered (these will be biophysical and socio-economic) and these 'descriptors' will need to be designated in terms of what is actually to be measured (i.e. the 'indicators' themselves). For example, soil fertility, plant cover and labour productivity might be three kinds of descriptors chosen to monitor a particular agroforestry system, and continuing measurements of the weight of soil biota (worms, soil insects, etc.), estimates of above-ground standing biomass and assessments of labour inputs for different tasks the actual indicators. Sometimes it may be desirable to place 'thresholds' below or above which the level of an indicator should not move (e.g. soil pH).

A major problem in measuring sustainability in agroforestry is in sampling. As has been noted, agroforestry systems are invariably 'patchy' so that even a simple case of sampling soil for organic matter fractions will need to be done in many more places than it would when dealing with an agricultural crop, or even an agricultural crop mixture. Because a 'suite' of indicators will usually be needed to cover the main feature of any agroforestry system, this problem is compounded. The quest for accuracy might have to be traded-off against the need to secure easily obtained, low-cost indicator measurements. For example, changes in complex plant canopies can be assessed using relatively crude, but easily replicated, methods such as the use of 'fish-eye' photography. Further discussion on sustainability indicators for agroforestry is to be found in Huxley (1995).

CONCLUSIONS

Agroforestry is a way of using land that, in the right circumstances, can help to increase productivity and it may prolong sustainable production. Its immediate benefits are to provide a range of products and services that can enhance the well-being of those actually occupying the land. There may also be some economic benefits such as reducing risk and possibly increasing income and/or stabilising it over a longer period. As with all land-use practices involving woody perennials, agroforestry practices involve an 'investment' of resources over time.

Some indigenous peoples have, in the past, evolved very suitable agroforestry practices through the use of local knowledge, and painstaking trial and error over a period of time. However, for any particular set of objectives for current agroforestry development thrusts, finding answers to questions such as 'What multipurpose trees to use', 'How many are needed?' and 'How best should they be arranged and managed?' will not be acheived most effectively by haphazardly trying out selections of woody and non-woody plant components and including some animals if they are needed. These apparently simple questions involve complex issues both biophysical and socio-economic that can only be solved by bringing practitioners and technologists together; the former to provide indigenous knowledge and experience, the latter to tap the wealth of scientific understanding that can help clarify how such systems can function. Although there are some topics where insufficient is known (e.g. for below-ground situations), there is a huge amount of information existing as the product of detailed research in allied fields of study such as forestry, agriculture, horticulture, ecology, soil science, etc., but much of it needs extending and elaborating before it can be immediately useful in agroforestry situations. This is, therefore a challenge for agroforesters!

Finally, agroforestry should be treated as part of a whole spectrum of sensible, potentially productive and sustainable ways of using land, as described in other chapters in this book. It holds promise only if we realise its limitations as well as its potentials.

FURTHER READING

There is now a substantial literature on agroforestry. Many journals publish papers on various aspects of the subject, but *Agroforestry Systems*, and *Agroforestry Abstracts* (both produced in cooperation with ICRAF) are dedicated to agroforestry alone. However, from what has been said in the text, much of what we need to know may not directly mention 'agroforestry' at all.

The topics covered in this chapter are dealt with in more detail in *Tropical Agroforestry* (Huxley, in press for Blackwell Science). General reading, mainly descriptive, is provided by *An Introduction to Agroforestry* by Nair (1993), which gives brief descriptions of some common tropical MPT species (plus references to this literature also). Useful specific accounts are to be found in Rocheleau *et al.* (1988) and Tejwani (1994), for Africa and India, respectively, and more generally in Gholtz (1987), Steppler and Nair (1987), Avery *et al.* (1991), MacDicken and Vergara (1990), Jarvis (1991). Bentley *et al.* (1993) and Tewari (1995). Soil aspects are covered by Young (1997).

10
Tropical Crops and their Improvement

N.W. Simmonds

This chapter is intended to give a bird's-eye view of the economic botany and breeding of tropical crops. Because the total list is a long one, the treatment is superficial but will it is hoped give the reader a broad picture of the marvellous diversity of botany and uses that the tropical agriculturist may expect to encounter. The conspectus in the following section therefore sets the scene by listing many species and identifying a fairly arbitrary selection of 18 crops that are deemed worthy of more detailed treatment because of their socio-economic importance. Other authors would no doubt have given different lists but choices had to be made. In the following section, very brief sketches of the chosen 18 crops are presented, along with collective treatments of the far more numerous minor crops.

The third and last section of the chapter presents a summary of plant breeding explicitly built upon tropical crops. Plant breeding principles are universal but the socio-economic circumstances in which they apply are very diverse. Hence an exclusively tropical concentration seemed appropriate for this book. The reader will find that there are many natural cross-connections between this and other chapters in this book; in general, the links are not identified because they are mostly obvious and much cross-referencing tends to obstruct clear exposition.

Leading references to explicit crops or subjects are given below, in context. There are a few general references that are given in the further reading section at the end of this chapter.

Conspectus of tropical crops

This is a list of the crops that are at least mentioned in the text. Those treated explicitly in the following section are named in **BOLD CAPITALS**. The order of listing is arbitrary, by use, families being arranged alphabetically within sections.

The following abbreviations are used to indicate area of origin, habit and life cycle: Am (America), Af (Africa), As (Asia), H (herbaceous), W (woody), A (annual), P (perennial).

(1) STARCHY FOOD GRAINS

Amaranthaceae: grain amaranths (*Amaranthus* spp., Am, H, A)
Chenopodiaceae: grain chenopods (*Chenopodium* spp., Am, H, A)
Gramineae, grass cereals: **RICE** (*Oryza* spp., As, Af, H, A), **SORGHUM** (*Sorghum bicolor*, Af, H, A), **MAIZE** (*Zea mays*, Am, H, A), **BULRUSH MILLET** (*Pennisetum americanum*, Af, H, A), finger millet (*Eleusine coracana*, Af, H, A), teff (*Eragrostis tef*, Af, H, A)

(2) STARCHY TUBERS

Araceae, tuberous aroids: giant taro (*Alocasia indica*, As, H, A/P), taro, eddo, dasheen (*Colocasia esculenta*, As, H, A/P), giant taro (*Cyrtosperma chamissonis*,

As, H, A/P), tannia, yautia, cocoyam (*Xanthosoma sagittifolium* et spp., Am, H, A/P)
Basellaceae: ullucu (*Ullucus tuberosus*, Am, H, A)
Cannaceae: Queensland arrowroot, achira (*Canna edulis*, Am, H, A)
Convolvulaceae: **SWEET POTATO** (*Ipomoea batatas*, Am, H, A)
Dioscoreaceae: **YAMS** (*Dioscorea* spp., Am, Af, As, H, A)
Euphorbiaceae: **CASSAVA** (*Manihot esculenta*, Am, W, A)
Leguminosae: yam bean (*Pachyrhizus tuberosus*, Am, H, A)
Marantaceae: arrowroot (*Maranta arundinacea*, Am, H, A), topee tambu (*Calathea allouia*, Am, H, A)
Oxalidaceae: oca (*Oxalis tuberosa*, Am, H, A)
Solanaceae: potato (*Solanum tuberosum*, Am, H, A)
Tropaeolaceae: isañu, mashua (*Tropaeolum tuberosum*, Am, H, A)
Umbelliferae: arracacha (*Arracacia xanthorrhiza*, Am, H, A)

(3) STARCHY (AND/OR OILY) PERENNIALS

Moraceae: breadfruit (*Artocarpus altilis*, As, W, P)
Musaceae: **BANANAS**, plantains (*Musa* cultivars, As, H, P), inset (*Ensete ventricosum*, Af, H, P)
Palmae: sago (*Metroxylon sagu* et spp.), pejibaye, pupunha, peach palm (*Guilielma gasipaes*, Am, W, P)
Lauraceae: avocado (*Persea americana*, Am, W, P)

(4) PROTEIN SEEDS

Leguminosae, legumes, pulses: various **BEANS** (*Phaseolus* spp., Am, H, A), **GROUNDNUT**, peanut (*Arachis hypogoea*, Am, H, A), cowpea, bodi bean (*Vigna unguiculata*, Af, H, A), chickpea (*Cicer arietinum*, As, H, A), lentil (*Lens esculentum*, As, H, A), bambarra groundnut (*Voandzeia subterranea*, Af, H, A), pigeon pea (*Cajanus cajan*, As, W, P), winged bean (*Psophocarpus tetragonolobus*, As, H, A)

(5) OIL PLANTS

Compositae: Niger seed (*Guizotia abyssinica*, Af, H, A)
Euphorbiaceae: castor (*Ricinus communis*, Af/As?, W/H, P)
Palmae: **COCONUT** (*Cocos nucifera*, As, W, P), **OIL PALM** (*Elaeis guineensis*, Af, W, P)
Pedaliaceae: sesame, sim sim (*Sesamum indicum*, As/Af?, H, A)

(6) SUGAR PLANTS

Gramineae: **SUGAR CANE** (*Saccharum* cultivars, As, H, P)
Palmae: sugar palm, gomuti (*Arenga saccharifera*, As, W, P), palmyra (*Borassus flabellifer*, As, W, P)

(7) FRUITS AND NUTS

Anacardiaceae: mango (*Mangifera indica*, As, W, P), cashew (*Anacardium occidentale*, Am, W, P)
Annonaceae: soursop, sweetsop, sugar apple, cherimoya, etc. (*Annona muricata* et spp., Am, W, P)
Bombacaceae: durian (*Durio zibethynus*, As, W, P)
Bromeliaceae: pineapple (*Ananas comosus*, Am, H, P)
Caricaceae: papaw, papaya (*Carica papaya*, Am, W, P)
Cucurbitaceae: musk melon (*Cucumis melo*, Af, H, A), watermelon (*Citrullus lanatus*, Af, H, A)
Guttiferae: mangosteen (*Garcinia mangostana*, As, W, P)
Moraceae: jackfruit (*Artocarpus heterophyllus*, As, W, P)

Myrtaceae: guava (*Psidium guajava*, Am, W, P)
Passifloraceae: passion fruits (*Passiflora edulis*, *P. quadrangularis*, Am, H, P)
Proteaceae: macadamia nut (*Macadamia* spp., Australia, W, P)
Rutaceae: citrus fruits such as orange, grapefruit, tangerine, lime, etc. (*Citrus* spp., As, W, P)
Sapindaceae: akee (*Blighia sapida*, Af, W, P), rambutan and litchi (*Nephelium lappaceum*, *N. litchi*, As, W, P)
Solanaceae: tree tomato (*Cyphomandra betacea*, Am, W, P), pepino, lulo, etc. (*Solanum muricatum*, *S. quitoense* et spp., Am, H/W, A/P), goldenberry (*Physalis peruviana*, Am, H/W, A/P)

(8) VEGETABLES

Cucurbitaceae: several squashes (*Cucurbita* spp., Am, H, A), cucumber (*Cucumis sativus*, As, H, A)
Malvaceae: okra (*Abelmoschus esculentus*, Af/As?, H, A)
Solanaceae: eggplant, aubergine (*Solanum melongena*, As, H, A), sweet and hot peppers, annual and perennial (*Capsicum* spp., Am, H/W, A/P), tomato (*Lycopersicon esculentum*, A, H, A)

(9) BEVERAGE PLANTS

Camelliaceae: **TEA** (*Camellia sinensis*, As, W, P)
Rubiaceae: **COFFEES** (*Coffea arabica*, *C. canephora* (*robusta*), *C. liberica*, Af, W, P)
Sterculiaceae: **COCOA**, cacao (*Theobroma cacao*, Am, W, P)

(10) SPICES, FLAVOURINGS

Lauraceae: cinnamon (*Cinnamomum zeylanicum*, *C. cassia*, As, W, P)
Myristicaceae: nutmeg, mace (*Myristica fragrans*, As, W, P)
Myrtaceae: clove (*Eugenia caryophyllus*, As, W, P), pimento, allspice (*Pimenta dioica*, Am, W, P)

Orchidaceae: vanilla (*Vanilla fragrans*, Am, H, P)
Palmae: betel nut (*Areca catechu*, As, W, P)
Piperaceae: pepper (*Piper nigrum*, As, W, P), betel leaf (*Piper betle*, As, W, P), kava (*Piper methysticum*, As, W, P)
Zingiberaceae: ginger (*Zingiber officinale*, As, H, P), cardamon (*Elettaria cardamomum*, As, H, P), turmeric (*Curcuma longa*, As, H, P)

(11) DRUGS, STIMULANTS

Erythroxylaceae: coca (*Erythroxylon coca*, Am, W, P)
Rubiaceae: quinine (*Cinchona* spp., Am, W, P)
Solanaceae: tobacco (*Nicotiana tabacum*, Am, H, A)
Sterculiaceae: cola nut (*Cola nitida*, *C. acuminata*, Af, W, P)

(12) INDUSTRIAL PLANTS, FIBRES

Agavaceae: sisal (*Agave sisalana*, Am, H, P), henequen, cantala (*Agave* spp., Am, H, P), Mauritius hemp (*Furcraea gigantea*, Am, H, P)
Bombacaceae: kapok (*Ceiba pentandra*, Af/Am?, W, P)
Malvaceae: New World **COTTONS** (*Gossypium hirsutum*, *G. barbadense*, Am, H/W, A/P), Old World cottons (*G. arboreum*, *G. herbaceum*, Af/As, H/W, A/P), kenaf (*Hibiscus cannabinus*, Af, H, A), roselle hemp (*Hibiscus sabdariffa*, Af, H, A)
Musaceae: Manila hemp (*Musa textilis*, As, H, P)
Tiliaceae: jute (*Corchorus capsularis*, *C. olitorius*, As, H, A)
Urticaceae: ramie (*Boehmeria nivea*, As, H, P)

(13) INDUSTRIAL PLANTS, OTHER

Compositae: pyrethrum (*Chrysanthemum cinerariifolium*, Mediterranean, H, P)
Euphorbiaceae: **RUBBER** (*Hevea brasiliensis*, Am, W, P)

Leguminosae: derris (*Derris elliptica* et spp., As, W, P)

(14) PASTURES, FODDERS AND COVERS

Gramineae, pasture/fodder grasses: about 60 genera (Bogdan, 1977)

Leguminosae, pasture/fodder legumes: about 30 genera (Bogdan, 1977)

Legume cover crops/green manures: *Calopogonium mucunoides* and *C. caeruleum* (Am), *Canavalia ensiformis* (Am), *Centrosema pubescens* (Am), *Crotalaria juncea* et spp. (As), *Mucuna* (*Stizolobium*) spp. (Am/As), *Pueraria* spp. (As), *Sesbania* spp. (As)

Crop botany
(1) STARCHY FOOD GRAINS

RICE, *Oryza sativa* (Gramineae)

General botany. The rices are tillering grasses, generally annual though some can be 'ratooned'. They are mostly 1–1.5 m tall but some deepwater rices may be several metres long. They are mostly day-length sensitive and flower 'on programme' if sown at a specific time of year but day-length neutrality has become commoner in recent decades. Typically, all the rices are wetland plants, ideally grown under irrigation in flooded, bunded paddies but much is grown in rain-fed fields and much also as 'dryland' rice, with no water control at all. The crop is entirely seed-propagated, either broadcast or drilled or, ideally, hand-transplanted to the paddy from a densely sown nursery. The inflorescences are terminal, non-shattering panicles bearing numerous grains that are essentially non-dormant and enclosed in tough glumes. Milling of 'rough rice' gives 'brown rice' which is nearly always further milled ('polished') to yield 'white rice'. Grain sizes, shapes, textures and flavours are very diverse. The Asian rices (*O. sativa*) dominate the crop, the West African *O. glaberrima* attaining only very local importance. Both species have troublesome local weed relatives that tend to have shattering panicles and dormant seeds.

History. The Asian rices are diploids ($2n = 24$) probably domesticated several millenia ago in a great tract of land from north-central India to southern China. In cultivation, several 'races' became differentiated, broadly towards lowland tropical adaptation on the one hand (*indica*) and summer-season temperate cropping on the other (*japonica*). There is an immense variety of more or less inbred land-race populations spread through the lowland tropics, with the greatest diversity in Asia. Though ancient in Asia, the crop reached West Africa and the New World only after European travels began.

Reproductive system. The rices are inbreeders and tolerant of prolonged selfing so are bred as inbred lines. The use of hybrids has been much studied; opinions as to their potential vary but, in practice, the necessary seed technology is difficult and the method has had only local uptake (in China). The wild rices are more outcrossed than the cultivated kinds so there is some reciprocal introgression between the populations. The International Rice Research Institute (IRRI) in the Philippines has the strongest breeding programme in the world but there are many other good ones, of which the All-India scheme is probably the most powerful. IRRI was responsible for the dwarf, day-length, neutral rices of the Green Revolution.

Agroecology. As indicated above, the tropical rices show an immense range of adaptation to day-length, altitude (up to 3000 m), temperature and water supply (rain-fed to deep flooding). There is some tolerance, too, of acid-aluminiferous land and salty soils. Probably in few crops has local selection taken local adaptation quite so far. Socially, rain-fed and dryland crops are quite as important as the much higher yielding irrigated crop.

Production. World production of rice is estimated as well over 500 Mt, of which about 90% is from Asia. China, India and Pakistan are the biggest producers.

Diseases. As might be expected of an inbred species grown in large contiguous stands, rice suffers from many and diverse pests and diseases. The former include borers, bugs and hoppers and the latter are due to fungi, bacteria and viruses. Breeding has had some successes but has also suffered some disastrous vertical resistance (VR) failures. Insecticide use on the crop has, rightly or wrongly, grown enormously as the dwarf rices grown at high inputs have spread since the 1960s.

References. Jennings *et al.* (1979) (breeding); Grist (1986).

MAIZE, *Zea mays* (Gramineae)

General botany. Maize is a robust grass, 2–3 m tall, with little or no tillering. Flowers are unisexual, male ones borne in a terminal panicle, female ones in a dense cyclindrical, lateral spike which becomes the 'ear' or 'cob' of grain. The latter is tightly enclosed in bracteate husks or 'shucks' which persist and protect the grains from bird damage. All maizes are day-length sensitive and vary widely in time to maturity. The crop is entirely seed-propagated and the grains are flattened and large, commonly 10–15 mm long; they are very diverse in colour (from white and yellow to diverse reds, purples and patterns) and also in texture (from flinty to 'floury', dependent upon starch chemistry).

History. Maize is montane-tropical (probably Central) American in origin and ancient. It still crosses with and has been introgressed by its relatives *Tripsacum* and *Euchlaena* (teosinte). There is an immense variety of local land-races throughout tropical America, grown from sea level to about 3000 m. The crop was early adapted to warm temperate summers in North America and, later, elsewhere. It was rapidly spread round the tropics in post-Columbian times and has long been an important food wherever it can be grown.

Reproductive system. Maize is diploid ($2n = 20$) and is genetically by far the best understood crop plant. It is wind pollinated, outbred (protogynous) and suffers severe inbreeding depression if selfed. Most tropical maizes are still open-pollinated land-races but hybrid cultivars (HYB cvs) have been immensely successful in the United States, Europe and, increasingly elsewhere (e.g. Kenya). The strongest tropical breeding programme is that of Centro Internacional de Mejoramiento de Maiz y Trigo (CIMMYT) in Mexico which has mostly worked on open-pollinated populations (OPP) but hybrid (HYB) breeding will surely increase. However hybrid cultivars (HYB cvs) are essentially local in adaptation.

Agroecology. Maize will not stand waterlogging or freezing but will take considerable heat and drought. It is less tolerant of the latter than sorghum but has tended to displace it in recent decades. The crop responds well to irrigation but is nearly always grown under natural rainfall. The grain is a staple carbohydrate for many and is often fermented. As a component of mixed cropping, maize is very important in providing support for climbing companions such as beans or yams.

Production. Maize, with a world production (mostly temperate) of more than 450 Mt, is one of the three great cereals, preceded by rice and wheat.

Diseases. Maize has diverse fungal, bacterial and virus diseases which have, in general, been very well controlled in both temperate and tropical agricultures by breeding horizontal resistance (HR). Two transient disasters (southern leaf blight, *Helminthosporium maydis*, in the United States and rust, *Puccinia polysora*, in Africa) were quickly negated by breeding. In Africa, streak virus is very damaging in some seasons but the International Institute of Tropical Agriculture (IITA) has recently done a fine job of breeding resistance. Pests are rather numerous, especially stem- and ear-borers.

Reference. Jugenheimer (1976).

SORGHUM, *Sorghum bicolor* (Gramineae)

General botany. Sorghum is a robust grass, often up to several metres tall, lightly or not at all tillered. It is a near relative of maize but the inflorescence is strictly terminal, a panicle that varies from open and drooping in habit to dense and erect. The species is very variable and scores of botanical names have been proposed but only one biological species is represented. Cultivated races are generally recognised, of which the most important are Durra, Kaffir and Bicolor; intermediate hybrid populations abound. There are also several interspecific hybrid derivatives, of only local interest, e.g. in China and the Americas. Like maize, sorghums are generally day-length sensitive. Grain matures in 3–7 months and the threshed seeds vary in colour from yellow to blackish; darker seeds are usually bitter-tanniniferous. Vegetative parts of the plant contain cyanogenic glycosides potentially toxic to stock.

History. The crop originated in tropical East/Central Africa from wild *S. arundinaceum* (both are diploid, with $2n = 20$). It has probably been domesticated for about 5000 years and became highly diverse in West Africa as well as becoming established in India quite quickly. The move to India may well have been by both sea and land. The crop went to the New World soon after discovery and later became a locally important component of temperate hot-summer agricultures in the Americas and South Africa.

Reproductive system. Sorghum is mostly self-pollinated and inbred lines are a common and feasible means of breeding. However, there is some crossing, so that cultivated and wild populations tend to introgress (cf. rice); also, there is much yield heterosis and a good cytoplasmic male sterility (CMS) system is available, so the most advanced temperate sorghums are all hybrids. Hybrid (HYB) varieties have also done well in India in recent years and sorghum is one of the several crops in which semi-dwarf plants grown at high density have excelled. However, the practical reality for the vast majority of small farmers who grow the crop is of variable local land-races, tall plants which are sometimes valued under mixed cropping for holding up yams or beans as well as for their grain. The International Crops Research Institute for the Semi-Arid Tropics (ICRISAT) is an important breeding centre and has an excellent collection.

Agroecology. The crop is drought-tolerant and will do well on a good start from highly seasonal rainfall. It is characteristic of the semi-arid tropics of the Sahel and Peninsular India, occupying a rainfall niche between the more demanding maize and the even less demanding millets. Sorghum is very much a small-farmer crop, providing diverse starchy food products from the coarse grain and is an important material for brewing.

Production. Sorghum is fifth in importance among the world's cereals, with production over 400 Mt. Africa and India are the principal tropical producers but yields and crops in some temperate countries are higher.

Diseases. The crop has diverse pests and diseases, some troublesome. Probably far more serious economically are the witchweeds, *Striga* spp. in Africa and India, and birds, especially *Quelea* in Africa. A little progress has been made in breeding resistance to *Striga* but the seeds are very long-lived in soil; in practice, *Quelea* is partly controllable at great expense but ineradicable.

Reference. Doggett (1988).

BULRUSH MILLET, *Pennisetum americanum* (Gramineae)

General botany. Bulrush or pearl millet, bajra in India, is a robust annual grass little or somewhat tillered. Inflorescences are dense spike-like panicles (hence one common name) and flowers are mixed hermaphrodite and male. The grains are small, variable in colour and free-threshing.

The crop, adapted as it is to dry places, is generally quick to grow and mature.

History. The crop (despite its Latin name) originated in western tropical Africa and still crosses with wild relatives in that area. It was long ago (1000 BC?) taken to India and became well adapted and variable in the semi-arid zone of the peninsula. The Sahel and India remain the main centres of production.

Reproductive system. The crop is a regular diploid (with $2n = 14$) and reproduction is entirely by seed. Flowering is markedly protogynous and the crop, like maize, is predominantly outbred, so land-race populations are variable. The crop, however, is fairly tolerant of inbreeding and inbred lines have been used, especially in India, both for the production of (successful) hybrids (HYB) and composite/synthetic (SYN) populations. Most African populations, at least, are still land-races. ICRISAT, at Hyderabad, and latterly also in the Sahel, is the main organisation for millet breeding research.

Agroecology. Millet is the most important cereal of poor, sandy, semi-arid lands, being more drought-tolerant (or drought-escaping) even than sorghum. It is very much a small-farmer crop, nearly always grown in mixtures, with varieties carefully chosen as to maturity to match the farmers' expectations of weather, crops and food needs. The grain stores well.

Diseases. As for sorghum, birds are very troublesome but *Striga* is not so damaging. There are several fungal diseases of which downy mildew, *Sphacelotheca graminicola*, is damaging and has been the object of some resistance breeding.

Reference. Burton (1983).

Starchy food grains, miscellaneous

There are five crops or groups of crops that deserve brief treatment. Two, the Amaranths and Chenopods, are dicotyledenous cereals and the other three are grasses. The three relevant species of *Amaranthus* (*caudatus*, *cruentus*, *hypochondriacus*) are all American and come from Central America to the Middle Andes; they are rarely seen now but one species has attained some local importance in India. They are mostly day-length sensitive, wind pollinated and outbred. Many are brightly coloured in reds and purples. The Andean Chenopods, *Chenopodium quinoa* (quinoa) and *Ch. pallidicaule* (cañahua) have survived rather better but have never been seriously grown outside their native Andes. Quinoa, the more important, is very variable in stature, maturity, colours and breeding habits; its fruits are usually loaded with more or less toxic saponins which must be washed out before the grain is eaten.

Of the grasses, the African rice, *Oryza glaberrima* is a West African analogue of the Asian *O. sativa*, but of no more than local importance. It is related to the annual *O. barthii* and its probable perennial progenitor, *O. longistaminata*. The species has never spread outside a narrow native range of seasonally flooded land around the Niger in which the seed is broadcast in due season. The crop is day-length sensitive and some cultivars will tolerate deep water.

There are several millets which are more subtropical than tropical. *Eleusine coracana*, the African finger millet, is truly tropical and is of importance at middling altitudes in (mostly) semi-arid areas of Central Africa and southern India (where it is called 'ragi'). The taxonomy of the group is confused. The cultigen is generally inbred.

Finally, teff (*Eragrostis tef*) is one of the several unique Ethiopian crops, the leading cereal of the country but cultivated nowhere else. It is a small, densely tillering grass with tiny seeds, very variable and self pollinated. It is mostly grown around 2000m and is made into a spongy/acid fermented food called 'injera'. The straw is palatable to stock. Teff is a major constituent of the remarkable permanent upland agricultural system of the country.

References. National Research Council (1976, 1983, 1989) (amaranths and chenopods); Risi and Galwey (1984) (chenopods); Seetharam et al. (1990) (*Eleusine* and the 'small millets'); Wanous (1990) (millets).

(2) STARCHY TUBERS
CASSAVA, *Manihot esculenta* (Euphorbiaceae)

General botany. Cassava is a straggling shrub with palmate leaves and (like many Euphorbiaceae) milky latex. It bears one to several, elongate, starchy root-tubers, which are the economic product. Rate of maturity varies widely from less than a year to several years. A considerable socio-economic attraction of the plant is that tubers last well if left in the soil undisturbed, though they must be processed/consumed quickly if lifted. It is therefore a valuable 'famine reserve' and has been widely grown as such. Cassava flowers regularly but not copiously or conspicuously and male and female flowers are separate; seeds are borne (potentially) in threes in trilocular fruits (cf. rubber, a relative). All parts of nearly all cassava plants contain cyanogenic glucosides which yield HCN on hydrolysis. So the plant is toxic and HCN must be very thoroughly washed out before consumption. Despite this feature, leaves are widely consumed as spinach, especially in West Africa. Some non- or low-cyanogenic clones are known, often under the name 'sweet cassava' (contrast 'bitter cassava'). Local traditions of washing, preparation, cooking and storage are often very elaborate.

History. The genus *Manihot* is northeastern South American, whence cassava as a crop originated. The taxonomy of the genus is very confused. At least one relative (*M. glaziovii*) was a minor rubber (Ceara rubber). Cassava was introduced to Africa soon after the discovery of the New World and quickly became a local staple; it was widespread in Asia by the early nineteenth century and has long been the most important tropical tuber crop.

Reproductive system. Cassava, a regular diploid ($2n = 36$), is easily (and universally) propagated by stem cuttings, so clones predominate. Many clones flower sparsely and natural seeding is uncommon; but most are fertile enough to use in breeding programmes. The International Institute of Tropical Agriculture (IITA) in Nigeria and Centro Internacional de Agricultura Tropical (CIAT) in Colombia both have strong breeding programmes and good genetic collections.

Agroecology. Cassava is, on the whole, adapted to the lowland wet tropics but has the reputation of surviving drought and infertility better than the other tubers. It will certainly survive on poor, sandy soils and has often been regarded as a famine reserve, to be planted and left alone if not needed but consumed in emergency. It is rarely seen in pure stand, being, like most tropical food crops, a relayed or intercropped constituent of shifting systems. Often, it enters the end of the cycle when the fertility has been most depleted. Economically, the crop is attractive in making rather low labour demands and in offering flexibility of timing of labour inputs.

Production. There are no reliable statistics. It is estimated that roughly one-third each of the world crop of about 140 Mt is produced in the three continents. Though overwhelmingly a small-farmer product for local consumption, there is some export of crude starch for stockfeed (e.g. from Thailand to Europe where the product is marketed as cereal replacement pellets (CRP)).

Diseases. Cassava has several very damaging diseases, the worst occurring in Africa. There are several viruses, of which mosaic in Africa is especially damaging; this disease and bacterial blight are starting to be reduced by resistance breeding. Cassava mealy bug appeared in West Africa in about 1980 but has been at least partly controlled by biological methods.

References. Byrne (1984); Cock (1985); Jennings and Hershey (1985); de Bruijn and Fresco (1989).

SWEET POTATO, *Ipomoea batatas* (Convolvulaceae)

General botany. *Ipomoea* is a large tropical genus of sprawling/twining herbs with cordate or palmately lobed leaves. The sweet potatoes are clonal sprawlers which cover the ground well and bear root-tubers. Only one species is involved but it probably has a hybrid ancestry and is extremely variable. Tubers vary widely in size, shape and colour of skin and flesh; the degree of sweetness of the flesh also varies. Flowering in most clones is sparse and erratic and seed rarely sets naturally.

History. The crop originated in lowland tropical central/northern South America. It is mostly hexaploid ($6x = 90$) but has close $2x$ and $4x$ relatives. Wild ancestors have not been certainly identified and ploidy status is uncertain (? auto- or allo-). Very probably, Pacific voyagers took material to Polynesia in pre-European times (a subject of prolonged argument); *easterly* spread to Africa followed the discovery of the New World, mostly in the hands of European travellers.

Reproductive system. Clonal propagation by rooting stem cuttings is very easy. There is, in addition to weak flowering and much sterility, also an erratic self-incompatibility. (Analogies with the *Solanum* potato are many and striking.) Breeding is feasible but has not been strongly pursued in the tropics, despite the local social importance of the crop. The strongest programme is probably that of IITA in Nigeria. An imaginative mass-selection programme has been developed in the United States aimed at adaptation and multiple disease resistance; the principle has wide potential (Simmonds, 1993).

Agroecology. In the moister lowland tropics, the crop grows and can be harvested more or less continuously. It grows up to about 3000 m. It is intolerant of cold and drought and more demanding than cassava. In subtropical and temperate climates, short-cycle cultivars do well (given water) and Polynesian peoples developed such clones in New Zealand a very long time ago. Tubers keep quite well in the ground but do not store well once lifted. As a typical tropical small-farmer crop, it is mostly grown in mixtures. The dense foliage cover smothers weeds well. The foliage makes an attractive spinach (cf. cassava) and the crop is occasionally grown specifically for this use. (The related *I. aquatica* is purely a spinach or feed crop.)

Production. More than 100 Mt, much of it in temperate countries.

Diseases. There are several troublesome viruses and insects (e.g. two weevils and a moth borer).

References. Martin and Jones (1986); Woolfe (1992).

YAMS, *Dioscorea* spp. (Dioscoreaceae)

General botany. The yams, members of a large genus spread throughout the tropics, are climbing (twining) herbs with tuberous rhizomes. The leaves are mostly cordate, with reticulate venation and plants hardly look monocotyledenous. Species have been domesticated in all three continents and the most important are: *D. alata* (As), *D. cayenensis* (Af), *D. rotundata* (Af) and *D. trifida* (Am). Though naturally perennial, cropping is annual. Wild species often contain toxic alkaloids and/or steroids, the latter being of some pharmaceutical importance.

History. Several early domestications, diffusely spread through the tropics must be presumed. All the important species became widespread in post-Columbian times but virtually nothing is known of dates or processes.

Reproductive system. Yams are clonal, usually propagated by bits of rhizome. Plants are dioecious, flowering is generally sparse at best and seeds set rarely or not at all. Chromosome numbers are often highly polyploid and erratic; outbreeding habits may be presumed. There has been very little breeding, to which sterility is an obstacle, though not usually an insuperable one.

Agroecology. The yams are all plants of the lowland wet or subhumid tropics, grown and liked nearly everywhere but attaining much importance only in West Africa, where they were once locally staple. They have the reputation of being more demanding of soil and labour than other roots (e.g. cassava and sweet potato) and are now generally in decline. They yield best if staked but most small-farmer crops are grown up sorghum or maize stalks or are even allowed to sprawl. Mixed cropping is the rule, of course. Tubers, which are lumpy and ill-shaped, store fairly well.

Production. Production is estimated very roughly as about 20 Mt per year.

Diseases. As clonal crops, yams might be expected to have many disease problems but do not, in fact, seem to be too badly troubled. There are several more or less insidious viruses (which are yielding to mericulture methods), several damaging insects and at least one serious leaf-spot.

Reference. Coursey (1967).

Starchy tubers, miscellaneous

As will be seen from the conspectus above, the minor tubers, all clonally propagated, are fairly numerous and spread over diverse families (nine listed). Several of the aroids are Asian but all the rest are American in origin. Of the latter, *Ullucus*, *Oxalis*, *Tropaeolum*, *Arracacia* and *Solanum* are Andean. Only the 'true' potato, *Solanum tuberosum*, is of any great importance and that mostly because it has been adapted to summer growth in temperate latitudes; in the tropics it is widely but locally scattered at high altitudes and, because popular, may be slowly descending to lower levels as adaptation by breeding proceeds. The other American tubers, *Canna*, *Pachyrhizus*, *Maranta*, *Calathea* and *Xanthosoma* are, broadly, from lower altitudes and the last is by far the most important of them. Tannia, under various names, has gone round the lowland moist tropics and is nearly everywhere, a significant component of mixed smallholder food cropping; it has characteristic sagittate leaves and yields starchy corms. The Asian analogues of tannia (species of *Alocasia*, *Colocasia*, *Cyrtosperma*) were/are important elements of Southeast Asian and Pacific food cropping. The botany of all the tuberous aroids is confused; there is much sterility and virtually nothing is securely known of their genetics or breeding habits. *Colocasia* is the most important of them; it probably originated in India and spread anciently to the Pacific and West Africa, and from the latter to the New World soon after discovery. It has peltate leaves and will withstand wet, even swampy conditions. The corm-tubers of both *Xanthosoma* and *Colocasia* are often harvested selectively, leaving the mother-plant to be earthed-up and 'ratooned' as a perennial.

Reference. National Research Council (1989) (Andean tubers).

(3) STARCHY (AND/OR OILY) PERENNIALS

BANANAS, *Musa* cultivars (Musaceae)

General botany. The bananas (including plantains, see below) are large to gigantic stooling herbs. Aerial stems are 'pseudostems' composed of densely packed leaf sheaths. The 'true' stem is thrust up the heart of a pseudostem and emerges at the top, bearing the terminal inflorescence. This, on emergence, is said to 'shoot'; basal flower clusters are female and collectively they form the fruit 'bunch'; distal clusters are (potentially) male and they and the bracts that subtend them are (usually) deciduous. Bananas are cytogenetically complex and many clones are polyploid and/or interspecific hybrid in constitution. Nearly all are more or less seed and pollen sterile, many completely so. The fruits of wild bananas are inedibly full of black stony seeds; those of cultivated kinds are said to be 'vegetatively parthenocarpic' because they grow a mass of edible pulp without the need of any stimulus by pollination. There are hundreds of clonal cultivars; one group can properly be called 'plantains'. Many bananas have fruits more or less starchy-acid at maturity. The 'plantains' are among these but there are others. *All* bananas can be cooked unripe and green as starchy vegetables. 'Banana' is a good general word for the whole crop and the word 'plantain' should only be used in the strict sense (though all too often it is mis-used). From planting, the first ('plant') crop usually takes about a year and thereafter, depending upon site, weather and management, a regular succession of bunches ('ratoons') is produced. In principle a banana patch may be immortal but in practice it is not. The underground parts of the plant (rhizomes or, perhaps better, corms) bear lateral buds which continue vegetative growth or may be planted as 'suckers'.

History. The wild bananas are nearly all Southeast Asian and the crop originated there. Polyploidy (especially triploidy, $2n = 3x = 33$) and interspecific crossing were involved in its evolution. Cultivars spread out into the Pacific and westwards to Africa long before European travellers arrived. The crop became staple in upland East/Central Africa, which is still the greatest centre of cultivation. It was taken by early European travellers from West Africa to the New World and was quickly adopted there too, though not as profoundly as in Africa. Commercial cultivations for export of fruit to Europe and North America developed during the nineteenth century. They were mostly in tropical America, to a lesser extent in West Africa.

Reproductive system. As we have seen above, the crop is very nearly totally sterile and therefore obligately clonal. Breeding of new export types has been attempted but with little success. However, the scientific bases for improvement are fairly well understood and breeding for the local food crop is just starting. Banana breeding is probably the most complex and refractory plant breeding problem ever tackled.

Agroecology. Wild bananas are jungle weeds, pioneers in the succession to rainforest. They are intolerant of much shade and require highish, well-distributed rainfall, good drainage and high temperatures. They do quite well, though growth slows down, up to about 1500 m altitude and are intolerant of frost. Some cultivar groups (those with *balbisiana* (B) genomes) are more tolerant of drought and will stand long dry seasons in monsoon climates fairly well. Bananas are intolerant of weed competition and responsive to mulching; their own trash, well spread, as in upland East Africa, is good mulch and some African cultivators used to bring in cut grass/bush from outside the banana patch. Bananas are most abundant as backyard plants but large areas are grown as more or less pure, long-lived stands in upland East/Central Africa. In plantation agriculture (e.g. cocoa) they make excellent nurse plants. In shifting systems, they are sometimes planted with tree crops, such as citrus, at the end of the cycle, to die out as the forest regenerates.

Production. Annual production is thought to be about 63 Mt of fruit of which about 90% is consumed locally. In the moister tropics, the crop is everywhere important socio-economically and it is a locally dominant staple in upland East Africa. For many people, the crop is a starchy tuber grown up in the air but with great advantages of seasonally well-distributed production and low labour demands. Locally, especially in Africa, ripe fruit is fermented to make 'beer', stems go for stock feed and the leaves provide wrappings and fibre.

Diseases. Bananas, in large-scale, monoclonal commercial cultivations, have been severely beset by diseases, especially wilt, Panama disease (*Fusarium oxysporum cubense*), leaf spots (*Mycosphaerella* spp.), bunchy top virus and nematodes (especially *Radopholus similis*) (this is a very short and over-simplified list). On the whole, highly heterogeneous small-farmer cultivations have been less troubled by diseases than the commercial ones but eelworms leaf spots, now seem to be increasing and viruses in Africa.

References. Rowe (1984); Stover and Simmonds (1987); Rowe and Rosales (1996).

Starchy and/or oily perennials, miscellaneous

An interesting feature of these crops is that, though not numerous, all five are *locally* very important socio-economically. The parthenocarpic breadfruit, *Artocarpus* (including its seedy, non-parthenocarpic form, the bread-nut) is a major food plant of the tropical Pacific, associated with aroid tubers and coconut. It was made famous by Bligh's voyages of 1787–93. There seem to be few clones and the crop, though widespread, is patchy and unimportant elsewhere. The inset (*Ensete*) is strictly confined as a crop to upland Ethiopia; wild *Ensete* spp. are widespread, from Central Africa to New Guinea, but nowhere else domesticated. The marvellous inset culture of Ethiopia is based on an ingenious method of clonal propagation by induced suckering. Pseudostems are cut before flowering and the sheath starch removed by pounding and washing to yield a fermented bread ('injera'). The sago palm, of Southeast Asian swamps, is analogous in that starch is beaten out of a stem felled before flowering and then split. *Metroxylon* spp. are, like dates, suckering palms (which is unusual), so they can be vegetatively propagated. The crop is more gathered from wild or protected stands than actually planted. It is thought to be highly productive but is almost untouched by research.

The last two crops are both fruits. The pejibaye palm (*Guilielma* or *Bactris*) is a small, suckering, spiny palm that yields heavy bunches of fruits, a staple food for some Indian peoples of lowland Central and northwestern South America. After boiling, the starchy mesocarp keeps well and the oily endosperm is also edible. There is much variability and no doubt that highly productive clones (including spineless ones, if wanted) could be bred. The 'heart-of-palm' makes an attractive vegetable. The crop is virtually unknown outside tropical America, a sorry state of affairs in view of its nearly non-seasonal production and evidently vast potential for lowland wet-tropical development. The avocado, *Persea*, has long been of much socio-economic importance only in Central America though widespread in tropical gardens elsewhere. It is very variable, clonally propagated, outbred and little researched. Fruits are drupes with a predominantly oily pulp and cropping is somewhat seasonal. Subtropical cultivations exporting fruits to temperate markets have thrived in recent decades.

References. Simmonds (1958) (*Ensete*); National Research Council (1976), Clement (1988); Ruddle *et al.* (1978) (sago); Stanton and Flach (1980).

(4) PROTEIN SEEDS

BEANS, *Phaseolus* spp. (Leguminosae)

General botany. There are four bean species of which the first-named is by far the most important: *P. vulgaris* (common bean, etc.), *P. coccineus* (scarlet runner bean), *P. lunatus* (Lima bean), and *P. acutifolius* (Tepary bean). All are herbaceous twiners (less often bushy plants) with trifoliate leaves. All are American and species that used to be placed in Asiatic *Phaseolus* have long been disposed in other genera, especially *Vigna*. They are all annual, day-length sensitive and seed propagated but some wild relatives are perennial. They are variable, conspicuously so in seed characters (sizes, colours, patterns). Though edible seeds are the principal economic product, there are many forms with much reduced pod fibre, selected for green vegetable production. Like many, indeed nearly all, legumes, mature seeds tend to contain diverse toxic factors (alkaloids, trypsin inhibitors, etc.) which must be eliminated by very thorough cooking.

History. Domestication (possibly multiple) was in central and northern South America and the species spread thence after the discovery of the New World, eastwards and westwards to Asia and Africa. Adaptation to temperate summers accompanied spread.

Reproductive system. All four species are regular diploids with $2n = 22$. *P. coccineus* tends to be cross-(bee-)pollinated but the other three are inbreeders (like nearly all the annual legume cultigens). Some interspecific hybrids have been made but have not been useful. Breeding, naturally, is by inbred lines (IBL) and Centro Internacional de Agricultura Tropical (CIAT) in Colombia, holding an excellent genetic collection, has been prominent in this.

Agroecology. The beans are mostly grown as small-farmer crops, at middling to high elevations, as components of rain-fed mixed cropping systems. Many tropical American societies can properly be described as maize–bean or maize–bean–squash cultures. Nutritionally, bean proteins complement lysine-deficient cereal proteins very well and beans are far more important socially than their comparatively small production would suggest. In the Americas, they are commonly relayed into maize to use the dead stalks as supports for twining. As pure-stand crops, the bush habit is probably more 'efficient' physiologically but, for the small farmer, indeterminate twining is thought to be more reliable.

Production. Probably about 15 Mt of *P. vulgaris*.

Diseases. The beans have their troubles, caused by diverse fungi (e.g. anthracnose, *Colletotrichum*), bacteria, several viruses and insects (including weevils in store). Strict inbreeding and a generally narrow genetic base are probably partly responsible.

References. Smartt (1976, 1990); Summerfield and Bunting (1980); Summerfield and Roberts (1985); Schoonhaven and Voysert (1991); Singh (1992).

GROUNDNUT, Arachis hypogoea (Leguminosae)

General botany. The groundnut or peanut is a major oilseed rather than a pulse/protein plant. The plant is a low, bushy or sprawling herb with compound leaves. It has characteristic variation in branching habit leading to horticultural types under such names as Valencia, Spanish, Bunch, Runner. Flowers emerge singly as 'pegs' which penetrate the soil, flower cleistogamically and set few-seeded, indehiscent legume pods a few centimetres underground. There is wide variation in maturity and seed dormancy, which must both be matched to seasonal rainfall.

History. The crop is American, probably centred on western Brazil. Numerous wild relatives survive in the area, mostly diploid, while the groundnut itself is allotetraploid, with $4x = 40$. It was spread both eastwards to Africa and India and westwards to Asia soon after discovery. West Africa became an important centre both of variability and cultivation.

Reproductive system. The crop is an obligate inbreeder (it could hardly be otherwise) and is bred as such. The numerous wild relatives are yet virtually unexploited. Much breeding has been done in the southern United States and Central Africa (now Zambia). ICRISAT has a substantial breeding programme for the semi-arid tropics of India and Africa and has a major collection.

Agroecology. The groundnut is adapted to light lands at low altitudes in the drier tropics. It will also do well in hot temperate summers, as in the United States. The crop has a particular need for calcium as an essential for pod development so tends to be intolerant of acid soils. In tropical agriculture it is a small-farmer crop grown, usually in mixtures, for home consumption and sale (e.g. in West Africa) as an oil-seed. The oil has excellent properties for industrial-food purposes and the cake is valued for stock feed. The notorious aflatoxin episode of the early 1960s was attributable to secondary fungal infection of damp produce in store rather than to any inherent property of the crop. The 'groundnut scheme' disaster of the late 1940s was due to a mixture of bureaucratic-political stupidity and technical errors.

Production. World production is over 14 Mt.

Diseases. There are two widespread and damaging cercosporoid leaf spots and a rust (*Puccinia arachidis*) which has spread from America to Africa. There are also several soil pathogens and insect pests, the latter including the aphid vector of several damaging viruses (e.g. rosette).

References. Smartt (1976, 1990); Summerfield and Bunting (1980); Weiss (1983); Summerfield and Roberts (1985); Institute of Biology (1990).

Protein seeds, miscellaneous

There are many minor legume crops of which but few are treated here. The cowpea or bodi bean is African in origin, a regularly inbred diploid with $2x = 22$, and an important seed and vegetable crop in semi-arid Africa and India. *Vigna* is near systematically to *Phaseolus* and *Dolichos* and the status of closely related crops such as the several Indian gram seeds is far from clear. Also tolerant of semi-arid conditions and capable of maturing under drought on residual rainfall, are the lentil (*Lens*) and chickpea (*Cicer*), both crops from the eastern Mediterranean and both still rather more important in the subtropics than in the tropics proper. Chickpea yields dhal in India where the crop has been long established and small-seeded (desi) and large-seeded (kabuli) forms are distinguished. *Voandzeia* is a sort of African analogue of *Arachis*, but very local. The perennial pigeon pea, *Cajanus*, is a short-lived shrub, rarely a small tree and, unlike most of the annual pulses, is an outbreeder. It is very probably Indian in origin (old arguments in favour of Africa notwithstanding) and related to *Atylosia*. It is drought-resistant and a significant source, for small farmers, not only of nutritious seed but also of sticks and firewood. The winged bean (pea) *Psophocarpus*, an Asian cultigen, has recently enjoyed something of a vogue (it is not clear why) as a multipurpose legume yielding seeds, green pods, shoots and tuberous roots.

Many other legumes might have deserved mention, both herbs and woody plants. Certainly, they have diverse uses in tropical agriculture but perhaps one should recall that most legumes are, in various degrees, toxic, some very highly so. Indeed some yield insecticides and fish poisons and many food crops need careful preparation before consumption. Collectively, however, their importance is likely to grow as constraints on fertiliser nitrogen increase.

References. Smartt (1976, 1990); National Research Council (1979); Summerfield and Bunting (1980); Singh and Rachie (1985) (cowpea); Smithson (1985) (chickpea); Summerfield and Roberts (1985); Nene *et al.* (1990) (pigeonpea); Dana and Karmakar (1990); Hall *et al.* (1997) (*Vigna*).

(5) OIL PLANTS

OIL PALM, *Elaeis guineensis* (Palmae)

General botany. The oil palm is a large, unbranched palm up to 20 m in height at maturity. The leaves are pinnate, with hooked spines on the petioles, and inflorescences are tightly packed into the axils. Male and female flowers are separated into different inflorescences in an erratic ratio of male to female. Pollen is very abundant and insect-borne; fruit bunches are dense, weighing about 5–20 kg each, and fruits are shiny, 2–5 cm long, and crimson-purple-blackish in colour. The main economic product is the yellow-orange coloured edible oil from the fleshy pericarp; the kernel oil is also extracted and all parts of the fruit except the exceedingly stony endocarp (shell) yields residues for stock feed.

History. The oil palm is native in tropical West Africa but its only near relatives (*E. oleifera* and *E. odora*) are northeastern South American.

Though anciently separated, they will hybridise. In the wild, the plant grows in wet places usually around savanna/forest margins but not in high forest. Fruits have long been gathered in West Africa (and still are) and the palm entered commercial cultivation only around 1900, first in West Africa, and shortly afterwards in Southeast Asia, especially in Malaysia and Sumatra. Early plantings in Asia originated from material earlier introduced to Java by way of Mauritius. It is therefore one of the most recent plantation crops, even though anciently gathered in Africa (as many other palms are locally gathered throughout the tropics, especially in tropical America).

Reproductive system. The oil palm is a regular diploid ($2n = 32$), outbred and intolerant of inbreeding. Good commercial populations are formed from seed derived from hand-pollination of selected parents. In general, parents are *dura* (homozygous *ShSh*, thick-shelled) crossed by *pisifera* males (homozygous *shsh*, more or less shell-less, sterile) to give *tenera* types (heterozygous *Shsh*, thin-shelled) progeny. Such families are a form of synthetic (SYN) and breeding has been outstandingly successful. Increasing use is being made of the *dumpy* mutant to reduce the size of the plant, ease harvesting and permit higher populations (cf. coconut). Since the plant is unbranched, clonal propagation by conventional means is impossible. Some 30 years of research on *in vitro* clonal multiplication has had some experimental, but yet no significant practical, success.

Agroecology. The oil palm is strictly lowland-tropical, thrives under high well-distributed rainfall and is tolerant of acid, aluminiferous soils. Well managed, it is capable of sustained yields of 6 t/ha/yr of high quality oil (contrast soya bean which does well to produce a crop of 1 t/ha). Though a plantation crop *per excellence*, much is locally grown by smallholders and much, too, is produced for village use. Commercial production demands a substantial infrastructure of transport and factories (which have to be provided for smallholder schemes).

Production. Production is about 8 Mt of oil per year. The main area of production, in Southeast Asia, is increasing and the crop is being extended in tropical America.

Diseases. The oil palm is relatively little troubled by pests and diseases. In West Africa, wilt (due to *Fusarium oxysporum*) is a major nuisance and would undoubtedly present problems if it spread into Asia. There are several 'mysterious' diseases of palms in Asia and America which are to be feared.

References. Hardon *et al.* (1985); Hartley (1989) (general); Soh (1990) (breeding).

COCONUT, *Cocos nucifera* (Palmae)

General botany. The coconut is a handsome, non-suckering, pinnate-leaved ('feather') palm growing up to 30 m tall in a lifetime of about 100 years. Inflorescences are large, axillary, monoecious, with distal male spikes and basal female flowers. Pollination is by both wind and insects and the flowers lend themselves well to bulk hand-pollination. The fruit is large, with a fibrous pericarp (which yields 'coir' fibre), a stony endocarp ('shell') and a bulky, oily endosperm ('meat', 'copra'). It matures in about a year. The endosperm is at first watery ('milk'). The fruit is tricarpellary but one-seeded and there is no dormancy. The plumule emerges and adventitious roots establish in the coir and the ground, the young seedling being meanwhile nourished by a fleshy 'haustorium' in the endosperm. Growth is rather slow, as the cylindrical stem builds up. Tall palms start to fruit in about 7 years, dwarf ones more quickly.

History. The coconut has pan-tropical relatives but is undoubtedly Southeast Asian in origin. It spread anciently to eastern Africa and far out across the Pacific to the west coast of America in the hands of human travellers helped, perhaps, by some floating, sea-borne nuts. Spread to West Africa and the Atlantic side of America was probably in the hands of European travellers. There is no evidence of hybridisation en route and 'wild ancestors' have never been identified.

Reproductive system. The palm is a regular diploid with $2n = 32$ and is outbred. It exists in variable local land-race populations some of which must be somewhat inbred by reason of 'founder effects'. The vast majority of coconuts are talls but several dwarf mutants are known: they tend to fruit earlier to have smaller nuts and to be somewhat inbred. The heterozygote tall × dwarf is morphologically intermediate and has attractive characters for commercial plantation. What little systematic coconut breeding has been done has been virtually confined to testing tall × dwarf combinations. Success is not clear because there have been considerable disease problems, yet unresolved. Attempts to propagate the plant clonally *in vitro* (cf. *Elaeis*) have so far failed. If a locally good cross can be identified, the techniques of hand-crossing and seeding nursery management are well established.

Agroecology. The coconut is a plant of the lowland wet tropics, intolerant of cold and drought but also of waterlogging. It is characteristic of coastal sands so long as fresh groundwater is within root-reach and palms will stand extreme exposure to wind and salty spray. Though commonest near coasts, however, it is by no means confined to them. No plant is more useful to man and many Pacific cultures are based upon it, accompanied by breadfruit, bananas and tuberous aroids. Coconuts provide food and drink, fibre, thatch and timber, besides combining well with very diverse crops, woody or herbaceous, beneath their shade, or grazing animals. At the plantation level, huge areas were planted in Southeast Asia and the Pacific during the late nineteenth/early twentieth century to provide oil for fats and soaps in temperate markets. Those old plantations are now running out and generally not being replaced; as an oil-producer, *Elaeis* is much superior. But there still seems to be a commercial place for coconuts shading cocoa or undergrazed by cattle and the crop remains of great social importance.

Diseases. Coconuts have some very troublesome pests of which Rhinoceros beetle is perhaps the best known but others (weevils, borers and eelworms) may be more damaging. Fungal diseases are not conspicuous but more or less mysterious maladies attributed to mycoplasms, viruses and physiological disorders are locally serious (e.g. 'lethal yellowing', 'cadang-cadang').

Reference. Child (1974).

Oil plants, miscellaneous

Of the better known annual oil seeds, *Arachis* is treated here as a legume (qv), while soya bean, sunflower, the Brassicas, safflower and jojoba are all essentially temperate or subtropical. The Shea butter tree (*Butyrospermum paradoxum*) is of local importance, both domestically and for export, in its native area, the West African savanna zone, and there are several other very minor tropical species.

Three species are worth slightly fuller mention. Castor oil, *Ricinus*, is a small tree, reduced locally by selection to a robust near-herb; it is a regular diploid with $2n = 20$, variable, basically monoecious but a useful male sterility is known. The crop is ancient in India but may be of either Indian or African origin. The seeds, borne in spiny three-seeded capsules, are highly toxic but the oil (merely unpleasant rather than poisonous) has long been used as a purgative as well as for diverse social and industrial purposes. In breeding, dwarf, annual, hybrid varieties are feasible. By contrast, the other two minor oil seeds listed, Niger seed and sesame, are both annual herbs. Niger seed (*Guizotia*), a characteristic crop of its native Ethiopia (where it is called 'noog'), is reminiscent of sunflower and more cultivated in India than Ethiopia. Sesame (*Sesamum*), an ancient crop in India and the near East, may be either of Asian or African origin. It is widely (but sparsely) grown in semi-arid areas for its excellent edible oil, produced, however, at poor yield. Flowers are borne in open spikes and seeds are small and numerous in the capsules (which tend to dehisce at maturity, a primitive character surprising in an old cultigen).

Reference. Weiss (1983).

(6) SUGAR PLANTS

SUGAR CANE, *Saccharum* cultivars (Gramineae)

General botany. The sugar canes are robust, stooling, perennial grasses. Stalks are thick, with a sweet-juicy pith and hard rind; they are cut off annually at ground level and milled to produce a juice from which sucrose is extracted, leaving molasses from which no more sucrose can be crystallised. The dry matter is 'bagasse', mostly burned to fire the factory but occasionally, made into hardboard. The first crop after planting is called the 'plant' crop, later ones 'ratoons'. All sugar canes are day-length sensitive and tend to flower ('arrow') at a time dependent upon latitude; but flowering stops growth and reduces sugar yield so must be prevented by appropriate breeding. The 'arrows' are large panicles of tiny flowers bearing minute, hairy, wind-borne seed ('fuzz'). The crop is planted by pieces of stem bearing two nodes or, more usually nowadays, by laying of whole stalks in furrows, followed by chopping into shorter pieces.

History. The crop originated in Southeast Asia, probably in the New Guinea area, from wild *S. robustum*, a non-sweet grass of river banks, which yielded the sweet 'noble' canes (*S. officinarum*) to human selection. Extensive hybridisation with related wild species and other genera accompanied geographical spread. Erratic high polyploidy as well as hybridity are characteristic and nomenclature is very confused. The crop reached Africa and Europe rather late and was quickly adopted in the New World after discovery (where it became one of the foundations of the slave trade).

Reproductive system. All sugar canes are vegetatively propagated and the crop is entirely clonal. Though, in practice, nearly non-flowering, most clones can be made to flower sufficiently to get crosses. The crop, which is outbred, has been outstandingly successfully bred and a complex interspecific hybrid genetic base has been greatly extended since pioneering studies in Indonesia and the West Indies around the early years of the twentieth century. Chromosome numbers are erratically high (>100). Disease-resistance breeding has been singularly successful. No matter what cytogenetic complexities, the ordinary routine of generating variable families and selecting between clones prevails. There are about 20 breeding programmes spread round the tropics and subtropics and international collaboration and exchange is outstandingly good.

Agroecology. The crop is well adapted in the tropics and subtropics at low altitudes. It needs abundant moisture for growth but a dry (and preferably cooler) season for harvest. Economical sugar contents (roughly >10%) are only attained when growth has been checked and dry weather also favours field transport. The trend is towards mechanisation (diverse harvesters are available) but much cane is still cut by hand. Ratoons commonly number 3–5 (rarely 10 or more). Burning trash before harvest is often done but is generally agreed to be agriculturally bad practice. Producers may be any size from large estates to smallholders. Heavy capital investment in factory and transport are essential. The 'nuclear estate' concept applies very well and the crop has many attractive features for smallholders, not least of which is that it rotates excellently with food crops (as the example of Barbados has shown for many decades).

Production. World sugar production of sugar in the later 1980s was about 105 Mt of which 66 Mt (63%) came from cane, the rest from beet.

Diseases. Sugar cane has numerous diseases caused by diverse fungi, bacteria and viruses. Some 15–20 are potentially devastating but all are well to excellently controlled by breeding. Insect pests (e.g. moth borers) are always present but not very damaging.

References. Blackburn (1984) (general); Heinz (1987) (breeding); Ricaud *et al.* (1989) (diseases).

Sugar plants, miscellaneous

Almost any palm, the coconut among them, will yield a sugary juice from tapping the stalk of an inflorescence. The juice may be fermented to yield 'toddy' or used as a source of more or less crystalline sucrose ('jaggery' in India). Two palm species characteristically used thus are the monocarpic feather palm, *Arenga* (gomuti), and the fan palm, *Borassus* (palmyra), both Asian, though the latter has African relatives. The date palm (*Phoenix dactylifera*) might almost be classified as a sugar-producing plant but is grown rather as a staple carbohydrate fruit and is essentially subtropical. Plants other than palms which are grown for sweet, fermentable juices from inflorescence stalks are members of the Agavaceae (e.g. *Agave* spp. in Mexico, which yield pulque, tequila and related drinks).

(7) FRUITS AND NUTS

The fruits and nuts listed are but a small sample of a very numerous group of useful plants which are spread through the tropics more as backyard plants than as field crops. There are, however, substantial commercial plantings of mango, cashew, pineapple, papaya, melon, passion fruit, macadamia and citrus fruit. Of these, the last are the most important but essentially subtropical.

Most are trees or shrubs, often long-lived, but pineapples, melons, passion fruits and the *Solanum* fruits are herbaceous. Propagation is varied, from strictly clonal (pineapple, citrus) to strictly sexual (papaya, Solanaceous fruits) but with a growing tendency to adopt clonal methods whenever possible. Several species have reproductive derangements such as polyembryony (in mango, mangosteen and citrus), seed-sterility associated with parthenocarpy (in pineapple and some citrus), and deranged sex determination systems (in papaya). The tropics of all three continents have provided entries and a longer list would be equally diverse as to sources. Macadamia is exceptional in being a very recent domesticate and from an Australian wild species, the only example from that continent.

Many, maybe nearly all, of the crops listed are old, some probably ancient, but we have no reliable dates for origins. The botanical classification of most of them, however, is fairly straightforward, suggesting no great evolutionary complications; citrus, however, is an exception because it is outstandingly confused by the description of up to 150 bad species from what is merely an array of old hybrid cultivars much complicated by somatic mutations. Serious breeding efforts have been devoted to very few of the fruits (even pineapple and citrus have been neglected) but there can be little doubt that something more than mere attempts to pick a few good clones would be rewarding. Papaya has had some serious breeding (by seedling populations) in Hawaii and South Africa and enough has been done on the two nuts, macadamia and cashew, to be assured of good potential for progress. Cashew seems to be a peculiarly attractive proposition for poor, sandy soils in semi-arid areas.

(8) VEGETABLES

The tropical vegetables are much more diverse than the short list given in the conspectus would suggest but they are not nearly so numerous as the fruits. There are probably a score or two of spinaches, some very local, and also such products as the nopalitos of upland Mexico, which are phyllodes of *Opuntia* scraped clean of prickles. Of those listed, the squashes (cucurbits) and peppers (*Capsicum*) are of considerable socio-economic importance in tropical America. Both are botanically complex, with several species each, and both are integral to the maize–bean cultures of middle America. Squashes and peppers are widespread in the tropics but nowhere else do they attain such importance. The *Capsicum* peppers are the principal source of pungency in tropical diets but there are also many bland ones (sweet peppers). The tomato is much more important in temperate than in tropical agriculture but it is, in fact, of tropical origin so appears in this list; it is not well adapted in the lowland wet tropics. Besides the squashes, the Cucurbitaceae have also provided the Asian cucumber, the melons (see Fruits and nuts), the great African gourd, *Lagenaria*, and diverse minor vegetables. Okra (*Abelmoschus* or *Hibiscus*), of which the immature fruits are eaten, has the bland mucilaginous character typical of the

Malvaceae, a family generally better known for its fibres (qv).

References. Herklots (1972); Tindall (1983).

(9) BEVERAGE PLANTS

TEA, *Camellia sinensis* (Camelliaceae)

General botany. Tea is an evergreen shrub or small tree that, in cultivation, is kept down to about a metre in height by pruning and harvesting leafy shoot tips ('plucking'). The crop is variable, consisting of 'hybrid swarms' of types with varying adaptations, leaf sizes and commercial properties. Diverse bad species have been proposed but phrases such as 'Assam-type' or 'China-type' are more often applied in practice. Some teas are dried immediately after plucking ('green tea') but most leaf is very lightly milled ('cut' or 'rolled') and fermented before drying, to make the familiar 'black tea'. Tea technology and quality assessment is a complex and subtle business. Though mostly used now as a beverage, the crop was, for many people, more of a chewing plant than a drink.

History. The wild progenitors of the crop are not known for sure but were (are?) fairly certainly wild in continental Southeast Asia. The crop is probably ancient in China and was an important article of trade for centuries before widespread cultivation in southern India, Sri Lanka, Indonesia (and later elsewhere) began in the late nineteenth century following the devastation of coffee plantations by the rust, *Hemileia*. The teas that spread into the tropics were very generally hybrid derivatives of the more temperate, small-leaved China teas and the tropical, large-leaved Assam types.

Reproductive system. Until the last few decades, tea was propagated by seed, more or less random as to origin, or from seed-gardens of (presumptively) superior parents. Apart from a few polyploids, teas are diploids with $2n = 30$. They are outbreeders which exhibit sterility on selfing. Vegetative propagation by cuttings and buddings has been successfully developed in recent decades and clonal (CLO) breeding and planting now predominate. But older fields are still seedlings, little or not at all improved. Given good clones, improved seedling (synthetic – SYN, see later) populations should be feasible and the genetical structure of the crop will gradually approximate to that of rubber.

Agroecology. Tea is more a subtropical than a tropical crop. At lower latitudes it is best grown at considerable altitudes (2000m or more), demands high rainfall and acid soils and is intolerant of drought. It is densely planted to form a continuous, smooth 'plucking table' when in production. Traditionally grown under light legume shade, it is now increasingly grown unshaded but well fertilised. It is predominantly an estate crop that makes considerable demands for labour to pluck the leaf and on skilled management to process the product. The best quality black teas are grown at high altitudes and the shoots are plucked young ('two-leaves-and-a-bud'), though older tips give higher yields. Chemically, quality components include pigments, tannins, volatiles and stimulants (caffeine and theobromine).

Production. About 2 Mt of 'made tea' are produced annually, mostly black.

Diseases. There are several troublesome fungi of which blister blight (*Exobasidium vexans*) is the worst. Several damaging insects have, of necessity, attracted much research on control measures that minimise the use of insecticides.

References. Eden (1976); Institute of Biology (1990).

COFFEES, *Coffea arabica, C. canephora* (Rubiaceae)

General botany. Coffea is an equatorial African genus of shrubs and small trees characteristic of the undergrowth of moist forests. The two species named are important economically and a few others, on a small scale, (e.g. *C. liberica*) also have economic importance. The plants are evergreen and dimorphic, with orthotropic main shoots and plagiotropic (flowering) laterals. Flowering tends to occur in one or two seasonal bursts and fruits are two-seeded berries ('cherries'). The seeds, cleaned of fruit pulp, washed and dried and polished, are the economic product. Some coffees are dried 'in the cherry' however. Quality judgements by the trade are (as for tea and cocoa) organoleptic; aroma and contents of stimulants (caffeine and theobromine) are main criteria.

History. The use of wild plants in tropical Africa is probably ancient. Arabica cultivation was taken from upland Ethiopia, first to the Yemen and Arabia, thence to Asia, thence (via European botanical gardens) to Martinique and continental America. Asian cultivations at fair altitudes in southern India and Indonesia thrived but were killed by leaf rust (*Hemileia vastatrix*) during the mid-nineteenth century (to be replaced largely by tea, qv). *Coffea canephora* (commonly called robusta) has long been cultivated in tropical Africa (at lower elevations than arabica) and was widely distributed in Asia and America during the twentieth century as a (somewhat) rust-resistant substitute for arabica. It now forms about 20% of the total crop, much of it destined for 'instant' coffee in which arabica quality would be wasted. The plants which founded the New World arabica crop in the early eighteenth century all traced back to a single seedling; success depended upon the freakish chance that arabica coffee is that *very* rare event, an inbred tree crop.

Reproductive system. The diploid ($2n = 22$) coffees (e.g. robusta, liberica) are self-incompatible outbreeders which suffer severe inbreeding depression; arabica is a self-compatible, inbred allotetraploid ($2n = 44$). Hence traditional populations were, respectively, variable open-pollinated (OP) populations and pure lines (IBL). But vegetative propagation has long been known (e.g. by African farmers) to be possible and the longer-term future for all coffees must surely lie with clones. *In vitro* meristem methods are feasible, which may be important because it is essential to propagate from (the rather scarce) orthotropic shoots (Simmonds, 1995b). Hybrid varieties have been proposed (e.g. in Ethiopia) but hand-pollinated seed would be excessively expensive; as a transitional compromise, F_2 hybrid varieties have been suggested. Good SYN (synthetic) varieties from self-incompatible robusta clonal parents should be feasible (cf. cocoa). Dwarf mutants are known and their use will no doubt become universal. Diverse interspecific hybrids are available and arabica genomes introgressed by robusta for disease resistance seem certain to be increasingly valuable.

Agroecology. The coffees are plants of the moist tropics and subtropics. Arabica does best at fair altitudes and will stand some cold (but little frost). Robusta is adapted to lower altitudes. Most arabica is produced in Central and South America (down to southern Brazil) with some in upland East Africa. Traditional planting practice was to grow at low inputs, with some shade; nowadays, fertilised, high-density, unshaded planting of dwarf bushes is increasing. Producers range from plantations (or, at least, sizeable farms) to smallholders and good management of processing is critical (though not always achieved).

Production. In the later 1980s, world production was about 6 Mt of 'green' coffee. In total it is one of the most valuable agricultural exports in the world. Prices are extremely variable and subject to international agreements and quotas.

Diseases. *Hemileia vastatrix* caused one of the classic epiphytotics, in Asia in the nineteenth century, but in the Americas only lately (1970s). As usual, vertical resistance (VR) failed but horizontal resistance (HR) breeding is in hand and well-timed sprays, if economic, can be helpful. Another serious fungus disease, coffee berry disease, caused by *Colletotrichum coffeanum*, which emerged in East Africa in the 1920s, is also yielding to HR breeding. There is no doubt that intelligent breeding is the correct approach.

References. Medina-Filho *et al.* (1984); Wrigley (1988); Institute of Biology (1990).

COCOA, *Theobroma cacao* (Sterculiaceae)

General botany. Cocoa (or cacao) is a small undergrowth tree of wet places in the rainforests of lowland northern South America. The whole genus is from this area. The tree is dimorphic, with orthotropic 'chupon' shoots and plagiotropic 'fans'. Flowers, which are midge-pollinated, are borne in huge numbers, in clusters ('cushions') on woody parts, so the tree is cauliflorous. Leaves are simple, ovate, evergreen and are produced in characteristic bursts or 'flushes'. The fruits ('pods') are more or less ovoid-cylindrical and leathery/woody, containing relatively few, large, cotyledenous seeds ('beans'), embedded in sweet-acid mucilage. The beans are variously 'fermented' in heaps (or boxes), then slowly dried. The product contains some 50–60% fat, which is partly separated in processing to give 'cocoa butter', of use in cosmetics as well as in confectionery. Cocoa, though listed here as a 'beverage', is admittedly rather more used as a solid confectionery. But it does contain the same stimulant alkaloids (caffeine and theobromine) as its relative, kola, and the unrelated tea and coffee.

History. The crop was carried northwards and grown in the Caribbean lowlands of Central America, by the great Indian civilisations of the area. 'Criollo' (pale-beaned) types were favoured. From the sixteenth century, cultivation spread in tropical America as the product became popular in Europe. 'Forastero' types, with smaller, darker beans (the common form among wild cocoas), were widely grown and locally, as in Trinidad, apparent hybrid swarms developed ('trinitario'). West African and Asian cultivations developed strongly in the nineteenth century and, until quite recently, Ghana dominated world production. The bulk cocoas are all of vigorous, red-beaned 'forastero' type; the 'criollos' and 'trinitarios' meet small, specialised markets.

Reproductive system. Virtually all cocoas are still seed-propagated even though clonal propagation by cuttings and buddings has been known for decades. The crop has a self-incompatibility system (a genetically peculiar one) but self-compatibility is fairly common, indeed general in the rather uniform and narrowly based 'amelonado' populations of Brazil and West Africa. Fairly extensive collecting has been done and good collections are held in Trinidad and Costa Rica. Recently planted populations have sometimes been based on using the self-incompatible system to produce synthetic or near-hybrid seedling populations that exploit the considerable vigour characteristic of crosses between 'amelonado' types and Upper Amazon collections; but manual pollination is more reliable. Clonal propagation has so far been inefficiently exploited but must increase and, indeed, shows signs of doing so as the 'centre of gravity' of production moves from West Africa to Southeast Asia.

Agroecology. The crop is strictly tropical and native to moist shady places but will withstand a moderate dry season. Cropping tends to be seasonal or bi-seasonal depending on climate. Traditionally a small-farmer crop grown at low inputs under shade (*Erythrina*, *Inga*, etc.) it is increasingly grown at higher yields, unshaded but with fertilisers (cf. coffees and tea). Processing equipment is fairly simple but a wooden drying floor sheltered from rain is virtually essential. Ghanaian small-farmer production has declined seriously under disease and mismanagement and Côte d'Ivoire and Cameroon now dominate production. Southeast Asian production develops apace, on a plantation scale and tending to grow modern clones under high inputs. An effective plantation practice is to grow cocoa under light coconut shade, but legumes such as *Gliricidia* are still seen.

Production. Production was recently about 1.6 Mt.

Diseases. The crop has been beset by a very damaging virus in West Africa (swollen shoot), by several fungi (notably *Phytophthora* spp., causing black pod, etc. and *Crinipellis perniciosa* (*Marasmius perniciosus*), causing witch broom) and by several insects. Breeding has made a little progress but vertical resistance (VR) against witch broom failed and the total effort has been small. An attempt to eradicate swollen shoot by cutting out in Ghana failed, despite heroic efforts. Disease resistance will have to be a substantial component of any serious breeding programmes and several good horizontal resistances have been identified.

References. Wood and Lass (1985) (general); Kennedy *et al.* (1987) (breeding).

(10) SPICES, FLAVOURINGS

The crops listed are all old, some probably even ancient and nearly all are Asian in origin. Several (e.g. clove, cinnamon, nutmeg, pepper) are of great historical importance because of the stimulus they gave to early European travel and trade and subsequent competition and conflict. Most are trees but the three Piperaceae are woody-shrubby or climbing and the Zingiberaceae are, of course, herbs. All are plants of moist wet areas at lower altitudes. Various parts of the plants provide the economic products, thus: bark of cinnamon: seed and aril ('mace') of nutmeg; dried flower buds of clove; prepared fruits of vanilla, pimento, pepper and cardamom; leaf of *Piper betel* (an adjunct to betel nut, *Areca*); endosperm of betel nut; roots or rhizomes of kava, ginger and turmeric. Several are somewhat narcotic (betel, kava) and one (turmeric) is an important colouring agent in curry powders (though it is an indifferent dye-stuff for cloth). The areca nut/betel leaf combination (with several adjuncts) is perhaps the commonest masticatory product in tropical Asia and responsible for the red-stained teeth, lips and spit of habitual consumers.

Most are seed propagated but peppers and gingers are clonal. None has been seriously bred. Vanilla is exceptional in requiring that flowers be hand-pollinated if plants are grown in the absence of their natural pollinator (as they normally are because the species is Central American and the main cultivation is in Malagasy).

Reference. Purseglove *et al.* (1981).

(11) DRUGS, STIMULANTS

The crops listed are very diverse. Coca (*Erythroxylon*) and its relatives are Andean plants, small trees and shrubs, of which the leaves yield the product. Leaves, chewed with lime, are an important local masticatory and the source of the cocaine alkaloids that (mostly illegally) enter international trade. The quinine (*Cinchona*) crop yields the most valuable natural drug which is still produced at about 300–500 t/yr of alkaloids. Traditionally the major treatment for malaria, it has not been entirely displaced by others and has other medical (and beverage) uses. Several ill-defined species are involved. Andean in origin, from middling altitudes, it was known as 'Peruvian bark' until a substantial agriculture was based upon it in the nineteenth century in Indonesia and India. There has been some breeding and the best cultivars are clones. Trees are generally coppiced and the shoots skinned to yield the alkaloid-bearing bark. Tobacco (*Nicotiana*) is an inbreeding, South American allotetraploid herb but, though tropical in origin, it is, like the tomato, also important as a summer crop in temperate agriculture. It has been very successfully bred by inbred lines and the object of many and important cytogenetic studies. Tobacco use (smoking, chewing, snuff), though nearly universal, is now under adverse social pressures in many countries. The alkaloid, nicotine, is addictive and other components are carcinogenic. Nicotine is an acute stomach poison. The related allotetraploid, *N. rustica*, is a source of industrial nicotine for use as an insecticide. Some related self-incompatible diploids are ornamental. There are several cola-nut (*Cola*) species, all small trees of the West African rainforest zone. Cultivated trees are mostly planted at the end of a cycle of shifting cultivation for their subsequent yield of 'nuts' which (like the related, but American, cocoa) contain the stimulant alkaloids, theobromine and caffeine. The nuts are chewed rather than eaten.

References. Akehurst (1981) (tobacco); Institute of Biology (1990) (quinine).

(12) INDUSTRIAL PLANTS, FIBRES

COTTONS, *Gossypium hirsutum, G. barbadense* (Malvaceae)

General botany. The wild cottons are pan-tropical, with species on all the continents but only two are of much economic importance. They are small trees or shrubs, drought-tolerant or even desert plants; cultivated forms have all been reduced in size and life cycle to robust, annual, shrubby herbs. Leaves are simple, palmate. Flowers are large and showy, yellow with a central 'spot'. Fruits ('bolls') are leathery, dehiscent capsules, often trilocular, which split at ripeness to reveal a white, fluffy cluster of linted seeds. A 'lint' hair is a single dead cell, dried, flattened, spirally twisted and therefore coherent and spinnable. There is also a fine 'fuzz' on the testa. 'Seed cotton' must be 'ginned' to tear the lint off the seed without too much damage. Cotton seed for planting needs acid treatment to germinate; most seed is milled for animal feed but must be fed with caution because it is toxic.

History. Of the wild species, only the African diploid, *G. herbaceum*, has spinnable lint. It, and its descendant, *G. arboreum*, which was developed in India, constitute the Old World diploids ($2n = 26$), still grown in India on a small scale, but very short-linted and of little economic importance. The New World tetraploids ($2n = 52$) are allotetraploids, with genomes from *G. herbaceum* and *G. raimondii*. In cultivation, all four species are old, with archaeological records (in India, Mexico and Peru) going back about 5000 years BP. Tetraploid perennials travelled widely from the New World eastwards to Africa and Asia soon after discovery. They were generally displaced by annuals as human selection pushed the crop to shorter seasons, especially in temperate summers. What emerged were the modern Uplands (*G. hirsutum*) and the high quality Sea Islands and Egyptians (essentially *G. barbadense*).

Reproductive system. The cottons are self-compatible inbreeders but with a tendency to outbreeding promoted by insect pollination. Inbred lines are nearly universal in both the important tetraploids; hybrid varieties would be attractive but the crossing technology is negated by the need to use insecticides which kill pollinators. However, hand-pollination is sometimes socio-economically feasible, as in India where hybrids are locally very successful. Whatever the breeding plans adopted, breeding objectives are maturity, lint yield, quality and disease resistance. Quality is complex and subject to an array of laboratory tests culminating in industrial spinning trials. There used to be very fine cotton research systems in tropical Africa, including excellent breeding.

Agroecology. The crop is characteristic of seasonally dry climates, even of deserts, if given some irrigation. The Egyptian/Sudanese ultra-fine *barbadense* crops are all irrigated (e.g. in the Gezira). The bulk of the world cottons, however, are 'middling Uplands' (*hirsutum*), as grown in the United States, the former USSR, Central Africa and India; their lint is shorter and coarser than that of *barbadense* but well adapted to industrial needs. The tropical cottons are mostly small-farmer crops grown under natural rainfall as a cash element in food cropping systems, as in India and the drier parts of a belt across Central Africa from Uganda to the west coast. Production must be highly organised as to seed supplies, ginning and marketing if a trade is to maintain a market niche.

Production. Cotton is still the biggest single clothing textile and production is about 18 Mt of lint, mostly temperate in origin. Production is expected to increase to about 24 Mt around 2000 AD.

Diseases. There are several troublesome diseases such as blackarm/angular leaf spot due to

Xanthomonas malvacearum (which is pretty well controlled by breeding). Insects are far more damaging and, indeed, cotton is probably the most insect-ridden (and most sprayed) of all crops. Pests include lepidopterans, weevils, sucking bugs and stainers. Insecticide use on the crop is now regarded by many as a major social hazard.

References. Prentice (1972); Munro (1987).

Other fibres, miscellaneous

The crops listed are very heterogeneous. One, kapok (silk cotton, *Ceiba pentandra*), is a large tree that produces a short, curly floss in the capsules, a material that used to be much valued for waterproof stuffing and insulation. Commercial production is mostly Asian. The Agavaceae listed, botanically a very complex group of tropical American plants, produce diverse leaf fibres of which sisal (*Agave sisalana*) is the most important as a coarse industrial fibre produced mostly in East Africa and Brazil. It is now on the decline in face of competition from synthetics. All are spiny, rosette-rhizomatous plants adapted to dry areas. A shoot flowers only after years of growth and then dies. If the main axis is cut out before flowering, sweet juice can be extracted and this is the basis of numerous Central American fermented/distilled drinks such as pulque and tequila.

The other plants all yield stem fibres. Manila hemp (ábaca, *Musa textilis*) used to be a major export from the Philippines as the principal marine cordage but has now virtually disappeared. The two jutes (*Corchorus*) yield bast-fibres, usually retted; they will stand varying levels of flooding late in crop. The crops are very recent in origin (mid-nineteenth century) and are the basic material of hessian/gunny/sacking fibres produced mostly in the Ganges/Brahmaputra delta. The two *Hibiscus* species, of which kenaf (Deccan hemp) is much the more important, are minor crops, mostly Indian, yielding bast fibres. Roselle hemp also has a shorter, branched form, with edible fleshy-aromatic calyces, locally called sorrel. *Boehmeria* is more a temperate Asian crop than a tropical one; ramie is an excellent, strong, lustrous bast fibre and the crop has the attraction of being a rhizomatous perennial that is clonally propagable.

Reference. Lock (1969) (sisal).

(13) OTHER INDUSTRIAL PLANTS

RUBBER, *Hevea brasiliensis* (Euphorbiaceae)

General botany. Rubber is a large tree of the Amazon rainforests. It has leathery palmate leaves which are erratically deciduous ('wintering') but quickly replaced by new flushes. Flowering is seasonal, in two bursts per year in a climate with bimodal seasons. Flowers are borne in loose panicles in vast numbers and very few set fruit. Fruits are trilocular capsules of which only three-seeded ones survive to maturity. Seeds are flattened-ovoid, characteristically patterned, oily/toxic and must grow quickly or die; they can be stored only briefly. All parts of the plant are laticiferous. In the bark, latex is secreted in a system of 'vessels' which lie in a spirally-oriented reticulum in the phloem. The art of 'tapping' is thus to cut the bark deeply enough to expose the cut vessels but not so deeply as to damage the cambium. Latex contains an emulsion of rubber particles and sundry organelles (which are imperfectly understood). Rubber is the terpenoid *cis*-polyisoprene, with variable molecular weight, 10^5 to 10^7. Chemically, it is a viscous fluid unless stabilised to make an elastic solid by cross-linking the polymer chains (e.g. with sulphur) to produce 'vulcanised rubber'. Latex concentrate may be stabilised with ammonia or rubber may be precipitated out of emulsion by various chemicals to produce 'lump', 'sheet', etc. The technology is very highly developed.

History. Rubber was a botanical curiosity derived from many different plants until late in the nineteenth century when all others were abandoned and industry and agriculture concentrated on the only good one, *Hevea*. The original stimulus was Goodyear's invention (1839) of vulcanisation and the use of the product to make waterproof fabrics and vehicle tyres. Seeds collected on the lower Amazon from 1876 onwards and taken to Southeast Asia were the basis of the new industry, a tale oft-told, and one with fascinating historical ramifications. Main developments were in Malaysia and Indonesia (evident in the 'boom' of 1910) but the crop became universally spread in the lowland wet tropics in a few decades. Some production later developed in West Africa but tropical American ventures were a total failure (and still are – see below).

Reproductive system. Rubber is an outbreeding diploid ($2n = 36$) showing severe inbreeding depression. Propagation of excellent clones by budding on seedling stocks is standard. Three-part trees, top-worked ('crown-budded') are feasible but yet little used. Standard clonal breeding plans are adopted. Clonal seed gardens provide superior synthetic populations of putative crosses between good parents but their structure and working are genetically ill-understood. In general, a combination of clones and good seedlings have converted a poor-yielding wild tree into a high-yielding cultigen in 100 years and about three sexual generations, a remarkable achievement. Wild relatives from the Amazon area have, so far, been useless.

Agroecology. Rubber demands high rainfall but will stand poor, acid, aluminiferous soils. High yields require good fertilising, though. 'Tapping' (i.e. paring thin layers of bark off the tapping 'panel') and collecting latex is done several times per week but must be stopped when trees are 'wintering' (out of leaf). Clonal propagation is an elaborate and demanding, but highly successful, horticultural procedure that varies a good deal in detail. Trees come into tapping in about 5 years and last for another 20–25 (roughly four panels of 5 years each). When a field is cut for replanting, the hard, durable wood has some value as small timber. Though naturally a big tree, cultivated ones are relatively small because tapping strongly inhibits growth, so much so that some clones in tapping are at risk of wind-damage. Rubber is important both in the estate and smallholder sectors and was a major factor in the economic success of Malaysia. Its capital needs for processing are modest but tapping has heavy demands for labour and wages so the estate sector is now in relative decline. Rubber, once established, must be nearly the ideal smallholder crop in terms of labour distribution and cash flow.

Production. Recent production has been about 5 Mt, the great bulk of it from Southeast Asia; synthetic rubbers (technologically complementary rather than competitive) totalled about 10 Mt.

Diseases. Outside tropical America, rubber is remarkably free of diseases and pests and a few fungal troubles are pretty well under control. In America, however, South American leaf blight (SALB, *Microcyclus ulei*) is devastating and has regularly destroyed all would-be plantations. Such breeding as has been done against it, has been based upon vertical resistance and has therefore failed catastrophically. A systematic programme based upon horizontal resistance and crown-budding would, however, have a good prospect of success; this has not been done because rubber breeding has now been sadly neglected.

Reference. Webster and Baulkwill (1989).

Other industrial plants, miscellaneous

Pyrethrum, a Mediterranean species by origin, became a tropical crop only in the 1960s and is now of some importance in upland Africa, especially Kenya. The pyrethrins are potent insecticides with low mammalian toxicity which lend themselves to chemical modification to give 'semi-synthetic' products and will probably be increasingly favoured for social/environmental reasons. Many *Chrysanthemum* spp. contain some pyrethrins but only the one species (*C. cinerariifolium*) is really useful. It is self-incompatible and clonally propagable, so excellent breeding progress is in prospect. For a time the crop seemed to have much promise for tropical small farmers in ecologically favoured localities but the emergence in recent years of strictly synthetic products must place this in doubt. The other crop listed, derris, is but the most prominent of a crowd of related woody legumes that yield rotenones (insecticides, but usually better known locally as fish poisons). Other genera are *Lonchocarpus* and *Tephrosia*. *Derris* is a woody climber propagated by cuttings and cultivated in Southeast Asia. Rotenones are extracted from the dried roots.

Crop improvement

The conspectus of crops given above lists over 100 crops, most of them long domesticated, some for several thousand years. They have all undergone some, often much, evolutionary change since domestication, under the normal evolutionary process of selection and fixation of genetic variation. In nature, 'Natural selection' rules; in cultivation, a mixture of natural and human selection operates. Farmers have been selecting effectively for millenia; plant breeding is simply a science-based extension of farmers' procedures, the current phase of crop evolution, so to speak. Plant breeding, as a systematic technology, effectively started in the nineteenth century and has been strongly developed in the tropics only during the twentieth century. There have been, as we shall see, some great successes but many crops are, as yet, untouched and there is still plenty of room for progress. This section is necessarily brief and superficial. A fuller treatment that includes tropical examples will be found in Simmonds (1979). Crop-by-crop sketches of evolution and current breeding programmes are given by Smartt and Simmonds (1995). Cytogenetic understanding is fundamental to crop improvement.

GENERAL PRINCIPLES

Objectives

Biomass, partition and yield

In any given well-managed environment, crop growth generally has a productive potential of x t/ha-yr of 'average' dry matter. This quantity is about 25 t in a long temperate summer; no general figure is available for the tropics but numbers around 60 t/ha-yr are quoted (e.g. for sugar cane). In practice, needless to say, few crops actually approach the limit.

A crop product is nearly always a fraction of total crop growth or biomass. The fraction is often called the harvest index, sometimes the partition ratio. Plant breeding can do little or nothing about enhancing photosynthesis, hence biomass *per se*, but it can and does: (1) protect biomass from losses due to diseases (by far the most important aspect of resistance breeding); and (2) enhance partition towards the desired product and away from unwanted or waste structures, such as stems and leaves. Yield is always an important objective of plant breeding, indeed is usually paramount, and improvement of partition is therefore a key element in making progress. Thus, yields of fruits, seeds, tubers, etc., may be favoured by reduction of the vegetative skeleton of the plant. Hence the tendency for many plants to become smaller in size and adapted to dense populations (e.g. rice, sorghum, cassava, banana, coffee, coconut and oil palm).

Since the product may be biologically more costly than crude biomass, it follows that increasing yields and enhanced partition may be accompanied by an actual decline in biomass. Rubber and cotton lint provide good examples. The matter deserves better research than it has received.

Quality

Next to yield, quality of product is sometimes important. Quality is inherent fitness for purpose and must be distinguished from condition. (Mangoes vary but, whatever the quality, a bruised mango is a bruised mango!) Quality is sometimes technologically defined (as for most fibres or for crops grown for a specific chemical constituent); it is sometimes organoleptically defined by smell, taste, texture, personal preference (as with most foods destined for direct human consumption). The plant breeder can often breed effectively for quality characters, whether estimated technically or subjectively. But it is well to remember than there are trade-offs; high quality for some people may not be high quality for others, there may be yield penalties or the market may not adequately reward the grower. Markets may also be given to practices which are superstitious and/or less than honest. So the cautious breeder will want to be convinced that breeding for quality is really worthwhile. Enhanced yield is always a sensible objective; quality features may not be, despite local conventional wisdom.

Socio-economic adaptation

Yield and quality are not everything. Varieties must be adapted to the farmer's environment, whether the farmer is a smallholder or a plantation company. Much can be covered by the rather clumsy word 'growability', implying ease of propagation, management and harvest; marketability is an extension of the same thing. Furthermore, secondary uses may be very important. Thus, maizes in tropical America and maizes and sorghums in Africa are very often used as supports for climbing crops relayed into them, for example, *Phaseolus* beans or yams. Cereal yield is valued but not at the expense of the relay crop. So big, strong sorghum and maize plants tend to be preferred; small, weak hybrid varieties, designed for high yields at high densities, don't work. In African cassava, root yield is not all-important because leaf spinach quality also influences acceptability. Breeders with temperate-country preconceptions often miss this important point.

Genetical features

Oligogenic and polygenic

The plant breeder sometimes sees Mendelian segregations (3:1, 1:1, 9:7, etc.) but most work is concerned with polygenically controlled characters. These can generally be modelled well enough in terms of several to many independent loci with additive alleles. We need not be concerned here with the many departures and complexities, sometimes partially interpretable by reference to dominance, epistasis, etc. (But we shall touch on dominance in connection with heterosis.) Starting from a simple additive model, with an arbitrary mean of zero and each allele having effect ± 1 we supposed this kind of pattern:

Genetic score	-3	-2	-1	0	$+1$	$+2$	$+3$
1 locus	–	–	1	2	1	–	–
2 loci	–	1	4	6	4	1	–
3 loci	1	6	15	20	15	6	1

Clearly, the genetic range widens as the numbers of loci increase. This pattern derives from Pascal's triangle which defines a symmetrical binomial distribution. Obviously, a bell-shaped distribution curve is building up and, in fact, this tends to the statistically 'Normal', which is the basis of statistics and the starting point of quantitative genetics. We never know how many loci are involved in a polygenic character, nor that alleles are truly equal and additive; it is enough that, in practice, Normal approximations work fairly well.

Genotype and phenotype

All gene expression is at least a little affected by environment, even 'good' Mendelian genes of large effect. Polygenic characters, however, are much affected which is why quantitative inheritance has to be treated statistically. Dodging many complications, we nearly always assume (indeed have to assume) that environmentally caused variation is additively superimposed upon genetic variation, that both are statistically more or less Normal and that total variability is measured by summing the variances. Figure 10.1 illustrates the idea. The figure also shows that we can think in terms of a physical outcome, a phenotype (P) determined by adding a genetic (G) component to an environmental (E) one. We can also think of fractions of phenotypic variance

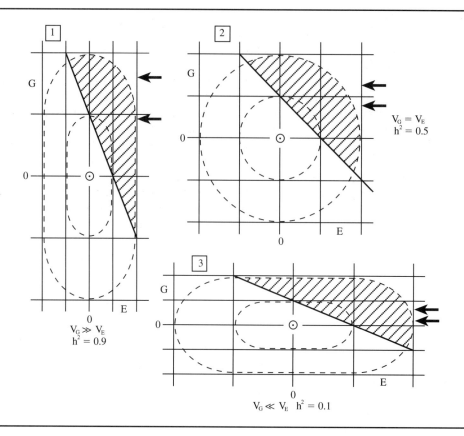

Fig. 10.1 Genotype, phenotype, environment and heritability. Genotypic values (G) plotted against environmental values (E) in the form of bivariate frequencies bounded by ellipses. Phenotypic values (P) are defined by diagonals. Scales are standardised about means of zero. The three cases are markedly contrasted, ranging from (1) high genetic variance and high heritability to (2) equal environmental and genetic variance to (3) low genetic variance and low heritability. The diagonal lines show the sectors cut off by phenotypic selection and the bold arrows mark the genetic advances, R, achieved by selecting at two levels. The higher the heritability the greater the advance; also, and obvious to inspection, the more intense the selection, the greater the advance.

attributable to genetic constitution; such a fraction is called a 'heritability' and

$$h^2 = V_G/V_P = V_G/(V_G + V_E)$$

Numerically, heritabilities are often inaccurate and rarely matter much but the idea is important because it lies at the heart of the problem of response to selection, the plant breeder's central preoccupation.

The reader should be warned that the preceding is grossly simplified. For a somewhat fuller treatment see Simmonds (1979) and, for really serious analysis, see Wricke and Weber (1988).

Inbreeders and outbreeders

Broadly, plants are either inbreeders or outbreeders (though there are some intermediates). Inbreeders are mostly self-pollinated and populations are largely free of deleterious recessive genes because they have been selected out. Pure, homozygous lines tend to dominate but some heterozygosity nearly always persists. There is little inbreeding depression or heterosis. By contrast, outbreeders are cross-pollinated (and have mechanisms that promote crossing), populations carry many deleterious recessive genes and individual inbred lines show inbreeding depression, often to the point of lethality; crosses between inbreds, however, show dramatic restoration of vigour, called 'heterosis'. These phenomena are generally attributed to, respectively, the exposure of homozygous deleterious recessives and the suppression of recessive expression by heterozygosity in crosses. This is the genetic basis of breeding hybrid varieties (HYB – see below).

Broadly, perennial plants are outbreeders and strong inbreeders are nearly all annuals. Therefore clonal crops (perennial by definition) are all outbred, are damaged by inbreeding and individual genotypes tend to be highly heterozygous. Furthermore, perennial crops are best bred as clones whenever it is practicable to do so.

Response to selection

Any population that varies genetically in some respect interesting to the plant breeder can be selected in the desired direction. The population may be a variable population, an array of inbred lines or an array of clones. Details will differ widely but the principles are general. Mendelian major genes present no problems; our real concern is with polygenic characters such as yield or chemical constitution or disease resistance. Response to selection depends upon an equation which we shall give but only with a very superficial explanation. The equation is:

$$R = i \times h^2 \times \sigma_P = i \times \sqrt{h^2} \times \sigma_G$$

where R, the genetic response, is proportional to a statistical measure of intensity of selection (i), to the genetic fraction of total variability (h^2) and to the amount of variability (σ_P, σ_G phenotypic and genetic standard deviations). Shorn of statistical and genetical trimmings (which are nearly endless) this seems and, indeed is, almost commonsensical. To make progress there must be appropriate genetic variability (covered by σ_G), it must be to some extent separable from environmental variability (h^2) and the breeder must work hard (i). If there is one fundamental principle of plant breeding, this equation is it.

G, E and GE effects

The heading refers to *genotypic* (G), *environmental* (E) and *interactive* (GE) effects. The need for this distinction arises from the universal experience that different genotypes (e.g. clones, lines, cultivars) respond differently to contrasted environments. Sometimes, the basis is obvious, as when cultivars differ in resistance and environments differ in disease potential. More often it is not. A common observation is that some cultivars do well in rich, high-input environments but not in poor, hard environments, and *vice versa*. Such interactions (using the word in the statistical sense) are revealed by appropriate analyses of variance and/or regressions (Table 10.1). There is a vast literature on GE interactions but the breeder must face the fact that they are, in effect, unpredictable and that the only practical way to achieve local adaptation is to select and test in the chosen place (Simmonds, 1991b). To try to breed elsewhere is almost to guarantee failure.

Table 10.1 Experimental rice yields (t/ha) in Taiwan and India. Old and new varieties (G_1 and G_2) at high and low fertiliser levels (E_1 and E_2). Δ are differences. New varieties are clearly more responsive to fertiliser. GE effects are about as big as G and E effects. Means of sets of experiments, based on Simmonds (1981).

	G_1	G_2	Δ
Taiwan			
E_1	4.47	4.90	0.43
E_2	5.03	5.86	0.83
Δ	0.56	0.96	1.39

$\Delta_{GE} = 1.39 - 0.56 - 0.43 = 0.40$
and $= 0.96 - 0.56 = 0.40$

	G_1	G_2	Δ
India			
E_1	2.60	3.10	0.50
E_2	3.55	4.73	1.18
Δ	0.95	1.63	2.13

$\Delta_{GE} = 2.13 - 0.95 - 0.50 = 0.68$
and $= 1.63 - 0.95 = 0.68$

PRACTICAL MATTERS

Breeding plans, general

There are four fundamental populations which follow upon the distinction that we drew above between inbred and outbred crops. The populations are as follows:

(1) *Inbred lines (IBL)*: These are feasible only in annual inbred species tolerant of prolonged selfing. Lines are, in effect, homogeneous and homozygous. Variability is generated only by crossing lines, whence new lines are isolated by selection while generation-wise selfing proceeds.
(2) *Open-pollinated populations (OPP)*: These are feasible only in seed-propagated outbred crops whether annual or perennial. Since inbreeding is forbidden because of inbreeding depression, heterozygosity and heterogeneity are maintained and such populations are quite variable.
(3) *Hybrid varieties (HYB)*: These are characteristic of a few outbred, annual crops which can be sufficiently inbred to give viable/usable inbred parent lines. Crosses between such lines are then highly heterozygous but phenotypically homogeneous, indeed, in the limit, as uniform in appearance as IBL or clones (CLO). In practice, however, effective HYB can sometimes be produced from less than perfectly inbred parents.
(4) *Clones (CLO)*: These are feasible in perennials which can be vegetatively propagated whether by tubers, cuttings or graftings. All are outbred, so a clone is highly heterozygous and will not breed true (even if fertile) but it is genetically uniform. Any cross between two different clones will generate genetic variability in the first generation, upon which selection may be practised.

These, then, are the four fundamental populations and their features are further explored in succeeding sections.

The four fundamental populations

Inbred lines (IBL)

IBL are found in rice, wheats, sorghum, most of the pulses, cottons and tobacco. Arabica coffee (an allotetraploid) is the quite exceptional inbreeder among perennials and it, too, is usually highly inbred. Sorghum and the cottons are good examples of plants which have something of the outbreeders about them and, though tolerant of selfing, also do well as HYB (see below). In IBL crops, the breeder crosses chosen parents, expects to find (and does find) a uniform F_1 family and then a variable, segregating F_2. The breeder may select in the F_2 but this is futile for characters of low heritability. Usually, bulks are run to F_3–F_5 and serious selection then starts *between lines, not between plants*. By F_7, essential homozygosity is attained and the best lines will enter final trials. There are endless variations of detail. The hope/intention is to isolate lines transgressive of the best parent in an array of significant economic characters. New IBL thus produced then become, in turn, parents of the next cycle.

Open-pollinated populations (OPP)

These are characteristic of strong outbreeders such as maize, many grass and legume forages and, among perennials, tree crops such as coconut and oil palm, Robusta coffee (outbred) and rubber seedlings. Variability is generated by bulk crossing of populations from different sources. Mass-selection, whereby bulks of phenotypically attractive parents are randomly intercrossed is one common procedure. Another is to rely on progeny testing of potential parents before making the vital decisions. The latter is slower and more laborious but more accurate, so there are trade-offs and definition of efficient plans may be difficult. At the limit, well chosen, progeny-tested parents (that can be maintained as clones to produce seed) generate narrowly based OPP which may be called synthetics (SYN). If only two parents are used and selfing is forbidden, then the SYN, though somewhat variable, is near a HYB. Rubber, oil palm and cocoa provide excellent examples of narrowly based SYN populations.

Hybrid varieties (HYB) (Stuber, 1994)

HYB cultivars depend upon the exploitation of heterosis in outbred, seed-propagated annual species. Inbred lines lose vigour by exposure of deleterious recessives but crossing restores vigour because of the action of complementary dominants: as a very simple model we might have parents aabbCCDD × AABBccdd giving HYB AaBbCcDd. In real life, of course, there are endless complications and only very rare IBL × IBL combinations are truly excellent. Note that heterosis is inexplicable on additive gene action. HYB cultivars are obviously heterozygous but homogeneous and new seed has to be produced for every generation. Successful HYB programmes are very few. Sorghum and maize in the United States are historically outstanding; in recent decades in temperate countries. HYB vegetables and ornamentals have multiplied and will go further. But HYB seed technology is not simple (we shall not explore it here) and tropical applications have been few and local, for example: maize in East Africa, bulrush millet and sorghum in India and some cottons in India. HYB cultivars in the tropics have been the subject of more enthusiastic verbiage than actual use. But HYB maize, at least, is virtually certain to increase (where the crop is grown in pure stand but maybe not in intercropped mixtures).

Clones (CLO)

Clonal crops are, obviously, all perennials even if we actually grow some of them (e.g. the tuber crops) as annuals. They are outbreeders which maintain crossing regimes by very diverse mechanisms but some, grown for vegetative products, have deranged systems of sexual reproduction. Such derangements can present formidable problems (as in the bananas and sugar canes) but, in general, if sexual progeny can be produced at will, CLO breeding is rather simple in principle, if not in practice. Cross any two superior CLO parents and a segregating family prospectively containing good new CLO will result. Examples of CLO crops are the tubers (cassava, sweet potato, potato, aroids), rubber, bananas, sugar cane, many (indeed most) fruits from date and mango to citrus and avocado, onwards. Other tree crops which have not yet been efficiently cloned (such as oil palm, coconut and coffee) would surely benefit from being so developed. If CLO propagation can be achieved, CLO breeding is genetically the most efficient process available (Simmonds, 1996).

The 'backcross method'

The phrase used here recurs in the plant breeding literature, especially from the United States. It refers to the practice (that goes back to the 1920s) of introducing a foreign gene into an inbred line or population by crossing the 'recurrent' and 'donor' parents and then backcrossing the products cyclically to the 'recurrent' parent as often as necessary. In principle, the final product is very near the genotype of the recurrent parent but with the foreign gene substituted. The technique (it should not be called a 'method') is useful for transferring vertical resistance (VR) genes (see below); it has long been a prominent element in temperate-country cereal breeding but declines as the use of VR declines. It is rarely relevant to tropical plant breeding but will have occasional uses when, for example, a dwarfing gene or a monogenic quality character must be transferred.

SPECIAL FEATURES

Disease resistance (Simmonds, 1991a, 1994)

Kinds of resistance

There are diverse pathogens of plants such as fungi, bacteria, viruses, insects and nematodes. Much plant breeding is devoted to developing resistance to attack by them, the economic object being, as we have seen above, to protect biomass and hence yield. With a few qualifications, for which we have no space here, the same genetical principles apply to resistance to all classes of pathogen. 'Resistance' means a state of 'less disease' which is usually economically sufficient; a state of 'no disease' is better called 'immunity' which does occur but is, in practice, uncommon. Evading diverse complications and special cases, we may say that there are two main kinds of resistance/immunity. They are: (1) vertical or pathotype-specific resistance (VR) that generates all-or-nothing effects due to major (Mendelian) genes, large and plainly classifiable as to activity; and (2) horizontal, general or field resistance (HR) which is pathotype-non-specific and polygenically controlled (nearly always with a large additive element) (Simmonds, 1991a).

Breeding strategy

The two main kinds of resistance, hereafter VR and HR, have contrasted characteristics and uses. VR, being monogenic, is easy to breed and may be transiently highly effective but is vulnerable to the evolution of new pathotypes. It tends to fail (and has generally failed) against mobile pathogens such as airborne fungi, when success can only be maintained by repetitive introduction of new VR genes at least a little faster than the pathogen can evolve. This has been done for temperate country wheats but is an acceptable strategy only if relentlessly efficient breeding is possible, which is not usually true in the tropics. However, VR genes may work quite well against immobile or non-adaptable pathogens, such as some soil fungi. By contrast, HR breeding, because polygenic, is harder and slower (though perfectly feasible) but is effective against all pathotypes and is therefore permanently effective ('durable'). An economically potent degree of resistance is always less than total so reduction of infection, not immunity, is sought. Most 'minor' diseases are minor because of HR.

Strategic decisions as to how to manage disease resistances in a breeding programme may be difficult and may involve issues too complex to touch here. Useful discussions in tropical agricultural contexts will be found in FAO (1986a, 1986b) and the genetics of HR is reviewed exhaustively by Simmonds (1991a).

Cytogenetic elements

Haploidy and polyploidy

Haploid plants can, in principle, be generated either by suitable foreign pollination, which stimulates an egg cell to growth without fertilisation, or by the *in vitro* culture of young pollen grains (often mis-called 'anther culture') followed by embryogenesis. In either case, the haploid plant will have to be 'diploidised' either by natural accident or by the use of colchicine. Such 'doubled haploids' are, in principle, absolutely homozygous and have found uses in the very rapid generation of inbred lines in inbred diploids such as barley and rice. They therefore offer economy of time spent in inbreeding and have had some practical use. By contrast, haploids out of diploid outbreeders are depauperate, indeed usually inviable (as expected). Dihaploids out of outbred tetraploids (e.g. potato) may have considerable potential value, though.

Polyploidy, sometimes natural, sometimes induced by colchicine, has been of very little use or interest in diploid crops, except as an occasional aid to cytogenetic tricks. Hopes for enhanced vigour and/or seed size have never been realised.

Mutation induction

Like polyploidy, high hopes used to be held out for the utility of mutations induced by physical or chemical means. In practice, apart from a few horticultural objects, the products have been useless. The genetic limitations as to what can sensibly be attempted and the practical problems of managing the treated populations efficiently have been overwhelming. If mutation induction has any uses in ordinary plant breeding, they are very scarce indeed.

Wide crossing

The great bulk of plant breeding operates within the bounds of the cultivated forms themselves. Occasionally, though, wide (i.e. interspecific, even intergeneric) crosses are attempted, usually with the object of introducing a disease resistance gene from the wild relative. The process demands repeated backcrossing to cultivars. Since the foreign gene usually turns out to be a VR, most such programmes (e.g. in wheats, and potatoes) have failed. In general, wild species have made little contribution to plant breeding but there are exceptions (such as rice) and there are polyploid crops such as sugar cane, sisal and bananas (all with more or less suppressed seed fertility) in which manipulation of whole genomes in wide crosses may have much to offer. Sugar cane, indeed, is the best example extant of such procedures.

In vitro *methods*

In vitro culture of embryos, meristems, tissues and cells on nutrient media in the laboratory go back many years, indeed decades. Useful outcomes include: (1) 'rescue' of precious hybrid embryos that would otherwise have died; (2) isolation of virus-free clones; (3) safe international transmission of healthy plantlets; (4) clonal multiplication *in vitro* ('mericulture') for experimental and/or commercial purposes; (5) generation of haploids from young pollen grains (see above). These are techniques of well-established value in specific circumstances.

If 'genetic engineering' or 'biotechnology' (by whatever name) is ever to be of any practical use in plant breeding (which is yet uncertain), then established *in vitro* methods will be an essential step in the process. Thus 'genetic transformation' or 'transgenosis' will nearly always require the intervention of an *in vitro* stage of free-cell culture whereinto the foreign DNA is introduced preliminary to embryogenesis and multiplication. However, the extravagant predictions of the 1980s have yet to be fulfilled, even though some very elegant experiments have been recorded. 'Biotechnology' in plant breeding is still in the bandwagon stage.

Genetic resource conservation

The response-to-selection equation briefly discussed above showed that progress depends on the availability of genetic variance. All past progress tends to reduce variance, so future prospects are automatically constrained. Hence, any well-found programme takes very seriously the question of conservation of variability (i.e. genetic resources) for future use. But this is a time of 'genetic erosion', so nothing less than systematic, conscious efforts to preserve genetic variability of our crops *in perpetuity* will suffice to meet the long-term need.

In the late 1970s very few people had seen the point, but times have changed and one of the Consultative Group on International Agricultural Research (CGIAR) institutes (see below) has been charged to promote genetic resource conservation (GRC) work on a worldwide basis. This is the International Board for Plant Genetic Resources (now IPGRI, International Plant Genetic Resources Institute) (see, for example, Anderson *et al.* 1988). Much has been done but much remains to do. Perhaps surprisingly, tropical crops have, on the whole, been better served than temperate ones. This is largely due to the fact that the CGIAR system saw the point and acted accordingly; thus the rice collection at the International Rice Research Institute (IRRI) is the best crop collection in the world (and probably the best-used too). But those in charge of research and development on the great tropical perennial cash crops have not been backward; thus substantial collections of bananas, sugar canes, rubbers, oil palms and cocoas have been variously assembled, disseminated and used. In the long-run, this is a profoundly important activity. The technologies of managing collections are very diverse. Holden and Williams (1984) give a valuable summary. For utilisation in relation to the 'genetic base', see Simmonds (1993).

Economics

Costs and returns

To grow a crop of anything incurs costs. In cash economies, these will be measured in money; in subsistence economies, they will be judged by farmers on personal scales of utility. Likewise, the product will be similarly valued. The net outcome is profit, measured in cash or farmer-satisfaction, according to circumstances. The plant breeder's primary objective is to increase growers' profits because, unless this is done, the breeder's varieties will not succeed. The breeder may do this by enhancing yield and/or quality, by protecting the crop against disease losses, by reducing costs of production and so on. But the economic gains due to a good new variety will not long remain with the growers (at least in market economies). As the variety spreads, so will the price fall and consumers as a whole will receive the later benefits. Therefore, *societies rather than farmers are the ultimate beneficiaries of plant breeding*. It is widely believed that plant-breeding is very attractive in social cost-benefit (and environmental) terms and so it very probably is.

Social issues

The Green Revolution in wheat and rice production was achieved by the exploitation of a *GE* effect (see above) on a gigantic scale: new, input-responsive, semi-dwarf varieties were grown under high inputs of water, fertilisers and other chemicals, with spectacular success. The same varieties are unremarkable at low inputs and most tropical food-crop agriculture, even now, remains virtually untouched by such technology. Thanks to IRRI breeding and *GE* interactions, yields of irrigated rices have risen, say, five-fold but a multitude of smallholders growing rain-fed or upland rices remain unchanged: they still grow traditional land-race cultivars at 1 t/ha or less because those are still the best available to them. It remains a serious question as to what, if anything, plant breeding can do for the small farmer on poor land who cannot afford enhanced inputs. Even if improved varieties are available, sustained supply of good quality planting material may present acute problems.

Lipton and Longhurst (1989) give an authoritative account of the wider socio-economic issues involved; a summary of the plant breeding component is given by Simmonds (1990).

EXAMPLES
Classified by breeding plans
Inbreeders (IBL)

Inbred crops of great importance in tropical agriculture include rice, most of the grain legumes (groundnut, cowpea, beans, chickpea, lentil), sorghum, the cottons, tobacco and (quite exceptional among trees crops) Arabica coffee. The breeding objectives of the food crops are, essentially yield and quality, with flavour and 'cookability' as features of the latter. Rice for high-input irrigated agriculture has been a major success story; other rices have not (see above). Disease resistances are a major feature of much breeding. In sorghum, bird resistance and stalk strength are substantial considerations. The most important tropical cottons are annual Uplands, essentially all developed in the twentieth century and a classic example of orderly IBL breeding and well-managed seed production systems.

HYB breeding methods are attractive in sorghum and cotton and have had some success in India. Similarly, heterosis has been reported in crosses of Arabica coffee in Ethiopia but the use of CLO in that crop would seem to be essentially more attractive than HYB; the necessary 'mericulture' technology is already available but may not be socio-economically appropriate.

Outbreeders (OPP and HYB)

The outstanding outbred annual food crop in the tropics is maize, with bulrush millet running it second. The great majority of both are represented by local OPP not yet much affected by scientific breeding. HYB maize has done well in parts of East Africa, however (e.g. Kenya), and HYB millets in India. More and more HYB maize, at least, must be expected but will be technically feasible only where there is a competent local seed industry. Of the legumes, pigeon pea is also a variable outbred OPP and no HYB technology is available.

Several tree crops are also grown as OPP

or as rather narrowly based synthetic (SYN) populations verging on HYB, though nowhere near so uniform because parents must, of necessity, be outbred. Thus, rubber, tea, cocoa, coconut and oil palm all have good seedling populations generated in seed gardens from selected clonal parents (the first three crops) or grown from hand-pollinated crosses of parents drawn from rather distinct sources (the two palms). Many relevant details are fascinating but beyond our scope.

Clones (CLO)

In the two preceding sections we have mentioned several crops that are either partly clonal (rubber, tea, cocoa) or that could with advantage be developed as CLO (Arabica coffee). Of these, rubber is already predominantly clonal, perhaps the outstanding example of mass-grafting in the history of agriculture, especially if development towards the three-clone tree proceeds as expected (rootstock, trunk and crown).

These crops are trees. There are also herbaceous (or shrubby) clones such as the tuber crops (cassava, yams, sweet potato, aroids), a fibre (sisal), a starchy fruit (bananas) and sugar cane. It could be argued that the tuber crops have not had the attention that their socio-economic importance in the wetter tropics would justify (de Vries *et al.*, 1967) but cassava and sweet potato are evidently making some progress. As low-input crops, initial progress is perhaps most likely to be by way of disease resistance (as seems to be true of cassava). The great success story of this list is undoubtedly sugar cane, arguably one of the most remarkable achievements in the history of plant breeding. Since the 1920s, the crop has been effectively reconstructed on a wide interspecific hybrid base and has a record second to none of controlling epidemic diseases by HR. The reason, of course, is plain enough: it is a great industrial crop backed by competent research and development at all levels. The breeding not only has a remarkable record of practical success; it is also an outstanding example of international collaboration and the solution, by breeding, of diverse biological and economic problems.

Agencies and outcomes

Introduction

In enquiring into what agencies think (or have thought) tropical crop breeding to be worthy of support, we first have to ask what are the motives for supporting it. There are two, which overlap. First, there is the socio-economic argument which assumes that well-used agricultural research in general is broadly favourable for rural societies in the first instance and subsequently favours the urban sector as well. On this argument, crop husbandry and breeding are regarded as complementary (they are) and both are supported; both food and cash crops are regarded as appropriate objects of study because societies need foreign exchange and internal cash flows as well as food.

Second, there is the more sharply focused economic argument that attaches to the great export crops grown and traded in high volumes and often dependent upon an elaborate and costly infrastructure of machinery, factories and transport. The more profitable they are the better. Research on such crops is certainly in the public interest and some governments support it; but, often, it is both practical and sensible to leave the task to the industry itself, on the argument that it *should* be the best possible judge of its own interests, as, indeed, it often (but not always) is.

Examples of agencies

The preceding section will have made it clear that agencies are likely to have complex and overlapping roles in tropical plant breeding and so they do. Three kinds of agency are distinguished. First, national agricultural research systems (which have grown greatly, though very unevenly, in recent decades) commonly tackle the whole spectrum of crops important for their countries. The Indian Agricultural Research

Institute is the outstanding example of a large, competent, nationwide system that tackles everything from food crops to export crops. Several comparable but less powerful systems are in place elsewhere in the tropics; often, there is a tendency to delegate responsibility for export/cash crops to semi-autonomous agencies.

Second, the Consultative Group on International Agricultural Research (CGIAR) runs a body of very strong institutes spread through the tropics (Anderson *et al.*, 1988). This, the 'CG system', is only about 20 years old, is internationally funded and has devoted itself to annual food crops to the extent of not noticing either that perennials are fundamental to tropical agriculture or that cash matters socially (misconceptions that now seem likely to be corrected) (Simmonds, 1995b). The institutes tend to a mixture of specialisation by crops and regions; for local applications, they naturally depend upon links with national systems.

Third, commodity programmes, centred on the great cash crops and run essentially by the industries themselves, have long and highly successful histories going back to the end of the nineteenth century. They have rarely lacked governmental backing, of course. Thus sugar cane, rubber, oil palm, coffee, tea, cotton and others have all been the subject of good to outstanding research, both agronomic and genetic. The great strengths of such systems are that grower–researcher links are close, that objectives are generally clear and uptake virtually certain. One can rarely say that of tropical food crop research.

Outcomes

The outcomes of tropical plant breeding have been very uneven. At best, as we saw above, some of the export/cash crops have made outstanding progress and there is no doubt, in the author's view, that such progress will be maintained if, but only if, the industries continue to thrive economically; rich-country politics are likely to be more important than technical competence, alas.

As to the tropical food crops, many, the lesser or local ones, are virtually unknown scientifically and the prospects of any agency, national or international, taking any of them seriously seems remote. Whatever their agricultural and social potential, we are unlikely to exploit breadfruit, pejibaye and avocado to the full. There are many such crops, of course, as the conspectus earlier in this chapter will have shown.

Some of the annual staple food crops have been rather better served. As we saw above, irrigated rices bred by the CG institute, the International Rice Research Institute in the Philippines, have revolutionised the crop. The essential feature was exploitation of the high-input/new variety package, a GE effect on the grand scale. Farmers could afford the inputs, therefore the technology worked. But rain-fed and upland rices (well over half the total rice area) have been but little touched by plant breeding or any other new technology. The CG system institutes and local research systems have also made some progress with other cereals, pulses and tubers but, as we saw above, there is a limit to what plant breeding can do on its own, if low inputs pre-empt the exploitation of GE effects. This matter ought to be a central pre-occupation of policy makers but one may doubt whether it is. Agricultural advance is a far wider matter than plant breeding or even agricultural research and development.

FURTHER READING

For general botany, Leaky and Wills (1977) and Cobley and Steele (1977) give useful introductions while Purseglove (1981, 1987) presents a fairly comprehensive treatment, crop by crop. Evolution and breeding of all significant crops, temperate and tropical, are outlined in Simmonds (1976, revised in Smartt and Simmonds, 1995). Food crops are reviewed by Simmonds (1995a). Mixed cropping and the importance of tropical tree crops are treated by Watson (1983, 1990), Huxley (1983c), Francis (1986), Cannell (1989a). Crop physiology and ecology (on a crop by crop basis) are treated by Squire (1990) and by Norman *et al.* (1984), respectively. Composition and nutritional features of fruits are summarised by Nagy and Shaw (1980). Much useful information on minor, little-known crops is assembled in Ritchie (1979) and National Research Council (1976, 1979, 1983, 1989). There is no systematic work on the breeding of tropical crops but useful summaries of some perennial and clonal crops are given in Ferwerda and Wit (1969) and Abbot and Atkin (1987). A general critique of tree crops and the neglect of perennials in tropical agriculture is given by Simmonds (1995b).

11
Production of Animal Feed

P.N. Wilson

Very few tropical animals are fed on a single feed – even cattle grazing a sown grass pasture will vary their diet with hedgerow plants and weed species. Some tropical milking cows are likely to be fed on a concentrated diet, in addition to fresh or conserved grass, for about half of the year.

It is, therefore, important to appreciate the complementary nature of the various types of feed. In the case of ruminants the diet will often consist of a combination of a forage and a concentrate in different proportions according to the system of management employed. In the case of tropical pigs and poultry kept intensively, the diet usually consists of a concentrated mixture, but this itself will comprise many different feeds. If mixed on the farm, the diet may comprise five or six feeds roughly mixed together, but in a purchased compound the number of feeds employed by the miller will probably number 20 or more. The object is to produce a 'balanced diet' with the correct ratio of protein to energy and the required amounts of major minerals, trace elements and vitamins.

Feed is traditionally broken down by the chemist into its constituent chemical fractions as shown in Table 11.1. However, apart from water, these traditional chemical components are only useful in providing a framework for classification and have no particular significance *per se*. For example, the ash fraction can be very useful (if it provides the right major and minor elements in the right proportions) or it can be relatively useless if, for instance, it contains a lot of indigestible silica because of soil contamination.

Feeds are usually classified according to their origin, and the system chosen is of less importance than the correct understanding of the definitions employed. One of the commonest errors is to regard a single 'straight' feed (e.g. sorghum) as a 'concentrate' merely because the nutrients contained in sorghum are more concentrated than those contained in a forage, such as grass.

The three main classes of animal feed are:

(1) Grass and forage crops.
(2) Straights.
(3) Concentrates and supplements.

These will be considered briefly in turn.

GRASS AND FORAGE CROPS

A detailed study of tropical grasslands, and their management, is given in Chapter 12. It is important to realise that grass, and other plant species associated with grass, comprises the most important feed for livestock in the tropics, since the majority of other crops which are useful for feeding to animals are also valuable sources of food for man.

Certain fodder crops, capable of giving high yields when properly managed under fertile and moist conditions, are grown on a moderate scale in the tropics for stall feeding or ensilage. Some of them can also be grazed, but the majority do not persist well under long-continued grazing and give lower yields than when they are cut for fodder. The plants most commonly used are tall, stool-forming grasses, such as *Pennisetum purpureum* (Elephant, Napier or Merker grass),

Table 11.1 Components of different fractions in the analysis of feed.

Fraction	Components
Moisture	Water
Ether extractive (EE)	Fats, oils, waxes, organic acids, pigments, sterols, vitamins A, D, E, K
Crude protein (CP)	Proteins, amino-acids, amines, nitrates nitrogenous glycosides, B vitamins, nucleic acids
Crude fibre (CF)	Cellulose, hemicellulose, lignin
Ash	Essential elements and minerals, non-essential elements and minerals
Nitrogen-free extractives (NFE)	Cellulose, hemicellulose, lignin, sugar, starch, pectins, organic acids, resins, tannins, pigments, water-soluble vitamins

Panicum maximum (Guinea grass), *Setaria sphacelata* and *S. splendida*, but to a lesser extent more prostrate, stoloniferous grasses, such as *Brachiaria mutica* (Para grass), are also used. Legumes and grass–legume mixtures are also sometimes cultivated for fodder or silage, but pure stands of legumes are more usually grown for hay, common species for this purpose being lucerne (*Medicago sativa*), velvet bean (*Stizolobium aterrimum* or *Mucuna utilis*), and certain varieties of soya bean (*Glycine max*).

These crops can be valuable for the production of dry-season feed in regions of seasonal rainfall. In conjunction with the utilisation of natural grasslands in the drier regions, it is often possible to grow limited areas or fodder crops, either on dry land during the rains, or in favourable places that are seasonally swampy or flooded, or under irrigation. They are also useful in the moister parts of the seasonal rainfall regions where mixed farming, involving the use of natural or sown pastures, is practised. Here, rain-grown fodder crops can be cut for silage or hay during the wet season and, if manured after cutting in the latter part of the rains, will continue growing into the dry season to provide a certain amount of grazing or a late cut.

In the humid tropics, where a dry season is absent or slight, fodder crops can also assist the development of more intensive mixed farming, because the natural grasslands are limited and poor and, in most places, suitable species for sown pastures have yet to be found. For example, in many rice-growing areas cattle subsist on the grazing or cut fodder provided by limited areas of coarse natural grasslands during the rice-growing season and, during the off season, they depend mainly on the poor grazing afforded by the fallow rice fields. Butterworth (1962) has indicated the poor feed value of the rice stubbles, and in many places, after their period of fallow grazing, the animals are in poor condition for the work of preparing the fields for the next rice crop. In these humid zones, and also under irrigation, fodder grasses can give high yields; their cultivation on limited areas would greatly improve the supply of cattle feed.

Fodder grasses

Cultivation

Most species have to be propagated vegetatively from stem cuttings or rooted stool pieces (e.g. Elephant and Guatemala grasses) or from stolons (e.g. Para grass) but some, such as *Setaria sphacelata* and certain strains of *S. splendida*, can be established from seed. It is usually best to plant in rows in order to facilitate weeding during the establishment stage. The best spacing will obviously vary with the species, the soil and rainfall, whether the crop is to be irrigated or not and whether cultivation is to be mechanised.

In a humid climate or under irrigation, fairly close spacing in rows about 90 cm apart may be expected to give the best yields, provided that sufficient nitrogen is supplied, but spacing can be varied within the limits of 60–120 cm between rows without greatly altering yield. In a drier climate, wider spacings may be advantageous where cultivation can be mechanised, since they involve little or no loss of yield after the first year. In some circumstances it may be desirable to have close spacing in order to reduce the labour, or cost, of weeding by obtaining a maximum smothering effect from the grasses. Closer spacing may also be advisable to lessen wind or water soil erosion.

After planting, cultivation for weed control will be needed for a few months, or perhaps the first year, but thereafter it should not be necessary since, if the crop is well manured and managed, it should suppress weeds. On soils with average day content cultivation is not needed except for weed control, and deep tillage is unlikely to be beneficial.

The effects of cutting interval and season on yields and composition

When fodder grasses are cut regularly, increasing the intervals between cuts (thus allowing the grasses to reach a more advanced stage before utilisation) results in increased yields of green or dry fodder, increased percentage contents of crude fibre, carbohydrates and dry matter, but decreased contents of crude protein and ash. This is illustrated by the figures in Table 11.2 for the composition of three species in Tanzania at various stages of growth, by those of Table 11.3

Table 11.2 Variation in composition of fodder grasses with stage of growth, in Tanzania. (Sources: [1] and [2] Van Rensburg, 1956: ash figures = total ash; [3] French, 1943: ash figures = soluble ash.)

Species	Stage of growth (Height in metres)	Percentage of dry matter			
		Crude protein	Crude fibre	N-free extract	Ash (see above)
Panicum maximum[3]	0.3–0.5	9.2	31.2	—	7.4
	0.6–0.9	9.3	34.5	—	8.4
	1.8	5.6	41.8	—	5.6
Pennisetum purpureum[1]	0.3	20.2	27.4	28.7	20.1
	1.2 (silage stage)	11.2	33.4	38.2	14.3
	2.4	5.9	35.4	42.3	14.3
	3.3 (with bare stem)	2.5	37.7	39.3	18.3
Setaria sphacelata[2]	0.3	21.3	26.4	32.3	16.2
	0.6	15.3	35.2	32.0	14.6
	1.2 (flowering)	8.7	42.0	36.1	10.4
	1.8 (after flowering)	4.1	36.3	50.5	7.8

Table 11.3 Effect of cutting interval on dry matter and crude protein content of four fodder grasses in Trinidad. (Source: Paterson, 1936.)

Species	Percentage dry matter Cutting interval (days)				Percentage crude protein in dry matter Cutting interval (days)			
	45	90	120	180	45	90	120	180
Elephant grass	13.4	21.5	24.3	23.2	9.6	5.9	6.1	4.9
Guatemala grass	16.4	16.4	16.9	22.8	8.1	5.7	3.9	3.3
Guinea grass	16.2	19.3	24.3	31.8	8.4	5.4	5.9	3.7
Para grass	18.0	17.4	22.1	17.1	11.1	6.9	6.6	4.0

showing the effect of four cutting intervals on the dry matter and crude protein contents of four species in Trinidad, and the data presented in Table 11.4 showing the effect of four cutting intervals on Elephant grass in Nigeria. With Elephant grass more than about 3 weeks old the stems mature more quickly than the leaves, resulting in a rapid deterioration in the nutrient value of the plant as a whole but, since animals do not normally eat the mature stems, it is the nutrient content and yield of the leaves which are important.

A cutting interval must be selected which gives a satisfactory compromise between nutrient

Table 11.4 Effect of cutting interval on Elephant grass (unfertilised) in Nigeria. (Source: Oyenuga, 1959.)

	Cutting interval (weeks)			
	3	6	8	12
Whole plants				
Dry matter (%)	16.5	19.0	21.7	25.9
Yield, green fodder (t/ha) in 11 months	73.0	97.0	98.0	132.8
Yield, dry matter (t/ha) in 11 months	12.4	19.1	22.0	34.4
Leaves only				
Crude protein (% of dry matter)	14.7	11.4	10.5	9.6
Silica-free ash (% of dry matter)	10.2	8.9	7.4	6.0
Crude fibre (% of dry matter)	25.7	28.1	29.5	29.6
N-free extract (% of dry matter)	40.9	43.1	44.8	47.6
Yield, crude protein (t/ha) in 11 months	1.9	1.1	1.3	1.6
Yield, ash (t/ha) in 11 months	1.3	0.9	0.9	1.0
Yield, total carbohydrates (t/ha) in 11 months	8.5	7.7	9.2	13.2
Leaf:stem ratio, by weight	12.0	1.3	1.2	0.9

Table 11.5 Yield, composition and feeding value of fodder grasses in Trinidad. (Source: Paterson, 1939.)

	Guatemala grass	Para grass	Elephant grass
Yield (t/ha)			
Fresh weight	106.4	100.4	160.1
Dry matter	20.3	24.4	23.4
Crude protein	1.4	1.5	1.6
Percentage of fresh herbage			
Dry matter, wet season	18.6	21.6	13.6
Dry matter, dry season	27.7	32.6	19.5
Crude protein, wet season	1.3	1.3	0.9
Crude protein, dry season	1.9	1.7	1.3
Percentage of dry matter (average for whole year)			
Crude protein	7.0	6.1	6.7
Crude fibre	34.1	34.8	33.4
Carbohydrate and fat	48.0	48.8	45.0
Total ash	10.8	10.3	15.0
CaO	0.4	0.7	0.9
P_2O_5	0.8	0.9	1.2
Na_2O	0.03	0.26	0.04
Starch equivalent, wet season	10.6	12.9	7.8

value and palatability on the one hand and yield on the other. This will vary with the species and also with the weather, especially the rainfall, which will affect the rate of regrowth of the grass, so that intervals normally need to be longer in the dry season than during the rains. It should also be noted that the dry matter content may increase by as much as 50% in the dry season (see Table 11.5). The protein content of the fresh material also rises, with the result that a maintenance ration at this time is provided by about two-thirds of the weight of fresh grass required in the wet season. The effect of cutting frequency on the rate of regeneration and persistence of the grass must also be considered. Too frequent cutting will weaken the grass, reduce the rate of regrowth and result in lower yields and in death of a proportion of the stools. Suitable cutting intervals usually vary between 6 and 10 weeks, depending on species and seasons. Workers in Trinidad recommended intervals of 6–8 weeks for Para grass, 7–8 weeks for Elephant grass and 8–10 weeks for Guatemala grass, while in Malaysia 6-weekly cutting seemed to be best for Guinea grass (Keeping, 1951). From his results for the nutrient yields of the leaves only of Elephant grass (Table 11.4) Oyenuga (1959) concluded that cutting every 3 weeks gave the highest yields of crude protein and ash, 12-weekly cutting gave the maximum yields of carbohydrates and total nutrients, and 6- or 8-weekly cutting was not recommended.

Height of cutting

The height of cutting may also affect yields, rate of regrowth and persistence. In the early Trinidad experiments it was found that, for all four species (Elephant, Guatemala, Guinea and Para grasses), cutting at ground level reduced the rate of regeneration and gave lower yields than cutting at 10–15 cm or 23–30 cm. Cutting also tended to cause death of stools, especially if combined with frequent cutting at intervals of 45 days. In another early experiment Paterson (1938) compared cutting at 8 cm and 30 cm above ground with Elephant, Guatemala and Para grasses, and found that the higher cut gave better yields with Guatemala grass but lower yields with Para grass, while with Elephant grass there was no significant difference between the two heights. For most species except Guatemala grass, cutting at 8 cm above ground will be satisfactory, but occasionally it may be desirable to cut at ground level to remove trash and old stumps.

Manuring

Liberal manuring may be necessary in order to maintain yield over a period of several years under a system of fairly frequent cutting. The main need is usually for nitrogen and, with high-yielding grasses, responses to this nutrient are usually linear up to fairly high levels; for example, Elephant grass in Zimbabwe (then Rhodesia) responded linearly up to at least 230 kg nitrogen/ha, and in Puerto Rico up to 900 kg nitrogen/ha/yr. The lower-yielding Guinea grass gave good responses up to 150 kg nitrogen/ha/yr in Malaysia. Nitrogen will also increase the protein content; with Elephant grass in Zimbabwe an increase of 0.83 kg of protein was obtained for every kg of nitrogen applied over a fair range of applications. Phosphate is likely to be required on many soils; it is best to give a good application at planting time as subsequent top dressings are less effective. Potassium may also be needed on some soils especially for Elephant grass, which is a gross potash feeder, but other grasses may also need it, e.g. in Malaysia 375 kg/ha of sulphate of potash more than doubled the yield of Guinea grass. Interactions between the three major nutrient elements may be important. Phosphorus and potassium may fail to produce significant effects for want of nitrogen but may lead to increases in yield in the presence of liberal supplies of nitrogen.

Fairly frequent fertiliser applications are needed. Even if frequent small applications of fertilisers do not give higher annual yields than the same amount applied less frequently, they will usually give more uniform yields throughout the year provided there is no severe dry season.

Yields

Yields vary greatly with species, soil, fertiliser practice and moisture supply. Average yields obtained with Guatemala, Elephant and Para grasses in Trinidad, on fairly good soil with 750 kg/ha/yr of fertiliser, and under an average annual rainfall of 1780 mm with a moderate dry season of 4 months, are given in Table 11.5. At the other extreme, Kennan (1950) under low rainfall in former Rhodesia, obtained average fresh weight yields over 2 years of only 17.4 t/ha/yr for Elephant grass, 14.0 t for *Setaria sphacelata*, and 11.5 t for Guinea grass. High yields can be obtained with irrigation; thus Van Rensberg (1956) obtained average fresh weight yields per hectare per annum over 4 years of 258 t of Elephant grass, 243 t for *Setaria splendida* and 158 t for *S. sphacelata*. In experiments in which the grass was grown without limitation of water or nutrient supply, fresh weight yields of over 400 t/ha/yr (85 t/ha dry matter) have been obtained with Elephant grass (Vicente-Chandler *et al.*, 1959).

Species

Pennisetum purpureum (Elephant grass) is the most widely used fodder grass as it is easily propagated from stem cuttings and usually gives the highest yields of herbage and nutrients. It is succulent and palatable but has a high moisture content; about 35 kg of fresh herbage is required to provide maintenance for a 450 kg cow in the wet season. It needs cutting every 6–8 weeks in the wet season, or even more frequently at the height of the rains, when it quickly reaches the stage of forming 'canes' which cattle will not eat. It can be grazed, if managed carefully, and it is often convenient to do this at the beginning of the rains, followed by several silage cuts, and then to graze again early in the dry season. It is suitable for cultivation up to 2000 m. There are many varieties, some of which are undesirable because they are too hairy.

Tripascum laxum (Guatemala grass) is propagated by rooted stool pieces. It is not as high-yielding as Elephant grass but is more leafy and palatable, persists well under appropriate cutting and maintains its yield during the dry season compared with other species. It should be cut at 23–30 cm above ground usually at intervals of 8–10 weeks. It has been found that 27 kg of Guatemala grass provides a wet-season maintenance ration for a 450 kg cow. It does not persist well under grazing as cattle tend to uproot it.

Panicum maximum (Guinea grass) exists in a very large number of strains, varying in vigour and habit, some of which are more suitable for pastures than as a fodder crop. Types used for cutting are usually propagated as rooted stool pieces, although some produce viable seed rather sparsely. It is not as quick to establish itself and smother weeds, nor as high yielding, as Elephant and Guatemala grasses. It can be cultivated up to 2000 m and should be cut at 8 cm above ground level about every 6–8 weeks.

Both *Setaria splendida* and *S. sphacelata* are leafy and succulent grasses with higher protein contents than the previous species. They can be grown up to 2400 m, but require a fertile soil. The former has normally to be propagated vegetatively but the latter produces seed abundantly.

Fodder legumes and grass–legume mixtures

Legumes are normally of better nutritive value than grasses, having higher contents of crude protein, calcium and phosphorus. They are not grown as fodder crops in pure stand as extensively as grasses, mainly because they give lower yields but also because they are more difficult to ensile unless mixed with other crops. Nevertheless, in stock-keeping areas experiencing seasonal rainfall, a most valuable use for limited areas of irrigable land is for the production of protein-rich leguminous hay or green fodder. For example, in Tanzania irrigated lucerne gave 88–122 t/ha/yr of green fodder containing 20% of crude protein in the dry matter. In some irrigated areas occupied by smallholders legumes are extensively grown to provide green fodder for stock feeding. Thus, in Egypt about 20% of irrigable cropped land grows berseem (*Trifolium alexandrinum*); in the Punjab about 14% of irrigable land is under berseem or Bokhara sweet clover (*Melilotus alba*), and in the Gezira area of the Sudan *Dolichos lablab* is extensively grown.

To a limited extent legumes are also grown in pure stand for the production of hay without irrigation. In South Africa, Zimbabwe and many parts of South America, lucerne is the most widely used species for this purpose, but in Africa soya beans (*Glycine max*), cowpeas (*Vigna sinensis*), velvet beans (*Stizolobium deeringianum*) and *Dolichos lablab* are also grown. Figures for the composition of hay of these species are given in Table 11.6. In Zimbabwe certain varieties of soya bean (e.g. Hernon, Biltan) have been specially selected for hay and these are capable of giving higher crude protein contents than those indicated in the table, probably up to 19%.

Although fodder grasses can give very high yields when supplied with adequate water and nutrients, they have a relatively low protein content. It would be useful if this could be counteracted by growing them in mixtures with suitable legumes, which might also release some of their fixed nitrogen to benefit the growth of the grass. However, even when grown alone, legumes yield less than grasses and when the two are grown together the tall fodder grasses shade the legumes, restrict their growth and limit the contribution they can make to yield and nitrogen supply. It is, therefore, unlikely that mixtures will give dry matter yields as high as those obtainable from heavily fertilised fodder grasses. (This subject is discussed in Chapter 12 in the context of mixed grass/legume swards used for grazing.) The success of mixed swards used for fodder depends not only on finding compatible perennial grass and legume species but also on determining sowing times, spacings and cutting procedures which will enable a satisfactory balance to be maintained between the two components. Compared with pure grass stands receiving little or no nitrogenous fertiliser, such mixtures often give higher yields of dry matter and increased crude protein yields. Various mixtures have been found useful, as indicated by the following examples.

In Puerto Rico there has been considerable use of *Pueraria phaseoloides*, *Centrosema pubescens* and *Leucaena leucocephala* combined with Elephant, Para or Guatemala grass. The best combination seems to be Elephant grass with *Pueraria* and it has been reported that, without fertiliser, a hectare of this mixture yielded as much as a hectare of each of the components in pure stand. However, Caro-Costas and Vicente-Chandler (1956) noted that the unfertilised mixture gave much lower yields of dry matter and crude protein than pure Elephant grass receiving 336 kg/ha of nitrogen. In Kenya, mixtures of Elephant, Guinea or Guatemala grass with *Desmodium intortum* or *D. uncinatum* have given higher dry matter and crude protein yields than those obtained from the grasses alone (Kenya National Agricultural Research Station, 1970). In one trial the crude protein content of Elephant grass was raised from 6.6 to 7.1% and that of Guatemala grass from 6.6 to 10.0% by growing the grasses with *D. uncinatum*, the crude protein contents of the mixtures being respectively 9.8 and 14.0% (Suttie and Moore, 1966). In trials in Queensland, *Stylosanthes guianenis* and Guinea grass proved a good combination, yielding 5.8 t/ha of dry matter to which the legume contributed 47% while mixtures of *Macroptilium atropurpureum* or *Centrosema pubescens* with the same grass gave 4.4 and 4.0 t/ha of dry matter respectively (Grof and Harding, 1970). Mixtures

Table 11.6 Leguminous hays: percentage composition of dry matter. (Sources: Various; enumerated by Webster and Wilson, 1980.)

Species	Crude protein	Digestible crude protein	Crude fibre
Dolichos lablab, Zimbabwe	12.1	—	21.3
Glycine max, Zimbabwe	13.8	—	32.2
Medicago sativa, Kenya	18.9	13.6	28.9
Medicago sativa, Tanzania	16.5	10.7	31.4
Stizolobium deeringianum, Northern Nigeria	11.2	7.3	34.2
Vigna sinensis, Northern Nigeria	10.9	7.4	27.4
Vigna sinensis, Zimbabwe	11.1	—	28.7

of *Setaria sphacelata* with *D. intortum*, *M. atropurpureum*, *Stylosanthes guianenis* or *Lotononis bainesii* have been successful (Blunt and Humphreys, 1970) and also those of Elephant grass with *C. pubescens* or *Glycine wightii*.

The highest yields of nutrients and of animal products will generally be obtained from pure grass stands heavily fertilised with nitrogen, which are also easier to establish and manage than grass–legume mixtures. However, the high cost of fertilisers will often make intensive animal production based on fodder grasses unprofitable and the system is unlikely to be adopted by small farmers. There is, therefore, a strong case for the wider use of grass–legume mixtures which are likely to give lower, but more profitable, yields.

Ensilage

Good silage can be made from maize and several types of sorghum in the tropics, but ensilage of most tropical fodder grasses, or of the species used in sown pastures, is difficult and commonly results in a product differing from, and inferior to, that prepared from temperate grasses. The exclusion of air from the ensiled material is difficult because the coarse-stemmed tropical grasses are not readily compacted and the consequent aeration leads to large losses by spoilage. Owing to the relatively low sugar content of tropical grasses and the large losses of sugar under high ambient temperatures by respiration, and by aerobic decomposition, ensiling does not result in the high concentrations of lactic acid that assist the stability of silage made from grass in the temperate zone. The pH of tropical silage is usually higher than that of temperate-zone silage, and considerable losses of dry matter and nitrogen occur during its preparation (Catchpole and Henzell, 1971). Addition of molasses can increase lactic acid content and improve preservation, but only when much larger amounts (40–80 kg/t wet weight) are added than are needed for the same purpose (7–10 kg/t) with temperate herbage and, even then, losses of dry matter and nitrogen are not always reduced.

Elephant grass is probably the commonest grass used for silage, with dry matter losses of only about 9–12% but wastage can easily exceed 20% in the absence of proper precautions. Satisfactory silage, with acceptable dry matter losses of 9–16%, can also be made with Guinea grass or *Setaria sphacelata*, especially if molasses is added.

The nutritive value of the product primarily depends on that of the original material and, as tropical grasses are lower in feed value than temperate grasses, they produce a poorer quality silage. This is accentuated as ensiling tropical grasses generally reduces the digestibility of the dry matter, and especially that of the crude protein, to a greater extent than with temperate grasses. Semple *et al.*, (1966) noted that silage made from fresh Pangola grass with 11.5–13.3% crude protein had a crude protein content of 9.4% with a digestibility of only 38.9%. Other workers have also reported digestibilities of about 40% and it appears that tropical grass silages are generally low in digestible crude protein and both their feed value and their intake by ruminants are lower than those for temperate grass silage. Silage made from legumes has a higher crude protein content but, as legumes contain less sugar than grasses, its preparation is more difficult. Better silage can probably be made from mixtures of grass and legume species.

It follows from the above considerations that silage making is often a risky operation in the tropics, and the economics may be unattractive especially when the end product is of low nutritional quality due to factors outside the control of the farmer. This has led some workers (e.g. Humphreys, 1991) to suggest that silage making has little tropical application. If an acceptable product is to be made it certainly requires a careful choice of species, and technical skills of a higher order than demanded of silage making in temperate regions.

STRAIGHTS

These are single feedstuffs, of either animal or vegetable origin, which may have undergone some processing. Single straights rarely provide the complete nutritional requirements for farm livestock, although it is suggested that the production of new and genetically improved varieties, such as high-lysine maize, may change this situation. Straights may in turn be subdivided into cereals, Cereal by-products, pulses or legumes, oil-seed residues, root crops, fibrous materials, animal by-products, recycled animal waste, synthetic nutrients, liquid feed, and oils and fats.

Cereals (e.g. sorghum)

These are important constituents in animal feeds for all forms of livestock although much more so in temperate rations than with tropical diets and, although often regarded as 'energy feeds', cereals can provide the preponderance of protein in the diet for non-ruminants simply because the percentage of cereals is often high. In temperate regions barley and wheat are the main cereal crops employed as animal feed but in tropical areas the range is greater with maize, sorghum, millets and rice all being used in certain areas. Most cereal crops are used both for human food and animal feed and thus there is competition for their use between man and animal. The ratio of usage is constantly changing. In areas where there is a shortage of cereals for human food, and where the economy does not permit the importation of cereals from other countries, little if any cereals may be used for animal feed with the possible exception of cereal by-products, such as bran and cereal straws.

Cereal by-products (e.g. rice bran)

Where cereals are grown primarily for human food they are normally processed to some degree, such as by removing the outer husks or glumes. The resultant by-products are thus available for animal feed but it should be noted that there is a trend towards a greater use of 'whole cereals' for human food because the husks are rich in vitamins and fibre and therefore regarded as useful additions to a balanced human diet.

Some cereal grains are used as the substrate for beer production in the larger towns and the resultant 'spent grains' may then be available as useful animal feed. They can be supplied directly to farms in a wet condition where the farms are sited conveniently close to breweries, or they may be ensiled or dried so that they can be stored and subsequently incorporated into compound feed.

Pulses or legumes (e.g. groundnuts)

Although the seed from leguminous crops normally commands an attractive price as human food (peas and beans), nevertheless certain crops are grown partly for animal feed use. Additionally, legume seeds which do not come up to the standards required for export may be diverted into the animal market. Such materials are relatively high in protein but they may contain gums, resins and sometimes toxic substances which render them less suitable for animal feed (Makkar, 1993).

Oil-seed residues (e.g. soya bean meal)

Traditionally oil-seed crops were used for the production of edible (e.g. palm) oils. The term used for the entire oil-seed crop (without extraction of the oil) is 'full fat' (e.g. full-fat soya). In the past it was mainly oil-seed by-products which were used for animal feeding. Where the by-product was produced by crushing the seed to expel the oil, the resultant material was known as a 'cake' (e.g. copra cake) and where the oil was extracted by the use of a solvent, the residue was known as a 'meal' (e.g. soya bean meal). In recent times there has been an escalation in the growing of soya for use as animal feed as its prime purpose, and this subject has been reviewed extensively by Owen and Jayasuriya (1989).

Root crops (e.g. cassava)

Root crops are capable of producing high yields of energy per unit of land area. The difficulty is that many such crops, particularly cassava, are difficult to harvest because they can be deep-rooted and are often low in protein. Over the last quarter of the twentieth century large amounts of cassava have entered world markets, particularly from countries such as Thailand. Indeed, cassava has now become a major export crop in several tropical countries and is perceived by arable farmers in temperate cereal-growing areas as a major import threat. Because of its high value on export markets there is relatively little use of whole cassava by tropical livestock although some cassava by-products are used in certain areas.

Fibrous materials
(e.g. bagasse and cereal straws)

This category covers a large number of crop by-products ranging from processed cereal straws to specialised products, such as bagasse pith which is the sugar cane stalk after removal of the sugar, and de-linted contton seed, which is a by-product of cotton ginneries.

Bagasse in particular is a very under-utilised crop by-product. McDowell and Hernandez-Urdaneta (1975) reviewed some of the Caribbean work with this material, and reported that a ration with 40% bagasse could provide weight gains in beef cattle equivalent to those from fertilised improved grass pastures, and that up to 33% could be used for finishing cattle. These authors reported work in the Dominican Republic which showed that cattle consumed and utilised bagasse better after it had been fermented for 2 months. They further reported that the digestibility of bagasse could be increased from 20% to over 50% after treatment with sodium hydroxide. Machinery has been developed to remove the outer layer from freshly harvested sugar cane stalks, this exposing the sugar-containing internal portion for feed use. This product is known as 'comfifth' and trials on this material in Barbados were said to give encouraging results.

Straws are the stems and flowering part of cereals with the grain removed either by winnowing by hand or combining with machine. For every tonne of grain harvested in the tropics there are between 1 and 3 t of straw available for use. The high proportion of structured carbohydrate, such as cellulose and lignin, gives straw a relatively low feeding value and much of the straw produced is either collected and used as animal bedding, burnt *in situ* or used as a fuel. In some localised areas there may be a limited demand for straw as a packing material. The advent of big round bales, in the few tropical areas where harvesting machinery is employed, has made the collection and utilisation of cereal straws much easier and as a consequence an increasing proportion of the straw produced each year is likely to be used as an animal feed. It can be fed either long, chopped or chemically treated with a variety of materials (such as caustic soda) which effectively break down the lignified cell walls and release the cell contents. In its normal long form, straw is of low energy value (less than 6 MJ/kg of dry matter) and very deficient in protein. If treated, particularly with nitrogenous chemicals such as ammonia, the energy level can be elevated to 8 or 9 MJ/kg and in addition a useful source of non-protein nitrogen may be made available. By manipulation in this way, straw can be upgraded to the value of a low/medium quality hay.

The site of production of crop residues has a major bearing on usage. The conventional method of feeding crop residues in many situations is to graze them *in situ*. This method of feeding is normally associated with low utilisation rates caused by trampling, soiling, termite damage and fire. In Malawi, for example, although maize stover is the most abundant crop residue available for feeding ruminant livestock, less than half is currently being utilised (Jayasuriya, 1993). The bulky nature of crop residues, such as straws and stovers, makes them particularly expensive to transport even over short distances.

Animal by-products (e.g. meat-and-bone meal)

These materials are by-products of the meat and fishing industries and are high in protein and in essential amino acids. In certain countries, such as Peru, the national economy is partly dependent on fishing operations designed to produce fish meal for export as an animal feed rather than fish for human consumption. Products of animal origin must be carefully processed if they are to be safely used as animal feed. If the processing is deficient they may become the carrier for pathogenic bacteria (such as *Salmonella* and anthrax), or for viruses or 'prion proteins' which are thought to be the cause of diseases such as scrapie in sheep and bovine spongiform encephalopathy (BSE) in cattle.

Recycled animal waste (e.g. dried poultry manure)

Large quantities of these materials are available on-farm, especially on intensive pig and poultry enterprises. Various attempts have been made in temperate countries to develop plants capable of processing these manures to a low-moisture material which can be used as an ingredient in ruminant diets, where it can provide a source of non-protein nitrogen together with small amounts of energy and minerals. However, there is a reluctance on the part of society to accept such practices, on both aesthetic and health grounds, and it is unlikely that recycled animals wastes will, in the medium and long term, be processed into animal feed. It should be stressed, however, that nutritionally there is nothing wrong in using such materials which can provide cheap supplies of essential nutrients, providing the health hazards are removed.

Synthetic nutrients (e.g. methionine)

This group of materials comprises chemicals and fermentation products produced by the chemical and biotechnology industries in order to supply essential micronutrients, such as amino acids, for use in diets for intensively managed pigs and poultry. Although this class of material currently contains relatively few products, it is likely that the list will enlarge in future as more micronutrients are recognised as essential ingredients in animal feed.

New processes enable the cost of such materials to be progressively reduced. Thus the first synthetic amino acid to be marketed was methionine with a price, in its early days, of several thousand of pounds (sterling) per tonne. However, as world demand has increased and production technology improved, the costs of production have been progressively reduced so that synthetic methionine can now be purchased, in real terms, at approximately one-fifth of the price originally demanded.

Liquid feed (e.g. molasses)

The main material in this class is molasses, a by-product of the sugar-processing industry, which is high in sugar and thus in energy. About 300 kg of molasses are available for every tonne of sugar processed. Another liquid feed is whey, a by-product of the milk-processing industry. Whey can be dried into a powder but can also be supplied fresh, by tankers, on to pig farms where it is fed to fattening pigs.

Oils and fats (e.g. tallow)

These two terms are interchangeable, since when a fat is in the molten stage it becomes an oil. The chemist would classify both as 'lipids'. The chief material in this class is tallow derived from slaughterhouses. Tallow could also be regarded as an animal by-product, a consideration which also applies to fish oil. Each livestock class has a limit to the amount of fat or oil which can be added to the diet, usually less than 5%. If larger quantities are fed, then appetite can be adversely affected and the excess lipid can inhibit the digestion of other components of the diet, particularly fibrous components in the case of ruminants.

Crop residues and by-products: general considerations

Crop residues and by-product animal feeds are of great importance. Not only does the use of such materials lessen the competition between human foods and animal feeds but it has been calculated (Kossila, 1984) that they could, in theory, supply about 84% of the annual energy requirement of all livestock. For both these reasons it is probable that greater quantities of crop residues and by-products will be used for feeding tropical livestock in the future. However, crop residues have many limitations – they have low metabolisable energy (ME) contents (normally <7 MJ/kg of dry matter) and crude protein (normally <5%) and their uptake by livestock is limited (Jayasuriya, 1993). Additionally, their use as animal feeds must compete with other uses, such as fuel, in tropical countries.

Owen and Jayasuriya (1989) have suggested that a number of factors are responsible for the low uptake of new by-products for animal feeding in the tropics. The most important factors are:

(1) The absence of inventories of crop residues within countries, including when and where residues are produced.
(2) The difficulties of transportation and storage.
(3) The absence of appropriate technologies for processing the materials and the removal of toxic or anti-nutritional factors.
(4) The virtual absence of agricultural extension services in many countries to demonstrate the economic benefits of their use.

In spite of the overall low level of usage of crop residues and by-products, there are some examples of successful uptake in certain areas. Thus Jayasuriya (1993) reported that the Feedlot Company of Malawi operated a highly successful feedlot beef fattening scheme in the Shire Valley, in the southern region of Malawi. The availability of by-products, together with an assured supply of fattening cattle, were the main reasons for locating the operation in this region. Bagasse and molasses from the sugar cane factory of the Sugar Corporation of Malawi Ltd (SUCOMA) and other by-products such as cottonseed cake, cottonseed husk, rice bran, rice husk, wheat offals and hominy chop, available from the Agriculture Development and Marketing Corporation (ADMARC), have been used in the preparation of compound rations. These rations, when given to Malawi zebu steers and non-descript crossbreds aged 1.5 to 3 years, have produced daily weight-gains of approximately 1.0 kg

Table 11.7 Utilisation of by-products in feedlot in Malawi. (Source: Wide, 1987.)

Breed used	Non-descript (cross breds)
Total land area	60 ha
Land under cultivation	Maize, millet and sorghum (15 ha) Leucaena (5 ha) Natural grazing (40 ha)
Period of fattening	Average of 95 days (from 250 kg to 360–400 kg)
Composition of food	Molasses (500–600 g/kg) Bagasse (150 g/kg) Cotton seed cake and wheat offals (300 g/kg) Urea (18 g/kg) + maize silage when available
Live-weight gain per day	Approx. 1.0 kg
Food conversion ratio (g food per g growth)	10:1
Dressing percentage	520 g/kg
Average grades achieved	89% choice, 10% prime and 1% standard
Average annual production	2000 animals

with a food conversion ratio of 10:1 over an average feeding period of 95 days. Dressing percentages have been around 530 g/kg yielding good carcass grades (Table 11.7). According to Wide (1987) the diets used have been formulated to contain from 115 to 130 g/kg crude protein and less than 120 g/kg crude fibre.

CONCENTRATES AND SUPPLEMENTS

Concentrates are products specially designed for further mixing before feeding. They are incorporated into diets at an inclusion rate of 5% or more, with planned proportions of cereals and other feeding-stuffs either on the farm or in a compound mill. Protein concentrates contain blended high-protein content ingredients, such as a fish meal, fortified with essential nutrients such as minerals, trace elements and vitamins. Where the rate at which concentrates are used in the mix is as high as 50%, they will contain cereals or cereal by-products or cassava. Some protein concentrates are formulated so that, after further mixing with cereals, the resultant product is then suitable for balancing farm forages. They may be added to chopped forage or silage and then fed as 'complete mix' to cattle. In this system the ratio of concentrate:forage is predetermined and the animals cannot 'pick and choose'.

Supplements are products used at less than 5% of the total ration. They are designed to supply planned proportions of vitamins, trace minerals and non-nutrient pharmaceutical additives. To facilitate adequate mixing into the total ration, the active materials (e.g. drugs) are normally added to supplements with a diluent carrier, such as finely ground cereal or cassava.

Many people misuse the terms 'concentrates' and 'compounds'. For instance, many research workers report experiments in which kibbled maize is or is not added to the diet of cattle and describe the maize as a 'concentrate' whereas in reality it is a 'straight' ingredient. Again, many farmers talk of 'concentrates' when they mean 'compounds'. The term 'compound' should be used to describe a balanced mixture of ingredients, usually made and processed by a feed manufacturer. A 'concentrate' is a mixture of trace elements and vitamins which is then added to other bulky raw materials to ensure the total diet is balanced. The fact that compounds or straights may be fed to cattle alongside hay or silage does not make them 'concentrates'.

In tropical countries with a developed compound manufacturing industry there is often confusion about the terms 'specification' and 'formulation'. A 'specification' is a description of a given blended feed in nutritional terms. Some 30 or so nutritional parameters may be listed and over 50 different materials may be allowed by the specification either freely without constraint, or restricted to minimum or maximum inclusion levels. A 'formulation' is the actual list of straight raw materials present in a given blended feed, at fixed levels of inclusion. In other words, a 'specification' is an open formula but a 'formulation' is a fixed recipe. The formulation is usually worked out on a computer, to meet the specification at 'least cost' so as to make the diet as economic as possible (see also Chapter 16).

Processing of raw materials and concentrates

Relatively few raw materials are fed in concentrate mixtures in non-processed form. There are a wide variety of processing methods and Table 11.8 lists the types of processing to which straights may be subjected (Wilson, 1979). The objective of cereal processing is to increase the exposure of starch grains and improve their susceptibility to enzyme degradation. It is not possible to describe all these processes individually, but the terms employed are self-explanatory. Many authors have reviewed the effects of cereal processing and digestion (e.g. Hutton and Armstrong, 1976).

There are marked differences in response to processing between individual animals. Thus Wilson et al. (1973) demonstrated significant variation between cows in the ability to digest intact maize kernels. Clearly in cases where cows are unable to digest whole grain efficiently some form of physical processing will increase the digestibility. On the other hand cows that can digest whole cereals without difficulty will exhibit little or no response when fed the same cereals cracked or pelleted. Although quantification of the effects of processing is difficult, with many conflicting reports in the literature concerning the benefits of a given processing technique, the qualitative effects of processing can be categorised and may be summarised as follows:

- Reduction or alteration in particle sizes.
- Alteration in physical and/or chemical structure.
- Effect of processing on the degradability of protein in the rumen.
- Effect of processing on palatability and on feed intake.

Reduction or alteration in particle sizes

Most dry or wet processes modify the particle size of the raw material and this results in a

Table 11.8 (a) Different types of feed processing and (b) the different classes of animal feed to which they are commonly applied. (After Wilson, 1979.)

(a)

Dry processing		Wet processing		Chemical processing	
D1	Grinding	W1	Soaking	C1	Alkali treatment
D2	Pelleting	W2	Steam rolling	C2	Acid treatment
D3	Dry rolling or cracking	W3	Steam processing and flaking	C3	Formaldehyde and tannin treatment
D4	Extruding	W4	Pressure cooking	C4	Mould inhibition
D5	Micronisation	W5	Exploding	C5	Oil extraction
D6	Roasting			C6	Antioxidants addition
D7	Popping				
D8	Sterilisation				

(b)

Feeding stuff	Processing to which feed is subjected
Cereal grains	D1, D3, D5, D6, D7, W2, W3, W4, W5, C1, C4
Root crops	D1, D2
Pulses	D1, D2
Oil seed residues	D3, W2, C5
Fibrous materials	D1, D2, C1, C2
Animal by-products	D8, W4, C6
Oils and fats	C6
Synthetic nutrients	Not applicable
Recycled wastes	D8, C1, C3, C5
Liquids	W4, C6
Trace nutrients	Not applicable

change in retention time in the alimentary canal and a consequential change in degree of rumen fermentation and in subsequent true digestibility. A secondary effect is to increase the voluntary intake of feed.

In general, very fine grinding increases the rate of passage and decreases digestibility. In cases where whole grains or seeds are poorly digested, therefore, there is an optimal grist size, as measured by the modulus of fineness, which differs between individual cows and is dependent upon the other constituents of the diet, especially the forage fraction (Wilson et al., 1973). Meal often gives poorer animal performance than pellets, because animals find meal less easy to eat. This may be a problem in tropical dairy milking parlours but not with pigs or poultry.

Alteration in physical and/or chemical structure

Relatively minor processing treatments, such as coarse grinding or cracking, produce little effect on the physical structure and even less effect on the chemical nature of the raw material. The main aim of fine physical grinding at medium to high temperatures is to overcome the two major feeding constraints with fibrous materials: low intake and low digestibility. Although the subject of processing of fibrous raw materials has been extensively reviewed (Swan, 1974) it is difficult to be precise about the quantitative effects but, in general, the poorer the material and the more crystalline the fibrous fraction, the higher are the responses to processing (Owen, 1976).

Feed intake is not always increased by grinding and beneficial effects on intake appear to depend primarily on the quality of the processed material and sometimes on the amount of the concentrates given. Campling and Freer (1966), in work with dry cows, noted that these animals ate more fibrous material in the pelleted form when low quality material was used. Journet (1970) showed that diets of finely milled grass and lucerne hays with 60% concentrates led to the production by cows of milk of only 2.2% fat and that milled forage alone or with only 30% concentrates gave normal milk fat content.

More severe treatments, such as micronisation of cereals or alkali treatment of fibrous materials such as cereal straws, produce marked effects on both physical and chemical structure. The addition of an alkali results in both chemical and physical changes in the plant cell wall. In the chemical reaction alkali–labile lignin–hemicellulose and lignin–cellulose ester linkages are hydrolysed. This renders the hemicellulose fraction more soluble and the cell wall carbohydrates more fermentable. The hydrolysis of the lignin–cellulose linkage is believed to make more cellulose available for enzymic degradation (Wilson and Brigstocke, 1977). The highly ordered crystalline form of cellulose has prevented the use of many untreated lignocellulosic materials for ruminant feedstuffs (Millet et al., 1974).

Effect of processing on the degradability of protein in the rumen

Many processing treatments, such as formaldehyde and tannin treatment and all heat treatments, reduce the amount of protein degradation in the rumen (Chalupa, 1975). The most favourable processing conditions have been shown to give net increases in the flow of amino acids to the small intestine and in their absorption. In cases where high yielding dairy cows are deficient in undegraded protein all such processing treatments are clearly advantageous (Roy et al., 1977).

Effect of processing on palatability and on feed intake

In general, the more extreme the treatment the greater the effect on cow acceptability and hence on voluntary intake. Ground materials are usually consumed more readily than when offered whole especially when reformed as a compound pellet. Heat expanded materials are in general more palatable than the same raw materials in a more dense form. Chemical treatment of lignin and lignocellulosic materials results in substantial increases in acceptance of materials which are only marginally palatable in their unprocessed state. Although processing treatments increase feed intake it does not follow that such treatments necessarily increase digestibility. As mentioned earlier, the treatment may increase the rate of passage through the alimentary canal and therefore decrease digestibility. Thus the hot processing of most cereals may show little or no effect over cold processing. However, with some cereals, such as sorghum, the same heat treatment may significantly increase true digestibility (Wilson et al., 1973).

To summarise, the processing of concentrated feeds has many advantages. Intakes can be increased and digestibility enhanced. Ørskov (1976) noted that processes should be developed which achieve the minimum rupture of the material in order to avoid pathological changes in the rumen wall and to ensure an efficient digestion of forage. However, processing is energetically expensive, particularly in energy-deficient tropical countries, and the greater the degree of processing, such as by chemical treatment, the greater the amount of energy required. In some cases the energy cost would be expended anyway, as in the case of pelleting of compound feeds, and the beneficial effects of such treatment are an added bonus. In most cases, however the process is only undertaken for its nutritional benefit, and in these cases the cost:benefit ratio of the treatment must be evaluated objectively.

Many of the above observations are only applicable where animal feeds are processed by a commercial compounder, and therefore it may be considered that these points are of little relevance in tropical countries. However, there are many tropical regions where a feed compounding industry is either developed (South America, India, Pakistan) or is developing rapidly (Central America, Thailand, parts of Africa). In any tropical countries which possess modern poultry farms, compound feeds will be a necessary input to egg or broiler production. In other areas there is a transition from the export of unprocessed crops to fully processed crop products for use elsewhere, including use as animal feed materials. It is, therefore, important that those concerned with the production and export of such tropical crop products have some understanding of their suitability for animal feeding.

Complementary nature of concentrates and forages and/or grass

Since dairy cows cannot be successfully fed on concentrates alone, it follows that the interaction of concentrates with other dietary components is of great importance. In the case of 'housed' dairy cows, the concentrate part of the diet can be carefully controlled, often by weighing the concentrates to each cow, either in the parlour during milking or outside the parlour in hoppers or troughs. The forage portion can rarely be allocated with such precision, and only in experiments can forage be rationed to individual cows (Wilson, 1977). Complete feeding allows the concentrate:forage ratio to be determined accurately but the actual amount fed to each individual cow is seldom controlled since such feeding systems deal with cows on a group basis. If follows that the maintenance of an optimal ratio is difficult to achieve under practical farming conditions.

Workers have drawn attention to the need for changing the energy concentration of the diet (M/D where M is the metabolisable energy and D is the dry matter) and the forage:concentrate ratio as the lactation progresses (Broster and Johnson, 1977). This ratio affects feed intake, the microbial population and the function of the reticulo-rumen, the site and extent of digestion, the efficiency of utilisation of energy for milk production, the mobilisation of body fat and milk composition (Broster, 1972).

In general a high percentage of concentrates in the diet is associated with an increase in the proportion of propionate and butyrate, a decrease in the proportion of acetate and a decrease in pH in the reticulo-rumen. When the acetate:propionate ratio falls below 2.5–3.0:1 there is a reduction in milk fat production. The deleterious effect of an adverse ratio is not rapidly reversed when the situation is corrected. It has been noted that when cows on a high concentrate diet were transferred to a more equitable ratio, milk fat percentage took 2–3 weeks to return to normal.

In comparing experimental results of varying forage:concentrate ratios, care must be taken that the affects of the ratio are not confounded with differences in the type of forage on the one hand and the differences in the nutritional specification of the concentrate on the other. This is particularly important when Central and South American studies are compared with other tropical research results. In the Americas the forage is normally maize silage and the concentrate is usually based on maize/soya/urea mixtures. On the other hand, under tropical conditions elsewhere the forage most commonly grown for intensively managed cows is grass and the concentrates (where used) are usually commercial compounds comprising a wide range of different raw materials.

ANTI-NUTRITIONAL FACTORS IN FEEDS FOR LIVESTOCK

Both human food and animal feed may contain hazardous materials, and the risks associated with animal feed are clearly the greatest. It would normally be uneconomic to provide nutrients in a refined form by the removal of all naturally occurring impurities (e.g. all weed seeds in cereals). In the case of plant materials, the chief hazards are disease (e.g. ergot) and chemical residues (such as herbicides). The main hazard of animal materials is microbiological contamination (e.g.

Salmonella), whilst in the case of dried poultry manure an additional hazard is the presence of poultry medicaments in the poultry faeces which may be injurious to ruminants. The main hazard in minerals is the presence of a toxic element (e.g. fluorine in rock phosphate). In the case of synthetic and trace nutrients the risk is of toxicity when excess levels are administered. Thus urea can be toxic when the level in the compound exceeds 3% and selenium is toxic where the level in the total diet exceeds about 3 mg/kg.

Anti-nutritional factors can be divided into four groups:

(1) Factors affecting protein utilisation and digestion (protease inhibitors, tannins, saponins, lectins).
(2) Factors affecting metal ion utilisation, (oxalates, phytates, gossypols, glucosinolates).
(3) Anti-vitamins.
(4) Others (e.g. mycotoxins, mimosine, cyanogens, nitrates, alkaloids, photosensitising agents, isoflavone).

Detailed accounts of all these groups of factors are available in reviews of the subject (e.g. Chubb, 1982; Cheeke and Shull, 1985) but this chapter selects one particular anti-nutritional factor which is commonly found in tropical feeds, tannin.

Tannins

Tannins are generally defined as naturally occurring polyphenolic compounds of high molecular weight which form complexes with proteins. They are commonly associated with many tropical feedstuffs, and may be classified into two groups based on their structural type: (1) the hydrolysable tannins, and (2) the condensed tannins, which are flavonoid-based polymers.

A great deal of work has been done on the effects of tannins on rumen metabolism. However, information on the effects of tannins on other parts of the gut is scant. A development (Reddy and Butler, 1989) on the radioactive labelling of tannins may provide valuable information on this aspect.

Binding of food tannins to salivary proteins and the epithelium of the mouth makes the food unpalatable and depresses voluntary intake. A negative correlation exists between tannin levels in feeds and their intake (Van Hoven, 1984). Grazing animals do not normally consume plants having a tannin content of over 2% dry weight basis (Donnelly and Anthony, 1969). However, McNaughton (1987) reported that palatability and feed intake were affected only by tannin levels beyond 5%. Certain tannin-rich plants are readily ingested by goats (Provenza *et al.*, 1990). These differences could be due to the different nature of the tannins present (Clausen *et al.*, 1990). Other factors affecting intake are:

(1) Decreased digestion of food (see below) which slows the passage rate and results in higher rumen fill and rumen distension.
(2) Phenols in blood may stimulate receptors in the brain, thereby causing a conditioned food aversion and reduced food intake (Distel and Provenza, 1991).
(3) Changes in hormone levels (Barry, 1989).

During mastication by livestock, protein and tannins come into contact with each other and form complexes in the mouth (Woodhead and Cooperdriver, 1979). Monogastric animals acclimatise to tannin-rich diets by increasing the size of salivary glands and producing more salivary proteins especially rich in proline. Proline-rich proteins (PRPs) have high affinity for tannins and form the first line of defence against food tannins (Mehansho *et al.*, 1987) in both monogastrics and ruminants. Tannins cause toxicity to rumen microbes probably by impairing cell wall permeability. Tannins decrease ruminal volatile fatty acid production and protein synthesis (Kumar and Singh, 1984). The rumen appears to have the capability to degrade tannins (Murdiati and Mahyudin, 1985).

In the mid-gut, tannins cause loss of mucus, injury to the alimentary canal tissue and gastroenteritis (Spier *et al.*, 1987). Excess tannins causes increased excretion of endogenous nitrogen and also of calcium and sodium in the faeces (Shahkhalili *et al.*, 1990). Breakdown of tannins occurs as these flow down the gut of sheep (Fahey *et al.*, 1980). Intestinal microflora play an important part in this degradation (Jung and Fahey, 1983c). Ruminants, especially goats, have a higher tolerance to tannins than non-ruminants due to extra mastication, larger amounts of saliva and rumen fermentation (Singleton, 1981). The threshold of toxicity for tannic acid administered to the rumen was 3–5% in cattle and 8–10% in goats (Singleton, 1981). Bohra (1980) reported higher feed intakes and higher digestibilities of the dry matter and crude protein in goats than in sheep given *Prosopis cineraria* leaves containing a high concentration of tannins.

Tannins not only decrease the utilisation of proteins but also of carbohydrates, amino acids, minerals and vitamins (Makkar *et al.*, 1987). However, some tannins have beneficial effects. Condensed tannins of *Lotus* spp., sainfoin and *Acacia seyal* at low levels increased nitrogen retention in ruminants by protecting feed proteins from ruminal digestion and thus increasing the supply of amino acids to the small intestine. The concentrations and availability of tannins of some important by-products of the Indian sub-continent are given in Table 11.9. The tannin contents of *Acacia nilotica* pod, and *Panicum miliaceum* polish, are high (Barry, 1989). On the other hand, tannin levels in pods of *Hevea brasiliensis* and *Hibiscus cannabinus* seeds are low. These latter by-products are safe for incorporation in livestock feed subject to the absence of other deleterious factors (Makkar *et al.*, 1990).

The above section deals with one selected, anti-nutritional factor of major significance in tropical crop by-products and tropical browse trees and bushes. It illustrates the range of problems that may be encountered when large quantities of materials with a high tannin content are fed to livestock. The example indicates the importance of obtaining technical advice when the use of such products is under consideration in tropical animal feeding programmes. Left to

Table 11.9 Estimated availability and tannin levels in some unconventional agro-industrial by-products. (Source: Makkar, 1993.)

By-product	Availability (t × 10^3)	Tannin levels (mg/kg dry matter) (tannic acid equivalent)
Acacia nilotica pods	60	348
Bixa orellana spent-seed	300	10
Camellia sinensis waste	15	19
Cassia tora seeds	30	27
Garcinia indica cake	15	45
Guizotia abyssinica cake	100	22
Helianthus annus straw	200	15
Hevea brasilienis seed-cake	150	23
Hibiscus cannabinus cake	30	7
Madhuca indica seed cake	300	71
Mangifera indica seed-kernel	1000	75
Panicum miliaceum polish	na	153
Panicum miliaceum bran	20	14
Pithecolobium saman pods	na	7
Pongamia glabra cake	130	15
Prosopis juliflora pods	1000	29
Schleichera oleosa cake	30	7
Shorea robusta meal	4800	80–100
Theobroma cacao pods	30	39

Note: Some of these estimates reported by Makkar are on the high side and represent maximum availabilities.

themselves, animals will usually restrict the voluntary intake of such materials, but when fed as part of a contrived diet, excessive tannin levels can produce the deleterious effects described.

FURTHER READING

Butterworth, M.H. (1985) *Beef Cattle Nutrition and Tropical Pastures*. Longman: London.

Cheeke, P.R. and Shull, L.R. (1985) *Natural Toxicants in Feeds and Poisonous Plants*. AVI Publishing: Westport, CT.

Haresign, W. and Cole, D.J.A. (eds) (1988) *Recent Advances in Animal Nutrition*. Butterworths: London.

Morrison, F.B. (1959) *Feeds and Feeding*. Morrison Publishing: Ithaca, NY.

Wiseman, J. (ed.) (1987) *Feeding of Non-ruminant Livestock*. Butterworths: London.

12
Tropical Grasslands Used for Livestock

P.N. Wilson

About 22% of the total dryland surface of the earth is under natural grassland suitable only for grazing (Vandermaele, 1977).

Good management of these widespread natural pastures requires an understanding of the ecology of the grassland community with which the grazier has to deal. This is especially true of tropical grasslands, since most of these occur in regions where the natural climax vegetation is forest or woodland and thus they form relatively unstable subclimaxes. In these associations the maintenance of grass as an important component, in competition with trees and shrubs, is largely dependent on management. It also helps to know the flora of the grass cover itself, to appreciate its agronomic characteristics and to understand the changes in its botanical composition. The latter may arise from local variation in soil and other natural conditions, or may indicate trends towards improvement or deterioration resulting from management.

TYPES OF NATURAL GRASSLAND

In the following summary, some of the more important types of grassland are classified by reference to their relationship to the major climax formations described in Chapter 2. A detailed description of the natural pastures in different tropical regions and their ecological relationships is outside the scope of this book, but many reviews are available (UNESCO/UNEP/FAO, 1979; Huntley and Walker, 1982; Bouliere, 1983; Sarmiento, 1984; Tothill and Mott, 1985).

Grasslands derived from forest at low and medium altitudes

In the regions of lowland evergreen, semi-evergreen and deciduous forests prolonged or repeated cultivation, followed by grazing and burning, may result in the vegetation becoming degraded to a fairly stable association of coarse grass and scattered shrubs. Large areas of such grassland, dominated by *Imperata cylindrica*, occur in Malaysia and parts of Africa, examples being the 'tall grass' areas of Uganda (dominated by Elephant grass, *Pennisetum purpureum*), and the large stretches of former forest in West Africa dominated by *I. cylindrica* and species of *Ctenium* and *Andropogon*. Some of the lower 'patanas' of Sri Lanka are similarly derived. With the possible exception of the Elephant grass areas, these grasslands are of poor productivity, being coarse, unpalatable and of low nutrient value. They are not capable of much improvement by management; more intensive land use requires their replacement by planting improved pastures (or other crops) where it is practical to do so.

Grassland associated with, or derived from, broad-leaved woodlands

In the broad-leaved woodlands, which occur under seasonal rainfall in the range of 600–1250 mm, the presence of a continuous ground cover (mainly of grass) under the widely spaced trees permits the passage of fire during the dry season. As a result, large areas of these woodlands have become fire-subclimax savannas of grass with scattered, fire-resistant trees. An indication of the nature of these grasslands can be given by two distinct examples that are widespread in Africa.

The first, occurring in regions with roughly 850–1250 mm of rain per annum and with 3–5 dry months, includes the types originally described for West Africa as 'undifferentiated: relatively moist types', and known in East Africa as 'scattered tree grassland: low tree-high grass'. They consist of woodlands and savannas characterised by tall, relatively dense tussocky grass, in which species of *Hyparrhenia* are dominant, and by a varying density of small trees, mostly 3–9 m high, with isolated patches of taller trees, usually associated with exceptional groundwater conditions.

The second type, known as 'miombo' in East Africa, consists of woodlands and savannas characterised by trees of the genera *Brackystegia*, *Julbernardia* and *Isoberlinia*, also with a ground cover of widely spaced tussocky grasses with *Hyparrhenia* spp. dominant. This type occurs in regions with one severe annual dry season, usually of not less than 6 months' duration, and a rainfall usually between 600 and 850 mm. It is found in parts of West Africa (e.g. northern Nigeria) and covers huge areas in East and Central Africa, being known in Zimbabwe as 'tall grass *Hyparrhenia* veld' and Mopane woodlands in Zambia and Botswana.

In both these types the pasturage is commonly known in Africa as 'sour veld', meaning that the young grass is palatable but soon becomes rank, unpalatable and of low nutritional value. At the onset of the dry season, growth ceases and the grass dries out to form standing hay of poor quality. There is, however, a small proportion of finer, creeping grasses of better nutrient value and a sparse occurrence of legumes. The composition and productivity of the herbage can be somewhat improved by the controlled use of fire, and correct intensive grazing management, which may repress the taller grasses in favour of the better creeping species. These grasslands are only moderately productive and in the better rainfall areas their replacement by sown pastures of improved grasses will usually give a better return where the cost of establishment is reasonable. Over much of the less productive 'miombo' country, however, the length and severity of the dry season prevents the use of sown pastures, and relatively extensive ranching of the natural grassland is practised with consequent lower productivity.

Grasslands associated with, or derived from, thorn woodlands and thickets

These grasslands are very widespread in the tropics, with between 350 and 750 mm of seasonal rainfall. The vegetation varies from woodlands, with a continuous cover of grass under widely spaced trees, to thickets of small trees with sparse grass beneath. In West Africa they occur in the northern Guinea and Sahel savanna zones. In East Africa they are extensive from south of the Gulf of Aden through to Zimbabwe and South Africa. In this region they include the types first referred to by Edwards (1956) as 'scattered tree grassland and open grassland – *Acacia–Themeda*' and 'dry bush with trees: *Commiphora – Acacia* – Desert grass' and several of the types described by Rattray (1957) for Zimbabwe namely: (1) *Eragrostis* and other species veld, (2) *Aristida* and other species veld, and (3) *Cenchrus* and other species veld. Similar grasslands also occur in India, and in parts of Central and South America.

This group of grasslands all have certain important characteristics. They are unstable subclimaxes and there is a tendency for the spread of trees and shrubs which must be controlled by management, although some of the woody species make a contribution to stockfeed during the dry season. The herbage consists mainly of perennial grasses less than 1 m high and is usually 'sweet veld', though in some areas the presence of species that are palatable only when young justifies the term 'mixed veld'. Legumes are scarce. Owing to the low and unreliable seasonal rainfall, the carrying capacity is low and very variable. Water sources are unreliable and the provision of adequate artificial supplies is usually too expensive.

Despite these common features, there are important differences between the grassland types in this group, which can be illustrated by reference to the contrasting grazing areas which occur towards the extremes of the rainfall range. The 'scattered tree grasslands, *Acacia-Themeda*', originally described by Edwards (1956), and occurring at medium altitudes in Kenya under an average rainfall of 500–750 mm, form one of the most productive types. The flat-topped *Acacia* trees are widely spaced and the grass cover is relatively good. Where tsetse fly is a problem it can be controlled by limited clearing of trees and shrubs. Bush regeneration can be checked by a combination of grazing management and controlled burning although this is not such a common practice as formerly. The maintenance of the dominant grass, *Themeda triandra*, which is a nutritious perennial species, requires light rotational grazing and the periodic use of fire, in the absence of which it is replaced by inferior grasses. The carrying capacity of the grassland is about 1 livestock unit to 8 ha, but under good management this can be doubled.

The less productive grasslands associated with *Commiphora–Acacia* bush and thicket are very extensive in East Africa. These consist mostly of deciduous bushes 3–5 m high with more widely scattered taller trees and a somewhat sparse ground cover of tufts of perennial grasses, augmented after rain by annual grasses and herbs. In many places bush clearing to eliminate tsetse, and to promote grass development, is indicated but this operation is expensive, owing to the capacity of the trees to regenerate. At best the resultant pasturage is of low productivity. The grass cover is sensitive to grazing pressure and anything more than light stocking with cattle can lead to a marked reduction in the perennial grasses, and will encourage the spread of bush. Carrying capacity will rarely exceed 1 livestock unit to 8 ha and is usually less.

Grasslands derived from montane evergreen forests

Grasslands resulting from the clearing of montane evergreen forests occur at elevations of 1000–3000 m. Annual rainfall is between 1000 and 5000 mm. The dry season is usually moderate or virtually absent, as in the highlands of Sri Lanka, but in some areas it may be more pronounced, as in Zimbabwe.

In East Africa this type of vegetation has been called 'highland grassland and forest' and consists of a patchy distribution of small areas of forest in extensive areas of open grassland. The grass species are of short to medium height. In the wetter and more fertile parts of the Kenya Highlands the dominant species is Kikuyu grass (*Pennisetum clandestinum*), which makes a productive sward in association with the clover, *Trifolium semiphilosum*. Elsewhere *Themeda triandra* is maintained by periodic fire, but tends to give way to coarse species, such as *Pennisetum schimperi* and *Eleusine jaegeri*, if grazing is intensified and fire prevented. In Zimbabwe the type called 'short grass mountain grassland' is usually dominated by *T. triandra* on more fertile soils and elsewhere by *Loudetia simplex*. Similar grasslands exist in Sri Lanka, where they are known as 'wet patanas' usually dominated by *Chrysopogon zeylanicus*, and in parts of India and Myanmar and at higher altitudes in South America and Costa Rica.

The productivity of these grasslands is good but varies with rainfall, soil and botanical composition of the herbage. Dense swards of nutritious grasses, such as Kikuyu, growing in favourable areas, will carry 2.5 livestock units/ha. On the other hand the mountain grasslands of Zimbabwe are less productive owing to the more severe dry season, the dominance of less valuable grasses and a tendency for shrubs and bracken to invade the pasture; at best they will carry 1 livestock unit to about 3 ha.

High-altitude grasslands

On the higher mountain slopes, usually above 2700 m between the limits of the forest and the snowline, there are natural climaxes composed mainly of low-growing herbs and short grasses. Comparatively large areas of these 'alpine meadows' or 'paranas' exist in the mountainous parts of tropical South America where they are grazed by sheep, llamas and alpacas and, to a lesser extent, by cattle. Elsewhere in the tropics such grasslands are limited in area and, being inaccessible and exposed, they are little used.

Seasonal swamp grassland

Natural climax grasslands and savannas occur in areas of seasonal rainfall where the nature of the soil or the topography results in annual alternation of excessively wet and dry soil conditions, thus preventing or restricting the growth of trees. The Lake Chad basin, lower regions of the Venezuelan llanos and the Beni region of Bolivia are well-known examples of this type. These grasslands may contain scattered trees and shrubs and occupy a vast area of plains and low hills where the annual rainfall of 800–1750 mm falls in one wet season of 6 to 7 months' duration. During the long rains large tracts of land are inundated by flooding. During the dry season the surface layers of the soil become parched, the grass rapidly dries out, and is regularly burned before the rains to promote a growth of young herbage.

The utilisation of such grasslands is largely conditioned by the alternation of floods and droughts. During the rainy season only islands of higher land are available to the cattle, which often swim from one island to another. Abundant feed is available early in the dry season but the grass quickly dries out, becomes unpalatable and is burnt off. The continued use of fire tends to eliminate better species and leaves only those which are resistant to flood, drought and fire. Improvement depends upon the practicability of major flood-control measures, possibly accompanied by irrigation, but such grandiose schemes are rarely economic. The combined use of an extensive area of seasonally flooded llanos and a smaller, more productive, area in the mountain foothills is usually a more feasible proposition. Cattle may be bred and fattened on the drier pastures, grazing improved grasses under fairly intensive systems of management, and run as store cattle on the less productive llanos using an extensive system of ranching.

MANAGEMENT AND UTILISATION

Low productivity of the majority of tropical grasslands

Although the relatively high-quality grasslands derived from montane evergreen forests are of considerable importance in some mountainous countries, such as Kenya, their total extent is limited and they are increasingly being replaced by crops. By far the greater part of the natural grasslands are those of the drier broad-leaved woodland, thorn woodland and thicket zones, where climatic conditions limit productivity and pose management problems. Over many of these areas the low and unreliable rainfall, and the severe dry seasons, result in a sparse tufted sward of grass which, in the drier areas, may cover only 5% of the ground. Many of the grazings are dominated by coarse grasses that are palatable and nutritious only when young, and legumes are scarce. The carrying capacity is low, but it is greater during the growing season than in the dry season, when the grass dries out to form very poor-quality standing hay. This marked seasonal variation in the carrying capacity poses a major management problem. Furthermore, in some areas these grasslands are relatively unstable subclimaxes, and the grass tends to be invaded by trees and shrubs that are better adapted to the climatic and edaphic conditions. This tendency is greater in the drier areas and is encouraged by the exclusion of fire, or by anything that weakens the competitive power of the grass, such as overgrazing. Bush encroachment should be checked constantly by management, because it results in lower carrying capacity, increased bare ground and consequent risk of erosion. Most of the grasslands in the more arid and savanna areas are only suited for extensive grazing by ruminants which can convert the structural carbohydrates in the vegetation to meat and other animal products useful to man. They are not usually capable of supporting arable farming because of the unreliability of rainfall and, in many areas, the low fertility of the soil.

It is against the background of the above limitations, which apply to the majority of the more extensive grassland types, that the management and utilisation of natural grasslands is considered below.

Aims of pasture management and factors involved

The main objectives of pasture management can be summarised as follows:

(1) To provide, as far as possible, a uniform and year-round supply of herbage for the maximum number of stock.
(2) To utilise the herbage at a stage which combines good nutritional quality with high yield.
(3) To maintain the pasture in good productive condition by encouraging the best species and maximising the ground cover. This will protect the soil from the beating action of rainfall, thus preventing runoff and erosion.

In evolving methods of management with these aims in view, the following factors should be considered:

(1) The influence of seasonal growth and of grazing on the maintenance of the sward.
(2) The variation in the composition and feeding value of herbage with stage of growth.
(3) The value of certain trees and shrubs as browse plants.
(4) The need for bush control.

Interaction of seasonal growth and of grazing on the maintenance of the sward

Defoliation of grasses by the grazing animal involves a number of biological changes. These include a reduction in leaf area (and therefore the photosynthetic capacity of the plant) as well as a depletion of the carbohydrate reserve. All aspects of growth are controlled physiologically by the shoot apex. Removal of this apex by grazing during growth exerts an adverse effect on the tillering and regeneration ability of the plant and thus on its productivity. Defoliation can be harmful when too much vegetation is removed so that the vigour of the plant is reduced. This reduces competitiveness and leads to sward deterioration.

Too early, too heavy, or too frequent grazing can result in reduction in grass yield, death of some of the plants and consequent development of bare or weedy patches. In seasonal rainfall areas the perennial grasses, which form the most important component of the pastures, make rapid growth during the early rains and quickly reach the flowering and seeding stage. Translocation of carbohydrates for storage begins as soon as the leaves start forming and continues until flowering. At the onset of the dry season growth stops and the aerial parts of the grass dry out. At the beginning of the next rains new growth takes place at the expense of the food reserves, which are not replenished until new leaf has been produced. Defoliation by grazing or cutting necessitates translocation of reserve nutrients to the remaining meristems for the growth of new leaves and tillers, and thus there is a reduction in the weight and carbohydrate content of the roots which is related to the intensity or frequency of the defoliation. The potential of grasses for regrowth after defoliation is partially correlated with the soluble carbohydrate content of the food reserve tissues, hence depletion of carbohydrate reserves is responsible for the ill-effects of too early, or too frequent, defoliation.

Two other factors are connected with overgrazing. First, stoloniferous, rhizomatous and prostrate plants, with a high proportion of basal leaves below the grazing level, are more resistant to grazing than taller, tufted species. Such grasses predominate in some tropical pastures – for example Kikuyu grass (*Pennisetum clandestinum*) in parts of the Kenya Highlands – but they are not important components of grasslands which are sensitive to overgrazing. Second, since cattle are selective grazers, the more palatable species are most adversely affected by overgrazing and hence one of the adverse effects is a deterioration in the flora and the resultant quality of the herbage.

These ill-effects can be avoided by adjusting stocking rates to the carrying capacity of the pasture, by refraining from grazing *immediately* regrowth starts after the dry season, and by allowing the pasture to have a periodic rest during the growing season in order to replenish root reserves. Adequate rest cannot always be provided by the intervals between rotational grazing cycles, as is normally the case with temperate grasslands, but necessitates periodic closing of the pasture to grazing for a significant part of the growing season. It is usually unsatisfactory to rest land at the same time in each resting year, as this tends to encourage undesirable species which benefit from resting at that particular time. For example, in East Africa regular resting during the first half of the rains encourages annuals at the expense of perennials, hence it may be desirable to alternate between resting from mid-wet to mid-dry season and from mid-dry to mid-wet season. Grazing during the dry season does little harm provided that it is not so intense as to leave insufficient cover to prevent erosion at the onset of the rains.

There are three 'critical periods' in the growth of grasses which spread and reproduce by seeding. These are:

(1) The flowering period.
(2) The period of change from a seminal rooting system to a coronal rooting system (generally occurring about a month or so after the seed has germinated).
(3) The period of transference of carbohydrates from the aerial to the underground part of the plant, usually at the onset of the dry season.

Natural grasslands containing a high proportion of seeding grasses should be more lightly grazed, and preferably rested, when the more important species are in one of these critical periods.

Variation in the composition and feeding value of herbage with stage of growth

The age changes in the composition and feeding value of tropical pasture grasses are similar to those that occur in temperate species. However, most young tropical grasses are less nutritious than temperate grasses of the same stage of development, and their feeding value declines more rapidly over time. In the tropics young grass is usually lower in protein and higher in crude fibre than in the temperate zone. It was generally assumed that this lower protein content severely limited animal productivity in the tropics, but recent work suggests that it is energy which is the limiting factor and not protein (Payne, 1990) unless the protein content is very low. Earlier work by Wilson and Ford (1973) demonstrated that tropical grasses generally accumulate less soluble carbohydrate than temperate grasses. When compared at the optimal temperatures for growth, the differences were even greater. Within the tropical grasses Duble *et al.* (1971) reported that digestibilities of the five species studied were inversely related to the cell wall content, and there are many reports indicating that tropical grasses have significantly higher cell wall contents, and hence lower digestibilities, compared to temperate species. Rees and Minson (1976) noted the beneficial effect of calcium fertiliser applied to *Digitaria decumbens* on both grass intake and digestibility, and suggested that the effect was related to changes in cell wall components which allowed an increased rate of cell wall breakdown in the rumen and thus a greater utilisation of the cell wall contents by the grazing animal.

As grass grows to maturity the protein, phosphate and potash contents fall, the crude fibre rises, and the carbohydrate and calcium levels remain fairly constant. Very low contents of phosphorus and sodium are often found in dry tropical herbage on which animals have to subsist in the dry season. The total yield of herbage, and the percentage of dry matter, increase up to an advanced stage of maturity. The protein content of tropical grasses usually falls very rapidly in the first few weeks of growth, after which the decline is slower until the late flowering, or seeding stage. After seeding, and with the onset of the dry season, there is a further decline, partly due to the shedding of seed and desiccated leaves. The digestibility of the protein also declines with the protein percentage. Glover and French (1957) established the relationship:

$$\text{Digestibility coefficient} = \left(\log \% \text{ protein in dry matter}\right) - 15.$$

However, as shown in Chapter 16, an increase in protein digestibility *per se* may not necessarily be advantageous and it is metabolisable protein rather than digestible protein which is important.

In general, the variation in protein content between species is less than that between different stages of growth in a single species but there are, nevertheless, important differences between grasses. There is usually a preponderance of species of low protein content. For example, Dougall and Bogdan (1958) found that, of 58 species sampled at the early flowering stage, 34 had crude protein contents of less than 10% and only seven exceeded 15%. The increase in crude fibre is usually slight after the first 6 weeks of the growing season but there is generally a further rise in the dry season. Fibre digestibility decreases as the grass ages but the decline is less than is the case with crude protein.

From the nutritional standpoint grass should be grazed at a moderately young stage (but not during a critical period as defined earlier) but the yield is greater with mature swards. It is therefore necessary to achieve a compromise between feeding value and yield. This is done on productive temperate pastures by frequent rotational, zero or strip grazing, but such intensive utilisation is impracticable for most natural pastures in the tropics. Efforts to use the grass at the optimum stage usually meet with limited success. Grazing must be relatively light because stock numbers must be adjusted to the low dry-season carrying capacity. Hence, in the rains the grass grows away from the stock and soon reaches the more mature stage of poorer quality, while in the dry season the protein content of the standing hay may limit fibre digestion and hence energy intake, consequently the animals lose weight (see Chapter 15).

Fodder and browse from trees and shrubs

In most of the grasslands in seasonal rainfall areas, trees and shrubs make a major contribution to stock feed, especially in the dry season. Goats, cattle and, to a lesser extent, sheep can obtain an appreciable proportion of their feed by browsing the leaves, flowers, seeds, twigs and even the bark of numerous species (Reed *et al.*, 1990). In some places herdsmen lop branches off selected trees for their animals to eat in the dry season. In Africa about 75% of the trees and shrubs are browsed to some extent by domestic animals. Dougall and Bogdan (1958) listed 57 plant species which are eaten by cattle (although some only by goats) and chemically analysed the parts eaten. Much of this plant material is high in protein and minerals; 29 of these species have crude protein contents exceeding 15%. The trees and shrubs come into leaf before the rains, and the edible parts retain their nutritive value well into the dry season. Their fibre content may be rather high but not above that of dry-season grass, or hay, and early digestibility trails indicated that the feeding value of fodder from trees and shrubs is greater than that of grass in the dry season (Göhl, 1982).

In certain areas it may be desirable practice to plant browse species for use during the dry season, as such plants usually retain their leaves longer due to their deep-rooted nature. Even when the leaves are shed the dry dropped leaf is of good nutritional value and will be eaten by cattle. Thus the planting of browse species such as *Bauhinia rufescens*, *Cadaba farinosa*, *Acacia mellifera*, *Balanites aegyptica*, and *Maerua crassifolia* has been recommended in parts of Chad and the southern Sudan by Tubiana and Tubiana (1975), but they pointed out that protection during establishment by fencing or the use of thorn enclosures may be necessary. In the very arid areas cattle rely much more on browse, and goats often survive exclusively on browse, as there is little alternative feed available, apart from ephemeral grass growth, during and just after the short rainy periods. Browse production and utilisation by rumimant livestock is discussed by Le Houérou (1980).

Bush thinning

While acknowledging the part played by trees and shrubs in providing fodder, these plants should not be encouraged at the expense of grass. Goats thrive on a mixture of grass and browse, but cattle do better on grasses and herbs and stock-carrying and liveweight gain capacity of pastures is normally increased by reducing the number of trees to the minimum required for shade. Thus, in Zimbabwe, the carrying capacity of woodland–grassland, with widely spaced trees of species of *Brachystegia* and *Julbernardia*, is about one livestock unit to 8–12 ha, but when the trees are reduced to the minimum needed for shade the cleared veld will support one livestock unit to 3 or 4 ha. In many drier areas the presence of trees and shrubs not only reduces the carrying capacity but also leads to the development of bare land and erosion, since the rainfall is insufficient to maintain the grass in competition with woody vegetation better adapted to the semi-arid climate. For example, in Tanzania it was found that pasture in which trees and shrubs were reduced carried a livestock unit to 0.6–1.2 ha, whereas adjacent 'bushland' pasture, although only carrying one livestock unit to 5.6 ha on a deferred grazing system, showed signs of sheet erosion. An initial reduction in the stand of trees must be obtained by felling or poisoning; burning will be ineffective, since mature trees of most species are relatively fire-resistant.

Control of bush regeneration and encroachment by burning After the initial thinning it will be necessary in many types of grasslands, especially those in the woodland and thicket areas, to prevent the regeneration and spread of bush. Often the most practicable and economic means of achieving this is by fire, but it is important that burning should be minimised, done at the proper time, and combined with good management. As a rule the best time to burn, for the purpose of bush control, is at the end of the dry season when the grass is still dormant but the trees and shrubs have begun to produce a flush of young leaves and are therefore in a vulnerable condition. In many places a few showers, known as the 'grass rains', occur towards the end of the dry season and are followed by a short, dry spell before the main rains begin and a common practice is to burn the grass immediately after the first showers. This avoids the risk of a long dry spell after the burn due to the late arrival of the rains. Burning earlier in the dry season results in the grass making transient growth at the expense of its food reserves and this is often done by herdsmen to produce some keep for hungry cattle. It is a bad practice from the standpoint of sustaining a good grass sward because it encourages bush encroachment and weakens the grass so that it cannot grow vigorously when the rains come.

A fierce fire is needed to control bush; therefore pastures should be rested for some time before burning so that there is an accumulation of dry grass. In the drier areas the grass cover is usually sparse and it may be necessary to rest the pasture for nearly a year, through most of the growing season and all the dry season. Previous grazing intensity will affect the quantity of grass available; where there has been heavy overgrazing, resting for longer than a year may be needed to accumulate enough grass for a good burn. After burning, the pasture should not be grazed at the beginning of the rainy season but allowed to grow unchecked for a time in order to allow replenishment of the food reserves. The requisite period of rest varies but is usually 1 or 2 months.

For effective bush control, burning must be done regularly, so that tree seedlings which survive one fire will not have grown sufficiently to escape a second. The required frequency of burning will vary with location and with pasture condition. For example, in Matabeleland the veld may need to be burned once in 3 years until a stable open parkland has been established; thereafter, stocked correctly, it should not be necessary to burn more often than once in 5 or 6 years. Certain shrubs occurring in grasslands cannot be controlled by fire alone and mechanical or arboricidal treatment may be needed. Examples are *Tarconanthus camphoratus* (known in Kenya as 'leleshwa'), which has extensively invaded pastures in the Rift Valley, and *Acacia pennata*, which is common in semi-arid Kenyan pastures.

Burning may be useful for purposes other than bush control. In 'sour' or mixed veld, where it is not possible to graze uniformly every season without deterioration resulting from overgrazing, there is always a lot of grass left uneaten. If cutting is impracticable (as it usually is), it may be desirable to remove this unpalatable herbage by burning every 2 years or so. Fire may also be an easy way of removing undesirable species of grass which have invaded a pasture owing to inappropriate management. For example, a good burn at the end of the dry season is the most effective way of ridding Kenyan pasture normally dominated by *Cynodon dactylon* and *C. plectostachyum* of an invasion by an annual grass, *Aristida kewensis*. Fire also prevents the replacement of *Themeda triandra* by the useless grass *Pennisetum schimperi* in the highland grassland and forest zone of Kenya, or by *Digitaria abysinnica* in marginal *Acacia–Themeda* country.

Where fire is used for purposes other than bush control, it may be undesirable to burn at the end of the dry season. Fires at this time are hot and fierce, and there is a danger of them getting out of control unless due precautions are taken. In some places fires may kill out some of the desirable grasses and increase the proportion of coarse, unpalatable, fire-resistant species. Consequently, where bush control is not the main problem, an early light burn to remove unpalatable material is preferable. It should be noted that the use of fire, as a means of sward improvement, is controversial (Humphreys, 1991). Winter (1987) has demonstrated that, over the long term, burning rarely causes substantial changes in the balance of grass species although it has a large impact on bush and scrub.

Other means of controlling bush Removing trees and shrubs manually or mechanically is usually too expensive because of the need to extract the root systems from which many species are capable of regenerating. Ring-barking, or cutting down trees and slashing shrubs, will effect some improvement but many plants will grow again from their stem bases. Repeated slashing will weaken and eventually kill many young trees and shrubs, but is usually an expensive procedure. Cutting combined with burning will deal with some species, especially if the cut material is stacked over the rootstocks and burned *in situ*. Where practicable, regular mowing will prevent the regeneration of many shrubs and trees, but it does not eliminate them. Mowing is usually impracticable owing to the large areas involved, the uneven ground and the presence of ant hills and trees. Intensive browsing with goats may kill young regrowth, by repeated consumption of the leaves and stripping of bark, but it necessitates confining the animals and they may damage the land by excessive trampling.

Some use has been made of arboricides for initial thinning of bush, controlling regeneration of indigenous species or dealing with invasions of introduced plants, such as *Lantana camara* or *Opuntia* spp. Power paraffin or diesoline will kill many plants, including most *Acacia* spp. if applied liberally to the junction of stem and root, but this necessitates scraping away the soil and is slow and expensive. Although arsenicals are cheap and very effective they cannot normally be used owing to their high mammalian toxicity.

The most widely used chemicals have been 2,4-D and 2,4,5-T but these arboricides are now banned in many countries because of their toxicity. Alternative materials, such as silvex, paraquat, dicamba and picloram, are available but expensive and product licences are required in many countries and the legal position needs to be checked on an individual country basis. For these and other reasons arboricides are no longer used extensively for the control of bush in grasslands although there are exceptions, such as the aerial application of silvex in Hawaii to control *Melastoma* and *Psidium* (Motooka *et al.*, 1967). The high cost of chemical treatment will only be justified if it can raise the carrying capacity to a relatively high level, which is unlikely on most low potential grasslands in the drier areas.

Grazing management

It will be apparent that, in order to achieve the aims stated earlier in this chapter for the management of pastures in semi-arid areas with seasonal rainfall, the following requirements should be met:

(1) Adjustment of stocking rate to carrying capacity, which will usually mean relatively light stocking well within the average productivity of the pasture. This requirement is fundamental.
(2) Periodic rests from grazing during the critical growth periods (but note the reservations of Humphreys (1991) referred to earlier).
(3) Bush control, usually by fire and involving a rest before and after burning.
(4) Some special provision for dry-season feed.

These aims are usually achieved by the adoption of deferred rotational, or seasonal, grazing by which the use of part of the pasture is delayed until the grass has matured. The pasture is fenced into paddocks, each ideally provided with a water supply, which are grazed in rotation which allows periodic rests and burns, and ensures that sufficient grass or standing hay is available all the year. Details of the system will depend on local climatic conditions, the type of grassland and whether or not burning is needed for bush control (Whiteman, 1980). There are many such systems but only two will be described below by way of illustration.

It should be noted, however, that little critical work has been carried out on the comparison of these different management systems on grass, and especially grass/legume, pastures and the results that have been obtained have not been consistent (e.g. Bowen and Rickert, 1979). Indeed, some authors question many of the assumed tenets of good grassland management. Thus Humphreys (1991), in a wide-ranging review, questions the evidence for advocating rotational grazing compared to continuous stocking, the efficacy of resting periods, the economic justification for making hay or silage as dry-season feed except as a last resort, and the role of zero grazing for tropical dairying system. His questioning of the scientific evidence for many advocated grassland management practices is valid, as also is his contention that systems evolved under temperate conditions are not readily transferred into the tropical environment. However, his total dismissal of many tried and proven tropical grassland practices, on the grounds that they are under-researched, may not be entirely justified.

Three herd:four paddock, four-year system

In this system each paddock is grazed continuously for 3 years and in the fourth year grazing is stopped after the beginning of the rains, the accumulated grass is burnt off at the end of the dry season, and the paddock rested for a further 1 or 2 months at the beginning of the next rainy season. Owing to the long period of continuous grazing involved, and the need to keep some grass standing in the dry season, the stocking rate has to be rather modest.

One herd:four paddock, four-year system

This system is suitable when a higher carrying capacity justifies expenditure on fencing and water supplies. It is flexible and permits heavier stocking rates, because it provides ample reserves for drought and bad seasons. The rotational details are shown in Table 12.1. Each paddock is used for growing-season grazing in one year. In the next year it is rested for burning. After burning, at the beginning of the third year, it has a growing-season rest and is then grazed in the dry season. At the end of the rains the animals are moved into the dry-season paddocks for grazing in the fourth year.

The adoption of a suitable system of deferred rotational grazing usually improves the carrying capacity of the pasture compared with continuous grazing. However, in the poor grasslands of the drier areas there may be no appreciable increase in production from the system compared with continuous grazing. Indeed, there are areas of such low productivity that they do not justify expenditure on bush thinning and fencing. Thus

Table 12.1 One herd:four paddock system.

Paddock	First year	Second year	Third year	Fourth year
A	Graze growing season	Rest for burning	Burn, graze dry season	Graze dry season
B	Rest for burning	Burn, graze dry season	Graze dry season	Graze growing season
C	Burn, graze dry season	Graze dry season	Graze growing season	Rest for burning
D	Graze dry season	Graze growing season	Rest for burning	Burn, graze dry season

it is important to realise that 'more management' is not necessarily 'better management' unless rotational grazing leads to higher animal production so that the extra expense is more than covered by the additional income.

In both continuous grazing and rotational grazing systems stocking rate is a very important factor, but appropriate rates can only be determined from local experience. The optimum rate is that which gives the maximum animal production consistent with maintaining, or improving, the productivity of the pasture, which will be adversely affected by either over- or understocking. These factors are amenable to quantitative analysis, and a linear response model has been devised by Jones and Sandland (1974). There is usually a tendency to overstock, especially as heavy stocking in the rains initially increases the total liveweight gain per hectare but, as this leads to deterioration in the pasture and liveweight losses in the dry season, it is advisable to err in the direction of under-stocking.

The use of mixed grazing systems with cattle, sheep or goats

Owing to the differences in the grazing habits of cattle, sheep and goats, mixed grazing may be preferable to stocking with cattle alone. Sheep are selective in their grazing; they prefer the finer grasses and nibble the growing leafage, leaving stems and coarser grasses, so that a pasture grazed by sheep alone tends to become very rough. Cattle graze more uniformly, but reject certain unpalatable grasses.

In grasslands containing trees and shrubs, mixed grazing with cattle and goats may be advantageous. In contrast to cattle, goats obtain the greater part of their feed from the trees and shrubs, thus aiding bush control. Goats are not very selective as regards plant species, but have a preference for succulent shoots and leaves that are slightly above their normal head level. Given the opportunity, they will browse mainly on trees and shrubs. They do not normally graze low down and therefore they neglect the growing points near the ground and the shorter grasses. In early large-scale experiments in Tanzania and Kenya, goats kept down coppice and seeding growth of most bush species, with the result that a uniform ground cover was maintained. By contrast, grazing cattle alone concentrated on the grasses, thus relieving the bushes of competition and encouraging the formation of thicket with little ground cover (Bogdan, 1955).

The use of fertilisers

The response to fertilisers depends on the rainfall and the type of grassland, but may of the natural grasslands are in the drier areas, where the response of the sparse cover is marginal. Most pastures suffer from nitrogen deficiency, especially where the grass is periodically burnt and the nitrogen in the aerial parts is lost to the atmosphere. Grazing cattle usually have only a low nitrogen intake and return only small amounts to the land in their excreta. Most of this is concentrated in small urine patches from which the recovery of nitrogen by the herbage is low because most is lost by volatilisation. The dung has a low content of nitrogen which is not readily available. On *Hyparrhenia* veld in Zambia, Smith (1965) found that cattle excreted approximately 5 kg of urinary nitrogen per head per annum and that their dung contained only 2% nitrogen or less. He concluded that, because of the low value of the dung as a nitrogen fertiliser, the high gaseous loss of urinary nitrogen, and the uneven distribution of their excreta, cattle had little effect on the nitrogen cycle of the grassland.

Provided that the rainfall allows a growing season of at least 4 or 5 months, nitrogen fertilisers usually increase the dry-matter yield and crude protein content of the herbage during the rainy season. For example, at Frankenwald in South Africa, where an average annual rainfall of 780 mm falls in 5 months, grazing experiments showed that a substantial profit could be obtained from applying nitrogen to the veld. On pastures dominated by *Hyparrhenia dissoluta*, *H. filipendula* and *Heteropogon contortus* in Zambia, 45 kg nitrogen/ha applied three times during the growing season more than doubled the dry matter yield of herbage cut at the growing stage and increased its crude protein content from 6.1 to 9.5%, the response being 23 kg dry matter and 3 kg crude protein per kg of nitrogen applied (Smith, 1961). On these grasslands, nitrogen, either alone or in a combination with phosphate, may improve the botanical composition of the herbage by encouraging the more nutritious species. Thus, in Swaziland, nitrogen and phosphate changed a *Hyparrhenia* pasture over 5 years into a more productive sward dominated by *Eragrostis curvula* (I'ons, 1967).

As the increased growth of grass from the application of nitrogen occurs in the rainy season, the use of the fertiliser increases the difference between wet- and dry-season carrying capacity, especially as the lesser amount of dry-season herbage is also very low in crude protein, which results in low intake, depressed digestibility and poor animal performance. Attempts to improve the quantity and quality of herbage for dry-season grazing by the application of nitrogen towards the end of the rains have usually proved ineffective or uneconomic. As a rule, such dressings have little effect on the amount of herbage dry matter available in the dry season and, although crude protein content may be raised if high rates of nitrogen are given, it falls rapidly after the onset of the dry season due to shedding of seed and leaves. For example, Tergas *et al.* (1971) were able temporarily to raise the crude protein content of *Hyparrhenia rufa* to 7% (the level adequate for the growth of beef cattle) by applying 150 kg nitrogen/ha towards the end of the rains, but it rapidly fell below this level during the dry season. Any increase in protein obtained could be utilised if hay was made at the end of the rains but, as explained below, this is not often practicable. Applying nitrogen earlier in the rains, in order to improve dry-season feed, is ineffective unless uneconomically high rates are used; thus, Miller and Nobbs (1976) found that over 400 kg nitrogen/ha was required to increase the crude protein content of dry season herbage of *Brachiaria mutica*.

The effect of nitrogen fertiliser must increase animal productivity if economic benefits are to be demonstrated. Unfortunately most fertiliser trials are measured by the effect on grass growth and composition, as plot sizes are often too small to allow assay by the grazing animal. However, Caro-Costas *et al.* (1972) carried out a replicated grazing trial evaluating the use of different levels of nitrogen applied to Pangola grass pastures on the growth of heifers and the results are presented in Table 12.2. The responses were highly significant, at 1.88, 1.84 and 1.47 kg liveweight gain/kg nitrogen applied over the base level of 63 kg nitrogen/ha. Similar work was conducted by Caro-Costas and Vicente-Chardler (1972) with Napier grass pastures when the equivalent responses were 1.98 and 1.90 kg liveweight gain/kg nitrogen applied. These results are also given in Table 12.2.

It is often necessary to apply phosphate if the full benefit is to be obtained from nitrogen but, in contrast to nitrogen, a good dressing of phosphate should last for some years since it is immobilised in the top few centimetres of soil and, unless heavy hay or silage crops are removed, much of the phosphate taken up by the herbage is returned in the excreta of grazing animals. Applied phosphate is only absorbed by the herbage while the surface soil is moist; when the latter dries out the fertiliser ceases to be effective and,

Table 12.2 Stocking rates and liveweight gains from pastures receiving different levels of nitrogen fertiliser. (Sources: [1] Caro-Costas et al., 1972; [2] Caro-Costas and Vicente-Chardler, 1972.)

Pasture	Fertiliser rate (kg N/ha)	Stocking rate (270 kg steers/ha)	Liveweight gain (kg/ha)
Napier grass pastures[2]	250	5.5	1060
	440	7.1	1435
	625	8.8	1772
Pangola grass pastures[1]	63	1.9	398
	188	3.2	633
	376	5.0	976
	533	5.9	1088

if the subsoil is deficient in phosphate, the content of this element in the dry-season standing hay may be insufficient for the needs of growing stock. Phosphate manuring may improve the botanical composition of the grass flora and may be essential in order to introduce or increase a legume component in the sward. Potash fertilisers may also be needed if legumes form an appreciable proportion of the sward but are not generally required on natural grasslands.

Legumes

Although nitrogen is generally deficient in natural grasslands, fertilisers are expensive and may not be profitable or practicable for small farmers. Legumes are usually sparse or absent but, when practicable, it may be desirable to introduce them because they will fix nitrogen and improve the nutritive value of the sward. Experiments with sown pastures have shown that legumes can sometimes improve the yield and crude protein content of grasses grown in association with them. For instance, Balachandran and Whiteman (1975) conducted a trial comparing the dry matter and crude protein yield of plots of pure *Desmodium*, pure *Setaria*, and mixed swards of *Desmodium* and *Setaria*. The dry matter yields of the *Desmodium* plots totalled 12.1 t/ha, the *Setaria* plots only 3.1 t/ha, and the mixed swards 10.4 t/ha. The nitrogen yields were 380, 40 and 253 kg/ha respectively. However, it is uncertain whether this benefit would result from the introduction of legumes into natural grasslands, although it may do so in the higher-rainfall areas.

The effect of legumes on the performance of grazing animals is to increase the average liveweight gain/head, or the stocking rate, or both. The best way of comparing results is as total liveweight gain/ha, since individual gains and stocking rates are related. Table 12.3 indicates that, although responses to legumes are subject to wide variation, the average increase in liveweight gain/ha was 108 kg (+86%). In the few examples where little or no improvement has been demonstrated, the lack of response is usually due to poor establishment or some other limiting factor (Butterworth, 1985). Other beneficial effects include extension of the grazing season (Wesley-Smith, 1972), earlier maturity (Cohen and O'Brian, 1974) and improved reproduction and calf production (Silvery et al., 1978).

Various legumes have been successfully introduced into natural grasslands. Thus in Zambia good establishment was achieved by sowing uninoculated seed of *Stylosanthes guianensis* and *Glycine wightii* in *Hyparrhenia*-dominated grassland after burning at the end of the dry season, scratching the soil with a rake and applying superphosphate. Both legumes, especially *Stylosanthes*, markedly increased the yields of dry matter and digestible crude protein, and these increases were due to the contribution of the legume since the yield and crude protein content of the grass component were either un-

Table 12.3 Effect of legumes on the liveweight gains of grazed swards by beef cattle.

Country	Beef production (kg/ha)		Reference
	Grass alone	Grass + Legume	
Australia	131	219	Lowe et al. (1977)
Australia	33	155	t'Mannetje and Nicholls (1974)
Australia	11	40	Norman (1974)
Papua New Guinea	78	106	Chadhokar (1977)
Philippines	22	117	Siota et al. (1978)
Thailand	34	122	Falvey (1976)
Fiji	143	352	Partridge (1979)
Nigeria	82	119	Haggar (1971)
Swaziland	115	176	Jones (1974)
Uganda	655	975	Stobbs (1969a)
Mexico	106	180	Garza (1973)
Brazil	110	260	Vilela et al. (1977)
Unweighted mean	126.7	235.1	

changed or decreased (Smith, 1963). In India, improvement was effected by introducing *Alytosia scaraboides* and *Centrosema pubescens* into *Heteropogon* communitites (Dabadghao and Shankarnarayan, 1970).

Phosphate manuring will often be needed for the establishment of legumes (Skerman, 1977) and, although trace-element deficiencies are unusual for grasses, they may need to be corrected for legumes which require adequate copper, zinc and molybdenum for satisfactory growth and nitrogen-fixation. An additional benefit of introducing selected legumes into the herbage is that they tend to be deeper rooted than tropical grasses and thus they remain green longer into the dry season and grow faster under shade when the rains come. Butterworth (1985) has reviewed the use of legumes in improving the productivity of selected tropical grasslands.

Although the legume is capable of fixing the greatest quantity of nitrogen by means of the microorganisms contained in the root nodules, other herbage plants also contribute to varying degrees. Thus genera such as *Andropogon, Brachiaria, Cynodon, Digitaria, Hyparrhenia, Melinis, Panicum, Paspalum, Pennisetum, Saccharum* and *Zea*, in association with microorganisms such as *Azotobacter* and *Beijerinckia* spp., also have this capability although they do not possess the typical root nodules of legumes (Dobereiner and Day, 1974). They all possess the C4 carbon pathway in photosynthesis which is more efficient than the C3 pathway in temperate grasses. Wide variation exists in the nitrogen-fixing abilities of the various grasses, but estimates of 1.5 kg/ha/day have been made for *Digitaria decumbens* and *Paspalum notatum* (Dobereiner and Day, 1974). The *Paspalum notatum–Azobacter paspali* association has been the most intensively studied (Day et al., 1975), and it has been shown that the nitrogen-fixation occurs largely in the rhizosphere of grass roots with the greatest activity in the growing season. Weir et al. (1979) demonstrated that *Rhyncheletrum repens*, a common coloniser of arable lands, may contribute as much as 120 kg nitrogen/ha/yr and work by Venkateswarlu and Rao (1983) suggested that the early studies may have over-estimated the nitrogen-fixing capacity of tropical grasses. This subject is also discussed in Chapter 3. It is clear that this evidence of nitrogen-fixation within the root systems of tropical grasses with associated, but free-living, microorganisms has important implications in the nitrogen-economy of tropical grasslands if it proves to be fully substantiated. This subject is dealt with at length by Whiteman (1980).

Food resources of cut-and-carry systems (zero grazing) may be improved by intercropping grasses with legumes. Thus, Muinga et al. (1992) have shown that intercropping Napier grass (*Pennisetum purpureum*) with leucaena forage (*Leucaena leucocephala*) significantly increased dry matter intake, decreased daily liveweight loss and increased milk yield of crossbred Brown Swiss × Sahiwal cattle in Kenya, provided that the stage of harvesting of the grass/legume forage mixture was carefully controlled.

Economics of improving tropical grasslands

It is relatively easy to demonstrate that tree bush and shrub removal, fertiliser application and legume introduction can increase tropical grassland productivity. However, due to the lower value of tropical livestock, the high costs of imported agrochemicals and the relatively low stocking rates in many tropical areas it is more difficult to show clear economic benefits. In general, minimum stocking rates of about 1 livestock unit/ha are required if such costly inputs are to be justified.

In many cases, overstocking has so impoverished the land that stocking rates which potentially may be of the order of 1 livestock unit/ha are reduced to levels which agrochemical application alone cannot rectify. In such cases, resting the land may be the least-cost solution to the immediate problem, but destocking may not be a practical proposition to farmers whose livelihood depends upon the regular sale of livestock or livestock products. Livestock numbers are usually the major determinant of overall livestock output.

THE USE OF NATURAL GRASSLANDS BY TROPICAL PASTORALISTS

Features of pastoralism

Large areas of land in regions of low and unreliable rainfall, where it is impracticable to grow crops, are thinly populated by pastoralists. The grasslands of these areas are derived from, or associated with, the drier types of broad-leaved woodlands, thorn woodlands and subdesert scrub, and are characterised by a sparse cover of grass with scattered trees and shrubs. Their overall carrying capacity is low; under traditional tropical pastoralism it is commonly of the order of 1 livestock unit to 8–10 ha, but it may be much lower where deterioration has resulted from prolonged mismanagement. Seasonal variation in carrying capacity is marked, but during the dry season the pastures are usually overstocked. As the rainfall is unreliable, years of severe drought are common, during which overstocking may result in animals dying of starvation. Water supplies are often inadequate and ill-distributed. Apart from directly causing stock losses, water scarcity may result in poor pasture utilisation because concentration of animals near watering points in the dry season leads to localised overgrazing and erosion while other pastures, remote from water supplies, may not be fully utilised. In many African pastoral regions a further limitation may be imposed by the occurrence of tsetse flies, the vector for the diesease trypanosomiasis, which render large areas unsuitable for grazing for a part, or the whole, of the year. Most pastoralists are nomadic, or semi-nomadic, not because of inherited custom or tradition, but as a direct consequence of the limitations of their environment. The movements of the people and their stock are mainly dictated by the need to go where fodder and water are available and to avoid areas with disease hazards, such as those infested with tsetse, *Stomoxys* or other biting flies (see Chapters 15 and 17).

Generally the land used by pastoralists is owned and grazed communally. On the other hand, individual or family ownership of livestock is normal and each owner usually keeps as many animals as possible, irrespective of their quality or the availability of pasture. This is partly because livestock are regarded as wealth, and social position and prestige may depend to some degree on the number of stock rather than on money or other possessions. Also cattle are sometimes needed to fulfil certain tribal obligations, such as the payment of bride price, which is a feature of the social life of many pastoral peoples. The head of a family may not be free to sell his cattle, since he holds them in trust for the family. Large numbers of stock may also be kept as an insurance against years of drought and famine, on the assumption that the more cattle possessed the more will survive a bad year. Little attempt is made to co-operate by restricting the numbers of stock held by individuals to those which might be expected to survive on the available fodder and water resources.

Because of their desire to own as many animals as possible, many pastoralists are reluctant to sell their livestock. They subsist on the milk, blood and meat of their animals, and by hunting, by collecting food from wild plants and by selling or bartering milk and skins to obtain grain. However, some tribes, for example the Fulani of northern Nigeria, have long been accustomed to selling stock to meet a demand for meat from other communities, and there are many examples in Africa where traditional systems of grazing management are well suited to the difficult environmental conditions which are the major constraint (Niamir, 1990).

The combination of communal ownership and grazing of the land with unrestricted individual or family ownership of stock leads to many difficulties of pasture management. Stock numbers may not be adjusted to carrying capacity and available water supplies; insufficient effort may be made to organise rotational grazing and resting of pastures where this is desirable and there is inadequate control of bush encroachment. Matters are made more difficult by unrestricted setting of fires to facilitate hunting or to promote a flush of green herbage during the dry season, a procedure which often provides only a negligible amount of feed and may exhaust grass plants already weakened by overgrazing. These widespread fires are usually uncontrolled and destroy useful standing hay, leaving the ground bare and exposed to erosion at the onset of the rains.

The pastoralists' expertise in animal husbandry is often thought to be low although certain qualities of stockmanship are sometimes developed to a high degree. Indeed, some workers consider that traditional pastoral systems are more environmentally sustainable than some of the recommended replacements (e.g. Niamir, 1991). The environment seldom makes it possible to avoid animals losing weight during the dry season, but such checks may be reduced by better grassland management. Seasonal mating, so that most calves are dropped as soon as there is adequate grazing for the cows, may reduce calf mortality and consequently aid the growth of the calf. This is not often practised nor is it always practical, so that calf mortality is generally high and calves are often underfed; the cow often does not give enough milk to support a healthy calf and an excessive proportion of the milk may be sold for human consumption. Pasture management and careful herding could make it possible to keep the stock on the better grazing when weaning or fattening, but this is not always practical and is seldom done. Correction of mineral deficiencies of livestock is often needed but mineral licks, apart from rock salt, are not commonly used. Not all herdsmen take sufficient interest in the measures recommended, or facilities provided, for the control of disease. For example, dipping schemes against ticks can only be effective if the cattle are protected at regular intervals, and this may be deemed to be difficult or inappropriate, either because of perceived high costs or because the regular gathering of cattle to a central dipping point is considered to be impractical.

Failure to adjust stocking rate to carrying capacity results in slow growth with annual dry-season checks. Cattle may take from 4–7 years to produce a poor, light-weight carcass. In drought years, as a consequence of overstocking, large numbers of cattle may die of starvation and thirst. Sometimes the herdsmen, threatened with heavy losses, may offer immature beasts for sale but the sale and slaughter of unfinished cattle is wasteful and represents a loss to the stockmen. These and other technical constraints to livestock production in Africa have been extensively reviewed by von Kaufmann and Fitzhugh (1993).

Results of pastoralism

Primitive pastoralism, which continues to be widely practised, is thought to be responsible for damage to large areas of land in Africa and Asia, primarily due to overstocking. This occurs in certain primitive pastoral systems that combine communal grazing with unrestricted individual ownership of stock, and recently it has tended to be more prevalent. Both the pastoralists and their livestock have become more numerous as a result of the reduction in the incidence of major endemic diseases. At the same time the areas available to pastoralists have often decreased owing to the land requirements of increasing numbers of cultivators, who have taken to growing cash crops in addition to subsistence food crops, and also to desertification.

The sequence of events which often occurs when pastoralism is combined with increasing population pressure on the land is well known. Overgrazing, which often begins at the watering places (where animals congregate in the dry season), spreads out from them destroying the grass cover especially where movement of stock is restricted. The removal or reduction of competition from the grass increases the spread of trees and shrubs that are better adapted to arid conditions, and this encroachment restricts the reestablishment of grass cover. Rain falling on the bare ground between trees runs off and erosion occurs at an accelerating rate. Once the absorption and percolation of rainfall through the soil ceases, the streams no longer flow regularly and come down in flash floods for short periods. Springs dry up, water for stock and vegetation becomes progressively scarcer and the end of the process is the creation of barren conditions; one of the forms of 'desertification', the other form being climate induced. The rate and extent of denudation varies with local circumstances, being greater where human and stock population are dense, where topography favours erosion, or where the rate of regrowth of the vegetation is reduced by harsh climatic conditions.

The economic benefits from this method of land use are frequently poor. Quite apart from the reluctance of people to sell their stock, the productivity of the system is often low. On well-run ranches in Tanzania, on grasslands similar to those occupied by native herdsmen, the average calving percentage can be 70–80%, with low calf and herd mortality. Average liveweight increases of 0.25–0.5 kg/day are obtained and the annual off-take of mature cattle is about 20% of the total stock population. Equivalent figures for nearby pastoralists are less reliable but some estimates of calving rate may be as low as 50%, with 20% of the calves not surviving the first 6 months and a further 20% failing to reach maturity so that the average annual off-take maybe of the order of 7% (International Bank for Reconstruction and Development, 1961).

Possibilities for improvement of pastural systems

The rectification of the above problems is not easy. Pastoralists are usually conservative and it is difficult for them to change their ways. In some cases their management systems, although unsophisticated, are difficult to improve upon in view of the environmental constraints. In many places, denudation has gone so far that special measures must be taken to rehabilitate the land before improved management can achieve good results. All these things involve capital expenditure, and the potential of much pastoral land is so low that there is little prospect of its paying a reasonable return on the investment required.

The use of tropical grasslands – general considerations

Importance of stocking rate

In certain cases the first essential in improvement is to reduce stock numbers in order to adjust them to carrying capacity. In some places, where the existing stocking is in excess of the carrying capacity of the land on any system of management, a permanent reduction in numbers may be needed. More commonly, the problem is not overstocking in relation to the real potential of the land but overstocking under the present limiting factors of pasture management. Hence an initial reduction in stock numbers, if followed by good pasture management, may in time enable the land to carry more stock. The first difficult step, of adjusting stock numbers to carrying capacity, often demands a radical change to traditional practice. Stock owners must be persuaded to accept an initial reduction in the size of their herds and, thereafter, to endeavour to raise animals for profit, selling a suitable proportion of their stock annually.

More recent studies (e.g. Scoones, 1987) have indicated that there is often a conflict between technical policies aimed at maximising grassland productivity and individual animal productivity by reducing stocking rates to theoretically better (lower) stocking rates, and economic policies designed to maximise the exploitation of the total grazing and livestock resources.

To help increase the off-take of animals it will be necessary to establish markets in the pastoral areas, with access roads for buyers, and to provide more stock routes, furnished with watering points and fenced night-stops. This enables animals to be trekked without suffering from shortage of feed and water or from the depredations of wild game. Where large numbers of immature cattle are sold in dry years, it may be possible to reduce the waste resulting from their slaughter before maturity (and at the same time to even out the supply of meat to the market) by the provision of holding grounds, possibly operated in conjunction with canneries, where the cattle can be fattened. A common problem during the early stages of an improvement programme is the disposal of very poor quality cattle that are below the standard required for canning, let alone for fresh meat. Herdsmen are unwilling to bring stock long distances to market when the chances are the beasts will be rejected yet these are the very animals which it is necessary to eliminate. The problem was originally addressed in Kenya by the provision of mobile abattoirs and meat-processing plants, which travelled into the more remote overstocked or drought afflicted areas to purchase low-grade cattle and process them into biltong, meat meal, blood meal and bone meal. More recently government-organised schemes enable cattle to be purchased for slaughter and canning to supply meat markets for home consumption and export.

Effect of stocking rate on animal productivity under extensive systems (e.g. ranching)

Having stressed the practical importance of correct stocking in line with carrying capacity, some experimental work on the subject will now be reviewed. Many trials have demonstrated the important relationship between gains per beast, which remain constant or decline with increased stocking rate, and gains per hectare, which initially increase. These early increases in productivity can be misleading, since clearly they only apply when the work is carried out at or under the carrying capacity of the pasture. Typical of a large number of reported experiments is the work carried out in Uganda by Thornton (1970) on *Pennisetum purpureum/Imperata cylindrica* natural pastures in Buganda, presented in Table 12.4.

An interesting observation made in this trial was that both *P. purpureum* and *I. cylindrica* decreased under heavy grazing whilst *Cynodon dactylon*, *Setaria sphacelata*, *Brachiaria* spp. and *Panicum maximum* increased. This increase in the shorter growing species (*Cynodon*, *Brachiaria*) resulted in lower dry matter production but vegetation of higher nutritive value. Different results were obtained in trials at Serere by

Table 12.4 Liveweight gains at different stocking rates from natural Buganda pastures. (Source: Thornton, 1970.)

Stocking rate (beasts/ha)	Liveweight gain	
	(kg/head)	(kg/ha)
1.2	85	105
2.5	92	230
3.3	93	306

Note: It is important that the data presented are not extrapolated into higher stocking rates where livestock performance will decrease when carrying capacity is exceeded.

Stobbs (1969a) on mixed *Hyparrhenia rufa* and *Stylosanthes guianensis* swards. Small East African zebu steers were used at different stocking rates, and the results demonstrated that, in these more intensive conditions, *H. rufa* declined under heavy grazing but the unpalatable, grazing-resistant *Sporobolus pyramidalis* increased (Table 12.5). The effect of stocking rate on livestock performance was not recorded.

There appear to be common patterns in the response of natural grazings to heavy grazing which may be different from the results on improved pastures. Perennial species in natural grazings tend to be replaced by ephemeral pioneer species and by woody brush species (bush encroachment). A reduction in the amount of vegetation tends to reduce the intensity of fires which normally control bush encroachment. There may be a reduction in uniform cover thus exposing land to erosion (Edwards and Nel, 1973). There is a generalised pattern of species which increase under heavy grazing and those which decrease, as indicated in Table 12.5 and in more detail in Table 12.6. Many of the increasers are relatively good grazing species. Animals which are heavily stocked (but not overstocked) tend to select a more nutritious diet (Winter *et al.*, 1977) and spend a longer time grazing than more lightly stocked cattle, but in spite of this selective grazing the more vigorous increasers maintain or increase their ranking in the sward.

Complications arise in any discussion on stocking rates where mixed grazing is practised. Cattle, sheep and goats have different grazing behaviour and so it is not possible to equate grazing pressure with x cattle to grazing pressure with y goats. It is, therefore, necessary to establish some crude relationships and the most commonly accepted equivalents are those based on liveweight suggested by Moorhouse (1975) presented in Table 12.7.

Table 12.5 Percentage botanical composition at different stocking rates at Serere, Uganda. (Source: Stobbs, 1969a.)

Species	Stocking rate (beasts/ha)		
	Low (1.6)	Medium (2.5)	High (4.9)
Hyparrhenia rufa	70.8	68.5	25.7
Sporobolus pyramidalis	0.5	1.6	38.7
Stylosanthes guianensis	18.9	22.0	20.2
Other spp.	9.8	7.9	15.4
Total	100.0	100.0	100.0

Table 12.6 Separation of tropical grass species into those that tend to increase (increasers) and those that tend to decrease (decreasers) under heavy grazing conditions. (Source: Butterworth, 1985.)

Increasers		Decreasers	
Species	Country	Species	Country
Brachiaria decumbens	Australia	*Chloris gayana*	Australia
Brachiaria decumbens	Uganda	*Heteropogon contortus*	Australia
Brachiaria decumbens	Zambia	*Hyparrhenia* spp.	Uganda
Cynodon dactylon	Uganda	*Hyparrhenia* spp.	Zambia
Cynodon dactylon	Zambia	*Loudetia simplex*	S. Africa
Cynodon dactylon	Zimbabwe	*Paspalum dilatatum*	Australia
Digitaria decumbens	Australia	*Paspalum scrobiculatum*	Australia
Heteropogon contortus	Zambia	*Paspalum scrobiculatum*	Uganda
Imperata cylindrica	Uganda	*Pennisetum purpureum*	Uganda
Panicum maximum	Uganda	*Themeda australis*	Australia
Pennisetum clandestinum	Australia		
Setaria sphacelata	Uganda		
Setaria sphacelata	Zambia		
Sporobolus pyramidalis	Uganda		

Table 12.7 Relationship between classes of livestock and definition of animal units. (After Moorhouse, 1975.)

Cattle	Animal unit equivalents	Speep and goats	Animal unit equivalents
400 kg steer	1.0	40 kg wether	0.125
1–8 month calf	0.35	Lamb under 1 year	0.06
8–12 month weaner	0.40	Maiden female	0.125
12-month steer	0.85	Breeding female	0.2
Breeding/milking cow	2.0	Ram or Buck	0.2
Bull	2.0		

Example: Mixed stocking with 5 breeding cows, 5 breeding ewes and 8 wether goats on a 50 ha paddock

$$= \frac{10 + 1 + 1}{50} = 0.24 \text{ animal units/ha.}$$

Jones and Sandland (1974) established a highly significant linear relationship between stocking rate and liveweight gain, applicable primarily to more sophisticated ranching conditions, according to the equation:

$$Y = a - bx$$

where Y is a gain per animal unit, stocking rate is x, and a and b are constants. Using this equation these authors tested the validity of the model on 33 data sets and showed that, for beef cattle and sheep, the data all fitted the linear equation very uniformly ($r = -0.992$). The general application of this simple equation represents an important advance in the understanding of the stocking rate:animal productivity relationship and the interpretation and analysis of grazing trials. The model predicts that zero animal production is reached when the stocking rate is double that required for maximum gain per hectare.

Jones and Sandland (1974) suggested that this linear model has its best application under longer-term grazing studies and that there may be differences in predicting optimum stocking rates with short grazing periods on high quality pastures. However, as rotational grazing on good quality swards is rarely encountered with most tropical grasslands, this problem is less likely to be encountered in the tropics than in temperate grassland systems. The authors also suggested that another application of the linear equation would be to simplify the design of stocking rate trials in that even if only two rates were used they could nevertheless give a good indication of the optimum. The robustness of the Jones and Sandland equation is discussed at length by Butterworth (1985), and more recent work suggests that more emphasis should be placed on total dry matter (TDM) production per unit area as a determinant of carrying capacity of the land rather than on stocking rate trials *per se* (Leeuw and Tothill, 1990).

Pasture management

Pasture management normally involves the division of the land into defined grazing areas or 'ranges' (usually of relatively large size, and necessarily related to water supplies) and the introduction of some form of rotational grazing coupled with the controlled use of fire to prevent bush regeneration or encroachment. The rotational grazing system adopted will vary with local conditions, but it should be kept as simple as possible, minimising the number of ranges and the frequency of stock movements. In a simple system introduced in Kenya the land was divided into four blocks; in each year three blocks were grazed in rotation for 4 months each, leaving one block for resting and burning.

Fencing is most desirable but usually too expensive, and the grazing areas are usually defined by natural features, the stock being controlled by herding. Alternatively the boundaries may be demarcated by lines of felled thorn trees or with live hedges or fences. The latter take time to establish and may be unreliable due to breaks in continuity, or to developing an open base as they mature. In most areas a compromise is possible if the demarcation plan is carefully thought out. An area of land may be delimited by a river on one side and a mountain range on the other. Cheap transverse divisions may be possible by running barbed wire on live trees and bushes, selecting a traverse where the bush is thickest. When bush clearing is to be done, it is important to plan for the siting of range boundaries, as it may be possible to leave thin strips of bush which can be made impenetrable by a crude system of barbed-wire fencing, alternating with rows of heaped-up thorn scrub. Provided the stocking rate is adjusted to the carrying capacity of the land, and animals are moved on before the herbage is in short supply, the standard of fencing required in extensive grazing systems can be much lower than that needed for intensive cattle management.

In view of the many disadvantages of communal grazing, it may be thought that a change to individually owned holdings is desirable in order to obtain good pasture management. However, in many pastoral areas individual holdings are impracticable for several reasons. First, seasonal migration of stock is often essential because large areas of pasture can only be used for part of the year. Second, where water is scarce the provision of water to individual holdings is often impracticable or uneconomic. Third, individual holdings would often be unacceptable because few pastoral regions are uniform. Under individual tenure some people would get only poor grazing, whereas under the communal system of common grazing all have equal access to good and poor areas. Fourth, the fencing of pastures on individual holdings would be expensive and often impracticable. In some circumstances, it may be sensible to divide large communal grazing areas into smaller units that can be effectively managed collectively by small groups of people.

It has been mentioned earlier that mixed grazing with cattle and goats is often advantageous in grasslands containing shrubs and trees, but this involves certain practical difficulties. In order to improve the composition and productivity of the pasture it is necessary to adopt a suitable ratio of cattle to goats, and this ratio will not necessarily remain constant. For example, a proportion of 1 steer to 20 goats may initially be suitable on grassland containing much thorn bush, but after 10 years the goats may have virtually eliminated the thorn bush, and a ratio of 1 steer to 5 goats would then be appropriate. Goats are difficult to confine within a range, since they can leap over 1.5 m fences and scramble through hedges unless restrained in some way (see Chapter 13). Goats are also more susceptible to the depredations of wild animals and theft. Consequently, it may be necessary to provide secure shelter for goats in places where this is not necessary for cattle.

Rehabilitation of denuded areas

In many places deterioration due to prolonged mismanagement has gone beyond the stage where improvement can be effected by introducing appropriate stocking rates, rotational grazing and control of bush regeneration and these management techniques need to be preceded by special measures to rehabilitate the grazing areas. Thinning over-dense bush may be needed to permit improvement of the grass cover and perhaps also for tsetse control. Soil conservation measures may be required on a large scale, especially where gullying has occurred, and should form part of a co-ordinated rehabilitation programme. Where pastures have been severely overgrazed it may be necessary temporarily to close them to grazing to allow regeneration of the grass cover. This can usually be done readily if the areas concerned are small, as where small hills or steep slopes need localised soil conservation works and regeneration of the grass cover in order to check erosion which is affecting surrounding lower land. With more extensive areas, different sections will have to be closed sequentially over several years, usually accompanied by a general reduction in stock numbers.

The productivity of natural pastures largely depends on their content of perennial grasses. Annual species, although palatable and nutritious when young, rapidly become mature and unpalatable and die after producing seed. Where the perennial grasses have been largely eliminated by overgrazing, it will be necessary to re-establish suitable perennial species. The species and methods to be used in reseeding will vary with the degree of denudation and the climatic and soil conditions of the site, and the subject has been extensively reviewed by Whiteman (1980). Some preparatory cultivation is normally necessary, especially as the eroded soil surface is usually capped by an impermeable crust from which most of the rainfall runs off, carrying away seed and seedlings during heavy storms. It is desirable to do the minimum amount of cultivation necessary as cheaply as possible, but the amount required for satisfactory results usually increases as the expected rainfall decreases. Methods which have proved successful have ranged from digging small holes with hoes or contour scratch-ploughing at intervals of 1–4 m, through disc- or tine-harrowing contour strips, to full time cultivation or disc ploughing. In the Kitui district of Kenya, large areas of denuded grazing land were successfully rehabilitated by scratch ploughing at intervals and reseeding, after thinning the bush.

Covering the land with cut branches and brushwood after seeding has sometimes proved beneficial, especially where little cultivation was done, since it breaks the impact of rain drops on the soil surface, reduces soil and seed loss by wind and water and provides shade. Where patches of land require reseeding within a larger area which is not closed to grazing, the brushwood will also give localised protection against grazing and trampling. Another method is to allow the development of a cover of annual grasses before sowing perennials, usually in the following season, but this procedure involves delay and may allow the development of undesirable weeds.

The seed used may be a mixture collected from locally adapted, perennial grasses, but experience has shown that some species are usually more successful than others. Thus, in Kenya, where much reseeding has been done in areas with rainfall ranging from 600 to 900 mm/yr, *Cenchrus ciliaris* has proved to be most generally successful, other useful grasses being *Chloris gayana, Panicum maximum, Eragrostis superba* and *Bothriochloa insculpta* (Sands *et al.*, 1970). No legume proved reliably persistent, but some success was obtained with *Neonotonia ternata, Glycine wightii, Stylosanthes guianensis* and *S. humilis*. It is usually advantageous to treat seed with an insecticide such as BHC (benzene hexachloride) against the depredation of insects, especially harvester ants. Pelleting of grass seed with lime or rock phosphate, soil or dung may help to reduce seed loss especially on sites subject to severe wash or high ground-level wind velocity (Jones, R.M., 1975). Seed rates are usually about 12 kg/ha and sowing is best done at the beginning of the rains; dry seeding is only advisable where the time of the onset of the rains is predictable. Grasses which are established vegetatively, such as *Cynodon dactylon*, should not be planted until the soil is thoroughly wet, and vegetative propagation is usually less reliable and much more costly than seeding.

Water supplies

Often an essential prerequisite for improvement will be the provision of well-distributed water supplies, but such improvement does not necessarily lead to long-term sustainable increases in animal productivity. Ideally watering points should not be further than 10 km apart, to restrict the distance over which animals have to travel and to avoid erosion near a watering point by limiting the cattle using it to 1250–1500 head. Swift (1975) observed in Mali that vegetation is devastated within 20–35 km around permanent watering places owing to excessive human and livestock populations concentrating in the dry season. The cost of providing water supplies is important, particularly as there are large areas where the pasture is so poor that it will not pay for even a moderate outlay on water, or where the water is so inaccessible that its supply becomes too expensive.

For these and other reasons it is important that

new watering points should form part of a coordinated plan for improved land use. They must be sited in relation to the situation and carrying capacity of available pastures, and be accompanied by control of stock numbers and, where appropriate, the introduction of a system of rotational grazing. If this is not done, then more water supplies will merely make matters worse by allowing overstocking and overgrazing to continue with increased severity. Tanks, hafirs and small earth dams, collecting runoff from a catchment area during the rains and storing it for dry-season use, are the cheapest forms of water supply. Such reservoirs should be fenced, and the water piped from them to troughs, otherwise large numbers of cattle approaching the water's edge are liable to break down the banks and create swampy patches. It is important to avoid overgrazing of catchment areas, with consequent erosion and silting up of the reservoirs, but this is difficult to achieve because herdsmen are liable to keep cattle near the reservoir during the dry season as long as any water remains. Boreholes with pumps to bring underground water to the surface, although more expensive, are more satisfactory since the water can be turned off (with the consent of the herdsmen) when it becomes desirable to force the stock elsewhere in order to prevent overgrazing. The same applies to the supply of water by pipelines from perennial streams and springs. These can often be used to bring water from mountains, where the rainfall is good, to neighbouring lower and drier lands. Finally, it should always be remembered that there is often competition between the human and livestock populations for limited water supplies.

Animal husbandry, disease control and livestock development

The most important improvements in animal husbandry that are needed are culling and castration (to permit selective breeding), control of calving times, better dry-season nutrition and correction of mineral deficiencies. These matters are dealt with in Chapters 17 and 18. Possibilities for improving the provision of dry-season feed have been mentioned earlier in this chapter and are dealt with fully in Chapter 11, but they are strictly limited in the drier pastoral areas. Where seasonally flooded or swampy land is available, or where there are small irrigable areas, advantage should be taken of opportunities to grow fodder crops. The alternative, feeding protein supplements, is rarely a practicable proposition as such feed is not usually available locally at a price which pastoralists can afford.

Improved land use in the wetter pastoral areas

At the beginning of this chapter it was mentioned that, although the majority of the pastoralists live in drier zones, pastoral tribes are found to a more limited extent occupying grasslands derived from the wetter broad-leaved woodlands and montane evergreen forests. Usually the degree of denudation which has occurred under these latter conditions is less than in the drier regions. Improvements in these regions of better rainfall, which usually have fairly fertile soils and are suitable for crop production, should be in the direction of more productive land use than can be achieved by pastoralism or ranching, and might be along the lines of dairy farming, mixed farming or the production of perennial crops.

FURTHER READING

Bogdan, A.V. (1977) *Tropical Pasture and Fodder Plants*. Longman: London.
Butterworth, M.H. (1985) *Beef Cattle Nutrition and Tropical Pastures*. Longman: London.
Humphreys, L.R. (1991) *Tropical Pasture Utilisation*. Cambridge University Press: Cambridge.
Tothill, J.C. and Mott, J.J. (eds) (1985) *Ecology and Management of the World's Savannas*. Australian Academy of Science: Canberra.
Whiteman, P.C. (1980) *Tropical Pasture Science*. Oxford University Press: Oxford.

13
Classes of Tropical Livestock

P.N. Wilson

This chapter is confined to a brief consideration of those species of farm animals which are relatively unimportant in temperate agriculture, but which are of great importance in the tropics. Sheep, horse kind, pigs and poultry will not be discussed in detail, since these domesticated livestock are common to both temperate and tropical agriculture and they are adequately dealt with in many standard reference works. They will also be referred to in the appropriate sections of subsequent chapters, when some specific aspects of the management of these classes of livestock are considered in more detail.

The omission of sheep, horse kind, pigs and poultry should not imply that these species are unimportant in the tropics. Indeed, in many parts of South America, Africa and Asia sheep and goats are, together, of greater economic and nutritional importance than cattle. Also, in those areas of the tropics which are rapidly developing their agricultural economies, the most dramatic increases in the poultry and pig populations invariably precede parallel development and expansion of the populations of dairy and beef cattle.

The distribution of the different species of farm animal differs in the tropics from that in temperate regions. The Indian zebu (*Bos indicus*) cattle are more abundant than the non-humped, European cattle (*B. taurus*). The dual-purpose goat is generally more important than the single-purpose hair sheep. Draught animals, bovines as well as horse kind, are still of great importance in many areas and there is a large number of species of local importance employed for this purpose, ranging from the Indian elephant to the South American llama, including the important beast of burden of North Africa and the Middle East – the Arabian camel or dromedary. In Asia, the water buffalo ranks as a most important triple-purpose animal and, in the subtropical parts of North America, the wild American bison has been successfully crossed with European-type cattle to form a hybrid beef-type animal known as the cattalo or beefalo, which recently spread, albeit in a very limited manner, into other subtropical areas. Commercial claims as to the fertility and productivity of these hybrid beef animals should be treated with great care. There are at present insufficient independent scientific data to substantiate many of these claims (Mason, 1975).

Swine and poultry are becoming increasingly important in many tropical countries. However, as livestock management systems change from crude forms of 'backyard enterprise' subsistence farming to more intensive husbandry practices managed by progressive agriculturalists, so we find a swing away from the so-called 'indigenous' breeds to widespread use of imported temperate-type breeds and strains. Thus the hybrid-type broilers and laying hens found in Africa, Asia or Latin America today are genetically identical to their counterparts in Europe and North America, although their housing, feeding and management systems may differ. This chapter will deal primarily with those classes of livestock which are different, in some important manner, from their temperate equivalents. Readers seeking a more detailed knowledge of pigs, poultry and horse kind are advised to consult standard temperate literature, or refer to Payne (1990) for a tropical treatment of the subject.

CATTLE

The Food and Agriculture Organization of the United Nations (FAO) estimates that the world cattle population is about 1.3 billion. Of this total figure approximately 556 million are distributed in tropical countries and the remainder in temperate zones. A rough breakdown of this distribution by continents, ranked as either temperate or tropical, is shown in Table 13.1. It is of interest that, since the first edition of this book was written, the total world cattle population has increased by over 100 million, somewhat less than the increase in the human population.

No accurate figures for the numbers of European-type and zebu-type cattle are available, nor is the size of the crossbred population (*Bos taurus* × *B. indicus*) known. Rough estimates place the zebu-type cattle in a distinct majority in tropical regions with a population of the order of 400 million. European-type cattle probably number about 50 million in the tropics as a whole, with an increasing population of crossbreds currently numbering about 100 million. As we shall see later, it is almost certain that the number of crossbred cattle will increase during the next few decades, at the expense of the zebu-type cattle population, since crossbred breeding programmes are now being pursued in many tropical countries.

Table 13.1 World distribution of cattle (millions). (Source: FAO, 1986d.)

Temperate regions		Tropical regions	
Africa	22	Africa	155
Americas	207	Americas	232
Asia (except Russia)	207	Asia	162
Europe (except Russia)	132	Oceania	7
Russia	121		
Oceania	24		
Total	713		556

In spite of their numerical superiority, the zebu-type cattle found in tropical countries contribute about one-half of the total cattle production consumed by man. However, the production of meat from tropical countries is rising at a greater rate than that from developed countries. de Haan (1993), quoting data from the FAO, showed that the percentage rise in red meat production in developing countries over the period 1962–88 had been 23% per year, compared to 1.9% per year in developed countries. Comparable figures for 'white' meats (pig and poultry) were 6.8 and 3.1%. The total world production of all meats in 1988 was 155 Mt, of which 56 Mt originated in developed countries compared to 99 Mt in developed countries. Two factors are responsible for the relatively low productivity of zebu cattle. First, the present production capacity of tropical cattle, whether for milk or meat production, is less than that of European-type cattle. Second, a relatively small percentage of tropical cattle owners tend their stock in order to exploit them for commercial reasons. Each cattle-keeping community raises stock for a variety of purposes, and the importance given to each purpose varies widely, sometimes within the confines of a small geographical area. It is true that, in the tropics as elsewhere, cattle are essential as sources of food, but even when dealing with this basic reason for cattle-keeping, religious and social customs abound which severely constrain maximum exploitation. For instance, socio-religious principles prohibit the eating of cattle-meat by Hindus, and socio-economic customs encourage the drinking of cattle-blood by certain African nomadic tribes, such as the Karamajong and the Masai of East Africa. In this context Hall and De Boer (1977) reviewed acceptable methods of increasing ruminant productivity in Asia, and de Haan (1993) has reviewed the determinants of success in livestock development projects in developing countries.

It is important to understand and appreciate the significance which cattle, more than any other class of farm livestock, have to the majority of the inhabitants of Africa and, to a lesser extent, in Asia. Cattle are part of their way of life, an integral part of their religion and a dominating factor in their social organisation. In the Ankole district of Uganda, for instance, until recent times the whole basis of the social structure of the tribe revolved around the 'unit' of the cattle kraal. A workable number of cattle was between 100 and 200 head, and if the cattle numbers increased then a son would be encouraged to marry, leave home and start a new kraal, taking with him for this purpose about 100 head of cattle. If a herd owner died without issue, then near relatives of the deceased would be ordered to take over responsibility for that kraal. Human social behaviour, who and when to marry, how many cattle to tend and where to live were, therefore, largely dictated by the husbandry requirements of the cattle herd. It follows that any attempt to change systems of cattle husbandry in such communities would have interfered, to a lesser or greater degree, with the social customs and religious beliefs of the people. There is, therefore, a real need for every animal husbandry adviser and agricultural planner to have a fair working knowledge of the social customs of cattle-owning people if technical change is to be successfully brought about without encountering opposition. This knowledge is much more important in tropical countries than in temperate regions, since cattle play a greater part in the social life and religious beliefs of the people living in such areas.

In addition to the important part which cattle play in the social lives of many tropical peoples, cattle in Africa are generally prized more for their quantity than for their quality. A great deal of social prestige is attached to the ownership of large herds, irrespective of the breeding value and productive efficiency of the animals comprising the herd. This is not surprising when it is understood that cattle are often regarded as a 'bank account on the hoof'. Ten poor-quality, old scrub cows are, in the eyes of many tropical communities, worth double five healthy, well-bred, productive heifers. The practical reason for this is that, until comparatively recently (and currently in some areas), the cattle populations fluctuated widely as drought, epidemic diseases and natural predators decimated cattle numbers. The more cattle a person possessed before calamity struck the more survivors might remain when the conditions improved. Large cattle numbers were, therefore, a form of biological insurance, especially as large herds were often split up and distributed over a wide geographical area so that if disease broke out in one location, animals in an adjoining area were likely to escape unscathed.

In Asia and the Far East the attitude towards cattle is quite different from that in parts of Africa. Cattle in Asia are only kept for what they can produce, be it meat, milk or work, and for many farmers it is capacity to work on the land and to produce milk which are most important. Consequently, strength, conformation and docility under the yoke are of prime importance. In many areas agriculture, including the carting of produce to markets, could not exist without oxen to pull the tillage implements and the carts. Tractors do not meet the needs of small farmers, who predominate in Asia. There are problems of capital cost, expertise, spare parts and, very importantly, replacement when the tractor is worn out. Oxen (Fig. 13.1), on the other hand, are self-replacing at little cost to the farmer and that cost is spread over several years. And when too old to work cattle provide meat which a worn-out tractor will never do.

Evolution of cattle

All domesticated cattle belong to the Bovidae family. The genetic relationship of the various species of the genus is not known with any certainty, and the following account is merely a survey of the literature on the subject. A full treatment of the present state of knowledge of the evolution of domesticated cattle was given by Mason (1984). There is a need to develop further the genetic methods of 'blood grouping' and 'DNA finger printing' in order to study the diversified cattle population of the world in more detail, so that one may trace their origins and migratory movements with a greater degree of accuracy.

The Bovidae family, as represented today, may be conveniently subdivided into four sub-

Fig. 13.1 Ox ploughing at Kitale, Kenya. Cattle are a very important source of draught power in many tropical countries.

groups or genera, known as taurine, bibovine, bisontine and bubaline.

Taurine subgroup

This subgroup includes the two most important domesticated cattle species which exist today – *Bos taurus*, or European-type cattle, and *B. indicus*, or zebu-type cattle. The former are thought to be descendants from urus (*B. primigenius*), one of the first wild animals to be domesticated by man and exploited as a triple-purpose animal. The evolution of *B. taurus* cattle from *B. primigenius* is thought to have taken place in mid- and northern Europe and southwest Russia. Urus originated on the west coast of the Pacific Ocean and spread through Asia and Europe to the eastern coasts of the Atlantic. Urus cattle can be traced through about 1 Myr of geological time, from the Pleistocene period right up to 1627, when the last known living representative died in captivity in Poland. *Bos indicus* cattle are thought to be descendants of *B. nomadicus*, a native of Asia considered to have been domesticated later than *B. primigenius*. *Bos indicus* probably evolved from *B. nomadicus* in south-central Asia. However, some workers regard both *B. taurus* and *B. indicus* as descendants of *B. primigenius*. Another member of this taurine subgroup is *B. opisthonamus*, a non-domesticated species once native to North Africa. *Bos opisthonamus* was the only taurine representative indigenous to the African continent, but it is now extinct, having been displaced by various migrations of other cattle species from the east. No satisfactory explanation has been given of the ancestry of zebu-type, non-humped shorthorn cattle which are found today in parts of the north Mediterranean coastline, China and Mongolia. These cattle form a quite distinct species, classed by some authorities as *B. brachyceros* (Jeffreys, 1953).

Bibovine subgroup (genus Bos bibos *or* Bibos)

This group includes several semi-domesticated species found in the Far East, in particular the gayal (*Bos bibos frontalis*) of Assam and parts of Burma (Simoons, 1984) and the banteng (*Bos bibos javanicus*) of Sabah and Indonesia. A review of banteng cattle in Sabah has been provided by Copland (1974). Other representatives of this subgroup are the gaur (*Bos bibos gaurus*) which is found in Assam, Burma, Bangladesh and parts of Malaysia, and the kouprey (*B. sauveli*) found in Cambodia. The gaur (*Bob bibos gaurus*) is not used for draught purposes and, until very recently, was never milked. It appears to have been domesticated exclusively for sacrificial purposes. As with cattle in many parts of the tropics, the gaur is currently used for paying bride prices, ransoms and fines, purchasing land or houses and as an item of barter.

The gaur possibly gives a clue to the history of domestication of other species of cattle. It is not known precisely where or when the first wild animals were captured, tamed and bred. We do know that man domesticated sheep and goats much earlier than cattle. The oldest remains of domesticated cattle have been found in Thessaly and date back to about 7000 BC (Protsch and Berger, 1973). Sheep and goats had already been domesticated and these, together with farm crops, supplied mans' principal nutritional needs. The ox would have been superior as a draught animal, but there is no evidence that animals were used in this way until about 3000 years later. However, the wild ox was often employed as the symbol of strength and virility and Mason (1976) has suggested that it was, therefore, for cultural and religious reasons that man first captured and domesticated, not only the gaur, but also other bovine species.

Bisontine subgroup

The only important member of this subgroup is the bison (*Bison bonasus*) which has been crossed with *Bos taurus* cattle to form the new hybrid, the cattalo or beefalo. Records of cattalo productive performance are not easily come by, and most are unsubstantiated, but it appears that cattalo calves are smaller at birth than purebred *B. taurus* calves, but make better gains from birth to weaning. Cattalo slaughter-stock apparently obtain lower grades than European stock, but are said to have a higher dressing-out percentage due to the lower weights of their hides and non-edible viscera. Another, non-tropical, member of the subgroup is the yak (*Poephagus mutus*) which is an important domesticated draught animal in Tibet.

Bubaline subgroup or Bovinae subfamily

This subgroup (or subfamily) is of great importance in India and other parts of the Near East and Southeast Asia. The nomenclature of the different species is still in dispute by systematic zoologists, but Payne (1990) recognised four distinct wild species and placed them in a separate genus, *Bubalus*. The most important domestic species is classified as *Bubalus bubalis*, commonly known as the water buffalo because of its habit of wallowing in the shallow water of rivers and swamps. Two species of lesser important are *B. mindorensis*, the tamarao or carabao indigenous to the Philippines, and the celebes buffalo, *B. depressicornis*, also known as the anoa or the dwarf buffalo of Borneo. Early writers include the African buffalo (*B. caffer*) in this bubaline subgroup, but later workers classify this species into a different genus, *Syncerus*.

It should be noted that various authorities have stated a case for a more systematic approach to the utilisation of the very large reservoir of wild game animals, present in Africa and other tropical regions, which belong to the Bovidae family. It is suggested that these animals could be 'ranched' in the wild state, and regularly 'cropped' as young stock reach slaughter-weights (e.g. Skinner, 1973; Young, 1973). If this proposal finds acceptance with those responsible for the policy affecting wild game, it may well be that *Syncerus caffer* (African buffalo) and possibly also *Bison bonasus* (European bison) will have to be regarded in the future as semi-domesticated livestock. It is also interesting to note that deer are now ranched in Europe and New Zealand as an extension to the current practice of extensive husbandry of beef cattle and sheep in these areas.

Differences between *Bos taurus* and *Bos indicus* cattle

There are many excellent textbooks describing the characteristics of European cattle (*Bos taurus*), and thus no specific description of this species will be given here. There is, in any case, prolific literature on the subject. A brief description of zebu (*Bos indicus*) cattle is, however, appropriate at this stage and those interested in a more detailed treatise should consult Mason and Maule (1960) and Payne (1990). Excellent bibliographies of the subject are provided in these two works.

Just as there are numerous well-defined breeds of European cattle, so there are also many different breeds and types of zebu cattle recognised in the tropics. However, the classification of the zebu cattle population is arbitrary, the same type of animal often being given a variety of separate names in different parts of the tropics. No attempt will be made to classify or define the various breeds of zebu in this brief survey, but it is sometimes useful to divide the species into two distinct subgroups on the basis of the anatomy of the hump (Milne, 1955).

(1) *Neck (or cervico-thoracic) humped zebu*
This subgroup has the hump placed in position anterior to the forelegs. The hump is composed of extensions of the rhomboideus muscle, with little intermuscular or intramuscular fat, and appears to function as an aid to the traction of the animal, giving a greater leverage to the action of the forelegs. This subgroup often possesses long horns, and for this reason is sometimes referred to as the 'long-horned zebu'. The best known and widely distributed example of this type is the Africander of South Africa, but other members are the Nilotic, and Setswana cattle of Northeast and Southeast Africa.

(2) *Chest (or thoracic) humped zebu*
This subgroup has the hump placed in position just over, or posterior to, the forelegs. The hump is composed of both muscular and adipose (fatty) tissue, extensions of the rhomboideus and trapezius muscles ramifying into a large quantity of subcutaneous fat. The structure appears to serve little or no tractive purpose, and is regarded by most workers as an energy-storage organ. This subgroup generally possesses short horns and is, therefore, sometimes referred to as 'shorthorn zebu'. There are numerous Indian and African breed names given to this type, some of the more important being the Sahiwal, Hallikar, Gir and Red Sindhi cattle of India (Fig. 13.2) and the Nkedi, Lugware and Nandi cattle (Fig. 13.3) of East Africa.

There is some evidence that when either of these two subgroups of zebu cattle is crossed

Fig. 13.2 An Indian zebu bull at Boa Vista, Rio Branco Province, Brazil. Notice the lack of vegetation in the background of the picture. The green leaves on the trees are inedible. The bull is about to be fed a mineral mixture, the only 'supplementary feed' provided during the dry season.

Fig. 13.3 A herd of Nandi zebus being demonstrated at the Baraton Livestock Station, Kenya.

with the humpless, hamitic longhorn cattle found in parts of the Middle East, North Africa and Eritrea, a stabilised crossbred is produced, known as the sanga. When neck-humped zebus are crossed with humpless longhorns, the resultant crossbred may be described as neck-humped sanga, the hump being much reduced in size compared to the zebu type. The best known example of the neck-humped sanga is the Ankole breed of cattle found in southwest Uganda. Similarly, when the cross is made between chest-humped zebus and humpless longhorns, the resultant crossbred is known as the chest-humped sanga, several examples of which may be found in Ethiopia.

The above classification may well be an oversimplification, and it should be noted that no two authorities agree on either the nomenclature of the subgrouping of the *Bos indicus* species or on the definition of the sanga group of cattle (see Payne, 1990; Maule, 1990). For a more detailed description of African and Indian *B. indicus* breeds the reader is referred to Mason and Maule (1960) and Payne (1990).

The points of comparison shown in Table 13.2 bring out the chief differences between *Bos taurus* and *Bos indicus* cattle. Certain of these differences have probably been produced as a result of artificial selection, particularly animal breeding by man, since the species was domesticated. Thus, items (7), (10), (16) and (19) have probably been modified, to a greater or lesser extent, by man. Some of the features noted are of little or no significance for productive performance or adaptation of the species to its environment, but the majority are of importance in one or other of these respects.

WATER BUFFALO (Bubalus Bubalis)

The domestic water buffalo is the third most important species of the bovine family, with a total world population (in 1982) of about 121 million (see Table 13.3). Unlike cattle, water buffaloes are restricted in their distribution to certain regions, principally in Asia. The present breeds are thought to be descended from the wild buffaloes which still inhabit the swamps and forests of India, Bangladesh, Pakistan and Malaysia. Indigenous buffaloes were probably domesticated long before non-indigenous zebu cattle moved into these regions from the west. The buffalo appears to have been domesticated

Table 13.2 The chief differences between *Bos taurus* and *Bos indicus*.

European (*Bos taurus*) cattle	Zebu (*Bos indicus*) cattle
(1) No hump.	Hump present in thoracic or cervico-thoracic region.
(2) Rounded ears, held at right angles to the head	Long drooping ears, pointed rather than rounded.
(3) Head short and wide.	Long and comparatively narrow head.
(4) Skin held tightly to body. Dewlap, umbilical fold and brisket small.	Skin very loose, often falling away from body in folds. Dewlap, umbilical fold and brisket extensively developed.
(5) Skin relatively thick, average thickness 7–8 mm.	Skin relatively thin, average thickness 5–6 mm.
(6) Large amounts of subcutaneous fat, especially in mature animals.	Relatively small amounts of subcutaneous fat, at all stages.
(7) Back line straight or relatively straight.	Back line high at shoulders, low behind hump, high over pin bones, sloping down markedly over tailbud.
(8) Hip bones wide and outstanding.	Hip bones narrow and angular.
(9) Thoracic ribs well sprung away from body.	Thoracic ribs poorly sprung, forming an angle with the vertebral column.
(10) Udder long, with a flat sole, well suspended between and behind the hind legs.	Udder more rounded with a curved sole, poorly suspended and carried in front of, rather than between, the hind legs.
(11) Hair fibres non-medullated, so that they are held limply on the skin surface.	Hair fibres usually medullated so that they tend to stand more erect and away from the skin.
(12) Hair relatively long, rough and double-coated. Seasonal differences in hair length. Average hair population density 800 follicles/cm^2.	Hair relatively short and smooth-coated. Little or no seasonal difference in mean hair length. Average hair population density 1700 follicles/cm^2.
(13) Hair and skin generally both pigmented or both non-pigmented.	Skin usually pigmented irrespective of colour of overlying hair.
(14) Legs short. Animal slow moving.	Legs longer. Animal faster moving.
(15) Skin and hair attractive to most cattle ticks.	Skin and hair much less attractive to cattle ticks.
(16) Fast maturing. Full mouth at or before 4 years.	Slow maturing. Full mouth rarely before $5\frac{1}{2}$ years.
(17) Milk yield, lactose content and nitrogen content drop when ambient temperatures reach or exceed 24°C.	Milk yield, lactose content and nitrogen content do not drop until ambient temperatures reach or exceed 35°C.
(18) 'Comfort zone' between 4°C and 15°C.	'Comfort zone' between 15°C and 30°C.
(19) Adult animals relatively large; fully grown bulls commonly reach 1000 kg liveweight.	Adult animals relatively small; fully grown bulls of most breeds rarely exceed 700 kg.

around 5000 years ago, as evidenced from the archaeological excavations of the Indus valley and the Deccan plateau in India. A cave painting discovered in central India revealed that the buffalo was known to Early Palaeolithic man, (Noorani, 1976). The domestication of the swamp buffalo is thought to have taken place in China about 1000 years later. The native word for buffalo in Malay, 'kerbau', indicates that these animals were well prized for their culinary properties, since 'bau' literally translated means 'superior meat'. A good general discussion on the husbandry and health of the domestic buffalo has been provided by Cockrill (1974), while

Table 13.3 World distribution of buffaloes (millions). (Source: FAO, 1983.)

Region	1965	1982
Africa	1.6	2.4
North, Central and South America	0.1	0.6
Russia	0.4	0.3
Asia	110	118
of which:		
China	28	18
India	52	62
Indonesia	3	3
Nepal	3	4
Pakistan	7	12
Philippines	3	3
Thailand	5	6
Vietnam	2	3
Total world	112	121

Bhattacharya has contributed a useful chapter on the characteristics, adaptability and nutrition of the buffalo in Payne (1990).

The domesticated species of water buffalo are now widely distributed throughout the northern tropical and subtropical regions of Asia, the countries bordering the eastern Mediterranean, Southern Africa, Australia and in isolated islands such as the Philippines. Furthermore, there are a number of countries where the buffalo has multiplied surprisingly fast. Such countries as Trinidad, Peru, Surinam, Guyana, Venezuela, Brazil and Australia deserve special mention. The largest concentration of water buffalo is found in India (about 62 million in 1982) and the second largest in China (about 18 million in 1982). It is not generally realised that the species is also found in Europe, some 800 000 buffaloes being distributed in southern European countries, particularly Italy and Romania. The FAO (1983) indicated that during the previous 17 years the worldwide buffalo population had increased from 112 to 121 million.

The water buffalo occupies an important place among the domestic animals of the tropics and may now be regarded as a truly triple-purpose animal. Meat, from young slaughter-stock, is of good quality and is palatable even to those who are mainly used to eating high-quality beef from European cattle breeds. Milk yields are comparatively high by tropical standards and the butterfat content is about double that of European milk breeds. The water buffalo is strongly built, achieves large mature liveweights and is capable of pulling very heavy loads through conditions which would prove quite impossible to equines, taurine cattle or light tractors. Buffaloes usually weigh heavier at birth (approximately 30 kg) than zebu calves and are slightly lighter compared to the heavier breeds of cattle of European origin. In the traditional systems of management, wherever these animals are raised, the average growth rate of buffaloes ranges from 170 to 400 g/day from birth to 15 months of age, which is usually the recommended age for slaughter in Western countries (Nagarcenkar, 1975).

In contrast to zebu cattle, the water buffalo is a large-boned massively built animal with a deep, low-set body, set on strong legs with large hoofs. All buffaloes have large horns of different shapes. Swamp buffaloes normally have back-swept horns while water buffalo horns are usually curled. Young buffaloes usually have soft grey-brownish hair which is 2–3 cm long and lies near the skin to provide a relatively complete cover. As the animal grows the hairs become more widely separated, so that adult hairs, 3–5 cm long, are sparsely scattered and provide little or no insulation. The skin is thick and invariably black, so that absorption of solar radiation is high. In direct sunlight the skin temperature may reach 50°C. Because of the melanin granules present in the coloured skin, ultraviolet light has little harmful effect on the underlying tissues and the animal is more sensitive to high temperature than to high radiation intensity.

The Egyptian water buffalo is usually dark grey in hair colour, whereas adult buffaloes found in Asia vary greatly from almost black to a pinkish off-white. There is also considerable variation in horn shape, length and weight, and differences in horn type are convenient aids in distinguishing between the various water buffalo breeds (Rife, 1962).

The skin area:body mass ratio of water buf-

Fig. 13.4 Buffaloes wallowing in a river in India. (Source: Indian Council of Agricultural Research, New Delhi.)

faloes is similar to that of cattle, but buffaloes have only about 20% of the sweat gland population density of *Bos taurus* breeds. The adult Murrah buffaloes have on average 87 sweat glands/cm^2 of body surface area, while the zebu breeds of cattle have ten to fifteen times more sweat glands per cm^2 (K. Gulshan, M.G. Govindaiah and R. Nagarcenkar, unpublished data). Because of their inefficient sweating mechanism, buffaloes rely primarily on respiratory evaporation and wallowing as a means of maintaining homeothermy. Buffaloes also have a higher haemoglobin content of the blood compared to cattle, although the significance of this physiological difference is obscure.

The water buffalo species may be conveniently divided into two major types, the river buffalo and the swamp buffalo. These two types have marked differences in anatomical characteristics and habit and crossbreeding between the two types is uncommon.

The *swamp buffalo* is the lighter animal and it possesses a small hump situated over the shoulder. It is chiefly found in Malaysia, where it is employed primarily for draught purposes and, to a lesser extent, for meat and milk (Hua, 1957). The swamp buffalo is a semi-aquatic, nocturnal animal which spends the hotter period of the day, from 10 AM to 4 PM, semi-submerged in natural swamps or self-made wallows. Owing partly to its physiologically relatively inert skin, the swamp buffalo is unsuitable for work when exposed directly to the sun. In order to maintain body condition, it requires a wallow for wetting the skin so as to increase the evaporative cooling as a means of decreasing the skin temperature (Fig. 13.4). Attempts to impose a different diurnal behaviour pattern on swamp buffaloes, or to deny them access to swamps or wallows, are usually unsuccessful. Such action can lead to a number of ill-effects, particularly cessation of breeding, increased calf mortality, incidence of 'joint-ill' and a reduction in growth rate. The usual management of swamp buffaloes is designed to conform to their normal behaviour pattern as closely as possible and they are commonly used for such tasks as puddling rice paddies and hauling timber during the early and late parts of the day. They are used for lighter work at any hour of the day, but when exposed to direct sunlight in the course of such work, a good stockman will reduce the heat load by giving the buffaloes a quick shower or a water bath at regular intervals.

The *river buffalo* is a larger and more versatile animal, with a less rounded body conformation. It is used primarily as a milch animal and used for draught or meat purposes to a much lesser extent than the swamp buffalo. Most of the important domesticated breeds of buffalo belong to this type. The hump is not so well defined as in the case of the swamp buffalo. The behaviour pattern is similar to that of taurine cattle and the provision of swamps and wallows, although greatly appreciated, is not as necessary for the physical well-being of these stock. The major breeds of riverine water buffalo may be distinguished by their horn characteristics and facial profile and to a lesser extent by their size and the colour of their hair. The five most important buffalo breeds are as follows:

(1) *Murrah (or Delhi)*

The most important Indian breed and the most efficient milking breed. The home of this buffalo type is in the Punjab and Delhi, but purebred herds are also found in Uttar Pradesh. The breed has a deep, massive frame with a short, broad back and comparatively light neck and head. The colour is normally jet black with white markings on the extremities. It has characteristic short, tightly curled horns and the udder is well developed. Milk yields are more variable than in other breeds, ranging from 1000 to 2500 kg but average about 1500–2000 kg per lactation (Amble *et al.*, 1970).

(2) *Nili and Ravi*

These types are found in the valley of the rivers Sutlej and Ravi in West Punjab and the Sahiwal district of Pakistan. There is no essential difference between the two types and they are now officially treated as a single breed, sometimes named the Nili-Ravi. These buffalo possess a medium-sized deep frame with an elongated, coarse head. The horns are small with a high coil and the neck is long and fine. The commonest hair colour is black with white markings, but brown animals are also found. Average lactation yields quoted are 1500–2250 kg (Houghton, 1960), although work in Pakistan has set more precise average yields at 1555–1971 kg with a heritability of 0.18 and a repeatability of 0.37 (Chaudhry and Shaw, 1965).

(3) *Surti*

This breed is mainly found in Gujarat State between the Mahi and Sabarmati rivers. Animals are well proportioned and of medium size, with short legs. The head is long and broad. The colour is usually black or brown, with white collars on the jaw and brisket. The horns are medium length and sickle-shaped. This breed has an exceptionally straight back line, and the udder is well formed with squarely placed teats of medium size. It is very early maturing, but is not regarded as a good breed for draught purposes.

(4) *Jaffrabadi*

This is the largest breed in body size, selected bulls weighing as much as 1250 kg. The home of the breed is in the Gir forest of Gujarat State, where large numbers are bred primarily for ghee production. The most notable feature of the breed is the large head with very prominent forehead and heavy horns which droop on each side of the head and which turn up at the tips. The dominant colour is black. The breed is relatively late maturing, maximum weight being reached about one year later than in the case of the Surti breed.

(5) *Nagpuri*

This breed is found mainly in central and southern India. The most noticeable feature is the characteristic formation of the long, sweeping, angular horns. The animals are of light build and possess comparatively fine limb bones and small feet. Some authorities regard this breed as being poor for milk production and it is mainly used for draught. Average milk yield is recorded at 1200 kg (Kalkini and Paragaonkar, 1969).

Other regionally differentiated types of buffalo are recognised by some authors, such as Bhattacharya (1990), the most notable being the Italian, Egyptian, Bulgarian, Thai, Chinese, Carabao and Vietnamese.

Kartha (1959) drew attention to the fact that the water buffalo has a remarkable capacity to adapt itself to extreme climatic conditions. He stated that the better-class water buffaloes were mostly confined to the Punjab where summer temperatures reach 46°C, and winter temperatures drop to 4°C, or even lower. In India the poorer types of water buffaloes are found in the wetter areas where the temperature variation is not so great. In such places the water buffaloes are preferred to taurine cattle because of their better milking capacity and their remarkable performance for draught in areas where bullocks would find it difficult or impossible to work efficiently. Kartha pointed out, however, that the buffalo cannot stand abrupt changes of temperature and that the species requires time to acclimatise itself to varying climatic conditions. It is of interest to note that there have been no major importations of water buffalo into Africa from Asia, although most African breeds of zebu cattle originated in western Asia. On the other hand Bhattacharya (1990) suggested that the wide distribution indicated that buffaloes are adaptable to a relatively large range of environmental conditions. Thus in Bulgaria buffaloes are even used for pulling snow-ploughs(!) (Cockrill, 1968). It would, however, appear that the swamp buffalo is less adaptable than the river buffalo, because of their need for unlimited amounts of water for cooling purposes in hot climates.

In all parts of the Hindu world, where there is a strict religious taboo against cow slaughter, the buffalo is acceptable as a meat-producing animal. However, in the Indian sub-continent and Southeast Asian countries attention has not been paid to exploiting the meat potential of this species. There is scope for producers to evaluate the buffalo as a source of meat production. Wilson (1961b) showed it had comparable or superior quality to beef. Polikhronov *et al.* (1971) found that under experimental conditions up to 1.03 kg liveweight gain/day could be obtained by raising buffaloes on a feeding regime with 65% of requirements met from high-quality forages.

Work reported by Charles and Johnson (1972) in Australia showed that with age at slaughter between 18 and 21 months, buffaloes had a dressing percentage of 55.2, a meat:bone ratio of 3.97:1 and a tissue composition of 68.6% meat, 10.6% fat and 17.3% bone, not dissimilar to genetically unimproved breeds of cattle. Meat produced from water buffaloes has a distinct bluish tinge and the fat is pure white in colour. Young buffalo steers slaughtered for beef at about 2 years of age yield very acceptable carcasses. A series of 'palatability dinners' reported by Wilson (1961b) revealed that fresh young water buffalo meat was preferred to the meat of deep frozen high-quality beef steers on the grounds of better flavour and more attractive colour of fat. The muscular development of the water buffalo is particularly good and the hindquarter is better developed than that of *Bos indicus* cattle and compares favourably with that of *B. taurus* animals of similar weight. Water buffalo meat is not well marbled, and fat is laid down sparingly in both the intramuscular and intermuscular sites; this presumably renders the meat more difficult to cook evenly without desiccation, although Pinkas and Hristov (1972) have reported that the buffalo meat had better flavour quality due to its higher juiciness and tenderness. It should, however, be remembered that meat in tropical countries is usually boiled rather than roasted, roast beef being very much a British and North American dish. The mature buffalo fattens very rapidly and large quantities of subcutaneous and abdominal fat are laid down. The skin of the water buffalo is thicker than that of cattle and can be tanned into a very tough and durable leather, known in the trade as 'hog hide' or 'hog leather'.

No critical studies of the basal metabolism of the water buffalo have been reported, but temperature, pulse and respiration rate are all lower than the equivalent figures for taurine cattle. Thus the body temperature is 37°C compared to 38.6°C for cattle, the pulse rate is 40/minute compared to 50–55 for cattle and the respiration rate is 16 inhalations/minute compared to 20–25 for cattle. Controlled field studies indicate that the thermoregulatory mechanism of the buffalo is more efficient than that of cattle, when the speed of recovery from heat stress is taken as an index (Mullick, 1960).

Milk yield has been shown to be correlated to sweat gland density and sweat volume; the more freely sweating buffaloes producing more milk. The morphology of the sweat glands exhibits reasonable heritability, ranging from 0.4 to 0.7, indicating the possibility of genetic selection for heat tolerance and thus for higher milk production (Sethi and Nagarcenkar, 1981; Sharma and Nagarcenkar, 1981).

The reproductive efficiency of buffaloes is related to climate. Semen quality is lower in the hotter seasons and there is a more marked seasonal breeding pattern than is the case with cattle. There is a lowered incidence of oestrus during the hottest months and oestrus is less pronounced during the adjacent periods.

Buffalo bulls are sexually mature at 2 years of age although even earlier reproductive activity is reported. However, bulls are not usually used for breeding before 3 years. Cows usually breed at around 40 months of age when they weigh about 500 kg, depending upon the breed (FAO, 1976), and continue to calve to 20 years of age or more. There are reports from Egypt that the age of first calving of Egyptian buffaloes can be as early as 29 months (Oloufa, 1968). Mediterranean buffalo breeds and swamp buffaloes appear to calve earlier compared to the riverine types of the Indian sub-continent. The higher age at first calving in Indian buffaloes is mainly due to poor nutritional regimes and experimental evidence has shown that rearing buffaloes on higher nutritional planes can reduce the age at first calving in Murrah buffaloes.

The calving interval is usually long (about 15 months), although Buvanendran et al. (1971), in work with Sri Lankan animals, found a wide range of calving intervals from 14 to 20 months. Similar findings have been reported for Nili and Ravi animals in India by Amble et al. (1970). The calving interval is mainly a function of management but seasonal differences in calving interval have been reported. Long calving intervals in buffaloes are mainly due to failure to detect oestrus, especially during summer and autumn months in tropical countries. This in turn leads to seasonal calving and is the reason for lean and flush periods of milk production in India, where the buffaloes are the mainstay of the dairy industry (Nagarcenkar, 1974). The gestation period is long, at about $10\frac{1}{2}$–11 months, that for swamp cows being about 2 weeks longer than that for riverine buffaloes. The oestrus cycle is 21 days in length and duration of oestrus is between 20 and 28 hours with less pronounced heat symptoms compared to cattle. Bulls frequently become infertile by the seventh year of age, even though libido remains and muscular strength increases well past that time.

A preference for mating occurs during the night, early morning or at sunset. This is particularly evident in buffaloes living in a hot, high-humidity environment. Buffalo bulls exposed to very high temperatures exhibit marked reduction in libido and mating does not take place when air temperatures exceed about 40°C. However, normal mating behaviour returns if the affected animals are cooled by bathing or water sprinkling. Female buffaloes are also affected and exposure to marked climatic stress results in unprotected cows being in constant anoestrus. Detecting oestrus in buffalo cows is difficult during the midday period. Behavioural signs of heat may be absent and 'silent heats' are common. Nagarcenkar (1974) stressed the importance of teaser bulls in helping to detect heat. In general, both sexes of riverine buffalo possess a greater degree of libido than is found in swamp buffalo types.

Water buffaloes are relatively slow maturing and growth frequently continues up to the tenth year of age, although it is comparatively slow and often only seasonal after the fifth year. The average birth weight ranges from 24 to 38 kg for female calves and 27 to 41 kg for males. Mature weights for swamp buffaloes are around 700 kg for males and 500 kg for females. River buffalo bulls have an average adult weight of between 550 and 1100 kg, varying according to breed. River buffalo cows average between 450 and 700 kg.

Sreedharan (1985) reported on the growth rates of Murrah buffaloes from birth to 48 months. Average gain/month was 17 kg to 6 months, 13 kg to 9 months and about 10 kg up to 27 months. From 27 to 48 months growth had slowed down to around 6 kg/month. Factors that affect the birth weight and growth of cattle also influence the development of the buffalo (Tomar and Desai, 1965). It has already been mentioned that the recorded milk yields of water buffaloes compare favourably with those of zebu cattle maintained under similar conditions. Maule (1953–54) presented data obtained from a survey in seven cattle and buffalo breeding areas in India, and his findings are summarised in Table 13.4, and presented in more detail in Table 13.5. It will be seen from Table 13.5 that the Murrah breed from India has the largest variability and what appears to be the greatest scope for improved genetic potential, although most studies on this trait have been done in Egypt, Italy and India. A study conducted on the data obtained from military farms with Murrah buffaloes has shown that a selection index based on weight and age at first calving, along with first lactation yield, could bring about marked genetic progress and economic improvement of this type of animal (Gokhale, 1974).

Maule (1953–54) provided further data showing the results obtained in well-managed government-owned herds for some of the major breeds of river buffalo. The Nili and Ravi breed tops the list with average milk yields of 2060 kg/lactation, followed by Murrah (1610 kg) and Surti (1370 kg). Exceptional yields, as high as 3500–4500 kg/lactation, have been recorded in

Table 13.4 Data for mean lactation yield of various breeds of water buffalo and zebu cattle in India. (Source: Maule, 1953–54.)

Item	Water buffalo (*Bos bubalis*)	Zebu cattle (*Bos indicus*)
Number of records	6160	4310
Mean daily yield (kg)	3.6	1.7
Estimated lactation yield (kg)	982	429
Average length of lactation (days)	300	264
Calving interval (months)	18	18.2

Table 13.5 Milk production in buffaloes.

Breed	Country	Lactation milk yield mean or range (kg)	Heritability	Repeatability	Reference
Nili/Ravi	India	1586–1855	—	—	Amble et al., 1970
Nili/Ravi	Pakistan	1555–1971	0.18	0.37	Chaudhry and Shaw, 1965
Bulgarian	Bulgaria	1290–1737	—	—	Polikhronov, 1969
Nagpuri	India	1200	—	—	Kalkini and Paragaonkar, 1969
Bhadawari	India	1111	—	0.35	Singh and Desai, 1962
Egyptian	Egypt	1110–2035	0.20–0.49	0.36–0.54	Alim, 1967
Caucasian	Former USSR	1110–1724	0.21	—	Agabeili et al., 1971
Murrah	India	1031–2565	0.17–0.53	0.17–0.62	Amble et al., 1970
Italian	Italy	1000–2025	—	0.17–0.38	Deshmukh and Roychoudhury, 1971
Marathwada	India	960	—	—	Hadi, 1965
Brazilian	Brazil	945–1113	—	—	Do Nascimento et al., 1970
Russian	Former USSR	595–872	—	—	Gadzhiev and Yusupov, 1971
Malaysian	Malaysia	226	—	—	Marsh and Dawson, 1948

some instances and lactation yields of 2700 kg are not uncommon. These reported yields must be seen in context. It is probable that the average (unrecorded) yields are as low as 500 kg milk/yr, but the average of unrecorded typical zebu cattle could be even lower at about 200 kg.

Ganguli (1981) has reviewed the information on buffalo milk quality. The most important feature is the comparatively high butterfat percentage which averages over 6.7%, but may exceptionally reach 15%. Work by Schalitchev and Polikhronov (1969) has shown that, for East European breeds of buffalo, the butterfat percentage is usually in the range 7–8. The solids-not-fat percentage averages between 9 and 10, while the mineral content is similar in most respects to that of the milk of zebu cattle, although possibly a little higher. Usually the total solids in river buffalo milk is within the range 16 to 19%, whereas swamp buffaloes (Carabaos and Chinese) have higher total solids ranging from 21 to 24% (Webb and Johnson, 1971).

More recently animal breeders have considered the prospects for genetic improvement of the productive traits of the buffalo, such as growth rate, milk yield and milk quality. Bhattacharya (1990) considered that the selection index approach could be successfully adopted. Because of their long generation interval, attempts have been made to accelerate breed improvement programmes by selection based on part-lactation yields. Results have indicated that the fifth month's yield or the first 150 days' yield are reasonably correlated to total lactation milk yield (Gokhale and Nagarcenkar, 1979, 1981).

Maule (1953–54) referred to the widespread use of water buffaloes in India in 'town dairies' such as those supplying milk to Bombay and Delhi. In the Bombay scheme, which is still continuing although relocated out-of-town, 15000 buffaloes are maintained in units averaging 300 cows each. Each unit has about 20 ha of grass for exercise and provision of some forage, but the buffaloes are mainly intensively fed. Their milk is processed at a central milk depot where it is 'diluted' with skimmed cows' milk to give a standard product averaging between 3.5 and 4.0% butterfat and 9% solids-not-fat.

Many visiting experts recommend that the large populations of river buffalo in Asia should be replaced with cattle, particularly imported *Bos taurus* cattle, on grounds of greater productivity. It is doubtful whether such advice is wise. The water buffalo has important attributes which result in it being well adapted to the hot, tropical environment. It is probable that the digestive system of the buffalo enables it to deal with a much higher level of crude fibre, because of the high population of protozoa and cellulose-degrading bacteria in the rumen compared to cattle (Panjarathinam and Laxminarayana, 1974). Buffaloes also synthesise more protein in the rumen and excrete less nitrogen through the urine. Various workers have also reported that the water buffalo is better able to utilise the fibrous by-products of Indian agriculture, such as rice straw, than either *B. indicus* cattle or imported *B. taurus* animals (Barsaul and Talapatra, 1970). More work is needed on this subject since insufficient comparative digestibility experiments have been carried out.

Basu (1985) has speculated that, had the water buffalo received as much attention from the geneticist and animal breeder as the 'improved' breeds of *Bos taurus* and *B. indicus* cattle, then its milk yield might well be more than double that of contemporary zebu cattle. In respect of meat production traits, it is interesting to note that, since the demonstration that the riverine breeds of water buffalo imported to the West Indian island chain are efficient meat producers, there has been a considerable interest in developing these animals for meat in the mainland of South America, particularly in Venezuela. Much of the Amazon Basin is flooded during the annual rainy season, and it would appear that these regions might be much better exploited by water buffaloes than either 'indigenous' criollo cattle or imported 'exotic' taurine breeds.

The use of buffalo hides for making thick, tough leather has already been referred to, but there are many other novel by-products. Strips of buffalo hide softened with fat are woven to make reins, lassos or ropes. Durable water buckets for use with irrigation equipment can be made from hides, which can also be split to make strong thin sheets (Bhattacharya, 1990). The large buffalo horns are used for the production of a wide range of fancy and decorative articles and may be a partial replacement for cottage industries formerly based on ivory.

In summary, the water buffalo is potentially a most important tropical bovine species, especially in very hot areas where rivers and swamps abound. It is probably capable of marked improvement in productivity, given a requisite amount of attention by the applied scientist. It can compete very successfully with other bovine species and any attempt to 'phase out' the buffalo and replace it with expensive exotic importations should be treated with caution. The water buffalo, as a triple-purpose tropical animal, has no obvious rival in those areas to which it is climatically adapted.

DROMEDARIES, LLAMAS AND ALPACAS

These three types of domesticated livestock belong to the family Camelidae. The camels belong to the genus *Camelus* while the smaller, South American llamas and alpacas are collectively known as Llamoids. There are two species of camel, the two-humped, temperate Bactrian camel (*C. bactrianus*) and a single-humped tropical camel, more correctly known as the dromedary (*C. dromedarius*) which may have evolved from the Bactrian camel which possesses a vestigial second hump (Nawito *et al.*, 1967).

The llamas and alpacas of South America are mainly confined to the 'altiplano' of the Andean mountain range, at altitudes averaging over 3000 m. They are small-bodied animals, the trunk and legs roughly resembling large, long-legged goats, but they possess a long neck so that the head is borne about 1 m away from the shoulders. Llamas are used as pack animals (they are rarely used to pull wheeled implements or carts) and the wool is spun and woven into blankets and wearing apparel. The animals are capable of withstanding very great extremes of diurnal variation in temperature (ambient temperatures ranging from below freezing to above 27°C in the short space of 24 hours) and they can maintain themselves on sparse, xerophytic vegetation high in fibre and low in digestible nutrients (Fig. 13.5). The total world population of llamas and alpacas is not known with certainty but probably does not exceed 7 million (see Table 13.6).

The dromedary is an important domesticated animal of the arid and semi-arid regions of the world and its wide distribution and larger total population (about 17 million in 1985) suggests that it should receive due attention in any textbook on tropical agriculture. Unfortunately this demand is seldom met, primarily due to the dearth of scientific literature on the subject even though its temperate relative, the Bactrian camel, has received a great deal of attention from Russian workers.

Two distinct types of dromedary may be recognised (Wilson, 1984), although there are no true 'breeds' in the normal sense of this term, and the various related herds of camel are often named after the tribes which own them. The first type is the heavy, thick-boned, slow-moving 'baggage camel' which is used as a pack animal. Second there is the sleeker, longer-legged, fine-boned 'riding camel'. Both types are less heavily built and possess longer legs than the Bactrian camel, and both have a much softer and thinner hairy coat. Dromedaries are extremely well adapted for hot, even very arid, climates, but they rapidly lose condition in the humid tropics and are particularly susceptible to attacks by

Table 13.6 Approximate numbers and distribution of South American Llamoids (thousands). (Source: Sumar, 1983.)

Country	Alpaca	Llama	Vicuna	Guanaco
Argentina	n.a.	100	10	500
Bolivia	300	2050	2	n.a.
Chile	n.a.	85	10	17
Ecuador	n.a.	3	—	—
Peru	3290	900	65	5
United States	—	3	—	—
Total	3590	3141	87	522

n.a. = not available.

Fig. 13.5 Vicuñas grazing natural pasture. (Source: Payne, 1990; Photo: S. Fernández-Baca.)

biting flies which are common in the wetter regions of the tropics and subtropics. Most of the world's population of dromedaries is to be found in Africa (see Table 13.7) and has increased significantly over recent times.

In Northeast Africa the southern limits of the breeding area for the dromedary is said by Mason and Maule (1960) to be the 15°N parallel, although herds are common in the Northern Frontier Province of Kenya, adjacent to Somalia. The distribution in the Horn of Africa is limited to the Somalia area, approximately east of the 41st meridian. In most of this area the mean annual rainfall is less than 350 mm.

The dromedary originally was thought to have evolved in Southwest Asia, possibly in Iran or Arabia, but it is now believed to have originated in North America, crossing the land bridge that is now the Bering Straits in the late Pliocene period. There are abundant accounts of the species in very early Egyptian, Jewish and Greek writings. Its spread along the North African coast is due to the Arab migrations across the deserts to the west coast of Africa, and southwards into what is now northern Nigeria (Fig.

Table 13.7 Appropriate numbers and distribution of camels (millions). (Source: FAO, 1986d.)

Continent/Region	1961–65	1985
Africa		
Tropics	7.4	12.2
Other regions	1.0	1.0
Asia		
Tropics	1.1	1.0
Other regions	2.7	2.9
Former USSR	0.3	0.3
Total tropics	8.5	13.2
Total other regions	4.0	4.2
Total world	12.5	17.4

13.6). No feral dromedaries exist today although there are still some wild Bactrian camels in the Gobi Desert (Payne, 1990).

The dromedary is a fatty-humped, hornless ruminant. The upper lip is divided and is a very sensitive and mobile organ, capable of performing amazing feats of labial dexterity. The dentition is most abnormal, the dental formula being:

Fig. 13.6 Camel (*Camelus dromedarius*) in a market in northern Darfur, Sudan. (Source: Payne, 1990.)

$$\frac{1:1:3:3}{3:1:2:3}$$

The young camel calf has three temporary incisors on the upper jaw, but only the third of these is replaced by a permanent tooth when the milk-teeth are lost. The canine teeth are long and pointed and are used with great effect as an offensive weapon. As it is often of importance to the camel owner to know the ages of individual animals, it is useful to estimate age by examination of the teeth. At birth the central incisors are already erupted, at 1 month the laterals appear and at 2 months the third pair break through. By 1 year the milk-teeth are mature but in the second year they show signs of wear. At 3 years they are well worn and by 4 years they become loose. The first pair of mature incisors erupt in the fifth year and the second pair in the sixth year. By the seventh year the camel has a full mouth of adult teeth (Payne, 1990).

The dromedary, even when young, grows horny callosities instead of skin under the sternum and at the elbows, knees and stifles. This fact once gave rise to much debate among biologists, as it was quoted as partial evidence for the hypothesis of the inheritance of acquired characters. The feet consist of two digits united by a single, horny 'sole' common to both. The feet are very large in area, a fact which enables the dromedary to travel across open deserts without the feet sinking into the loose sand. Camels have no gall bladder, a curious anatomical deficiency which this genus shares in common with the horse.

Male dromedaries possess a paid of 'poll-glands' which emit a strong-smelling liquid in the rutting season. The opening of the sheath is very small and points to the rear, so that the male animal urinates backwards between the hind legs. It is not wise to stand directly behind any large domestic animal, and this is a second good reason for not standing close behind a male dromedary!

The breeding season of camels exhibits geographical variation (Wilson, 1984). The male shows little sexual behaviour outside the 'rutting' season, when the testes increase in size, the soft palate lengthens and an offensive odour is emitted. Control of the camel in rut is very difficult, and the rutting season may last for up to 5 months.

Males can breed at 3 years of age but are not fully sexually mature until 5–6 years. They remain sexually active until about 20 years and when fully mature a male can serve up to 70 females during the rut (Matharu, 1966). Female camels are normally sexually mature at 3 years and are commonly first served at 4 years of age. They can continue to breed until over 20 years old.

The usual ratio of male dromedaries (stallions) to females is 1:70 (Burgemeister, 1974). In Ethiopia the number of females/breeding stallion is smaller. The loss of mobility in some nomadic tribes has caused a high incidence of inbreeding in their herds of dromedaries, and this is said to have resulted in the production of undesirable recessive genotypes, especially deformed foals (Burgemeister, 1974).

The dromedary cow has an udder divided into four roughly equal quarters, each bearing a teat, similar to cattle. The gestation period varies between 350 and 380 days, but cows normally breed every second year, so the calving is strictly seasonal and usually biennial. The lactation period varies according to the nutrition of the animal, but commonly extends for as long as 18 months (Wilson, 1984).

Dromedary milk is consumed in many parts of the world, especially by nomads. In some regions of Somalia and the lowlands of Ethiopia, dromedary milk is one of the main components of the diet. There is a great lack of information concerning the milk yields of dromedary cows. Table 13.8 presents limited data compiled by Knoess (1976), which indicate that average yields probably lie between 1000 and 3000 kg/yr, although yields of 3600 kg have been recorded in Pakistan.

Camel milk is white as it contains little or no carotene but it is rich in vitamin C, a factor of great nutritional significance in the diet of transhumant camel herders with little access to alternative vitamin C sources (Kon, 1972). In most regions, butter, cheese and ghee are not made from dromedary milk, which is instead soured and stored as curds. Table 13.9 presents data on the composition of dromedary milk. The limited information available indicates that the milk is high in protein and fat, as well as vitamin C.

The dromedary is essentially a bush-browser, but stock will graze on grasses if no shrubs or trees are available. The usual behaviour pattern for the dromedary is to feed and rest during the day, and travel at night. In some countries working dromedaries are fed green fodder and concentrate feeds such as millet, oats, beans, cotton

Table 13.8 Milk production of the dromedary. (Source: Knoess, 1976.)

Country	Lactation length (day)	Milk yield (kg)	Nutritional status
Ethiopia (Awash Valley)	365	2442	Good
Pakistan	270–540	2700–3600	Good
Pakistan	270–540	1350	Poor
Pakistan	540	3000	Good
Pakistan	270	1700	Desert conditions

Table 13.9 Composition of dromedary milk. (Sources: 1 – Harbans Singh (1962); 2 – Leupold (1967); 3 – IFTV (1973); 4 – Knoess (1976); 5 – FAO (1968b).)

Constituent	Reference				
	1	2	3	4	5
Moisture (%)	87.6	—	—	85.6	88.5
Ash (%)	0.8	—	0.8	0.9	0.7
Protein (%)	3.9	—	3.7	4.5	2.0
Ether extract (%)	2.9	2.9	4.2	5.5	4.1
Lactose (%)	5.4	4.0	4.1	3.4	4.7
Solids-not-fat (%)	10.1	—	8.7	8.9	—
Calcium (mg/kg)	—	—	—	4	9
Phosphorus (%)	—	—	—	14	9
Vitamin C (mg)	—	5.6	5.6	2.3	—

seed, maize bran and grain. Different cereal straws are also fed in a chaffed form. Dromedaries, therefore, are capable of utilising a wide range of agricultural by-products and waste materials as well as more traditional feedstuffs.

Although intensive rearing of dromedaries is rarely carried out, under experimental conditions it has been found that they are capable of high levels of productivity when grazed on fields of alfalfa or on pastures of *Panicum maximum*. The highest intake of fodder recorded is 50 kg of alfalfa/day expressed on a dry-matter basis by a dromedary weighing 360 kg. Details of the types of browse eaten by African dromedaries are given by Le Houérou (1980). In Iran some dromedaries are fed in feedlots (Kazem-Khatami, 1970). The ration usually consists of 15–20 kg of a low-cost mixed feed comprising straw, sugar-beet pulp, silage, molasses and barley. Dromedaries have been grazed experimentally on sugar-beet tops (Kazem-Khatami, 1970) which are capable of supporting liveweight gains of 0.95 kg/day for females and 1.4 kg for males.

The nutritional requirements of camels are not known with any accuracy but calculated energy and protein requirements are presented in Table 13.10.

Desert caravans of dromedaries often move during the coolest part of the night, from 2 AM to 8 AM. The ability of the dromedary to go for long periods without water is well known. In Somalia, dromedaries travelling over the desert routes are generally provided with water once every 3–7 days; in the Sudan stock are watered once every 3–6 days; in Arabia once every 3–4 days, while Algerian dromedaries are generally provided with water once every 2–3 days under similar conditions. A well-documented and classic endurance record of the dromedary was set in Australia in 1891–92, when a troop of soldiers travelled with a caravan of dromedaries across 537 miles of desert in 34 days without water. A quarter of the caravan survived this ordeal! Payne (1990) stated that the camel can tolerate the loss of 27% of its body weight during periods of intense dehydration and can drink, at any one time, water equivalent to 30% of its body weight.

During the winter or rainy season, the moisture content of the trees and bushes browsed by the dromedary is quite sufficient to supply the total water requirement and water may not be drunk at all during this period. In the dry season the vegetation is desiccated and extra water, over and above the winter requirement, is required for heat regulation. Adult working stock require on average between 10 and 40 litres of water/day under dry-season conditions.

A misconception about the hump of the camel being a water-storage organ is so commonly quoted that it is necessary to state that the hump of the dromedary, like the hump of zebu cattle, is composed of fatty tissue and is an energy store. Fat provides more energy per unit weight than any other feedstuff or reserve tissue and hence fat is the most economic form of energy reserve. The amount of fat contained in the hump of a dromedary averages about 20 kg, and on complete oxidation this could produce about the same weight of oxidation water. The total fat contained in the hump would, therefore, only be sufficient indirectly to supply a dromedary's water requirements for a relatively short period. The thermal regulation of the dromedary, and the physiological characteristics which result in such a high level of water economy, have not received sufficient critical scientific study. The

Table 13.10 Calculated energy and protein requirements of adult camels. (Source: Wilson, 1984.)

Function	Daily theoretical requirement	
	Energy (MJ/ME)	Protein (g DCP)
Maintenance		
500 kg male or castrate	54	300
400 kg breeding female	45	260
Production		
per litre of milk	5	50
per hour of work of a pack or draught camel	8	negligible

ME = metabolisable energy; DCP = digestible crude protein.

following are the most important factors which make the dromedary the most efficient desert animal yet domesticated by man:

(1) *Insulation.* The thick, wool-like hair, the thick hide and the deep layer of subcutaneous fat effectively insulate the underlying body tissues from the radiant heat load received from the sun.
(2) *Body temperature range.* The dromedary has an extremely wide diurnal variation in temperature. Normal early morning temperatures average about 35°C and normal midday temperatures approximately 41°C. The dromedary thus has a daily temperature range of about 6°C, compared to about 1°C in most other mammals including cattle. When deprived of water the range may exceed 6°C.
(3) *Dehydration of body tissues.* The dromedary can lose up to about 40% of its body water before pronounced physiological disturbance results in an 'explosive' increase in body temperature. Most mammals, including man, normally die after the loss of body water has reached a critical value of about 20%.
(4) *Water-drinking capacity.* When most mammals have experienced severe thirst and are given access to unlimited quantities of water they can only drink a small amount comparatively slowly or they suffer a physiological disturbance known as 'water intoxication'. The dromedary, under similar conditions, is able to drink large quantities of water very rapidly without ill-effect. There is a record of a dromedary, 300 kg in liveweight, drinking 103 litres of water in less than 10 minutes. The water drunk in this manner is evenly distributed in the body tissues in less than 2 days. Most mammals take several weeks to rehydrate their body tissues after periods of severe desiccation, since when water is drunk in large quantities after dehydration most is excreted in the urine.
(5) *Maintenance of appetite during conditions of thirst.* Most mammals lose their appetites when experiencing severe thirst. The dromedary does not do so, and is capable of maintaining its feed intake at a reasonable level, providing feed is available, in spite of pronounced thirst and dehydration of the body tissues.
(6) *Low respiration rate.* The normal respiration rate of a resting dromedary is 5 to 8 respirations/minute, one of the lowest recorded respiration rates in any mammal. The consequence is that the evaporative water losses from the buccal cavity and upper respiratory tracts are reduced to a minimum.
(7) *Ability to reduce faecal and urinary water loss.* When dehydrated, the camel reduces water output in the urine and faeces. In addition, the urine flow is restricted (Siebert and Macfarlane, 1971), the plasma volume does not fall rapidly and the packed cell volume (PCV) does not rise as rapidly as in other species (Payne, 1990).

The dromedary calf is exceptionally delicate in comparison to the hardiness of the adult animal. Large numbers of calves die before attaining the age of 3 weeks. Although no data are available, many stock owners consider that the rich colostral milk is responsible for early digestive disorders and it is a common practice to 'ration' the calf to only one quarter of the udder, the other three teats being milked by hand and the milk used for human consumption.

Traditionally, dromedaries are used as beasts of burden, for riding and for many operations in agriculture and transport. They are still used in great numbers for powering water-wheels, sugar cane crushers and different types of mill. They are also used for pulling various types of cart. Male dromedaries in Ethiopia have been tested for their suitability for ploughing (Knoess, 1976). They were harnessed to reversible mouldboard ploughs, worked 7 hours a day and were able to plough 1 ha of land to a depth of about 16 cm in 20 hours.

The meat of the dromedary is eaten by Arabs, that of young castrated males being very fine in texture and flavour. The carcass of a male dromedary in Iran weighs about 300–400 kg, while that of a male Bactrian camel might weigh up to

650 kg (Kazem-Khatami, 1970). The carcass weight of female dromedaries is around 250–350 kg. The meat of young dromedaries is similar in taste and texture to beef.

Dromedary wool and hair are of economic importance and command good prices on world markets. In particular, the undercoat of calves has very high commercial value when collected during the moulting season. The calf hair possesses qualities of durability, is exceptionally fine and light and is used for weaving blankets and cloths. The dromedary hide forms a rather poor-quality thick leather, which is usually used for making saddlery. The long leg-bones are used for making tent pegs and the dung of the dromedary, which is exceptionally dry, is used as a fuel.

It follows from the above that the full economic potential of dromedaries is relatively under-utilised. Countries in the drier zones of the world should reconsider the role of the dromedary if they are to exploit the unique potential of these animals for draught purposes, to produce protein (both milk and meat) and other animal products at relatively low cost from desert and steppe lands and from farm waste. An evaluation of the role of the dromedary populations of the world is urgently needed. This evaluation should include technological studies of various aspects of dromedary husbandry, including extensive and intensive management practices, optimal stocking rates, disease prevention and control and the effect of different stocking rates on the natural vegetation. Without adequate regard to such considerations, any increase in the world population of dromedaries might result in overgrazing and consequently in soil erosion in certain areas.

It must be a cause of much regret to all those interested in animal production in the tropics that such a well-adapted animal has been so neglected by the agricultural scientist. The vast majority of the world population of dromedaries are still maintained according to traditional techniques which were evolved some 5000 years ago. Given skilled attention to such major limiting factors as nutrition, disease control, breeding and management it is possible that the world population of about 17 million dromedaries (in 1985) could make a much greater contribution to meat and milk production in the arid regions of the tropics, while continuing to supply an important source of draught power and an impressive range of by-products.

GOATS

The FAO (1984) estimated that the tropical and subtropical goat population was of the order of 460 million, of which approximately 430 million were distributed in the tropics. A very rough breakdown of the geographical distribution is shown in Table 13.11. It will be noted from Table 13.11 that the largest concentration of goats is to be found in Africa and Asia. and that this large goat population represents approximately 15% of the total world population of grazing (ruminant) domestic animals.

Domestic goats belong to the genus *Capra*, but the systematists are undecided as to whether all goats belong to the same species or whether different species and subspecies should be recognised. Ellerman and Scott (1951) divided the genus into five currently living subspecies:

Table 13.11 Approximate distribution of tropical and subtropical goat population (millions). (After FAO, 1984.)

Region	Number	Percentage of total population	Percentage of total grazing ruminants*
Africa	151	32.9	29.1
North, Central America and Caribbean	14	3.1	6.3
South America	20	4.3	7.0
Asia and the Pacific	255	55.5	23.8
Europe	13	2.7	4.3
Former USSR	7	1.4	2.4
Total	460	100.0	15.3

*Goat population as percentage of total population of buffaloes, cattle, goats and sheep.

(1) *Capra aegagrus hircus.* True goats.
(2) *Capra aegagrus ibex.* The ibex.
(3) *Capra aegagrus caucasica.* Typified by the Caucasian tur.
(4) *Capra aegagrus pyrenaica.* The Spanish ibex.
(5) *Capra aegagrus falconeri.* The markhor.

All these subspecies of the wild goat, *C. aegagrus*, were previously considered as distinct species but as they freely interbreed they are now classified as a single species.

Classification of goats

At least four methods of classifying domestic goats have been used, namely:

(1) Origin (Webster and Wilson, 1980).
(2) Body size (Devendra and Burns, 1983).
(3) Ear shape and length (Mason and Maule, 1960).
(4) Function (Williamson and Payne, 1965).

Each of the above methods has both strengths and limitations. The classifications of Williamson and Payne (1965) on the basis of function (meat, milk and wool) and that of Webster and Wilson (1980) in the previous edition of this book, based on country of origin, are somewhat broader in scope. The inherent difficulty in classifying goats by country of origin is that the geographical source of many breeds lacks proper documentation at the present time. However, on balance this method is to be preferred, and will be employed in the present work. Such a system of classification recognises at least five major goat types: European, Oriental, Asiatic, African and South American.

European type

Most of the European-type goats originated in Central Europe, particularly in the mountainous regions of Switzerland and Austria. Three of the important individual breeds, organised into recognised breed societies, are:

(1) *Toggenburg*

The home of this milk breed is in northeast Switzerland. It is a large goat, distinguished by its fawn or chocolate colour with white or cream stripes and markings. The ears are generally dark in colour but are marked with characteristic white edges. Both sexes are polled. In common with other European breeds, the ears are borne erect and many individuals possess a pair of tassels – small pendant outgrowths of skin – below the lower jaw at its junction with the neck.

The Toggenburg is probably the least successful European breed to have been introduced into the tropics. This may or may not imply a criticism of the breed *per se*, but more probably its lack of success in the tropics is fortuitous, since many introductions were made without ensuring that the requisite animal husbandry skills were available to ensure that the exotic animals were maintained in good health, with adequate nutrition and by stockmen exhibiting a sufficient level of expertise. In spite of this generalisation, there have been a few success stories reported, notably in Kenya and in Brazil.

(2) *Saanen*

This milk breed originated in west Switzerland. The goats are white, light grey or cream in colour and the breed is often named by its colour in countries other than Switzerland, such as 'Netherlands White' in Holland and 'White German' in Germany. Saanens have a short coat and are generally polled. Ears are erect and point forward. This breed is very popular in tropical and subtropical countries and has been widely distributed. They are good milkers and have been successfully employed for this purpose in Puerto Rico, the West Indies, Fiji, Ghana, Kenya, Malaysia and Australia (Devendra and Burns, 1983). In Israel, favourable comments have been made on the prolificacy of the breed (Epstein and Herz, 1964), but elsewhere it is reported that hermaphroditism can be a problem with polled Saanen goats, and for this reason it is suggested that only horned bucks should be employed (Devendra and Burns, 1983).

(3) *Alpine*

This is a very handsome milking goat, primarily black in colour and bearing white, cream or fawn markings which enable it to be very readily recognised. It is a highly developed milk breed, widely distributed throughout Europe although its home is in the Swiss and Austrian Alps. Breed societies have been formed in many countries and it is variously described as the French Alpine, Italian Alpine, British Alpine, etc. The Alpine has been introduced in the West Indies, Guyana, Madagascar, Mauritius and Malaysia. It is said to acclimatise very well. In the West Indies milk yields of up to 4.5 kg/day have been reported for Alpine goats in their second and third lactations.

Oriental type

This type of goat originated in the eastern Mediterranean and the Middle East. It is more suited to drier, arid conditions and for this reason it has now become widely distributed throughout the drier tropics. It has been extensively crossed with European breeds, the heat tolerance of the Oriental goat combining with the productive capacity of the European to form a well-adapted, tropical, dual-purpose animal much sought after in many continents. The two most important Oriental breeds are:

(1) *Nubian*

This breed is associated with the eastern Mediterranean and Northeast Africa, but especially the Sudan. The Nubian is mainly polled, but horns may be present in both sexes of certain strains. It is a large, long-legged breed and the ears are long and pendant or 'drop-eared'. The head has a convex facial profile, with a pronounced Roman nose, especially evident in the buck. The colours are various, black being perhaps the most common, but dark and light markings are frequently found on the same animal. It is a very hardy breed, capable of standing up to very harsh conditions and of living on fibrous, xerophytic vegetation for long periods. It would appear that the genetic potential of this breed may be much greater than previously recognised. It milks well, and the large average weight means that the surplus kids are valuable for meat production.

(2) *Angora*

This breed originated in Central Asia and was brought to Anatolia in Turkey during the thirteenth century. It has now been widely spread into other dry, tropical areas but especially South Africa, where it is known as the Sybokke, and also into Texas, where it is called the Mohair goat. The breed is white in colour, and bears long lateral horns in both sexes. The hair or wool is very valuable as it is strong and takes dyes very readily, thus commanding a high price as a top-quality yarn. The fleece, averaging 30.5 cm in length, hangs in separate and well-defined ringlets. It may be regularly combed out (in which case the annual yield of mohair is between 1.5 and 5 kg/yr) or it can be shorn off, in which case the fleece varies more in length, but the yield can be increased to a maximum of about 7 kg/yr. The average mohair yield in Texas, where very large herds of Angora goats are kept, is about 3 kg/yr.

Although this breed is kept mainly for the valuable mohair, meat and milk are subsidiary products in Turkey where the average carcass weight is about 13 kg. Angora goats are very valuable for crossbreeding purposes. Crossbreeding has been practised in South Africa, Lesotho, India, Madagascar and Fiji. In the Indian experiments Angora goats were crossed with the Gaddi hill breed, and the three-quarter Angoras were reported to produce about three times as much mohair as the local goats (Lall, 1968).

Asiatic type

Over half the total world population of goats is found in Asia, divided almost equally between the tropical part of the continent (especially India and Pakistan) and the temperate area to the north. The most important breed is the Kashmir or Kashmiri, but numerous other Asiatic breeds are recognised, many taking district or provincial names which they share in common with breeds of cattle and water buffaloes. Two of the most important Asiatic breeds will be briefly described:

(1) *Kashmir or Kashmiri*

This breed originated in Central Asia, around the mountains of Tibet. It is usually white in colour, occasionally black and white, and exceptionally black. The breed is prized for its long, fine hair, known as 'cashmere'. The breed is double-coated, the outer coat containing hairs up to 12–13 cm in length, and an inner coat of very high quality with shorter hairs averaging 2–5 cm in length. It is the inner coat alone which yields the cashmere hair (usually spoken of as cashmere wool); average crops being between 125 g and 200 g/goat/yr.

Because much of this highly prized cashmere fibre has been marketed via India and Pakistan, the goats that produce it are thought to comprise a fully tropical breed. In fact they are adapted more to cold environments rather than to high temperatures, and are traditionally farmed in the high mountainous areas of Central Asia, including Tibet, Inner Mongolia, Iran, Turkey, Kurdistan and the southern parts of the former USSR. This breed, therefore, thrives in arid regions at high altitudes and provides meat, skins and hair as well as cashmere wool. The goats are also used for draught, and represent one of the very few breeds of goat still used for this purpose.

(2) *Jumna Pari*

This breed, also known as the Etawah, is primarily a milch goat, probably the most widely distributed milk-producing goat in India and Southeast Asia. It is a large animal with a characteristic convex face, a somewhat fore-shortened muzzle and long lop ears. It originated in the region bounded by the Ganges, Jumna and Chambal rivers in India, and possibly also in the Etawah of Uttar Pradesh, from whence is derived its alternative name.

The udders of the females are well developed. Average milk production in India is approximately 3.8 kg/day, and a maximum lactation yield as high as 562 kg has been recorded (Fig. 13.7). The butterfat content is high, averaging 5.2%. In spite of its excellence as a milk producer, the breed is also used for meat in certain areas. The average liveweight of the male is 70–80 kg and that of the female 40–60 kg. The killing-out percentage is stated by Srivastava *et al.* (1968) to be about 45%.

Because of its good milking capacity and growth rate, the breed has been widely used for crossing on to smaller indigenous goats in countries as far apart as the West Indies, Malaysia and Indonesia.

African type

Mason and Maule (1960) suggested that a useful subdivision of the African-type goats may be based on the relative length of the ear. They classified the goats of East, Central and South Africa as being either long-eared or short-eared. Other authors place more importance on the size of the animal, and recognise the African dwarf goats as forming a separate group from the normal-sized types. In this chapter we will merely note four of the more distinctive African-type breeds without attempting a classification based on ear or body size. Many African breeds are dual-purpose.

(1) *Beneder*

This breed is found in the southern half of Somalia, north of the River Juba. They are large in size and diverse in colour, but red- or black-spotted goats are common and white

Fig. 13.7 Small East African goats at the Government's Stock Farm, Serere, Eastern Province, Uganda. Notice the excellent 'condition' of the goats and the marked colour variation exhibited by this breed. The pasture grass is Rhodes grass (*Chloris gayana*), and the goats are being given a protein and mineral supplement.

goats are rare. The coat is usually short and smooth but sometimes long and coarse. The ears are long and pendant with turned-up tips. The horns are so placed that they point backwards and run parallel along the back of the animal.

(2) *Galla or Somali*

This breed is found in the Northern Frontier Province of Kenya and adjacent areas, particularly southern Somalia and Ethiopia. It is white or off-white in colour and is a meat-type breed. It does not thrive in wetter regions and is rarely found in the more temperate areas adjacent to Mount Kenya. The goats have fine short hair and thin skins which are of good quality. The breed is used for both meat and skin production but some local strains are occasionally milked.

(3) *Small East African*

This breed is said to be a genetic strain possessing two genes for recessive pituitary hypoplasia. The possession of this gene results in the adult goat weighing about 15–30 kg at maturity, instead of the normal liveweight for an African goat of 45 kg or more. The breed is distributed over a wide range in East Africa, where its main products are meat and skins. It has small horns and is generally short-haired, the hairs varying greatly in colour (Fig. 13.8). The does breed at any time, but twinning has been recorded in only about 10% of total births. Mason and Maule (1960), however, did not consider the breed should be termed 'dwarf'. Wilson (1958b; 1960) has described the nutrition of this type of goat in detail.

(4) *Boer*

This South African breed was evolved by the Dutch settlers from local Bantu goats, possibly with the addition of some European, Oriental and Asiatic blood. Improved Boer goats are white with red head markings, and when well managed produce meat of excellent quality. The skins are also highly valued,

Fig. 13.8 Boer goat in South Africa. (Source: Payne, 1990.)

fertility is high and the milk yield moderate. The breed bears some resemblance to Nubian Goats, having long lop-ears. The horns project backwards from the head. The meat is coarse-grained but soft and palatable, and in some areas the Boer is also an important milk producer in spite of its relatively modest yield. Twin or triplet births are usual, and under good husbandry conditions the kids reach liveweights of 40 kg by about 12 months of age and 100 kg when mature (Joubert, 1973). Ueckermann et al. (1974) reviewed the milking capacity of Boer does.

South American type

About one-half of the goat population of South America is found in Brazil and this country possesses two separate recognisable breeds:

(1) *Moxoto or Black Back*
This goat is found in the hot arid areas of northeast Brazil and thought to be descended from the Portuguese Charnequeiro. The goat is of medium size, adult females weighing about 30 kg. The colour is light brown or fawn, and the goat typically has a black stripe along the back and on the belly, a black face and black legs. Although the breed is a reasonable milk yielder, it is mainly bred for its meat and, to a lesser extent, for its skin.

(2) *Marota*
Devendra and Burns (1983) stated that this breed was found only in Bahia State and resembled the Saanen breed, being of similar size and colour. These goats are kept mainly for their skins, but the does are sometimes milked, producing modest yields of up to 0.5 kg/day.

Many types of goat now found in the tropics are crossbred, formed by breeding highly productive European-type goats with heat-tolerant breeds, such as the Nubian. The best known example of a stabilised crossbreed is the Anglo-Nubian, thought to contain Nubian, British and also possibly Jumna Pari blood. This cross was first made in the nineteenth century and Anglo-Nubians are now very widely distributed in the tropics and subtropics. In the United States they are highly favoured and the term employed for the breed in that country is 'Nubian' which must not be confused with the true Nubian breed of Oriental type, already described. Anglo-Nubians may be of almost any colour. They possess short, fine coats, a convex facial outline and the long drop-ears of their Nubian ancestors. Most Anglo-Nubian goats are extremely good milkers, yielding milk of consistently high quality, with a high butterfat content.

Since there are about 300 breeds of goats, the majority being found in the tropics, it is impossible to describe more than a small selection in this book. For a fuller treatment of the subject, the reader is referred to Mason (1981) and Devendra and Burns (1983). The latter authors have produced a useful check list of the main breeds employed for different productive purposes, and this guide to breed usage is presented in Table 13.12.

Goats as meat and milk producers

Tropical goats are multi-purpose animals, producing meat, milk, skins and hair. Their primary function is meat production, although in temperate countries milk has become of greater importance during the twentieth century. Goat meat is highly regarded in all countries where there is a tradition for meat consumption from both sheep and goats, and in Hindu countries the goat is the major supplier of meat to the human population.

Where these traditions apply, there can be no good case for planners to try to substitute beef from imported exotic cattle for the indigenous production of preferred goat meat. In such circumstances, a better policy would be to increase the local production of goat meat by proper attention to feeding and breeding goats for higher yields of meat per breeding female and per unit area of land. Table 13.13 presents data for the contribution by goats to the production of meat and milk in some tropical and subtropical countries. It will be noted that certain countries, such as Iraq and Libya, produce over half their total milk requirements from milch goats, while some, such as India and Libya, produce over one-third of their total national meat requirement from goats.

Goat meat is traditionally preferred to mutton in the Sudan, many other parts of Africa and some Asian countries. In Malaysia the demand for goat meat is such that it commands a higher price per kilogram than other local meats. Devendra (1987) suggested that the preference for goat meat may be related to special features which make it organoleptically different from that of mutton. In the latter meat the fat is distributed all over the body, while in goats most of the fat is deposited in the abdominal regions (Devendra and Burns, 1983). Thus, on a weight basis, goat meat has a higher lean-meat content than mutton, and there is therefore less wastage in cooking.

Some typical killing-out percentages have been given in the section describing the various breeds of goat, and Devendra and Burns (1983) have summarised the available information on this subject. The highest published killing-out percentage is 55.4, obtained with crossbred Boer females and indigenous Tanzanian castrates, reported by Hutchinson (1964). The lowest percentage, 43.5, was obtained with Small East African goats by Wilson (1958b). The average killing-out percentage of all breeds listed was 48%.

Three different types of goat meat are produced (Devendra, 1981):

(1) Meat from kids (goat-veal or cabrito, 8–12 weeks of age).
(2) Meat from young goats (prime meat) (1–2 years of age).
(3) Meat from old goats (over 2 years old).

The first type is popular in Central and South America; the latter type is, as would be expected, tough and equivalent to 'cow beef'.

Goats' milk, like meat, is not only enjoyed but even preferred by many people. In the Mediterranean region, goats comprise approximately one-quarter of the overall livestock population, and most of these goats are kept for their milk which is consumed either fresh or preserved as yoghurt or cheese Goats as producers of milk serve a useful purpose in supplying animal protein to tropical rural communities. This supply is

Table 13.12 Improved breeds in the tropics and subtropics. (Source: Devendra and Burns, 1983.)

Speciality	Breed	Country of origin
Milk		
High yield	Alpine	Switzerland; temperate, wet
	Anglo-Nubian*	UK; temperate, wet
	Saanen*	Switzerland; temperate, wet
	Toggenburg	Switzerland; temperate, wet
Medium/low yield	Barbari	India; tropical, dry
	Beetal	India; tropical, dry
	Black Bedouin	Israel, Egypt; subtropical, very dry
	Damani	Pakistan; tropical, dry
	Damascus*	Syria, Lebanon; subtropical, dry
	Dera Din Panah	Pakistan; tropical, dry
	Jumna Pari	India, Southeast Asia; tropical/subtropical, dry
	Kamori	Pakistan; subtropical, dry
	Kilis	Turkey; warm temperate, dry
	Malabari	India; tropical, humid
	Marwari	India, tropical, dry
	Sudanese-Nubian	Egypt and Sudan; subtropical/tropical, dry
	Zaraiby*	Egypt; subtropical, dry
Meat	Anglo-Nubian*	UK; temperate, wet
	Boer	South Africa; subtropical, dry
	Fijian	Fiji and Pacific Islands; tropical, humid
	Jumna Pari	India, Southeast Asia; tropical/subtropical, dry
	Katjang	Indonesia, Malaysia; tropical, humid
	Ma 'Tou	China; subtropical, humid
	Sirohi	India, tropical, dry
	Sudan Desert	Sudan; tropical, very dry
Prolificacy	Barbari	India; tropical, dry
	Boer	South Africa; subtropical, dry
	Black Bengal	India; tropical, dry
	Criollo	South America; tropical, subtropical, dry
	Damascus*	Syria, Lebanon; subtropical, dry
	Katjang	Indonesia, Malaysia; tropical, humid
	Malabar	India; tropical, humid
	Ma 'Tou*	China; subtropical, humid
	Sudan Desert	Sudan; tropical, very dry
	West African Dwarf	West Africa; tropical, humid
Mohair	Angora	Turkey; warm temperate, dry
Pashmina (Cashmere)	Kashmiri	Central Asia; high mountains, cold
Skins	Black Bengal	India; tropical, dry
	Maradi (Red Sokoto)	Niger and Nigeria; tropical, dry
	Mubende	Uganda; tropical, fairly dry
Desert survival (milk, meat)	Black Bedouin	Israel, Egypt; subtropical, very dry

*Indicates that the breed is polled.
Note: Most other (non-listed) breeds are dual-purpose.

Table 13.13 Contribution by goats to the production of meat and milk in selected countries. (After Devendra and Burns, 1983.)

Country	Goat population (millions)	Contribution by goats to		Mean milk yield (kg goat/yr)
		Milk (% of total production)	Meat (% of total production)	
Brazil	17.3	5	2	44
Cyprus	0.3	50	14	109
India	67.2	3	35	18*
Indonesia	7.8	—	11	—
Iraq	1.8	58	12	135
Libya	1.6	50	36	42
Morocco	7.6	33	16	26
Pakistan	11.4	26	—	90
Sudan	6.9	23	—	64
Turkey	18.1	28	16	67

*This very low average figure must be regarded as somewhat suspect. However, not all goats are milked.

Table 13.14 Mean composition of goats milk compared to the milk of Indian and European cows and water buffaloes (%). (After Devendra and Burns, 1983 and Rao and Nagarcenkar, 1977.)

Species	Fat	Protein	Lactose	Ash	Solids-not-fat	Total solids
Bos indicus cows	4.8	2.8	4.6	0.7	8.1	13.5
Bos taurus cows	3.7	3.4	4.8	0.7	8.9	12.7
Buffalo cows	6.8	3.9	4.9	0.8	9.6	—
Goats	4.9	4.3	4.1	0.9	9.3	14.2

valuable to vulnerable groups, subsistence peasants, pregnant and nursing mothers and young children, who could not afford to buy meat or milk and who may be too poor to own cattle or buffaloes. The composition of goats' milk compares very favourably with that of cattle and water buffaloes. The average butterfat content is about 5.0%, higher figures being obtained towards the end of the lactation. The casein (milk protein) percentage averages 4.3, which compares very favourably with about 3.0% for cows, and the albumen content is about 0.7%, compared to 0.4% for cows. Further data concerning the comparative composition of milk from cows, buffaloes and goats are presented in Table 13.14.

The milk of the goat is in some respects closer in composition to human milk than the milk of the cow, and for this reason goats' milk is an excellent baby food. It is, however, slightly alkaline in reaction while cows' milk is slightly acidic. The fat globules in goats' milk are small in average size, and the proportion of very small globules greater, so they are digested in the human stomach about four times as fast as are the large globules in cows' milk. Lastly, the milk from the goat has a much higher degree of initial cleanliness than milk from dairy cows. This is due to the fact that the faeces of the goat are pelleted and hence there is much less contamination of the udder with coliform bacteria and also because tropical goats are not so commonly infected with tuberculosis as are tropical dairy cattle. More recent studies have indicated that the overall nutritive value of goats' milk does not differ as much from that of cows' milk as the data presented in Table 13.14 would suggest. The milk from both species has a relatively low content of the essential fatty acids (linoleic, linolenic and arachidonic), and both milks contain approximately 66% by weight of saturated fatty acids in the lipid fraction. In addition, the amino acid composition of both milks is also similar as is the calcium and phosphorus content in the mineral fraction. Goats' milk has a lower content of vitamins B (particularly B_{12}) and C, and does not contain any precursors of vitamin A, unlike the milk from both cattle and water buffaloes (Jenness, 1980).

Goats lactate well in the tropics when compared to cattle or water buffaloes on a weight-for-weight basis, but there is a very great difference in yield between breeds and between strains within a breed. The highest lactation yields are attained by European-type goats in the subtropics and by crossbred goats, such as the Anglo-Nubian, in the hotter tropics.

Devendra and Burns (1983) have summarised the records for milk yields and lactation lengths of goats of different types and their data are summarised in Table 13.15. A comparison of the milk production of tropical milch animals, including goats, on the basis of their average liveweights is presented in Table 13.16.

Table 13.15 Lactation milk yields and lactation lengths of indigenous goats in the tropics and subtropics. (Source: Devendra and Burns, 1983.)

Breed	Country	Lactation yield (kg)	Daily yield (kg)	Lactation length (days)
Non-seasonal breeders				
Barbari	India, Pakistan	150–228	1.6	180–252
Black Bedouin	Israel	n.a.	1.3–2.0	n.a.
Black Bengal	Bangladesh, India	25–30	n.a.	n.a.
Boer	South Africa	n.a.	1.3–1.8	n.a.
Chapper	Pakistan	75	0.7	105
Chegu	India	40	0.4	100–110
Criollo	Venezuela	60	0.2–0.6	n.a.
Damani	Pakistan	104	1.0	105
Dera Din Panah	Pakistan	200	1.5	130
Ganjam	India	50	0.5	100
Jakharana	India	122	1.0	115
Kamori	Pakistan	228	1.8	120
Kashmir	India	20	0.2	100
Katjang	Malaysia	90	0.6–0.8	126
Maradi	Niger	75	0.5–1.5	100
Nubian (Sudanese)	Sudan, Egypt	70	1.0–2.0	n.a.
SRD (Creoule)	Brazil	n.a.	0.1–1.0	n.a.
West African Dwarf	Nigeria	38	0.3	126
Seasonal breeders				
Angora	Turkey	35–38	0.5	123–164
Beetal	India, Pakistan	140–228	1.2	208
Damascus	Cyprus	500–560	2.0	190–290
Gaddi	India, Pakistan	40–50	0.8	90–100
Jumna Pari	India	200–562	1.5–3.5	170–200
Kilis	Turkey	280	1.0	260
Malabari	India	100–200	1.0	181–210
Mamber	Israel	350–450	1.5	n.a.
Marwari	India	90	0.9	106
Najd	Iran	250	1.0	250
Sirohi	India	116	0.9	134

n.a. = not available.

Although the figures of Table 13.16 have been rounded, and can be criticised as not being applicable to all tropical conditions, it is clear that many tropical breeds of goat produce about as much milk per unit liveweight as European crossbred cattle and considerably more than most zebu cattle or water buffaloes. On the same basis, European or European crossbred goats are superior to all types of larger ruminants. As feed intake is closely related to liveweight, it follows that the goat is a comparatively efficient converter of feed into milk in the tropics. In areas where feed is in short supply and livestock must fend for themselves, the goat is often able to live and lactate in areas where there is an insufficiency of feed for larger types of animal. Milking facilities can be simple and unsophisticated because of their small size and cleaner life styles compared to cattle (see Figs 13.9 and 13.10).

Table 13.16 Comparison of milk yield of tropical milch animals on the basis of yield per unit liveweight.

Type of stock	Mean lactation yield (kg)	Mean liveweight (kg)	Yield (kg)/10 kg liveweight
Goat			
Tropical type	200	40	50
European crossbred	400	50	80
European type	600	60	100
Cow			
Unimproved zebu	400	300	13
Improved zebu	1000	350	29
European crossbred	2000	400	50
Water buffalo			
Unimproved breed	1000	500	20
Improved breed	1500	500	30

Fig. 13.9 A milking parlour for goats at Maracai, Venezuela. Note the overhead vacuum line to which the milking machine is attached. The goats stand on the small wooden platforms while being milked, and a feed trough can be seen at each standing.

Fig. 13.10 Intensively reared milking goats at Maracai, Venezuela. These imported Saanen goats are fed on forage and compound diets inside these rotationally grazed goat paddocks.

The lactation yield of goats, as in the case of other milking ruminants, usually increases from the first up to the third lactation, and thereafter remains fairly constant for the next three or four lactations, after which a steady decline sets in. The maximum daily yield is not reached until between the eighth and twelfth weeks of the lactation, which results in the lactation curve of the goat being flatter when compared to that for dairy cows. Cattle lactation curves have sharper peaks, with the maximum daily yield between the fourth and eighth weeks.

The gestation period of the goats is about 5 months in duration. That for European-type goats is generally given as 150 days, but slightly shorter gestation periods of 147 days have been recorded for smaller, African-type goats. Mating takes place early, sexual maturity being reached soon after the sixth month of age. The oestrus cycle is 3 weeks in duration, and the oestrus periods last for about 48 hours. The female goat will accept service when pregnant, and this action occasionally leads to abortion. It is therefore advisable to separate pregnant does from the bucks whenever this is practical (Wilson, 1957). Some tropical breeds of goat are reported to have a restricted breeding season, while others breed freely throughout the year (Devendra and Burns, 1983). The presence or absence of a pronounced breeding season does not seem to depend primarily on seasonal variation in day length associated with latitude; it appears that other factors are involved, one of which is probably the time of onset of the rains which determines seasonal feed supply.

In addition to their value as meat and milk animals, goats are prized for the production of hair and skins. The skin of the West African Red Sokoto goat is outstanding for its superior quality and hence the premium which it commands in world markets (Robinet, 1967). As has already been noted, the Kashmir breed is primarily kept for the high-value cashmere wool, or pashmina. The hair of common goats is used extensively in the carpet trade and also in the making of ropes and bags.

In spite of the obvious importance of the goat in many parts of the tropics, there is controversy as to whether the goat is beneficial or detrimental from the standpoint of its effect on tropical ecology. On the one hand early writers such as Maher (1945) condemned the goat as being the major destroyer of vegetation and hence causing erosion. Examples of drastic ecological change are quoted in support of this viewpoint from East Africa, Cyprus, southern Italy, St Helena, Ethiopia and Israel. On the other hand, workers such as Wilson (1957) and Knight (1965) have shown that the goat, because of its browsing habit, can be extremely useful in certain tropical regions by preventing or restricting bush encroachment. The encroachment of bush into grassland areas is a highly undesirable ecological change and it is primarily brought about by continuous overgrazing by cattle. Reporting on an early study, Hornby and Van Rensburg (1948) stated, 'Whenever in bushland country there is some sort of ground cover of grasses and herbs which can be expected to extend given a chance, such an area may be improved even by heavy goat browsing, whereas the same area would be damaged by anything more than the lightest grazing by cattle'.

Staples *et al.* (1942) reported the results of a classic experiment in which goat-browsing, cattle-grazing and the effects of mixed stocking by both cattle and goats, were compared over a 4-year period. It was concluded that goats were not so responsible for soil erosion as cattle, since the primary cause of erosion is removal of the ground cover of grasses and herbs rather than a reduction in the population of trees and bushes. After cattle have removed the ground cover by extensive overgrazing, the vegetation can no longer support anything except light stocking by cattle, but it can continue to carry many browsing goats. The result is that the goat is the last animal found on land which is eroding for lack of ground cover, and the observer wrongly concludes that the goat has been primarily responsible for the whole ecological change. The tropical goat is thus made the 'scapegoat' for the consequences of overgrazing by cattle. Campbell *et al.* (1962) reviewed the situation in regard to this controversy in West Africa, and have suggested that the goat could be an important animal in tree and shrub savanna regions of the world.

When land which has been overgrazed by cattle or by cattle and sheep shows signs of soil exposure, it should not be grazed by goats and the resultant soil erosion blamed on the latter species. Specialists in the agricultural development of land resources often exaggerate the destructive habits of goats, even though other contributing factors such as overpopulation, uncontrolled fires and damage to the vegetation by pests and diseases are recognised. The arguments frequently advanced against the practice of mixed stocking by cattle and goats are that goats are difficult to control in numbers and more difficult to confine than cattle by the use of fences and hedges. The former is a criticism of the stock owner rather than the goat *per se*, and the latter point can be partly met by the technique of placing a 'collar' round the necks of the goats, formed by a triangle of long sticks, which effectively prevents the wearer from scrambling through, or jumping over, normal cattle fences.

Milch goats are best kept under more intensive conditions, such as by running them in yards with simple protective shelters. In cases where it is necessary to allow the goats access to pasturage or browse as the major item in their diet, it should be noted that the total daily intake of nutrients can be obtained in approximately 2–3 hours of browsing. This contrasts markedly with cattle, since these larger animals need to graze for a minimum period of about 8 hours each day. Allowing for walking and idling time, the minimal periods at pasture for cattle cannot be reduced much below 10–12 hours without restricting herbage intake and hence productive performance. The goat, on the other hand, can obtain its daily requirements in 4–5 hours, inclusive of walking time.

Finally, tropical goats can contribute towards improving the standard of nutrition of the human population, particularly by providing protein both in the form of meat and of milk. The short generation interval of goats makes it possible to increase production more rapidly than from cattle, and goats have the important advantage over sheep of generally higher levels of fertility. Most pertinent of all, goats are able to live in arid areas on sparse vegetation of no value for human nutrition, of limited use by other domesticated livestock and usually of little value for arable farming. The small size and relatively low individual values of goats brings them well within the capacity of low-income peasants. It is not without good reason that the goat is often known, in tropical areas, as 'the small man's cow'.

FURTHER READING

Cockrill, Ross W. (1976) *Buffalo in China.* FAO: Rome.

Gattenby, R.M. (1986) *Sheep Production in the Tropics and Sub-tropics.* Longman Group: London.

Payne, W.J.A. (1990) *An Introduction to Animal Husbandry in the Tropics*, 4th edn. Longman: London.

Smith, A.J. (ed.) (1985) *Milk Production in Developing Countries.* Centre for Tropical Veterinary Medicine, University of Edinburgh: Edinburgh.

Wilson, R.T. (1984) *The Camel.* Longman Group: London.

14
Adaptation of Livestock to Tropical Environments

P.N. Wilson

Many attempts to export temperate breeds of livestock to the tropics have met with dismal failure. After a short time in the tropics, the productivity of many breeds of exotic stock decreased, their condition deteriorated and they became susceptible to tropical diseases. It was only later that agricultural scientists devoted attention to the subject of the adaptation of livestock to hot climates. The environmental physiology of farm livestock, and the effects of solar radiation and heat stress on their productivity in tropical and subtropical areas, have been well documented. A comprehensive review by Johnson (1987) is available on this subject. This chapter reviews some of the more important findings which have emerged, with particular reference to the adaptation of bovines to hot climates. The basic principles of thermal adaptation are common to all vertebrates, although the different types of external covering (hair, wool feathers and bristles) and the different types of underlying skin structure, give rise to variations in the mechanism of heat loss and the maintenance of homeothermy. The reason for focusing attention on bovines is that much of the early critical work has been carried out on them and because the successful thermal adaptation of cattle is probably of greater importance that the adaptation of any other classes of livestock, since cattle are kept mainly in the open.

A temperate animal taken to a hot climate is affected in two distinct ways: directly by the influence of high temperature and intense radiation and, possibly, humidity on the animal itself, and indirectly by the effect of heat on the animal's environment, including shelter and housing.

DIRECT EFFECTS OF HEAT ON TROPICAL LIVESTOCK

Brody, in a classical review of the energetics of livestock (1945), stated that: 'An animal's main task is to maintain its internal environment constant'. All domestic animals are homeotherms. In the case of vertebrate farm animals this means maintaining the internal temperature of the animal constant, or nearly constant. The 'normal' diurnal variation in the body temperature of most farm animals is of the order of 0.6–1.2°C. Rises in temperature of more than about 1.2°C are indicative of ill health or poor adaptation to hot conditions. One notable exception to this general rule is the dromedary which has a 'normal' diurnal variation in body temperature of about 6°C, (see Chapter 13 for a detailed description of the dromedary). The maintenance of an almost constant internal body temperature is known as *homeothermy*, and it is necessary for the efficient functioning of the brain tissues. When these delicate tissues are overheated, many cells are destroyed and the animal eventually dies.

Every vertebrate animal has a particular range of environmental temperature to which it is adapted and in which it is able to live most efficiently at a metabolic rate which is independent of environmental temperature. This temperature range is spoken of as the 'comfort zone' and it varies widely from species to species and between different breeds. *Bos indicus* cattle for example, have a comfort zone of 15–27°C while *B. taurus* cattle perform best at lower temperatures, with a 'comfort zone' of 0–20°C. When animals are kept at temperatures below or above their comfort zones their metabolic rate is increased, either to keep the animal warm (shivering) or to assist in heat dissipation (panting). It should be noted that the comfort zone varies with age, younger animals performing best at the higher end of the range. This point is illustrated in Fig. 14.1.

In order to follow the different steps in the maintenance of homeothermy, let us consider the reactions of a cow (whose normal body temperature is approximately 38.6°C) exposed to direct tropical sunlight and to an air temperature of about 40°C. We will assume that the cow is unable to alleviate the heat stress by seeking shade. In practice some of the following events occur simultaneously:

(1) The skin temperature will rise. This is a direct result of the external surface of the animal receiving direct sunlight, and, therefore, radiant heat energy. Because the cow has a layer of subcutaneous fat underneath its hide, there will be a heat gradient between the outer surface of the animal and the underlying tissues. External skin and blood temperatures will differ markedly at this initial stage.

(2) The rise in temperature of the skin will have been detected by the nerves which penetrate into the hide of the animal, and messages to this effect will be communicated to the brain. This will set up a chain reaction, involuntarily controlled by the brain, each step of which is an attempt to reduce the heat stress of the animal. The first involuntary response will be that the cow reduces her activity, thereby lowering her metabolic rate. Walking will slow down or stop; grazing will be reduced or cease.

(3) As a reduction in the activity in the animal can only reduce the metabolic heat load and will not enable it to dissipate the extra heat received from the sun, a more complicated series of responses will now take place. The most important of these is that the animal will transpire. Transpiration can take place in two distinct ways. First, sweat can be secreted in the apocrine sweat glands of the animal, and pass up the sweat

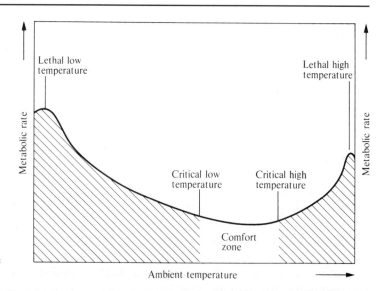

Fig. 14.1 The relationship of the basal metabolic rate to the ambient temperature in homeotherms, showing the position of the 'comfort zone'.

ducts to the skin surface, from whence the aqueous fraction of the sweat can be evaporated. Second, water can be diffused through the epidermal layers of the animal on to the surfaces, from whence it can be evaporated. These layers can be outside the animal (skin), or inside the animal (buccal cavity and respiratory tracts). In practice, it is difficult to determine the relative importance of evaporative losses of 'diffusion water' and 'sweat'. Sweating is the primary and most important heat-regulating mechanism in all farm animals other than poultry (Finch *et al.*, 1982). It can be clearly recognised with animals with eccrine sweat glands secreting large droplets of sweat, such as equines, but it is not so readily apparent in animals having small apocrine sweat glands, such as the cow or goat. (In the case of cattle which are adapted to a hot climate, the chain reaction will normally stop at this stage.)

(4) Assuming that the evaporation of water and sweat from the surface of the body has not restored homeothermy, the cow will evaporate large quantities of water from the upper part of the respiratory tracts and the buccal cavity. This is assisted by a rise in the respiration rate, which is achieved by increasing the number of respirations per minute with a corresponding decrease in the tidal volume. At this stage, therefore, the amount of air entering and leaving the lungs per unit time is not significantly altered.

(5) After a period of high respiration rate, the cow will be tired and depressed. She will probably lie down, but where there is no shade this will serve merely to increase the heat stress on the animal by exposing a greater body surface area to the direct sunlight. From now on each progressive reaction of the animal, instead of alleviating heat stress, will have the opposite effect.

(6) The heart rate will rise. This rise in the heart rate is caused by the increased respiration rate. The muscles concerned with respiration will require an enhanced oxygen supply because of their increased activity. This will be achieved by increasing the blood supply through increased pumping by the heart. The higher pulse will increase the muscular heat produced by the cow and this will lead to a further warming of the blood. The animal is now overheated both by external and internal causes.

(7) The true body temperature will commence to rise. The true body temperature, sometimes known as the deep body temperature, is generally taken to be the temperature of the blood leaving the heart.

(8) A little later, due to the lagging effects of the contents of the alimentary canal, the rectal temperature of the body will also rise. Generally, the rectal temperature will be between 0.1 and 0.3°C less than that of the true heart temperature measured in the dorsal aorta.

(9) The cow is now in a condition of marked distress. The true body temperature and the rectal temperature continue to rise. Eventually the 'tidal volume' will increase. In other words, the amount of air entering and leaving the lungs per unit time will become greater than normal. This is sometimes spoken of as 'deep panting' or 'second-phase breathing'.

(10) This second-phase breathing will result in a marked increase in body temperature due to the increased muscular activity of the respiratory system. This in turn will result in yet warmer blood being pumped to the brain. Eventually, damage of a permanent nature will be done to the brain tissues and, when this has occurred, the animal will go into a coma and death will normally result.

Evaporative cooling

We must regard the evaporative losses of water from the body surfaces as being the most important of all the processes concerned with heat regulation. For every kilogram of water evaporated from the animal, either from the external skin surface or from the mouth and upper respiratory system, 2.4 MJ of excess heat will be lost (Johnson, 1987).

At low environmental temperatures, non-evaporative cooling (i.e. loss of heat due to conduction, convection and radiation) is responsible for more heat loss than evaporative cooling. Thus at freezing point, 75% or more of the heat lost is through non-evaporative cooling. As the environmental temperature rises, the proportion of evaporative cooling rises, and the proportion of non-evaporative cooling falls. At environmental temperatures around 32°C, about 80% of the heat loss is due to evaporative cooling and only 20% to all other factors. These changes in proportions of non-evaporative and evaporative cooling are illustrated in Fig. 14.2.

The principal effect of sweating is to limit the rise in skin temperature, and since skin temperature largely governs respiration rate, a high sweating rate is usually associated with a relatively low respiration rate (Gatenby, 1986).

Early workers were sceptical about the importance of sweating in the heat-tolerance mechanism of the bovine but early studies by Dowling (1958) and Taneja (1959), served to establish sweating as one of the major avenues of heat dissipation in the heat-stressed animal. Brook and Short (1960a, 1960b), in a classical experiment with a group of Merino sheep, which were congenitally lacking sweat glands, compared the temperature response of abnormal and normal ewes in a psychometric chamber at an ambient temperature of 40°C. It was found that the temperature of the ewes without sweat glands was about 1°C above the corresponding temperature of the control group. However, evaporation of moisture from the skin is less important in sheep than in cattle. Insulation (by the fleece) and evaporation from the respiratory tract are the main ways in which sheep maintain homeothermy (Johnson, 1987). Table 14.1 shows the rate of water loss from the skin of various animal species.

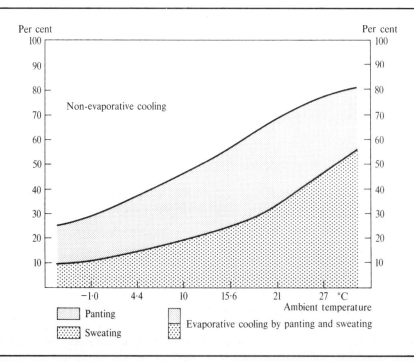

Fig. 14.2 The proportion of non-evaporative to evaporative cooling in cattle, at different ambient temperatures.

The structure, number and volume of sweat glands

The sweating mechanisms of most farm animals are relatively inefficient when compared to those of man. This is illustrated in Table 14.2 in which statistics relating to the efficiency of the sweat glands of sheep, European cattle and man are compared.

A diagrammatic sketch of an apocrine sweat gland of a cow is shown in Fig. 14.3. Sheep possess similar apocrine glands. Each sweat gland is relatively deep-seated and is closely associated with a hair fibre. The duct of the apocrine sweat gland is relatively long and narrow and in general the structure does not seem to be a very efficient means of secreting a liquid capable of evaporation, and conveying it to the surface of the animal. The histology of the apocrine sweat gland has been studied and three distinct types of gland have been identified:

(1) Tubular coiled glands, of variable length, with small diameters (found usually in European cattle).
(2) Baggy glands, of variable length with large diameters (found usually in zebu cattle).
(3) Club-shaped glands, of more constant length with a wide, lower end and a narrow, semi-coiled, upper end (found usually in crossbred *Bos taurus* × *B. indicus* cattle).

It was once assumed that the ranking of different species and breeds of cattle for heat tolerance agreed closely with their ranking for mean sweat gland volume. Baggy glands have the greatest volume, followed by club-shaped glands, with the tubular coiled glands having lowest volumes. However, the classical study of 14 different breeds of *B. indicus* cattle by Walker (1960) suggested that the rate of sweat secretion is a better index of heat tolerance than the volume of the sweat gland. This point may be illustrated by

Table 14.1 Rate of water loss from the skin of some ungulates. (Source: Whittow, 1971.)

Species	Air temperature (°C)	Cutaneous evaporation (g m^2/hr)	Reference
Buffalo (*Syncerus caffer*)	40	240	Robertshaw and Taylor (1969)
Camel (*Camelus dromedarius*)	57	240	Schmidt-Nielsen *et al.* (1957)
Eland (*Taurotragus oryx*)	40	200	Taylor and Lyman (1967)
Goat (*Capra hircus*)	40	50	Allen and Bligh (1969)
Horse (*Equus caballus*)	40	100	Allen and Bligh (1969)
Ox (*Bos taurus*)	40	145	McLean (1963)
Pig (*Sus scrofa*)	35	24	Ingram (1964)
Sheep (*Ovis aries*) with sweat glands	>30	63	Brook and Short (1960b)

Table 14.2 Approximate values for efficiency of sweat glands of sheep, cattle and man exposed to high temperatures. (Source: Brook and Short, 1960a.)

Animal	Sweat Glands		Rate of sweating		Ratio of sweat output gland/hour to gland volume (mg hr/mm^3)
	No./cm^2	Volume of secretory part (mm^3)	Sweat/gland (mm/hr)	Total sweat (g m^2/hr)	
Cattle (*Bos taurus*)	1000	0.010	0.06	588	6.0:1
Man	150	0.003	1.3	2000	433:1
Sheep	290	0.004	0.01	32	2.5:1

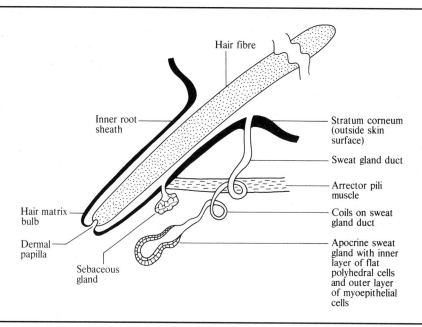

Fig. 14.3 Diagram of a section through the skin of a crossbred *Bos taurus* × *B. indicus* cow, showing the relationship of a hair follicle with its associated apocrine sweat gland.

comparing the similar sweat gland volumes, but vastly different sweating rates, of sheep and man in Table 14.2. Early studies by *Pan et al.* (1969) showed that sweating rate was positively correlated with sweat gland density, but negatively correlated with sweat gland volume. However, later studies by Amakiri and Mordi (1975) revealed that neither sweat gland volume nor density has a direct effect on sweating rate.

The total number of sweat glands is determined at birth. As the animal grows, so the number of sweat glands per unit surface area diminishes. At low environmental temperatures the rate of evaporation over the body is roughly uniform, but as temperatures rise locational differences in evaporation rates appear. Evaporation is greatest from the back and forequarters of the animal, i.e. those regions which naturally receive the greatest incidence of solar radiation (Wilson, 1989).

The sweat glands of cattle differ in structure and functional ability between breeds and species. Data relating to the rate of evaporation from the skin of pure and crossbred cattle reveal some startling differences. When Brahman and Shorthorn cattle were equally stressed by heat, Brahmans had the higher rate of cutaneous evaporation. In general, the sweat rate of *Bos taurus* cattle is less than that of *B. indicus* cattle, particularly at high air temperatures (Amakiri and Mordi, 1975).

Differences in the number, size and volume of sweat glands between *Bos indicus* and *B. taurus* cattle have long been subject to controversy. Early workers considered that *B. indicus* cattle possess more sweat glands per unit area of skin than *B. taurus* cattle, but this has been disputed by Shafie and El-Tannikhy (1970).

Amakiri (1974b), observed that the Friesian had a larger sweat gland volume than indigenous Nigerian breeds. The White Fulani (*Bos indicus*), which possessed the lowest sweat gland volume, also had the highest rate of evaporation. This lends support to the suggestion made by Jenkinson and Nay (1973) that the amount of moisture evaporated from the skin is related more to the functional ability of the sweat glands than to their distribution over the body.

The dromedary has a much thicker skin than taurine cattle and deeply embedded sweat glands. It has an average of 200 glands/cm² which is about one-third of the number found on cattle. Camel sweat contains about four times as much potassium and sodium and also has large amounts of bicarbonate (Yagil, 1985). Camels do not lose their appetite to the same extent as other species and thus do not lose weight and condition to the same extent as cattle or buffaloes exposed to arid climates and high ambient temperatures (Lewis, 1978).

The sweat glands of goats are morphologically similar to those of cattle. Most goat breeds have a lower sweat-gland density than cattle (100–300 glands/cm² compared to 1000 glands/cm²) but the output of sweat per gland is comparable (Robertshaw, 1982). However, certain tropical goat breeds, such as the Black Bedouin, have a much greater output of sweat which can be as high as 140 g/m²/hr, which is about four times as great as *Bos indicus* cattle. It is possible that most tropical goat breeds likewise have a greater output of sweat but detailed comparative studies of the many different breeds have not been undertaken.

It has already been shown that, when cattle are exposed to high temperatures, the chief cause of heat loss from the animal is the evaporation of water from the internal and external body surfaces. However, water evaporation is not the only means by which animals may lose heat. The following equation, often spoken of as the 'homeothermic equation', indicates all the various ways by which an animal may lose or gain heat from its environment:

$$M - E + F - Cd - Cv - R = 0$$

where M is the metabolic heat produced by the animal; E is the heat lost through evaporation of water from the body surfaces; F is the heat lost or gained by the consumption of feed which is either hotter or colder than the body temperature; Cd is the heat lost by conduction (i.e. the transfer of heat energy from particle to particle by increased molecular activity); Cv is the heat lost by convection (i.e. the transfer of heat energy by a circulation of heated materials, usually air); and R is the heat lost through radiation (i.e. the transfer of energy across space without heating the space through which it passes).

The metabolic heat production (M) depends on:

(1) Basal heat production for maintaining essential body processes.
(2) Digestive heat production that varies with the type of digestive system of the animal.
(3) Muscular heat production, which is much higher in working draught animals.
(4) Productive activities, such as growth, milk production or reproduction.

Where the environmental temperature is above the normal body temperature, a cow clearly cannot lose heat to its environment by convection, conduction or by radiation. In fact, the animal will be gaining heat by the radiant energy received from the sun, by the convection currents of warm circulating air and by conduction of heat to the animal from its hot surroundings. Convection heat loss increases when the animal is exposed to breezes and livestock housing should therefore be constructed so as to encourage air movement around the animals. Artificial cooling methods may sometimes be economically beneficial but only when levels of production are high. In this connection the use of low-cost wind and water sprays has been reviewed by Johnson *et al.* (1987), and Salah *et al.* (1992) have demonstrated that when bulls were cooled by water sprinkling their semen quality was significantly improved.

The animal may possibly lose heat if, during the period of heat stress, it ingests large quantities of water at a lower temperature than the body of the animal, such as river water. Water which has been standing in an exposed trough or shallow pond may be at or near body temperature in the hot tropics. River or well water will, however, invariably be below body temperature and therefore is a preferable source of supply for tropical livestock (Baker *et al.*, 1987).

It is pertinent to comment at this stage on the various ways in which the animal is affected by the radiation received from the sun. Radiation may be divided into infrared or heat radiation (long wavelengths); visible light (medium waves); and ultraviolet radiation (short waves). The proportion of long and medium waves in the total radiation will increase as one moves from the poles towards the equator. The proportion of short waves present in the radiation will increase as one moves from sea level to high altitudes. It follows that the greatest radiation intensity is encountered by farm animals kept at a high altitude on the thermal equator.

Solar radiation not only increases the heat load on the animal but can adversely affect the skin, causing skin cancer. However, most of the infrared radiation received by the animal will be absorbed by the coat. The characteristics of skin colour, hair length and hair colour will influence this absorption since light, sleek hairs will enable a greater proportion of the infrared radiation to be reflected. The visible light falling on the animal can be more readily reflected than infrared radiation, the amount also depending on the physical characteristics of the coat surface.

Definition of heat tolerance

Now that we have considered the various ways by which an animal receives a heat load from its environment, and also the chain of the events which takes place when this heat load is recorded by the peripheral nervous system of the animal, we can attempt to define the term 'heat tolerance'. An old definition of a heat-tolerant animal was 'one which eliminates large amounts of excess heat and allows productive processes to continue at a high level at high air temperatures'. In other words, emphasis used to be placed upon the *elimination* of excess heat by animals kept in hot environments.

This definition is now considered to be too narrow, and current workers would define a heat-tolerant animal as 'one which has a high efficiency of energy utilisation and allows productive processes to continue at a high level *without the production* of excessive amounts of heat'. In other words, emphasis is now placed more upon the first term (M) in the homeothermic equation. The greater the amount of metabolic heat produced by the animal itself, the greater the need to eliminate this heat from the animal by some means or other under hot conditions. Conversely, at lower basic metabolic rates, less heat is produced and the result is a greater adaptation of the animal to its environment.

It should be noted that both definitions place emphasis upon the importance of productive processes. The animal is clearly not heat tolerant if it lives in a tropical environment but fails to produce calves, milk or liveweight gain. Unfortunately, most attempts to measure heat tolerance have usually failed to take into account the parallel measurement of the animal's production. There is a need to devise a new index, easily measured, which incorporates an objective assessment of productivity.

Heat-tolerance indices

The dated, but widely used, test for heat tolerance has been the *Iberia Heat Tolerance Test*, evolved by Rhoad (1944). This is a simplified field test, only the rectal temperature being measured, with no account taken of the animal's productivity. The animals on test are kept in direct sunlight, with environmental temperatures between 29.4 and 35°C between the hours of 10 AM and 3 PM. The heat-tolerance index is then defined as follows:

$$\text{Heat-tolerance index} = 100 - 10(t_{3\text{PM}} - t_{10\text{AM}})$$

where $t_{3\text{PM}}$ is the rectal temperature of the animal at 3 PM and $t_{10\text{AM}}$ is the rectal temperature of the animal at 10 AM.

Thus if an animal whose morning body temperature is 38°C is exposed to a high air temperature for 5 hours so that its temperature at 3 PM has risen to 41°C then the heat-tolerance index would be:

$$\text{Heat-tolerance index} = 100 - (10 \times 3) = 70$$

An animal which showed no increase in rectal temperature during the 5-hour exposure period would be regarded as well adapted and would have a heat-tolerance index of 100. Rhoad calculated the heat tolerance of different cross-breeds and obtained the following results:

Purebred zebu = 89
½ zebu ½ Angus = 84
Santa Gertrudis = 82
½ Africander ½ Angus = 80
Jersey = 79
¼ Zebu ¾ Angus = 77
grade Hereford = 75
purebred Aberdeen Angus = 59.

Payne *et al.* (1952) working in Fiji showed that the range of heat-tolerance index found within a breed is extremely large and that the difference between breeds may be less than the differences of individual animals within breeds. Payne showed that the range within purebred Nellore (Zebu) was 70–84; within grade Friesians 58–78; and within a group of grade Jerseys 74–86. As with other aspects of animal husbandry, it is the characteristics of the individual animal that are important rather than the mean characteristics of the breed from which it is drawn.

The Rhoad's Heat Tolerance Test has been criticised as it merely measures the ability or inability of the animal to reflect solar radiation efficiently. Dowling (1956) proposed the use of another index, which measures directly the ability of the animal to recover from heat stress after the stress has been removed.

In Dowling's test, known as the *Rainsby Test* (after the station at which it was devised), a group of animals are vigorously exercised for a period of at least 1 hour, after which the rectal temperatures of the animals are measured throughout the subsequent cooling period. The results are then presented in graphical form (see Fig. 14.4). An animal which is heat tolerant will begin to revert to normal body temperature as soon as the heat stress has been removed. Conversely, animals which are ill-adapted to a hot climate will continue to increase in temperature for some time after the heat stress has been removed and will return to normal much later. It must be appreciated that both these early field tests fail to take into account the important measurement of the productivity of the animal without which no test can be said to be really meaningful in agricultural terms.

In 1955, McDowell *et al.* devised a laboratory test called the *Six-hour Hot Room Test* which has since been widely used in breed comparisons in the United States, particularly at the Iberia Livestock Experimental Station, where it replaced

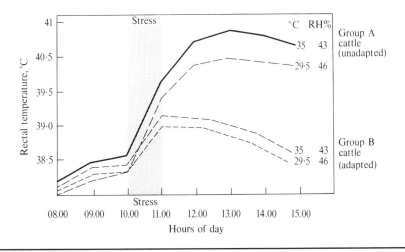

Fig. 14.4 Diagram illustrating the concept of the 'Rainsby Heat Tolerance Test'. Cattle are subjected to thermal stress for a short period (10.00–11.00 hours), and the patterns of their rectal temperature response are compared.

the earlier Rhoad's Heat Tolerance Test. It involved the measurement of rectal temperature and respiration rate over a 6-hour exposure period. Analysis of test results of groups of heifers, dry and lactating cows, revealed that location, season, environmental and rectal temperatures prior to the day of the test, stage of lactation and level of production influenced the response of all breed groups in both parameters measured.

Johnson (1987) regarded these early heat tolerance tests as being only partly useful. He considered the most serious criticism of the indices to be that they failed to take productivity into consideration and that, at best, they served largely as indicators of lack of ability to maintain heat balance. Available indices may suffice for general selection on the basis of homeothermy only, but where interest lies in the levels of productive performance attainable, all the indices are limited in application.

Coat characteristics

The role played by the hair coat in thermal balance in a hot environment is two-fold:

(1) It affords a certain degree of protection against radiant heat from the sun.
(2) It interferes with the dissipation of heat from the animal's body surface.

The number of hair follicles per unit of skin area (follicle density) is fixed at birth. Thus, as the animal grows older, or more specifically as body weight increases, follicle density becomes lower. Follicle density is greater in male than in female calves and is about 20% higher in Brahman crosses than in European-type cattle.

Early work stressed the importance of coat colour in reflecting or absorbing solar radiation. Hair fibres which are light in colour reflect more solar radiation than hairs which are dark. The amount of sunlight reflected can be as much as 50% in the case of a white-haired animal, but considerably less with a dark-coloured beast. The importance of coat colour is greater during the season of high light intensity than during the period of lower light intensity (Finch and Western, 1977; Finch et al., 1980).

Attention has also been focused on the texture of the coat. Rectal temperatures and respiration rates of woolly-coated animals were invariably higher than those of animals with fine, glossy coats. Coat texture as well as colour affected the mean absorption coefficient, i.e. the heat produced at the surface of the body by conversion of radiant solar energy into heat energy, expressed as a percentage of the theoretical maximum. White-haired zebus had a mean absorption coefficient of 49%, red Africander cattle 78% and black Aberdeen Angus 89%. Later work indicates that the effect of coat colour in this context is relatively small compared with that of coat texture. It has been suggested that medullated hair fibres might be more effective in reflecting infrared radiation from the sun than non-medullated fibres.

Much attention has been devoted by workers to the role of the hair coat as a barrier to heat dissipation. The total insulation of the hair coat of dairy calves is proportional to the hair weight per unit area of skin, and at moderate to low hair densities, coat insulation increases with increasing fibre density. The nature of the hair also has a bearing on its insulative properties. The hair fibres of cattle exhibit marked seasonal variation in the length, diameter and weight of hair per unit area. Large and significant changes are also found in the incidence and degree of medullation of the hair at different seasons of the year. The presence of a central medulla in the centre of a hair fibre enables the hair to stand more rigidly and therefore assists in the evaporation of surface water, since a semi-erect coat allows freer air movement on the skin surface than a tangled, limp coat. The hair medulla must therefore be regarded as an important characteristic of the coat in the regulation of heat dissipation. There is a significant correlation between the presence of medullated hair fibres and the maintenance of near-normal rectal temperatures. This point is borne out in Table 14.3. Thus, a coat of short, thick medullated hairs is sleek and smooth in texture and has a minimal effect on heat loss, while thin, long, non-medullated hair forms a

Table 14.3 Percentage medullation and rectal temperature of different types of cattle in Australia. (Source: Dowing, 1959.)

Type of cattle	% medullation	Rectal temperature (°C)
Brahman × Shorthorn	100.0	38.5
Illawara Shorthorn	79.9	38.6
Australian Shorthorn, group A	62.3	38.8
Australian Shorthorn, group B	46.4	39.2
Long-coated Shorthorn	29.5	40.9

woolly coat which presents a serious barrier to heat flow whilst on the other hand acting as an insulation layer which reduces heat absorbtion.

The type of hair coat is partly genetically determined, *Bos indicus* cattle having shorter and lighter hair than that of *B. taurus*. However, by far the most effective coat as far as insulation is concerned is that of sheep. Under extreme conditions, the solar heat load impinging on a sheep may be five to seven times its heat production (MacFarlane, 1964). The fact that most of this heat is not absorbed by the animal is indicative of the high insulative property of the sheep's coat.

Earlier work by Turner and Schleger (1960) showed that the coat characteristics of cattle are directly related to productive performance. A large number of imported cattle in Australia were typed for 'coat score' by visual observation and manual handling. 'Coat score' was an attempt to define the length, thickness and texture of hair in quantitative terms without actually measuring the hairs in the laboratory. When these coat scores were plotted against subsequent growth it was found that they had a low, but statistically significant, negative correlation with liveweight gain during the period 16–28 months of age ($r = -0.095$). They also established a statistically significant negative correlation between coat score and skin temperature, and showed that the correlation of coat score to succeeding liveweight gain was greater than the correlation of gain in the second year of life to gain in the first year. They therefore suggested that an evaluation of coat score would give a better indication of meat production, as measured by growth, than would an examination of liveweight gain data during the first year of life.

If these conclusions of Turner and Schleger can be verified with other breeds of cattle in other parts of the tropics, then we are nearer to the goal of defining certain factors or characteristics which are directly related to productive performance at high temperatures. Coat scores are not only of greater repeatability than measurements of body temperature, but they can be assessed at any time without the need to conduct field or laboratory experiments to obtain measurements during periods of actual exposure to heat stress.

There is evidence that the hair coat may also be used as an indicator of the condition and productive performance of the animal. Some of the qualities of the coat are related to the physiological status of the animal, notably with regard to endocrine function, which in turn has a bearing on the animal's response to heat stress.

It is possible that a component of coat type, not necessarily related to the insulative properties, may be correlated with liveweight gain and thus may be significant as an indicator of growth potential. Later work by Turner and Schleger (1970) suggested that this component is the total production of new hairs, and this is supported by the fact that unthriftiness strongly depresses new hair production. There is general agreement that a short, light-coloured coat with a smooth and glossy texture is best for minimising the adverse effects of solar radiation on livestock.

Pigmentation of the skin

The skin of farm animals can be either pigmented or non-pigmented. Usually, the breeds of domesticated livestock which have been evolved in the tropics, such as zebu cattle, possess pigmented skins irrespective of hair colour, while those evolved in temperate regions usually have skins which are non-pigmented. In some breeds, such as the Friesian, both pigmented and non-pigmented skins are found on the same animals. Black hairs usually overlie the pigmented areas, while white hairs usually overlie the non-pigmented areas.

A pigmented skin is most desirable in the tropics, since it is less susceptible to sunburn and photo-sensitivity disorders. A good example of this is provided by white-faced Hereford cattle, which are prone to epithelioma due to the high photo-sensitivity of their unpigmented eyelids. Similarly, non-pigmented breeds of pigs, such as the Large White breed, are particularly susceptible to sunburn. Many instances are on record of white-skinned pigs dying after relatively short periods of exposure to direct tropical sunlight.

Skin thickness

Thickness of skin (or hide) varies between and within breeds, and both thick and thin skins may be associated with any combination of milk or beef, and early or late maturing, temperate cattle. However, Amakiri (1974a, 1975) has shown that *Bos taurus* cattle usually possess thicker skins and more subcutaneous fat than *B. indicus*, and that the thinner skin typical of tropically adopted cattle is an adaptive attribute in a hot climate. Thus the average skin thickness in Holstein-Friesian cattle was 6.2 mm, compared to 4.9 mm in White Fulani animals and 4.8 mm in the N'Dama breed. The papillary and reticular layers were also thicker in the Holstein-Friesian, the comparative figures being 1.24 and 4.77; 0.92 and 3.57; and 0.85 and 3.56 mm respectively.

The effect of high temperatures on poultry

Poultry are very adaptable domesticated livestock. Although efficiency of production is affected by very hot climates, there are only a few locations in the tropics where climatic conditions make poultry keeping impossible (Johnson, 1987).

The 'normal' body temperatures of birds are generally much higher than those of mammals, averaging between 39 and 44°C. The upper lethal body temperature is about 47°C. In a hot environment, a high body temperature is advantageous because it diminishes the heat load and therefore the use of water in heat dissipation. Birds have no sweat glands, but they are able to lose heat through evaporative cooling by panting, and in addition a little water diffuses through the skin and feet from whence it can be evaporated, provided that feathers do not impede air movement. Panting starts in fowls in still air at ambient temperatures in the range 27 to 33°C. Well-adapted breeds of poultry are capable of maintaining an extremely high respiration rate, up to 170 respirations/minute, without apparently suffering from thermal stress (Siegel, 1968). Unadapted breeds of poultry are incapable of achieving or maintaining high respiration rates of this order and it is probable that this essential difference in respiration rates is the chief factor determining tropical adaptation of poultry. Heat production of poultry varies markedly according to the state of productivity and activity of the animal, and the basal (metabolic) heat production of laying hens is greater than that of non-laying birds.

Both feed (metabolisable energy) intake and metabolic heat production by individual laying hens vary considerably, and are markedly affected by environmental temperatures. Figure 14.5 illustrates the curvilinear relationship between environmental temperature and daily metabolisable energy intake by laying hens. Energy intake decreases at an ever-increasing rate as environmental temperature increases within the range of 21 to 38°C. As ambient temperature reaches the body temperature of the bird, feed intake appears to cease (Smith, 1971).

Many workers have reported detrimental effects of high environmental temperatures on egg production (e.g. Deaton *et al.*, 1982), and egg weight (Smith, 1974). The decline in productivity of laying hens under high ambient temperatures above 32°C may be caused by a combination of heat stress *per se* and insufficiency of certain nutrients in the daily feed intake (Smith, 1971). Therefore, diets in use in hot climates should be designed to take into consideration the fact that, under tropical conditions, the feed intake of hens declines. Consequently more protein, minerals and vitamins should be incorporated per kilo-

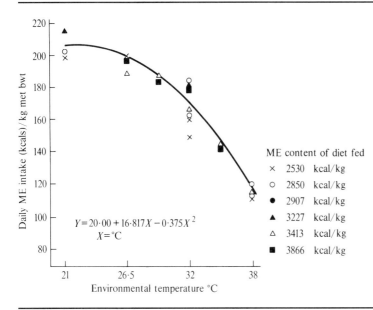

Fig. 14.5 The relationship between environmental temperature and daily metabolisable energy intake by laying hens per unit of metabolic body weight. (Adapted from Smith, 1971.)

gram of feed of ensure that birds consume sufficient quantities of these nutrients. In other words, diets of higher nutrient density are required. A comprehensive review of the effect of high environmental temperatures on the productivity of laying hens has been published by Smith (1973).

As birds are being intensively managed in the tropics on an increasing scale, it follows that a great deal of attention must be devoted to the construction of suitably designed poultry houses. The essential requirements are shelter from sun and rain, and the maximum air flow through the poultry house. The enclosed 'sweat box' type of deep litter house, designed to rear and maintain poultry through a temperate winter, is totally unsuited for the tropics. Instead, open-sided houses, with due regard to adequate roofing or walling along the side facing the prevailing wind, are desirable. Preferably, houses should be either built with relatively tall roofs or else they should be constructed with roofing materials which reflect, and do not absorb, solar radiation. Corrugated aluminium is good but expensive, and similar effects can be obtained by painting less costly corrugated iron roofs with several coats of glossy white paint. The provision of adequate drinking water, and water-trough space sufficient for the needs of each bird housed, are essential points, since birds can go for several days without feed, but usually die in hot climates if denied access to water for more than 24 hours (see Fig. 14.6).

Chicks are more tolerant of high temperatures than adult birds, but there are dangers of transporting day-old chicks in chick boxes where temperatures can often exceed 35°C. With no access to water, chick mortality can be high during this critical stage.

INDIRECT EFFECTS OF HEAT ON TROPICAL LIVESTOCK

Effect of tropical climate on animal behaviour

The behaviour patterns of farm animals, both in temperate areas and in the tropics, has received a great deal of study. The behaviour pattern is usually determined by watching a group of animals continuously for a 72-hour observation period, and recording the behaviour of each animal at regular time intervals. It has been found that the

Fig. 14.6 An intensive hen battery house, fitted with equipment for the automatic feeding and watering of the hens and for removal of the droppings. The eggs roll forward out of reach of the hens, and must be collected by hand. (Source: Zambia Information Services.)

grazing behaviour of temperate cattle in a temperate area is similar to that of tropically adapted livestock in a tropical area. However, when unacclimatised temperate livestock are observed in the tropics, large differences from the normal behaviour can be detected.

Cattle normally spend between 7 and 9 hours grazing in every 24-hour period. This grazing activity is broken up into three separate periods, one between dawn and mid-morning, a long period between midday and dusk, and a further, shorter, period around midnight. Unadapted cattle generally exhibit a behaviour pattern with a very short grazing period and an abnormally long resting period. Moreover, unadapted livestock exhibit less inclination to graze during the main afternoon period when ambient temperatures and solar radiation intensity are greatest. This point is illustrated by the figures shown in Table 14.4 from early work in Trinidad by Wilson (1961a) for mean grazing, ruminating and idling times of purebred zebu heifers and high-grade Holstein heifers, observed simultaneously in the same pasture. The data show that the less-adapted animals spent approximately 2 hours less in grazing, and 1 hour less in ruminating, than the well-adapted purebred zebus.

Experimental work, reviewed by Payne (1990), has shown the importance of providing tropical livestock with good night grazing. This is contrary to common practice in many tropical countries where herds of livestock are often placed into a night corral, completely devoid of any vegetation (Wilson, 1961c).

The early experiment conducted in Trinidad (Wilson, 1961a) has shown that it is possible to alter the diurnal grazing pattern by varying the quality of the pastures offered to the cattle by day and by night. The system in which the best pastures were provided by day and worst pastures by night resulted in 4.5 hours of day grazing and 2.9 hours of night grazing. Conversely, when

Table 14.4 Difference in grazing behaviour of zebu and crossbred cattle.

Breed	Grazing hours	Ruminating hours	Idling hours	Hours spent drinking
Pure-bred zebus	8.8	9.3	5.8	0.1
High-grade Holsteins	6.8	8.4	8.7	0.1

the best pastures were offered by night and poorer quality swards offered during the day the herd grazed for 3.1 hours by day and 4.2 hours by night. Maust *et al.* (1972) showed that increases in night grazing times during periods of heat stress significantly increased feed intake thus ameliorating the depression in feed intake noted when cows in hot climates are denied access to feed at night. Since the heat stress on an animal in the tropics is much greater during the day it follows that it is preferable to induce an animal to adopt a system of behaviour in which the maximum night-time grazing is practised; indeed Payne (1990) suggested that the availability of night grazing is critical for *Bos indicus* cattle when forage supply is short. Such systems should be more widely adopted in the tropics than they are at present.

Many behavioural patterns are means of alleviating the heat load on the animal; i.e. they are thermoregulatory. Examples of this kind of behaviour include the wallowing of pigs and buffalo in muddy water under hot conditions, thus achieving an evaporative cooling rate comparable to that of man, and the use of natural shade. The benefits of utilising shade and water cooling in beef and milk production have been studied by Adeyemo *et al.* (1979) in Nigeria and more fully discussed in Chapter 15. The beneficial effect of water sprinkling on bull fertility has already been mentioned (Salah *et al.*, 1992).

Effect of hot climates on growth, meat and milk production

Livestock growth rates are highest when the animals are kept within their particular comfort zone. When livestock are maintained at higher ambient air temperatures the appetite is depressed, feed intake is reduced and consequently the animals grow more slowly. In a classic experiment conducted by Hancock and Payne (1955), eight pairs of identical twin calves were split, half being reared in New Zealand (temperate environment) and half sent to Fiji (tropical environment), with similar conditions of feeding and management. An appreciable depression in growth rate occurred only when the temperature in Fiji was at its highest (above 29.4°C). At calving, the New Zealand heifers were approximately 10% heavier than those reared in the tropics. The growth retardment of the tropically raised heifers was reasonably uniform, affecting equally all body measurements with the exception of belly girth. The greater belly girth of the Fiji animals was attributed to their significantly greater water intake.

An assessment of the way in which a hot climate affects growth and beef production is complicated by the difficulties encountered when attempting to distinguish direct from indirect climatic effects and those effects that are not connected with climate. Many experiments have measured the extent of reduction in feed intake as a result of thermal stress. The subject has been extensively reviewed by Yousef (1985). A long-term experiment by McIlvain and Shoop (1971) quantified the effects of temperature and humidity on weight gain. They found that high temperatures in conjunction with high humidity levels substantially reduced weight gain of beef steers.

The effect of higher ambient temperature is to decrease both the quantity and quality of the milk produced by *Bos taurus* cattle. Since there is a decrease in voluntary feed intake with rising ambient temperature, it is to be expected that a hot environment, in addition to affecting the level of milk production directly, will also cause changes in milk composition comparable to those caused by under-feeding. In general, far more work has been conducted in this field with B. taurus than with B. indicus cattle, probably because decline in milk yield with rising temperature starts at a lower temperature in temperate breeds than in tropical cattle. Maust et al. (1972) noted that the response of milk yield to high temperature was complex and dependent on heat tolerance, the stage of lactation and on the level of milk production. Cows giving a high milk yield in early lactation were more depressed by heat stress than cows giving a low milk yield. McDowell et al. (1976b) also showed that high temperatures had the greatest depressing effect on milk yield during the first 60 days of lactation. A combination of high humidity and high ambient temperature can have a profound effect on milk production. Thus Johnson et al. (1987) found that the milk yield of Holstein cows was seriously depressed by high humidity above an air temperature of 27°C.

The classic experiment carried out by Payne and Hancock (1957) indicated that the effect of climate on milk quality is more marked on the butterfat constituents of milk than on the solids-not-fat. These workers showed that the average milk production of the New Zealand cows was 44% higher than that of their co-twins in Fiji, and that the total quantity of butterfat produced was 50% greater in the temperate climate. Under hot-room conditions in the United States, it has been shown that the butterfat yields of Holstein cows fell at temperatures above 27°C, while workers studying crossbred cows in Tasmania noted a fall in fat percentage at temperatures around 28°C. High ambient temperatures also result in a rise in the chloride content of milk and a fall in the milk-sugar and total nitrogen content (Rodriquez et al., 1985).

Decreases in the solids-not-far percentage of the milk of cows of various breeds have been found to occur at temperatures higher than 30°C. Decreased protein percentages in the humid tropics have been reported. The values range from 3.9% in Sindhi × Brown Swiss to 3.6% in criollo cattle. Maust et al. (1972) showed that the effects of temperature on both butterfat yield and butterfat percentage were greater than the effect on milk yield *per se*, and the subject of the effect of climate on milk composition has been well reviewed by Johnson (1985).

The effect of climate on the growth and reproduction of sheep and goats has not been as fully studied as is the case for cattle. Payne (1990) stated that temperate-type sheep in the Australian tropics exposed to high temperatures exhibited low lambing percentages and produced small lambs with high post-natal mortality. The young lamb or kid, like the young calf, is less well adapted to very high ambient temperatures than adult animals.

Similar considerations apply to pigs. Young piglets do not possess efficient heat-regulating mechanisms and are unsuited for either very hot or very cold ambient temperatures. However, as they thrive best at temperatures above 32°C for their first few days of life, there are normally fewer heat-related stress problems with young piglets in the tropics than in temperate regions, provided adequate housing is provided (Johnson, 1987). As the pig ages the optimal ambient temperature for efficient liveweight gain falls, and in most tropical climates (where the mean annual temperature varies between 25 and 35°C) pigs weighing 30–70 kg are probably being reared in near optimal conditions. Very high ambient temperatures affect embryo survival in sows and also have an effect on oestrus (Curtis, 1985).

Effect of tropical climate on reproduction

The most important obstacle hampering both beef and diary production in tropical areas is a low reproductive rate. Among others, Venter et al. (1973) have identified the following factors as contributing to this low reproductive rate:

(1) Poor/no demonstration of oestrus behaviour. Lactational anoestrus has also been shown to be more marked in tropically productive breeds (Seebeck, 1973).
(2) Poor conception rates resulting from lowered male fertility.
(3) High mortality from birth to weaning.

Apart from variations in management practices and standards, the most important factor influencing reproduction in the tropics is climate. At low ambient temperatures the testes are held close to the abdominal wall so that their temperatures can approximate to the true body temperature of the animal. As temperature increases, the testes descend lower into the scrotal sac. When the body temperature of the animal is slightly above normal while the ambient temperature is less than body temperature, the temperature of the testes will be reduced. However, when ambient temperatures are such that the temperature of the surface of the animal exposed to solar radiation is higher than 38.6°C, the testes will be above the true body temperature of the animal. As spermatogenesis is adversely affected by high temperatures, the effect of high ambient temperatures on non-adapted *Bos taurus* bulls is to decrease their efficiency; Salah *et al.* (1992). At a temperature of 40°C it is found that as little as 12 hours' exposure can prove critical to optimum spermatogenesis, and a decline in semen quality occurs after a week at this temperature. The more heat-tolerant Africander is less adversely affected by high temperatures than temperate breeds. Field studies by Venter *et al.* (1973) showed that, at temperatures greater than 18°C, 11 out of 12 Shorthorn bulls were culled either on the basis of poor semen quality or abnormalities of the genitalia, while only one of ten Africander bulls was culled. Juma *et al.* (1971), working with sheep, showed that the scrotal cutaneous receptors are capable of influencing the general body temperature, thus giving a much broader thermoregulatory role to the scrotum than the purely local one previously considered. Nakayama *et al.* (1991) have suggested that injection of testosterone in bulls in hot environments would enhance their reproductive performance.

The effect of climate on the reproductive efficiency of the female is apparent in several ways:

(1) *Growth and reproductive development of heifers.* Hafez (1967) state that heifers maintained at temperatures of 10°C reached sexual maturity at 10 months, while those kept under temperatures of 27°C were later maturing at 13 months, possibly as a result of reduced growth rate at high temperatures. Delayed onset of puberty, particularly in *Bos indicus* cattle, constitutes a major limiting factor in the breeding of yearling heifers for beef production. Baccari *et al.* (1983) showed that growth hormone output, and consequently growth rates, were depressed at high temperatures.

(2) *Conception rate and foetal development.* Bonsma *et al.* (1972) found that 2-year-old *Bos taurus* heifers, when imported from a temperate region to a subtropical region, suffered an overall drop in calving percentage from 80 to 43%. The birthweight of calves born to unadapted European breeds following a summer pregnancy in the tropics is lower than that of indigenous breeds. These workers found that, with calves born following a summer gestation in the subtropics, 33% were classified as dwarfs. The problem of the low fertility in the adult female has been widely recognised (e.g. Rosenberg *et al.* 1977). This low fertility arises from a failure of conception, or from early embryonic mortality resulting from high body temperatures (Johnson, 1987).

(3) *Calving index of adult cows.* Finally, several workers have suggested that the reproductive efficiency of cows may decrease at an earlier age in a tropical environment compared to a temperate environment (e.g. Thatcher, 1974). An examination of the lifetime records of many European cattle maintained in the tropics supports this hypothesis. The increase in calving index of such cattle at each subsequent lactation is a very common feature of the records of dairy herds kept in a tropical environment. For a more detailed account of the factors affecting fertility the reader is referred to Payne (1990).

Effect of seasonal variation in day-length

In temperate regions certain classes of livestock, especially sheep and goats, exhibit seasonal reproductive behaviour. A decrease in day-length and a corresponding increase in night-length results in the onset of oestrus. Sheep and goats can be made to breed at abnormal times of the year by artificially reducing the day-length, although hormone treatment is also necessary in order to ensure high levels of fertility out of season. The effect of photoperiod on reproduction has been reviewed by Reiter (1976) and Thwaites (1985).

In the tropics, especially near to the equator, seasonal variation in day-length is a matter of minutes rather than of hours. Many workers have therefore assumed that such small differences would be insufficient to cause a seasonal reproductive pattern, and it is well known that most breeds of livestock can reproduce at any month of the year in tropical latitudes. However, careful analysis of the parturition data for different species of livestock in the tropics usually reveals a seasonal incidence of conception and, therefore, of parturition (Simplicio et al., 1982). It has yet to be shown whether these seasonal rhythms of reproductive behaviour are dependent upon the small differences in day-length or whether they are imposed by other factors, such as seasonal differences in nutritional quality of feed. It is, however, important for the tropical animal husbandman to realise that some seasonal variations in reproductive behaviour do exist, as it may well be advisable to plan seasonal breeding programmes which are coincident with the normal rhythm. In this connection it is interesting to observe that as little as 10 minutes difference in day-length can determine whether a plant does or does not flower. Payne (1990) suggested that the relatively small seasonal variation in day-length affects the seasonal shedding of woolly coats by some temperate-type cattle imported into the tropics.

It is well known that the artificial provision of extra lighting for intensively managed livestock, such as poultry, results in a increase in production in temperate areas. Limited work carried out in the tropics indicates that the provision of artificial lighting may prove beneficial, even in those areas where the length of day does not vary much from an average figure of 12 hours. Poultry houses constructed with low hanging eaves, designed to keep out driving tropical rain, effectively cut out the early morning and late evening sun. The effective day-length inside such houses may be as little as 8 hours. The provision of artificial light in such instances may prove economic and lead to a significant increase in productivity.

The effect of high temperature is to increase the water requirement of all forms of livestock in the tropics compared to their normal requirements in temperate areas. Part of the increased water intake is required to replace water lost from the body by evaporation, and the subject is discussed in detail by King (1983). The increased water requirement can be as high as twice or more of the need of the same type of stock in a temperate climate. For instance, whereas the daily intake of drinking water of temperate dairy cows is approximately 5 litres/100 kg liveweight, the same type of animal maintained in the tropics will have a free-water intake approximating to 10 litres/100 kg liveweight. If the tropical forage is much drier than the temperate forage, then this difference will be increased still further. In the case of intensively managed livestock with simple stomachs, fed on dry meals, the extra water consumption is rarely in excess of 50%. However, it should be emphasised that water consumption under tropical conditions, particularly of cattle, varies widely from one animal to another (Coppock et al., 1988).

In a review of ruminant nutrition in the tropics, Payne (1990) gives a detailed breakdown of the chief components and effects of the tropical slimate – ambient temperature, humidity, solar radiation and climatic stress – and examines the effect of each on water consumption. A more extensive review of the water requirements of ruminants is given by the Agricultural and Food Research Council (1980). The amount of water drunk is governed mainly by the severity of the heat, and the amount of dry matter eaten. An extensive early account of the relationship between water consumption and dry-matter intake, and how this changes with rising temperature, has been provided by Payne (1963) who confirmed that not only do *Bos indicus* cattle consume less water in general than *B. taurus* cattle, but they also consume less water per unit of dry-matter intake. Thus, it has been concluded that high heat tolerance (as exhibited by *B. indicus* breeds) is associated with low water consumption, and conversely low heat tolerance with a high water consumption.

The composition of the grass or forage can also affect water consumption. Payne (1963) found that raising the crude protein content in feed caused both *Bos taurus* and *B. indicus* cattle to increase their consumption of water. The water content of the feed can also affect water intake. Table 14.5 presents data relating to the seasonal differences in the water intake of zebu heifers in Uganda and crossbred zebu-Holstein cattle in Trinidad. Such differences are common in the tropics, where there are pronounced wet and dry seasons. The differences between the two estimates are large and must be ascribed to differences in breed, climate and water content of the herbage.

When the figures are expressed as total water intake (both free-water drunk and feed-water consumed) the seasonal difference is greatly reduced. In Trinidad, for instance, in the experiments referred to in Table 14.5 the *total* daily water intake per 100 kg liveweight was 12.1 kg/cow in the wet season, and 13.1 kg/cow in the dry season. In places, such as Trinidad, with a climate mainly dependent upon moist easterly air currents, the chief climatic differences are found in the rainfall rather than in the ambient temperature. The heat stress on the animal is therefore reasonably constant throughout the year, with day temperatures almost invariably in the upper twenties (°C). The total water requirement of livestock is therefore relatively constant from one season to another. Under these conditions, the differences in the amount of drinking water which the farmer must provide for his grazing animals is dependent mainly upon the water content of the herbage, and this is itself mainly dependent upon the seasonal variation in rain precipitation and thus in seasonal grass growth.

Indirect effect of climate on tropical livestock

The most important indirect effect of climate on tropical livestock has already been mentioned during the above discussion on water requirement and feed intake. The quantity and quality of feed on offer to tropical livestock is primarily dependent upon the climatic factors influencing, and possibly limiting, plant growth. A much fuller treatment of this climatic effect on tropical vegetation is to be found in Chapter 2.

The second most important indirect effect of climate on farm animals is its influence on the distribution of the major pests and diseases, and the arthropod vectors which are responsible for their spread. A good example of this is given by the disease trypanosomiasis, which is mainly spread by the tsetse fly (*Glossina* spp.) in Africa. The distribution of the tsetse fly is directly related to the presence or absence of suitable breeding sites, and these are themselves influenced by the climate of that region. If the climate or the vegetation pattern can be modified so that the tsetse fly is denied suitable breeding grounds, then trypanosomiasis could be eliminated from tropical areas where it is endemic. This subject will be discussed further in Chapter 17.

Table 14.5 Seasonal differences in free-water intake. (Sources: Wilson, 1961c; Wilson *et al.*, 1962.)

Type of stock	Wet-season intake		Dry-season intake	
	Water drunk cow/day (kg)	Water drunk/100 kg liveweight (kg)	Water drunk cow/day (kg)	Water drunk/100 kg liveweight (kg)
Zebu heifers at Serere (Uganda)	10.1	5.1	20.3	8.8
Crossbred Holstein × zebu cows at Centeno (Trinidad)	8.4	1.9	37.0	8.1

Adaptation

Acclimatisation is the term used to describe the process by which an animal adapts itself to its environment. It is defined by Bligh and Johnson (1973) as 'A change which reduces the physiological strain produced by a stressful component of the total environment'. This change may occur during the lifetime of an organism or be the result of genetic selection; if the stresses imposed by climate are too great it will not acclimatise and it will become unthrifty and may die. Adaptation to tropical environments depends upon the animal increasing its heat loss, reducing its heat production or increasing its tolerance to higher, or more variable, body temperature. The opposite applies to adaptation to cold temperatures.

Acclimatisation can be either short- or long-term. Permanent acclimatisation may be due to changes in animal behaviour (e.g. more night-time grazing) or to changes in physiological reactions (e.g. an increase in the diurnal temperature variation of the animal). In the very long term there may be natural or artificial selection for morphological characters which may assist heat tolerance (e.g. selection for lighter hair colour or pigmented skin).

FURTHER READING

Bligh, J., Cloudsley-Thompson, J.L. and MacDonald, A.G. (eds) (1976) *Environmental Physiology of Animals*. Blackwell: Oxford.

Johnson, H.D. (ed.) (1987) *Bioclimatology and the Adaptation of Livestock*. World Animal Science Series, B5. Elsevier: Amsterdam.

Wilson, R.T. (1989) *Ecophysiology of the Camelidae and Other Desert Ruminants*. Springer-Verlag: London.

15
Cattle Management in the Tropics

P.N. Wilson

Usually, in an agricultural textbook, the section on cattle management is a description of the systems which are best suited to the production of milk and/or meat on farms of different sizes in defined locations. The systems described are determined by two factors: (1) the commodity or commodities for sale, and (2) the scale of farming operation. In the tropics this distinction is invalid. Cattle are not kept solely for the production of saleable commodities; they are maintained, to a large degree, because of their intrinsic social value or for work. This point has already been referred to earlier. Thus a rigid distinction between 'beef production systems' and 'dairy production systems' is not helpful. Before tropical cattle can be efficiently and effectively managed for the production of saleable commodities, such as milk and beef, it is important to have an appreciation of the characteristics of indigenous systems of livestock management in the tropics.

Although some progress has been made in recent decades to assist cattle-owning peoples to move from a subsistence economy (based on livestock) to a cash economy (in which stock are exploited as a means of economic production) it will take time before the traditional practices disappear. Customs, such as the payment of the 'bride price' with so many head of stock, continue to be practised in many rural areas. Temple and Reh (1984) have noted that promoting efficient livestock production is more difficult than increasing the productivity of other sectors of primary industry. Joubert (1975), referring to Southern Africa, stressed the important role of international agencies in promoting improved livestock productivity.

The social importance of cattle will now be illustrated by reference to the system of cattle management once practised by the Abahima of Ankole district, Uganda. This example has been chosen since the traditional system of management of the Abahima has been well documented (Mackintosh, 1938), and it provides an illustration of the interdependence of livestock husbandry and socio-anthropology. It should be noted that this particular system of cattle-keeping once practised by the Abahima, and the customs associated with it, is not typical of cattle-keeping in the tropics. Nevertheless a study of the Abahima illustrates the symbiotic relationship which certain peoples in Africa have with their animals.

The Abahima managed their large herds of sanga-type, Ankole cattle on the inland plains bordering the northwest shores of Lake Victoria. The altitude varies between 1000 and 3000m, and the mean annual rainfall is about 110mm. The vegetation is open grassland, with light thorn scrub, and the dominant species is red oat grass, *Themeda trianda*.

The Abahima considered four factors to be of importance when judging the perceived value, or 'desirability', of an animal. The preferred colour was dark red, and calves with other colours were usually slaughtered at birth. The horns were required to be long and white, with their tips bent forward. Breeding for this character resulted in stock with massive horns, weighing as much as 70kg. This heavy weight often resulted in the head of the animal being held down so that movement was slow and the gait somewhat erratic. The hump was required to be large, and selection for this factor resulted in many stock with humps which toppled over, giving the animal an asymmetrical appearance. Lastly, the animals were selected for fatness, obese bulls being highly prized. It will be noted that none of these factors has any direct economic significance. On the contrary, continuous artificial selection for large horns, humps and excessive amounts of fat resulted in a lowering of the efficiency of the stock for production and reproduction.

The herdsmen were always present at the birth of a calf. The tribal magician (*Omufumu*) was called in to assist with cases of difficult parturition, and charms were fixed to the unborn calf and its dam. In cases of difficulty the tribal doctor (*Omuzuzi*) would be called upon to dissect the calf *in utero* so as to save the cow. Female calves were especially prized and were kept in the owner's hut for the first week, before being placed in the communal calf-house. Calves were not allowed to run with their dams, but were allowed to suckle before milkings, representing a system of calf management common to many parts of the tropics. The young calves were grazed near the kraal and herding was the responsibility of children. The calves were housed during the six hottest hours of the day to protect them from the sun and biting flies such as *Stomoxys*. Calves did not join the main herd until after they were about 8 months of age. Heifers were not given their adult name until after the birth of their first calf, and different nouns were used to describe the various stages of growth through calfhood to first calving. Each of these names was a precise description of the animal in question.

The organisation of the Abahima tribe was largely determined by the need to group a certain number of people at each kraal, sufficient to carry out the various duties connected with tending cattle. There was an optimum ratio of men to the number of cattle. Thus the division of a herd into two parts, when it had grown too large to be managed at a single kraal, was accompanied by the division of the family group into two sections, one for each of the newly formed herds. If a stock owner died, leaving a small herd of, say, 100 cattle and four sons to inherit his property, the four sons remained together living in a communal kraal tending a common herd, until natural increase allowed the herd to be subdivided. On the other hand, if a herd reached a size of, say, 800 cattle then the owner would have expected his sons to marry, leave home and each form a new kraal with a section of the herd, since 800 cattle based on one site would have led to overstocking and greater disease risk. The social organisation of the Abahima therefore revolved around four factors:

(1) The number of cattle at each kraal.
(2) The carrying capacity of the land and the availability of water.
(3) The need to spread disease risk by cattle dispersal.
(4) The size of the cattle owner's family.

The ceremony of marriage could not take place without the physical transfer of cattle from the groom's family to that of the bride. Cattle formed the 'bride price', but the term is misleading since this payment was a form of insurance against the risk of marriage failure. The bride price was returnable should the marriage fail through some alleged fault on the part of the bride. Before the wedding took place, the bride was fattened on a diet rich in milk, since obesity was regarded as a sign that milk was plentiful, and that the bride came from a family owning many cattle. The chief rite in a marriage ceremony was 'milk spitting' between bride and groom. Milk was regarded as the symbol of union, equivalent to the ring still used, although of pagan origin, in Christian marriage ceremonies.

The dung of cattle was regarded as semi-sacred, because nothing associated with cattle was regarded as unclean. For this reason the Abahima buried their dead in the dung-hill which was the central feature of each cattle kraal. It is therefore clear that attempts by zealous agricultural officers to persuade the Abahima to use cattle manure for crop production purposes have met with conspicuous lack of success!

The corpse was prepared for burial by smearing it with butter on the face and body and with dung on the arms and legs. Special milk from the 'leading cow' was poured into the mouth of the corpse, so that the gods might observe that the deceased was well fed up to the time of death, a symbol that he owned many cattle and therefore came from a 'good' family.

This brief glimpse at a few of the traditional social customs of the Abahima indicates that, to cattle-owning tribes of this type, livestock were not merely regarded as a means of production but were an integral part of social and religious life. Tribal beliefs, superstitions and taboos were closely related to cattle. Thus any alteration in the system of cattle management usually had side-effects on human social organisation. It is therefore important that advisers have a full appreciation of the consequences of attempting to change livestock systems. This is not to say that changes cannot be made. Cattle are so important to livestock owners that any new ideas which are beneficial will be implemented, provided that they do not disrupt the social organisation of the tribe or infringe deeply held beliefs.

At the present time the social organisation of many tropical communities is undergoing drastic change. The impact of education and improved communications will modify traditional social customs and may increase the capacity to absorb new ideas and embrace more technically advanced livestock systems. However, many of the cattle-owning peoples of India, Africa and South America will probably be the last to be influenced by modern technology, and nomadic peoples last of all.

The adviser should remember that modern systems of livestock husbandry are not necessarily better than traditional practices. Modern technologies, such as disease prevention and control, obviously are improvements on primitive cures and remedies, but the traditional standards of basic husbandry and attention to detail are often of a high standard and the systems of management well adapted to local ecological and climatic conditions. Nomadic pastoral systems and small-scale mixed farming economics are relatively resilient, and often well adapted to the local environment and the tropical climate. Droughts and floods do not usually affect all regions within a country equally, and the nomadic pastoralist is well placed to avoid local catastrophes. Similarly small-scale, mixed farming enterprises have an inbuilt balance of the risks of animal disease and crop failure. Thus, the replacement of nomadic pastoral systems by ranching systems, and the modification of small-scale mixed farming systems to a larger-scale monoculture, can introduce a new and unacceptable level of risk. These factors must be taken into consideration before planning major changes in the traditional livestock management systems in tropical areas. It is also salutary to note that, according to Sweet (1991) 'No range management system has yet been perfected that combines the sociological and organisational factors associated with the use of common grazing land'.

With this note of caution, five different systems of cattle management in the tropics will now be considered:

- *Extensive systems*:
  ○ Nomadic pastoralism.
  ○ Ranching.
- *Intensive systems*:
  ○ Small-scale mixed farming (peasant agriculture).
  ○ Medium-scale cattle farming (yeoman agriculture).
  ○ Large-scale cattle farming (estate agriculture).

EXTENSIVE SYSTEMS OF CATTLE MANAGEMENT

Nomadic pastoralism

The classic nomadic pastoralists are to be found in Africa, the best known examples being the Fulbe (or Fulani) in West Africa and the Masai, Suk, Turkana and Karamajong in East Africa. Further north, nomadism is practised by the Borana and Somali, who keep mixed herds of cattle and dromedaries. Many nomadic people also tend large flocks of sheep and goats, and these small stock are usually the responsibility of the womenfolk. Nomads often depend on a diet in which milk and, occasionally, blood figure predominantly. Where blood is drunk, it is obtained by puncturing the jugular vein with a sharp instrument.

The chief reason for nomadism is the seasonal grazing requirement of the cattle herds and sheep and goat flocks. Some historians claim that many nomadic tribes were once sedentary, but an increase in cattle numbers, coupled with a decrease in grazing quality and a reduction in the availability of drinking water, forced the tribes to migrate each year to 'dry-season grazing land' where there was a more dependable supply of both grass and water. This system of seasonal migration is known as transhumance (or semi-nomadic herding), when the herdsmen retain a 'base village' where limited subsistence ephemeral cropping may be practised. Whilst these systems may appear to make efficient use of available grazing because some resting of pastures is involved, this is not always the case. Nomadism is a high-risk system because of the drought hazards involved. Also, as larger numbers of animals are bred, but retained for prestige and insurance purposes rather than sent to market, the problems of overstocking result in rapid degradation of the range resources on which the system is dependent. The effect on the quality of the natural grasslands of the area may be disastrous, a point already made in Chapter 12. Most authorities regard nomadic pastoralism as being on the decrease, and state that this system of cattle management will eventually die out. It is true that many tribes, previously completely nomadic, are now passing through the transitory stages of transhumance and permanent settlement. These stages may be briefly summarised as follows:

(1) A seasonal grazing system in which wet- and dry-season grazing lands are far apart (usually by a distance of at least 100 km) and are separately located each year. In each grazing ground, the stock are rotated around the various watering points in such a manner that it is common for the animals to be driven 10–20 km away from water for one day (overnight) and then grazed back towards the water supply on the following day. Occasionally, when stocking rates are heavy, this system may be extended so that the animals go for 3 days without water. With dromedaries the period without water can be further extended, up to about a week (see Chapter 13). Under this system, the wet- and dry-season grazings are continually being shifted to new locations, both within and between seasons. There is therefore, no possibility for permanent settlement, no village life and consequently the nomads live in tents or temporary hutments. Trading is reduced to a minimum, the chief commodities purchased being clothing, ornaments, cooking utensils and salt for both human and cattle consumption. Examples: the Jie, Turkana and Somali of East and Northeast Africa.

(2) A seasonal grazing system in which the wet- and dry-season grazing lands are separate, usually by a distance not exceeding 100 km, but with the wet-season grazing area permanently located in one place. This system allows permanent houses to be built in the wet-season grazing area, with the result that some form of village life can develop. Trading consequently takes place throughout the duration of the wet season, when the pastoralists may sell surplus milk and buy a range of consumer goods. During the dry season these semi-nomadic pastoralists often migrate to agricultural areas in order to graze the stubbles of cereal crops such as millet and sorghum. Provided the herds of cattle are kept under control, the arable farmers concerned often encourage this practice, since they value the manure which the cattle leave behind them, and the partial removal of stubble by the livestock makes subsequent cultivation easier. Semi-nomadic pastoralists are usually partially integrated into the social and political life of their country. They may pay taxes and some may even take part in elections and have their representatives in the government. This sys-

tem is known as 'transhumance'. Example: the Fulbe (Fulani) of West Africa and the Masai of Kenya.

(3) A system involving more or less permanent settlement in the wet-season grazing area. The problem of the dry season is overcome by conserving herbage or growing fodder crops wherever possible for use when fresh grass is unobtainable. There may still be some movement in and around the wet-season grazing area, and each kraal or camp may have an effective life of between 5 and 10 years, after which the centre may be shifted a few kilometres to an area where grass is more abundant and overstocking less acute. Example: the Abahima and Karamajong of Uganda.

It is likely that some nomadic pastoralists will progress through these stages and eventually become permanently settled, combining the growing of crops with the tending of livestock where soil and climate permit. A good example of a tribe which has reached this last stage is the Iteso tribe of eastern Uganda, which is now permanently settled and which combines the growing of cotton and food crops with the management of large cattle herds and goat flocks. When this final stage has been reached it is quite common to find the tribe employing herdsmen from other, less advanced, communities to take over the responsibility of cattle herding (Wilson, 1958a).

As might be expected, the knowledge of livestock and the standard of husbandry practised by nomadic pastoralists is sometimes of a high order. Thus, many nomadic tribes carry out crude vaccination of their animals against diseases, such as bovine pleuro-pneumonia and rinderpest. This is done by inoculating virulent material from infected animals into healthy stock and then restricting the severity of the local reaction by cauterisation. Unfortunately, the crude technique of cauterisation with hot irons may be injurious or even fatal. The common 'cure' for East Coast fever is to cauterise the inflamed lymphatic glands and many animals are permanently damaged by this crude procedure. Such practices are carried out with the best intentions, but nomadic peoples are willing to accept new ideas relating to their livestock providing they do not run counter to accepted beliefs and provided the efficacy of the new technique can be demonstrated. For instance, it is unwise to introduce modern vaccination programmes (which always result in the death of a few old and unhealthy stock) without warning the cattle owners concerned that unfortunately vaccination cannot protect all animals immediately and that some will die. Once the trust of nomadic peoples has been lost, much time and effort must be expended before confidence can be restored.

Although nomadism may be described as an agricultural system, nevertheless nomads are often regarded as aimless wanderers, and their way of life is frequently opposed by politicians and planners. As Frazer-Darling and Farvar (1973) stated: 'The memory of Genghis Khan has produced the warped civilised judgement that nomadism is a practice to be discouraged or even wiped out'. It is, therefore, salutary to note that the Food and Agriculture Organization of the United Nations (FAO) reported (1972) that nomadism, correctly practised, is not an aimless wandering of tribal people with their livestock. It represents a highly rational adaptation of human life to a severe and adverse arid environment. This point is illustrated by a study in the Sudan undertaken by Hunting Technical Services and quoted by Chalmers (1976). This study compared the performance of cattle in migratory herds with those in sedentary herds. The data are presented in Table 15.1, from which it will be seen that, in all the factors examined, the well-managed migratory herds were superior. Obviously the data have been selected to make the case that most systems, well suited to local conditions, can be made to work whereas alternative systems, sometimes regarded as superior, can often produce inferior results. It follows that well-meaning attempts to settle nomadic pastoralists have often been unsuccessful. Frazer-Darling and Farvar (1973) state that the settling of the Bakhtiara and Basseri tribes in Iran lead to mortality in sheep flocks as high as 75%. Heady (1973) reported that the installation of

Table 15.1 Comparative performance of cattle in migratory and sedentary herds in the Sudan. (Source: Chalmers, 1976.)

Parameter	Migratory herds (total 546)	Sedentary herds (total 149)
Percentage gain or loss in herd	+8	−35
Percentage deaths before 6 months	11	40
Percentage deaths, all ages	15.5	31.7
Percentage calving rate	64.8	39.5
Percentage cows calving before 4 years	65	39

permanent watering points in Saudi Arabia without controlling stock numbers was a mistake, which resulted in overgrazing of an area around the watering points up to 75 km in radius. Similar effects were noted in West Africa by Swift (1973) who commented: 'The problem is how best to help, how to prevent outside intervention from aggravating the situation, how to improve traditional nomadism so that nomads can help themselves in the ways they know best'.

In summary, a rational case can be made for nomadic pastoralism in those arid areas of the tropics which are subject to extremely erratic rainfall. In such circumstances, nomadism is a sustainable system of rotational grazing and water conservation. It maximises the output from grazing areas of low potential yield and minimises the risks to the livestock population. As Chalmers (1976) has pointed out, it is a system closely related to the behavioural pattern evolved over many centuries on the plains of Africa by indigenous wild game, such as the wildebeest.

Ranching

Ranching is carried out on a large scale in South America, Central America, in parts of East and Central Africa, particularly Zimbabwe, South Africa, Australia and the Pacific Islands and on a more limited scale in specific locations in Southeast Asia, particularly the Philippines. In most of these areas ranching has not evolved as an indigenous system for cattle management in the tropics. It is a system introduced from Europe and modified to suit tropical conditions.

The greatest influence on the spread of ranching techniques has been Spanish, since ranching was carried out on a large scale in Spain for many centuries and has spread from the areas colonised by the Spaniards from the fifteenth and sixteenth centuries, particularly Central and South America. Ranching is usually considered to be a system of extensive beef production, but in many areas of Africa, South America and Puerto Rico herds of dairy stock are also ranched. The milking operation is usually carried out on a portable bail system which is moved around the open ranges with the cattle. Much of the fresh milk supply for Caracas, Venezuela, is produced in low-rainfall ranching areas adjacent to Lake Maracaibo, from whence it is conveyed 400 km by refrigerated milk tanker. Milk ranches usually consist of at least five self-contained herds of dairy cattle, each milked separately on the open range. The milk is brought to a central depot, cooled and transported to the fresh milk market. The dairy cattle found on South American milk ranches are mainly of the criollo type (improved crossbred Spanish longhorn) but crossbred Friesians and Brown Swiss cattle are also used for this purpose.

Annual milk production averages in ten traditional milking ranches within the central area of Veracruz, Mexico were reported by Johnson (1987) and are presented in Table 15.2. An average of 29 cows were milked throughout the year in each ranch, which represent only 31% of the total cows. In these environments there was a strong influence of climatological conditions on milk production, especially of rainfall. In this zone 80% of the rainfall falls between June and October. Milk production from June to November was 35% higher than from December to May (4.3 vs 2.8 kg/day/cow), in spite of greater thermal stress. Lactation lengths are quite variable in traditional milk-ranching systems in the tropics. They range from a few days to several

Table 15.2 Average daily milk production (June 1982–May 1983) in ten commercial ranches in the Mexican humid tropics. (After Johnson, 1987.)

Month	Total females*	Total cows	Milking cows	Daily milk yield (kg)
June	168	89	32	4.1
July	168	94	32	4.5
August	168	96	27	4.5
September	156	90	29	4.1
October	156	90	28	4.7
November	156	90	32	4.0
December	156	90	31	3.5
January	177	99	31	3.0
February	176	98	29	2.5
March	174	96	25	3.0
April	159	89	25	2.3
May	168	94	28	2.4
Average	165	93	29	3.5

*Including heifers over 1 month of age.

Table 15.3 Overall means of milk production in a traditional ranch in the Mexican humid tropics. (Source: Johnson, 1987.)

Variable	$X \pm SE$
Observations	267
Average age parturition (months)	49.0 ± 1.7
Milking days	168.0 ± 3.8
Milk/lactation (kg)	762.0 ± 23.5
Milk/day (kg)	4.5 ± 0.1
Calving interval (days)	419.0 ± 7.4
Milk consumed/calf/day (kg)	1.8 ± 0.1

months depending on the productive potential. Milking is generally stopped when cows decrease their production to less than 0.5 kg/day, which may occur soon after calving or when cows are pregnant and drying-off. Low producers are left in the pastures with calves until the next lactation period. Information on productive performance during several years in a traditional milking ranch in the Mexican tropics is presented in Table 15.3. The average length of lactation was 168 days with a total milk production of 762 kg, demonstrating the low levels of cow productivity achieved in these extensive ranching systems.

Most of the ranches in tropical Africa, Asia and Australia are beef ranches, the cattle generally being of the zebu type (as in the case of the Africander cattle ranches in South Africa and Boran ranches in Kenya), but increasing numbers of crossbred cattle (*Bos taurus* × *B. indicus*) are being raised on the more progressive ranches. The beef ranches in the Americas used to be stocked with Spanish longhorn cattle but, during the twentieth century, the impact of improved stock has been very great and most American ranches are now stocked with crossbred animals, such as the Santa Gertrudis, or with purebred European breeds, as in the case of Shorthorns and Herefords in Argentina. (For fuller descriptions of the crossbred cattle mentioned here, reference should be made to Chapter 13).

Ranching is often the commercial alternative to nomadic systems of cattle management. For this reason most of the land at present ranched is either low in fertility, as in the case of the Rupununi savannas of Guyana, or land of very low rainfall, such as the cattle ranches in parts of Kenya and South Africa. The result is that the stocking rates are low and quoted in terms of cattle to the square kilometre rather than beasts per hectare. In the Rupununi, for instance, in spite of a low overall stocking density the number of cattle required for an efficient cattle ranch is very large and capital also has to be expended in the erection of fences, stockades, slaughter-houses, and races, cattle dips and crushes and also in the provision of water and the improvement of communications. In many parts of South America ranching is in the hands of large companies, able to supply capital and technical knowledge when required and equipped to export the finished beef animals economically and efficiently from the remote areas of production. In ranching the profit per beast is very low and a large annual throughput of slaughter-stock is required if the unit is to be viable. This makes it difficult for the nomadic pastoralists to change over to a modern ranching system. Various authorities, such as the FAO, have suggested that co-operative ranching might be possible with several cattle owners combining their resources, including their herds of cattle, in order to make the required capital investment possible. Up to the present, however, there are few examples of large-scale, successful co-operative ranching, but the situation may change in the future. Also, because so much capital is tied up in breeding stock, it is difficult for individual ranchers to obtain agricultural credit, since cattle are not usually regarded as adequate security for the granting of loans on normal terms.

The major technical objectives of modern ranch management are to decrease annual fluctuations in cattle numbers and seasonal fluctuations in liveweight, to maximise reproductive performance and minimise mortality while at the same time improving the quality of the ranges and of the finished stock.

Some of the major limiting factors of cattle ranching in the tropics are as follows:

(1) *Lack of stratification*

In advanced ranching systems, the different phases in the management of cattle (such as breeding, rearing and the production of meat or milk) are carried out on separate, specialised farms. In the tropics such a stratified system of cattle management is unusual and most ranches carry out all these operations on the same holding. Furthermore, there is little or no stratification within the ranch itself. Thus, instead of the best ranges being set aside for breeding herds and fattening units, while the poorer ranges are used for the production of store cattle, the whole ranch tends to be utilised by all classes of stock more or less indiscriminately (see Chapter 12).

An example of a ranching country which could adopt a much greater degree of stratification is Venezuela. Many of the ranch lands are adjacent to the rivers draining into the Amazon Basin, and this river system is typified by seasonal flooding, followed by periods of relative drought. A few hundred kilometres to the north the land is better drained and hence more fertile, and capable of the production of high quality forage, both for direct grazing and conservation. If the breeding and finishing herds could be concentrated in the latter areas, while the store animals were allowed to utilise the low-lying llanos to the south, a much better overall utilisation of resources would be possible. Such a system would necessitate the periodic movement of large numbers of animals from higher to lower land and vice versa, and this particular limiting factor will be considered next. Creek (1984) has described a similar problem, and its possible solution by stratification, in Kenya.

(2) *Poor communications*

Many ranching areas in the tropics are geographically remote and can be reached only by air or river. Examples of such areas are the Rupununi savannas in southwest Guyana (reached by air since the closing of the long cattle-trail in the mid-1950s) and the El Beni Department of Bolivia (reached mainly by air but also connected with the Atlantic Ocean by some 2000 km of river via the Amazon). The consequence of such poor communications is that the freight rates are high and this results either in a low net return to the producer or else the necessity to reduce bulky livestock by meat and milk processing. Thus cattle may be marketed in the form of canned meat or meat extract and milk may be sold as butter, ghee or cheese.

(3) *Lack of good local markets*

Many tropical ranching areas are situated in countries with a relatively small human population (e.g. north Australia, northern Kenya, Argentina and Bolivia) with the result that the local market is limited and production must be aimed at the export trade. Production for export is usually more complex than production for local consumption. For example, meat must usually be deep-frozen and the disease-control requirements of most meat-importing countries are very demanding. Furthermore, the export trade in meat tends to be managed by large shipping companies which own cold-storage facilities and hence control of the market tends to pass from the producer to the merchant-exporter.

A further difficulty is the problem of forecasting world supply and demand for meat. Although supply and demand patterns tend to be cyclical, as illustrated in Fig. 15.1, it is difficult to estimate accurately the actual length of each phase of the cycle. Since beef production on a ranching system is essentially long term (4 or 5 years from conception to consumption) it follows that the market must be 'read' several years in advance, if supply and demand are to be kept in balance. Experience indicates that economists are not able to predict the phased pattern of supply and demand with the required degree of accuracy in respect of timing.

(4) *Marked seasonality in cattle growth and productivity*

Most of the ranching areas in the tropics are situated where either climate, soil or both are limiting for crop production, and where grass and animal growth are often markedly seasonal. (The implications of seasonality of grass growth have been fully discussed in Chapter 12.) As indicated in Fig. 15.2, it is usual for tropical ranch animals to gain in weight for about 8 months of the year during the grass growing season and to lose weight for the remaining dry period. After a prolonged drought the growth and productivity of the cattle in the first part of the subsequent rainy season is often better than at any other time in the year. This enhanced growth is spoken of as 'compensatory growth' and has been discussed in detail by several workers, such as Wilson and Osbourn (1960).

(5) *Low availability of crop by-products, forage crops and concentrated feed*

As has been pointed out already, tropical ranching areas are often situated on lands unsuited for arable cropping and, as such regions often lack good communications, it is very difficult to supplement the natural range vegetation with alternative livestock

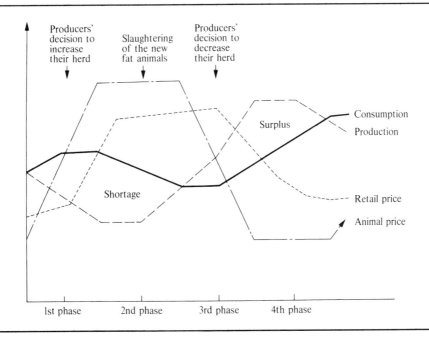

Fig. 15.1 Phases of a meat price cycle.

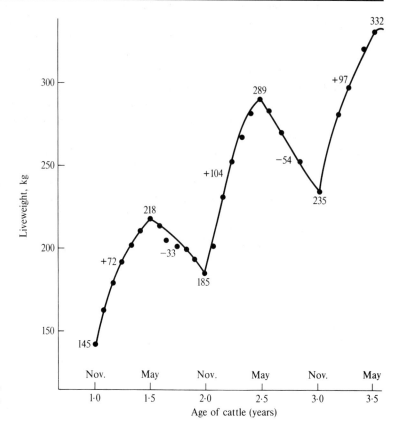

Fig. 15.2 Seasonal liveweight changes of cattle, 1–4-year-old, grazing native pasture at Katherine, northern Australia. (After Norman, 1966.)

feed. The making of hay and silage is not usually practicable, and the freightage on imported livestock feeds from other regions is generally prohibitive. For this reason, fodder conservation in most tropical ranching areas merely consists of closing up one or more ranges for a season. Such crude conservation measures enable a supply of fibrous energy feeds to be maintained throughout a dry season in the form of 'standing hay' (forage), but this invariably results in a deficiency of protein in the conserved material. Under such conditions it may be advisable to introduce a seasonal breeding policy, so that cows calve during those months of the year when the best-quality herbage is available. These seasonal shortages of feed supply can be overcome by removing the cattle from the open range and concentrating them into intensive feedlots adjacent to roads or railways, so that quantities of conserved feed may be transported at reasonable cost. Such an intensive system can no longer be recognised as 'ranching', although in many instances a viable stratified system can be evolved in which the breeding stock are ranched while the fattening stock are more intensively fed (the provision of supplementary feed is discussed fully in Chapter 11).

(6) *Difficulty in giving individual attention to ranch stock*
The point has been made that the viable size of a tropical ranch is of necessity large, due to the low profit per beast and the low carrying capacity of the range land. The result is

that stock management is rendered difficult, and cattle can only be given the minimum of attention. A stockman can only get to know his animals individually when they are in groups of 100 or so or less. On many ranches the units (or mobs) of cattle number 1000 head or more. On many ranches stock are only seen at close quarters once or twice each year, when they are rounded up for dipping, culling, vaccinating and selection for slaughter. Animals on open range tend to be widely dispersed and it is rare for a large mob to keep close together except in areas where wild predators abound. It is consequently difficult to locate and treat stock which become injured, which have calving difficulties or which are in need of individual attention. One major consequence of this is that diseases may not be correctly diagnosed until a large number of cattle are infected and possibly past the time of optimal treatment. It is, therefore, vital that the stockmen should be on the lookout for symptoms of the major diseases of the area and be able to introduce the required treatment without undue delay.

(7) *Wild temperament of ranch cattle*
Stock which are infrequently handled tend to be wild compared to stock which are in regular contact with man. Skill is required in managing ranch stock if damage to the cattle, and injury to the stockmen, are to be avoided. Good stockmen handle cattle in such a way that the natural fear which most animals have for man is minimised. This can be achieved by curtailing the use of the stock-whip and stick to the minimum and by encouraging the animals to approach the stockmen by dispensing 'range cubes' (pelleted compound rations), molasses, salt or minerals during visits to the mobs of cattle on the open range. Even with these elementary measures of good stockmanship, the construction of efficient facilities for confining cattle is important. Kraals (corrals), camps and stockades must be constructed with strong materials, and proper cattle races and crushes are essential for restraining animals while they are being vaccinated, branded, spayed or castrated. There are many instances of heavy losses of stock due to inadequate security measures taken in the kraal (Fig. 15.3) or stockade. Ranch stock are excitable and panic easily, so that a loose gate or defective fence may result in a cattle charge and consequent mortality due to the stock piling up and trampling each other while trying to escape. The details of the construction of suitable cattle enclosures and crushes have been described by Payne (1990).

(8) *Shortage of water*
In most ranching areas supplies of drinking water for the stock are deficient and steps have to be taken to provide additional water by conserving rainwater in dams during the wet season, or by drilling boreholes. Each range on a ranch ideally should have its own supply of drinking water and large ranges require several drinking sites if they are to be efficiently utilised, without overstocking in the area surrounding the watering places. The amount of water required will depend upon the average weight of the cattle, the water content of the herbage, the mean maximum ambient temperature and relative humidity and the evaporation rate of water from the surface of the dam or trough (see Chapter 16). A rough calculation can be made by assuming that tropical ranch stock will require about 5 litres of water/day for every 50 kg liveweight in the dry season and about half this amount in the wet season. It follows that a mob of 500 store cattle averaging 400 kg liveweight would require roughly 500×40 litres = 20 000 litres of water/day in the dry season. These estimates have been substantiated by Pandey *et al.* (1989). If stock have to be watered from troughs instead of from river banks or dams, there is a major problem in providing such large quantities of water in a short time. Troughs should be so constructed that they are long and narrow, so that many stock can

Fig. 15.3 Cattle kraal and sorting yards in the Rupununi savannas, Guyana. The cattle are crossbred *Bos taurus* × *B. indicus* cattle. The influence of the Hereford bull may be noted in the white-faced animal in the background.

drink simultaneously. If water can only be provided slowly from its source, such as by wind-pump from a borehole, then it is important to construct storage troughs large enough to provide at least half of the required daily amount of water at any one time. In the example cited above, the capacity of storage troughs should be at least 10 000 litres (see also Chapter 11).

It can be seen from the above summary of ranching systems that the principal feed of range cattle is the natural vegetation of the range. However, judicious supplementary feeding of animals at critical physiological stages can often result in tangible benefit. In Table 15.4, the response of ranched breeding cows to a small daily amount of supplementary feed is quantified in terms of increased fertility.

However, although supplementary feeding may be beneficial, the efficient husbandry and utilisation of range grasses is vital to efficient ranching. As ploughing and reseeding of the range is usually impracticable and uneconomic, the rancher must develop grazing systems which

Table 15.4 Response of breeding cows on range to supplementary feeding during the winter/dry season.

Author	Treatment	Calving percentage
Bauer (1965)	Nil	77
	420 g protein concentrate	88
Bembridge (1963)	Nil	63
	454–908 g cotton seed	76
Elliott (1964)	Nil	58
	908 g cotton seed	82
Ward (1968)	Nil	60
	908 g groundnut	75

Note: This table shows that very modest daily supplementation with a high protein concentrate increased calving percentage by amounts varying between 11 and 24%.

lead to a favourable ecological succession. As the application of fertiliser on a vast scale is usually uneconomic, the rancher must use the resting period and the tropical legume rather than the manure distributor to maintain fertility and to increase grassland productivity. However, there may be certain range areas, such as those in the Northern Territory of Australia, where the provision of deficient trace minerals may produce such a dramatic increase in grassland productivitity that the cost:benefit ratio is highly favourable. (See Chapter 12 for further information on grassland management.)

The most important factor under the direct control of the rancher is the number of livestock carried on the land. Overstocking is the most common mistake made. However, it is often difficult to determine the actual carrying capacity of a new area other than on a 'trial-and-error' basis or, occasionally, as a result of experimentation. Thus, the rancher does not often realise that he is overstocking the range until the detrimental effects of overstocking become apparent in terms of sward deterioration and reduced livestock productivity. For these reasons, the build-up of cattle numbers on a new range should proceed with caution and steps should be taken to de-stock as soon as it is apparent that the optimum stocking rate has been exceeded.

A useful technique which enables a reduction in stock numbers, without the disposal of growing and breeding stock, is to spay (an operation in which the ovaries are removed) all the elderly breeding cows and cull cows. Spaying, by preventing further pregnancies, will enable these animals to gain in weight in such a manner that they can be marketed for beef at some profit to the rancher. The resultant reduction in the calf crop will, in time, stabilise or reduce livestock population to a safer level. This spaying technique has been employed with success in parts of South America, particularly the Rupununi savannas of Guyana, and is a method which can be recommended to ranching areas faced with problems of acute overstocking. The actual operation of spaying can be quickly learnt, rapidly performed in the field by qualified operators under veterinary supervision, and the resultant mortality from post-operative infection should not exceed about 5%.

The foregoing account of some of the major limiting factors of large-scale cattle ranching may appear to be pessimistic. While it would be unwise to minimise the difficulties which confront the tropical rancher, the future prospects of efficient cattle ranching in areas without problems of high human population pressure are promising. The world's demand for, and the price of, beef follows closely the increase in numbers and standards of living of the human population. The labour requirement in cattle ranching, per beast or per hectare, is usually lower than in any other form of tropical agriculture. Land suitable for cattle ranching is usually unsuitable for other forms of land use and is, therefore, available at relatively low prices (as low as US$10/km^2 in parts of South America!). Once the original capital investment in terms of land and breeding stock has been amortised, the recurrent costs of cattle ranching are relatively low. Finally, it is probable that the productivity of ranch-type cattle, particularly crossbred cattle, will be markedly increased in future by the application of improved breeding techniques (see Chapter 18). It is also likely that the efficiency of long-distance transport, both of live cattle and processd meat products, will become more efficient and relatively less expensive.

INTENSIVE SYSTEMS OF CATTLE MANAGEMENT

Small-scale mixed farming

The successful introduction of mixed farming techniques in temperate agriculture has led many authorities to recommend this system in appropriate areas of the tropics. In certain areas, such as the Kipsigis district of Kenya, intensive mixed farming was successfully introduced in a relatively short period. In many other parts of the tropics, attempts to introduce fully integrated mixed farming into peasant agricultural systems have been unsuccessful, in spite of the fact that 80% of tropical livestock are found on farms which also grow some crops (IDRC, 1988).

Table 15.5 shows that this system embraces about 290 million bovine cattle, or 46% of all tropical cattle. If systems including ley farming

Table 15.5 Estimate of the number of bovines reared in major systems of tropical agriculture. (Source: Payne, 1976.)

System	No, of bovines (millions)	No. of bovines as % of total	Type of bovine
Sedentary subsistence	290	46	Cattle and buffaloes
Commercial ranching	180	29	Cattle
Transhumance	63	10	Cattle
Nomadic herding	32	5	Cattle
Sedentary shifting cultivation	30	5	Cattle and buffaloes
Perennial crop cultivation	15	2	Cattle and buffaloes
Regulated ley farming	15	2	Cattle
Migratory shifting cultivation	6	1	Cattle

and perennial crop cultivation are added, then this type of farming system affects about 50% of all tropical cattle. Some of the most important managerial difficulties of small-scale mixed farming with cattle will now be examined.

Size of field or paddock

The size of most tropical fields, or paddocks, has been determined more by the amount of land a family can cultivate with a hand hoe than by the area which can be cultivated with the aid of a draught animal or tractor. The smaller the area of land planted with a grass ley or permanent pasture, the greater the proportionate cost of providing stock-proof fences or hedges per head of stock. Apart from cost, intensive grazing techniques become difficult when field size and number are so small that herd size is severely limited and rotational grazing becomes impractical. In general, fields of less than 0.5 ha in size are unlikely to repay the capital and recurrent costs of confining cattle to them. A series of at least five paddocks is required to enable adequate grass regrowth and to reduce reinfestation with cattle parasites to manageable proportions. It follows that an area of at least 2.5 ha of grass is about the minimum required for small-scale intensive cattle management, but there are many areas of the tropics where the present average farm size is much less than this.

It may be undesirable for governments to encourage, by subsidisation, the planting of pasture grasses on small farms where the area for this form of land use falls below a certain minimum figure. In those countries where, for political reasons, a maximum size of agricultural holding is prescribed by law, it is important that the restriction does not prohibit efficient tropical livestock production. In India, maximum farm sizes of between 20 and 30 ha have been suggested, and these would certainly allow efficient livestock management to take place. In other tropical countries, such as certain South American republics, 'land reform' has allocated large blocks of land, formerly in private hands, on an equal-area basis, and in certain instances the resultant holdings are of an uneconomic small size. In Trinidad, land made available on former wartime air-bases was allocated to small-scale dairy farmers, with the average size of holding being about 10 ha. Although holdings of that size were just viable at the time of land distribution, escalating variable farm costs have since rendered such farms marginal for efficient milk production. It is, therefore, extremely important that planning authorities should take a long-term view of future economic trends when determining minimum, mean or maximum size of agricultural holdings.

High cost of fencing

Live hedges are uncommon in the tropics. Where suitable species are available they are difficult to maintain in a stock-proof manner for long periods (see also Chapter 12). Fencing materials are relatively expensive, since few local timbers are capable of being termite-proofed for reasonable periods and such timbers usually command high prices. Wire, whether barbed or plain, is usually imported and carries high transport costs. In some areas the use of electric fencing has been successfully introduced at great saving in capital costs partly, if not mainly, off-set by higher recurrent labour costs incurred in moving the fence. Even where electric fencing is economically used, permanent fencing is desirable round the perimeter of the grassed area, particularly where pastures are adjacent to neighbouring arable fields. In certain areas where improved, but disease-susceptible, breeds of cattle are intensively kept double perimeter fencing is necessary and often obligatory. This further increases the costs and also leads to a slight reduction in the area available for grazing.

In certain areas Australian 'slung fencing' has been successfully introduced. This consists of stretching a high-tensile strength wire between two large 'king posts' erected at approximately 100 m intervals. A lightweight wooden fence is then hung from the tightly stretched wire, so that the fence is suspended with the lower part of the fence touching or just clearing the ground. Such fencing can be erected at a lower cost than traditional fencing, since the posts need to be sunk into the ground only at 100 m intervals instead of every 3 m. If properly erected, they will effectively hold cattle, even semi-wild range cattle, but they are often ineffective for small ruminants which are able to crawl under, or jump over, the fence at the mid-way position between two 'king posts'.

The high cost of fencing has led many workers to consider the prospect of zero grazing systems, in which the grass or forage is cut and carried to the cattle which are kept in yards or sheds. Cattle are only allowed out on to grass pastures for the purpose of exercise, and there is no grazing in the normal sense. Indeed, in many tropical countries, such systems were introduced before the advent of suitable species of improved pasture grasses for grazing. The traditional name for such a system was a 'cut-and-carry' or 'soilage grass' system (see Chapters 11 and 12). Such techniques become uneconomic as labour rates rise and the cost of cutting and carting forage become prohibitive, but there are many areas of the tropics where they are still practised, especially where the work is performed by family labour as in parts of Asia.

An early example of such a zero grazing system is the technique introduced into Cuba by Preston and Willis (1969) in which whole sugar cane, sometimes supplemented with molasses and urea, was cut and carted for feeding to cattle, usually beef cattle. The machinery involved in such systems is expensive and requires a minimum area, usually in excess of 5 ha, for economic operation. The capital costs, and minimum areas, are usually prohibitive for small-scale peasant operations, although there may be prospects for sharing machinery, and hence costs, on a co-operative farming basis.

Provision of productive cattle

Intensive livestock production is not possible with ranch-type cattle suitable for more extensive systems. Productive breeds are essential, and in most cases this means that tropically adapted breeds of *Bos taurus* cattle, or crossbred *B. taurus* × *B. indicus* cattle, must be employed. The breeding, selection and multiplication of productive cattle of the required standard is usually, if not invariably, outside the scope of the small-scale producer. The most that the small farmer can do is to cull his own stock on performance as well as on appearance. Even this requires a minimum standard of herd recording, such as for breeding, milk yields and calving interval, which is more than can be reasonably expected from an untrained producer. The government, or some livestock producers' co-operative, should organise the breeding and supply of improved livestock on a national basis. Useful adjuncts to this work are regulations to limit the indiscriminate breeding by scrub bulls and the development of a nationwide artificial insemination (AI) service which can spread the use of the country's best proven bulls to the smallest producer owning only a few cows. This point will be dealt with in greater detail in Chapter 18.

The production levels required in tropical

cattle will vary from place to place and according to the different costs of production and the market values of the end products. A common mistake in the past has been to think in terms of existing levels of production, and ways by which they might be slowly increased, instead of dealing with the longer-term economics of the situation from the start. Such considerations should enable a proper evaluation of the minimum acceptable production levels for future economic forms of land use with cattle. Where the problem has been tackled from the former standpoint, figures of 500–1000 litres of milk/cow/yr, or 0.1–0.2 kg liveweight gain/head/day, have been quoted for milk and beef production respectively. Such figures are easily realised with *Bos indicus* cattle. However, where the problem has been approached from the economic standpoint, it is apparent that such low targets are, or will shortly become, unacceptable and result in the farm running at a loss. This is particularly so if a proper economic valuation is placed upon land and interest charges are levied on the theoretical capital investment. As long ago as 1962 Maule suggested the following very general objectives for intensive cattle production levels in the tropics.

- *Beef production.* Animals should be able to subsist without supplementary feeding and make reasonable good growth without excessive seasonal fluctuations, and be fit to kill at about 4–5 years with a dressing-out percentage of about 50. Good breeding cows should rear three calves in 4 years.
- *Milk production.* Cows should be able, with good management, to withstand difficult climatic conditions and give a yield of at least 2000–3000 litres of milk/yr and to breed a calf every 12–14 months starting from the age of 3 years.

However, these are examples of generalised recommendations and there are situations where they would be inapplicable. For instance, Maule's standards could not be met where disease ruled out the use of European or crossbred cattle for milk and his target for beef would be inappropriate in areas with a marked dry season and consequent wide variations in the quantity and quality of herbage.

Supplementary and dry-season feeding

In many tropical countries the seasonality of grass growth is such that some degree of supplementary feeding is needed for optimum productivity, particularly for dairy cattle, during the latter part of the dry season and possibly also during the onset of the rains. This need for extra feeding becomes more acute as breeds of cattle with higher productive potential are used. Unfortunately, many of the traditional dry-season feeds are coarse roughages, low in digestible protein. Taller silage-type grasses, whose deep-rooting systems allow a degree of dry-season growth, also come into this category with digestible crude protein levels generally below 5%. In many parts of the tropics the dry season is not so much a period of total shortage of herbage as of acute shortage of protein. It is relatively easy to conserve 'foggage', or standing hay, but such feeds do little to rectify this basic nutrient imbalance. This point is illustrated by the data presented in Table 15.6, which compare the nutritional quality of a pasture grass in the wet and dry seasons in Trinidad. Table 15.7 shows that, whereas a high-quality Pangola grass pasture can provide enough digestible crude protein and total digestible nutrients for the needs of an average weight low-yielding dairy cow in the wet season, the intake of protein in the dry season is about 25% below requirement.

McDowell (1981) has shown that, on average, tropical forages can supply about $1.8 \times$ maintenance-energy requirement, sufficient for the energetic needs of low yielding cows but not for medium or high yielders. It is therefore important that the small-scale farmer understands the need to supply some form of protein supplement at this time. In certain areas this can be provided by judicious use of local by-products, such as copra meal, cotton seed meal, palm kernel meal, spent grains, etc., which are all relatively high in crude protein (Jayasuriya, 1993) (see Chapter 11.)

Table 15.6 Seasonal variation in nutritional quality of Pangola grass (*Digitaria decumbens*) pastures. (Source: Butterworth *et al.*, 1961.)

Season	Dry matter (%)	Composition expressed as % of dry matter						
		Crude protein	Crude fibre	Ether extract	Ash	Nitrogen-free extractives	Digestible crude protein	Total digestible nutrients
Wet	23.4	11.1	30.5	3.0	8.4	47.0	6.8	59
Dry	39.3	6.8	29.5	2.1	7.8	53.8	3.2	62

Note: For an explanation of the nutritional terms used in this table refer to Chapter 16 and the section dealing with improved nutrition.

Table 15.7 Intake of nutrients (kg/day) from Pangola grass pastures compared with requirements of a 450 kg heifer yielding 7 litres milk/day. (After Butterworth *et al.*, 1961.)

Item	Digestible crude protein	Total digestible nutrients
Wet season	0.9	8.0
Dry season	0.6	10.4
Requirements for maintenance plus 7 litres of milk	0.8	5.4

Housing

Although the need for housing is much less in warm tropical climates than in colder, temperate regions of the world, nevertheless some cattle buildings are necessary for efficient small-scale production. It is often traditional to house, or kraal, all stock at night and although this is not essential for the well-being of the cattle it may be necessary to reduce losses through theft or by predatory animals. In mixed farming areas, the danger of stock breaking out from night pastures and damaging neighbouring arable crops is sufficiently great to induce the owner to invest in night housing. Wherever possible, however, the aim should be to encourage night-grazing systems in order to maximise the voluntary intake of herbage. Wilson (1961a, 1961c) has shown that up to 40% of the total grazing time may occur during the 12 'night' hours. If cattle are denied access to night-grazing, the lost feed intake is not fully compensated by more intensive day-grazing.

There is less need to house calves in the tropics than is generally perceived by small-scale farmers in mixed farming areas, where the 'calf house' is often regarded as an integral part of the homestead. If the various dangers of night-grazing detailed above can be overcome for adult cattle, there is usually no good reason why calves should not also be kept out at night. On the other hand, the need of the calf for night-time grazing is less than that of the adult, as can be seen by the data presented in Table 15.8 which show that whereas adult cows spend 4.4 night hours grazing and walking, the time spent by immature heifers and calves is much less, 1.0 and 0.6 hours respectively.

The chief need for some form of cattle buildings in small-scale mixed farming systems is for suitable milking accommodation for small dairy herds. Often the type of buildings unfortunately recommended are miniature milking sheds or parlours suitable for temperate climates. In the tropics milking accommodation has to meet two requirements. First, to provide minimum protection for the milker during the rainy season; second, to provide reasonable facilities for the storage of milk and milking utensils, and for cleaning and sterilising the milking equipment. A simply constructed lean-to shed, sited on well-drained land with protection from the prevailing wind and driving rain, is quite sufficient. If cheap local materials are used, then the structure can be rebuilt on a different site every few years so as

Table 15.8 Comparison of grazing behaviour of female cattle at different stages of growth (hours). (Source: Wilson, 1961a.)

Behaviour	Calves		Heifers		Cows	
	Day	Night	Day	Night	Day	Night
Grazing + walking	6.0	0.6	6.1	1.0	4.9	4.4
Ruminating	0.7	3.5	1.9	3.9	2.5	4.0
Idling	5.3	7.9	4.0	7.1	4.6	3.6

Fig. 15.4 An open-air milking shed at Kabete, Kenya. There are facilities for feeding and watering the cows and the milking utensils, after washing, are sterilised by exposure to the sun.

to prevent excessive fouling of the land and a build-up of parasites in the immediate vicinity. During the dry season there is no reason why milking should not be carried out in the pasture itself, with the cows tied to temporary stakes driven into the ground (Fig. 15.4). In this way building costs are reduced to the minimum and the cattle benefit by reduced disturbance to their normal grazing routine. The dairy should have good facilities for keeping the milk clean and cool and there should be a small room, preferably with a concrete floor and walls which can be painted or regularly washed, in which the milking utensils can be cleaned, sterilised and stored. The tropical sun is a most effective sterilising agent and, after washing thoroughly, it is a good hygienic practice to expose buckets, churns and other metal utensils so that they are dried and sterilised by the sun. The dairy should be supplied with running water if possible and care must be taken to see that flies are effectively excluded or else killed with an efficient safe insecticide (Fig. 15.5).

It should be stressed that tropical dairy buildings do not have to be constructed in expensive materials in order to be efficient. A dark, damp concrete structure with cracked walls and a broken floor, situated in a valley subject to occasional flooding, is far less desirable than a cheap, open wooden shelter sited on well-drained land.

Fig. 15.5 A herd of imported pedigree Guernsey cattle being milked in a modern milking shed at Dar-es-Salaam, Tanzania. This herd is being managed intensively as a 'town dairy'.

The same general principles apply to calves. Newly born calves should be reared in individual pens, but again there is no need for expensive brick or concrete divisions between calf pens. Simple free-standing wooden hurdles, tied together into squares each containing a calf, are required. Such accommodation has several merits; it is cheap, it can easily be replaced and after each crop of calves the hurdles can be dismounted, sterilised and repainted (with a non-toxic material) so that the prospect of transmission of disease from one group of calves to another is minimised. The open nature of the simple wooden hurdles enables ample air movement in and out of the calf pens. Thus, two main factors causing the spread of calf disease – poor ventilation and high humidity – are reduced to a minimum.

Particular attention should be given to the roofs of animal buildings including cattle buildings. The purpose is to keep out heavy rain, but a secondary purpose should be to reduce to a minimum the impact of solar radiation on the backs of the animals (see also Chapter 14).

Sylvo-pastoral systems

Three major types of integrated systems are defined by Payne (1990) as follows:

(1) Grazing and/or browsing in natural forest.
(2) Grazing or harvesting forage grown under planted trees, including those used for the production of timber, firewood, nuts, fruit and industrial products.
(3) Browsing and/or harvesting tree forage.

Although there are differences between sylvo-pastoral systems in wet and dry regions, most have a number of biological advantages, the chief of which have been identified by Payne (1990) as:

- Available solar energy is used rather efficiently due to the vertical stratification of the vegetative components of the system.
- The soil is protected from severe erosion by two or more plant storeys.
- There is usually vertical stratification of the root systems of the different plant species.

- The tree crop ensures some recycling of soil nutrients.
- Where leguminous trees are used, or where legumes are included in the sward, there are beneficial long-term effects in soil fertility.
- There is a reduction in the annual cost of weeding under the tree crop.
- There is an increase in total product and revenue output per unit area of land.

However, in spite of the above advantages, such systems are not widely adopted in the wetter tropics as they tend to be more managerially complex, and some of the technical problems encountered have not been properly investigated nor administratively resolved. In the drier regions sylvo-pastoral systems are located in relatively fragile ecosystems, easily destroyed by a combination of fire and continuous overgrazing. They are essentially subsystems of pastoral systems so that the problem is to stop and, if possible, reverse the degradation that is occurring.

Payne (1985) reviewed the possibilities for integrating cattle with tree crop production in both the wetter and drier regions of the tropics. These systems offer considerable opportunities for the expansion of cattle production in the humid tropics. To date, the cattle–coconut system has been more developed than other integrated systems and this has been described as a possible model for the development of other cattle/tree crop systems (Payne, 1990).

Cattle/coconut systems The Fats and Oil Team, FAO (1979), estimated that the total world area of coconuts was 7.2 Mha, of which 46% was in the Philippines and 78% in Asia. Using these data, Payne (1985) estimated that the area of coconuts in the humid tropics suitable for cattle/coconut operations was about 6 Mha, but that cattle/coconut system production systems only used about one fifth of this area.

These systems are easily established in those regions where total annual rainfall is 2000 mm or more, and is seasonally well distributed. Soils should be well drained. The coconut varieties should be planted at optimal spacing for maxi-

Table 15.9 Effect of grazing cattle and buffaloes in oil palm plantations on the yield of fresh fruit bunches (t/ha). (Source: Devendra, 1989.)

Year	Grazed	Non-Grazed	Difference
1980	30.6	25.6	5.0
1981	17.7	15.9	1.8
1982	25.1	23.0	2.1
1983	23.5	18.3	5.2
Mean	24.2	20.7	3.5

mum nut yield. Shade-tolerant grass and legume species should be undersown. *Brachiaria brizantha*, *B. miliiformis* and varieties of *Panicum maximum* appear to be suitable grasses and *Centrosema pubescens* is a legume used in many regions. Fertiliser application to palms and forage plants is usually essential in the long term.

Productive cattle should be rotationally grazed beneath the trees once these have grown sufficiently to escape damage from the animals. Adequate supplies of minerals and water should be available for the cattle. Stocking rates under good managerial conditions will vary from 1.5 to 3.0 Isu (international stocking unit)/ha (Whiteman, 1977).

Other systems Cattle/oil palm, cattle/rubber and cattle/fruit-tree crop systems are being developed, particularly in Southeast Asia. These systems and others have been described by Payne (1985). Integration of cattle and other livestock with timber trees is developing rapidly in the American tropics (Budowski, 1980). Devendra (1989) has suggested that small ruminants, especially sheep, are most suitable for grazing under rubber and coconut plantations, and that both cattle and buffaloes can be grazed under oil palm. This worker reported that there was a beneficial effect of the livestock manure on the yield of fresh oil palm fruit in Malaysia, as shown in Table 15.9.

Milking procedures and calf rearing

The commonest system of milking practised in the tropics by small-scale producers is twice-a-day milking, with calf at foot. The calf is usually separated from the dam during the night and allowed to suckle for several minutes prior to the morning milking, which generally takes place within an hour of dawn (Little *et al.*, 1991). The cows and calves usually graze separate pastures by day and the calf is allowed its second daily feed from its dam immediately prior to the afternoon or evening milking.

This traditional milking system is often used with purebred zebu cows, which let down their milk more readily in the presence of their calves. Unfortunately, the system often results in too little milk being drunk by the calf, so that calf growth is unduly restricted. The calf is thus weak when it is exposed to tick- and fly-borne diseases and to internal parasites. Part of the high incidence of calf mortality in the tropics is due to the poor nutrition of the calf, which is itself related to the system of milking with calf-at-foot. (Further information on the nutritional requirements of calves is given in Chapter 16.)

The system of milking with calf-at-foot is not essential for the majority of *Bos indicus* breeds, and there are many recorded examples of successful complete milking-out of zebu cattle, coupled with bucket feeding of the calves. However, in any unselected herd of zebu cattle, there is generally a small percentage of cows which are difficult, if not impossible, to milk-out completely without the presence of the calf (Preston, 1989). In such cases, the normal endocrinological reflex action, by which milk is released from the milk alveoli into the milk sinus, cannot be modified so that the let-down stimulus is provided by the act of milking. In order to achieve an efficient milking system these 'difficult' cows, which do not readily let down their milk, must be culled. This undesirable characteristic may be inherited, and if so it may be desirable to eliminate any bulls which may be transmitting this undesirable character to their daughters.

With crossbred, or *Bos taurus*, cattle the normal milking procedures adopted in temperate regions, and described in temperate textbooks on dairying, can be employed. In the case of zebu cattle it is highly desirable that calf-at-foot milking should be replaced by complete milking-out, but this changeover will require higher culling rates for the first few cow generations. More attention must be paid to the nutrition of the growing calf than is usually practised and a system should be aimed at in which about 250 litres of whole, skimmed or reconstituted milk, or milk substitute, is fed to each calf during its first 3–4 months of life. It is usually helpful to adopt a system in which milk, or milk substitute, is gradually replaced by a balanced concentrate, which is itself progressively replaced by grass. The total period between birth of the calf and full grazing should normally not be less than 5, and not more than 8, months. The period between birth and weaning from a liquid diet should normally not be less than 5, and not more than 12, weeks. Calves should not be weaned off liquid milk before they are eating a minimum amount of 0.8 kg head/day of concentrates and until they are observed to be ruminating, or chewing the cud. Although systems of once-a-day feeding of milk substitute can be successful under temperate conditions, nevertheless twice-a-day feeding systems are normally recommended for tropical areas, particularly with *Bos indicus* cattle.

Zebu calves should be expected to grow at a rate not less than 0.3 kg/head/day for their first year of life, crossbred calves at a rate not less than 0.4 kg/head/day and *Bos taurus* calves at not less than 0.5 kg/head/day. Rates such as these are rarely realised in management systems in which cows are milked with calf-at-foot, with little if any attention given to the nutritional needs of the calf during its first 6 months.

An alternative milking and calf-rearing system is 'once-a-day milking'. In this system the calf is separated from its dam at night and the latter is then milked out by hand as completely as possible in the morning. The calves may be 'at foot' at the morning milking, but are only allowed a few sucks, sufficient to stimulate milk let-down. After the morning milking the calves are allowed to run with their dams for the rest of the day, being separated again at dusk. It is calculated that, under such conditions, the milker obtains about 55% of the milk yield and the calf about 45%. This ensures that the calf has an adequate milk supply, but with cows giving over 750 litres of milk it is wasteful in that the calf will obtain more milk than it requires and too small a proportion will be available for sale (Little *et al.*, 1991).

The act of hand milking is often carried out under unhygienic conditions in the tropics and a large proportion of the milk produced is unacceptable because of adulteration, high bacterial count or contamination with solid matter. It is difficult to enforce higher standards of personal hygiene in the dairy than is practised in the home, and the problem of clean milk production must be tackled by a concerted approach by all interested parties – milk buyer, agricultural officer, health inspector. One undesirable milking practice found in both African and Asian countries is 'wet milking', during which spittle is conveyed by the milker to the teats of the cow which is then milked by a stroking action between the thumb and forefinger instead of by the more usual 'squeeze and release' action with the teat held more firmly in the palm of the hand. The use of the strip-cup to detect mastitis and the use of a strainer, such as butter-muslin, to separate solid matter from the liquid milk before it is poured into churns or bulk tanks, requires active encouragement. It may be desirable for the milk purchaser to take the initiative by providing farmers with these materials free and to pay a slightly lower price for the milk.

One of the disadvantages of small-scale milk production is that operations of this order of size do not justify the large capital expense of installing milk-cooling apparatus. It is therefore essential that the milk is sold off the farm soon after milking, twice daily deliveries being desirable. The data presented in Table 15.10 show how the bacterial count escalates logarithmically when milk is left at high temperature for a 12-hour period, as it would be if only one delivery per day was made and the milk was not cooled.

Medium-scale cattle farming

Although the average size of cattle holding in the tropics may indicate that there are a large number of medium-scale cattle farms, in fact they comprise a small proportion of the total

Table 15.10 The relationship between temperature and bacterial count in milk 12 hours after milking. (After Williamson and Payne, 1965.)

Temperature (°C)	No. of bacteria/ml (thousand)
4.4	4
7.2	9
10.0	18
12.8	38
15.6	453
21.1	8 800
26.7	55 300

number of enterprises. The reasons for this are important to understand. Under temperate systems of manual cultivation, the size of holding is limited to that area which can be conveniently worked by family labour, with the minimum of employed assistance. Whereas the tractor has taken over from the draught animal, the average size of holding has increased, to create a predominance of medium-scale farms ranging between 20 and 100 ha in size.

In the tropics this evolution in size of holding has not taken place to the same extent. The laws of inheritance often prevent the amalgamation of holdings, so that farm size, instead of increasing, has decreased as the size of family has increased. On the other hand, estates controlled by foreign companies or individuals have usually been extremely large, sizes being measured in square kilometres rather than by hectares. Over recent years these large holdings often have been appropriated by government agencies, and the land reallocated to peasant farmers, with the consequence that the size has been drastically reduced, usually to holdings of about 20 ha or less. It therefore follows that there has been no steady evolution of medium-scale cattle farms in the tropics corresponding to that in developed temperate regions.

Possibly the co-operative movement, now being encouraged by governments in many parts of the tropics, will enable medium-scale cattle farms to develop, the land and the stock being held conjointly by the members of the co-operative society. If such changes come about, it is likely that the co-operatively owned medium-scale farms may rapidly adopt the technologies of beef and milk production which have been evolved elsewhere. Dairy cattle may, therefore, be fed partly on natural or conserved forage and partly on manufactured feedstuffs, possibly produced by feed mills owned by the co-operative or by a group of co-operatives. Improved breeding techniques may be used, based on artificial insemination (AI), with the dairy cows being milked through herringbone or rotary milking parlours, based upon the designs evolved in temperate countries. Beef production may also be intensified, with supplementary feed provided during the dry season so that the cattle gain in weight throughout the year.

Co-operative movements along these lines are currently found in South America, Africa and Asia. Whether or not such co-operatively run, medium-scale cattle farms will become a stable feature of tropical livestock production, or whether they are examples of a transitional phase in a long-term evolutionary process, only time will tell. Whatever the outcome, it is important to realise that the chief limiting factors of medium-scale cattle farming in the tropics are usually of a political or socio-economic nature, rather than technological.

There are also regions of the tropics with favourable climatic conditions where European or crossbred cattle have been successfully employed on medium-sized family farms. Thus at this higher end of the production scale conducted at Paso del Toro, Veracruz, the reproductive performance of Holstein, Brown Swiss and Jersey cows, under a permanent barn system, was compared (Johnson, 1987). Feeding consisted of sorghum or corn silage, plus concentrate which was supplied according to the production level. Milking was by hand, twice daily, and calves were separated from their mothers two days after birth. Milk production was higher in Holstein than in Brown Swiss and Jersey cows and higher in Brown Swiss than in Jersey cows (Table 15.11). No evidence of interaction between breed-year, breed-season and breed-year-season was detected. Costs of milk production were lower for Holstein cows. Later the Jersey breed was eliminated and Holstein and Brown Swiss-Zebu crossbred cows were added to the programme. These cows were managed under a rotational grazing system of Pangola grass (*Digitaria decumbens*), plus some native and introduced legumes, plus a molasses–urea supplement. Cows were milked twice a day without the presence of the calf. Lactation length was greater and milk production better in Holstein-Zebu cows than in Brown Swiss-Zebu cows as shown in Table 15.12.

Table 15.11 Milk yield means of Holstein, Brown Swiss and Jersey cows in an intensive milk production system in a tropical climate. (After Johnson, 1987.)

Variable	Holstein	Brown Swiss	Jersey
Cows	58	56	18
Lactations	123	127	46
Average parturition age (months)	47	46	54
Lactation length (days)	325	315	318
Milk yield (kg/lactation)	3534	2821	2537

Table 15.12 Milk yield F_1 cows (Holstein × Zebu and Brown Swiss × Zebu) on grazing management. (Source: Becerril et al., 1981.)

Variable	Holstein × Zebu	Brown Swiss × Zebu
Lactations	62	25
Average parturition age (months)	47	43
Lactation length (days)	214	173
Milk yield (kg)	2149	1302

This work demonstrates that the genetic potential for milk production on medium-scale dairy farms can be rapidly increased by the use of either *Bos taurus* cattle (where environmental conditions permit) or by crossbred cattle. The levels of productivity achieved are greater than when unimproved *B. indicus* stock are used (cf. Tables 15.2 and 15.3).

Large-scale cattle farming

As stated previously, the proportion of farms falling into this category is small, especially if the tropical parts of America and Australia are excluded. However, large areas of the tropics are suitable for this form of extensive land use and it is likely that the numbers of such farms will increase. However, political pressures may prevent large farms being owned privately, and may determine that they are either state-owned and managed or run by large farming co-operatives.

Some regions of the tropics, such as the wetter, medium-altitude areas of Africa, are capable of intensive large-scale grassland production which can rival the best temperate grasslands in terms of productivity. At present this potential is relatively little exploited, since the fertile soils of the tropics have been used for cash crops rather than for grassland production. For instance, many parts of the West Indies have been widely used for the production of sugar cane for export rather than for forage crops for cattle, since the profitability of sugar production, until recently, has exceeded that of livestock production. There are, however, signs that in certain parts of the tropics this pattern may change. One factor which may bring about this change is the rising cost of labour which is operating against high-labour intensive crops, such as hand-harvested sugar cane, coffee and cocoa. Another factor is the increasing world demand for livestock products, particularly meat, which are comparatively less labour intensive and which command relatively high prices on world markets. For these and other reasons the last few decades have seen an increase in grass area at the expense of cane in the Caribbean. Another recent development has been an increase in grass production, and a corresponding decrease in maize production, in certain parts of South America and Africa.

These examples illustrate an important difference between the development of large-scale cattle enterprises in the tropics and in temperate regions. Temperate systems are economically efficient in terms of output per man and profitability on invested capital. Unfortunately, they rely very heavily on the availability of imported cereal grain, for instance the typical feedlot developed in the United States. Not only are cereal grains being increasingly used for direct feeding to man, but the cost of imported cereals is usually higher than the cattle feed market can bear. This increase has been partly due to the increasing cost of energy inputs into the cereal grain production cycle, such as the price of fuel and the high support energy cost of producing fertilisers. In addition certain tropical food crops are capable of out-producing cereal crops by a large margin, such as sugar cane, cassava and other starchy root crops.

Most of the large-scale cattle enterprises in temperate regions are highly specialised, producing either milk or beef, but rarely the two products simultaneously. Good examples are the large dairy enterprises in California and the large beef feedlots in Texas. Some workers, such as Preston (1976), argue that a dual-purpose approach would be better suited to the needs of developing tropical countries. In places where there is an unsatisfied demand for both beef and milk, it is wasteful to discard the majority of bull calves in large-scale dairy systems and, similarly, it would not be sensible to use cows capable of giving 2000 litres of milk/cow/yr on a single-suckling system in which the nutritional needs of the calf are much less than the potential milk output of the cows.

In temperate regions, especially the UK, although beef and dairy cattle systems are usually concentrated onto separate specialised farms, most home-produced beef comes from the dairy herd. The calves leave the farm of origin at about 1 week of age and are reared and fattened on specialist beef enterprises. It is possible that similar stratified systems could be developed in many tropical areas. Clearly, the specialist dairy enterprises would tend to be near the towns where the demand for fresh milk was greatest. The beef-rearing farms, on the other hand, could be in remote areas, especially in places where there was plenty of land available for grass rearing of beef stock, or else concentrations of crop by-products suitable for intensive fattening of beef (Jayasuriya, 1993).

Whether the enterprises are specialised or dual-purpose, it is likely that all large-scale cattle farms in the tropics will be stocked in future mainly with *Bos taurus* cattle or with crossbreds. The choice between these two alternatives will be primarily dependent on the climate and the disease risk prevailing in the country in question. Some recent estimates of the comparative performance of indigenous *B. indicus* cattle on the one hand and crossbred cattle on the other, when evaluated in terms of liveweight gain, are presented in Table 15.13 and this subject is also considered in Chapter 18.

The techniques of cattle management on large-scale farms in the tropics are very similar to those employed on farms of similar size in temperate agriculture and for this reason a detailed description will not be given in this book. Attention will be drawn to a few important points when considering this form of tropical land use.

Capital The capital required for intensive grassland farming is comparatively large. The cost of cattle, fencing materials, buildings and equipment are relatively more expensive in the tropics and this high cost is not always counterbalanced by the reduced cost of land. Data on the initial investment of capital required for intensive large-scale cattle farming are difficult to come by, although some useful indicator figures have emerged from a few tropical countries. Ayre-Smith (1976) gave figures for the ratio between the approximate cost of establishing holdings calculated to produce a net income of US$2500/yr. These data are presented in Table 15.14. It will be noted that, although the capital investment for cattle enterprises is less than that required for cropping enterprises, nevertheless the costs are still high. They would be out of reach of all except large corporations, rich individual farmers or co-operatives backed with government finance.

Levels of intensification The systems of cattle management at present employed on large-scale cattle farms in the tropics are usually insufficiently intensive for maximum efficiency and

Table 15.13 Breed differences in daily liveweight gain under feedlot conditions in tropical countries. (After Preston, 1976.)

Country	Ration	Zebu (kg/day)	Crossbred (kg/day)	Percentage difference (%)	Reference
Brazil	Sugar cane	0.87	1.13	30	Preston, 1974
Cuba	Ear maize	0.88	1.14	29	Willis and Preston, 1970
Cuba	Molasses	0.80	0.93	17	Preston, 1974
Kenya	Maize silage	0.88	1.06	20	Creek *et al.*, 1973

Table 15.14 Approximate cost of establishing holdings to provide a net annual income of US$2500. (Source: Ayre-Smith, 1976.)

Enterprise	Location	Net investment cost/ha (US$)	Ratio of income: investment
Tree crops and vegetables	Eastern Caribbean	2250	1.11
Bananas	Central America	2000	1.25
Milk (and beef)	Eastern Caribbean	1250	2.00
Beef (and milk)	Eastern Caribbean	625	4.00
Beef	Central America	275	9.09

Note: Costs given are those prevailing in 1973.

profitability. One of the criticisms of existing livestock estates and ranches is that they are often understocked and their general technical efficiency is often surprisingly low. In marked contrast to nomadic pastoral systems overstocking is rare but understocking so commonplace that there is often a significant positive correlations between liveweight gain per hectare and stocking rate and between economic output per hectare and stocking rate (e.g. Nestel and Creek, 1964). In some early surveys (SMA, 1963), on 1000 ha extensive beef farms in Jamaica, it was shown that the average stocking rate was only about 1 beast/ha whereas the carrying capacity of many of these farms was thought to be in excess of 2 beasts/ha and in many parts of the wetter tropics it can exceed 4 beasts/ha.

Productivity of tropical pastures This topic has been dealt with in Chapters 11 and 12 but the point needs reinforcing that the response of tropical pastures to the application of nitrogen fertiliser is often underestimated. There is a tendency for temperate concepts of grassland productivity to be transported into the tropics by estate managers and consultants. In most temperate countries, grass growth is limited by light, temperature and soil water. In the wetter tropics these factors are of lesser significance, or operate for shorter periods, and hence nutrient supply is often the chief limiting factor to maximum grass growth. Economic response to nitrogen fertiliser applications of 1 t/ha/yr of sulphate of ammonia have been recorded, but the average fertiliser application in large-scale intensive farms is much less. Tropical grassland farmers are aware that water is the limiting factor to growth in the dry season, yet few realise that nitrogen is the chief limiting factor for the remainder of the year, specially when grass is grown in the absence of a legume.

In many cases the high energy input costs for the production of nitrogenous fertilisers would alter the economics of applying nitrogen to tropical grassland, in which case a greater emphasis on grass–legume mixtures will result. The point is that whether the nutrients are provided from the fertiliser bag or from the symbiotic bacteria in legume nodules, the response of most tropical grasslands to extra nitrogen is dramatic. Where prolonged dry seasons are not encountered and soils are good, stocking rates can often be increased to figures in excess of normal temperate standards.

Seasonal production In areas where rainfall is limiting, and there are one or more severe dry seasons each year, climatic limitations will require a planned programme of seasonal calving, seasonal milk production and seasonal beef marketing. A good temperate parallel is the seasonality of grass growth in New Zealand which has forced the New Zealand dairy farmer into becoming a summer milk producer, with the whole herd dry for the two severest winter months. At present, most tropical grassland farmers tend to breed all the year round and attempt to even out the seasonal fluctuations in grass growth by irrigation or, more commonly, grass conservation. More attention should be devoted to the advantages of seasonal breeding and seasonal productivity, not only of dairy and beef cattle but also of other tropical livestock such as sheep and goats. This subject is dealt with more fully in Chapter 16.

FURTHER READING

Butterworth, M.H. (1985) *Beef Cattle Nutrition and Tropical Pastures*. Longman: London.
Crotley, R. (1980) *Cattle, Economics and Development*. CAB: Farnham Royal.
Payne, W.J.A. (1990) *An Introduction to Animal Husbandry in the Tropics*, 4th edn. Longman: London.
Smith, A.J. (ed.) (1985) *Milk Production in Developing Countries*. Centre for Tropical Veterinary Medicine, University of Edinburgh: Edinburgh.

16
Livestock Improvement by Feeding and Nutrition

P.N. Wilson

GENERAL CONSIDERATIONS

The productivity of tropical livestock can be improved in two ways: by improving their feeding and management or by breeding better animals. Better management may effect improvement relatively quickly. For instance, more skilful milking techniques or more balanced feeding programmes can be adopted overnight and milk production thereby increased in a matter of days. Genetic improvement is of necessity a much slower process, but in practice both methods should be carried out simultaneously. If genetic improvement is completely neglected, than a stage will soon be reached where little further advance can result from better management and improved nutrition. Although improvement in the environment and improvement in the animal's genotype are considered separately in this book the reader should remember that in practice both methods of raising animal productivity, wherever possible, should proceed together.

The purpose of the farm animal is to convert feed into some form of animal product. The range in feed conversion efficiency of animals within a breed is relatively limited, although quite small differences can be of economic importance in the case of farm animals which are intensively fed on purchased feeds, such as chickens or pigs. The range of feedstuffs which the tropical farmer can offer to his livestock may be wide, but it is vital that the right feeds, in the right proportions, are fed to his animals. A deficiency of one item in the diet may cause ill-health and hence low productivity. A surfeit of one ingredient may result in similar disadvantages due to reciprocal antagonisms between dietary components. Indigenous livestock are fed upon a wider collection of feedstuffs in the tropics than is usually found in temperate countries, although the digestive efficiency for a particular feedstuff does not differ greatly between temperate and tropical livestock. Temperate livestock have been bred to consume more feed and partition its eventual utilisation more towards 'productive' processes, such as milk or egg production, but the amounts of nutrients passing through the gut wall from the same quantity and quality of feed appear to be fairly similar in tropical livestock.

The nutritional requirements of farm animals can be divided into two distinct types: first, the maintenance requirements to keep the animal alive and in good health, and second the production requirements to enable the animal to grow or to produce milk, eggs, fibre or offspring. It would appear that the maintenance needs of well-adapted tropical stock are somewhat less than those of temperate animals, because the former are more lethargic and spend less energy in maintaining constant body temperature; on the other hand the temperate standards for production requirements can be taken as a reasonable working guide for the tropics.

Under systems of small-scale subsistence agriculture the cost of feed is low, since little concentrate feed is bought and minimal expenditure is incurred in pasture improvement or the growing of specific forage crops. Under systems of nomadic pastoralism, for instance, the feed costs can be equated to the cost of buying rock salt as a mineral supplement. However, as land for nomads becomes more limited and as large- and small-scale tropical agriculture becomes more efficient, the cost of properly balanced livestock feed must increase and will inevitably become a more significant item of expenditure. It is, therefore, important that livestock feeding programmes should be efficiently planned and that the maximum use should be made of local feedstuffs, correctly supplemented where necessary. The balanced rations thus devised must be fed in amounts appropriate to the liveweight and production level of the animals in question. This exercise requires a working knowledge of the basic principles of animal nutrition. This subject can only be dealt with in general outline in this book and the reader is referred to the standard text books on the subject, such as McDonald *et al.* (1988) and Payne (1990) which describe some of the practical aspects of livestock rationing.

Before dealing with the main nutrients present in animal diets it is important to note that the overall efficiency with which feed is converted into animal products is relatively low. Indeed, if the object is to maximise human food production from scarce resources (see Chapter 1) then there is clearly more merit in using land to grow crops for direct human consumption rather than to grow fodder crops for feeding to farm animals. However, the situation is not quite as simple as it may appear. Livestock, especially ruminants, can be supported on land unsuitable for arable cropping but which can supply rough grazing and browse for buffaloes, cattle, goats, sheep and dromedaries. They are also able to utilise a wide variety of waste products unsuitable as food for man (Muller, 1980; Jayasuriya, 1993). In addition, many livestock products, such as milk and eggs, provide food which is more balanced and higher in nutritional quality, especially important for young children. Lastly, even where the land in question is potentially suitable for arable farming, mixed farming systems may represent more sustainable systems of land use, such that the total output from the area is greater than if the area was used only for crop production. Animal manures provide useful and low cost plant nutrients and 'ley farming' systems enable soil fertility to be restored between cropping cycles. It follows that the general low feed conversion efficiencies even of very productive livestock, indicated in Table 16.1, have to be interpreted with caution and with knowledge of the optimal land resource management for the region in question.

Bearing in mind these important considerations, a brief account will now be given of the main nutrients that are found in most feedstuffs and their requirements by farm animals.

FOOD COMPONENTS

The feed eaten by livestock comprises the same basic ingredients that form their own animal

Table 16.1 Conversion of food protein to product protein in productive farm livestock. (Source: Wilson, 1973.)

Livestock class	Production level	Annual protein consumed (kg)	Annual protein produced (kg)	Efficiency of conversion (%)
Cow	6800 litres	586	222	38
Hen	300 eggs	7.3	2.3	31
Broiler	1.8 kg carcass	3.6	1.1	31
Fish	0.4 ha carp pond	3400	680	20
Rabbit	4 litters of 10	48	8	17
Porker	$2\frac{1}{4}$ litters of 12	840	123	15
Lamb	2 litters of 3	125	11	9
Steer	300 kg carcass at 1 yr age	568	34	6

Protein = N × 6.25 = crude protein

tissues and products. The relative proportions of the various elements vary, but both feed and animal body constituents comprise water, protein and other nitrogenous compounds, carbohydrates, fats (or lipids) minerals and vitamins. Of all these components water is often the chief limiting factor, especially in the hot dry tropics.

Water

Water is the main component of animal bodies, the percentage varying from about 75% in young animals to 50% in fat adults. The average animal contains between 55 and 65% of water. This water requirement is met partly in the form of drinking water and partly as water ingested with the feed. A small amount of 'metabolic water' is obtained when fat is oxidised within the animal's body, a point already touched upon in Chapter 13 in the discussion on the role of the fatty hump of the camel.

The water demand of livestock in the tropics varies according to breed, age, size, productivity and climate. Many workers have shown that *Bos indicus* cattle require less drinking water than *B. taurus* in the same environment (Payne, 1990; Pandey et al., 1989).

One of the problems of grazing cattle in many parts of the tropics is the seasonal effect of climate on herbage dry matter. For most of the year in arid areas the grazing consists mainly of low-quality herbage, which is essentially 'standing hay' or 'foggage'. This herbage has a dry matter content, which can be as high as 93%, a high fibre level and a low digestibility. In contrast, herbage in the humid tropics has a high water content which makes it difficult for the animal to obtain sufficient dry-matter intake. Data relating to this point are limited, but Butterworth *et al.* (1961), working with dairy cattle in the West Indies, reported a dry-matter intake of 3.7% of liveweight during the dry season, compared with 3.0% in the wet season. On the other hand, the water content of certain grasses during the wet season may be so high that the grazing animal will obtain most of its water requirements through the feed.

A major feature of water deprivation for ruminants is its effect on liveweight gain. The short-term effect is marked, due to a reduction in body water. In studies on the reaction of different breeds to drought in Australia, Frisch (1972) showed that zebu crossbred heifers were much more tolerant of drought conditions and managed to achieve target body weight at a younger age than exotic heifers. Some wild animals in hot dry climates seem capable of surviving without regular access to drinking water. They presumably manage to obtain their water requirement from dew and from the small amount of water present in the dry vegetation, and to a lesser extent on metabolic water (Payne, 1990). As already shown in Chapter 13, the camel is an outstanding example of an animal which can go for long periods without free water.

Clearly, livestock fed intensively on cereal-based diets have a large requirement for drinking water and the supply of sufficient quantities of clean water is extremely critical. Tropical poultry raised on deep litter, for instance, will die if denied water for more than 24 hours, whereas they can live for several days with no feed without suffering much more than a temporary setback to their growth or productivity.

The water intake of poultry increases with increased ambient temperature but there are large breed, strain and individual bird differences. The relevant work has been reviewed by Sykes (1977) and the data have been used to construct the graphs presented in Fig. 16.1. Up to an ambient temperature of 25°C the mean water intake/bird is about 200 g/day and above this basal level there is an average increase of approximately 10 g/day for each additional degree of temperature. Thus at 30°C the mean intake is about 250 g/day. The water requirements have also been examined by measuring the various forms of water loss from chickens, including evaporative, urinary and faecal losses as well as the losses of water in the egg. These losses are shown in Fig. 16.1 in line B, while line C represents the data for evaporative losses separately. It will be noted that up to about 25°C the non-evaporative losses predominate, but after 25°C the sharp rise in evaporative loss forms the greater part of the

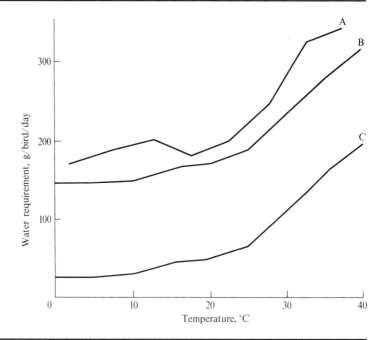

Fig. 16.1 Water intake and water loss of laying poultry in relation to ambient temperature (°C). Line A = mean water intake at 5°C intervals; line B = estimated total water loss at 5°C intervals; line C = evaporative water loss at 5°C intervals. (After Sykes, 1977.)

total water loss. It can be seen that lines A and B show a very reasonable agreement and this work on water loss substantiates the recommended water requirements of Sykes (1977).

Water requirements of tropical livestock, especially grazing ruminants, are therefore difficult to assess since they depend on the water content of feed, prevailing climatic conditions, breed and type of animal and physiological state. Thus, milking cows require more water than dry cows and laying hens more than poultry which are broody or moulting. Recent studies have also shown that cattle show a peak demand for both water and feed during the early morning and again during early evening, serving to emphasise the importance of ensuring that both water and grazing are freely available during these periods.

Carbohydrates

Carbohydrates provide the main source from which energy is derived, and fats are elaborated, although proteins can also be broken down to provide energy, albeit somewhat inefficiently. Since larger amounts of energy-producing feeds are required than any other class of nutrient, the provision of sufficient carbohydrates in a diet is very important. Carbohydrates are supplied as sugars, starches, cellulose, hemi-cellulose or constituents of fibre, such as lignin. Animals with simple stomachs mainly utilise carbohydrate supplied as sugar and starch, although they are able to digest a very limited amount of fibre and even that which is not digested assists in stimulating digestive processes. The main carbohydrate supplies for pigs and poultry are cereal grains and by-products. Cereals contain between 60 and 80% of starch, and tropical root crops such as yams, sweet potatoes, tannias, cassava, contain 65–95% of starch in the dry matter. With all livestock a small proportion of indigestible fibre can be of benefit, by assisting the movement of feed through the gut.

The main carbohydrate supplies for ruminant farm animals are derived from grass and other 'fodder crops' such as clovers, and edible leaves of tropical trees such as *Mellotus philippensis* (Bhargava *et al.*, 1977) and *Leucaena leucocephala* (Adeneye, 1979), and various products resulting from the drying and processing of these crops, such as hay, straw, silage and dried grass. These materials are usually high in fibre, especially when they are mature, and variable proportions of the fibrous and non-fibrous carbohydrates are reduced to volatile fatty acids by the action of the microorganisms present in the alimentary canal. The major volatile fatty acids formed by this process are acetic, butyric and propionic acids. Since these volatile fatty acids do not themselves occur either in milk or animal tissue in any appreciable quantity, but are the precursors of animal products, they are termed 'intermediate metabolites'. They are to be found either in the gut or in the bloodstream. A small proportion of the feed carbohydrates escape rumen fermentation, and pass into the mid-gut relatively unchanged. From there they can be absorbed, usually as a sugar such as glucose, into the bloodstream.

The microflora of the rumen varies according to the type of feed being offered. The flora are composed of two main types of microorganism, bacteria and protozoa, and also substantial numbers of anaerobic fungi. The protein derived from protozoa has a superior biological value in terms of amino acid balance and future research may discover ways in which the rumen ecology could be modified to provide a more useful microbial population and a higher ratio of protozoa to bacteria. However, some nutritionists advise against disturbing the normal microbial population. The changeover from one diet to another may necessitate a major change in the composition of the rumen microflora, and so it is usually advisable to alter diets slowly, over a period of days, rather than abruptly.

The value of carbohydrate feeds is related to the amount of energy they can supply. This value may be expressed in various ways, and there is no common view on the best definition of energy to use. All methods should express the energy value on a dry-matter basis, since the water content is variable and makes no contribution towards energy requirement. The simplest method is to deduct from the dry matter the mineral content of the feedstuff, and to designate the remainder the 'organic matter'. This organic matter percentage is related to the energy content, but only roughly so. A further extension of this method is to make a further deduction for nitrogenous matter and crude fibre, so as to calculate the 'nitrogen-free extractive' or NFE.

The best approach is to express the energy value of the feed at different points in its digestion and utilisation by the animal. The starting point is the gross energy of the feed as consumed, measured as the total heat it will produce when ignited in a bomb calorimeter. This method provides the 'calorific value' of feed, which can be expressed in terms of calories, therms or joules. A calorie is defined as the amount of heat required to raise the temperature of 1 g of water 1°C, confined to a narrow range of temperature increase, normally from 14.5°C to 15.5°C; 1000 calories equals 1 kilocalorie (kcal), sometimes written as a K calorie or a Kal; 1000 kilocalories equals 1 megacalorie or 1 therm. The universal unit for energy is the joule (J). It is that force which, if applied to a mass of 1 kg, gives it an acceleration of 1 m/s. One million joules equals 1 megajoule or 1 MJ. Medical dieticians employ calorific values for assessing the energy value of foods, but animal nutritionists use the joule which is the internationally accepted SI unit of energy. Fortunately, calories and joules are related, and 1 calorie equals 4.184 joules.

Proceeding from the starting point of gross energy, a system has been developed which enables deductions to be made to allow for the inefficiency of the animal's digestive system. The animal never utilises all the energy fed to it as some will be lost or voided as faeces, urine or methane. The difference between the gross energy input and output in the faeces is the digestible energy. The digestibility is measured by conducting a gross energy determination on the feed and a second gross energy determination on the faeces, the difference between the two being the digested fraction of the feed. Strictly speaking, this digestible energy should be called 'apparent' digestible energy, since no allowance is made for the extra energy contained in the faeces due to the presence of discarded gut wall in the digesta; the so-called 'endogenous loss'.

Not all the digestible energy is available for the animal's use. Two more corrections must be made before we arrive at the metabolisable energy (ME), which is the energy available for carrying out essential bodily functions. The first correction is for the energy absorbed from the gut but subsequently voided in the urine, while the second is for the gases, particularly methane in the case of ruminants, produced during the fermentation process. Only a proportion of the metabolisable energy of the feed is available for productive purposes, such as milk or egg production. Some of it will be required by the animal to repair worn tissues and to maintain body temperature. These two functions may together be considered as 'maintenance'. More energy is lost during conversion of metabolisable energy into a form suitable for maintenance and production and this loss is described as the 'heat increment' of the feed. The residual value is known as the net energy of the feed, and as Armsby was responsible for calculating some of the earliest net energy values, this term is sometimes spoken of as the 'Armsby net energy value'. Figure 16.2 shows how all of these different energy terms may be related to each other in the form of a family tree.

The net energy content of the diet is used by the farm animal in one of two ways. First, to keep the animal alive and well; second to enable it to produce a product such as work, milk, eggs or wool. The first part of the net energy is called the maintenance requirement and the latter is designated the production requirement. Thus the net energy estimate can be divided into the net energy required for maintenance (NEM) and the net energy required for production at a defined level (NEP). The total net energy, for both maintenance and production, is referred to as NEM + P. Further complications arise since a single feed will have different values for NEM and NEP, and these will vary at different levels of energy concentration in the diet. This fact must be borne in mind when tables listing net energy values are being used.

Another method, now not so widely used, is to calculate a value relating the energy level of the feed to the net energy value of pure starch, taken as a standard fattening feed. This relationship is the starch equivalent (SE) of the feedstuff, where starch equivalent is defined as the number of kilograms of pure starch which would provide as much net energy for a fattening steer as 100 kg of the feedstuff in question. It should be noted that the definition of SE strictly applies to fattening steers. The same system can be used, with modification, for use with lactating cows, or for non-ruminants, but it should be recognised

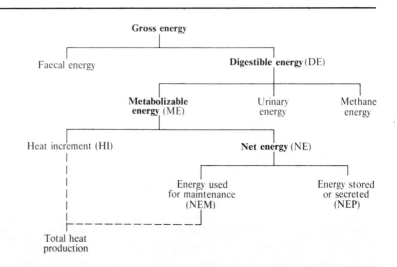

Fig. 16.2 Partitioning of feed energy within the animal.

that its original purpose was to compare different feeds with pure starch according to their ability to lay down fat in a steer. Starch equivalents are linked with the name of Kellner (1905) who first used this method of comparing the energy values of different feeds. A number of other methods are employed in other countries, such as the total digestible nutrients (TDN) system of the United States and Scandinavian feed unit (SFU) system of Norway. It is not possible to detail all these other systems in this book.

It should be stressed that the Armsby net energy values calculated for temperate cattle are not wholly applicable to zebu cattle in tropical conditions, nor does it follow that a Kellner starch equivalent worked out for temperate fattening steers will apply, say, to feed fed to fattening tropical meat goats. This is because the efficiency of utilisation of digested energy for productive purposes will differ according to many factors, including climatic conditions and the species of animal.

Unfortunately nutritionists working with the different species of livestock tend to use different systems of energy determination. Thus, ruminant nutritionists favour the metabolisable energy system, as defined by the Agricultural and Food Research Council (1980) and MAFF *et al.* (1975). Pig nutritionists tend to use the digestible energy system, as suggested by the Agricultural Research Council (1967), while poultry nutritionists are wavering between the metabolisable energy system and the net energy system. In general, the metabolisable energy system is the best one to describe the different classes of feedstuffs for a given livestock species, while the net energy system is the most relevant to employ when predicting standards of animal performance. Within a given class of animal kept under the same conditions, metabolisable energy tends to be utilised with the same degree of efficiency as net energy for a given productive purpose, so it is possible to move from one energy term to another, as illustrated in Fig. 16.2, by the use of conversion coefficients, without undue error.

Since the tropical animal husbandman is often confronted with nutritional requirements for energy quoted in one unit, and a list of energy values of different tropical feeds employing another unit, it may be necessary to translate values, such as from Scandinavian feed units to total digestible nutrients. This exercise can be dangerous, since different energy units are not strictly convertible for all ages and breeds of animal, and for all feeding systems. However, approximate conversions can be made. For instance, a crude conversion of net energy (NEM + P) into starch equivalent may be obtained by the use of the formula $9.86 \, \text{MJ} \simeq 1 \, \text{kg}$ of SE.

Fats and oils

Fats and oils are collectively known as lipids. They are similar to carbohydrates as they are primarily composed of carbon, hydrogen and oxygen. They differ since they have relatively less oxygen in their molecules than carbohydrates, hence more oxygen is needed for combustion and consequently about two and a half times as much energy is liberated per unit weight on complete oxidation. Lipids act as electron carriers, as substrate carriers in enzyme reactions, as components of cell membranes and as energy stores.

The lipids present in most animal feeds are compounds of glycerol with fatty acids. Fats and oils are broken down during the digestive process into these major constituents, which are then absorbed and either built up in the body of the animal to form a specific animal fat, usually a triglyceride, or oxidised to gaseous end products if the animal uses its fat as an energy source. This oxidation process occurs every dry season in the case of livestock, such as tropical ranch cattle, which lose weight during periods of feed shortage by depleting their fat reserves in order to provide their energy requirements. It also occurs each lactation in the case of high-yielding dairy cows, which usually lose weight for the first 4–10 weeks after calving.

The quantity and quality of fats laid down in a farm animal can be modified by manipulating the diet of the animal. It follows that diets for meat animals have to be so designed that the right sort of fat is laid down in the animal, namely a hard light fat. Within narrow limits, the amount and quality of butterfat excreted in the milk of cows and goats can also be controlled in this way. Certain tropical peoples place great store on a high fat content of slaughter-stock, but the market trend is to pay premium prices for animals which have a reasonable, but not excessive, covering of subcutaneous fat, the minimum of abdominal fat and a reasonable amount both between and within the muscle bundles. This latter fat, known as intramuscular fat or 'marbling fat', is partly responsible for determining the eating quality of meat, especially where the meat is roasted.

There is no problem in preventing excess fat deposition in ranch cattle. On the contrary, it is sometimes difficult to obtain a good 'finish' with an even covering of subcutaneous fat. However, there are problems of excess fat deposition with intensively fattened tropical livestock. These difficulties are due to the relatively low maintenance requirements of such animals in hot climates which enable them to utilise more of their energy intake for the deposition of fat. If the energy requirements of such animals are fully provided by carbohydrate feeds, the lipid content of the diet may be primarily employed for fat deposition and so carcass grade will deteriorate. This problem is especially acute in the case of swine since the improved breeds of pigs, when imported into the tropics, lay down more fat when fed standard pig diets than they would do in a cooler climate.

It is known that a proportion of the carbohydrate ingredients in the diet can be replaced by certain fats and oils without serious effects on carcass quality. As a general rule, however, it is wise to avoid diets high in lipid in the case of tropical livestock, especially pigs, for which the problem of excess fatness has not been solved. Feedstuffs containing fats or oils with a high proportion of oleic or other fatty acids associated with soft fat should be avoided, particularly during the finishing stages of the fattening process. The hardness of fat can also be affected by the rate of growth. Quick-growing stock tend to have softer fat than slow-maturing animals, so that when high planes of nutrition are adopted, with consequent high rates of liveweight gain, careful consideration should be given to the fat content of the diet.

Fat colour depends upon the presence or absence of pigments in the diet. Feedstuffs high in carotene, such as yellow maize, produce yellow-coloured animal fat. This yellow pigment is not depleted at the same rate as the fat tissue during periods of inadequate nutrition. Animals fattened over a long period on feedstuffs containing carotene tend to yield carcasses with an excessively coloured fat. For this reason the majority of ranch cattle, which can take up to 7 years to reach slaughter weight, often contain a highly coloured fat. The colour of the fat is also determined by species and, to a lesser extent, by breed. Thus, grass-fed cattle will contain fats with some degree of yellow colour. Water buffaloes, on the other hand, fed under similar conditions, will contain fat which is almost pure white.

Proteins and non-protein nitrogenous substances

Proteins are complex chemical compounds, containing about 16% of nitrogen in addition to the elements common to carbohydrates – carbon, hydrogen and oxygen. Most proteins, especially the more important ones, contain some sulphur. On hydrolysis proteins break down into peptides and other nitrogenous substances including amino acids, and the value of a particular protein for monogastrics may be expressed by detailing the amino acids released. A typical amino acid composition of some livestock feeds is shown in Table 16.2. It will be noted that the levels are low when the amino acids are expressed as a percentage of the total dry matter of the feed. It is, therefore, sometimes more convenient to express the amino acid composition as a proportion of the total protein content.

Proteins are found in all living cells, and play a vital role in most of the chemical reactions that take place there. Enzymes and hormones are also made up of complex proteins. Proteins are the building blocks for all living tissues, hence the supply of the raw materials which allow proteins to be synthesised in the animal body is of

Table 16.2 Amino acid composition of selected tropical feeds (% of dry matter). (After McDonald et al., 1988.)

Feedstuff	Crude protein	Arginine	Cystine	Glycine	Histidine	Leucine	Isoleucine	Lysine	Methionine	Phenylalanine	Threonine	Tryptophan	Tyrosine	Valine
Barley	11.0	0.5	0.4	0.4	0.4	0.6	0.3	0.3	0.2	0.4	0.3	0.1	0.3	0.4
Maize	8.4	0.4	0.3	0.3	0.2	0.9	0.3	0.2	0.2	0.4	0.3	0.1	0.3	0.4
Rice (polished)	7.7	0.6	0.1	0.6	0.2	0.5	0.3	0.2	0.1	0.3	0.3	0.1	0.4	0.5
Sorghum	7.9	0.3	0.1	0.3	0.2	1.0	0.4	0.2	0.1	0.4	0.3	0.1	0.3	0.5
Soya bean meal	45.0	3.0	1.0	1.7	1.1	3.0	1.7	2.5	0.7	2.0	1.5	0.5	1.5	1.9
Groundnut meal	47.2	5.1	1.0	2.4	1.0	2.7	1.4	1.5	0.5	2.3	1.2	0.3	1.8	1.9
Rapeseed meal	28.1	2.2	1.4	1.7	0.9	2.3	1.3	1.9	0.7	1.3	1.5	0.2	1.0	1.7
Meat and bone meal	46.0	3.1	0.5	6.8	0.7	2.5	1.1	2.1	0.6	1.4	1.4	0.2	0.9	1.9
Fish meal	65.0	3.7	0.6	4.7	1.3	4.1	2.4	4.4	1.4	2.6	2.3	0.6	2.0	2.8

great importance with stock which are actively growing. Proteins are not absorbed as such but, after enzymic breakdown, as amino acids. Therefore it is not necessary to feed protein to farm animals in exactly the form it will subsequently take as part of the animal's body or its secretions, such as the casein in milk.

Non-ruminant farm animals obtain most of their protein by eating plant or animal proteins, breaking them down by a variety of enzymes and absorbing the various amino acids thus formed. It is, therefore, of importance that non-ruminants are provided with feeds with a proportion of the essential amino acids which the animal needs and cannot synthesise in its body. There must also be an adequate proportion of non-essential amino acids to provide nitrogen for synthesis of more complex metabolites. Because it is difficult and expensive to conduct routine amino acid analyses of feeds, it is usual to obtain amino acid composition of different feedstuffs from standard tables.

In the case of ruminants a different situation occurs. Although ruminants must eventually depend upon essential amino acids provided by their 'feed', these amino acids come from two different sources. The first is direct from the diet as fed to the animal, some of which will escape fermentation in the rumen and will arrive in the mid-gut as undegraded protein, which can be digested and absorbed through the gut wall into the bloodstream. The second is from protein and other nitrogenous material which is fermented in the rumen by the same microorganisms which have transformed the carbohydrate fraction of their feed into volatile fatty acids. In this case the end products of the fermentation process are simple nitrogenous compounds, especially ammonia. Having reduced a proportion of the feed protein to ammonia, the microorganisms use this simple chemical as a building block for their own body proteins. The dead microorganisms pass down the gut with the digesta. The animal then absorbs the amino acids, derived from the breakdown of the microbial protein by proteolytic enzymes, in the same way as for amino acids from the non-degraded protein obtained from the feed itself.

This nutritional phenomenon can be exploited by the farmer, since under certain circumstances it is possible to provide a non-protein nitrogen (NPN) source to the ruminant animal by feeding it simple nitrogenous compounds, such as urea or ammonium salts. This allows the microorganisms in the rumen to build up these simple compounds into protein within their cells, which in turn can be digested by the host ruminant in the middle gut. However, there is a limit to the amount of degradable protein and NPN in the feed which can be absorbed by the microorganisms. The total amount of NPN which can be utilised in this way is about 30 g of nitrogen/kg organic matter fermentable in the rumen. Any NPN in excess of this cannot be utilised by the microorganisms and, therefore, is of limited use to the host animal. Indeed, this excess NPN may prove toxic under certain circumstances since the ammonia formed will pass into the blood, where it can cause 'ammonia toxicity' if present at too high a level. Low levels will be converted by the liver into urea which can then be recycled back into the gut by means of the saliva secreted from the mouth or by direct absorption from the blood supply to the gut. Thus the NPN that has been altered by the liver in this way has a second chance of being utilised by microorganisms and transformed into microbial protein. A diagram showing the various pathways in which nitrogen is utilised by ruminants is presented in Fig. 16.3.

It follows that the ratio of degradable protein to non-degradable protein is very important. This ratio determines the ability, or inability, of the animal to deal with different amounts of NPN in its diet. Table 16.3 presents data on the degradability values of some common animal feeds. As the actual nitrogen, or crude protein, percentage of feedstuffs varies widely, it is usual practice to carry out a Kjeldhal analysis to determine the nitrogen content of the feedstuff which can vary between plant varieties, stages of growth and different growing seasons. The nitrogen content thus determined is then multiplied by 6.25, since proteins contain on average about 16%N. However, the mean of 6.25 encompasses a wide range of conversion factors from different

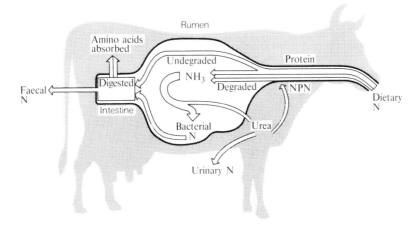

Fig. 16.3 Diagrammatic representation of nitrogen metabolism in the ruminant.

Table 16.3 Approximate protein degradability of selected numinant feeds (%) illustrating wide variation between sources.

Feedstuff	Degradability (%)
Groundnut meal	85
Grass silage	82
Wheat	82
Barley	80
Barley straw	80
Grass (fresh)	80
Hay	80
Rapeseed meal	78
Maize silage	60
Maize	55
Whitefish meal	34

Table 16.4 Factors for converting nitrogen content of selected tropical feedstuffs to protein content. (After McDonald et al., 1988.)

Food protein	Nitrogen (g/kg)	Conversion factor
Barley	172	5.83
Cottonseed	189	5.30
Eggs	160	6.25
Maize	160	6.25
Meat	160	6.25
Milk	157	6.38
Oats	172	5.83
Soya bean	175	5.71
Wheat	172	5.83

feedstuffs, and is more relevant for animal-derived feeds than for feeds of plant origin (see Table 16.4). It follows that this crude protein percentage is always somewhat higher than the true protein percentage. It will include an allowance for NPN, which as we have noted is a useful addition in the case of ruminants but may introduce a source of error with non-ruminants unable to utilise NPN to any appreciable extent. For this reason, a standard parameter known as the 'protein equivalent' (PE) has been devised. This is defined as a percentage of true protein, plus half the NPN × 6.25. Tables based on the work of British nutritionists sometimes give values in terms of protein equivalent. Tables compiled by American workers frequently give the crude protein percentage (CP) often misleadingly abbreviated to protein percentage. It is therefore important never to confuse protein equivalents from British textbooks with crude protein percentages from American works.

The difficulties occur when applying these basic principles of protein metabolism in practice. There is less of a problem with pigs and poultry, where the skill of the livestock feeder is to ensure that the amino acid balance of the diet is about right and that the total protein supplied meets normal requirements. In ruminants, however, the situation is more complex. The ruminant requires a proportion of undegradable protein which will be passed to the hind gut for normal digestion and absorption but it also requires a supply of degradable protein (or alternatively NPN) to enable the rumen microorganisms to produce microbial protein. Not only this, but that protein which is degraded in the rumen is itself present in different forms depending on whether the protein breakdown is slow or fast. Slowly degraded protein (SDP) is efficiently 'captured' by the rumen microorganisms, but protein which is rapidly degraded (FDP) is partially lost and so less effectively captured. In order to take these complex considerations into account a new ruminant protein system has been devised (Agricultural and Food Research Council, 1992; Webster, 1992), known as the Metabolisable Protein System. The scheme is complex and cannot be described in detail in this book, but as it is likely to be widely implemented in future some reference to it cannot be omitted. For this reason the tables of standard requirements presented later in this chapter include a column for the metabolisable protein (MP) requirement.

It has already been stated that proteins are made up of amino acids, the nature and proportions of the amino acids varying in different materials. Certain feeds, especially those of plant origin, have a ratio of amino acids which differs from that required in the synthesis of protein by the farm animal. Livestock may, therefore, require a large amount of feed in order to meet their requirements for the most limiting amino acid. In extreme cases, the feed offered may be completely deficient in a certain amino acid. Proteins in feeds accordingly have a different 'biological value' (BV). As a general rule, animal proteins (such as fish meal) have a higher BV than vegetable proteins (such as grass).

The plant breeder has made attempts to increase the BV of improved varieties by selecting for high levels of certain essential amino acids, such as lysine. An example of this is the production of 'high lysine corn' as a result of the plant breeder selecting the 'opaque 2' gene in maize. With the addition of this gene the proportion of lysine is doubled, and the percentage of total protein is increased from about 11 to 14% or more. It is likely that other plant species, used as animal feeds, may be improved by similar genetic manipulation in the future.

Minerals

Although the mineral matter, or 'ash', present in the animal body forms a small proportion of the total dry matter, certain quantities are essential for normal growth and production. The proportion of minerals in the diet is therefore important, and the actual levels may be critical for stock, such as milking cows or laying hens, which are excreting large quantities in their milk or eggs. As the loss of certain minerals, such as sodium, is much greater for tropical animals (that are sweating or salivating profusely) than for temperate animals, the standard temperate requirements for certain minerals are inadequate for the tropics. Unfortunately, the mineral requirements of stock kept in tropical environments have not been precisely determined. It is therefore a sound practice to provide free access to a balanced mineral mix so that animals may regulate their own intake to some extent.

The ability of animals to balance their own diet by varying the intake of different feeds is not confined to minerals, although it is most pronounced in the case of mineral imbalance, such as sodium or copper deficiency. If animals are over fat, they will tend to select feed of higher protein and lower energy concentration. Similarly if stock are lean and thin, they will tend to select high energy feeds in preference to high protein diets. This remarkable ability sometimes referred to as 'nutritional wisdom', can sometimes be used by the farmer to his advantage, as in the case of poultry which can be offered two contrasting diets, one relatively high in energy and the other in protein, thus allowing the birds to exercise their 'nutritional wisdom' in selecting the correct blend of the two. This ability is demonstrated to a lesser degree by pigs and ruminants.

Minerals may be divided into two categories according to their requirement by the animal and the concentration in the body. Elements which are required in relatively large quantities are referred to as 'major' or 'macro' elements; these include calcium, phosphorus, magnesium, sodium, potassium, sulphur and chlorine. Minerals required in smaller quantities are classed as 'micro' or 'trace' elements; included in this category are iron, zinc, copper, manganese, iodine, cobalt, molybdenum, chromium and selenium. These trace elements are present in very low concentrations and are therefore generally expressed in terms of milligrams per kilogram (mg/kg), (equivalent to parts per million or ppm).

As more research is conducted the number of recognised essential minerals is increasing, when 'essential' means that, in the absence of the element, deficiency symptoms or disease are exhibited. Such deficiency diseases can only be corrected by administering the required quantity of the mineral concerned. McDonald et al. (1988) suggested that 40 or more elements may have essential animal functions to perform. However, these newly discovered extra essential minerals are either required in such small amounts or else are so widely distributed in animal feed and drinking water, that deficiencies are rarely seen and thus the elements are only of academic importance. A list of the nutritionally and practically important minerals is presented in Table 16.5, together with their mean concentration in animal tissues.

Although minerals such as calcium and phosphorus are stored in the body (e.g. in the form of skeletal deposits) it is unwise to consider these deposits as 'reserves' on which the animal can draw for long periods. Depletion of the bone stores of calcium and phosphorus, for instance, can lead to a weakening of the skeletal structure, giving rise to deformation or to fragile bones which are easily fractured. Other elements, such as sodium and chlorine, are not stored in any appreciable quantity. Consequently, in places where minerals are in short supply, supplementation must be regularly provided by some

Table 16.5 Selected nutritionally important essential mineral elements and their approximate concentrations in animal tissues. (Source: McDonald et al., 1988.)

Major		Trace	
Element	g/kg	Element	mg/kg
Calcium	15	Iron	20–80
Phosphorus	10	Zinc	10–50
Potassium	2	Copper	1–5
Sodium	1.6	Molybdenum	1–4
Sulphur	1.5	Iodine	0.3–0.6
Chlorine	1.1	Selenium	1–2
Magnesium	0.4	Manganese	0.2–0.5
		Cobalt	0.02–0.1

means or other. Unfortunately, mineral excess can at times be as detrimental as mineral deficiency, hence the correct balance of the minerals in a ration is very important. A notable example of mineral excess is the occurrence of fluorosis (a condition typified by fragile bones) in Tanzania, caused by too much fluorine in the soil and water supplies and hence in the feed.

In the case of intensively managed livestock, which are fed mainly or exclusively on bought-in concentrates, the mineral supplementation of the ration should have been taken care of by the manufacturer. The farmer must check that this has been done, especially if he is purchasing feed from a small local mixing plant which may not have the knowledge or facilities required. Even where the feed is correctly supplemented, there may still be a need for the farmer to take additional steps to ensure that an essential mineral is never limiting production. This is important when unacclimatised stock are kept intensively in the tropics. For instance, it has been estimated that at temperatures of 40°C, *Bos taurus* cattle may lose up to 0.7 g of sodium and 0.9 g of chloride/100 kg liveweight/day in dribbled saliva and sweat. This mineral loss must be regularly replaced.

Sometimes a mineral may be of especial importance to young stock Thus in the case of young pigs, where iron is often a limiting element in sow's milk, the farmer may need to take measures to prevent a deficiency of iron in the piglet even though the sow's diet may have been adequate in this mineral. The piglets can be given supplementary iron either in the form of injections or by painting the teats of the sow with some suitable form of iron so that, as the piglets suckle, the iron is ingested.

Browsing animals, such as goats and camels, are usually better able to meet their mineral requirements than selective grazers, such as imported European cattle. The leaves of certain weeds and the leaves and pods of many tropical trees have a high mineral content, and livestock will obtain quite high intakes of minerals from such sources. Animals which mainly eat grasses and fibrous fodder crops usually have reasonably high intakes of calcium but such diets are likely to be low in phosphorus. Conversely, stock which are maintained primarily on cereal diets, such as feedlot beef cattle, will tend to have a good supply of phosphorus but calcium levels may be limiting.

In temperate agriculture, gross deficiencies or excesses in the mineral content of the herbage are often corrected by appropriate fertiliser treatment to the soil. In the tropics such corrections are only economically possible under intensive farming conditions. Under extensive ranching conditions, little or no modification can generally be made to the mineral balance of the soil, and the situation must be remedied by supplementing the diet of the animals. In the hotter parts of Australia, however, the correction of trace element deficiencies of the soil has now become an economic proposition, even in some extensive grazing areas, although problems arise with the unevenness of distribution. For example in eastern Australia where molybdenum supplementation has been attempted, this has resulted in the 'plum pudding effect' in that some areas are oversupplied and others missed out.

Vitamins

The vitamin requirements of extensively managed tropical livestock are met fairly readily, since the vitamins are either present in the herbage or else they are synthesised in the alimentary canal of the animals themselves. The term 'vitamin' is derived from 'vitamine' or 'vital amine', although it is now known that not all vitamins are amine derivatives.

As more intensive methods of animal husbandry are adopted in the tropics, especially intensive pig and poultry production, more attention to the amount and potency of the vitamin content of concentrate rations will be necessary. One important factor is that many vitamins are less stable under hot, moist tropical conditions than they are in temperate climates. Some vitamins have their potency reduced, or destroyed, by heat or by chemical action such as oxidation. Feed stores in the tropics are usually very hot, often being constructed of corrugated iron which is a very effective conductor of heat. Supplies of feed often have to be kept for long periods, and hence the risk of vitamin degradation is more pronounced. The vitamins of importance to animal nutrition are vitamin A, members of the related B complex which extends to vitamin B_{12}, and C, D, E and K. The vitamins are listed in Table 16.6, together with their chemical names.

Vitamin A

This is a fat-soluble vitamin synthesised in livestock from carotene obtained from green feedstuffs. It is destroyed by oxidation, so hay, straw and silage (which undergo some degree of oxidation during the curing process) supply very little vitamin A precursor compared to fresh grass. Vitamin A is stored in the liver, and so the supply need not be continuous, except in the case of animals such as milking cows and laying hens which are regularly excreting large amounts of vitamin A in their milk and eggs. The livers of fish, birds and mammals are a rich source of vitamin A and cod-liver oil and halibut-liver oil are commonly fed as vitamin A supplements.

Vitamin A is manufactured synthetically and can be obtained in pure form as retinol. Because of the loss of potency of this vitamin through oxidation, it is necessary to 'stabilise' vitamin A in concentrates by adding an antioxidant. Thus

Table 16.6 Vitamins important in animal nutrition. (Source: McDonald et al., 1988.)

Vitamin	Chemical name
Fat-soluable vitamins	
A	retinol
D_2	ergocalciferol
D_3	cholecalciferol
E	tocopherol*
K	phylloquinone†
Water-soluble vitamins	
B_1	thiamin
B_2	riboflavin
	nicotinamide
B_6	pyridoxine
	pantothenic acid
	biotin
	folacin
	choline
B_{12}	cyanocobalamin
C	ascorbic acid

* A number of tocopherols have vitamin E activity.
† Several naphthoquinone derivatives possessing vitamin K activity are known.

the declared percentage of non-stabilised fish oil (or other vitamin A source) originally added to the feed may be misleading. Vitamin A deficiency has been reported in poultry fed a diet containing as much as 3% of non-stabilised cod-liver oil. Wherever practicable, the practice of hanging up bunches of fresh green leaves inside intensive poultry houses is commendable. This precaution not only ensures that poultry have access to a potent vitamin A precursor, but it also lessens problems due to feather-pecking and cannibalism.

Vitamin B complex

This omnibus title covers a group of water-soluble vitamins as shown in Table 16.6. A complication in nomenclature arises from the fact that in the United States vitamin B_1 is known as vitamin F_1, and B_{12} as vitamin G. Vitamin B_{12} was previously known as the animal protein factor (APF). It is involved in protein synthesis, possibly as a co-factor in protein chain initiation. It is a very complex organic compound and its action is closely associated with copper metabolism of the animal. Most members of the B complex are not stored in animals in appreciable amounts and so a regular supply is needed. It is a common practice to add vitamin B_{12} to a large range of animal feedstuffs, but the other members of this B group are fairly readily synthesised in the alimentary canal of ruminants and are present in varying amounts in most common ingredients of pig and poultry rations, so deficiencies should not often be encountered. Meat and bone meal, whey and dried yeast are rich sources of B vitamins, and incorporation of small percentages of one or more of these ingredients in concentrate rations will usually take care of normal requirements. The recent concern that scrapie disease of sheep may give rise to bovine spongiform encephalopathy (BSE) in cattle, and that BSE may in turn give rise to Creutzfeldt-Jacob disease (CJD) in man, may inhibit the future use of meat and bone meal in livestock diets. It is important to note that meat and bone meal is now a prohibited substance for use in animal diets in many countries, especially those within the European Union, and that this prohibition may eventually extend to tropical countries.

Vitamin C (ascorbic acid)

This vitamin, once spoken of as the antiscorbutic vitamin, is synthesised by farm livestock and is therefore comparatively unimportant in animal nutrition. It is widely believed that there is a requirement for a dietary supply of vitamin C in the tropics for certain classes of livestock, especially laying poultry. However, the evidence, reviewed by Kechik and Sykes (1974), is conflicting. Some trials reported in the literature show a marked response to vitamin C addition while others reveal no significant effect. There is some evidence that ascorbate metabolism is affected by some forms of stress, such as starvation (Caudwell and Sykes, 1975). Extreme temperatures will cause stress but it does not follow that temperatures within the range normally tolerated should create an additional nutritional requirement.

Vitamin D complex

This group of vitamins is associated with calcium and phosphorus metabolism, and deficiency of vitamin D indirectly gives rise to rickets. Vitamin D is not present in plant material to any appreciable extent; however it is synthesised in the skin of farm animals through the action of solar ultraviolet radiation, either direct or reflected. For this reason vitamin D deficiency is rare in animals which are exposed to sunlight in the tropics. Vitamin D deficiency is most likely to occur with intensively kept pigs and poultry, housed in buildings with low-pitched roofs which effectively exclude all ultraviolet radiation. In such cases, supplementation of the diet with sources of vitamin D_3, such as animal fat, meat and bone meal (see above) or fish-liver oils, may be necessary. Vitamin D_2 is less readily utilised by poultry than vitamin D_3, consequently vitamin D supplementation of poultry diets should always be in the form of D_3. Vitamin D is fat-soluble, and it will be noted that many substances of animal origin which are rich in vitamin D are also good sources of vitamin A.

Vitamin E

This group of vitamins, all chemically related to the tocopherols, are collectively known as the anti-sterility vitamins since a deficiency gives rise to degeneration of the gonads and eventually to sterility. However, their main role is to prevent damage to membranes from dietary oxidised fatty acids or 'peroxide-free radicals'. They are present in green forage and in many cereal grains and hence the occurrence of avitaminosis E among tropical livestock is rare even though the poor storage conditions referred to above may increase the levels of free radicals in the diet. Grass stored as hay is low in vitamin E. Vitamin E is stored in the animal's body, so non-breeding stock can withstand quite long periods on a vitamin E deficient diet. Meat meal and animal fat can be used in concentrate diets to ensure adequate vitamin E intake. Vitamin E is associated metabolically with the trace element selenium. It is important that a sufficient supply of vitamin E and selenium is maintained since a deficiency can result in muscular dystrophy in young animals or various other membrane-associated symptoms, such as liver or heart necrosis or exudative diathesis in poultry.

Vitamin K

The absence of this vitamin delays the time taken for blood to clot. Its original Dutch name was 'Koagulation Factor', abbreviated to 'Vitamin K'. It is synthesised by the microorganisms of the alimentary canal of ruminant animals. Livestock with simple stomachs must obtain their requirements from one of the many common plant sources, such as cereal grains. Vitamin K deficiency is found in tropical poultry and the condition can be rectified by adding menadione (synthetic vitamin K) if good sources of naturally occurring vitamin K are unavailable.

FEEDING STANDARDS

While the previous sections have listed the different nutrients required by livestock for productive purposes, and described the reasons why each nutrient is necessary, it is essential that the farmer, and the feedstuff manufacturer, should have information on the the quantity of each nutrient to feed to different types of livestock at each stage of growth. Unfortunately, as stated previously, precise information applicable to tropical livestock is not available. Temperate requirements have been carefully assessed and adapted, but much more work is needed before tropical standard requirements can be presented with confidence. This consideration applies especially to pigs and poultry which, even in the tropics, are usually kept intensively.

Careful analyses of the diets available to many extensively managed tropical stock reveal nutrient intakes which are so low that, were temperate data wholly applicable, the stock in question would all be dead! Clearly, therefore, any attempt to lay down tropical feeding standards must be made with great care and the only claim which can be made about the data which follow is that they will eventually have to be revised.

The reader should bear these points in mind when using the subsequent tables.

It is important to define the term 'requirement'. Strictly speaking, the term 'nutrient requirement' should be reserved for the biological need of a given animal for a certain nutrient at a point in time. Since the farmer is normally feeding a group of animals, such as a herd or a flock, and not individual animals, it follows that 'average requirements' have to be applied rather than 'true requirements'. As the true requirements of similar individuals vary, a decision has to be made as to whether to aim at the mean requirement of the herd or flock or the average requirement of the better stock, say the top 20%, in order that those animals which are producing most are not penalised by suboptimal nutrition.

Again, due to the errors inherent in mixing feeds and feeding them on the farm, it may be desirable either to over-formulate nutrients into feeds or to produce feeding scales which are biased on the generous side. A slight oversupply of a nutrient is unlikely to be deleterious; indeed it may increase productivity by a small increment. Similarly, if one nutrient is marginally under-supplied the animal will not die, or production cease; in practice a slight lowering of productivity will be the result.

However, 'requirements' are best interpreted dynamically, in terms of response curves, rather than mechanistically in terms of a single value. Most nutrients follow the classic dose–response curve indicated in Fig. 16.4. As the supply of the limiting nutrient is increased from low levels, the response is almost linear over a wide range of nutrient intake. As the level of nutrient reaches the 'biological requirement' (defined as that level which will produce the maximum response) the curve bends over, each additional increment of nutrient resulting in a proportionally lower response. If the biological requirement is exceeded, the response decreases, and over this latter part of the curve the excess nutrient may be regarded as having a deleterious effect on the animal.

It is often assumed that the 'correct' addition of the nutrient in question is the biological requirement at the maximum point of the dose–response curve. In biological terms this may be so, but in economic terms the position may well be different. The last few increments of nutrient prior to the point of maximum response are usually expensive, since a small increase in production results from these last but expensive nutrient additions. The economics of livestock production require that the cost of the input

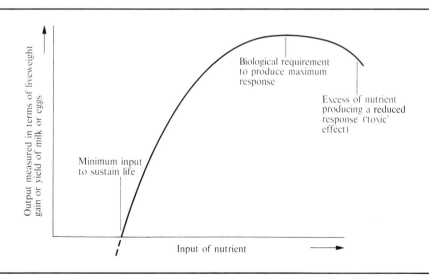

Fig. 16.4 Input–output response to the increasing supply of a given nutrient.

should never exceed the value of the resultant output. If the dose–response curves are interpreted in financial terms, it will be understood that the 'economic requirement' is set below the biological requirement. This is made clear in Fig. 16.5, which indicates that the economic formulation level of a nutrient will always be somewhat below the biological requirement.

In the tropics, the cost of nutrients may be much greater than in temperate countries and the value of the livestock products correspondingly less. Where this is so, it may be inadvisable to formulate feeds to temperate economic requirements, but to be content with lower levels of productivity at less cost. Thus the following feeding standards for tropical livestock should be treated with added caution, and interpreted according to the economic conditions which prevail.

Many feeding standards express the energy, protein and mineral requirements as percentages of the diet. The amount of nutrient consumed is therefore dependent on the quantity of feed eaten. As feed intakes vary it is preferable to express feeding standards as actual quantities (by weight) of each dietary component. With certain nutrients, such as the trace elements, the errors involved in formulating on a percentage basis instead of a weight basis are small and in such cases it does not matter whether the nutrients are calculated on a percentage basis, or an intake basis, by weight. However, with the major nutrients, such as protein and energy, the errors involved, and the economic consequences of not getting the sums right, are significant.

SCHEDULE OF NUTRIENT STANDARDS FOR LIVESTOCK

Nutritional requirements of cattle

Water requirements of cattle

Because of the various factors affecting the water demand of tropical livestock, the data tabulated in Table 16.7 are just a general guide to minimal water requirements, expressed in terms of total water needed (i.e. both free drinking water and water contained in wet feeds).

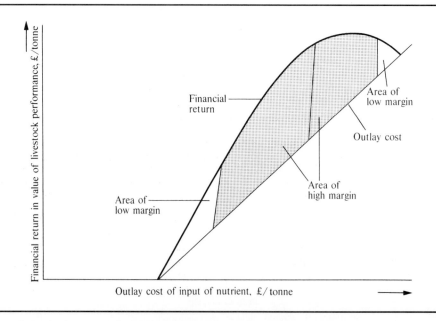

Fig. 16.5 Economics of response. The area between the two lines represent the margins defined as return–outlay.

Where the feed consumed is low in dry matter, as in the case of lush grass, the estimates for free drinking water can be reduced by more than one half, but where the feed is high in dry matter, as in the arid tropics, the data in Table 16.7 represent the minimal daily need for drinking water. When the drinking water contains salt, as in the case of tidal rivers and estuaries, the water intake is increased by up to 50%. Cattle can tolerate up to about 1.25% of salt in water, but with other species the tolerances differ. Thus camels can tolerate as much as 5.5% salt; sheep and goats up to about 1.6% (Payne, 1990).

Salinity is not the only potential problem in respect of the mineral content of drinking water. It has been suggested that high magnesium concentrations may have adverse effects on performance (Addison, 1968) but more recent work has failed to substantiate this (Saul and Flinn, 1985). Similarly, sulphate is more toxic than chloride (Squares, 1988). On the other hand it would appear that brackish water, containing a mixture of minerals at a concentration of 3600 mg total dissolved solids (TDS), may sometimes lead to better performance (in terms of milk yield of cattle) than freshwater with less than 500 mg TDS (Bahman *et al.*, 1993). Clearly much will depend on the actual amounts of the different mineral components of the TDS.

Calves

Calves must receive colostrum, preferably from their own dam, within the first 12 hours of life and also for the first 3 days. Colostrum contains antibodies which may protect the calf against many of the infections to which the dam has previously been exposed. For 12 hours or less, these antibodies can be absorbed through the gut wall. When the calf is a day old, the antibodies can no longer be absorbed in adequate amounts and thus are unable to offer the calf immunity. However, colostrum is also invaluable as an excellent high density feed. The importance of the calf receiving adequate amounts of colostrum early in life cannot be overstressed. The data presented in Table 16.8 indicate the effect of early colostrum intake on the health of the calf.

Having ensured the provision of adequate colostrum very early in life, calves may be easily weaned. In such cases it is advisable to rear calves on whole cows' milk for about the first month, before changing over to a cheaper milk substitute or skim milk. The replacement should be made over a 2-week changeover period, and the milk substitute or skim milk should be fed preferably twice a day. During the milk-feeding period the calves should have access to fresh clean drinking water and long roughage,

Table 16.7 Estimates of the water requirements of tropical livestock. (Sources: Baudelaire, 1972; Agricultural and Food Research Council, 1980; King, 1983; Payne, 1990.)

Type and class of livestock	Mean liveweight (kg)	Average requirement (litres/day)	Dry season requirement (litres/day)	Dry season frequency of drinking
Buffaloes	500	80	100	Daily
Camels	500	30	60	4–5 days or longer
Cattle				
Low Yielding	350	40	60	
	500	80	100	1–2 days
High Yielding	350	80	100	
	500	120	140	Daily
Sheep and goats				
Non-lactating	35	5	6	1–2 days
Lactating	35	10	12	Daily

Table 16.8 Relationships of colostrum test to calfhood disease and mortality (data from 2206 calves reared).

Colostrum test result	Percentage of calves requiring treatment for disease	Percentage of calfhood mortality
Little or no colostrum consumed	50.0	7.4
Inadequate colostrum consumed	29.2	2.8
Adequate colostrum consumed	21.1	1.3

Table 16.9 Total milk replacer and concentrate intake by calves (0–12 weeks). (Source: Kilkenny and Stollard, 1971.)

System	Amount of milk replacer (kg)	Amount of concentrates (kg)
Bucket, twice daily, single pens	12.7	156
Bucket, twice daily, group pens	12.2	159
Bucket, once daily, single/group pens	13.2	152
Automatic machine	16.8	125

preferably good quality dry hay or straw. Although the calf will not consume much roughage during its first weeks, the amounts ingested help to stimulate the development of the rumen and assist the development of the correct rumen microflora. In addition to long roughage the calf should be allowed access to a dry concentrate feed.

The calf will be ready for weaning from milk-type diets when the rumen has sufficiently developed to deal adequately with solid feed. It is easy to check when this point has been reached since the calf will ruminate when the rumen size and function have developed sufficiently. The calf should not be weaned until rumination activity has been observed, and until the intake of dry feed has reached a level of at least 0.5 kg/day.

When no good quality milk substitutes or dry calf feeds are available it is advisable to rear calves on whole milk for at least the first month and then to feed skim milk for at least a further 3 or 4 months. The quantities of milk replacer, and concentrate dry calf feed, required for successful early weaning systems are detailed in Table 16.9. Calves left running with their dams will wean themselves naturally over an extended period, and will not be independent from their dams until they have reached 6 months or so. It is usual for beef cows to become pregnant while still suckling their calves.

Growing and fattening cattle

The requirements of growing cattle are related to their liveweight and to their expected weight gain. Table 16.10 allows the reader to interpolate intermediate values if required. Energy requirements have been expressed in both TDN and ME at slightly lower rates than those normally advocated for temperate stock. The reason is that the maintenance requirements for growing cattle in the tropics are less than the requirements for stock maintained in cooler climates. Calcium and phosphorus intakes are important to the growing ruminant animal and approximate daily requirements have been included in Table 16.10. It should also be noted that a column has been included in this table for rumen degradable protein (RDP) and also for metabolisable protein (MP). No undegradable protein (UDP) will normally be required for moderate growth. A more detailed description of tropical beef cattle nutrition is provided by Butterworth (1985) and McDowell (1985).

Dairy cows

The requirements for dairy cows are presented in two parts. First, their requirements for maintenance are set out in Table 16.11, based on the liveweight of the cow. Second, the additional production requirement, for cows in milk, is shown in Table 16.12. In order to calculate total requirements, the maintenance and production

Table 16.10 Approximate daily nutrient requirements for growing cattle (assuming 0.5 kg daily liveweight gain). (Sources: MAFF et al. 1975; Agricultural and food Research Council, 1992.)

Liveweight (kg)	Daily intake of dry matter (kg)	Daily nutrient requirements						
		ME* (MJ)	TDN (kg)	DCP (g)	RDP[†] (g)	MP (g)	Calcium (g)	Phosphorus (g)
150	3.7–4.3	35	2.3–2.7	315–350	330	220	14	12
200	5.0–5.5	42	2.8–3.1	340–380	340	240	16	14
250	6.3–6.5	50	3.2–3.6	365–410	385	260	18	16
300	7.5–8.0	58	3.7–4.9	390–445	430	280	20	18
350	9.0–10.0	66	4.3–5.5	410–500	470	300	22	20
400	9.5–10.5	73	4.8–6.0	435–525	510	320	24	22
450	10.0–11.0	80	5.5–6.5	460–550	555	340	26	24

*ME = 8.3 + (0.091 × LW) + allowance for 0.5 kg gain/day, where LW is the liveweight.
[†]Roy et al. (1977).
ME = metabolisable energy; TDN = total digestible nutrients; DCP = digestible crude protein; RDP = rumen degradable protein; MP = metabolisable protein.

Table 16.11 Approximate maintenance requirements for selected nutrients of dairy cows. (Sources: MAFF et al., 1975; National Research Council, 1978; Agricultural and Food Research Council, 1992.)

Liveweight (kg)	Daily intake of dry matter (kg)	Daily intake of nutrients				
		ME* (MJ)	TDN (kg)	DCP (g)	RDP[†] (g)	MP (g)
315	6.0–7.5	37	2.3–2.5	180–275	290	220
365	6.6–8.0	41	2.5–2.9	225–280	320	240
415	7.5–8.6	46	2.9–3.1	275–320	360	260
455	8.0–9.5	50	3.1–3.4	300–350	390	280

*ME = 8.3 + (0.091 × LW), where LW is the liveweight.
[†]Roy et al. (1977).
ME = metabolisable energy; TDN = total digestible nutrients; DCP = digestible crude protein; RDP = rumen degradable protein; MP = metabolisable protein.

requirements must be added together. For each litre, or kilogram, of milk produced, the cow will require 14 MJ of ME or 0.3 kg of TDN, and 60 g of DCP or 45 g MP, the actual amount depending on the butterfat content of the milk. It will be noted from Table 16.12 that, for each 1% increase in butterfat, the energy requirement rises by 0.2 kg of TDN and 3.0 MJ when at a low yield level and up to 0.9 kg TDN and 13 MJ when at a reasonably high milk yield. The calcium and phosphorus levels in the feed of milking cows are extremely critical, as large quantities of these minerals are excreted in the milk. It may be necessary to add these minerals to the diet in the form of bone meal, dicalcium phosphate or defluorinated rock-phosphate, or else by means of purchased mineral supplements.

Working oxen

The maintenance and production requirements for mature working oxen are presented in Table 16.13. These two requirements are also additive. Thus an ox weighing 550 kg and working an 8-hour day would require 74 MJ/day and 438 g DCP/day.

Table 16.12 Approximate milk production requirements for selected nutrients of dairy cows (assuming a liveweight of 365 kg with a solids-not-fat in the milk of 8.6%).

Daily milk yield (kg) and butter fat (%) (BF)	Daily nutrient requirements							
	ME (MJ)	TDN (kg)	DCP (g)	RDP* (g)	UDP* (g)	MP† (g)	Ca (g)	P (g)
5 at 4% BF	26	1.6	275	208	—	475	10	8
5 at 5% BF	29	1.8	275	234	—	475	10	8
10 at 4% BF	52	3.2	550	412	138	700	20	16
10 at 5% BF	58	3.6	550	463	97	700	20	16
15 at 4% BF	78	4.9	825	615	278	925	30	24
15 at 5% BF	88	5.5	825	692	217	925	30	24
20 at 4% BF	104	6.5	1100	819	418	1150	40	32
20 at 5% BF	117	7.4	1100	921	335	1150	40	32

*Roy *et al.* (1977).
†Agricultural and Food Research Council (1992).
ME = metabolisable energy; TDN = total digestible nutrients; DCP = digestible crude protein; RDP = rumen degradable protein; UDP = undegradable protein; MP = metabolisable protein.

Table 16.13 Approximate total production** and maintenance requirements for selected nutrients of working oxen. (Sources: MAFF *et al.*, 1975; National Research Council, 1978; Agricultural and Food Research Council, 1980, 1992.)

Liveweight (kg)	Daily maintenance requirements						Production requirements per working hour				
	DM (kg)	ME* (MJ)	TDN (kg)	DCP (g)	RDP† (g)	MP (g)	ME (MJ)	TDN (kg)	DCP (g)	RDP† (g)	MP (g)
250	5.0–7.0	33	2.2	136	180	147	2	0.12	13.6	16	15
350	6.0–8.0	42	2.7	181	230	190	2	0.14	13.6	16	15
450	7.0–9.0	50	3.3	227	280	228	2	0.15	13.6	16	15
550	8.0–10.0	58	3.8	272	330	264	2	0.16	13.6	16	15

*ME = 8.3 + (0.091 × LW), where LW is the liveweight of the oxen.
†Roy *et al.* (1977).
**'Total production' requirement is the sum of the daily maintenance requirements + the production requirement per working hour × number of hours worked/day.
DM = dry matter; ME = metabolisable energy; TDN = total digestible nutrients; DCP = digestible crude protein; RDP = rumen degradable protein; MP = metabolisable protein.

Nutritional requirements of other ruminants

Lactating sheep and goats

The maintenance requirements are set out in Table 16.14 and the production requirements in Table 16.15. These requirements are additive, in similar fashion to the requirements of dairy cattle. Tropical sheep and goats are usually suckled and rarely milked by hand, hence suitable nutritional requirements for hand-reared lambs and kids are not tabulated. Strictly speaking, the maintenance requirements of both sheep and goats are increased towards the end of pregnancy, particularly when the females are carrying litters rather than single offspring. However, since it is rarely possible, without the use of scanning techniques, to know whether the females are carrying singles or litters it is not possible to differentially feed according to bio-

Table 16.14 Approximate maintenance requirements for selected nutrients for sheep and goats. (Source: MAFF et al., 1975; National Research Council, 1978; Agricultural and Food Research Council, 1990.)

Type	Approximate liveweight when mature (kg)	Daily requirements				
		ME* (MJ)	TDN (kg)	DCP (g)	RDP† (g)	MP (g)
Temperate breeds	70	8.6	0.75	58	67	65
Crossbred	45	6.3	0.55	45	49	50
Tropical	30	4.8	0.45	30	37	40

*ME = 1.8 × 0.1 liveweight.
†Roy et al. (1977).
ME = metabolisable energy; TDN = total digestible nutrients; DCP = digestible crude protein; RDP = rumen degradable protein; MP = metabolisable protein.

Table 16.15 Approximate production requirements for milking sheep and goats (assuming a liveweight of 50 kg and milk solids-not-fat of 11.5%).

Daily milk yield (kg)	Butterfat (%)	ME* (MJ)	TDN (g)	DCP (g)	RDP† (g)	UDP† (g)	MP (g)
3	4	31	960	210	194	136	171
	5	33	1080	240	209	123	180
5	4	43	1600	350	291	251	257
	5	48	1800	400	316	231	270

*ME = 1.8 × 0.1 liveweight.
†Roy et al. (1977).
ME = metabolisable energy; TDN = total digestible nutrients; DCP = digestible crude protein; RDP = rumen degradable protein; UDP = undegradable protein; MP = metabolisable protein.

logical need. Usually the female will be able to support the multiple pregnancy by drawing from body reserves, and so the greatest need for extra feeding is after parturition during the suckling period. It is for this reason that the nutrient requirements are set out in greater detail *post-partum* rather than *pre-partum*.

Water requirements of other ruminants

Detailed estimated water requirements for cattle have already been presented in Table 16.7, together with approximate standards for other ruminants. It should be remembered that the needs for lactating stock are much greater as the water secreted in the milk must all be replaced by extra drinking water, approximately on a litre-for-litre basis.

Nutritional requirements of pigs

The water requirements for pigs are set out in Table 16.16. Temperate standards have been increased, but even greater amounts of water will be needed where pigs are kept in conditions of high temperature and low relative humidity in the drier tropics.

Where pigs in the tropics are allowed to act as scavengers and are fed on waste products their nutritional requirements cannot be regulated but where they are fed intensively or semi-intensively in the tropics the composition of their diets and their feeding regime may be critical. In temperate agriculture it used to be common practice to express pig rations in terms of 'pig meal' required at different growth stages, and to indicate the 'nutrient ratio' of the meal at each

Table 16.16 Water requirements of pigs. (Source: McDonald *et al.*, 1988.)

Type	Daily water intake (kg/head)
Growing pigs	1.5–2.0 at 15 kg liveweight, increasing to 6.0 at 90 kg liveweight
Non-pregnant sows	5.0
Pregnant sows	5.0–8.0 (increasing as pregnancy progresses)
Lactating sows	15.0–20.0 (highest after farrowing; lowest at weaning)

Table 16.17 Guide to amino acid composition of protein appropriate to the diets of growing pigs, pregnant, and lactating sows.

Amino acid	Percentage of the protein		
	Growing pigs	Pregnant sows	Lactating sows
Histidine	1.5	2.1	1.9
Isoleucine	3.5	3.7	4.5
Leucine	5.0	7.6	6.4
Lysine	5.5	3.5	3.8
Methionine + cystine	3.1	2.5	2.5
Threonine	3.2	2.8	2.6
Tryptophan	1.0	0.8	0.8
Trysine + phenylalanine	3.5	6.3	6.3
Valine	3.5	4.4	4.6

stage. Thus, young piglets were weaned on to a high-protein diet (nutrient ration one part protein:four parts carbohydrate) and could be fattened for pork or bacon on meals of lower protein content (nutrient ration one part protein:six or seven parts carbohydrate). Such simple standards are applicable where 'wholemeal' feeding is the rule and pigs are kept semi-intensively. In some parts of the tropics pigs are raised in this way and the problem is to balance a bulky diet, consisting of vegetable waste and by-products, with a high protein concentrate.

Where pigs are intensively reared for high levels of productivity, their feeding standards are as critical as those commonly applied to poultry (see later). The economic targets of higher rates of liveweight gain, better feed conversion efficiencies and more lean meat production require that pig nutrition, and swine-feeding systems, become more sophisticated. As the pig is a monogastric it is unable to obtain protein from the breakdown of bacteria which have utilised simpler nitrogenous substances. It is therefore important to ensure that proteins fed to pigs are of a high biological value, containing the essential amino acids in the correct ratio. This balance is most readily achieved by including high-protein feeds of animal origin, such as fish meal or meat and bone meal. Such materials are likely to be higher in limiting essential amino acids, such as lysine and methionone, than plant protein. Table 16.17 provides a guide to amino acid requirements for growing pigs, pregnant and lactating sows. It should be stressed that pigs can usually survive without supplementary nutrients, but at much lower levels of production.

The main nutrient requirements for pigs of different weight and type are detailed in Tables 16.18 and 16.19. These tables have been modified from work published by the Agricultural Research Council (1967) and McDonald *et al.* (1988). Tropical research work in pig nutrition has also been taken into consideration. In interpreting the vitamin requirements in Table 16.19 (and also Table 16.22) it is important to make allowance for the 'background levels' of vitamins present in the normal, unsupplemented animal feeds. Although an oversupply of vitamins can be regarded as a useful insurance policy, certain vitamins are expensive and it is wasteful to supply more than are actually required by the stock in question. Although the nutritional requirements of pigs are best specified in nutrient terms, nevertheless it is important for the pig farmer to have some ready means of calculating feed requirements on a total tonnage basis.

Whether pigs are reared on *ad libitum* system, or on a restricted feeding system, the quantities of feed consumed should approximate to those tabulated in Table 16.20. As nutritional density increases, feed intake/day will drop and, conversely, as the diet offered is of a lower nutritional density the total amount of that diet

Table 16.18 Energy feeding standards for pigs (MJ of digestible energy). (Sources: Agricultural Research Council, 1967; McDonald et al., 1988.)

(1) Growing pigs

Liveweight (kg)	Expected liveweight gain per day (g)	Daily feed intake (kg dry matter)	DE (MJ)
20	500	0.9	13
30	625	1.3	18
40	750	1.6	23
50	790	1.9	28
60	790	2.2	31
70	790	2.4	35
80	790	2.6	38
90	790	2.8	40

(2) Pregnant sows

Initial liveweight (kg)	DE requirement (MJ)	
	Weeks 1–12	Weeks 14–16
135 and under	23	25
160	25	27
180	28	30
205	30	32
230	32	34

(3) Lactating sows

Litter size	DE requirement according to liveweight (MJ)		
	135 kg	180 kg	225 kg
5	59	65	71
6	63	70	76
7	68	74	80
8	72	78	85
9	76	82	89
10	80	85	93

consumed each day will rise. Table 16.20, taken from Whittemore and Elsley (1976), provides five possible alternative feeding scales. Scale A represents a generous scale of *ad lib* feeding to slaughter. Scale B, *ad lib* to 50 kg followed by a moderately liberal (B1) or restrictive (B2) regime. Scale C represents a much simpler system whereby pigs are fed to appetite until they are eating a set amount of 2.5 kg daily, after which they are restricted to that amount until slaughter. Scale D is a restricted regime up to 75 kg, after which it approximates to Scale B1. The more restricted scales produce leaner, and slower-growing pigs.

The majority of tropical pigs will be fed diets relatively high in fibre. Such fibrous diets can contribute up to about 30% of energy requirements, but there will be a resultant reduction of both the overall digestibility of the feeds and possibly significant reductions in energy intake. Sows are better able to utilise fibre than younger pigs. If appropriate care is taken there is much potential for the incorporation of fibrous materials into feeding strategies for pigs in tropical countries (Frank *et al.*, 1983; Close, 1993).

Nutritional requirements of poultry

The water requirements for poultry have been discussed earlier in theoretical terms. However, in practice water has to be provided to poultry in troughs or drinking devices, and in these circumstances a large proportion of the water supplied is inevitably spilt and therefore wasted. Thus it is more practical to specify requirements in terms of water trough space or of drinkers needed for a group of birds, and the data in Table 16.21 are presented in this practical fashion. The estimates are for medium-sized breeds; lighter or heavier breeds will need correspondingly smaller or larger quantities.

Chickens are normally grain-feeders, and the basic ingredient of poultry diets is usually some form of cereal or cereal by-product. Energy is rarely a limiting factor in traditional poultry diets and other nutritional components, particularly the amino acid balance and the mineral and vitamin inclusion levels, assume the greatest importance.

For these reasons energy tends to be supplied *ad libitum* and it is assumed that requirements are satisfied by voluntary intake. Where ambient temperatures are low, the birds healthy and the feed supplied is of reasonable nutrient density, the birds will eat more feed as the nutrient density is reduced and less feed when the nutrient

Table 16.19 Feeding standards for pigs (values expressed as composition of dry matter of diet). (After McDonald *et al.*, 1988.)

Nutrient	Growing pigs					Pregnant sows	Lactating sows
	1.4–4.5 kg	4.5–9.0 kg	9–20 kg	20–50 kg	50–90 kg		
Crude protein (%)	32	26	20	18	15	14	16
Digestible energy (MJ/kg)	16	15	14	13	13	13	13
Amino acids (%)							
Histidine	0.5	0.3	0.2	—	—	—	—
Isoleucine	1.3	0.9	0.7	0.75	0.75	1.0	1.0
Leucine	1.4	1.0	0.7	0.6	0.5	0.3	0.8
Lysine	2.2	1.6	1.3	1.0	0.7	0.4	0.7
Methionine + cystine*	1.3	0.9	0.6	0.5	0.4	0.3	0.4
Threonine	0.9	0.8	0.7	0.6	0.5	0.4	0.5
Tryptophan	0.3	0.2	0.15	0.15	0.1	0.1	0.1
Tyrosine + phenylalanine[†]	1.0	0.7	0.5	—	—	—	—
Valine	1.0	0.7	0.5	0.4	0.4	1.0	1.0
Vitamins							
A (IU/kg)	10 000	9000	8000	7000	6000	8000	8000
B_{12} (µg/kg)	18.0	18.0	18.0	10.0	10.0	15.0	15.0
D (IU/kg)	1000	1000	900	800	700	1000	1000
E (mg/kg)	15	15	15	15	15	15	15
Choline (mg/kg)	1000	1000	1000	1000	1000	1500	1500
Nicotinic acid (mg/kg)	20.0	20.0	15.0	15.0	15.0	15.0	15.0
Pantothenic acid (mg/kg)	10.0	10.0	10.0	10.0	10.0	10.0	10.0
Pyridoxin (mg/kg)	2.5	2.5	2.5	2.5	2.5	1.5	1.5
Riboflavin (mg/kg)	2.5	2.5	2.5	2.5	2.5	3.0	3.0
Thiamine (mg/kg)	2.5	2.5	2.0	1.5	1.5	2.0	2.0
Minerals							
Calcium (%)	1.0	1.0	0.9	0.8	0.6	0.9	0.9
Phosphorus (%)	0.7	0.7	0.7	0.6	0.5	0.7	0.7
Potassium (%)	0.25	0.25	0.25	0.25	0.25	0.25	0.25
Sodium (%)	0.1	0.1	0.1	0.1	0.1	0.1	0.1
Copper (mg/kg)[‡]	≤250	≤250	≤250	≤250	4	6	6
Iodine (mg/kg)	0.15	0.15	0.15	0.15	0.15	0.1	0.1
Iron (mg/kg)	60	60	59	58	57	60	60
Magnesium (mg/kg)	350	350	300	250	225	300	300
Manganese (mg/kg)	11	11	11	11	11	11	11
Selenium (mg/kg)	0.15	0.15	0.15	0.15	0.15	0.15	0.15
Zinc (mg/kg)	60	60	55	50	50	50	50

* Cystine may supply up to half the methionine requirement.
† 0.14% tyrosine can replace an equal weight of phenylalanine.
‡ 5 mg/kg is the basal requirement. Additional copper up to 250 mg/kg acts as a growth promoter.

density is increased. However, as temperature rises voluntary feed intake, and hence energy intake, falls steeply as illustrated in Fig. 16.6.

Figure 16.6 is based on data derived from small-scale energy balance trials on carefully selected birds (Davis *et al.*, 1973), calorimetric estimations over 4-day periods (Van Es *et al.*, 1973), and longer-term laying trials under practical conditions (Emmans *et al.*, 1975). The range of the observations from these trials is shown in

Table 16.20 Ration scales for growing pigs (in kg meal/day). (Source: Whittemore and Elsley, 1976.)

Weight of pig (kg)	Amount of feed				
	Scale A	Scale B1	Scale B2	Scale C	Scale D
20	1.0	1.0	1.0	1.0	0.8
25	1.2	1.2	1.2	1.2	1.0
30	1.4	1.4	1.4	1.4	1.2
35	1.6	1.6	1.6	1.6	1.4
40	1.8	1.8	1.8	1.8	1.5
45	2.0	2.0	2.0	2.0	1.7
50	2.2	2.2	2.2	2.2	1.8
55	2.4	2.3	2.3	2.4	2.0
60	2.5	2.4	2.4	2.5	2.1
65	2.7	2.5	2.4	2.5	2.3
70	2.8	2.6	2.5	2.5	2.4
75	3.0	2.7	2.5	2.5	2.5
80	3.1	2.8	2.6	2.5	2.6
85	3.3	2.9	2.6	2.5	2.7
90	3.4	3.0	2.7	2.5	2.8
95	3.5	3.0	2.7	2.5	2.9
100	3.6	3.1	2.8	2.5	3.0
105	3.7	3.1	2.8	2.5	3.1
110	3.8	3.2	2.8	2.5	3.2

Note: For definition of scales A–D see text.

Table 16.21 Water trough requirements for different classes of poultry in the tropics. (After Payne, 1990.)

Type	Age (weeks)	Length of water trough per 100 birds (m)	No. water fountains and capacity (litre)
Fowls	1–4	1.0	
	5–8	1.8–2.0	4 × 2.5
	9–18	2.5	
	Laying	2.5	
Turkeys	1–4	1.0–1.3	
	5–8	2.5	4 × 3.0
	9–16	3.0	
	Breeders	3.0–3.8	

Note: The allowances are generous by temperate-zone standards, as under normal tropical conditions birds need to drink more frequently than they would in the temperate zone.

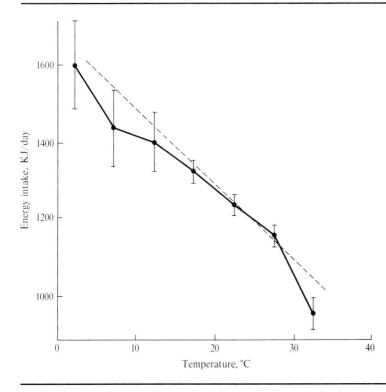

Fig. 16.6 Effect of ambient temperature on intake of energy (kJ) in chickens. This figure presents the mean results from nine trials at 5°C intervals in ambient temperature for a bird of 1.5 kg liveweight. The dotted line represents the calculated regression. ME = 1690 − 20.1°C ($r = 0.80$). (Source: Sykes, 1977.)

Table 16.22 Summary of selected nutrient requirements of chickens expressed as dietary concentrations on a dry matter basis. (Sources: Agricultural Research Council, 1975; McDonald et al., 1988.)

	0–4 weeks	4–8 weeks	Growing pullets	Laying pullets	Breeding pullets
Assumed feed intake (g/day)	—	—	75	110	110
Assumed ME content of diet (MJ/kg)	12	11.5	11	11	11
Crude protein (%)	20	15	12	16*	16*
Arginine (g/kg)	11	7	7	5	5
Glycine + serine (g/kg)	14	10	8	?	?
Histidine	5	4	3	2	2
Isoleucine (g/kg)	9	6	5	5	5
Leucine (g/kg)	15	10	8	7	7
Lysine (g/kg)	11	8	7	7	7
Methionine (g/kg)	5	4	3	4	4
Methionine + cystine (g/kg)	9	7	5	6	5
Phenylalanine + tyrosine (g/kg)	16	11	8	7	7
Threonine (g/kg)	7	5	4	4	4
Trytophan (g/kg)	2	2	1	1	1
Valine (g/kg)	10	7	5	5	5
Calcium (g/kg)	12	9	8	35	33
Chloride (g/kg)	1	1	1	1	1
Copper (mg/kg)**	4	4	4	4	4
Iodine (mg/kg)**	0.4	0.4	0.3	0.3	0.3
Iron (mg/kg)**	75	75	75	75	75
Magnesium (g/kg)**	0.4	0.4	0.4	0.4	0.4
Manganese (mg/kg)**	100	100	100	100	100
Non-phytin phosphorus (g/kg)	5	4	4	4	4
Total phosphorus (g/kg)	7	6	5	5	5
Potassium (g/kg)	3	3	3	3	3
Sodium (g/kg)	1.5	1.5	1.5	1.5	1.5
Zinc (mg/kg)**	50	50	50	50	50
Vitamin A (IU/kg)**	2000	2000	2000	6000	6000
Vitamin B_{12} (mg/kg)	0.02†	—	—	0.002	0.002
Vitamin D_3 (IU/kg)**	600	600	600	800	800
Vitamin E (IU/kg)**	25	25	25	25	25
Vitamin K (mg/kg)**	1.3	1.3	1.3	1.3	1.3
Biotin (mg/kg)	0.15	—	—	—	—
Choline (mg/kg)	1300	—	—	1300	1100
Folic acid (mg/kg)	1.5	—	—	0.3	0.3
Linoleic acid (g/kg)	10	10	10	12‡	2§
Nicotinic acid (mg/kg)**	28	28	28	28	28
Pantothenic acid (mg/kg)**	10	10	10	10	10
Pyridoxine (mg/kg)	3.5	—	—	2.0	4.0
Riboflavin (mg/kg)**	4	4	4	4	4
Thiamin (mg/kg)	3	—	—	2	2

*These protein and amino acid requirements are for an egg output of 50 g/hen/day.
†Assumes negligible reserves at hatching.
‡Requirements for maximum egg weight for pullets reared on conventional diets.
§Requirement for maximum hatchability.
**Added as supplement to diet.

Fig. 16.6 by vertical lines, and the dotted line shows the calculated effect of temperature on energy intake measured in kilojoules (kJ) of ME.

Until recently, the protein requirements of poultry were often expressed in terms of the overall protein percentage of the diet, or the percentage of amino acids contained in the total diet. Such standards have limited use since it is the daily intake of the different amino acids that is important. Overall protein levels can be kept relatively low providing a full range of amino acids is present in approximately the correct proportion (Agricultural Research Council, 1975). Table 16.22 provides data for the requirements for the chief amino acids, vitamins and minerals of intensively managed poultry, together with an indication of the energy density of the diet, expressed as MJ/kg.

The protein content of the diets for laying pullets recommended in Table 16.22 would provide 18 g/day of crude protein. This is slightly higher than the previously accepted standard of 15 g protein/day but work by Morris (1975) has shown that there is a response, especially in egg weight, to increasing protein intake of typical amino acid content up to 18 g/day within the ambient temperature range of 15 to 27°C. For reasons given earlier in this chapter, the economic optimal level of protein may be less than 18 g/day; for instance when egg weight is unimportant the optimal level of the chief limiting amino acids may be correspondingly less. On the other hand where it is important to achieve a minimum egg weight, higher levels may be more profitable.

Since the maintenance metabolism of poultry falls as ambient temperature rises it might be expected that the protein requirements would also fall. However, experiments at the North of Scotland College of Agriculture and Gleadthorpe Experimental Husbandry Farm (G.C. Emmans, unpublished data) suggest that at temperatures up to 24°C response to protein intake remains virtually unchanged. Thus, in the absence of other data, temperate standards can be employed in the tropics where birds are housed appropriately.

Table 16.23 Feed consumption guide for birds with target weight at 18 weeks of less than 1.5 kg.

Week	Per bird		Per 1000 birds	
	Week (kg)	Cumulative (kg)	Week (kg)	Cumulative (kg)
1	0.06	0.06	64	64
2	0.13	0.19	127	191
3	0.18	0.37	186	377
4	0.24	0.61	236	613
5	0.28	0.89	282	895
6	0.32	1.21	322	1217
7	0.35	1.56	354	1571
8	0.37	1.93	372	1943
9	0.38	2.31	386	2329
10	0.40	2.71	399	2728
12	0.42	3.54	422	3557
14	0.45	4.43	454	4450
16	0.49	5.39	490	5412
18	0.53	6.43	526	6446

It must be remembered that as temperature rises the level of egg output and fertility will fall, and this aspect has already been considered in Chapter 14. This fall may in part be caused by the reduction in energy intake at high temperature (above 24°C). Hence a higher energy density diet may be of benefit, as temperatures above 24°C are common in the tropics.

Table 16.23 gives an approximate guide to the total quantities of feed required for poultry flocks. The data are presented in terms of requirement/bird/day and requirements per 1000 birds/day are also tabulated.

Wherever poultry are fed on diets containing unground cereal or other grains and seeds, it is important to ensure that they are provided with insoluble grit *ad libitum* to allow the birds to grind the feed particles in their gizzards. Birds given access to free range will pick up their own grit, but birds kept intensively, and fed on unground cereals, must have grit provided. Where laying birds can be fed a calciferous grit separately from the main diet an advantage may be obtained since the birds will regulate their calcium intake to suit their needs.

FURTHER READING

Butterworth, M.H. (1985) *Beef Cattle Nutrition and Tropical Pastures*. Longman: London.

Haresign W. and Cole, D.J.A. (eds) (1988) *Recent Advances in Animal Nutrition*. Butterworths: London.

McDowell, L.R. (1985) *Nutrition of Grazing Ruminants in Warm Climates*, Academic Press: Orlando, FL.

McDonald, P., Edwards, R.A. and Greenhalgh, J.F.D. (1988) *Animal Nutrition*, 4th edn. Longman: London.

Payne, W.J.A. (1990) *An Introduction to Animal Husbandry in the Tropics*, 4th edn. Longman: London.

17
Livestock Improvement Through Health and Hygiene

P.N. Wilson

GENERAL CONSIDERATIONS

Animal health is a relative term, and is defined by Tyler and Lee (1990) as 'the condition of an animal that enables it to attain acceptable levels of production within the farming system in which it is maintained'. Thus an animal which might be considered healthy by a Masai pastoralist would be considered unthrifty if kept in a feedlot in Venezuela. It is therefore important to differentiate between clinical disease, displaying one or more signs of ill health, and subclinical disease, where there are no overt signs of disease but where there is measurable production loss.

Not long ago the main cause of loss in ranching operations in Brazil was said to be death of suckling calves as a result of attack by wild jaguars (Ellis and Hugh-Jones, 1976). In certain other parts of the tropics, more animals are carried away by floods each year than are killed by the major enzootic diseases. The main 'production disease' among tropical ruminants is starvation, and the major limiting nutrient in the tropics is water. It follows from these opening remarks that disease (dis-ease) is but one, albeit important, aspect of a wide spectrum of environmental hazards which confront domesticated livestock in the tropics. The prevention and cure of specific animal diseases must be viewed within this wider context. Figure 17.1 attempts to illustrate this point by showing a series of limiting factors which prevent livestock from living and performing well. Although disease is of great importance, it is not usually the first limiting factor, nor can it be considered in isolation from other aspects of good management, such as adequate nutrition (Putt and Hanks, 1993).

Maintaining his stock alive and healthy must be primarily the responsibility of the farmer. Most environmental hazards can be avoided by good management, and many tropical diseases can be prevented on the farm with an elementary knowledge of animal hygiene. There are, however, certain conditions, especially contagious diseases and diseases which are spread by common vectors (such as birds and flies), which the farmer cannot combat on his own. For these conditions a national policy must be enforced, such as a tsetse eradication campaign or a rinderpest vaccination programme. Even though the farmer is unable to tackle such country-wide problems single-handed, it is important that he should be familiar with the signs of all disease conditions which fall into this category. He can then take steps to notify the authorities of outbreaks on his farm and thus assist the government veterinary officers in carrying out their professional tasks. However, to encourage farmers to report disease outbreaks some form of compensation for losses involved is desirable.

For various reasons the tropics are more amenable to the development of animal diseases than temperate climates. To start with, most temperate parasites are also found in the tropics. Many of these require to survive outside their host for part of their life cycle, during which time they are exposed to the external environment. During these critical periods, humid tropical climates can be more benign to the parasite than temperate climates.

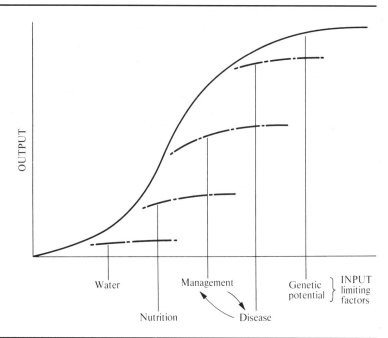

Fig. 17.1 Relationships between various inputs and livestock production. (Source: Ellis and Hugh-Jones, 1976.)

Second, traditional features of livestock management in the tropics often result in land adjacent to kraals, watering points, farms and homesteads becoming increasingly fouled, thus producing ideal conditions for the development of the infective phases of a wide variety of tropical parasites and pathogens.

Finally, chronic malnutrition can be associated with the rapid development of parasitic diseases. Well-fed animals are usually more resistant to infections than badly managed, poorly fed stock. The seasonal productivity of many tropical grasslands often results in a large proportion of tropical livestock being chronically underfed for many months each year. Unless seasonal migration is practised, this will inevitably result in large concentrations of ill-fed animals existing near to dwindling feed stocks and dried-up watering points, producing ideal conditions for the rapid spread of infectious diseases.

THE ECONOMICS OF ANIMAL DISEASE

The cost of animal diseases in the tropics is difficult to assess because there are at present many tropical animal diseases which exert a depressing effect on livestock health and productivity while remaining undiagnosed and unrecorded. Most estimates are likely to err on the conservative side because of this fact. Bearing this point in mind, it is salutary to note that Jordan (1986) reported that the trypanosomes carried by the African tsetse fly were once responsible for the loss of livestock valued then at over £100 million in West Africa alone each year, a figure which now would be much greater, and even more so if calculated for the whole of Africa.

There is an urgent need to quantify data on tropical diseases, and their resultant cost to the producer, on a co-ordinated basis. Sample surveys should be undertaken to establish more accurately the pattern of disease problems under different types of production systems in different regions of the world. Disease patterns vary markedly from herd to herd and from country to country, and it is only by collecting data on a systematic basis that the total pattern will emerge. In the absence of such work, it is not possible to distinguish between the complete absence of a specific disease from a given area on one hand and the unrecorded presence of the disease on the other. Again, without veterinary survey work, it is not possible to differentiate between true absence of the disease and its presence in a mildly chronic or modified form (Woods, 1985).

In spite of these important generalisations, certain interesting comparisons may be made. Thus East Coast fever (theileriosis) which has killed millions of cattle in East Africa, does not appear to have made any significant impact in West Africa, where the tick vector does not occur even under similar climatic conditions (Irvin et al., 1981). Dermatophilosis causes severe losses in West Africa, but appears to be less important elsewhere on that continent (Ellis and Hugh-Jones, 1976). Again, bovine pleuropneumonia, trypanosomiasis and East Coast fever have never been recorded in Latin America, in spite of the fact that these diseases were probably introduced from Africa, before quarantine regulations were set up. Lastly, foot-and-mouth disease is a very important health problem in South America but relatively less so in Africa, despite the fact that many different forms of the foot-and-mouth virus have been found in Africa but only three in South America. When chronicled in this way, the pattern of disease incidence in different tropical countries appears to be random. However, by compiling data of the different countries, the various breeds and strains of livestock present and the different nutritional and climatic factors of significance, it should be possible to identify the major causal factors determining the presence or absence of a given tropical disease. Indeed, it may be possible to go a stage further and produce a simulation model capable of predicting the sort of diseases likely to be encountered when management practices change, or when the climate is modified by man, such as by widespread forest-clearing which can lead to desertification.

As a clearer understanding of the dynamics of the situation is obtained, it may be possible to move into the important area of economic evaluation of different disease control measures. Thus, the technique devised for evaluating different socio-economic programmes can be adapted for animal health plans (Ellis, 1972). Studies along these lines have been conducted in different parts of the world, such as the work of Putt et al. (1987) on the economics of controlling tsetse flies and cattle trypanosomiasis. Other similar projects are in progress in South America.

Practical experience, and more sophisticated studies such as those referred to above, suggest that the cost of entirely eliminating a major disease from a large continental country bordering several other states is often prohibitive. For this reason, it is sometimes more realistic to 'live with the disease' and build up livestock populations which are immune or resistant, either naturally or following vaccination (e.g. haemorrhagic septicaemia), rather than to embark upon a costly and difficult programme of disease eradication.

As might be expected, indigenous livestock are more resistant to the local diseases of the area than imported exotic stock, and for this reason many veterinary departments in the tropics in the past have restricted the entry of 'susceptible exotic stock'. Such restrictions are based on good scientific foundations and several workers, such as Spooner et al. (1973), have shown that it is possible to identify the gene or genes responsible for resistance to a specific disease. Unfortunately, although such a policy may be scientifically justified from the veterinary standpoint, it may not be economically viable. Thus, it may be thought financially more desirable to have, say, 1000 head of stock producing 0.5 kg liveweight gain per day with a relatively high mortality rate averaging 20% than to have 1000 head of local stock with a low mortality rate of 5% but only giving a productive return of 0.2 kg liveweight gain per day. The former policy, in spite of greater disease hazards, yields 146 t of liveweight gain per annum while the latter and safer policy only yields 58.4 t liveweight gain each year. With economic and political considerations being as important as they are, many tropical countries will accept the greater disease risk of importing exotic stock in order to maximise the gross national product from which government revenue is derived. For this reason, many tropical countries which previously legislated to prevent the importation of susceptible animals are now changing their policy and allowing importation, if not of livestock, at least of exotic semen and/or embryos.

This conflict between what may be scientifically preferable and what is economically desirable exists at the level of the farmer as well as at the level of national policy. For instance, it is generally preferable to feed animals on non-medicated feedstuffs which allow disease outbreaks to be detected as and when they occur, rather than to supply animals with feeds with added antibiotics and other anti-bacterials which mask the early manifestations of disease and hence delay detection and which lend to the risk of lowering natural immunity. However, the addition of such biologically active materials to pig, poultry and calf rations undoubtedly increases livestock performance and liveweight gain, and supplementation of diets with these products is becoming as common in some intensive animal industries in the tropics as it is in North America and Europe.

The ratio of qualified veterinary surgeons to the number of livestock present in the country is usually very wide in tropical areas. In many places, the only veterinarians are those employed in government service and such officers are administratively unable to deal with cases of non-infectious diseases, or injury to individual animals. This fact, coupled with the generally lower value of tropical livestock compared to their temperate counterparts, means that the tropical farmer needs to become more self-reliant regarding animal health matters. Unfortunately, due to a lower general standard of education, the tropical producer is often less informed and less capable of dealing with animal health problems, even with the aids of modern technology, than temperate area stockmen.

It is a common tenet of temperate animal husbandry that a successful farmer is usually one who devotes a great deal of personal attention to his stock, treating each animal as an individual. The same applies in tropical areas, and in certain cattle-owning tribes (such as the Abahima described briefly in Chapter 15) the standard of traditional stockmanship is often high, even allowing for many malpractices due to ignorance or tribal custom. This tradition has already broken down in the temperate poultry industry, where attention to the individual is impracticable and the smallest unit dealt with is a group of birds. In this respect certain specialised extensive systems of animal production, such as ranching, are similar to large-scale poultry farming. It is uneconomic to devote too much attention to the individual and the sickly cow or sheep identified at pasture should if possible be disposed of humanely and hygienically since it will rarely repay the expense of isolation, treatment and possible cure. It is an unfortunate fact that the relatively low profit margins per head of stock force the large-scale tropical producer to consider his animals in groups rather than as individuals. There are naturally exceptions to this general rule, the most important being in connection with elite breeding stock which economically justify regular individual attention.

DEFINITIONS OF DISEASE

There are several different types of disease agent and these are briefly described later. There are also different methods of spreading disease, which may be defined as follows:

- *Infectious disease.* These diseases involve the entry and development of the agent in the body of the host. Spread of infectious diseases is more rapid when the host animals are in close proximity, such as the rapid spread of salmonellosis in an intensive piggery.
- *Contagious disease.* These diseases are spread by direct contact, usually very close contact, as in the cases of sexually transmitted diseases such as vibriosis.
- *Epidemic disease.* This term refers to the occurrence of a disease above its usual frequency in the population. Rinderpest is a good example.
- *Endemic disease.* These diseases are present at a fairly constant level in the population at all times. Mange in camels is a good example.
- *Pandemic disease.* This term refers to an epidemic affecting a very large area in a short time. The outbreak of rinderpest in the late

nineteenth century, which swept across most of the African continent, is a good example.
- *Acute disease.* These diseases flare up quickly in individual animals, which may either die or recover. Good examples are blackleg and anthrax.
- *Chronic disease.* Chronic diseases have much longer time-scales. They may take a long time to show clinical signs and then persist for months or even years. Examples are bovine pleuro-pneumonia and ringworm.

THE INFLUENCE OF LIVESTOCK HOUSING ON HEALTH AND HYGIENE

Livestock buildings in the tropics, especially those constructed for pigs and poultry, have to provide shelter from driving rain and possibly from direct sunlight but having achieved this attention should be devoted to making the buildings cool, hygienic and easy to maintain in a clean condition. A slightly sloping concrete floor and an efficient system of drainage are of greater importance than high walls and expensive roofs. Direct sunlight is an extremely effective sterilising agent and for this reason wide open-sided structures, which allow entry of the early morning and the evening sun, are far preferable to the box-like buildings suitable for keeping temperate livestock warm through the winter. Careful attention should be given to the efficient disposal of effluent. Manure and compost heaps properly constructed on well-drained sites are not health hazards, but pools of stagnant urine and manure heaps constructed in low-lying places give rise to major fly problems and are a constant source of irritation and infection to man and beast alike.

There are many unfortunate examples of badly designed, wrongly constructed livestock buildings in the tropics. Elaborate equipment, designed for temperate conditions, such as tower silos, rotary parlours, effluent treatment plants, etc., rarely justify their high initial cost unless they are used and maintained by qualified staff, possessing appropriate mechanical skills. The runoff from slopes may be extremely rapid and livestock buildings should not be so sited that they act as traps for storm water. Black walls and roofs absorb solar radiation and become excessively hot, as does the air contained within them. On the other hand, white-washed walls reflect solar radiation so efficiently that they produce an intense amount of reflected radiant heat and glare to man and animals unfortunate enough to be near to them. They may provide a more acceptable environment inside the animal house but if stock are free to move outside, the environment close to such housing may be stressful. The optimum would be white roofs and walls of any colour that does not reflect sunlight onto the animals. Attention to such elementary details can make all the difference between success and failure in building design. The construction of suitably adapted livestock buildings has a markedly beneficial effect on tropical animal health and hygiene and hence on livestock productivity. The worst possible situation is the construction of badly designed permanent buildings, expensive to modify and difficult to maintain.

THE AGENTS OF TROPICAL ANIMAL DISEASE

Although there are a number of specific animal diseases which are peculiar to the tropics, such as East Coast fever and trypanosomiasis, by and large most types of livestock disease are roughly similar in both temperate and tropical regions. For this reason, the basic textbooks of veterinary science, such as Urquhart *et al.* (1987), give adequate coverage of the major diseases of tropical livestock. In this chapter only the principal types of diseases will be described, namely those caused by bacteria, viruses, protozoa and worms. Finally a brief section will be devoted to animal health problems due to 'production disease'.

Diseases caused by bacteria and viruses

The most dreaded diseases in temperate and tropical areas alike are epidemics caused by diseases which can spread rapidly between animals within a herd, and between herds within large regions of a country. Such diseases are usually caused through the invasion of the host animal by single-celled microorganisms, such as bacteria, or by viruses or mycoplasma. Although bacteria invade the organs of the body, they normally live outside living cells, such as in the gut or in the bloodstream. Viruses multiply inside living cells which they injure or kill in the process. The host animal is commonly invaded through its natural orifices – mouth, nose or the genital tract – or else through external injuries to the skin surface. These external injuries may, in turn, be made by larger parasites of the host animal, such as biting flies and ticks, which then introduce the smaller parasites through their saliva.

There are two different ways of treating diseases caused by bacteria and viruses. First, the disease organism may be attacked with pharmaceutical products capable of weakening or killing the pathogen. Many bacterial diseases may be treated in this way, and in this sense antibiotics may be regarded as pharmaceutical products designed to kill or weaken bacteria. Unfortunately, viruses are seldom killed by drugs and thus they are often more difficult to control than diseases of bacterial origin.

The second mode of attack is to assist the host animal in preparing an effective defence mechanism against the invading parasite. This defence may be sufficiently effective that the invader is overcome immediately, or else is unable to multiply quickly enough to damage the host animal. This reaction is known as the immune response, and immunity may be either genetically bestowed on the host (i.e. natural resistance) or may be the result of protection acquired as a result of a former invasion by the same type or parasite. Animals have the ability to recognise 'foreign bodies', including pathogens, which are collectively known as antigens. The response to an antigen is to produce an antibody, which is capable of attacking and, where successful, destroying the invading antigen. This antigen–antibody response is the essential component of the immune reaction and it confers protection to the animal which, in certain cases, will persist throughout life. The response is highly specific to a specific antigen and it is capable of being measured, in the form of a 'titre', which can diagnose an infection and assess its virulence. A special form of immunity is often given to newborn animals, which obtain the immune-response antibodies from their immune dam. Some of these antibodies may be present in the bloodstream at birth while others pass from the mother to the offspring through the milk (see Chapter 16). For the first 12 or so hours of life the young animal can absorb these antibodies from its gut, but soon afterwards the gut wall 'closes', effectively preventing any further passage of antibodies. This passively acquired immunity lasts for only 2–3 weeks. Thereafter a young animal deprived of colostrum is left in a vulnerable state until it acquires its own immunity, which may take one or more weeks.

It is this ability of the host to react to invasion that is the basis of the protection offered by vaccines and related pharmaceutical products. These materials may be manufactured in one of three ways. First the infective agent may be attenuated by biological or physicochemical means (e.g. heat treatment), so that when injected into the host animal it stimulates the immune response reaction without causing disease. Second, the parasite may be completely killed but, when injected as a vaccine, it is still capable of producing antibodies in the host animal. Third, the livestock can be protected by injecting a product consisting of antibodies themselves, obtained from the serum of other animals which have successfully reacted to the disease and produced a reservoir of antibodies in their blood serum. This is known as passive immunity.

Veterinary skill is required in deciding upon the best type of vaccination programme to combat a specific disease. The same disease may be caused by one or more different strains of related viruses, or by related serotypes of bacteria. Unless the correct strain or serotype is used in the preparation of the vaccine, the programme may be ineffective, with the result that money may be wasted. Again, it is very important that laboratories manufacturing tropical vaccines are under professional supervision. Thus if an attenuated (modified) virus has not been correctly weakened, the vaccination programme, instead of protecting the livestock population, may become the means of spreading disease. For this reason, it is often good sense for tropical countries to rely on major pharmaceutical companies for the manufacture of vaccines, tailor-made to their specific requirements. Not only are the chances of mistakes lessened, but international pharmaceutical companies of repute normally carry insurance cover so that, in the event of mistakes, adequate financial compensation can be made.

Diseases caused by protozoa

The livestock population in the tropics suffers from a greater range and variety of protozoan parasites than is found in temperate regions. Many of these parasites are conveyed to the animal through arthropod vectors, especially biting flies, such as the tsetse (*Glossina* spp.) and *Stomoxys*, or ticks. The control of biting flies is a difficult and expensive operation, usually dependent on a major modification of the vegetation so as to remove bushes and trees which provide the fly with suitable breeding grounds (Jahnkw, 1974). For this reason, more attention is now being given to the control of fly-transmitted diseases by the administration of drugs to the host animal, such as Antrycide or Dimidium bromide in the case of trypanosomiasis (human sleeping sickness) (Jordan, 1986), Fig. 17.2. There are over 20 different species of tsetse fly and certain groups have specific environmental requirements, the knowledge of which can aid their eradication or control. Thus, riverine species such as *G. palpalis* and *G. tacinoides* prefer cool humid conditions associated with shaded river banks. In the case of these particular flies removing the vegetation from the river banks may be a suitable control measure, although the area may be difficult for mechanical equipment to reach and the cost of tree clearance may be prohibitive. Another species, *G. morsitans*, prefers drier, open savanna with scattered bush vegetation. Early attempts to clear such terrain, using pairs of heavy tractors dragging a long heavy chain between them, have proved to be ineffective since regrowth occurred relatively quickly. Complete eradication of bush,

Fig. 17.2 Picture showing the emaciating effect of trypanosomiasis on cattle. The white cow in the background is unaffected, but is of equal age to the diseased black cow in the foreground. (Source: Shell.)

by the use of herbicides, is more effective but usually prohibitively expensive.

Recent work has examined the effects of trypanosome infection on the metabolism of cattle (Trail *et al.*, 1995) and goats (van Dam *et al.*, 1996). These workers have demonstrated that infected animals had lower feed intakes, increased heat production and increased requirements of metabolisable energy. This illustrates the important principle that a single discrete host:parasite relationship has numerous side-effects which collectively effect animal production in a variety of different ways.

Ticks transmit many of the more serious viral and protozoan diseases and they are such important vectors of tropical diseases that tick control is one of the first prerequisites of more intensive forms of animal husbandry (Irvin *et al.*, 1981). Ticks may be controlled on the individual farm by dipping or spraying the livestock at frequent intervals (Fig. 17.3). In certain areas, efficient programmes of this sort have led to complete tick eradication and, therefore, to the successful elimination of such diseases as tick paralysis, redwater fever (babesioisis), gall sickness (anaplasmosis), spirochaetosis, East Coast fever

Fig. 17.3 Cattle entering a cattle dip in Zambia. The dip contains an insecticide against African cattle tick. (Source: Zambian Information Services.)

(theileriosis), biliary fever and heartwater. In other places programmes have been less successful due to a build-up of a population of ticks resistant to the acaricide used in the dip and sprays. It is therefore prudent to adopt a tick control programme in which the different basic chemical types of acaricide (chlorinated hydrocarbons and organo-phosphorus compounds) are used in rotation. Full details of the construction of suitable dip and spray races for use in the tropics are given by Payne (1990). It should be noted that dipping and spraying will also reduce the population of other ectoparasites, and may also temporarily repel attacks from troublesome biting flies. Most of the chemicals in dips or sprays are poisonous to man and should only be used by trained stockmen taking due caution.

Some knowledge of the life cycle of the various types of tropical ticks is important for those planning tick control or eradication programmes. A very brief summary will therefore be given of the biology of tropical ticks. The engorged female tick lays her eggs on the ground in protected crevices in a single cluster which may contain several thousand eggs. The eggs hatch after a few weeks and the resultant larvae or 'seed ticks' emerge from the crevices and migrate on to the vegetation growing in areas selected for grazing by cattle, sheep or goats. The different species of tick have different life cycles. The one-host tick spends all its time on the same host animal, passing through larval, nymphal and adult stages on one animal. The females drop from their host to lay their eggs only when they have engorged themselves with their final blood meal. An example of a one-host tick is *Boophilus decoloratus*, or blue tick, which is the major vector of redwater and gall sickness. The two-host ticks spend about 2 weeks on one host animal in their larval stage and a shorter period of 5 or 6 days on the second host. Three-host ticks have yet shorter periods at each stage of their life cycle, less than 1 week being spent attached to each of their three host animals. It follows from these considerations that the interval between dipping or spraying treatments is governed by the length of the different stages of the ticks' life cycle. Thus, with the one-host tick, spraying or dipping at 2-week intervals will be effective, while for the two-host and three-host ticks the interval between treatments should be less than 1 week, although there may be severe difficulties in achieving this.

Diseases caused by worms (helminths)

Three important groups of animal parasites come into this category. They are the roundworms or nematodes; the flukes or trematodes; and tapeworms or cestodes. Few, if any, tropical livestock are completely free of some degree of infestation with worms from one or more of these groups, and many stock remain in an unthrifty, unproductive state because of chronic infestation by parasitic worms. Farmed fish are also very susceptible to worm infestation. The two main methods of dealing with helminth parasites are by preventing their transmission from one host to the next or by directly attacking the worm by administering an anthelminthic drug.

Under wet tropical conditions, grass growth is so rapid that the optimal speed of grazing rotation may be such that the herd is repeatedly reinfesting itself with worms which have been hatched from the eggs excreted in a previous grazing cycle. There may well be conflicting needs from the point of view of good grass management on the one hand and good animal hygiene on the other. Practical considerations of space or of the utilisation of the herbage result in the pasture being regrazed while it is still heavily contaminated. Occasionally it may be possible to increase the mortality rate of the larvae by harrowing to break up the faecal pats, so increasing the rate at which the faecal material dries out (Sewell, 1976). However, harrowing is an expensive operation not often practical in tropical pasture management. Australian work (CSIRO, 1973) has suggested that alternate grazing by cattle and sheep may greatly assist in reducing the numbers of larvae on the pasture. This procedure might well be used, together with rotational grazing, in order to allow an increasing stocking rate and better nutritional utilisation of the pasture on the one hand and a reduction in the helminth burden on the other. In some areas, where labour costs are low, hand feeding or 'zero grazing' may be economic and effective.

Helminth diseases manifest themselves in four main ways. First, some diseases, such as haemonchosis, may give rise to acute disease outbreaks, leading to such high levels of mortality that the disease may appear to be epidemic in nature. The parasites associated with such acute outbreaks include *Haemonchus*, *Fasciola*, *Mecistocirrus*, *Bunostomum*, *Trichostrongylus* and *Oesophagostomum*.

Second, and much more common, there are the less dramatic chronic conditions which result in impaired growth and hence low levels of animal productivity. A good example of helminths falling into this group are the large liver flukes such as *Fasciola*. It is known that the reduced growth rate is mainly caused by a loss of plasma protein through the damaged walls of the gut or bile duct (Murray, 1975). As a result of this plasma-protein loss, a greater proportion of the available amino acids in the bloodstream are used to repair the damaged tissues with the consequence that less are available for muscular growth. The result is a lowered rate of liveweight gain and a reduction in lean meat production.

Third, certain helminth parasites, such as *Taenia saginata* and *Echinococcus granulosus*, are of zoonotic importance, since these parasites of farm animals can also cause disease conditions in man. For this reason, economic loss occurs due to condemnation of the affected organs during meat inspection. Naturally such parasites are of major importance to countries with an important export trade in meat.

Lastly, there are helminths, such as *Stilesia hepatica*, which cause parts of the carcass to be condemned on visual inspection purely for aesthetic reasons since there is no risk of spreading these animal diseases to man. Cattle livers affected with flukes are another example, although in this case there is physical damage to the liver which, besides making it unsightly, would reduce the quantity of edible tissue for the consumers.

Many of the helminths can be controlled by the administration of appropriate prophylactic or therapeutic drugs. With some of these care must be taken to see there are no untoward side-effects due to photosensitisation, in which the breakdown products of the drug render the animal abnormally sensitive to solar radiation. Anthelminthics are usually administered orally, but occasionally some of the absorbed drug finds its way to the skin where it is broken down into a potentially toxic substance. Because of this danger it is important to keep stock out of direct daylight for 1 or 2 days after such a drug has been administered. Some workers advise the minimal use of anthelminthics on the basis that a healthy animal can cope with a 'normal' infestation of helminths and is, therefore, capable of keeping the level of parasitism in check. The use of anthelminthics, it is argued, makes the animal more susceptible and thus prone to subsequent reinfestation. This view is not always supported by critical research, but it is true that a helminthicidal programme, once embarked upon, must be maintained long term if reinfestation is to be avoided.

One of the older anthelminthics employed is phenothiazine, or one of the related chemical compounds. The use of this drug illustrates the difficulties which sometimes surround the practice of veterinary medicine in the tropics. Phenothiazine is broken down in the body of the farm animal and eliminated as a red dye, and the sight of the blood-red urine, when unexpected, is alarming. The drug was formerly introduced into certain tropical countries without sufficient warning being given to the herdsmen about these side-effects. The sight of red urine, pink milk and a discoloured fleece or hair has led the stockmen to believe that their livestock were being poisoned. Great harm can be done to the relationships between the veterinarian, livestock officer and herdsmen if the implications of such procedures as dosing with phenothiazine are not made clear from the start. Another potential hazard is the use of molluscicides to control the other host of many tropical trematodes – the freshwater snail. Some of the earlier molluscicides were not only toxic to snails but also poisonous to freshwater fish, and since the habitat of the snail is low-lying swampy land adjacent to rivers, it followed that freshwater fish were inadvertently killed. The advent of N-tritylmorphiline has helped matters in this respect since this product is a potent molluscicide which is relatively non-toxic to fish and other forms of wild life. It is, however, relatively costly and so its use in the tropics is limited.

Molluscicides are most effective if they are used to kill infected snails just before they shed the cercariae, which is that stage in the trematode life cycle ingested by the ruminant. However, such an operation requires detailed knowledge of the life cycle of the trematode in its snail host, which varies between species (Ollerenshaw, 1971). Such knowledge is not readily available in tropical countries, so the control treatments are less selective and hence less effective. The reproductive capacity of the snails is such that any survivors will rapidly repopulate the habitat, especially in warm or hot climates.

The control of helminths provides a good illustration of the importance of assessing the cost-effectiveness of alternative techniques of control. Clearly this cost must be considered relative to the deleterious effects of the parasites, the productivity of the animals and the market value of their products. A knowledge of these factors is necessary to enable correct decisions to be made with respect to different control techniques. There are many unfortunate instances of parasite or disease control in tropical countries where the cost of treatment has outweighed the benefit to the livestock producer. Decisions have to be made as to whether to assist the local livestock population to build up a relative immunity to a given disease, to reduce the disease level or opt for a complete disease-free policy. Every case must be considered on its merits, and it is unwise to translate experience from one country to another where different conditions apply.

A good example of this problem is foot-and-mouth disease, which is kept out of the UK by a strict 'detect and slaughter' policy. However, as has been mentioned, foot-and-mouth disease is widespread throughout Africa and Asia and mortality of zebu cattle from this disease is comparatively low. There are undoubted losses in productivity, as measured by milk production and liveweight gain, but the disease is not severe enough to justify the cost of an elimination programme. On the other hand, where a meat export trade is being developed, the importing countries may demand that foot-and-mouth disease is eliminated or such imports will be prohibited.

Diseases caused by fungi

A wide range of fungi can affect farm animals, especially their more exposed surfaces, such as the skin, the mucosal surfaces of the respiratory tract and the urogenital system. Ringworm in tropical cattle is a well known obvious example. External fungal diseases can usually be controlled by the application of fungicides.

Control of disease vectors

As has been described earlier, many important tropical diseases are transmitted by, and develop in, arthropod vectors. Early control programmes concentrated on direct attacks on the vectors (as in the case of dipping against ticks) or on the environment (such as the clearing of bush to remove the breeding sites of tsetse flies).

It is now increasingly difficult to justify such expensive routine measures, and in future new approaches will need to be developed. Animals can be bred with some resistance to attack by vectors. The vectors themselves can be diverted from their normal livestock hosts by the use of odour-baited, insecticide-impregnated traps and target screens. Finally, the vectors themselves may in future be controlled by disrupting their normal reproductive process, such as by the introduction of sterile males into the vector population.

An interesting classic development in biological control has been successful in eliminating an important ectoparasite of cattle, the screwworm (*Callitroga hominivorax*), from the semi-tropical, southeastern region of the United States. This technique consists of releasing large numbers of laboratory reared, sterilised male flies during the peak mating season. These sterilised males mate with normal female screwworm flies and, provided the programme is so planned that the sterilised males are in the correct ratio to the total screwworm population, a complete eradication is possible within a few generations. The technique was first developed on one of the small Dutch islands in the Caribbean and was adopted on a large scale in Florida. It has also been applied recently in the extensive control programme mounted by the Food and Agriculture Organization of the United Nations (FAO) for the successful eradication of screwworm in Libya (Gillman, 1992). It is possible that a similar approach will prove successful with other major disease vectors and ectoparasites of farm animals as a result of future research, although such methods are more readily carried out on discrete islands than on whole continents.

Production diseases

For purposes of convenience, the major nutrients have been discussed in Chapter 16, dealing with feeding and nutrition, while the major pathogens causing livestock disease have been dealt with in the present chapter. However, when major nutrients are in very short supply they not only limit livestock productivity, such as by reducing milk yield or liveweight gain, they also cause other serious disturbances such that the animal in question exhibits clinical signs of 'dis-ease' (Payne 1990). The avoidance of 'production diseases' has been complicated, since they fall between the disciplines of animal nutrition and veterinary science. In practice close collaboration is needed in order to solve production disease problems.

Many production diseases can become so acute, that the nutrient imbalances are best rectified by injection rather than by adding the missing nutrient to the feed. It is therefore reasonable to consider acute production diseases as being similar to diseases caused by pathogens. However, the long-term solution to production diseases must be found by improving the quality of the feed supply to tropical livestock. Unfortunately, many important mineral deficiencies and/or toxicities do not result in clear-cut clinical signs specific only to a given nutrient. Thus behavioural abnormalities, such as depraved appetite or pica, can be caused by the deficiencies of several minerals, such as sodium and phosphorus. Therefore in order to determine which nutrients are involved it is necessary to carry out chemical analyses of the animal feed, the soil on which the feed was grown and/or the affected animal tissues.

Soil surveys, such as those described in Chapter 3, may indicate which particular minerals are deficient or in excess. Unfortunately, the pattern in the soil is not usually reflected by the pattern in the plants growing thereon, since different plants absorb minerals selectively and not uniformly. Forage analyses are a more reliable guide than soil analyses, but different animals are able to absorb differential amounts of soil minerals from the same herbage plant, and an excess or deficiency of a particular mineral in plant tissue does not necessarily mean that the animal consuming it will exhibit a mineral imbalance. Therefore attention has been given to examination of the animal tissues themselves rather than concentrating solely on the data from soil or herbage analyses. Since the various nutrients are not distributed evenly throughout the animal body, it is important to select the correct organ or tissue for an analysis of a particular nutrient. Thus the most relevant organ in respect of copper deficiency is the liver, since copper is selectively stored in this organ. If fluorine is thought to be in excess, then a chemical analysis of bones is indicated, since this particular element is preferentially stored in the skeleton. If imbalances between calcium, phosphorus and magnesium are suspected, then the levels circulating in the blood give an indication of the balance of these three elements. Table 17.1 indicates which analyses are thought to be of most significance in assessing the specific deficiencies and toxicities of 18 minerals.

It has already been stressed (see Chapter 16) that many nutrients interact, either synergistically or antagonistically, with each other. For this reason it is advisable to investigate problems of mineral deficiency or toxicity by looking at the

Table 17.1 Detection of specific mineral deficiencies or toxicities in cattle. (After McDowell, 1976.)

Element	Analyses for assessment of mineral status	Critical level	
		Deficiency	Toxicity
Calcium	Diet	0.18–0.60%	
	Blood plasma	8 mg/100 ml	
Magnesium	Blood serum	1–2 mg/100 ml	
	Urine	2–10 mg/100 ml	
Phosphorus	Blood plasma	4.5 mg/100 ml	
	Diet	0.18–0.43%	
Potassium	Diet	0.60–0.80%	
Sodium	Saliva	100–200 mg/100 ml	
	Diet	0.1%	
Sulphur	Diet	0.1%	
Cobalt	Liver	0.05 mg/kg	
	Diet	0.05–0.1 mg/kg	
Copper	Liver	25 mg/kg	700 mg/kg
Iodine	Milk	300 µg/day	
Iron	Haemoglobin	10 g/100 ml blood	
	Transferrin	13–15% saturation	
Manganese	Liver	10 mg/kg	
	Diet	20–40 mg/kg	1000–2000 mg/kg
	Hair		70 mg/kg
Selenium	Liver	0.25–0.5 mg/kg	5–15 mg/kg
	Diet	0.05–0.10 mg/kg	5 mg/kg
Zinc	Blood plasma	0.04 mg/100 ml	
	Diet	10–50 mg/kg	
Fluorine	Bone		4500–5500 mg/kg
Molybdenum	Diet		6–20 mg/kg

whole mineral profile, rather than by investigating each suspect mineral individually.

Payne (1975) adapted a diagnostic technique frequently used in human medicine, known as the 'metabolic profile test'. This test investigates a number of different blood metabolites taken from a random sample of animals from the population under investigation. When applied to milking cows, the test requires samples of blood from 21 cows, 7 of which are dry, 7 near to peak milk yield and 7 lower yielding cows, towards the end of their lactation. These 21 blood samples are analysed for glucose (an indicator of energy status), urea, albumin and globulin (which, taken together, act as indicators of protein status), packed cell volume (an indicator of dehydration), and a range of individual minerals, usually including magnesium, calcium, phosphorus, iron, copper and zinc. Since there are likely to be sampling and laboratory errors regarding the level of any one particular metabolite in the blood taken from a single cow, the test is evaluated statistically in terms of mean values for each metabolite examined, together with an indication of the statistical variation exhibited between the seven animals comprising one of the three herd groups. The statistical parameter used is the standard deviation, and a nutrient is deemed to be deficient or in excess if it is two standard deviations or more from the mean value of the factor in question.

Unfortunately, tests of this nature are expensive to conduct and skill is required in interpreting the results in a way which takes into account the practical considerations involved with the herd. It is clearly a step in the right direction, since it is the animal's actual tissues which are under test, rather than samples of the feed it eats. The test is statistically analysed on a computer, and the result printed out in terms of tables, scatter diagrams and histograms. It is the computerised histogram which lends the name 'profile' to the test, as illustrated in Fig. 17.4.

CONCLUSION

The discussion on metabolic profiling emphasises the point that an increasing number of disease conditions of farm livestock are due to several untoward factors occurring simultaneously. Thus an infection of a healthy animal with bacteria may produce no clinical signs of disease and have little effect on the animal's productivity. However, the same level of infection suffered by an animal with chronic protein deficiency may result in clinical disease and, possibly, in death. Again, other diseases may only manifest themselves when two or more pathogens are present in the same animal host. Thus salmonellosis, a very common infection of cattle, may have little effect on a healthy adult cow, but if the same animal is simultaneously suffering from liver fluke infection then the animal may develop clinical signs of salmonellosis and may die as a consequence.

It is possible that most of the main diseases caused by the presence of a single major pathogen have been identified, though not all can be cured. On the other hand, probably only a relatively small number of the many disease conditions which are multi-factorial in origin have been properly investigated. Obviously such complex conditions are more difficult to identify, and their prevention and/or cure must depend upon simultaneous treatment of all the causal factors involved. Thus, in the case of animals suffering from infection by *Salmonellae* and *Fasciolae*, it will be necessary to treat both the liver fluke and the bacterial parasites, and it will probably be necessary to dress the pastures with a chemical to control the fluke vector – the water snail. Furthermore, if the animals in question are undernourished, it will also be necessary to improve the level of feeding.

For these reasons, the control of the diseases of multi-factorial origin requires a team effort, involving the veterinary surgeon, animal nutritionist and stock owner. Only when all concerned fully understand the complexity of the problem, and are committed to rectifying the various causes, will advances in animal productivity result.

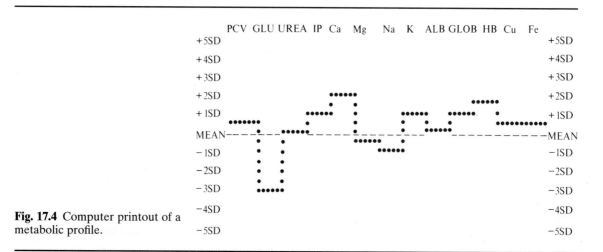

Fig. 17.4 Computer printout of a metabolic profile.

FURTHER READING

Losos, G.J. (1986) *Infectious Tropical Diseases of Domestic Animals*. Longman in association with IDRC, Canada: London.

Ristig, M. and McIntyre, I. (eds) (1981) *Diseases of Cattle in the Tropics*. Martinus Nijhoff: The Hague.

Urquhart, G.M., Armour, J., Duncan, J.L., Dunn, A.M. and Jennings, F.W. (1987) *Veterinary Parasitology*. Longman: London.

18
Livestock Improvement Through Breeding

P.N. Wilson

INTRODUCTION

The term 'animal breeding' may be used in one of three different ways. It may be defined as 'sexual reproduction', and assessed by the efficiency of reproduction in terms of conception and birth rate or generation interval. In this sense the term 'good breeding' is synonymous with 'high fertility'. Second, 'animal breeding' may be used to denote animal multiplication, such as the mass mating of an improved breed of ranch cattle. Lastly, the term may be used to denote genetic improvement, when it is properly applied to systems in which every mating is designed either to assess the breeding value of an untested animal, or else to multiply the progeny of animals which have already been tested and shown to be genetically superior. Unfortunately, the animal geneticist has developed a sophisticated terminology so that a communication gap exists between him and the practical 'animal breeder'. Indeed, much of the fundamental work in animal genetics is developed not with domesticated animals but with laboratory creatures, such as *Drosophila* (fruit flies), which appear to the practical farmer to have little relevance to the breeding of farm livestock. However, all these aspects of animal breeding are pertinent to the theme of livestock improvement in the tropics, and each will be briefly considered.

SEXUAL REPRODUCTION AND FERTILITY

When male animals are run freely with breeding females, an understanding of the specific characteristics of the oestrus cycles of tropical livestock is unnecessary. However, as soon as seasonal breeding, controlled mating or artificial insemination (AI) is introduced, a knowledge of such matters is basic. The length of the oestrus cycle, the duration of oestrus and the time of ovulation in relation to the onset or termination of oestrus all vary from animal to animal within a breed. It is therefore necessary to ascertain the characteristics of the individual animal when strictly controlled mating is practised.

With cattle, the onset and duration of oestrus can be traced experimentally by taking small samples of the vaginal and cervical mucus and measuring its viscosity with an oestrometer. There is an abrupt change in viscosity as oestrus commences, preceded by a slow but steady rise in viscosity during the last half of anoestrus. The time of ovulation can also be detected by this means, thus enabling service or insemination to take place at the optimal time for conception. The approximate parameters for the various phases of the reproductive cycles of tropical livestock are given in Table 18.1. It should be noted that the range is very wide and that about three-quarters of the individuals within a species will conform to the 'norm' while the rest will have cycles which lie outside the limits shown.

The chief variable in the reproductive rhythm of the female is the length of the oestrus cycle. This parameter can be modified by subjecting the animal to some change in its environment, such as by altering the level of feeding or the management routine. The actual duration of oestrus, and the time of ovulation in relation to the beginning or the end of oestrus, is much less

Table 18.1 Average parameters of the reproductive cycles of domestic livestock.

Species	Length of oestrus cycle (days)	Duration of oestrus (hours)	Average time of ovulation in relation to oestrus	Average Length of gestation (days)	Range of gestation period (days)
Alpaca	—	—	—	343	—
Buffalo					
River	21–22	12–36	18–42 hours after onset	308	302–313
Swamp	21–22	12–36	18–42 hours after onset	332	301–343
Camel	21–28	72–96	Not known accurately	373	360–390
Cow					
Tropical	20–24	12–18	9–14 hours from end	286	284–288
Temperate	20–24	12–18	9–14 hours from end	280	273–282
Ewe (tropical)	14–19	24–36	24–48 hours after onset	147	140–160
Goat (tropical)	18–21	24–36	30–40 hours after onset	146	144–148
Llama	—	—	—	330	—
Mare	21–26	120–168	24–36 hours from end	336	330–342
Sow	21–22	48–72	35 hours after onset	114	110–117

variable. It used to be thought that the duration of oestrus was shorter in tropical cattle than in temperate stock. It is now realised that signs of oestrus may be of shorter duration in the tropics but that the female is on heat, and will accept service, for periods of similar length to those recorded for temperate breeds (Rollinson, 1955).

The consequences of low fertility or reproductive failure in the tropics are far-reaching. Many tropical cows produce only two live calves by the age of 5–6 years. Thus where the tropical environment is harsh it is quite common for cows to calve only in alternate years, and the consequent losses of productivity are immense (Donaldson, 1971). Ward (1987), in reviewing the production performance of *Bos taurus* and *B. indicus* beef cows in Zimbabwe, showed that the superiority of the latter was largely due to stress-related low reproductive performance, and the inability to exploit the genetic growth potential, in exotic stock.

The measurement of reproductive efficiency, or fertility, is not simple and many of the factors used in this measurement leave a lot to be desired. Measurements are of two types: those which are applicable under natural mating conditions and those which pertain when reproduction is controlled. In the former instance, the usual measure employed is the calving rate or lambing percentage. This factor can be defined in several ways, for example:

(1) The number of offspring born expressed as a percentage of the number of breeding females which are mated.
(2) The number of offspring branded, sheared or marketed expressed as a percentage of the number of breeding females.
(3) The number of offspring that reach weaning age expressed as a percentage of the number of breeding females.

Another parameter is the calving or lambing interval, defined as the number of days (or months) between successive parturitions. Unfortunately, this factor measures only females that produce offspring, since the calving or lambing interval of barren females, being an infinite quantity, escapes the calculation and produces a bias in the result. This factor assumes that there is an ideal reproductive interval (normally taken as 365 days in the case of cattle), but under certain circumstances there may be good management reasons for making the interval shorter or longer than normal. Thus, if a cow breeds out of season, it may be good practice to hold her back from service so that she next calves at the required time.

Another factor employed in the controlled mating situation is to calculate the number of services per conception. Unfortunately, successful conceptions do not invariably result in an equal number of successful parturitions, since embryos may be reabsorbed or aborted during pregnancy. As with other parameters considered, this factor has the disadvantage that it cannot take into account barren females, since the number of services can only be recorded once conception has been achieved.

Payne (1970) suggested that, under tropical conditions, the number of services per conception for tropical cattle normally varies between 1.3 and 1.6, equivalent to about 60 to 70% of the cows conceiving and calving to their first service. However, where disease or chronic malnutrition prevail, the number of services per successful conception may rise steeply, to 3 or more. Seebeck (1973) has shown that there are significant differences in fertility between different breeds and crossbreeds in the tropics and that these often have very important economic implications.

Another parameter, applicable in the controlled breeding situation, is the 'non-return rate' defined as the percentage non-return to first insemination taken at a fixed point in time after that insemination, normally at 60 days or 90 days. This index is usually employed when the efficiency of an artificial insemination service is being assessed. However, it has a more general application and can be calculated whenever matings are controlled, and recorded, in flocks and herds.

Seasonal breeding

The reproduction of certain temperate livestock, particularly sheep, is mainly governed by the season of the year. A decrease in daylight during the autumn causes the ewe to ovulate and it is usually only by controlling the length of day by artificial means that sheep can be induced to breed at other times of the year. However, in most parts of the tropics there is little variation in day-length, and it is not yet known how much effect this reduced variation has on seasonal breeding patterns. Within the latitude limits of 20°N and 20°S sheep can breed all the year, but even so there is still a tendency for seasonal reproductive activity. The mechanisms which operate the seasonal change in reproductive behaviour are not known with certainty. They may be linked to the rainfall distribution which in turn influences grass growth and hence animal nutrition, or they may be linked to seasonal changes in parasitism and animal health which affect reproductive efficiency. The practical point is that the natural seasonal incidence of parturition can be used as a guide for controlled breeding practice. It is in the interest of the tropical livestock keeper to work with these patterns rather than to plan against them. When there is not a steady demand for animal products the year round, it may be preferable to adopt seasonal breeding programmes for both milk and meat animals, especially if by so doing reproductive efficiency is enhanced and productivity increased. There is some evidence that such seasonal calving programmes may be beneficial. Thus Tomar and Mittal (1960) reported that June-calving Indian cattle lactated for an average of 317 days compared to an average lactation length of only 268 days for November calvers.

Artificial insemination (AI)

This aid to selection can now be used, with varying degrees of success, for most species of domestic livestock. The technique has been fully described by Payne (1990). AI will not by itself solve the problem of low reproductive efficiency in the tropics. Payne stressed that improvements in cattle health and management must be a prerequisite of livestock improvement in tropical countries before AI can be fully exploited but, if these conditions are met, the technique is capable of raising productivity (milk yield) by 1% per year (White *et al.*, 1981). However, other workers consider that breeding improvement is not necessarily dependent on a prior improvement of the environmental constraints. Since costs are high, satisfactory levels of fertility from an AI service must be obtained, and Table 18.2 indicates that often fertility rates achieved in

Table 18.2 Fertility results from tropical artificial insemination (AI) schemes. (After Payne, 1970.)

Place	Year	No. of first services	Fertility to first service (%)	Remarks
East Pakistan	1959–60	?	51.2	Semen ex Malir Livestock Experiment Station
India (Punjab)	1964–65	46 000	51.2	51 000 cows inseminated of which 46 000 followed up and 28 000 pregnant
Jamaica	1950–56	7 244	37.6–74.8 (average 61.1)	Beef and dairy cattle, average 1.6 services/pregnancy
Jamaica	1958–59	2 900	68–69	Semen from Jamaica Hope bulls inseminating approx 2000 cows/yr
Kenya	1966–67	1 734	21.7–62.5 (average 45.5)	Data from Kenya Ministry of Agriculture
Malaysia	1959–60	506	50–79	
Rhodesia and Malawi	1960–61	2 870	70	Mainly milk recorded cows
Uganda	1965	5 868	51.6–62.1	52% of inseminations from frozen semen

practice in AI schemes are too low to make good economic, or even good genetic, sense.

AI is capable of benefiting animal breeding programmes in many ways. It allows widespread use of superior sires which can produce tens of thousands of offspring instead of the few hundred possible by natural service. By markedly widening the bull:cow ratio it can save overall animal breeding costs in areas where male animals are not in demand for draught purposes. It minimises the spread of venereal disease and it can enhance the rate of genetic improvement by increasing the selection pressure on a breeding population. AI is currently mainly used in tropical cattle breeding schemes and attempts to introduce it to other classes of livestock have been relatively unsuccessful. It is probable that the technique will eventually be used in tropical pig improvement programmes, but to date the results with swine have been disappointing because of the difficulty of efficiently detecting oestrus in gilts and sows and the problems associated with the dilution and storage of boars' semen.

The exception to this general rule is the turkey. Improved breeds of turkey have so altered the conformation of the breast muscles that natural mating is not possible, and AI is a prerequisite of reproduction in improved turkey strains. Whether or not such a development is in the best interests of tropical turkey producers is another matter. There is a simple choice, either to use the traditional breeds of smaller breast size or else to import improved breeds, together with the necessity to inseminate turkey hens artificially with all the consequential difficulties entailed.

It should not be assumed that AI only has a part to play in intensive farming areas, and not under extensive conditions. There are several reports of successful AI techniques employed in connection with tropical cattle ranching, and Butterworth (1985) has reported that advanced methods, such as routine pregnancy diagnosis, have also made a contribution to the economic management of tropical beef cattle. Successful AI depends upon successful heat detection by the stockman. Although partially effective artificial aids to heat detection are available (such as dye-containing pouches which can be attached to the hind-quarters and are ruptured when the cows come into oestrus and are mounted by other cows), nevertheless it is the efficient stockman's eye, and good stock record-keeping, which are essential to an efficient AI service. It is vital that the stockmen employed on cattle farms using AI instruct their staff in the detection of 'standing heat' of cows in oestrus. The riding of other cows by a sexually excited cow is not by itself a sign that the cow is in oestrus, nor is the fact that she may be occasionally mounted by other cows within the herd. A 'correct' standing heat (in which the cow in question raises no objection to repeated mounting by other cows, and will stand quietly for that purpose) must be differentiated from other sexual behavioural patterns. Since 'standing heats' may only last for 12 hours, several observations must be made by the stockman if successful AI results are to be achieved. In some cases, the desired results were obtained only after the stockmen were encouraged by the payment of bonuses for cows conceiving to AI. In other cases a successful technique is to gather the cows daily in a corner of their paddock where they can be observed for standing heats continuously for a period of about 3 hours. As each standing heat is checked, the cow in question is marked by a blob of paint using a long-handled brush, and at the end of the 3-hour period the designated cows can be removed for immediate insemination.

Where AI is used to introduce *Bos taurus* blood into tropical cattle populations, temperate experience on the techniques of AI, such as semen collection, can be drawn upon with confidence. However, where *B. indicus* bulls are used, difficulties in semen collection are sometimes encountered as zebu bulls are often 'shy breeders'. For this reason, electro-ejaculation techniques may be desirable in place of the more normal use of live 'teaser' animals or dummy cows.

Although AI is a very useful technique in breeding programmes, it will only effect improvement if the semen is derived from genetically superior bulls. Unfortunately, the 'prestige value' of AI schemes has often led some tropical countries to adopt large-scale programmes without ensuring that the bulls employed were genetically superior. Indeed, there is some evidence that certain schemes are detrimental to livestock improvement, since the yields of daughters bred by AI are lower than those of the dams milking alongside them. It is most important that AI schemes should be linked with performance testing or, progeny testing programmes (see below), and that they be introduced with great care (Bondoc *et al.*, 1989).

It is sometimes assumed that, since the tropical breeds of livestock are generally slow maturing compared to temperate breeds, sexual maturity is also delayed. It is correct that the average age at first calving of zebu cattle is often a year or more later than that of European stock, but there are nevertheless instances of zebu cattle coming into oestrus and conceiving at less than one year of age and of tropical goats reaching sexual maturity at an age of a few weeks, and tropical sheep within 3 months. The reason why reproduction normally takes place much later is due to environmental constraints, particularly malnutrition and disease, rather than any innate inability of tropical livestock to mature early. For these reasons, it is necessary to separate the sexes of animals soon after weaning if promiscuous breeding is to be minimised.

Embryo transfer (ET)

This relatively new technique is fully described by Wilmut (1980). Essentially it consists of the removal of an embryo from the oviduct of a donor animal and its placement in the uterus of a recipient female at the correct stage of her oestrus cycle. As with AI, ET is mainly used in cattle but it can be employed in other domestic livestock. The technique permits the rate of genetic improvement to be greatly accelerated as it enables the more widespread use of superior dams, in exactly the same way that AI allows the prolific use of highly selected sires (Seidel, 1984). In the tropical context, ET enables the importation of complete, albeit embryonic, diploid animals whereas AI alone only permits the import of haploid semen, which must then be used to fertilise native female animals which may be of lower genetic potential (Smith, 1988). A further development in reproductive biology has been the discovery that the fertilised ovum can be split, either once or several times, at an early stage of its development, so that one or more identical twins, or clones, can be produced. This enables genetic improvement via the female line to make yet more rapid progress as the results of a single, successful mating can be manifest by the production of several superior offspring, instead of the normal single calf (Polge, 1991; Chalack, 1993).

ANIMAL MULTIPLICATION

Owing to the low productivity of many classes of tropical livestock, it may be desirable to replace or cross them with animals of superior potential. The 'grading up' of flocks and herds on individual farms is an important aspect of this process, but it is usually a good practice to hasten the replacement period by multiplying superior breeds and strains on stations under direct or indirect government control, subsequently releasing the resultant stock for use on private farms. The need for such multiplication centres is greatest in those tropical countries where there are few, if any, alternative commercial sources of superior stock. In many parts of Africa and Asia there are no 'pedigree breeders' whose chief livelihood is obtained by raising and selling better breeding-stock to farmers with more limited facilities. There is a growing international trade in tropical breeding-cattle but this is often unorganised and there is little guarantee that the stock thus obtained are superior to those which the farmer could raise himself. Also, once a tropical producer has established a local reputation for his breeding-stock, it is likely that demand will exceed supply and there is a danger that the unsuspecting buyer may be purchasing 'culled' animals which are of no better value than the average of his own stock.

In many parts of the tropics, particularly in India, there are excellent 'elite herds' maintained at government research centres, and there is a regular inflow of superior bulls, or semen, to such centres. In other words, the top of the breeding pyramid is complete. The deficiency is in the second section of the pyramid, which should be occupied by the 'multipliers' who receive the results of the genetic improvement programme conducted in the elite herds, and multiply large numbers of improved (but not necessarily pedigree) stock for dispersal to the livestock producers at the base of the pyramid.

For every elite herd there should be at least 10 or 20 multiplying herds extending the benefits of genetic improvement to the individual livestock producer. Without this essential step there is a danger that the elite herds will disperse their superior stock direct to the grass-roots producer, in trivial numbers which cannot do justice to the genetic improvement programme. The correct ratio of elite herds to multiplying herds is fundamental to any national livestock improvement programme and it is unfortunate that this ratio is often distorted to 1:1 instead of to a figure nearer to 1:10 or 1:20 which is required.

Three important factors should be considered in connection with state-controlled or co-operatively run animal multiplication schemes:

(1) *Stock must be of a consistent high standard*
Even with attention to the choice of genetically superior sires and elite groups of breeding females, some below-standard progeny are bound to be born and these animals should be culled, and preferably slaughtered, and not passed on to the commercial producer. Rigid control of animal health is also necessary, since it is a tragedy when a government-sponsored livestock improvement scheme unwittingly becomes the means by which disease is spread to private farms. It is often useful to couple multiplication schemes with herd-book registration and the formation of breed societies so that the identity and parentage of all stock produced can be recalled. The use of suitable brand marks is preferable to ear-tagging, since it is easier to prevent unauthorised branding than illicit ear-tagging. It is most desirable to associate the producers with the administrative aspects of multiplication schemes, so that farmers realise that it is in their own interests to maintain quality and to prevent unauthorised registration of poor quality stock.

(2) *The future breeding programme must be agreed from the very start*

It is easy to take one step forward followed by two steps backward in livestock breeding. The advantage gained by the supply of superior stock to farmers will soon be lost if these stock are mated with average or below-average animals. For this reason, long-term planning is essential from the start. Two alternatives are possible:

(a) The producer may be encouraged to refrain from breeding replacement stock on his own and to return repeatedly to the multiplication centre for his future requirements. This policy is normally adopted with poultry, since the birds supplied by the multiplication centre are often the product of crossing in-bred lines and further breeding from such birds would result in regression and a marked increase in genetic variability. Therefore poultry keepers are usually encouraged to return for fresh supplies of day-old chicks at regular intervals and to refrain from practising breeding on their own farms. Such a policy is less feasible with larger stock, since the multiplication centres would seldom be able to meet the demands which such a policy entails, especially where the number of such centres falls short of real need.

(b) The producer may be assisted with a breeding service, such as an AI scheme or stud-loaning scheme, so that the improvement achieved from the original group of animals from the multiplication centre can be maintained or increased. Grave problems arise when crossbred stock, such as $\frac{1}{2}$ *Bos taurus* × $\frac{1}{2}$ *B. indicus*, are released from the multiplication centre. These stock will exhibit hybrid vigour (heterosis) and as a consequence they will be unusually uniform. In the next generation, whatever animals they are bred to, genetic segregation will occur and the variation will be relatively wide so that heavy culling will be necessary. For this reason, it is advisable for the multiplication centre to retain half-bred animals for a further generation of breeding so that the problem of segregation can be properly addressed (Syrstad, 1990).

Where improved stock are distributed among commercial farmers care must be taken to see that appropriate proven sires are available to prevent undesirable mating with scrub males or with males of the wrong genetic constitution. For example, the multiplication centre may be producing seven-eighths bred stock ($\frac{7}{8}$ *Bos taurus* × $\frac{1}{8}$ *B. indicus*) for distribution to farmers. There may be good grounds for attempting to stabilise the population at this level, since a higher proportion of European blood may result in increased mortality, and a higher proportion of zebu blood may lead to lower productivity. In this case, it is important that $\frac{7}{8}$ bulls are available (for purchase, for loan or via an AI service).

(3) *Steps must be taken to ensure that stocks distributed from multiplication centres are well cared for and used for a specific purpose*

There are many instances of government multiplication schemes selling breeding stock at subsidised prices with a result that, instead of finding their way to farms, they end up in the slaughterhouse. Perhaps the purchaser of improved stock, made available to farmers at government expense, should enter into a written undertaking that he will keep the stock according to the tenets of good husbandry and not dispose of them without permission. It may be unwise to subsidise the price at which breeding-stock are released from multiplication centres, since this may result in the producer assuming that the animals are inferior. It is important to inculcate a proper sense of values into tropical livestock producers' minds and it is right that superior breeding-stock should command higher prices than slaughter-stock. A sense of economic balance must be preserved. Superior stock must not be 'priced out' of the reach of the progressive producer, neither should they be so heavily subsidised that they become devalued and thereby encourage non-productive exploitation.

With regard to the need to insist that the stock from government multiplication centres are well managed, the early experience of the Kenyan government during the grade-Sahiwal multiplication programme is of interest. This programme made available $\frac{3}{4}$ and $\frac{7}{8}$ bred Sahiwal × East African zebu stock to African peasant farmers, for intensive milk production on small farms. The demand for these stock exceeded supply and producers could only obtain them on the recommendation of the local agricultural officer. Successful farmers were made to sign a firm undertaking not to dispose of the animals without written permission, and to carry out five 'good husbandry practices':

(1) To spray the stock regularly against ticks for a minimum period of 9 months.
(2) To erect a perimeter fence around the farm and maintain this in stock-proof conditions.
(3) To record milk yields in a record book supplied by the government.
(4) To make adequate provision for a minimum quantity of dry-season feed (usually in the form of Elephant grass, *Pennisetum purpureum*).
(5) To inoculate or vaccinate all cattle regularly against anthrax, blackwater disease, brucellosis and septicaemia.

If these conditions were not fulfilled, the farmer forfeited his right to any further government help. The number of defaulters was small, most producers being willing to comply with the regulations because they were anxious to benefit by receiving superior breeding stock.

This case study illustrates another point in tropical livestock improvement. The main limiting factor in many tropical countries is stockmanship and livestock management. This fact is often recognised, but the wrong steps are taken to correct the situation. Too much emphasis is often placed on increasing the size and staffing of bureaucratic headquarter institutions rather than producing a cadre of well-trained and properly motivated stockmen. It is important to ensure that the correct infrastructure of milk collection and processing is in place before breed improvement schemes for milk production are designed for small-scale farmers. Without this provision, small farmers may not be able to benefit from these genetic improvement programmes; Gryseels and Boodt (1986).

GENETIC IMPROVEMENT

Two different types of factor are responsible for the differences between individual animals within a breed. First, there are the environmental factors (E) such as climate, nutrition, health and management, which have been considered in preceding chapters. Second, there are genetic factors (G) which are due to the genes received from the two parental gametes – the female ovum and the male spermatozoon.

The genes themselves, responsible for the 'biological specification' of the individual animal in detailed biochemical terms, are polymer molecules of deoxyribonucleic acid (DNA). The genes are arranged in roughly linear fashion along the chromosomes and the 'genetic coding' is spelt out by the spatial arrangement of the amino acids along the length of the strand of DNA.

The environmental (E) and genetic (G) factors interact, so that the total variation between animals is equal to the sum of the effects of all the E and G factors, and also to the interactions between them (Syrstad, 1990). Thus:

$$\text{Total variation} = E + G + EG$$

With a single individual, it is not possible to separate the effects of the various components in the above equation and to estimate how much of the productive level is due to each factor. With groups of livestock, however, estimates of the relevant importance of E, G and EG can be obtained.

One cannot rapidly improve livestock productivity by genetic means for a particular character if most of the variation of that character is environmentally determined. Genetic improvement of a trait can only take place if that trait is reasonably highly heritable. It is, therefore, important that the relative heritabilities of the various production traits should be known, in order that breeding programmes may concentrate attention on those factors of highest heritability. Obviously the amount of selection pressure exercised by the breeder is of great importance in influencing the rate of genetic improvement.

Heritability (h^2) may be defined as the proportion of those phenotypic differences in a trait that can on average be passed on from parent to offspring. It is sometimes described as the strength or intensity of inheritance. Unfortunately, heritability coefficients are difficult to calculate and so in many cases an indirect assessment is made by measuring the 'repeatability' of a character which occurs more than once in the lifetime of an individual. Examples of 'repeatable' factors are milk yield (in successive lactations) and litter number (in successive parturitions).

A further complication arises since the heritability of the same factor differs according to the methods employed and the circumstances in which the heritability is measured or estimated. Thus Mahadevan (1970) has shown that the heritability of weaning weight may vary from 0.1 to 0.8, with the most common value being about 0.3, on a scale of 0 to 10. Harsh environments gave rise to a greater degree of environmental variation and consequently to lower estimates of heritability. However, although the heritability coefficients differ according to the population studied for the purposes of calculation, the order of ranking of the heritabilities of different characters is generally similar, providing the estimates are obtained from a sufficiently large animal population. Again, although individual estimates of heritability of *Bos taurus* cattle may differ from estimates for the similar trait in *B. indicus* cattle, mainly due to the fact that the environments in which the two populations are maintained will differ, nevertheless those factors which are strongly inherited in temperate cattle are also strongly inherited in tropical cattle.

It will be seen from Table 18.3 that carcass characteristics are highly heritable. Production factors, such as weight gain and feed conversion, are moderately highly heritable. Milk and egg production are lowly heritable and reproductive factors, such as litter size and fertility, are very lowly heritable. Therefore genetic improvement programmes with pigs generally can make faster progress in improving carcass conformation that in improving litter size; in poultry usually more progress can be made in increasing egg numbers than egg hatchability. With dairy cattle, it is easier to make progress in selecting for cows with higher butterfat than breeding for cows with regular reproductive performance.

As Payne (1990) has pointed out, the most effective breeding plan for improvement of highly heritable traits would be by means of mass selection. Thus, in the case of improvement in growth rate, a large number of animals should be examined for this trait, and those with a high growth rate (all other things being equal) should be used for breeding the next generation while those with a low growth rate should not. On the other hand if the heritability is low, as with most characteristics of reproduction which are only evident in one sex, then more sophisticated techniques, such as progeny testing, are likely to be of value. Progeny testing will be considered in detail later in this chapter.

Although the heritability of milk production is similar to that of butterfat production, nevertheless many estimates of the heritability of milk yield in tropical cattle are probably artificially low because the animals were not fed and managed in such a manner that the genetic potential for milk yield could be expressed.

Unfortunately, the meat animal, which lends itself more readily to genetic improvement because of the reasonably high heritabilities of liveweight gain and meat characteristics, has received much less attention in the tropics than the dairy animal. There are exceptions to this especially where between-breed differences are marked. For instance, a measure of success has attended the efforts of the American breeders to

Table 18.3 Representative heritability estimates for different livestock species. (Source: average of many estimates especially Gowe and Fairfull, 1980; White et al., 1981; Mitchell et al., 1982; Dalton, 1985; Nicholas, 1987; Payne, 1990.)

Species	Item	Heritability coefficient
Beef cattle	Area of eye muscle (L. dorsi)	0.70
	Post-weaning growth rate	0.50
	Final weight	0.50
	Birthweight	0.40
	Carcass grade	0.40
	Efficiency of feed utilisation	0.40
	Pre-weaning growth rate	0.27
	Weaning weight	0.23
Dairy cattle	Butterfat percentage	0.55
	Protein percentage	0.50
	Body conformation	0.30
	Milk production	0.30
	Butterfat production	0.27
	Reproductive performance	0.05
Pigs	Carcass length	0.65
	Thickness of back-fat	0.50
	Weight gain: weaning – 80 kg	0.45
	Efficiency of feed utilisation	0.40
	Dressing percentage	0.30
	Number of functional nipples	0.20
	Litter size at birth	0.20
	Weaning weight of litter	0.17
	Litter size at weaning	0.15
	Weight gain: birth to weaning	0.10
Poultry	Egg weight	0.65
	Mature pullet weight	0.55
	Weight of day-old chick	0.50
	Thickness of egg shell	0.40
	Egg yield (number)	0.25
	Egg hatchability	0.10
	Fertility	0.03
Sheep	Liveweight gain	0.55
	Yearling birthweight	0.40
	Clean fleece weight	0.40
	Birthweight	0.30
	Weaning weight	0.30
	Fleece quality	0.20
	Number of lambs born	0.12

Notes:
(1) Value of 1.0 = 100% heritability; value of 0 = no evidence of any heritability.
(2) Different methods of assessing heritability result in different values but the *ranking* (within a group) should be very similar.

introduce *Bos indicus* characteristics, of Brahman type, into the beef ranches of the southern United States, particularly Florida and Texas. Greater attention should be devoted to the improvement of *B. indicus* cattle for beef purposes in the tropical parts of Africa. The buffalo is also a candidate for genetic improvement for its beef characteristics in Asia and those countries to which it has been successfully exported, such as parts of the West Indies, South America and Australasia.

In any discussion on heritability it is important to clarify whether one is talking about the heritability of differences between individuals within a herd, differences between herds within a breed, or differences between breeds. In most of the data discussed in this section, estimates of heritability will have been obtained between individuals within a herd or within a group of related herds. Only few large-scale studies have been conducted in which the heritability of differences between species, or between temperate and tropical breeds, have been examined. Mahadevan (1970) has drawn attention to the fact that the differences between breeds have a high heritability because different breeds are synonymous with different genetic populations. On the other hand, differences within and between herds of a similar breed tend to have a very low G component. For instance, where AI is practised, the same bull will tend to be used on many separate herds to about the same extent, and the G variation between them is consequently less. On the other hand, the E variation between herds will be large compared to the within-herd E variation, since the different herds will be fed and managed differently while the animals within a herd will be fed and managed in a similar way (Syrstad, 1990).

As a result of these considerations (in which the E components swamp the G components in between-herd comparisons) whereas G components dominate the E components in within-herd comparisons, heritability estimates indicate that only about 10% of the herd-to-herd variation in milk production is genetic, whereas the differences between individual cows in the same herd are estimated to be about 30% genetic in origin.

The heritability coefficients presented in Table 18.3 are mainly derived from within-herd studies and thus tend to be on the high side of the range. Thus it is important to know whether quoted heritability coefficients have been derived from within-herd or between-herd studies. In the case of AI schemes, in which the data from a large number of herds served by the same AI bulls are used, the heritabilities will be based on a mixture of within- and between-herd data, and the estimates thus obtained will tend to be intermediate.

Systems of breeding

The three main questions that confront the tropical livestock breeder are:

(1) 'What is the selection goal required?'
(2) 'What breed should be improved?'
(3) 'What breeding system should be adopted to maximise the rate of improvement?'

With regard to the second question, four different approaches, given below, are possible.

(1) Selection from within the indigenous tropical breeds

As has been mentioned in Chapter 15, this approach is safe but slow. Disease risks are minimised and there is no problem of acclimatisation (Peters, 1993). On the other hand, the time taken to reach the desired goal is often excessive and national development plans often dictate that steps are taken to raise animal productivity in the short term. The merits and demerits of this approach have been discussed in detail by Trail (1981), and the existing information on the genetic components for heat tolerance has been reviewed by Horst and Mathur (1990).

(2) Mass importation of temperate livestock

This approach is likely to provide rapid results but it is expensive, often prohibitively so, and it suffers from major risks in respect to tropical diseases and acclimatisation. When this approach is adopted, as with the importation of American Holstein cattle into Puerto Rico, a large capital and recurrent sum must be earmarked for the control of ticks and other vectors of major tropical diseases. It may be necessary to subsidise the importation of breeding stock over long periods. In Puerto Rico, for example, female replacements were, until recently, still being imported and most stud bulls used for AI were progeny tested in the United States. Many other tropical countries are currently following this system (Fig. 18.1). Unfortunately, unless management factors are adjusted to take account of the specific needs of susceptible temperate livestock, many of these expensive livestock movements will fail (Bondoc *et al.*, 1989; Vaccaro de Pearson, 1990). On the other hand, where management is not a limiting factor, there is evidence that sires that perform well in temperate environments will also produce offspring that perform above average in tropical climates (McDowell *et al.*, 1976a), but caution is advised by Baptist (1990) who demonstrated that exotic cattle may sometimes be less productive, in lifetime overall economic terms, despite higher milk yields.

(3) Mass importation of tropically adapted cattle from other tropical countries

This approach has the merit of combining many of the advantages of (1) and (2) above, while minimising most of the disadvantages other than cost. It enables certain tropical countries to benefit from the experience of, and advances made in, other tropical areas (Fig. 18.2). A good example of this approach is the importation of Sahiwal cattle (Fig. 18.3) from Pakistan to Kenya. Other examples are provided by the wide distribution of Santa Gertrudis cattle from Texas to many tropical countries and the spread of improved goat breeds, such as the Anglo-Nubian, from the eastern Mediterranean to the tropics. The major difficulty preventing widespread adoption of this policy is that most tropical countries are extremely short of improved stock and, therefore, have few available for export purposes.

Fig. 18.1 Jamaica Brahman prizewinning cow at the Jamaica Annual Livestock Show. This breed of Indian cattle was imported into the West Indies for draught purposes, but is also used for beef.

Fig. 18.2 Kenana heifer at the Government Stock Farm, Entebbe, Uganda. This breed of cattle was imported into Uganda from the southern Sudan in 1955.

Fig. 18.3 An imported Sahiwal bull in the Kipsigis area of Kenya. Several importations of Sahiwal bulls have been made into Kenya for crossing on to East African Shorthorn zebu cattle, with a view to forming a stabilised crossbred population.

(4) Crossbreeding of temperate × tropical livestock and the evolution of stabilised or semi-stabilised crossbreeds

This system is being adopted in the tropics on an increasing scale. The results in many different parts of the tropics are very promising, especially in terms of longevity, and it is likely that the proportion of the world's livestock population falling into this crossbred category will increase significantly in future (Hocking and McAllister, 1986; Vaccaro de Pearson, 1990).

The critical policy decision when implementing a programme of crossbreeding, followed by breed stabilisation, concerns the proportion of temperate and tropical genotype which is deemed desirable in the end product. Wherever possible, the optimal ratio should be established by experiment, or by drawing upon the experience of other tropical countries with similar environmental conditions. This is of fundamental importance. If this is not done, the breeding programme will be *ad hoc*, and the results will repeat the mistakes of earlier introductions when temperate males were top-crossed to indigenous tropical females until health and performance declined. This decline was rectified by changing the policy to one of back-crossing to tropical sires in order to re-introduce the genes responsible for tropical adaptation. In time this resulted in a regression to the indigenous type and a fall back to the original levels of production. This aimless system of crossbreeding has been described as 'zigzag' breeding and most tropical countries can furnish examples of such pointless programmes. It is, therefore, most important that the long-term objective should be clearly defined from the start (Syrstad, 1990).

Extremely hot and arid countries will probably require a stabilised crossbred with the greater contribution of the genotype coming from the tropical parental breed (Vaccaro de Pearson, 1990). Areas in which the mean maximum temperature rarely exceeds 40°C but often exceeds 30°C, and in which humidities tend to be high, will probably require a population stabilised at between $\frac{5}{8}$ and $\frac{3}{8}$ of temperate parental type. As elevation, or a decrease in humidity, lessens the climatic stress so the livestock population will perform best with an increasing proportion of temperate genotype. The disease situation and the standard of management at any one place will modify the ratio as it has been demonstrated that there are significant breed differences in disease tolerance and susceptibility (Seifert, 1971; Trail *et al.*, 1988; Paling, 1990), and obviously much will depend on the choice and excellence of the breeds used to form the new crossbreed. Broad generalisations as to the best combination of the two species for optimal performance in different tropical environments are presented in Table 18.4.

Table 18.4 A guide to the combination of *Bos taurus* and *B. indicus* for performance in the tropics.

Region	Proportion B. taurus (%)	Proportion B. indicus (%)
Very arid tropics	0–40	100–60
Hot humid tropics	30–60	70–40
Hot dry tropics	40–70	60–30
Warm humid tropics	60–90	40–10
Warm dry tropics	75–95	25–5
Subtropics and high-altitude tropics	90–100	10–0

Table 18.5 Average production of milk of crossbred cattle in a hot, dry region of India.

Breeding of cow	Mean milk yield (kg)
$\frac{1}{8}$ *Bos taurus* $\frac{7}{8}$ *B. indicus*	2199
$\frac{1}{4}$ *Bos taurus* $\frac{3}{4}$ *B. indicus*	2719
$\frac{1}{2}$ *Bos taurus* $\frac{1}{2}$ *B. indicus*	3171
$\frac{5}{8}$ *Bos taurus* $\frac{3}{8}$ *B. indicus*	3175
$\frac{3}{4}$ *Bos taurus* $\frac{1}{4}$ *B. indicus*	3029
$\frac{7}{8}$ *Bos taurus* $\frac{1}{8}$ *B. indicus*	2809

A careful analysis of the production records of animals of varying proportions of tropical and temperate genotypes would enable a rough prediction of the desirable genetic proportions for that particular environment without the need to conduct an expensive, long-term experiment. Thus the data presented in Table 18.5, taken from a hot, dry region of India, would indicate that optimal performance in this particular environment might be obtained by stabilising a crossbred population of cattle between the limits of $\frac{1}{2}$ and $\frac{5}{8}$ *Bos taurus*.

Care must be taken to see that performance records are unbiased. Thus, if it were shown that the sample of cattle used to provide the data in Table 18.5 were heavily culled for high milk yields in the $\frac{1}{2}$ *Bos taurus* and $\frac{5}{8}$ *B. taurus* categories, but relatively unselected in the other categories, then the optimal proportion of *B. taurus* blood might be $\frac{3}{4}$ or higher. It is also important that factors other than the production of saleable products are taken into account and that reproductive efficiency and longevity are given due attention (Vaccaro de Pearson, 1990). This point is borne out by reference to the data presented in Table 18.6. Examination of the first two columns might indicate that the purebred *B. taurus* type was the most suitable type for this particular environment. However, a study of the last two columns would show that it would be most unwise to proceed beyond the $\frac{3}{4}$ *B. taurus* stage for purposes of forming a new stabilised crossbreed. After the $\frac{3}{4}$ *B. taurus* stage is reached, calving intervals lengthen and longevity is reduced. As Syrstad (1985) demonstrated, the delicate balance between genetic performance ability and adaptability is determined by the degree of exotic inheritance and the respective production conditions.

Although it is important to define the end points of a cross breeding population, it is unwise to place too much emphasis on 'fractional breeding' as an end in itself. There is a distinction between aiming at a *mean* proportion of tropical and temperate genotypes in a crossbred population, and aiming at a *fixed* proportion of tropical and temperate genotypes so that stock which do not exactly conform are excluded. The designation '$\frac{7}{8}$ *Bos taurus*: $\frac{1}{8}$ *B. indicus*' can be given to a population to describe the average contribution to the genotype derived from each contributing species. The same designation is less meaningful when given to an individual, since in the course of the two back-cross generations following the original cross between a European bull and a zebu cow, the random assortment of chromosomes at each meiotic division will have given rise to individuals which possess either more or less genes derived from the temperate ancestor than the fraction '$\frac{7}{8}$' indicates. For instance, an animal which theoretically may be described as '$\frac{7}{8}$ *B. taurus*' may in fact have considerably more than seven-eighths of its total genes contributed by the temperate ancestor. Indeed, this is likely to be the case if heavy selection has been practised for production factors which have been introduced into the genotype by the repeated use of European bulls. The only generation in a crossbreeding programme in which the proportional genetic contributions from *B. taurus* and *B. indicus* is known with precision is the F_1, or hybrid, generation in which precisely half the genes are derived from each species. In subsequent generations the proportions will be only approximate and to exclude productive stock would be unjustified. This point is illustrated by the data presented in Table 18.7 which refer to the composition of a crossbred herd being stabilised at the $\frac{7}{8}$ *B. taurus* level. The only types not represented in this interbreeding population are purebred *B. taurus* and purebred *B. indicus*. It will be noted that, whereas the highest mean yield (2836 kg) is obtained from $\frac{7}{8}$ *B. taurus* cows, certain animals within this $\frac{7}{8}$

Table 18.6 Data illustrating the importance of taking reproductive efficiency and longevity into consideration when planning programmes for stabilising crossbreeds of cattle.

Breeding of cow	Mean milk yield (kg)	Age at first calving (months)	Calving intervals (days)	Total no. of lifetime calvings
$\frac{1}{2}$ *Bos taurus*	1673	36	380	5
$\frac{5}{8}$ *Bos taurus*	1736	32	397	5
$\frac{3}{4}$ *Bos taurus*	1977	31	456	3
$\frac{7}{8}$ *Bos taurus*	2577	30	534	$2\frac{1}{2}$
$\frac{15}{16}$ *Bos taurus*	2886	29	611	2
Purebred *Bos taurus*	3173	28	660	$1\frac{1}{2}$

group exhibit the lowest recorded yield (2123 kg). This topic has been extensively reviewed by Peters (1993).

The term 'stabilisation' should not be interpreted to mean that the crossbreed in question has become a homozygous, true-breeding population with standard breed characteristics. There are few, if any, new tropical crossbreeds which would meet such a definition and it is unimportant that they should do so. The term is used to denote that interbreeding is taking place in a crossbred population in which the majority of individuals are well adapted to their environment and capable of high levels of productivity, and in which the average proportion of temperate and tropical genotype is fixed. Where particular breed characteristics, such a colour, size, horn type and shape, can be incorporated and stabilised simultaneously with adaptability and productivity this may be an advantage in terms of breed recognition, but undue attention to superficial breed points will result in less attention being given to major factors of economic importance.

In the first generation (F_1) after the crossbreeding of two different species, breeds or strains, hybrid vigour, or positive heterosis, is often exhibited Rege *et al.*, (1994). Hybrid vigour is defined as the phenomenon in which the offspring are better than the mean of both parents. In many cases, the difference is sufficiently large that it is readily discernible, in terms of greater growth rate, higher milk yields or larger litters, and does not rely on detailed measurements for demonstration. The genetic differences between parents must be wide to obtain significant heterosis, and it should be remembered that, because of unavoidable environmental limitations, not all the offspring will demonstrate the average superiority of their crossbred generation. It must also be understood that hybrid vigour is confined to the first generation. Although it may be stabilised thereafter by the use of suitable breeding programmes, the second generation (F_2) will not exhibit a further degree of heterosis over and above that already obtained. Indeed, in the second generation there will be a need to use a much higher degree of selection pressure than normal, as some of the F_2 offspring will revert, by genetic regression, to the characteristics of their grandparents. For this reason, the genetic stabilisation of crossbreeds is a long drawn out business. Two examples of new tropical crossbreeds will now be considered, one developed for intensive milk production and the other for extensive beef production. These breeds are the Jamaica Hope and the Santa Gertrudis.

Jamaica Hope This is a crossbreed combining *Bos taurus* blood derived from the Jersey (and to a lesser extent, the Friesian) and *B. indicus* blood from the Sahiwal imported from India. It was formed in Jamaica, and has spread through the West Indian islands into South America. The breed was organised into a closed herd-book and stabilised at about the $\frac{7}{8}$ to $\frac{5}{16}$ *Bos taurus* level by Lecky in the early 1950s, but credit for the pioneering work must be given to Cousins (Lecky, 1962) who investigated crosses between Sahiwal, Jersey, Guernsey and Friesian breeds. The early days of the development of this breed were marked by the 'zig-zag' breeding system already discussed. By 1930 the Guernsey crossbred progeny were eliminated, but the desirable proportions of temperate and Sahiwal blood were not known and both purebred European and zebu bulls were in use.

The grounds on which the Jersey was eventually chosen as the major contributor of *B. taurus* blood were:

Table 18.7 Types of crossbreed represented in a tropical dairy herd stabilised at level of $\frac{7}{8}$ *Bos taurus*.

Type of crossbreed	Number of cows	Mean yield (kg)	Range (kg)
$\frac{63}{64}$ *Bos taurus*	2	2618	2527–2705
$\frac{31}{32}$ *Bos taurus*	7	2677	2459–3195
$\frac{15}{16}$ *Bos taurus*	19	2773	2318–4000
$\frac{7}{8}$ *Bos taurus*	73	2836	2123–4214
$\frac{3}{4}$ *Bos taurus*	21	2646	2268–4009
$\frac{1}{2}$ *Bos taurus*	3	2268	2186–2423
$\frac{1}{4}$ *Bos taurus*	1	2291	2291

(1) Greater heat tolerance.
(2) Higher fat-corrected milk yields.
(3) Better reproductive performance and earlier breeding.
(4) Hardiness and more efficient utilization of rough grazing.

The recent use of Holstein blood reflects the decrease in the importance of high butterfat and the increase in emphasis on protein and total milk yields for the Jamaican milk market. Milk yield data show that, under conditions of good feeding and management, the Jamaica Hope (Fig. 18.4) can consistently maintain herd averages of 2500 kg and yields of over 4000 kg are not uncommon (Roache *et al.*, 1970; Wellington *et al.*, 1970) as shown in Table 18.8. On the other hand the latter authors suggested that the different breeders of Jamaica Hope cattle varied as much in their efficiency of cow management as their cattle varied in productivity.

As Payne (1970) pointed out, the Jamaica Hope breed has earned a reputation for efficient milk production under a wide range of tropical conditions, and the potential demand for this breed outstrips the supply. Payne suggested that the situation would be altered if the large-scale importation of Holstein cattle from the United States were top-crossed on to the Jamaica Hope

Table 18.8 National milk records of Jamaica Hope cattle (1965–73) based on the Dairy Herd Improvement Scheme. (After Wellington, pers. comm.)

Year	Average lactation milk yield (kg)	Average lactation length (days)	Number of herds	Number of lactaions
1965	2102	268	34	1228
1966	2265	258	47	2055
1967	2319	262	49	2571
1968	2390	271	35	1639
1969	2544	272	32	1468
1970	2603	269	33	1575
1971	2620	268	42	1812
1972	2571	266	52	2277
1973	2551	266	46	1958

Fig. 18.4 Jamaica Hope show-winning cow. This is one of the few crossbred breeds of milking cattle to be developed in the Americas. It is a stabilised cross between imported European Jerseys and imported Indian Sahiwals. The cow shown is $\frac{15}{16}$ Jersey.

breed so that, eventually, a stabilised Holstein × Jamaica Hope crossbred would be evolved. Such a policy would permit the breed to be not only of maximum benefit to the commercial milk producers of Jamaica but would also enhance the prospects for the export of this breed to other tropical countries. One suggested method would be to mate all Holstein females imported into Jamaica to Jamaica Hope sires and to follow this by a rigorous programme of grading-up to the Jamaica Hope. Such a programme would eventually permit the resultant progeny to be registered in the Jamaica Hope herd-book.

Santa Gertrudis This tropical breed was developed on the King Ranch in Texas. The history of the ranch stretches back to 1851 when Richard King introduced Spanish longhorn cattle, known as Texas Longhorns. Between 1870 and 1910 various crossbreeding programmes with improved *Bos taurus*-type bulls were conducted, but it was not until crossbreeding work with Shorthorns and Brahman cattle commenced in 1910 that notable progress was made. In 1910, the King Ranch acquired a half-bred Shorthorn × Brahman zebu bull, which was mated to purebred Shorthorn cows. The comparison of the progeny of the half-bred bull and his son with those of contemporary progeny from purebred Shorthorn and Hereford heifers was so favourable to the crossbreds that a breeding programme was set up, under the leadership of Kleberg, resulting in the production of an outstanding bull, Monkey, who weighed 500 kg at 1 year of age. Monkey was the result of a chance mating between a Brahman bull and a high-grade Shorthorn cow and he was widely used for breeding with an elite herd of crossbred Brahman × Shorthorn cows. Monkey sired many outstanding bulls including Santa Gertrudis, who eventually gave his name to the new breed.

The breeding plan was unusual in that attention was focused on the herd rather than on individual animals. In various single- and multiple-sire herds an intensive line-breeding programme to the bull Monkey was developed. Each of the single-sire herds was made up of animals of a particular grade, while simultaneously the multiple-sire herds were built up on a mass-selection basis ignoring their particular breeding lines or strains. Selection was on a 'total score' basis and relatively high levels of culling were practised. There was considerable flexibility during the early stages of the programme, but in 1940 a new breed, roughly $\frac{3}{8}$ Zebu and $\frac{5}{8}$ Shorthorn, had been developed and the Santa Gertrudis was officially recognised as a new breed.

The Santa Gertrudis breed is noted for its early maturity, hardiness and its propensity to 'finish' when provided with sparse grazing. It is also capable of excellent performance when maintained under feedlot conditions and has topped the official American '140 day-gain performance test'. In 1977 an animal from the King Ranch on one such test gained an average weight of 1.93 kg/day, which is an exceptionally high figure for ranch-type cattle.

The Santa Gertrudis is now spread worldwide. Its breeding programme is still influenced by the King Ranch in Texas and breeding bulls from this centre fetch high prices. The breed society has adopted an open herd-book policy and four top-crosses with registered Santa Gertrudis bulls enable animals from other breeds to qualify for registration as pedigree Santa Gertrudis cattle. The breed society considers liveweight gain and feed conversion efficiency as essential selection requirements, but it has also endeavoured to stamp the breed with easily recognisable characteristics, such as a deep-red hair colour, pigmented skin, loose hide and a small hump in the male.

The Jamaica Hope and the Santa Gertrudis are good examples of crossbreeds which have been successful in achieving a high degree of adaptation to a tropical environment while incorporating the high productive capacity normally associated with European-type cattle. Each breed is still evolving and neither breed has yet achieved pure-breeding 'stability' which is generally associated with well-established temperate breeds, such as the Friesian and the Hereford. Other examples of successful starts to the formation of new tropical crossbreeds are provided by the Bonsmara (Africander × Shorthorn); the Beefmaster (Shorthorn × Hereford × American Brahman); the Bradford (Brahman × Hereford); the Jamaica Red (Red Poll × Zebu); the Nelthropp or Senepoll (Senegal zebu × Red poll); the Achiote (criollo × Shorthorn); the Jamaica Black (Aberdeen Angus × zebu); the Charbray (Charolais × American Brahman); the Australian Brangus (Aberdeen Angus × Brahman) and many others. The essential feature of the long-term stabilisation programme of these new crossbreeds is that the animals are mated *inter se* in a closed, or relatively closed, population once sufficient genetic variation has been introduced from the original foundation stock. These stocks can be derived from two sources, as with the Bonsmara, or may be the result of triple or quadruple crosses, as with both the Jamaica Hope and the Santa Gertrudis breeds. Once sufficient numbers of animals of approximately the desired genotype have been obtained, further breeding work consists of regular genetic selection within the interbreeding population, with only exceptional outcrossing with animals from other genetic sources. This fixation of the breed generally requires relatively high levels of inbreeding and/or several generations of *inter-se* breeding to reduce the excessive heterozygosity and to allow the recognition of undesirable recessive genes, followed by heaving culling, in order to ensure that as much of the population as possible consists of superior breeding-stock.

In the majority of crossbreeding programmes the *Bos taurus* genotype is introduced via the male parent and *B. indicus* via the female. In certain tropical countries, it is impractical to import purebred European bulls because disease or climatic stress would severely curtail the effective breeding life of any 'exotic' stock. The use of AI and/or ET enables crossbreeding programmes to take place without importing animals of temperate breeds into the tropics, and also substantially reduces costs. The usual technique of AI, in which semen is obtained from stud bulls once or twice a week and stored for limited periods after suitable dilution with media such as egg yolk, is inapplicable. A 'bank' of semen from temperate bulls is required and the stock of semen held requires to be replenished at relatively long intervals. This requirement can be met by the technique of freezing semen and maintaining it for several years, if necessary, in a deep-frozen state. The two most commonly employed freezing materials are liquid nitrogen, which will store semen at $-193°C$, and solid carbon dioxide, which will maintain the semen temperature at $-79°C$. Semen can be air-freighted in special containers, packed in liquid nitrogen or solid carbon dioxide inside a thermos flask, and can reach most remote tropical countries in less than 2 days. The cost of sending 100 ampoules of semen by this means from, say, the UK to Fiji would be about £200 (or £2 per straw). Semen exports have now been made successfully to most tropical countries to allow them to embark upon crossbreeding programmes.

A further advantage in introducing exotic semen by air-freight is that the semen can be obtained from older, progeny-tested sires. The breeding value (BV) of the sire in respect to reproductive characters is (or should be!) known (McDowell *et al.*, 1976c). However, it must be realised that the progeny tests of the bull in his native country will have been carried out with different cows and in a dissimilar environment from that in which his tropical offspring will perform. Sometimes there may be a 'bull × place' interaction so that although the bull showed promise in his homeland his tropical progeny may sometimes be disappointing. It is therefore prudent to use the semen of imported bulls sparingly in the first instance, and to 'bank' the rest of the semen until the performance records from his tropical progeny become available. This latter test may be better described as an 'adaptability test' and it can be incorporated into a breeding programme with very little extra expense or effort, other than the loss of 2 or 3 years at the start (Peters, 1993).

The role of breed societies

The original concept underlying the formation of breed societies was to maintain the 'purity' of the breed. For this purpose, a register of animals within the breed was drawn up, and all further entries to the register were confined to the progeny of this elite breeding group. In a closed herdbook, both parents must be on the register for the progeny to qualify. In an open-herd book, non-registered animals can gain entrance to the register after a certain number of top-crossings with male animals already on the register. The number of such top-crosses demanded varies between breed societies, but is normally not less than three.

The founding members of the breed societies assumed that, by restricting breeding to an elite group of animals, breed improvement would necessarily follow. If the original foundation stock were genetically superior to their contemporaries, and if the founder members were fortunate in having sires capable of passing on high levels of productivity to their offspring, then this exercise may well have been successful. Conversely, if the original collection of breeding animals were phenotypically superior, but genetically only average, the formation of the society would not lead to any perceivable progress. For this reason, it was normal for such societies to be formed only after the breed had demonstrated its potential merits for the production of meat, milk or other factors of economic importance.

Unfortunately, many of the 'standards of excellence' originally drawn up by breed societies related to minor points of no economic significance, such as colour, horn type and shape and other unimportant factors, such as the relative size of the 'milk vein', which were thought, erroneously, to be correlated to factors of economic importance, such as milk yield. Breed societies of the tropics should avoid the early mistakes made by their temperate counterparts. In many cases, undue stress on breed type and too little emphasis on good lifetime performance has led to the decline of several well-known breeds. In temperate countries breeders are still prepared to pay very high prices for show winners possessing all the essential breed characteristics, but not necessarily superior from the economic standpoint. More recently, the role of world breed societies has been reappraised. The more progressive breed societies now provide a first-rate advisory service primarily aimed at assisting commercial breeders to make the correct choice of sire in their commercial breeding programme, with economic performance the main criterion.

This new approach, which focuses attention on productivity characteristics rather than show points, is commendable and should guide similar organisations in the tropics. In different tropical countries, various organisations will assume responsibility for national breeding policies. Often such breeding policies will be determined by governmental bodies and within this policy framework the breed societies will make their specific contribution. In certain tropical countries governments may let the breed societies take responsibility for breed improvement, assisting them with finance or making available resources such as state-controlled AI stations. Whatever the situation, it is clearly in the best interests of the livestock producers that the breed societies and governmental institutions should work in close harmony (Simm *et al.*, 1994), and it is also important that large-scale programmes are mounted on a global basis, with support from international agencies such as the Food and Agriculture Organization of the United Nations; (FAO) (Cunningham, 1993).

METHODS OF SELECTION

Although the theory of animal genetics is complicated, the question confronting the animal breeder is simple. Essentially it is how best to choose between two similar animals for future breeding. We shall, therefore, consider the various methods which are available, and consider the pros and cons of each.

Individual selection and performance testing

Individual selection is the practice of selecting future breeding stock on the basis of their own individual performance, without any consideration of their pedigree, progeny or the performance of their near relatives. It is the most simple method of selection, and primarily applicable where the factor in question is strongly inherited. It is also a valuable selection technique when the production factor in question can be observed in both sexes, such as liveweight gain. However, in practice the degree of selection pressure exerted may often be insufficient to secure any measurable genetic improvement as the need to sell the maximum number of pedigree breeding females results in unacceptably low culling rates (Roache et al., 1970). Dairy cattle performance testing in Africa has been reviewed by Peters and Thorpe (1988), who pointed out that many of these reported tests were short term, implemented solely 'on station' and did not include important traits required for the assessment of overall production merit. As Peters (1993) has indicated, fitness traits as well as production parameters, are very important.

A refinement of individual selection is the 'performance testing' of meat animals in order to select suitable breeding males. Performance testing provides a good estimate of the potential of sires for the main attributes of meat production. It has the additional advantage that the test can be carried out either on an individual farm or at a central performance testing station. The prime requirement for performance testing is the keeping of objective, accurate records, such as those related to liveweight, health and feed consumption. Unfortunately, such record-keeping is expensive, and therefore performance testing is normally confined to the larger breeding companies and breed societies and to stations run by government and para-statal organisations.

Where meat is supplied to a critical market, as in the case of tropical countries exporting meat to Europe and the United States, factors connected with carcass and meat quality may be as important as factors concerned with the performance of animals on the farm. The difficulty is that there is obviously a need to correlate suitable records of performance on the live animal, such as ultrasonic measurement of tissue composition, with attributes of carcass or meat quality which can usually only be obtained on the dead animal. It is in the interests of tropical livestock breeders supplying meat to the export market to obtain factual information on the meat and carcass attributes of the beasts they send to slaughter, and to compare sire performance accordingly.

Pedigree selection

This is defined as a method of selecting breeding-stock on the basis of the performance of their ancestors rather than on their own individual performance. The method is clearly applicable when there are no records of the individual themselves, but where ample data exists for the records of their ancestors.

The difficulty with this technique is that there is likely to be some disparity between the accuracy, and recording techniques, related to the various ancestors of the animals in question. For instance, a bias could well be introduced if disadvantageous data were not recorded as such, but were simply left out of the records as 'missing data'. Again, the various ancestors may well have been reared and managed in widely differing environments and so the environmental effects may mask the genetic characteristics of the ancestors in the different parts of the pedigree.

One of the temptations in pedigree selection is to rely too heavily on the presence of one or two outstanding ancestors, without due weight being given to the numerous other ancestors of average or below-average merit. Although it is true that certain outstanding sires, such as Monkey in the Santa Gertrudis breed considered earlier, have left their mark on the performance of the breed for several generations, generally such animals are in a minority and most pedigrees are not unduly dominated by such outstanding individuals. In most cases, therefore, undue weight should not be placed on the 'best' ancestors in a pedigree but attention should be paid to the weighted genetic merit of all ancestors, compared with the weighted genetic merit of their similarly managed contemporaries.

Selection on the basis of sib performance (family selection)

Sib testing is the practice of choosing future breeding-stock on the basis of the performance of their close collateral relatives, such as their brothers and sisters or half-brothers and half-sisters. With families consisting of four or more full sibs which are not highly inbred, a faster rate of genetic improvement can be achieved from sib testing than from individual selection, where the factors in question have a heritability of less than about 0.35. Sib testing, therefore, has a major role to play in poultry breeding where no difficulty is encountered in assembling a large number of full sibs as the basis for selection. The technique also has a role in pig breeding, but with cattle the production of four full sisters must take at least 4 years and on average about 5 years, with the result that the time scale is drawn out and the annual rate of genetic improvement is correspondingly reduced. The term 'collateral relatives' embraces not only full brothers and sisters but also other relatives, such as cousins. However, selection by sib testing is mainly relevant where close relationships are involved, and the data becomes of lesser importance as the relationship extends beyond the immediate close relatives of the animal in question.

Selection through progeny testing

The maximum rate of genetic improvement can only be achieved when breeding-stock are selected on their ability to beget offspring which are above the average of their contemporaries in performance. This ability cannot be estimated objectively by any other means than close observation of the actual progeny of the animals in question. Objective tests must be made in order to estimate an animal's breeding potential.

As male animals are capable of producing many times more offspring than females, more attention is devoted to the genetic testing of sires than to that of dams, but it should be remembered that both sire and dam are of equal importance in determining the genotype of their offspring, since each parent contributes precisely half the genes present in the fertilised ovum.

The classic method of assessing the genetic value of a male animal is by means of a progeny test. These tests take many forms, but the basic feature is that each male is mated to a random selection of females and the resultant offspring are reared under comparable conditions so as to enable an objective estimate to be made of the relative performance of offspring sired by the different males on test. If males are being progeny tested for milk production, then their daughters must be bred before any data can be collected on their milking performance. The male to which the daughters are bred is irrelevant to their sire's progeny test result. Progeny testing for milk yield, twinning percentage of sheep and goats, litter size in pigs and other factors associated with the reproductive performance of the offspring of the male on test, is essentially a long-term process. There is, therefore, a real danger that the male may be old, or even dead, by the time his progeny test results are known. This point is very important in the tropics, where the turnover of livestock generations is longer due to slower growth rates, but where longevity is less and effective breeding lives are shorter (unless semen or embryos are stored) (Vaccaro de Pearson, 1990).

Another difficulty operating against the success of progeny testing in tropical countries is that there are few herds large enough to provide sufficient females for mating to the various bulls on test. A minimum of 10 (although 25 is far preferable) daughters is required to progeny test a bull for milk yield and many more for factors of low heritability. It will probably be necessary to mate the bull to at least 25 females in order to provide a minimum of 10 'acceptable' daughters, i.e. daughters giving rise to at least one fully recorded lactation. If there are 10 bulls on test at any one time, then 250 females and 100 fully recorded daughters (a total of 350 head of female stock) will be required in order to accommodate one objective progeny test. If this programme is repeated each year, with consequent overlap of the rearing periods for the test daughters, then the size of farm required will be very large indeed and about 1000 head of stock will need to be maintained.

In some countries there is increasing use of AI for extensive progeny testing purposes. In this way, several thousand recorded cows can be incorporated into the test result of bulls which show promise at a single testing centre. Unfortunately, this method is only applicable when a large number of farmers using the AI service keep accurate livestock records. This situation is uncommon in the tropics where the majority of herds are unrecorded. There are, however, limited opportunities for utilising the resources of recorded herds attached to such institutions as universities, farm schools, prisons and missions in conjunction with an AI programme, so that a compromise version of this scheme could be used.

There are three different approaches to the progeny testing of bulls for milk yield. First, the bulls can be evaluated by comparing the average performance of their daughters. This method is less rigorous when the daughters are dispersed, in different numbers, over a range of different farms and stations. Second, bulls can be evaluated by comparing their daughters' performance with that of their dams which are milking alongside them over corresponding periods of time. Such a test is known as a 'daughter–dam comparison', but it is limited in its application as it is important that both daughter and dam should be in milk together. It also necessitates the use of correction factors to allow for the expected differences in yield between the daughters' heifer lactation and their dams' adult lactations. Lastly, the bulls can be evaluated by simply comparing their daughters' yields with those of their contemporaries milking alongside them for the same lactation period. This test is known as the 'contemporary comparison test', and it has been widely used.

In the case of beef cattle, the progeny test provides the best method of assessing the sires' transmission ability in respect of characters exhibiting low heritabilities, such as weaning weight. Beef cattle have an advantage over dairy cattle in that progeny testing can be applied to both sexes. The progeny testing of livestock other than cattle follows the same basic principles and the minor variations to suit the particular needs of the different species will not be discussed here. However, an interesting new development with regard to progeny testing of pigs is worthy of mention and has already been incorporated into certain tropical breeding programmes with swine. This technique consists of inseminating sows with mixed semen, part from a control boar of known breeding potential bearing a distinctive feature (such as a colour-mark) and part from a boar on test. The offspring of the two boars will have an identical uterine environment during pregnancy and the differences in their post-natal growth and development can be attributed mainly to the genetic differences in their respective sires.

Progeny testing must be regarded as the most reliable method for assessing the genetic value of an animal, particularly where the characters in question are not strongly heritable. Unfortunately there are various factors which mitigate against the widespread adoption of progeny testing in the tropics. Livestock recording is still in its infancy, and the result is that there is usually only a small number of recorded progeny available for each sire on test. Again, the interpretation of progeny records can be misleading if the mates of the sire on test are significantly better or worse than the national breed average, or the herd average. It is vital to arrange progeny testing programmes so that the sires on test are mated to an unbiased sample of females. Failure to ensure this results in an invalid test, or the difficulty of 'correcting' the progeny test data to allow for the mates being atypical.

For many years research workers have sought in vain for a clear-cut correlation between some easily identifiable morphological character and economic production factors, such as milk yield or carcass gain. Many links between visible factors and hidden economic factors have been suggested, thus Ankole stockmen considered that excessive horn growth in bulls predisposed to high milk yields in their daughters (see Chapter 15). Unfortunately, most of the suggested correlations have proved to be valueless. The productive factors of greatest economic importance are under the control of many genes so that the chances of correlating some or all of the genes responsible for these factors with visible characteristics are remote.

A more promising prospect is to identify physiological characteristics of the young animal which may be correlated to its own future productivity when adult, or even to its genetic ability to transfer high levels of productivity to its offspring. Several decades ago Rendel (1960) demonstrated definite relationship between the B locus and the fat percentage of milk such that cows with the genotype known as $B^O1^Y2^D$ produce milk with 0.16 unit of extra butterfat compared to cows which do not carry this particular allele. Further studies have indicated that useful correlations between blood groups and productive indices also exist in swine (Brucks, 1964). If this work is confirmed with tropical breeds, it may be possible in the future to predict the breeding value of a male or female animal even for characters, such as milk yield, litter size or meat quality, which are sex-limited in their expression, merely by typing their blood, or protein components of the blood, such as haptoglobins. Even more promising would be the demonstration that the secretion rate of important hormonal and endocrinological secretions is highly heritable. It has already been shown (Chapter 14) that the output of sweat from the sweat glands is correlated to heat tolerance, and it may well be that the output of growth hormone by the pituitary could be used as a genetic maker for growth rate or milk production. Further research work in this field should be closely followed by all those with a real interest in the improvement of livestock, but a warning should be sounded that the prospect of any such shortcuts being found for the majority of production characteristics of tropical livestock is probably remote.

There is currently (in the 1990s) an international attempt to map the genomes of cattle and pigs, and this exercise is being carried out by a group of co-operating laboratories in Europe and the United States, with each laboratory examining the gene sequences of a different group of chromosomes. If the individual genes can be related to biochemical or physiological traits it should then be possible to 'test' DNA material for the presence of desirable, productive genes and also for the presence (or absence) of deleterious genes, and this should enable animals to be 'screened' for their suitability for different productive requirements and for different environments.

It is likely that the genomes of *Bos taurus* and *B. indicus* cattle will be similar in most respects, and thus, although the work is being carried out on European-type cattle, it could theoretically be exploited for use with tropical cattle. Tropical pigs are normally of the same species as temperate pigs so there the temperate pig genome map should be of immediate relevance to most tropical pig populations.

However, the production of the first complete 'genome maps', and the linking of the different genes to the various traits of economic importance, will take some time to complete, and the use of this technology then has to be assessed, on economic grounds, in practice. It is likely that it will mainly be of relevance in temperate livestock breeding programmes, but it may enable more objective judgements to be made of the suitability, or otherwise, of temperate breeding stock for export to tropical countries. It is a development of great potential importance which will need careful monitoring over the next few decades.

Selection for several factors simultaneously

It is important that, in selecting for one important factor (such as milk yield) other equally important factors (e.g. heat tolerance and disease resistance) are not neglected. Unfortunately the rate of genetic improvement for more than one factor is much slower than for only one, and for many unrelated factors it will be very slow indeed. However, if the different selection factors are combined into an index, and especially if the factors are weighted by their economic importance so that the resultant index is predominantly financial rather than biological, the problem is largely overcome. By the use of such an index, high merit in one trait can compensate for deficiencies in another trait. The use of such multi-trait indices is complex and requires mathematical skills and a computer to manipulate the numerous data involved, but they are increasingly employed by pig and poultry breeding organisations supplying improved stock for export to the tropics. One of the main problems of using a breeding index is that scientists cannot easily explain to breeders how the index works. For instance, poultry breeders currently employ selection indices with up to 16 items for egg production alone, and pig breeders use indices which include up to 9 traits to describe growth and body confirmation. There must therefore be a degree of trust between the livestock producer purchasing superior livestock bred by this means and the company or organisation selling them. Such trust is really nothing new – the purchaser of a pedigree cow from a well-known breeder must have faith in the breeders' ability before paying premium prices for his stock. The difference is that many of the 'qualities' contained within a complex index cannot be readily perceived until several generations of offspring have been bred and their overall performance observed (Miller, 1993).

When performance records are used as factors in a selection index the records need to be adjusted to remove all known sources of environmental bias which would otherwise distort the data. For instance, performance in different seasons of the year will be influenced by seasonal effects, and performance of reproductive females of different ages will be affected by age effects. It is now possible, with the aid of powerful computers, to take all these complex interactions into account and thus derive estimated breeding values (EBVs) for sires (and also for females producing many offspring such as sows and hens) by use of a statistical method known as BLUP. BLUP stands for Best Linear Unbiased Prediction, and has been fully described by Henderson (1973) who developed it and by Nicholas (1987) who has presented an example of the methodology involved. Its application in practical animal breeding has been reviewed by Wray *et al.* (1993).

Regular genetic improvement will be brought about by the objective testing of accurately collected and collated livestock records, and basing breeding decisions upon the analysis of such records by performance testing and progeny testing procedures. At the end of the day the biblical saying is still pertinent to the animal breeder: 'By their fruits ye shall know them'.

FURTHER READING

Brachett, B.G., Seidel, G.E. and Seidel, S.M. (eds) (1981) *New Technologies in Animal Breeding.* Academic Press: London.

Cundiff, L.V. and Gregory, K.E. (1977) *Beef Cattle Breeding.* USDA Dept. of Agriculture Research Service: AGR 101: Washington, DC.

Lasley, J.F. (1978) *Genetics of Livestock Improvement*, 3rd edn. Prentice Hall: Englewood Cliffs, NJ.

Nicholas, F.W. (1987) *Veterinary genetics.* Clarendon Press: Oxford.

Willis, M.B. (1991) *Dalton's Introduction to Practical Animal Breeding*, 3rd edn. Blackwell Science: London.

References

Abbot, A.J. and Atkin, R.K. (eds) (1987) *Improving Vegetatively Propagated Crops.* Academic Press: London.

Abruna, F.R. and Vicente-Chandler, J. (1967) Sugar cane yields as related to acidity of a humid tropical Ultisol. *Agron.J.*, **59**, 330–32.

Acquaye, D.K., Maclean, A.J. and Rice, H.M. (1967) Potential and capacity of potassium in some representative soils of Ghana. *Soil Sci.*, **103**, 79–89.

Adam, A.V. (1986) International code of conduct for pesticides. *Span*, **29**, 94–5.

Adams, F. and Pearson, R.W. (1970) Differential response of cotton and peanuts to sub-soil activity. *Agron.J.*, **62**, 9–12.

Addison, J.M. (1968) Harmful levels of magnesium salts are not uncommon. *Vet.J.Agric.*, **66**, 468–9.

Adeneye, G.A. (1979) A note on the nutrient and mineral composition of *Leucaena leucocephela* in Western Nigeria. *Anim.Food Sci.Tech.*, **4**, 221–5.

Adepetu, J.A. and Corey, R.B. (1977) Changes in N and P availability and P fractions in Iwo soil from Nigeria under intensive cultivation. *Plant and Soil*, **46**, 309–16.

Adeyemo, O., Heath, E., Adadeuoh, B.K., Steinbach, J. and Olaloku, E.A. (1979) Some physiological and behavioural responses in *B. indicus* and *B. taurus* heifers acclimatised to the hot, humid seasonal equatorial climate. *Int.J.Biometeor.*, **23**(3), 231–41.

Agabeili, A.A., Guseinov, L.A. and Serdyuk, V.S. (1971) A new buffalo breed: The Caucasian. *Zhivotnovodstvo, Mosk.*, **33**(8), 61. (*Arum.Breed.Abstr.*, **40**, 76).

Agnew, C.T. (1982) Water availability and the development of rainfed agriculture in S.W. Niger. *Trans.Inst.British Geographers*, **7**, 419–57.

Agnew, C.T. (1983) *Pastoralism in the Sahel: U204 Third World Studies.* Open University: Milton Keynes.

Agnew, C.T. (1989) Sahel drought: meteorological or agricultural? *Int.J.Climatology*, **9**, 371–82.

Agnew, C.T. (1990) Spatial aspects of drought in the Sahel. *J.Arid Environ.*, **18**, 279–93.

Agnew, C.T. (1994) Evaporation and evapotranspiration. *In*: A. Goudie (ed.) *The Encyclopaedic Dictionary of Physical Geography*, pp. 191–6. Basil Blackwell: Oxford.

Agnew, C.T. (1995) Desertification, drought and development in the Sahel. *In*: A. Binns (ed.) *People and Environment in Africa*, pp. 137–49. John Wiley & Sons: Chichester.

Agnew, C.T. and Anderson, E. (1992) *Water Resources in the Arid Realm.* Routledge: London.

Agricultural and Food Research Council (1980) *The Nutrient Requirements of Ruminant Livestock.* CAB: Farnham Royal.

Agricultural and Food Research Council (1992) *Nutritive Requirements of Ruminants: Protein.* Tech.Com. on Responses to Nutrients, Rep.No.9. CABI: Farnham Royal.

Agricultural Research Council (1967) *The Nutrient Requirements of Farm Livestock. No.3. Pigs.* HMSO: London.

Agricultural Research Council (1975) *The Nutrient Requirements of Farm Livestock. No.1. Poultry.* HMSO: London.

Ahenkorah, Y. (1970) Potassium supplying power of some soils of Ghana cropped to cocoa. *Soil Sci.*, **109**, 127–35.

Ahlawal, I.P.S., Singh, A. and Saref, C.S. (1981) Effects of winter legumes on the nitrogen economy and productivity of succeeding cereals. *Exp.Agric.*, **17**, 57–72.

Ahmed, P. (1991) Agroforestry: a viable land use of alkali soils. *Agroforestry Systems*, **14**, 23–37.

Aina, P.O. (1979) Soil changes resulting from long-term management practices in Western Nigeria. *Soil Sci.Soc.of Amer.J.*, **43**(1), 173–7.

Akehurst, B.C. (1981) *Tobacco*, 2nd edn. Longman: London.

Akehurst, B.C. and Sreedharan, A. (1965) Time of planting – a brief review of experimental work in Tanganika, 1956–62. *E.Afr.Agric.For.J.*, **30**, 189–201.

Akobundu, I.O. (1987) *Weed Science in the Tropics.*

Principles and Practices. John Wiley & Sons: New York, NY.

Akunda, E. and Huxley, P.A. (1990) *The Application of Phenology to Agroforestry Research.* ICRAF Working Paper No.63. International Centre for Research in Agroforestry: Nairobi.

Alegre, J.C. and Sanchez, P.A. (1989) Central continuous cropping experiment. *In*: N. Caudle (ed.) *Tropsoils Technical Report, 1986–87,* pp. 86–7. Tropsoils Management Entity, North Carolina State University: Raleigh, NC.

Alim, K.A. (1967) Repeatability of milk yield and length of lactation of the milking buffalo in Egypt. *Trop.Agric.(Trin.),* **44**, 159–63.

Allan, T.G. (1986) Land clearing in African savannas. *In*: R. Lal, P.A. Sanchez and R.W. Cummings Jr. (eds) *Land Clearing and Development in the Tropics,* pp. 69–80. A.A. Balkema: Rotterdam.

Allen, T.E. and Bligh, J. (1969) A comparative study of the temporal patterns of cutaneous water vapour loss from some domesticated mammals with epitrichial sweat glands. *Comp.Biochem.and Physiol.,* **31**, 347–51.

Allison, F.E. (1966) The fate of nitrogen applied to soils. *Advances in Agron.,* **18**, 219–58.

Altieri, M.A., Javier Trujillo, F. and Farrell, J. (1987) Plant–insect interactions and soil fertility relations in agroforestry systems: implications for the design of sustainable agroecosystems. *In*: H.L. Gholtz (ed.) *Agroforestry: Realities, Possibilities and Potentials,* pp. 89–108. Martius Nijhoff and International Centre for Research in Agroforestry: Dordrecht.

Altieri, M.A. and Liebman, M. (1986) Insect, weed and plant disease management in multiple cropping systems. *In*: C.A. Francis (ed.) *Multiple Cropping Systems,* pp. 183–218. Macmillan: New York, NY.

Amakiri, S.F. (1974a) Seasonal changes in bovine skin thickness in relation to the incidence of *Dermatophilus* infection in Nigeria. *Res.Vet.Sci.,* **17**, 351–5.

Amakiri, S.F. (1974b) Sweat gland measurements in some tropical and temperate breeds of cattle in Nigeria. *Anim.Prod.,* **18**, 285–91.

Amakiri, S.F. (1975) The skin structure of some cattle breeds in Nigeria – studies in relation to hide production. *J.Niger.Vet.Med.Assn.,* **4**, 21–8.

Amakiri, S.F. and Mordi, R. (1975) The rate of cutaneous evaporation in some tropical and temperate breeds of cattle in Nigeria. *Anim.Prod.,* **20**, 63–8.

Amble, V.N., Gopalan, R., Malhotra, J.C. and Mehrotra, P.C. (1970) Some vital statistics and genetic parameters of Indian buffaloes at military dairy farms. *Ind.J.Anim.Sci.,* **40**, 377–88.

Anderson, D.M. (1984) Depression, dust bowl, demography and drought: the colonial state in East Africa during the 1930s. *Afr.Affairs,* **83**, 321–43.

Anderson, J.M. and Ingram, J.S.I. (1993) *Tropical Soil Biology and Fertility: A Handbook of Methods,* 2nd edn. CAB International: Wallingford.

Anderson, J.R., Herdt, R.W. and Scobie, G.M. (1988) *Science and Food. The CGIAR and its Partners.* Consultative Group on International Agricultural Research/World Bank: Washington, DC.

Anderson, L.S. and Sinclair, F.L. (1993) Ecological interactions in agroforestry systems. *Agroforestry Abstracts,* **6**, 57–91.

Andrews, D.J. (1972) Intercropping with sorghum in Nigeria. *Exp.Agric.,* **8**, 139–50.

Archibold, O.W. (1995) *Ecology of World Vegetation.* Chapman & Hall: London.

Arcoll, D.B., Goulding, K.W. and Hughes, J.C. (1985) Traces of 2:1 layer-silicate clays in Oxisols from Brazil, and their significance for potassium nutrition. *J.Soil Sci.,* **36**(1), 123–8.

Arnon, I. (1972) *Crop Production in Dry Regions.* Leonard Hill: London.

Asamenew, G. and Saleem, M. (1991) *The Concept and Procedure of Improved Vertisol Development in Ethiopia.* IBSRAM Workshop, Nairobi, 5–7 March 1991. International Board for Soil Research and Management: Bangkok.

Atampugre, N. (1993) *Behind the Stone Lines.* Oxfam: Oxford.

Aubert, G. and Tavernier, R. (1972) Soil survey. *In*: Committee on Tropical Soils, Agricultural Board, National Research Council (eds) *Soils of the Humid Tropics.* National Academy of Sciences: Washington, DC.

Avery, M.E., Cannell, M.G.R. and Ong, C.K. (eds) (1991) *Biophysical Research for Asian Agroforestry.* Winrock-Oxford & IBH Series.

Ayre-Smith, R.A. (1976) Principles for increasing the efficiency of livestock production in a developing country. *Trop.Anim.Hlth.Prod.,* **3**, 43–51.

Baanante, C.A., Bumb, B.L. and Thompson, T.P. (1989) *The Benefits of Fertiliser Use in Developing Countries.* Paper Series P-8, IFDC. International Fertilizer Development Center: Muscle Shoals, AL.

Baccari, F. (Jr.), Johnson, H.D. and Hahn, G.L. (1983) Environmental heat effects on growth, plasma T_3 and post-heat compensatory effects on Holstein calves. *Proc.Soc.Exp.Biol.Med.,* **173**, 312–18.

Bache, B.W. and Heathcote, R.G. (1969) Long-term effects of fertilizers and manure on soil and leaves of cotton in Nigeria. *Exp.Agric.,* **5**, 241–7.

Bahman, A.M., Rooke, J.A. and Topps, J.H. (1993) The performance of dairy cows offered drinking water of low or high salinity in a hot arid climate. *Anim.Prod.,* **57**(1), 23–8.

Baker, C.C., Coppack, C.E., Nave, D.H., Brasington, C.F. and Stermer, R.A. (1987) Effect of chilled water on lactating Holstein cows in summer. *J.Dairy Sci.,* **70**(Supplement 1), 113.

Baker, E.F.I. (1975) Effects and interactions of 'package deal' inputs on yield and labour demand of maize. *Exp.Agric.*, **11**, 295–304.

Balachandran, N. and Whiteman, P.C. (1975) Nitrogen yield and recovery by *Setaria* and *Desmodium* swards during the pasture phase and a subsequent crop phase. *Proc.Austr.Conf.Trop.Pastures, Townsville*, **2**(5), 5–15.

Balek, J. (1983) *Hydrology and Water Resources in Tropical Regions.* Elsevier: Oxford.

Baptist, R. (1990) The actuarial approach to evaluate breeding objectives for tropical livestock. *Proc.4th Wrld.Cong.on Genetics Applied to Livestock Prod.*, **14**, 410–13.

Barrau, J. (1959) The bush fallowing system of cultivation in the Continental Islands of Melanesia. *Proc.9th Pacific Sci.Cong.*, **7**, 53–5.

Barrett, E.C. and Martin, D.W. (1981) *Use of Satellite Data in Rainfall Monitoring.* Academic Press: London.

Barrow, C. (1987) *Water Resources and Agricultural Development in the Tropics.* Longman Scientific and Technical: Harlow.

Barrow, E.G.C. (1996) *The Drylands of Africa: Local Participation in Tree Management.* Initiatives Publishers: Nairobi.

Barry, T.N. (1989) Condensed tannins: their role in ruminant protein and carbohydrate digestion and possible effects upon the rumen ecosystem. *In*: J.V. Nolan, R.A. Leng and D.I. Demeyer (eds) *The Role of Protozoa and Fungi in Ruminant Digestion*, pp. 153–69. Penambul Books: Armidale, NSW.

Barsaul, C.S. and Talapatra, S.K. (1970) A comparative study on the determination of digestibility coeffients of feedingstuffs by different species of farm animals. *Ind.Vet.J.*, **47**, 348–55.

Barthélemy, D., Edelin, C. and Hallé, F. (1989a) Architectural concepts for tropical trees. *In*: L.B. Holme-Nielson and H. Baslev (eds) *Tropical Forests: Botanical Dynamics, Speciation and Diversity*, pp. 89–109. Academic Press: London.

Barthélémy, D., Edelin, C. and Hallé, F. (1989b) Some architectural aspects of tree ageing. *In*: E. Dreyer, G. Ausserac, M. Bonnet-Masimbert, P. Dizengrenel, J.M. Farre, J.P. Garrec, F. Le Tacon and F. Marta (eds) *Ann.Sci.For.*, **46**(Supplement), *Forest Tree Physiology*, pp. 194s–198s. Elsevier/Institut National de la Recherche Agronomique: Amsterdam.

Basu, S.B. (1985) *Genetic Improvement of Buffaloes.* Kalyani Publishers: New Delhi.

Bate, R. and Morris, J. (1994) *Global Apocalypse or Hot Air?* Institute of Economic Affairs: London.

Bationo, A., Christianson, C.B. and Mokwonye, U. (1989) Soil fertility management of the pearl millet-producing sandy soil of Sahelian West Africa: the Niger experience. *In*: *Soil, Crop and Water Management Systems for Rainfed Agriculture in the Sudano-Sahelian Zone*, Proceedings of an International Workshop, 11–16 January 1987, ICRISAT Sahelian Centre, Niamey, pp. 159–68. International Crops Research Institute for the Semi-Arid Tropics: Patancheru, India.

Baudelaire, J.P. (1972) Water for livestock in semi-arid zones. *Wrld.Anim.Rev.*, **3**, 1–9.

Bauer, M. (1965) Five years' study of ranch breeding stock, 1959–1964. *Rhod.Agric.J.*, **62**, 28–33.

Baumer, M. (1991) Animal production, agroforestry and similar techniques. *Agroforestry Abstracts*, **4**, 179–98.

Beaumont, P. (1989) *Environmental Management and Development in Drylands.* Routledge: London.

Becerril, P.C., Román-Ponce, H. and Castillo, H.R. (1981) Comportamiento productivo de vacas Holstein, Suizo Pardo y sus cruzas con Cebú F_1 en clima tropical. *Téc.Pecu.Méx.*, **40**, 16–24.

Begg, J. and Turner, N. (1976) Crop water deficits. *Advances in Agron.*, **28**, 161–217.

Behnke, R.H., Scoones, I. and Kerven, C.K. (eds) (1993) *Range Ecology at Disequilibrium.* Overseas Development Institute Publications: London.

Bell, A. (1988) Trees, water and salt – a fine balance. *Ecos*, **58**, 2–9.

Bellis, E. (1953) Nitrogen in rainfall. *1951 Annual Rep.of Dept.of Agric.Kenya.* Department of Agriculture: Nairobi.

Belsky, A.J., Amundsen, R.G., Duxbury, J.M., Riha, S.J., Ali, A.R. and Mwonga, S.M. (1989) The effects of trees on their physical, chemical and biological environments in a semi-arid savanna in Kenya. *J.Appl.Ecol.*, **26**, 1005–24.

Bembridge, T.J. (1963) Protein supplementary feeding of breeding stock proves profitable under watershed ranching conditions. *Rhod.Agric.J.*, **60**, 98–103.

Bene, J.G., Beall, H.W. and Côté, A. (1977) *Trees, Food, and People: Land Management in the Tropics.* International Development Research Centre: Ottawa.

Bentley, W.R., Khosla, P.K. and Seckler, R. (eds) (1993) *Agroforestry in South Asia: Problems and Applied Research Perspectives.* Winrock-Oxford & IBH Series.

Beran, M.A. and Rodier, J.A. (1985) *Hydrological Aspects of Drought.* Studies and Reports in Hydrology 39. UNESCO-WMO: Paris.

Bergersen, F.J. (1980) *Methods for Evaluating Biological Nitrogen Fixation.* John Wiley & Sons: Chichester.

Bhargava, B., Katiyar, U.C. and Saxena, R.P. (1977) A note on the nutritive value of Rainy (*Mellotus philippensis*) tree leaves as sole feed for sheep. *Ind.J.Anim.Sci.*, **47**, 594–5.

Bhattacharya, P. (1990) Buffalo. *In*: W.J.A. Payne

(ed.) *An Introduction to Animal Husbandry in the Tropics*, pp. 422–71. Longman: London.

Bibby, J.S. and Mackney, D. (1969) *Land Use Capability Classification.* Tech.Mono.1. Soil Survey of England and Wales, Rothamsted Experimental Station: Harpenden, UK.

Biot, Y. (1990) THEPROM: an erosion-productivity model. *In*: J. Boardman, I.D.L. Foster and J.A. Dearing (eds) *Soil Erosion on Agricultural Land*, pp. 465–79. John Wiley & Sons: Chichester.

Birch, H.F. (1958) The effect of soil drying on humus decomposition and nitrogen availability. *Plant and Soil*, **10**, 9–31.

Bird, G.W., Edens, T., Drummond, F. and Groden, E. (1990) Design of pest management systems for sustainable agriculture. *In*: C.A. Francis, C.B. Flora and L.D. King (eds) *Sustainable Agriculture in Temperate Zones*, pp. 55–110. John Wiley & Sons: New York, NY.

Bishop, J. (1992) *Economic Analysis of Soil Degradation.* London Environmental Economics Gatekeeper Series No. LEEC GK 92–01. International Institute for Environment and Development: London.

Blackburn, F. (1984) *Sugar-cane.* Longman: London.

Blaikie, P.M. (1989) Explanation and policy in land degradation and rehabilitation for developing countries. *Land Degrad.Rehabil.*, **1**, 23–37.

Blaikie, P.M. and Brookfield, H.C. (1987) *Land Degradation and Society.* Methuen: London.

Blair, G.J., Lefroy, R.D.B., Singh, B.P. and Till, A.R. (1994) Development and use of a carbon management index to monitor changes in soil C pool size and turnover rate. *In*: G. Cadisch and K.E. Giller (eds) *Driven by Nature: Plant Litter Quality and Decomposition*, pp. 273–82. CAB International: Wallingford.

Bligh, J. and Johnson, K.G. (1973) Glossary of terms for thermal physiology. *J.Appl.Physiol.*, **35**, 941–61.

Bloomfield, C. and Coulter, J.K. (1973) Genesis and management of acid sulphate soils. *Advances in Agron.*, **25**, 265–326.

Bloomfield, C., Coulter, J.K. and Kanaris-Sotiriou, R. (1968) Oil palms on acid sulphate soils in Malaya. *Trop.Agric.(Trin.)*, **45**, 289–300.

Blunt, C.G. and Humphreys, L.R. (1970) Phosphate response of mixed swards at Mt. Cotton, south-eastern Queensland. *Austr.J.Exp.Agric.Anim. Husb.*, **10**, 431–41.

Boa, W. (1958) Development of NIAE ditch cleaner. *J.Agric.Eng.Res.*, **3**, 17–26.

Boa, W. (1966) Equipment and method for tied-ridge cultivation. *Farm Power and Machinery Informal Wkg.Bull.*, **28**, FAO: Rome.

Bogdan, A.V. (1955) Bush clearing and grazing trial at Kisokon, Kenya. *E.Afr.Agric.For.J.*, **19**, 253–9.

Bogdan, A.V. (1977) *Tropical Pasture and Fodder Plants.* Longman: London.

Bohra, H.C. (1980) Nutrient utilization of *Prosopis cineraria* (Khekjri) leaves by desert sheep and goats. *Annals of Arid Zone*, **19**, 73–81.

Bolin, B., Doos, B.R., Jager, J. and Warrick, R.A. (1991) *The Greenhouse Effect, Climate Change and Ecosystems.* John Wiley & Sons: Chichester.

Bolton, J. (1968) Leaching of fertilizers applied to a latosol in lysimeters. *J.Rub.Res.Inst.Malaysia*, **20**, 274–84.

Bondoc, O.L., Smith, C. and Gibson, J.P. (1989) A review of breeding strategies for genetic improvement of dairy cattle in developing countries. *Anim.Breed.Abstr.*, **57**, 819–29.

Bonnier, C. and Segier, J. (1958) *Symbiose rhizobium-legumineuses en region equatoriale.* 2. Pub.Nat. Agron.du Congo Belge, Series Scientifique, 76. Institut National pour l'Etude de Agronomique du Congo: Brussels.

Bonsma, J.C., Badenhorst, J.F.G. and Skinner, J.D. (1972) The incidence of foetal dwarfism in Shorthorn cattle in the sub-tropics. *S.Afr.J.Anim.Sci.*, **2**, 19–21.

Bos, M.G. (1990) Research on acid sulphate soils in the humid tropics. *In: Workshop on Acid Sulphate Soils in the Humid Tropics*, 20–22 November, Bogor, Indonesia, pp. 1–9. Agency for Agricultural Research and Development: Jakarta.

Bose, S. (1984) Shifting cultivation in India. *J.Ind.Anthropolog.Soc.*, **19**(1), 55–65.

Boserup, E. (1965) *The Conditions of Agricultural Growth.* Allen & Unwin: London.

Boserup, E. (1981) *Population and Technological Change.* Chicago University Press: Chicago, IL.

Bouldin, D.R. (1986) The chemistry and biology of flooded soils in relation to the nitrogen economy in rice fields. *Fertilizer Res.*, **9**, 1–14.

Bouldin, D.R., Quintana, J. and Suhet, A. (1989) Evaluation of mineralization potential of legume residues. *In*: N. Caudle (ed.) *Tropsoils Technical Report 1986–87*, pp. 304–305. Tropsoils Management Entity, North Carolina State University: Raleigh, NC.

Bouldin, D.R., Mughogho, S., Lathwell, D.J. and Scott, T.W. (1979) *Nitrogen Fixation by Legumes in the Tropics.* Cornell International Agricultural Mimeograph 75. Department of Agronomy, Cornell University: Ithaca, NY.

Boulière, F. (ed.) (1983) *Tropical Savannas: Ecosystems of the World, 13.* Elsevier: Amsterdam.

Bowen, E.J. and Rickert, K.G. (1979) Beef production from native pastures sown to fine-stem Stylo in the Burnett region of SE Queensland. *Austr.J.Exp. Agric.Anim.Husb.*, **19**, 140–49.

Bowen, G.D. (1985) Roots as a component of tree

productivity. *In*: M.G.R. Cannell and J.E. Jackson (eds) *Attributes of Trees as Crop Plants*, pp. 303–15. Institute of Terrestrial Ecology: Abbots Ripton.

Boyer, J. (1972) Soil potassium. *In*: Committee on Tropical Soils, Agricultural Board, National Research Council (eds) *Soils of the Humid Tropics*, pp. 102–35. National Academy of Sciences: Washington, DC.

Bradley, R.G. and Crout, N. (1995) *The PARCH Model User Guide*. Tropical Crops Research Unit, University of Nottingham: Nottingham.

Bremner, J.M. (1968) The nitrogenous constituents of soil organic matter and their role in soil fertility. *In*: *A Study Week on Organic Matter and Soil Fertility*. North Holland Publishing Company: Amsterdam.

Brenner, A. (1996) Microclimatic modifications in agroforestry. *In*: C.K. Ong and P. Huxley (eds) *Tree-crop Interactions: A Physiological Approach*, pp. 159–87. CAB International/International Centre for Research in Agroforestry: Wallingford/Nairobi.

Brewbaker, J.L. (1987) *Leucaena*: a multipurpose tree genus for tropical agroforestry. *In*: H.A. Steppler and P.K.R. Nair (eds) *Agroforestry: A Decade of Development*, pp. 289–323. International Centre for Research in Agroforestry: Nairobi.

Briggs, D. and Smithson, P. (1992) *Fundamentals of Physical Geography*. Routledge: London.

Brinkmann, W.L.F. and Vieira, A.N. (1971) The effect of burning on germination of seeds at different soil depths of various tropical tree species. *Turrialba*, **21**(1), 78–82.

Brody, S. (1945) *Bioenergetics and Growth*. Reinhold: New York, NY.

Bromfield, A.R. (1972) Sulphur in N. Nigeria soils. 1. Effects of cultivation and fertilizers on total S and sulphate patterns in soil profiles. *J.Agric.Sci.*, **78**, 465–70.

Brook, A.H. and Short, B.F. (1960a) Regulation of body temperatures of sheep in a hot environment. *Austr.J.Agric.Res.*, **11**, 402–407.

Brook, A.H. and Short, B.F. (1960b) Sweating in sheep. *Austr.J.Agric.Res.*, **11**, 557–69.

Brookman-Amissah, J. (1985) *Forestry and Socio-economic Aspects of Modification of Traditional Shifting Cultivation through the Taungya System in the Subri Area, Ghana*. FAO Forestry Paper 50/1. FAO: Rome.

Broster, W.H. (1972) Effect on milk yield of the cow of the level of feeding before calving. *Dairy Sci.Abstr.*, **34**(4), 265–88.

Broster, W.H. and Johnson, C.L. (1977) A modern approach to feeding dairy cows. *ARC Res.Rev.*, **3**, 9–10.

Brubacher, D., Arnason, J.T. and Lambert, J.D.H. (1989) Woody species and nutrient accumulation during the fallow period of milpa farming in Belize. *Plant and Soil*, **114**, 165–72.

Brucks, R. (1964) Blood groups of the pig with special reference to the L system. *Z.Tierzücht.Zücht.Biol.*, **80**, 66–80.

Bruijn, G.H. and Fresco, L.O. (1989) The importance of cassava in world food production. *Netherlands J.Agric.Sci.*, **37**, 21–34.

Brunt, M. (1967) The methods employed by the Directorate of Overseas Surveys in the assessment of land resources. *Act.2e Symp.Inter.Photo-interpretation*, Vols 3–10: Paris.

Brunt, M. and Rees, C.C. (1965) *Malawi Land Use Survey*. Mimeo. Land Resources Division, Directorate of Overseas Surveys: London.

Bryson, R.A. (1974) A perspective on climatic change. *Science*, **184**, 753–60.

Bryson, R.A. (1989) Will there be a global greenhouse warming? *Env.Cons.*, **16**(2), 97–9.

Buddenhagen, I.W. and Persley, G.J. (1978) *Rice in Africa*. Academic Press: London.

Budowski, G. (1980) The place of agro-forestry in managing tropical forests. *Int.Symp.Trop.For.Util. Cons.* Yale University: New Haven, CT.

Bumb, B.L. (1989) *Global Fertilizer Perspective, 1960–1995: The Dynamics of Growth and Structural Change*. International Fertilizer Development Center: Muscle Shoals, AL.

Bumb, B.L. (1991) Trends in fertilizer use and production in Sub-Saharan Africa, 1970–1995: an overview. *Fertilizer Res.*, **28**, 41–8.

Burgemeister, R. (1974) *Probleme der Dromedarhaltung und Zücht in Sud-Tunesien*. Justus Liebig University: Griessen.

Burley, J. (1984) Global needs and problems of collection, storage and distribution of multipurpose tree germplasm: A background document. *In*: J. Burley and P. von Carlowitz (eds) *Multipurpose Tree Germplasm*, Proceedings of a Planning Workshop, pp. 43–221. Nairobi: International Centre for Research in Agroforestry.

Burnham C.P. (1989) Pedological processes and nutrient supply from parent material in tropical soils. *In*: J. Proctor (ed.) *Mineral Nutrients in Tropical Forest and Savannah Ecosystems. Special Publication No.9 for the British Ecological Society*, pp. 27–41. Blackwell Science: Oxford.

Burton, G.W. (1983) Breeding pearl millet. *Plant Breed. Rev.* **1**, 162–82.

Butai, P.C. (1987) Review of research on fertilizer and soil fertility in Zimbabwe. *In*: H. Ssali and L.B. Williams (eds) *Proc.East and Southeast African Fertilizer Management and Evaluation Network Workshop*. International Fertilizer Development Center: Muscle Shoals, AL.

Butterworth, M.H. (1962) The digestibility of sugar-cane tops, rice aftermath and bamboo grass. *Emp.J.Exp.Agric.*, **30**, 77–81.

Butterworth, M.H. (1985) *Beef Cattle Nutrition and Tropical Pastures.* Longman: London.

Butterworth, M.H., Groom, C.G. and Wilson, P.N. (1961) The intake of Pangola grass under wet- and dry-season conditions in Trinidad. *J.Agric.Sci.*, **56**, 407–10.

Buvanendran, V., Jalatge, E.F.A. and Ganesan, K.N. (1971) Influence of season on the breeding pattern of buffaloes in Ceylon. *Trop.Agric.(Trin.)*, **48**, 97–102.

Byerlee, D. and Heisey, P. (1993) Strategies for technical change in small-farm agriculture, with particular reference to Sub-Saharan Africa. In: N.C. Russell and C.R. Dowswell (eds) *Policy Options for Agricultural Development in Sub-Saharan Africa*, pp. 21–52. Centre for Applied Studies in International Negotiations/Sasa Kawa Africa Association/Global 2000: Mexico, DF.

Byrne, D. (1984) Breeding cassava. *Plant Breed. Rev.*, **2**, 73–134.

Calegari, A. (1995) *Leguminosas Para Adubaçao Verde de Verao no Paraná.* Instituto Agronômico do Paraná: Londrina, Brazil.

Camberlin, P. (1995) June–September rainfall in North East Africa and atmospheric signals over the tropics. *Int.J.Climatology*, **15**, 773–83.

Campbell, Q.P., Ebersöhn, J.P. and Broembsen, H.H. van (1962) Browsing by goats and its effect on the vegetation. *Herb.Abstr.*, **32**, 273–5.

Campling, R.C. and Freer, M. (1966) Factors affecting the voluntary intake of food by cows. 8. Experiments with ground, pelleted roughages. *Brit.J.Nutr.*, **20**, 229–44.

Cannell, M.G.R. (1971) Effects of fruiting, defoliation and ring-barking on the accumulation and distribution of dry matter in branches of *Coffea arabica* L. in Kenya. *Exp.Agric.*, **7**, 63–4.

Cannell, M.G.R. (1983) Plant population and yield of tree and herbaceous crops. In: P.A. Huxley (ed.) *Plant Research and Agroforestry*, Proceedings of a consultative meeting held in Nairobi, 8–15 April 1981, pp. 489–502. International Centre for Research in Agroforestry: Nairobi.

Cannell, M.G.R. (1985) Dry matter production in tree crops. In: M.G.R. Cannell and J.E. Jackson (eds) *Attributes of Trees as Crop Plants*, pp. 160–93. Institute of Terrestrial Ecology: Abbots Ripton.

Cannell, M.G.R. (1989a) Food crop potential of tropical trees. *Exp.Agric.*, **25**, 315–26.

Cannell, M.G.R. (1989b) Physiological basis of wood production: a review. *Scand.J.For.Res.*, **4**, 459–90.

Cannell, M.G.R., Mobbs, D.C. and Lawson, G.J. (1998) Complementarity of light and water use in tropical agroforests. II. Modelled tree production and potential crop yield in aid to humid climates. *For.Ecol.Manage*, **102**, 275–82.

Cannell, M. and Pitcairn, C. (1993) *Impacts of Mild Winters and Hot Summers 1988–1990.* Department of the Environment, HMSO: London.

Caro-Costas, R. and Vicente-Chandler, J. (1956) Comparative productivity of Merker grass and of a Kudzu-Merker mixture as affected by season and cutting height. *J.Agric.Univ.Puerto Rico*, **40**, 144–51.

Caro-Costas, R. and Vicente-Chandler, J. (1972) Effect of heavy rates of fertilization on beef production and carrying capacity of Napier grass pastures over 5 consecutive years of grazing under humid tropical conditions. *J.Agric.Univ.Puerto Rico*, **56**, 223–7.

Caro-Costas, R., Vicente-Chandler, J. and Abruña, F. (1972) Effect of four levels of fertilization on beef production and carrying capacity of Pangola grass pastures in the humid mountain region of Puerto Rico. *J.Agric.Univ.Puerto Rico*, **56**, 219–22.

Carr, M.K.V. and Stephens, W. (1992) Climate, weather and the yield of tea. In: K.C. Wilson and M.C. Clifford (eds) *Tea: Cultivation to Consumption*, pp. 87–136. Chapman & Hall: London.

Carr, S.J. (1989) *Technology for Small-scale Farmers in Sub-Saharan Africa: Experience with Foodcrop Production in Five Major Ecological Zones.* Wrld.Bank Tech.Paper No.109. World Bank: Washington, DC.

Castle, D.A., McCunall, J. and Tring, J.M. (eds) (1984) *Field Drainage: Principles and Practices.* Batsford Academic and Educational: London.

Catchpole, V.R. and Henzell, E.F. (1971) Silage and silage making from tropical herbage species. *Herb.Abstr.*, **41**, 213–21.

Catling, D. (1993) *Rice in Deepwater.* Macmillan: London.

Catling, H.D., Hobbs, P.R., Islam, Z. and Alam, B. (1983) Agronomic practices and yield assessments of deep water rice in Bangladesh. *Field Crops Res.*, **6**, 109–32.

Caudwell, F.B. and Sykes, A.H. (1975) Unpublished data quoted in Sykes, A.H. (1977) 'Nutrition-environment interactions in poultry'. In: W. Haresign, H. Swan and D. Lewis (eds) *Nutrition and the Climatic Environment.* Butterworths: London.

Caviedes, C.N. (1988) The effects of ENSO events in some key regions of the South American continent. In: S. Gregogy (ed.) *Recent Climatic Change*, pp. 252–66. Belhaven Press: London.

CFSCDD (1986). *Soil Conservation in Ethiopia: Guidelines for Development Agents.* Community Forests and Soil Conservation Development Department, Ministry of Agriculture: Addis Ababa.

Chadhokar, P.A. (1977) Establishment of Stylo (*Stylosanthes guianensis*) in Kunai (*Imperata cylindrica*) pastures and its effects on dry matter yield and animal production in the Markham Valley, Papua New Guinea. *Trop.Grassl.*, **11**, 263–72.

Chairuddin, G., Iriansyah, Klepper, O. and Rijksen, H.D. (1990) Environmental and socio-economic aspects of fish and fisheries in an area of acid sulphate soils: Palau Petak, Kalimantan. *In: Workshop on Acid Sulphate Soils in the Humid Tropics*, 20–22 November, pp. 374–92. Agency for Agricultural Research and Development: Jakarta.

Chalack, D.A. (1993) Increasing profitability by utilizing embryo technologies. *Brit.Cattle Breeders' Club Digest*, **48**, 21–4.

Chalmers, A.W. (1976) Advantages and disadvantages of nomadism with particular reference to the Republic of the Sudan. *In*: A.J. Smith (ed.) *Beef Cattle Production in Developing Countries*, pp. 388–97. Centre for Tropical Veterinary Medicine, University of Edinburgh: Edinburgh.

Chalupa, W. (1975) Rumen by-pass and protection of protein and amino acid. *J.Dairy Sci.*, **58**, 1198–218.

Chapman, E.C. (1975) Shifting agriculture in tropical forest areas of south-east Asia. *IUNC Pub.*, **32**, 120–35.

Charles, D.D. and Johnson, E.R. (1972) Carcass composition of the water buffalo (*Bubalus bubalis*). *Austr.J.Agric.Res.*, **23**, 905–11.

Charreau, C. and Nicou, R. (1971) L'emeliaration du profile cultural dans les sols sableux et sablo-argileux de la zone tropicale Seche Ouest Africaine et ses incidences agronomiques. *Agron.Tropicale, Paris*, **26**, 209–55.

Chase, R.G. and Boudouresque, E. (1989) A study of methods for the re-vegetation of barren, crusted Sahelian forest soils. *In: Soil, Crop and Water Management Systems for Rainfed Agriculture in the Sudano-Sahelian Zone*, Proceedings of an International Workshop, 11–16 January 1987, ICRISAT Sahelian Centre, Niamey, pp. 125–35. International Crops Research Institute for the Semi-Arid Tropics: Patancheru, India.

Chaudhry, I.A. and Shaw, A.O. (1965) Production traits of Pakistani buffaloes. *Pakist.J.Sci.*, **17**, 252–8.

Cheeke, P.R. and Shull, L.R. (1985) *Natural Toxicants in Feeds and Poisonous Plants*. AVI Publishing: Westport. CT.

Child, R. (1974) *Coconuts*. Longman: London.

Chin, S.C. (1985) Agriculture and resource utilization in a Lowland rainforest Kenyah community. *Sarawak Museum J.*, **35**, 56.

Chisholm, A.H. and Dumsday, R.G. (eds) (1987) *Land Degradation: Problems and Policies*. Cambridge University Press: Cambridge.

Christian, C.S. and Stewart, C.A. (1953) *Survey of Katherine-Darwin Region*. CSIRO Land Res. Services Bull. 1. Commonwealth Scientific and Industrial Research Organisation: Canberra.

Christian, C.S. and Stewart, C.A. (1964) Methodology of integrated surveys. *Proc.of UNESCO Conf.on Principles and Methods of Integrating Aerial Studies of Natural Resources*, Toulouse, France.

Chubb, L.G. (1982) Antinutritive factors in animal feedstuffs. *In*: W. Haresign (ed.) *Recent Advances in Animal Nutrition – 1982*, pp. 21–37. Butterworths: London.

CIAT (1974) *Annual Report 1973*. Centro Internacional de Agricultura Tropical: Palmira, Colombia.

CIAT (1984) Upland rice in Latin America. *In: An Overview of Upland Rice Research*, pp. 93–119. International Rice Research Institute: Los Banos, Philippines.

Clark, C. and Turner, B.J. (1974) World population growth and future food trends. *In*: M. Rechcigl, Jr (ed.) *Man, Food and Nutrition*. CRC Press: Cleveland, OH.

Clarke, W.C. (1971) *Place and People: An Ecology of a New Guinea Community*. National University Press: Canberra.

Clausen, T.P., Provenza, F.D., Burritt, E.A., Reichardt, P.B. and Bryant, J.P. (1990) Ecological implications of condensed tannin structure: a case study. *J.Chem.Ecol.*, **16**, 2381–92.

Clement, C.R. (1988) Domestication of the Pejibaye palm (*Bactris gasipaes*) past and present. *Advances in Econ.Bot.*, **6**, 155–74.

Cline, M.G. (1955) Soil survey of the territory of Hawai. *Soil Survey Bull., USDA*. United States Department of Agriculture: Washington, DC.

Close, W.H. (1993) Fibrous diets for pigs. *In*: M. Gill, E. Owen, G.E. Pollott and T.L.J. Lawrence (eds) *Animal Production in Developing Countries*, pp. 107–17. Brit.Soc.Anim.Prod.Occ.Pub.16. British Society of Animal Production: Edinburgh.

Cobley, L.S. and Steele, W.M. (1977) *An Introduction to the Botany of Tropical Crops*. Longman: London.

Cocheme, J. and Franquin, P. (1967) *An Agroclimatology Survey of a Semi Arid Area in Africa South of the Sahara*. WMO Technical Note 86. World Meteorological Organization: Geneva.

Cock, J.H. (1985) *Cassava. New Potential for a Neglected Crop*. Westview Press: Boulder, CO.

Cockrill, W. Ross (1968) The draught Buffalo (*Bubalus bubalis*). *The Veterinarian (Oxford)*, **5**, 265–72.

Cockrill, W. Ross (ed.) (1974) *The Husbandry and Health of the Domestic Buffalo*. FAO: Rome.

Cohen, R.D.H. and O'Brian, A.D. (1974) Beef production from three pastures at Grafton, N.S.W. *Trop.Grassl.*, **8**, 71–9.

Colfer, C.J.P. (1991) *Toward Sustainable Agriculture in the Humid Tropics: Building on the TropSoils Experience in Indonesia*. TropSoils Bulletin 91–02. Soil Management Collaborative Support Program, North Carolina State University: Raleigh, NC.

Collins, M. (1990) *The Last Rain Forests*. Mitchell Beazley and IUCN: London.

Collinson, A.S. (1988) *Introduction to World Vegetation*. George Allen & Unwin: London.

Conklin, H.C. (1957) *Hanunoo Agriculture*. FAO For.Dev.Paper No.12. FAO: Rome.

Cooke, G.W. (1967) *The Control of Soil Fertility*. Crosby Lockwood: London.

Copland, R.S. (1974) Observations on Banteng cattle in Sabah. *Trop.Anim.Hlth.Prod.*, **6**, 89–94.

Coppock, D.L., Ellis, J.E. and Swift, D.M. (1988) Seasonal patterns of activity, travel and water intake for livestock in South Turkana, Kenya. *J.Arid Environ.*, **14**, 319–31.

Corley, R.H.V. (1983) Potential productivity of tropical perennial crops. *Exp.Agric.*, **19**, 217–37.

Coulter, J.K. (1970) *Soils of Central Africa. A Review of Investigations on their Fertility and Management*. Mimeo. Committee on Tropical Soils: Washington, DC.

Coulter, J.K. (1972) Soils of Malaysia. A review of investigations on their fertility and management. *Soils and Fertilizers*, **35**, 475–98.

Couper, D.C., Lal, R. and Claasen, S.L. (1981) Land clearing and development for agricultural purposes in Western Nigeria. *In*: R. Lal and E.W. Russell (eds) *Tropical Agricultural Hydrology*, pp. 119–30. John Wiley & Sons: London.

Coursey, D.G. (1967) *Yams*. Longman: London.

Courtney, R.S. (1993) The end of the world is not nigh. *Coal Trans.*, **8**(3), 40–42.

Creek, M., Destro, D., Miles, D.G., Redfern, D.M., Robb, J., Schleicher, E.W. and Squire, H.A. (1973) A case study in the transfer of technology. *Proc.3rd Wld.Conf.Anim.Prod.*, Melbourne, Australia.

Creek, M.J. (1984) Stratification of the beef industry in Kenya. *In*: B.L. Nestel (ed.) *Development of Animal Production Systems*. Elsevier: Amsterdam.

Crill, J.P. (1982) *Evolution of the Gene Rotation Concept for Rice Blast Control*. International Rice Research Institute: Los Banos, Philippines.

Critchley, W. (1991) New approaches to soil and water conservation. *ILEIA Newsletter*, May, 51–2.

Critchley, W., Reij, C. and Willcocks, T.J. (1994) Indigenous soil and water conservation: a review of the state of knowledge and prospects for building on tradition. *Land Degrad.Rehabil.*, **5**, 293–314.

Critchley, W. and Siegert, K. (1991) *Water Harvesting: A Manual for the Design and Construction of Water Harvesting Schemes for Plant Production*. Report AGL/MISC/17/91. FAO: Rome.

CSIRO (1973) Sheep, cattle and worms. *Rural Res.Dec.1973*, 25–6.

Cunningham, E.P. (1993) Animal genetic resources – the perspective for developing countries. *In*: M. Gill, E. Owen, G.E. Pollott and T.L.J. Lawrence (eds) *Animal Production in Developing Countries*, pp. 33–6. Brit.Soc.Anim.Prod.Occ.Pub.16. British Society of Animal Production: Edinburgh.

Curtis, S.E. (1985) Physiological responses and adaptation in swine. *In*: M.K. Yousef (ed.) *Stress Physiology in Livestock. Vol. 11, Ungulates*. CRC Press: Boca Raton, FL.

Dabadghao, P.M. and Shankarnarayan, K.A. (1970) Studies of *Iseilema*, *Sehima* and *Heteropogon* communities of the *Sehima-Dicanthium* zone. *In*: *Proc.11th Int.Grassld.Cong.*, 1970, pp. 36–45.

Dagg, M. and McCartney, J.C. (1968) The agronomic efficiency of the NIAE mechanised tied-ridge system of cultivation. *Exp.Agric.*, **4**, 279–94.

Dahl, G. and Hjort, A. (1976) *Having Herds*. Stockholm Studies in Anthropology. University of Stockholm: Stockholm.

Dalton, D.C. (1985) *Introduction to Practical Animal Breeding*, 2nd edn. Collins: London.

van Dam, J.T.P., van der Heide, D., van der Hel., van den Ingh, T.S.G.A.M., Verstegen, M.W.A. Wensing, T. and Zwart, D. (1996) The effect of *Trypanosoma vivax* infection on energy and nitrogen metabolism and serum metabolites and hormones in West African Dwarf goats on different feed intake levels. *Anim.Sci.*, **63**(1), 111–21.

Dana, S. and Karmakar, P.G. (1990) Species relationships in *Vigna* sub-genus *Ceratropis* and its implication in breeding. *Plant Breed.Rev.*, **8**, 19–42.

Dancette, C. (1980) Water requirements and adaptations to the rainy season of millet in Senegal. *In*: *Proceedings of International Workshop on Agroclimatological Research Needs of Semi-Arid Tropics*, Hyderabad, November 1978, pp. 106–20. ICRISAT: Hyderabad.

Daniel, J.N. and Ong, C. (1990) Perennial pigeonpea: a multipurpose species for agroforestry systems. *Agroforestry Systems*, **10**, 113–29.

Davidson, D.A. (1992) *The Evaluation of Land Resources*. Longman Scientific and Technical: Harlow.

Davis, R.H., Hassan, O.E.M. and Sykes, A.H. (1973) Energy utilization in the laying hen in relation to ambient temperature. *J.Agric.Sci.*, **80**, 173–7.

Davy, E.G., Mattei, F. and Solomon, S.I. (1976) *An Evaluation of Climate and Water Resources for Development of Agriculture in the Sudano-Sahelian Zone of West Africa*. WMO Special Environment Report 9. World Meteorological Organization: Geneva.

Dawelbeit, M.I. (1991) On-farm comparison of different tillage systems in Vertisols of Rahad Project – Sudan. *Proc.12th Int.Conf., Int.Soil Tillage Res. Organ.*, Soil Tillage and Agricultural Sustainability, 8–12 July 1991, Ibadan, Nigeria, pp. 521–5. Ohio State University: Columbus, OH.

Day, J.M., Neves, M.C.P. and Dobereiner, J. (1975)

Nitrogenase activity on the roots of tropical forage grasses. *Soil Biol.Biochem.*, **7**, 107–12.

Deaton, J.W., McNaughton, J.L. and Lott, B.D. (1982) Effect of heat stress on laying hens acclimated to cyclic versus constant temperatures. *Poult.Sci.*, **61**, 875–8.

De Datta, S.K. (1981) *Principles and Practices of Rice Production.* John Wiley & Sons: New York, NY.

De Datta, S.K., Gomez, K.A. and Descalsota, J. (1988) Changes in yield response to major nutrients and in soil fertility under intensive rice cropping. *Soil Sci.*, **146**, 350–58.

De Datta, S.K. and Malabuyoc, J. (1976) Nitrogen response of lowland and upland rice in relation to tropical environmental conditions. *In: Climate and Rice*, pp. 509–39. International Rice Research Institute: Los Banos, Philippines.

De Datta, S.K. and Patrick, W.H. (eds) (1986) *Nitrogen Economy of Flooded Soils.* Martinus Nijhoff: Dordrecht.

Delwaulle, J.C. (1973) Resultats de six ans d'observations sur l'erosion au Niger. *Revue Bois et Forets des Tropiques*, **150**, Juillet-Aout, 15–40.

Denmmead, O.T. and Shaw, R.H. (1962) Availability of soil water to plants as affected by soil moisture content and meteorological conditions. *Agron.J.*, **54**, 385–90.

Dennison, E.B. (1959) The maintenance of fertility in the southern Guinea savanna zone of Northern Nigeria. *Trop.Agric.(Trin.)*, **36**, 171–8.

Dennison, E.B. (1961) The value of farmyard manure in maintaining fertility in Northern Nigeria. *Emp.J.Exp.Agric.*, **29**, 330–36.

Dent, D.L. (1986) *Acid Sulphate Soils: A Baseline for Research and Development.* Publication 39. International Institute for Land Reclamation and Improvement: Wageningen.

Deshmukh, I. (1986) *Ecology and Tropical Biology.* Blackwell Scientific Publications: London.

Deshmukh, S.N. and Roychoudhury, P.N. (1971) Repeatability estimates of some economic characters in Italian buffaloes. *Zert.fur Vet.Med.A.*, **18**, 104–107.

Deshpande, T.L., Greenland, D.J. and Quirk, J.P. (1968) Changes in soil properties associated with the removal of iron and aluminium oxides. *J.Soil Sci.*, **19**, 108–22.

Devendra, C. (1981) Meat production from goats in developing countries. *In:* Brit.Soc.Anim.Prod.Occ. Pub.4, pp. 395–406. British Society of Animal Production: Edinburgh.

Devendra, C. (1987) The role of goats in food production systems in industrialised and developing countries. *Proc.IVth Wrld.Conf. on Goat Prod.*, **1**, pp. 3–40. Brasilia.

Devendra, C. (1989) Ruminant production systems in developing countries: resource utilization. *In: Feeding Strategies for Improving Productivity of Ruminant Livestock in Developing Countries.* Panel Proc.Series. International Atomic Energy Agency: Vienna.

Devendra, C. and Burns, M. (1983) *Goat Production in the Tropics.* Comw.Bur.Anim.Breed.Genet.Tech. Comm. 19. CAB: Farnham Royal.

D'Hoore, J.L. (1965) *Soils Map of Africa. Commission pour Cooperation Technologique en Afrique*: Lagos.

Distel, R.A. and Provenza, F.D. (1991) Experience early in life affects voluntary intake of blackbrush by goats. *J.Chem.Ecol.*, **17**, 431–50.

Djokoto, R.K. and Stephens, D. (1961) Thirty long-term fertilizer trials under continuous cropping in Ghana. *Emp.J.Exp.Agric.*, **29**, 181–95; 245–57.

Dobereiner, J. and Campello, A.B. (1971) Non-symbiotic nitrogen fixing bacteria in tropical soils. *Plant and Soil (Special Volume)*, 457–70.

Dobereiner, J. and Day, J.M. (1974) *In: Soil Management in Tropical America*, Proc.Seminar, pp. 197–210. Centro Internacional de Agricultura Tropical: Colombia.

Dobereiner, J., Day, J.M. and Dart, P.J. (1972) Nitrogenase activity and oxygen sensitivity of the *Paspalum notatum-Azotobacter paspali* association. *J.Gen.Microbiol.*, **71**, 102–16.

Doggett H. (1988) *Sorghum*, 2nd edn. Longman: London.

Dommergues, Y. (1960) Mineralization de l'azote aux faibles humidities. *Trans.7th Int.Cong.Soil Sci.*, **2**, 672–8.

Dommergues, Y. (1963) Evaluation de taux de fixation de l'azote dans un sol dunaire reboise en filao (*Casuarina equisetifolia*). *Agrochemica*, **7**, 335–40.

Dommergues, Y.R. (1987) The role of biological nitrogen fixation in agroforestry. *In*: H.A. Steppler and P.K.R. Nair (eds) *Agroforestry: A Decade of Development*, pp. 245–72. International Centre for Research in Agroforestry: Nairobi.

Dommergues, Y. and Ganry, F. (1985) Biological nitrogen fixation and soil fertility maintenance in management of nitrogen and phosphorus fertilizers in Sub-Saharan Africa. *In*: A.V. Mokwunye and P.L.G. Vlek (eds) *Management of Nitrogen and Phosphorus Fertilizers in Sub-Saharan Africa*, pp. 95–115. Martinus Nijhoff: Dordrecht.

Donaldson, L.E. (1971) Investigations into the fertility of Brahman crossbred female cattle in Queensland. *Austr.Vet.J.*, **47**, 264–7.

Do Nascimento, C.N.B., Gui Maraes, J.M.A.B. and Gondim, A.G. (1970) Milk productivity factors in black water buffalo. *Ser.Estud.Bubalinos Int.Pesq.Exp.Agropeus Noyte.Belem Bras*, **1**(1). (Anim.Breed.Abstr., **40**, 1486).

Donnelly, E.D. and Anthony, W.B. (1969) Relationship of tannin, dry matter digestibility and crude protein in *Sericea lespedeza*. *Crop Sci.*, **9**, 361–3.

Doorenbos, J. and Kassam, A.H. (1979) *Yield Response to Water*. FAO: Rome.

Doorenbos, J. and Pruitt, W.O. (1975) *Crop Water Requirements*. Irrigation and Drainage Paper 24. FAO: Rome.

Dorm-Adzobu, C., Ampudu-Agyei, O. and Veit, P. (1993) Agroforestry by mobisquads in Ghana. *In*: N. Hudson and R.J. Cheatle (eds) *Working With Farmers for Better Land Husbandry*, pp. 165–9. Intermediate Technology Publications: London.

Dougall, H.W. and Bogdan, A.V. (1958) Browse plants of Kenya with special reference to those occurring in S. Baringo. *E.Afr.Agric.For.J.*, **23**, 236–45.

Doughty, L.R. (1953) The value of fertilizers in African agriculture: field experiments in E. Africa, 1947–1951. *E.Afr.Agric.For.J.*, **19**, 30–31.

Douglas, M. (1994) *Sustainable Use of Agricultural Soils: A Review of the Prerequisites for Success or Failure*. Development and Environment Reports No. 11. Institute of Geography, University of Berne: Berne.

Dowling, D.F. (1956) An experimental study of heat tolerance of cattle. *Austr.J.Agric.Res.*, **7**, 469–81.

Dowling, D.F. (1958) The significance of sweating in heat tolerance of cattle. *Austr.J.Agric.Res.*, **9**, 579–86.

Dowling, D.F. (1959) The medullation characteristic of the hair coat as a factor in heat tolerance of cattle. *Austr.J.Agric.Res.*, **10**, 736–43.

Dregne, H.E. and Chou, Nan-Ting (1992) Global desertification trends and costs. *In*: H.E. Dregne (ed.) *Degradation and Restoration of Arid Lands*. Texas Tech University: Lubbock, TX.

Dregne, H.E. and Willis, W.O. (eds) (1983) *Dryland Agriculture*. American Society of Agronomy No. 23. American Society of Agronomy: Madison, WI.

Drosdoff, M. (1972) Soil micronutrients. *In*: Committee on Tropical Soils, Agricultural Board, National Research Council (eds) *Soils of the Humid Tropics*, pp. 150–62. National Academy of Science: Washington, DC.

Druyan, L.M. (1989) Advances in the study of Sub-Saharan drought. *Int.J.of Climatology*, **9**, 77–90.

Duble, R.L., Lancaster, J.A. and Holt, E.C. (1971) Forage characteristics limiting animal performance on warm season perennial grass. *Agron.J.*, **63**, 795–804.

Dudal, R. and Moorman, F.R. (1962) *Characteristics of major soils of South-East Asia and considerations of their agricultural potential*. FAO, Rome.

Dufrene, E. and Saugier, B. (1993) Gas exchange of oil palm. *Functional Ecol.*, **7**, 97–104.

Dugdale, G., McDougall, V.D. and Milford, J.R. (1991) Rainfall estimates in the Sahel from cold cloud statistics: accuracy and limitations of operational systems. *In*: M.V.K. Sivakumar, J.S. Wallace, C. Renard and C. Giroux (eds) *Proceedings of Workshop on Soil Water Balance in the Sudano-Sahelian Zone*, 18–23 February, pp. 65–74. IAHS Publication 199. International Association of Hydrological Sciences: Wallingford.

Dunn, A.M. and Jennings, F.W. (1987) *Veterinary Parasitology*. Longman: London.

Dykes, T.A. and Nabors, M.W. (1986) Tissue culture in rice and its application in selecting for stress tolerance. *In*: *Rice Genetics*, pp. 799–810. International Rice Research Institute: Los Banos, Philippines.

Economist (1995) Reading the patterns: the evidence that greenhouse gases are changing the climate. *The Economist*, **335**(7908), 109–111.

Eden, M.J. (1985) Forest cultivation and derived savanna and grassland: comparative studies from southern Papua and south west Guyana. *In*: J.C. Tothill and J.C. Mott (eds) *Ecology and Management of the World's Savannas*, pp. 260–64. Australian Academy of Science: Canberra.

Eden, M.J. and Andrade, A. (1987) Ecological aspects of swidden cultivation among the Andoke and Witoto Indians of the Colombian Amazon. *Human Ecol.*, **15**(3), 339–59.

Eden, T. (1976) *Tea*, 3rd edn. Longman: London.

Edwards, C.E., Lal, R., Madden, P., Miller, R.H. and Haize, G. (1990) *Sustainable Agricultural Systems*. Soil & Water Conservation Society, St. Lucie Press: Delray Beach, FL.

Edwards, D.C. (1956) The ecological regions of Kenya. *Emp.J.Exp.Agric.*, **24**, 89–108.

Edwards, P.J. and Nel, S.P. (1973) Short-term effects of fertilizer and stocking rates on the Bankenveld. 1. Vegetation changes. *Proc.Grassl.Soc.S.Afr.*, **8**, 83–8.

Ellerman and Scott (1951) *Check List of Palaearctic and Indian Mammals: 1758–1946*. British Museum: London.

Elliott, R.C. (1964) *Some nutritional factors influencing the productivity of beef cattle in Southern Rhodesia*. Ph.D. thesis, University of London.

Ellis, P.R. (1972) *An Economic Evaluation of the Swine Fever Eradication Programme in GB*. Study No. 22. Department of Agriculture, University of Reading: Reading.

Ellis, P.R. and Hugh-Jones, M.E. (1976) Disease as a limiting factor to beef production in developing countries. *In*: A.J. Smith (ed.) *Beef Cattle Production in Developing Countries*, pp. 105–16. Centre for Tropical Veterinary Medicine, University of Edinburgh: Edinburgh.

Elwell, H.A. and Norton, A.J. (1988) *No-till Tied-ridging: A Recommended Sustained Crop Production System*. Institute of Agricultural Engineering: Harare.

Elwell, H.A. and Stocking, M.A. (1982) Developing a simple yet practical method of soil loss estimation. *Trop.Agric.(Trin.)*, **59**, 43–8.

Elwell, H.A. and Stocking, M.A. (1984) Estimating

soil life-span for conservation planning. *Trop. Agric.(Trin.)*, **61**, 148–50.

EMBRAPA (1984) Upland rice in Brazil. *In: An Overview of Upland Rice Research*, pp. 121–34. International Rice Research Institute: Los Banos, Philippines.

Emmans, G.C., Charles, D.R. and Dun, P. (1975) *Gleadthorpe Experimental Husbandry Farm Poultry Booklet*. Ministry of Agriculture, Fisheries and Food: London.

Empire Cotton Growing Corporation (1951–56) *Progress Reports from Experimental Stations; Namulonge, Uganda; Lake Province, Tanganyika*.

Enwezor, W.O. and Moore, A.W. (1966) Phosphorus status of some Nigerian soils. *Soil Sci.*, **102**, 322–8.

Epstein, H. and Herz, A. (1964) Fertility and birth-weight of goats in a sub-tropical environment. *J.Agric.Sci.*, **62**, 237–44.

Eriksson, E. (1952) Composition of atmospheric precipitation. 1. Nitrogen compounds. *Tellus*, **4**, 145–270.

Evanari, M. and Noy-Meir, I. (eds) (1986) *Ecosystems of the World. Vol. 12: Hot Deserts and Arid Shrublands*. Elsevier: Oxford.

Evans, J. (1992) *Plantation Forestry in the Tropics*, 2nd edn. Clarendon Press: Oxford.

Ewel, J.J. (1976) Litter fall and leaf decomposition in a tropical forest succession in eastern Guatemala. *J.Ecol.*, **64**(1), 293–308.

Ewell, J.J. (1991) Diversity: Yes we got some bananas. *Conservation Biology*, **5**, 423–5.

Ewell, J.J., Mazzarino, M.J. and Berish, C.W. (1991) Tropical soil fertility changes under monocultures and successional communities of different structure. *Ecological Applications*, **1**, 289–302.

Fahey, G.C., Al-Haydari, S.Y., Hinds, F.C. and Short, D.F. (1980) Phenolic compounds in roughages and their fate in the digestive system of sheep. *J.Anim. Sci.*, **50**, 1165–72.

Fakhri, A. and Fajer, E. (1992) Plant life in a CO_2 rich world. *Scientific American*, **266**(1), 18–24.

FAO (1968a) *FFHC Fertilizer Programme: Physical and Economic Summary of Trial and Demonstration Results*. Report LA. FFHC/68/14. FAO: Rome.

FAO (1968b) *Feed Composition Tables for Use in Africa*. FAO: Rome.

FAO (1972) *Production Yearbook 1971, Vol. 25*. FAO: Rome.

FAO (1974, revised 1978) *FAO-UNESCO Soil Map of the World. Vol. 1. Legend*. UNESCO: Paris.

FAO (1976) *Proc.Singapore Meeting on Expert Consultation of Buffalo Res.Needs*. FAO: Rome.

FAO (1979) *A Provisional Methodology for Soil Degradation Assessment* (with mapping of North Africa at a scale of 1:5 million). FAO: Rome.

FAO (1983) *Production Yearbook, Series No. 4*. FAO: Rome.

FAO (1984) *Production Yearbook 1983, Vol. 36*. FAO: Rome.

FAO (1985) *Guidelines: Land Evaluation for Irrigated Agriculture*. Soils Bull. No. 55. FAO: Rome.

FAO (1986a) *Breeding for Durable Disease and Pest Resistance*. FAO: Rome.

FAO (1986b) *Breeding for Durable Resistance in Perennial Crops*. FAO: Rome.

FAO (1986c) *Early Agrometeorological Crop Yield Assessment*. Plant Production and Protection Paper 73. FAO: Rome.

FAO (1986d) *Production Yearbook 1985, Vol. 39*. FAO: Rome.

FAO (1987) *African Agriculture: The Next 25 Years*. FAO: Rome.

FAO (1988) *FAO Fertiliser Yearbook 1987*. FAO: Rome.

FAO (1989a) *FAO Production Yearbook 1988*. FAO: Rome.

FAO (1989b) *Sustainable Agricultural Production: Implications for International Agricultural Research*. (Prepared by TAC.) FAO Research and Technical Paper No. 4. FAO: Rome.

FAO (1990) *FAO Fertiliser Yearbook 1989*. FAO: Rome.

FAO (1992) *Sustainable Development and the Environment. FAO Policies and Actions, Stockholm 1972–Rio 1992*. FAO: Rome.

FAO (1997) *FAOSTAT Database*. http://apps.fao.org/lim500/Agri_db.pl.

FAO/RAPA (1992) *Regional Strategies for Arresting Land Degradation*. FAO Regional Office for Asia and the Pacific: Bangkok.

FAO/UNDP/UNEP (1994) *Land Degradation in South Asia: Its Severity, Causes and Effects Upon the People*. World Soil Resources Report 78. FAO: Rome.

FAO/UNFPA/IIASA (1982) *Potential Population Supporting Capacities of Lands in the Developing World*. Tech.Rep. Project FPA/Int/513. FAO: Rome.

Farmer, G. (1989) *Rainfall*. IUCN Sahel Studies, pp. 1–25. International Union for the Conservation of Nature and Natural Resources: Nairobi.

Fassbender, H.W., Alpízar, L., Heuveldop, J., Fölster, H. and Enríquez, G. (1988) Modelling agroforestry systems of cacao (*Theobroma cacao*) with laurel (*Cordia alliodora*) and poro (*Erythrina poeppigiana*) in Costa Rica: III. Cycles of organic matter nutrients. *Agroforestry Systems*, **6**, 49–62.

Fassbender, H.W., Beer, J., Heuveldorp, J., Imbach, G., Enriquez, G. and Bonnemann, A. (1991) Ten years of organic matter and nutrients in agroforestry systems at CATIE, Costa Rica. *In*: P.G. Jarvis (ed.), *Agroforestry: Principles and Practice*, pp. 173–83. Elsevier: Amsterdam.

Fats and Oils Team, FAO (1979) *The World Coconut Situation and Outlook*. 5th Session, FAO Tech.

Working Party on Coconut Prod., Protection and Processing, Manila, Philippines. FAO: Rome.

Felix-Henningsen, L., Liu, L.W. and Zakosek, H. (1989) Pedogenesis of red earths (Acrisols) and yellow earths (Cambisols) in the central sub-tropical region of south-east China. In: E. Maltby and T. Wollersen (eds) *Soils and their Management: A Sino-European Perspective*, pp. 19–46. Elsevier Applied Science: London.

Felker, P., Cannell, G.H., Clark, P.R., Osborne, J.F. and Nash, P. (1983) Biomass production of species (mesquite), leucaena, and other leguminous trees grown under heat-drought stress. *For.Sci.*, **29**, 592–606.

Felker, P., Clark, P.R., Osborn, J.F. and Cannell, G.H. (1984) *Prosopis* pod production – comparison of North American, South American, Hawaiian, and African germplasm in young plantations. *Econ.Bot.*, **38**, 36–51.

Feller, C. (1979) Une methode de fractionment granulometrique de la matière organique des sols, application aux tropicaux, a texture grossières, tres pauvres en humus. *Cahiers Orstom Series Pedologiques*, **17**(4), 339–46.

Fernandes, E.C.M. and Nair, P.K.R. (1986) An evaluation of the structure and function of some tropical homegardens. *Agric.Systems*, **21**(4), 279–310.

Ferwerda, F.P. and Wit, F. (eds) (1969) *Outlines of Perennial Crop Breeding in the Tropics*. Veenman and Zonen: Wageningen.

Fillery, I.R.P., Simpson, J.R. and De Datta, S.K. (1984) Influence of field environment and fertilizer management on ammonia loss from flooded rice. *Soil Sci.Soc.of Amer.J.*, **48**, 914–20.

Finch, V.A., Bennett, I.L. and Holmes, C.R. (1982) Sweating response in cattle and its relation to rectal temperature, tolerance of sun and metabolic rate. *J.Agric.Sci.*, **99**, 479–87.

Finch, V.A., Dmi'el, R., Boxman, R., Shkolnik, A. and Taylor, C. (1980) Why black goats in hot desert? Effects of coat colour on heat exchanges of wild and domestic goats. *Physiol.Zool.*, **53**, 19–25.

Finch, V.A. and Western, D. (1977) Cattle colours in pastoral herds: natural selection or social preference? *Ecology*, **58**, 1384–92.

Flohn, H. (1987) Rainfall teleconnections in northern and eastern Africa. *Theoretical and Applied Climatology*, **38**, 191–7.

Floyd, C.N. (1991) *The Use of Crop Residues to Improve Soil Fertility in Sub-Saharan Africa*. Mimeo.Rep. Natural Resources Institute: Chatham, Kent.

Folland, C. (1987) Sea temperatures predict African drought. *New Scientist*, **116**(1580), 25.

Follett, R.F. and Stewart, B.A. (eds) (1985) *Soil Erosion and Crop Productivity*. American Society of Agronomy: Madison, WI.

Fones-Sondell, M. (1989) *Perspectives on Soil in Africa: Whose Problem?* Gatekeeper Series No. 14, Sustainable Agriculture Programme; International Institute for Environment and Development: London.

Forbes, J.C. and Watson, R.D. (1992) *Plants in Agriculture*. Cambridge University Press: Cambridge.

Fortmann, L. and Riddell, J. (1985) *Trees and Tenure: An Annotated Bibliography for Agroforesters and Others*. International Centre for Research in Agroforestry and the Land Tenure Centre: Wisconsin Nairobi.

Fox, R.L., Asghar, M. and Cable, W.J. (1983) Sulfate accretions in soils of the tropics. In: G.J. Blair and A.R. Till (eds) *Sulfur in South-East Asian and South Pacific Agriculture*. University of New England: Armidale, NSW.

Fox, R.L., Plucknett, D.L. and Whitney, A.S. (1968) Phosphate requirements of Hawaiian Latosols and residual effects of fertilizer phosphorus. *Trans.9th Int.Cong.Soil Sci.*, **2**, 301–10.

Francis, C.A. (ed.) (1986) *Multiple Cropping Systems*. Macmillan: New York, NY.

Frank, G.R., Aherne, F.X. and Jensen, A.H. (1983) A study of the relationship between performance and dietary component digestibilities by swine fed different levels of dietary fibre. *J.Anim.Sci.*, **57**, 645–54.

Frazer-Darling, F. and Farvar, M.A. (1973) Ecological consequences of sedentarization of nomads. In: T. Farvar and J.P. Milton (eds) *The Careless Technology*. Tom Stacey: London.

French, M.H. (1943) The composition and nutritive value of Tanganyika feeding stuffs. *E.Afr.Agric.For. J.*, **9**, 88–99.

French, M.H. (1950) The nutritive value of East African grass and fodder plants. *E.Afr.Agric.For.J.*, **15**, 214–23.

Fresco, L.O. and Kroonenberg, J.B. (1992) Time and spatial scales in ecological sustainability. In: *Land Use Policy*, pp. 155–68. Butterworth and Heinemann.

Frisch, J.E. (1972) Comparative drought resistance of *Bos indicus* and *Bos taurus* crossbred herds in Central Queensland. 1. Relative weights and weight changes of maiden heifers. *Austr.J.Exp.Agric.Anim. Husb.*, **12**, 231–5.

Furley, P.A. (1994) Tropical moist forest: transformation or conservation? In: N. Roberts (ed.) *The Changing Global Environment*, pp. 304–31. Basil Blackwell: Oxford.

Gadjah Mada University (1987) Tropical peat and peatlands for development. *Int.Peat Soc.Symp. 9–14. Conclusions and Recommendations*. Gaja Mada University and the Agency for Assessment and Application of Technology: Yogyakarta, Indonesia.

Gadzhiev, G. and Yusupov, Yu. (1971) Milk production in buffaloes. *Molochmyas Stotovod.Mosk.*, **16**(10), 26. (ABA **40**, 161)

Ganguli, N.C. (1981) Buffalo as a candidate for milk production. *Ind.Dairy Fed.Bull.*, **137**, 1–56.

Garnier, B.J. (1967) *Weather Conditions in Nigeria.* Climatological Research Series 2. Department of Geography, McGill University: Montreal.

Garrity, D.P. (1996) Tree–soil–crop interactions on slopes. *In*: C. Ong and P.A. Huxley (eds) *Tree-crop Interactions: A Physiological Approach*, pp 299–318. CAB International/International Centre for Research in Agroforestry: Wallingford/Nairobi.

Garza, T.R. (1973) Potencial anual del zacate Guinea fertilisado y bajo pastoreo rotocional en clima Am. *Téc.Pec.Méx.*, **21**, 26–32.

Gatenby, R.M. (1986) Exponential relation between sweat rate and skin temperature in hot climates. *J.Agric.Sci.*, **106**, 175–83.

Gholtz, H.L. (ed.) (1987) *Agroforestry: Realities, Possibilities and Potentials.* Martius Nijhoff and International Centre for Research in Agroforestry: Dordrecht, Netherlands.

Gibbon, D. and Harvey, J. (1977) *Subsistence Farming in the Dry Savanna of Western Sudan.* School of Development Studies Discussion Paper 18. University of East Anglia: Norwich.

Giller, K.E. and Wilson, K.J. (1991) *Nitrogen Fixation in Tropical Cropping Systems.* CAB International: Wallingford.

Giller, K.E., McDonagh, J.F. and Cadisch, G. (1994) Can biological nitrogen fixation sustain agriculture in the tropics? *In*: J.K. Syers and D.L. Risser (eds) *Soil Science and Sustainable Land Management in the Tropics*, pp. 173–91. University of Newcastle: Newcastle-upon-Tyne.

Gillman, H. (1992) *The New-world Screw-worm Eradication Programme, N. Africa, 1988–92.* FAO: Rome.

Gire, G. and De, R. (1979) Effect of preceding grain legumes on dry land pearl millet in N.W. India. *Exp.Agric.*, **15**, 169–72.

Glantz, M.H. (1987) Drought in Africa. *Scientific American*, **256**(6), 34–40.

Glantz, M.H., Katz, R.W. and Nicholls, N. (eds) (1991) *Teleconnections: Linking Worldwide Climate Anomalies.* Cambridge University Press: Cambridge.

Glover, J. and French, M.H. (1957) The apparent digestibility of crude protein in the ruminant: IV The effect of crude fibre. *J.Agric.Sci.*, **49**, 78–80.

Glover, N. and Beer, J. (1986) Nutrient cycling in two traditional Central American agroforestry systems. *Agroforestry Systems*, **4**, 77–87.

Göhl, B.C. (1982) *Tropical Feeds.* FAO: Rome.

Gokhale, S.B. (1974) *Inheritance of part lactations and their use in selection among Murrah buffaloes.* PhD thesis. Agra University: Agra, India.

Gokhale, S.B. and Nagarcenkar, R. (1979) Studies on part lactations and their use in prediction of total lactation in Murrah buffalo. *Wrld.Rev.Anim.Prod.*, **16**, 57–63.

Gokhale, S.B. and Nagarcenkar, R. (1981) Inheritance of part yields and their use in the selection of buffaloes. *Trop.Anim.Hlth.Prod.*, **13**, 41–7.

Goldstein, G. and Sarmiento, G. (1987) Water relations of trees and grasses and their consequences for the structure of savanna vegetation. *In*: B.H. Walker (ed.) *Determinants of Tropical Savannas*, pp. 13–38. IRL Press: Oxford.

Goldsworthy, P.R. and Heathcote, R. (1963) Fertilizer trials with groundnuts in Northern Nigeria. *Emp.J. Exp.Agric.*, **31**, 351–66.

Goldthorpe, C.C. (1994) An organisational analysis of plantation management. *The Planter*, **70**, 5–18.

Golley, F.B. (ed.) (1983) *Ecosystems of the World. Vol. 14: Tropical Rainforest Ecosystems.* Elsevier: Oxford.

Gong Zi-Tong (1985) Wetland soils in China. *In*: *Wetland Soils: Characterization, Classification and Utilization*, pp. 473–88. International Rice Research Institute: Los Banos, Philippines.

Good, R. (1970) *The Geography of Flowering Plants.* John Wiley & Sons: New York, NY.

Gotora, P. (1991) Adaptive no-till tied-ridging trials in small-scale farming areas of Zimbabwe. *Proc.12th Int.Conf., Int.Soil Tillage Res.Org.*, Soil Tillage and Agricultural Sustainability, 8–12 July 1991, Ibadan: Nigeria, pp. 410–16. Ohio State University: Columbus, OH.

Goudie, A. (ed.) (1994) *The Encyclopaedic Dictionary of Physical Geography.* Basil Blackwell: Oxford.

Gowe, R.S. and Fairfull, R.W. (1980) Some lessons from selection studies in poultry. *In*: *Proc.of Wrld.Cong. on Sheep and Beef Cattle Breeding*, Vol. 1, pp. 261–81. Dunmore Press: Palmerston North, New Zealand.

Gowing, J.W. and Young, M.D.B. (1996) Evaluation and promotion of rainwater harvesting in semi-arid areas. Final Technical Report. University of Newcastle: Newcastle-upon-Tyne.

Grainger, A. (1996) Forest environments. *In*: W. Adams, A.S. Goudie and A.R. Orme (eds) *The Physical Geography of Africa*, pp. 173–95. Oxford University Press: Oxford.

Grant, P.M. and Shaxson, T.F. (1970) The effect of ammonium sulphate fertilizer on the sulphur content of tea garden soils in Malawi. *Trop.Agric. (Trin.)*, **47**, 31–6.

Gray, B.S. and Siggs, A.J. (1994) Global perspective of the future of the plantation industry. *In*: Chee Kheng Hoy (ed.) *International Planters Conference on Management for Enhanced Profitability in Plantations, 24–26 October 1994*, pp. 1–20. Incorporated Society of Planters: Kuala Lumpur.

Grayson, B.T., Green, M.B. and Copping, C.G. (eds) (1990) *Pest Management in Rice*. Elsevier Science: Barking.

Greenland, D.J. (1958) Nitrate fluctuations in tropical soils. *J.Agric.Sci.*, **50**, 82–92.

Greenland, D.J. (1959) A lysimeter for nitrogen balance studies in tropical soils. *J.W.Afr.Sci.Assn.*, **5**, 79–89.

Greenland, D.J. (1965) Interaction between clays and organic compounds in soils. Part I. Mechanisms of interaction between clays and defined organic compounds. *Soils and Fertilizers*, **28**(5), 415–25. Part II. Adsorption of soil organic compounds and its effect on soil properties. *Soils and Fertilizers*, **28**(6), 521–32.

Greenland, D.J. (1984) Rice. *Biologist*, **31**, 209–25.

Greenland, D.J. and Nye, P.H. (1960) Does straw induce nitrogen deficiency in tropical soils? *Trans.7th Int.Cong.Soil Sci.*, **3**, 478–85.

Greenland, D.J. and Okigbo, B.N. (1983) Crop production under shifting cultivation and maintenance of soil fertility. *In: Potential Productivity of Field Crops under Different Environments*, pp. 505–24. International Rice Research Institute: Los Banos, Philippines.

Greenwood, M. and Hayfron, R.J. (1951) Iron and zinc deficiencies in cacao in the Gold Coast. *Emp.J.Exp.Agric.*, **19**, 73–86.

Gregory, P. (1989) Water use efficiency of crops in semi-arid tropics. *In: Soil, Crop and Water Management Systems for Rainfed Agriculture in the Sudano-Sahelian Zone*, Proceedings of an International Workshop, 11–16 January 1987, ICRISAT Sahelian Centre, Niamey, pp. 85–98. International Crops Research Institute for the Semi-Arid Tropics: Patancheru, India. Pradesh.

Gregory, P.J. (1965) *Rainfall over Sierra Leone*. Research Paper 2. Department of Geography, University of Liverpool: Liverpool.

Gregory, S. (1988) El Niño years and the spatial pattern of drought over India 1901–70. *In*: S. Gregoy (ed.) *Recent Climatic Change*, pp. 226–36. Belhaven Press: London.

Grewal, S.S. and Abrol, I.P. (1986) Agroforestry on alkali soils: effect of some management practices on initial growth, biomass accumulation and chemical composition of selected tree species. *Agroforestry Systems*, **4**, 221–32.

Griffiths, E. (1965) Micro-organisms and soil structure. *Biol.Rev.*, **40**, 129–42.

Griffiths, G.Ap. and Manning, H.L. (1949) A note on nitrate accumulation in a Uganda soil. *Trop.Agric. (Trin.)*, **26**, 108–10.

Griffiths, J.F. (1972) *Climates of the World. Vol. 10: Climates of Africa*. Elsevier: Amsterdam.

Grimes, R.C. and Clarke, R.T. (1962) Continuous arable cropping with the use of manure and fertilizers. *E.Afr.Agric.For.J.*, **28**, 78–80.

Grist, D.H. (1986) *Rice*, 6th edn. Longman: London.

Grof, B. and Harding, W.A.T. (1970) Dry matter yields and animal production of Guinea grass (*Panicum maximum*) on the humid tropical coast of N. Queensland. *Trop.Grassl.*, **4**, 85–95.

Gryseels, G. and Boodt, K.de (1986) Integration of crossbred cows on small-holder farms in the Debre Zeit area of the Ethiopian highlands. International Livestock Centre for Arid Regions: Addis Ababa.

Guha, M.M. and Watson, G.W. (1958) Effects of cover crops on soil nutrient status and growth of Hevea. 1. Laboratory studies on the mineralization of nitrogen in different soil mixtures. *J.Rub.Res.Inst. Malaysia*, **15**, 175–88.

Gunathileke, H.A.J. and Liyanage, M de S. (1996) Multiple cropping under coconuts. *In*: P.A. Huxley and H. Ranasinghe (eds) *Agroforestry For Sustainable Development in Sri Lanka*, Proceedings of a 3-Day Participatory Training Course, 9–11 September, 1994, Colombo, Sri Lanka, pp. 77–94. University of Wales/University of Sri Jayewardenapura: Bangor/Colombo.

Gupta, P.C. and O'Toole, J.C. (1986) *Upland Rice, a Global Perspective*. International Rice Research Institute: Los Banos, Philippines.

Gupta, S.C. and Larson, W.E. (1979) Estimating soil water retention characteristics from particle size distribution, organic matter percent and bulk density. *Water Resources Research*, **15**, 1633–5.

de Haan, C. (1993) Determinants of success in livestock development projects in developing countries: a review of World Bank experience. *In*: M. Gill, E. Owen, G.E. Pollott and T.L.J. Lawrence (eds) *Animal Production in Developing Countries*, pp. 129–34. Brit.Soc.Anim.Prod.Occ.Pub.16. British Society of Animal Production: Edinburgh.

Hadi, M.A. (1965) A preliminary study of certain productive and reproductive characters of Marathwada buffaloes of Maharashtra State. *Ind.Vet.J.*, **42**, 692–9.

Hafez, E.S.E. (1967) Bioclimatological aspects of animal productivity. *Wrld.Rev.Anim.Prod.*, **3**, 14–22.

Haggar, R.J. (1971) The production and management of *Stylosanthes gracilis* at Shika, Nigeria. 1. In sown pasture. *J.Agric.Sci.*, **77**, 427–36.

Hall, A.E., Singh, B.B. and Elders, J.D. (1997) Cowpea breeding. *Plant Breeding Reviews*, **15**, 215–74.

Hall, W.J. and De Boer, A.J. (1977) Increasing ruminant productivity in Asia. *Wrld.Rev.Anim.Prod.*, **13**(3), 9–16.

Hallé, F., Oldeman, R.A.A. and Tomlinson, P.B. (1978) *Tropical Trees and Forests: An Architectural Analysis*. Springer-Verlag: Berlin.

Hamilton, G.J. and Christie, J.M. (1971) *Forest Management Tables (Metric)*. Forestry Commission Booklet No. 34. HMSO: London.

Hancock, J. and Payne, W.J.A. (1955) The direct effect of tropical climate on the performance of

European-type cattle. 1. Growth. *Emp.J.Exp. Agric.*, **23**, 55–74.

Harbans Singh, A. (1962) *Handbook of Animal Husbandry for Extension Workers.* Ministry of Food and Agriculture: New Delhi.

Hardon, J.J., Rao, V. and Rajanaidu, N. (1985) A review of oil palm breeding. *In*: G.E. Russell (ed.) *Progress in Plant Breeding*, pp. 223–38. Butterworths: London.

Haresign, W. and Cole, D.J.A. (eds) (1988) *Recent Advances in Animal Nutrition.* Butterworths: London.

Hargrove, W.L. and Thomas, G.W. (1981) Effect of organic matter on exchangeable aluminium and plant growth in acid soils. *In*: *Chemistry in the Soil Environment.* Am.Soc.Agron. Special Pub.No. 40, pp. 155–66. American Society of Agronomy: Madison, WI.

Harris, D. (1980) Tropical savanna. *In*: D. Harris (ed.) *Human Ecology in Savanna*, Academic Press: London. pp. 3–27.

Harrison, M.N. (1970) Maize improvement in E. Africa. *In*: C.L.A. Leakey (ed.) *Crop Improvement in East Africa.* Tech.Comm.No.19. Commw.Bur. Plant Breed. and Genet. CAB: Farnham Royal.

Hartley, C.W.S. (1989) *The Oil Palm, Elaeis guineensis*, 3rd edn. Longman: London.

Hastenrath, S. (1985) *Climate and Circulation in the Tropics.* D. Riedel: Boston, MA.

Hayward, D.F. and Oguntoyinbo, J.S. (1987) *The Climatology of West Africa.* Hutchinson: London.

Heady, H.F. (1973) Ecological consequences of Bedouin settlement in Saudi Arabia. *In*: T. Farvar and J.P. Milton (eds) *The Careless Technology.* Tom Stacey: London.

Hedge, A.M. and Klingebiel, A.A. (1957) The use of soil maps. *In*: *Soil: The Yearbook of Agriculture, 1957.* United States Department of Agriculture: Washington, DC.

Heinz, D.J. (ed.) (1987) *Improvement of Sugarcane through Breeding.* Elsevier: Amsterdam.

Hellums, D.T. (1992) Role of nonconventional phosphate fertilizers in tropical agriculture: IFDC's research perspective. In: J.J. Schultz (ed.) *Phosphate fertilizers and the environment: Proceedings of an international workshop*, pp 89–98. International Fertilizer Development Center, Alabama, USA.

Henderson, C.R. (1973) Sire evaluation and genetic trends. *In*: *Animal Breeding and Genetics*, (Proc.of Symp. in honour of J.L. Lush), pp. 10–41. American Society of Animal Science and American Dairy Science Association: Champaign, IL.

Herklots, G.A.C. (1972) *Vegetables in South-East Asia.* Allen and Unwin: London.

Hess, P.G., Battisit, D.S. and Rasch, P.J. (1993) Maintenance of the ITCZ zones and the large scale tropical circulation on a water covered earth. *J.Atmospheric Sciences*, **50**, 691–713.

Heuveldop, J., Fassbender, H.W., Alpízar, L., Enríquez, G. and Fölster, H. (1988) Modelling agroforestry systems of cacao (*Theobroma cacao*) with laurel (*Cordia alliodora*) and poro (*Erythrina poeppigiana*) in Costa Rica. II. Cacao and wood production, litter production and decomposition. *Agroforestry Systems*, **6**, 37–48.

Hew, C.K. (1972) A comparison of coconut upkeep techniques on coastal clay soils. *In*: R.L. Wastu and D.A. Earp (eds) *Cocoa and Coconuts in Malaysia*, pp. 402–11. Incorporated Society of Planters: Kuala Lumpur.

Hocking, P.M. and McAllister, J. (1986) Factors affecting longevity in dairy cows. *Brit.Cattle Breeders' Club Digest*, **41**, 3–9.

Holden, J.H.W. and Williams, J.T. (eds) (1984) *Crop Genetic Resources: Conservation and Evaluation.* Allen & Unwin: London.

Holdridge, L.R. (1947) Determination of world plant formations from simple climatic data. *Science*, **105**, 367–8.

Holdridge, L.R. (1959) Determinations of world plant formations from simple climatic data. *Science*, **130**, 572.

Holdridge, L.R. (1971) *Forest Environments in Tropical Life Zones.* Pergamon Press: Oxford.

Holtkamp, R. (1990) *Small Four-wheel Tractors for the Tropics and Sub-tropics: Their Role in Agricultural and Industrial Development.* Deutsche Gesellschaft für Technische Zusammenarbeit (GTZ): Eschborn, Germany.

Honisch, O. (1974) Water conservation in three grain crops in the Zambesi Valley. *Exp.Agric.*, **10**, 1–8.

Hornby, H.E. and Van Rensburg, H.J. (1948) The place of goats in Tanganyika farming systems. 1. In deciduous bushland formation. *E.Afr.Agric.For.J.*, **14**, 94–8.

Horst, P. and Mathur, P.K. (1990) Genetic aspects of adaptation to heat stress. *Proc.4th Wrld.Cong.on Genetics Applied to Livestock Prod*, **14**, 286–7.

Hoogmoed, W.B. (1986) Analysis of rainfall characteristics relating to soil management from some selected locations in Niger and India. *Wageningen Agricultural University Report 86–3.* Wageningen Agricultural University: Wageningen.

Horsley, S.B. (1991) Allelopathy. *In*: M.E. Avery, M.G.R. Cannell and C. Ong (eds) *Biophysical Research for Asian Agroforestry*, pp. 167–83. Winrock and South Asia Books: Arlington, VA.

Hough, M., Palmer, S., Weir, A., Lee, M. and Barrie, I. (1996) The Meteorological Office rainfall and evaporation calculation system MORECS: version 2.0. *Met.Office Mem. 45.* Meteorological Office: Bracknell.

Houghton, J.T., Jenkins, G.J. and Ephraums, J.J. (1991) *Climate Change: The IPCC scientific assessment.* Cambridge University Press: Cambridge.

Houghton, T.R. (1960) The water buffalo in Trinidad. *J.Agric.Soc.Trin.and Tob.*, **60**, 339–56.

Howard, A. (1924) The effects of grass on trees. *Trans.Roy.Soc., Proc.B.*, **97**, 284–321.

Hsu, P.H. (1965) Fixation of phosphorus by aluminium and iron in acid soils. *Soil Sci.*, **99**, 398–402.

Hua, I.C. (1957) The use of Malayan swamp buffaloes and Murrah Malayan crossbreds for milk production in Taiping. *J.Malay Vet.Med.Assn.*, **1**, 141–3.

Hudson, N.W. (1971) *Soil Conservation*. Cornell University Press: Ithaca, NY.

Hudson, N.W. (1987) *Soil and Water Conservation in Semi-arid Areas*. Soils Bulletin 57. FAO: Rome.

Hudson, N.W. (1991) *A Study of the Reasons for Success or Failure of Soil Conservation Projects*. Soils Bulletin 64. FAO: Rome.

Hudson, N.W. (1992) *Land Husbandry*. Batsford: London.

Hudson, N.W. (1995) *Soil Conservation*, 3rd edn. Batsford: London.

Hudson, N.W. and Cheatle, R.J. (eds) (1993) *Working With Farmers for Better Land Husbandry*. Intermediate Technology Publications: London.

Hughes, C.E. (1994) Risks of species introduction in tropical forestry. *Commonw.For.Rev.*, **73**, 243–52.

Huke, R.E. and Huke, E.H. (1983) *Bangladesh: Rice Area by Season, Culture Type and Farm Size Distribution*. International Rice Research Institute: Los Banos, Philippines.

Hulme, M. (1996) *Climate Change and Southern Africa*. Climate Research Unit/World Wildlife Fund: Norwich.

Hulugalle, N.B. (1990) Alleviation of soil constraints to crop growth in the upland Alfisols and associated soil groups of the West African Sudan Savannah by tied ridges. *Soil Tillage Res.*, **18**, 231–47.

Humphreys, L.R. (1991) *Tropical Pasture Utilisation*. Cambridge University Press: Cambridge.

Huntley, B.J. and Walker, B.H. (eds) (1982) *Ecology of Tropical Savannas*. Springer Verlag: Berlin.

Hutchinson, H.G. (1964) *4th Annual Report 1963*. Livestock Research Division, Ministry of Agriculture, Tanganyika: Dar-es-Salaam. (*ABA* 33, 932).

Hutton, K. and Armstrong, D.G. (1976) Cereal processing. *In*: H. Swan and D. Lewis (eds) *Feed Energy Sources for Livestock*, pp. 47–63. Butterworths: London.

Huxley, P.A. (1983a) *Considerations When Experimenting with Changes in Plant Spacing*. ICRAF Working Paper No.15. International Centre for Research in Agroforestry: Nairobi.

Huxley, P.A. (1983b) Phenology of tropical woody perennials and seasonal crop plants with reference to their management in agroforestry systems. *In*: P.A. Huxley (ed.), *Plant Research and Agroforestry*, Proceedings of a consultative meeting held in Nairobi, 8–15 April 1981, pp. 501–25. International Centre for Research in Agroforestry: Nairobi.

Huxley, P.A. (ed.) (1983c) *Plant Research and Agroforestry*. Proceedings of a consultative meeting held in Nairobi, 8–15 April 1981. International Centre for Research in Agroforestry: Nairobi.

Huxley, P.A. (1983d) Some characteristics of trees to be considered in agroforestry. *In*: P.A. Huxley (ed.) *Plant Research and Agroforestry*. Proceedings of a consultative meeting held in Nairobi, 8–15 April 1981, pp. 3–12. International Centre for Research in Agroforestry: Nairobi.

Huxley, P.A. (1985) The basis of selection, management and evaluation of multipurpose trees: an overview. *In*: M.G.R. Cannell and J.E. Jackson (eds) *Attributes of Trees as Crop Plants*, pp. 13–35. Institute of Terrestrial Ecology: Abbots Ripton. (Expanded text in ICRAF Working Paper No.25.)

Huxley, P.A. (1986) The tree/crop interface, or simplifying the biological/environmental study of mixed cropping agroforestry systems. *Agroforestry Systems*, **3**, 252–66. (Previously ICRAF Working Paper No.13, 1983.)

Huxley, P.A. (1995) *An Agroforestry Sustainability Framework. Part 1: Introduction*. Draft document prepared for the Agroforestry Sub-Group. FAO: Rome.

Huxley, P.A. (1996) Biological factors affecting form and function in woody/non-woody plant mixtures. *In*: C.K. Ong and P.A. Huxley (eds) *Tree-Crop Interactions: A Physiological Approach*, pp. 235–98. CAB International/International Centre for Research in Agroforestry: Wallingford/Nairobi.

Huxley, P.A. (in press) Multipurpose trees: biological and ecological aspects relevant to their selection and use. *In*: F.T. Last (ed.) *Tree Crop Ecosystems*. Elsevier: Amsterdam.

Huxley, P.A., Darnhofer, T., Pinney, A., Akunda, E. and Gatama, D. (1989) The tree-crop interface: a project designed to generate experimental methodology. *Agroforestry Abstracts*, **2**, 127–45.

Huxley, P.A., and Greenland, D.J. (eds) (1989) Pest management in agroforestry systems: A record of discussion held at CAB International, Wallingford, UK, 28–29 July 1988. *Agroforestry Abstracts*, **2**, 37–46.

Huxley, P.A., Pinney, A., Akunda, E. and Muraya, P. (1994) A tree-crop interface orientation experiment with a *Grevillea robusta* hedgerow and maize. *Agroforestry Systems*, **26**, 23–45.

Huxley, P.A. and Van Eck, W.A. (1974) Seasonal changes in growth and development of some woody perennials near Kampala. *J.Ecol.*, **62**, 579–92.

ICIPE (1990) *Annual Report*. International Center for Insect Physiology and Ecology: Nairobi.

ICRAF (1989a) *Annual Report*. International Centre for Research in Agroforestry: Nairobi.

ICRAF (1989b) *Viewpoints and Issues on Agroforestry and Sustainability*. International Centre for Research in Agroforestry: Nairobi. (Compiled by P.A. Huxley, limited circulation.)

ICRAF (1996) *Annual Report*. International Centre for Research in Agroforestry: Nairobi.

ICRISAT (1984) *Agroclimatology of Sorghum and Millet in Semi-arid Tropics*, Proceedings of an International Symposium at ICRISAT, Hyderabad. International Crops Research Institute for the Semi-Arid Tropics: Andtora Pradesh, India.

ICRISAT (1988) *Annual Report: West African Programmes 1987*. International Centre for Research Institute for the Semi-Arid Tropics: Patancheru, India.

IDRC (1988) *Crop and Animal Production Systems Programme*. International Development Research Centre: Ottawa.

IFAD (1992) *Soil and Water Conservation in Sub-Saharan Africa. Towards Sustainable Production by the Rural Poor*. Report prepared by the Centre for Development Cooperation Services, Free University of Amsterdam. International Fund for Agricultural Development: Rome.

Igue, K. and Gallo, J.R. (1960) *Zinc Deficiency of Corn in São Paulo*. Bull.No.20. IBEC Res.Inst.: New York, NY.

IITA (1990) *Annual Report 1989–90*. International Institute of Tropical Agriculture: Ibadan, Nigeria.

Ikehashi, H. and Ponnamperuma, F.N. (1978) Varietal tolerance of rice for adverse soils. *In*: *Soils and Rice*, pp. 801–23. International Rice Research Institute: Los Banos, Philippines.

Ikombo, B.M. (1984) Effects of farmyard manure and fertilizers on maize in semi-arid areas of eastern Kenya. *E.Afr.Agric.For.J.*, (Special issue on dryland farming research in Kenya), **44**, 266–74.

Ingram, D.L. (1964) The effect of environmental temperature on heat loss and thermal insulation of the young pig. *Res.Vet.Sci.*, **5**, 357–64.

Inns, F. (1992) Field power. *In*: *Tools for Agriculture. A Guide to Appropriate Equipment for Smallholder Farmers*, pp. 9–18. Intermediate Technology Publications: London.

Institute of Biology (1990) *Exploited Plants*. Collected papers from *The Biologist*. Institute of Biology: London.

Institute of Soil Science Academia Sinica (1981) *Proc.of Symp.on Paddy Soils*. Springer Verlag: Berlin.

IFTV (1973) *Kamelmilch*. Merkblatt No.18. Institut Für Tropische Veterinärmidezin: Giessen.

International Bank for Reconstruction and Development (1961) *The Economic Development of Tanganyika*. John Hopkins Press: Baltimore, MD.

I'ons, J.H. (1967) The effects of fertilizers on subtropical *Hyparrhenia* veld in Swaziland. *Exp.Agric.*, **3**, 143–51.

IRRI (1963) *The Rice Blast Disease*. John Hopkins Press: Baltimore, MD.

IRRI (1974) *Annual Report for 1973*. International Rice Research Institute: Los Banos, Philippines.

IRRI (1978) *Soils and Rice*. International Rice Research Institute: Los Banos, Philippines.

IRRI (1979a) *Proceedings of the Rice Blast Workshop*. International Rice Research Institute: Los Banos, Philippines.

IRRI (1979b) *Rainfed Lowland Rice*. International Rice Research Institute: Los Banos, Philippines.

IRRI (1982a) *Drought Resistance in Crops, with Emphasis on Rice*. International Rice Research Institute: Los Banos, Philippines.

IRRI (1982b) *Proceedings Rice Tissue Culture Planning Conference*. International Rice Research Institute: Los Banos, Philippines.

IRRI (1983) *Weed Control in Rice*. International Rice Research Institute: Los Banos, Philippines.

IRRI (1984a) *An Overview of Upland Rice Research*. International Rice Research Institute: Los Banos, Philippines.

IRRI (1984b) *Organic Matter and Rice*. International Rice Research Institute: Los Banos, Philippines.

IRRI (1985) *Wetland Soils; Characterization, Classification and Utilization*. International Rice Research Institute: Los Banos, Philippines.

IRRI (1986a) *Progress in Rainfed Lowland Rice*. International Rice Research Institute: Los Banos, Philippines.

IRRI (1986b) *Progress in Upland Rice*. International Rice Research Institute: Los Banos, Philippines.

IRRI (1986c) *Small Farm Equipment for Developing Countries*. International Rice Research Institute: Los Banos, Philippines.

IRRI (1987a) *Azolla Utilization*. International Rice Research Institute: Los Banos, Philippines.

IRRI (1987b) *International Deepwater Rice Workshop*. International Rice Research Institute: Los Banos, Philippines.

IRRI (1988a) *Green Manure in Rice Farming*. International Rice Research Institute: Los Banos, Philippines.

IRRI (1988b) *Hybrid Rice*. International Rice Research Institute: Los Banos, Philippines.

IRRI (1989) *IRRI Towards 2000 and Beyond*. International Rice Research Institute: Los Banos, Philippines.

IRRI (1993) *Rice Research in a Time of Change*. International Rice Research Institute: Los Banos, Philippines.

Irvin, A.D., Cunningham, M.P. and Young, A.S. (eds) (1981) *Advances in the Control of Theileriosis*. Martinus Nijhoff: The Hague.

Irvine, F.R. (1969) *West African Agriculture. Vol. 2 Crops*. Oxford University Press: Oxford.

IUCN/UNEP/WWF (1991) *Caring for the Earth: A Strategy for Sustainable Living*. The World Conservation Union: Gland, Switzerland.

Ivory, D.A. (1990) Major characteristics, agronomic features, and nutritional value of shrubs and tree fodders. *In*: C. Devendra (ed.) *Shrub and Tree Fodders for Farm Animals*, Proceedings of a workshop in Denpasar, Indonesia, 24–29 July 1989, pp. 22–38. International Development Research Council: Ottawa.

Jacks, G.V. (1956) The influence of man on soil fertility. *Advances in Sci.*, **13**, 137–45.

Jacks, G.V. and Whyte, R.O. (1939) *The Rape of the Earth: A World Survey of Soil Erosion*. Faber & Faber: London.

Jackson, I.J. (1989) *Climate, Water and Agriculture in the Tropics*, 2nd edn. Longman Scientific and Technical: Harlow.

Jahnkw, H.A. (1974) *The Economics of Controlling Tsetse Fly and Cattle Trypanosomiosis in Africa with Special Reference to Uganda*. Economics Institute, University of Hohenheim: Hohenheim, Germany.

Jain, J.K. (1968) The climatic conditions of the arid region of Rajasthan, India. *In*: *Proceedings of Symposium on Arid Zones*, pp. 20–27. National Committee for Geography: Calcutta.

Jarvis, P.G. (ed.) (1991) *Agroforestry: Principles and Practice*. Proceeding of an International Conference, 23–25 July 1989, University of Edinburgh. Elsevier: Amsterdam.

Jayasuriya, M.C.N. (1993) Use of crop residues and agro-industrial by-products in ruminant production systems in developing countries. *In*: M. Gill, E. Owen, G.E. Pollott and T.L.J. Lawrence (eds) *Animal Production in Developing Countries*, pp. 47–56. Brit.Soc.Anim.Prod.Occ.Pub.16. British Society of Animal Production: Edinburgh.

Jeffreys, M.D.W. (1953) *Bos brachyceros* or dwarf cattle. *Vet.Rec.*, **65**, 393–6.

Jenkinson, D.Mc.E. and Nay, T. (1973) The sweat glands and hair follicles of Asian, African and South American cattle. *Austr.J.Biol.Sci.*, **26**, 259–75.

Jenkinson, D.S. (1988) Soil organic matter and its dynamics. *In*: A. Wild (ed.) *Russell's Soil Conditions and Plant Growth*, pp. 564–607. Longman: New York, NY.

Jenkinson, D.S. and Ayanaba, A. (1977) Decomposition of carbon-14 in labelled plant material under tropical conditions. *Soil Sci.Soc.of Amer.J.*, **41**(5), 912–15.

Jenkinson, D.S. and Rayner, J.H. (1977) The turnover of soil organic matter in some of the Rothamsted classical experiments. *Soil Sci.*, **123**, 298–305.

Jenness, R. (1980) Composition and characteristics of goat milk: Review 1968–1979. *J.Dairy Sci.*, **63**, 1605–30.

Jennings, D.L. and Hershey, G.H. (1985) Cassava breeding. *In*: G.E. Russell (ed.) *Progress in Plant Breeding*, pp. 117–164. Butterworth: London.

Jennings, P.R., Coffmann, W.R. and Kauffman, H.E. (1979) *Rice Improvement*. International Rice Research Institute: Manila.

Jensen, M.E. (ed.) (1974) *Consumptive Use of Water and Irrigation Water Requirements*. Report for American Society of Civil Engineers. American Society of Civil Engineers: New York, NY.

Jodha, N.S. (1979) *Intercropping in Traditional Farming Systems*. ICRISAT Progress Rep. No. 3. International Crops Research Institute for the Semi-Arid Tropics: Patancheru, India.

Johnson, H.D. (1985) Physiological responses and productivity in cattle. *In*: M.K. Yousef (ed.) *Stress Physiology in Livestock. Vol.II, Ungulates*. CRC Press: Boca Raton, FL.

Johnson, H.D. (ed.) (1987) *Bioclimatology and the Adaptation of Livestock*. World Animal Science, B5. Elsevier: Amsterdam.

Johnson, H.D., Li, R., Meador, N., Spencer, K.J. and Manalu, W. (1987) Influence of temperature, humidity, wind and water sprays above the thermoneutral zone on milk yields for lactating Holstein cattle. *J.Dairy Sci.*, **70**(Supplement 1), 122–9.

Jones, E. (1960) Contribution of rainwater to the nutrient economy of soil in Northern Nigeria. *Nature*, **188**, 432.

Jones, E., Nyamudeza, P. and Busangavanye, T. (1989) Rainfed cropping and water conservation and concentration on Vertisols in the SE lowveld of Zimbabwe. *In*: P. Ahn and C.E. Elliott (eds) *Vertisol Management in Africa*, pp. 132–42. International Board for Soil Research and Management: Bangkok.

Jones, H.G. (1992) *Plants and Microclimate*. Cambridge University Press: Cambridge.

Jones, M.J. (1975) Leaching of nitrate under maize at Samaru, Nigeria. *Trop.Agric.(Trin.)*, **52**, 1-10.

Jones, O.R., Allen, R.R. and Unger, P.W. (1990) Tillage systems and equipment for dryland farming. *Adv.Soil Sci.*, **13**, 89–130.

Jones, O.R. and Stewart, B.A. (1990) Basin tillage. *Soil Tillage Res.*, **18**, 249–65.

Jones, R.J. (1974) The relation of animal and pasture production to stocking rate on legume-based and nitrogen-fertilized subtropical pastures. *Proc.Austr. Soc.Anim.Prod.*, **10**, 340–43.

Jones, R.J. and Sandland, R.L. (1974) The relation between animal gain and stocking rate. Derivation of the relationship from the results of grazing trials. *J.Agric.Sci.*, **83**, 335–46.

Jones, R.M. (1975) Pasture establishment. *In*: Austr.Inst.Agric.Sci.Queensland Branch, *Refresher*

Course in the Management of Improved Pastures, pp. 68–80.
Jones, U.S., Katyal, J.C., Mamaril, C.P. and Park, C.S. (1982) Wetland rice nutrient deficiencies other than nitrogen. *In*: *Rice Research Strategies for the Future*. International Rice Research Institute: Los Banos, Philippines.
Jordan, A.M. (1986) *Trypanosomiasis Control and African Rural Development*. Longman: London.
Jordan, C., Caskey, W., Escalante, G., Herrera, R., Montagnini, F. and Uhl, C. (1983) Nitrogen dynamics during conversion of primary Amazonian rain forest to slash and burn agriculture. *Oikos*, **40**(1), 131–9.
Joubert, D.M. (1973) Goats in the animal agriculture of southern Africa. *Z.Tierzücht.Zücht.Biol.*, **90**, 245–62.
Joubert, D.M. (1975) The livestock situation in Southern Africa: A study in perspective. *S.Afr.J.Agr. Affairs*, **1**, 14–24.
Journet, M. (1970) Use of pelleted feeds for dairy cows. *Annals Zootech.*, **19**, 85–7.
Jugenheimer, R.W. (1976) *Corn Improvement, Seed Production and Uses*. John Wiley & Sons: New York., NY.
Juma, K.H., Gharib, F.H. and Eliya, J. (1971) A note on studies on heat tolerance in Anassi sheep. *Anim.Prod.*, **30**, 369–70.
Jung, H.G. and Fahey, G.C. (1983) Effects of phenolic monomers on rat performance and metabolism. *J.Nutr.*, **113**, 546–56.
Juo, A.S.R. and Fox, R.L. (1977) Phosphate sorption characteristics of some benchmark soils of West Africa. *Soil Sci.*, **124**, 370–76.
Kalkini, A.S. and Paragaonkar, D.R. (1969) Nagpuri buffalo. *Indian Dairyman*, **21**, 47–8.
Kamprath, E.J. (1967) Residual effects of large applications of phosphorus on high phosphorus-fixing soils. *Agron.J.*, **59**, 25–7.
Kamprath, E.J. (1970) Exchangeable aluminium as a criterion for liming leached mineral soils. *Soil Sci.Soc.of Amer.Proc.*, **34**, 252–4.
Kamprath, E.J. (1972) Soil acidity and liming. *In*: Committee on Tropical Soils, Agricultural Board, National Research Council (eds) *Soils of the Humid Tropics*, pp. 136–49. National Academy of Science: Washington, DC.
Kang, B.T., Wilson, G.F. and Sipkens, L. (1981) Alley cropping maize (*Zea mays*) and Leucaena (*Leucaena leucocephela*) in southern Nigeria. *Plant and Soil*, **63**(2), 165–79.
Kanitkar, N.V. (1944) *Dry Farming in India*. Sci.Mono.No. 15. Indian Council of Agricultural Research: New Delhi.
Kartha, K.P.R. (1959) *In*: J.C. Williamson and W.J.A. Payne (eds) *An Introduction to Animal Husbandry in the Tropics*, pp. 250–66. Longman: London.
Kass, D.C. and Drosdoff, M. (1970) *Sources of Nitrogen in Tropical Environments*. Agronomy Mimeograph No. 70. Department of Agronomy, University of Cornell: Cornell, NY.
von Kaufmann, R.R. and Fitzhugh, H.A. (1993) Technical constraints to ruminant livestock production in sub-Saharan Africa. *In*: M. Gill, E. Owen, G.E. Pollott and T.L.J. Lawrence (eds) *Animal Production in Developing Countries*, pp. 13–24. Brit.Soc.Anim.Prod.Occ.Pub.16. British Society of Animal Production: Edinburgh.
Kayombo, B. and Lal, R. (1994) Responses of tropical crops to soil compaction. *In*: B.D. Soane and C. van Ouwerkerk (eds) *Soil Compaction in Crop Production*, pp. 287–316. Elsevier: Amsterdam.
Kazem-Khatami (1970) *Camel Meat*. Ministry of Agriculture: Tehran.
Kechik, I.T. and Sykes, A.H. (1974) Effect of dietary ascorbic acid on the performance of laying hens under warm environmental conditions. *Brit.Poult. Sci.*, **15**, 449–57.
Keeping, G.S. (1951) A review of fodder grass investigations in Malaya. *Malay Agric.J.*, **34**, 65–75.
Kellner, O. (1905) *Die Ernahrung der Landwirtschaftlichen Nutztiere*. Paul Parey: Berlin.
Kennan, T.C.D. (1950) Preliminary report on a comparison of several heavy yielding perennial grasses for the production of silage. *Rhod.Agric.J.*, **47**, 531–43.
Kennedy, A.J., Lockwood, G., Mossu, G., Simmonds, N.W. and Tan, G.Y. (1987) Cocoa breeding; past, present and future. *Cocoa Growers' Bull.*, **38**, 5–22.
Kenya National Agricultural Research Station (1970) *Annual Report for 1970, Pt.2*. Govt.Printer: Nairobi.
Kerkhof, P. (1990) *Agroforestry in Africa: A Survey of Project Experience*. Panos Publications: London.
Kerr, J. and Sanghi, N.K. (1992) *Indigenous Soil and Water Conservation in India's Semi-Arid Tropics*. Gatekeeper Series No. 34, Sustainable Agriculture Programme. International Institute for Environment and Development: London.
Khush, G.S. (1980) Breeding rice for multiple disease and insect resistance. In: *Rice Improvement in China and Other Asian Countries*, pp. 219–38. International Rice Research Institute and CAAS: Los Banos, Philippines.
Kidson, J. (1977) African rainfall and its relations to upper air circulations. *Qrtly J. of Roy.Met.Soc.*, **103**, 441–56.
Kiepe, P. and Rao, M.R. (1994) Management of agroforestry for the conservation and utilisation of land and water resources. *Outlook on Agric.*, **23**, 17–25.
Kiepe, P. and Young, A. (1992) Soil conservation through agroforestry: experience from four years of demonstrations at Machakos, Kenya. *In*: H. Hurni

and K. Tato (eds) *Erosion, Conservation and Small-scale Farming*, pp. 303–12. Geographica Bernensia: Berne.

Kilkenny, J.B. and Stollard, R.J. (1971) *Calf Rearing Costs and Performance, 1970*. MLC Beef Improvement Service Records Rep. No. 25. Meat and Livestock Commission: Milton Keynes.

King, F.H. (1911) *Farmers of Forty Centuries*. Rodale Press: Emmaus, PA.

King, J.M. (1983) *Livestock Water Needs in Pastoral Africa in Relation to Climate and Forage*. ILCA Res.Rep.No.7. International Livestock Centre for Arid Regions: Addis Ababa.

King, K.F.S. (1987) The history of agroforestry. *In*: H.A. Steppler and P.K.R. Nair (eds) *Agroforestry: A Decade of Development*, pp. 3–21. International Centre for Research in Agroforestry: Nairobi.

King, K.F.S. and Chandler, T. (1978) *The Wasted Lands: The Programme of Work of ICRAF*. International Centre for Research in Agroforestry: Nairobi.

Kiome, R. and Stocking, M. (1993) *Soil and Water Conservation in Semi-Arid Kenya*. Research Bulletin No. 61. Natural Resources Institute and the Overseas Development Administration: Chatham and London.

Klingebiel, A.A. and Montgomery, P.H. (1961) *Land Capabiliity Classification*. Agric.Handbook No. 210. USDA Soil Conservation Service. United States Department of Agriculture: Washington, DC.

Knight, J. (1965) Some observations on the feeding habits of goats in the South Baringo district of Kenya. *E.Afr.Agric.For.J.*, **30**, 182–8.

Knoess, K.H. (1976) *Assignment Report on Animal Production in Middle Awash Valley*. FAO: Rome.

Kon, S.K (1972) *Milk and Milk Products in Human Nutrition*. FAO Nutrition Studies No. 10. FAO: Rome.

Konno, T. (1990) Buprofezin: A reliable IGR for the control of rice pests. *In*: B.T. Grayson, M.B. Green and L.G. Copping (eds) *Pest Management of Rice*, pp. 210–22. Elsevier Applied Science: London.

Koppen, W. (1931) *Die Klimate der Erde*. Berlin.

Kossila, L.V. (1984) Location and potential feed use. *In*: F. Sundstol and E. Owen (eds) *Straw and Other Fibrous By-products as Feed*. Elsevier: Amsterdam.

Kowal, J.M. and Kassam, A.H. (1978) *Agricultural Ecology of the Savanna: A Study of West Africa*. Clarendon Press: Oxford.

Kozlowski, T.T. (1968a) *Water Deficits and Plant Growth. Vol. 1: Development, Control and Measurement*. Academic Press: New York, NY.

Kozlowski, T.T. (1968b) *Water Deficits and Plant Growth. Vol. 2: Plant Water Consumption and Response*. Academic Press: New York, NY.

Kozlowski, T.T. (1972) *Water Deficits and Plant Growth. Vol. 3: Plant Responses and Control of Water Balance*. Academic Press: New York, NY.

Kozlowski, T.T. (1976) *Water Deficits and Plant Growth. Vol. 4: Soil Water Measurement*. Academic Press: New York, NY.

Kozlowski, T.T. (1978) *Water Deficits and Plant Growth. Vol. 5: Water and Plant Disease*. Academic Press: New York, NY.

Kozlowski, T.T. (1980) *Water Deficits and Plant Growth. Vol. 6: Woody Plant Communities*. Academic Press: New York, NY.

Kozlowski, T.T. (1983) *Water Deficits and Plant Growth. Vol. 7: Additional Woody Crop Plants*. Academic Press: New York, NY.

Kramer, P.J. (1983) *Water Relations in Plants*. Academic Press: Orlando, FL.

Krause, R., Lorenz, F. and Hoogmoed, W.B. (1984) *Soil Tillage in the Tropics and Subtropics*. Schriftenreihe de GTZ, No. 150. Deutsche Gesellschaft für Technische Zusammenarbeit (GTZ): Eschborn, Germany.

Kumar, Rao, J.V.D.K., Dart, P.J. and Sisky, P.V.S.S. (1983) Residual effect of pigeon pea on yield and nitrogen response of maize. *Exp.Agric.*, **19**, 131–4.

Kumar, R. and Singh, M. (1984) Tannins, their adverse role in ruminant nutrition. *J.Agric.Fd.Chem.*, **32**, 447–53.

Lal, R. (1985) A soil suitability guide for different tillage systems in the tropics. *Soil Tillage Res.*, **5**, 179–96.

Lal, R. (1990) Ridge-tillage. *Soil Tillage Res.*, **18**, 107–11.

Lal, R. (1991a) Current research on crop water balance and implications for the future. *In*: M.V.K. Sivakumar, J.S. Wallace, C. Renard and C. Giroux (eds) *Proceedings of Workshop on Soil Water Balance in the Sudano-Sahelian Zone*, 18–23 February, pp. 31–44. IAHS Publication 199. International Association of Hydrological Sciences: Wallingford.

Lal, R. (1991b) Mulch rate effects on maize growth and yield on an Alfisol in Western Nigeria. *Proc.12th Int.Conf.,Int.Soil Tillage Res.Organ.*, Soil Tillage and Agricultural Sustainability, 8–12 July 1991, Ibadan, Nigeria, pp. 612–25. Ohio State University: Columbus, OH.

Lal, R. and Okigbo, B. (1990) *Assessment of Soil Degradation in the Southern States of Nigeria*. Environment Working Paper No. 39. World Bank: Washington, DC.

Lal, R., Sanchez, P.A. and Cummings, R.W., Jr. (eds) (1986) *Land Clearing and Development in the Tropics*. A.A. Balkema: Rotterdam.

Lall, H.K. (1968) Cross-breeding for Indian-type Angora goats. *Ind.Fmg.*, **17**(12), 45–7.

Lancaster, W. and Lancaster, F. (1991) Limitation on sheep and goat herding in the Eastern Badia of Jordan: an ethno-archaeological enquiry. *LEVANT*, **XXIII**, 125–38.

Lancaster, W. and Lancaster, F. (1992) Tribal formations in the Arabian Peninsula. *Arabian Archaeology and Epigraphy*, **3**(3), 145–72.

Landon, J.R. (ed.) (1991) *Booker Tropical Soil Manual. A Handbook for Soil Survey and Agricultural Land Evaluation in the Tropics and Subtropics*. Longman Scientific and Technical: Harlow.

Lanly, J.P. (1984) Nature, extent and developmental problems associated with shifting cultivation; the overview. *In*: *Report of an Expert Consultation on Education Training and Extension for Shifting Cultivation in Developing Countries. December 1983*, pp. 15–23. FAO: Rome.

Lathwell, D.J. (ed.) (1979) *Crop Response to Liming of Ultisols and Oxisols*. Cornell Int.Agr.Bull. 35. Cornell University: Ithaca, NY.

Laudelout, H. (1989) Plant breeding and minimal input of Ca and Mg against Al toxicity. *In*: E. Maltby and T. Wollersen (eds) *Soils and their Management: A Sino-European Perspective*, pp. 313–20. Elsevier Applied Science: London.

Lawson, G. (1995) *Agroforestry Modelling Newsletter*, Issue 3, June 1995 and subsequent issues. Institute of Terrestrial Ecology, Bush Estate, Penicuik, Edinburgh.

Laycock, D.H. (1964) An empirical correlation between weather and yearly tea yields in Malawi. *Trop.Agric.(Trin.)*, **41**, 277–91.

Leakey, C.L.A. and Wills, J.B. (1977) *Food Crops of the Lowland Tropics*. Oxford University Press: Oxford.

Leakey, R.B., Mesen, F., Shiembo, P.N., Ofori, D., Nketiah, T., Hamzah, A., Tchoundjeu, Z., Njoya, C., Odoul, P., Newton, A.C., Dick, J. McP. and Longman, A. (1992) *Low Technology Propagation of Tropical Trees: Appropriate Technology for Rural Development*. Institute of Terrestrial Ecology: Edingurgh.

Lecky, T.P. (1962) The development of the Jamaican Hope as a tropical adapted dairy breed. *In*: *Proc.U.N.Conf.on the Application of Sci.Tech. for the Benefit of the Less Dev.Areas*. Agenda Item C5.2.

Ledig, F.T. (1983) The influence of genotype and environment on the dry matter distribution in plants. *In*: P.A. Huxley (ed.) *Plant Research and Agroforestry*, Proceedings of a consultature meeting held in Nairobi, 8–15 April 1981, pp. 427–54. International Centre for Research in Agroforestry: Nairobi.

Lee, H.A. and McHargue, J.S. (1928) The effect of manganese deficiency of the sugar cane plant and its relationship to Patiala blight of sugar cane. *Phytopath*, **18**, 775–86.

Lee, K.E. (1969) Some soils of the British Solomon Islands Protectorate. *Phil.Trans.Roy.Soc.B*, **255**, 211–57.

Leeds-Harrison, P.B. and Rickson, R.J. (1991) Field drainage in non-temperate climates. *In*: P. Smart and J.G. Herbertson (eds) *Drainage Design*, pp. 90–117. Blackie & Son: Glasgow.

Leeuw, P.N. and Tothill, J.C. (1990) The concept of rangeland carrying capacity in sub-saharan Africa – myth or reality? *Land Degradation and Rehabilitation Paper 29b*, pp. 1–16. Overseas Development Institute: London.

Le Houérou, H.N. (ed.) (1980) *Browse in Africa: the Current State of Knowledge*. International Livestock Centre for Africa: Addis Ababa, Ethiopia.

Le Mare, P.H. (1970) A review of soil research in Tanzania. Mimeo, Committee on Tropical Soils, Agricultural Board, National Research Council. National Academy of Science: Washington, DC.

Leng, R.A. (in press) *Trees – Their Role in Animal Nutrition in Developing Countries in the Humid Tropics*. FAO Animal Production and Health Paper. FAO: Rome.

Leupold, J. (1967) *Die wirtschaftliche Bedeutung des Dromedars*. Urban and Schwarzenberg: Munich.

Lewis, J.G. (1978) Game domestication for animal production in Kenya: shade behaviour and factors affecting the herding of eland, oryx, buffalo and zebu cattle. *J.Agric.Sci.*, **90**, 587–95.

Lindroth, A. (1993) Potential evaporation – a matter of definition. *Nordic Hydrology*, **24**(5), 359–64.

Lipton, M. and Longhurst, R. (1989) *New Seeds and Poor People*. Unwin Hyman: London.

Little, D.A., Anderson, F.M. and Durkin, J.W. (1991) Influence of partial suckling of crossbred dairy cows on milk offtake and calf growth in the Ethiopian highlands. *Trop.Anim.Hlth.Prod.*, **23**, 108–14.

Lock, G.W. (1969) *Sisal*, 2nd edn. Longman: London.

Lockwood, J.G. (1984) The southern oscillation and El Niño. *Prog.Phys. Geog.*, **8**, 102–10.

Lok, S.H. (1992) People's participation in conservation farming. *In*: S. Arsyad, I. Amien, T. Sheng and W. Moldenhauer (eds) *Conservation Policies for Sustainable Hillslope Farming*. Soil and Water Conservation Society: Ankeny, IA.

Loomis, R.S. and Connor, D.J. (1992) *Crop Ecology*. Cambridge University Press: Cambridge.

Louw, G. and Seeley, M. (1982) *Ecology of Desert Organisms*. Longman: London.

Loveless, A.R. (1983) *Principles of Plant Biology for the Tropics*. Longman: New York, NY.

Lowe, K.F., Filet, G.F., Burns, M.A. and Bowdler, T.M. (1977) Effect of pod-seeded *Siratro* on beef production and botanical composition of native pasture in SE Queensland. *Trop.Grassl.*, **11**, 223–9.

Lundgren, B. (1982) Introduction. *Agroforestry Systems*, **1**, 3–6.

Lundgren, L., Taylor, G. and Ingevall, A. (1993) *From Soil Conservation to Land Husbandry: Guidelines Based on SIDA's Experience*. Natural Resources Management Division, Swedish International Development Authority: Stockholm.

McCalla, A.F. (1994) *Agriculture and Food Needs to 2025: Why We Should Be Concerned*. Consultative Group on International Agricultural Research: Washington, DC.

McClearly, J.A. (1968) The biology of desert plants. *In*: G.W. Brown (ed.) *Desert Biology*, Vol. 1, pp. 141–94. Academic Press: London.

McClellan, G.H. and Northolt, A.J.G. (1986) Phosphate deposits of tropical Sub-Saharan Africa. *In*: A.Z. Mokwunye and P.L.G. Vlek (eds) *Management of Nitrogen and Phosphate Fertilizers in Sub-Saharan Africa*, pp. 173–223. Martinus Nijhoff: Dordrecht.

McClung, A.C., de Freitas, L.M.M., Mikkelsen, D.S. and Lott, W.L. (1961) *Cotton Fertilization on Campo Cerrado Soils, State of São Paulo, Brazil*. Bulletin No.27. IBEC Research Institute: New York, NY.

MacDicken, K.G. and Vergara, N.T. (eds) (1990) *Agroforestry: Classfication and Management*. John Wiley & Sons: New York, NY.

McDonald, P., Edwards, R.A. and Greenhalgh, J.F.D. (1988) *Animal Nutrition*, 4th edn. Longman: London.

McDowell, L.R. (1976) Mineral deficiencies and toxicities and their effect on beef production in developing countries. *In*: A.J. Smith (ed.) *Beef Cattle Production in Developing Countries*, pp. 216–41. Centre for Tropical Veterinary Medicine, University of Edinburgh: Edinburgh.

McDowell, L.R. (ed.) (1985) *Nutrition of Grazing Ruminants in Warm Climates*. Academic Press: Orlando, FL.

McDowell, R.E. (1981) Limitations for dairy production in developing countries. *J.Anim.Sci.*, **64**, 1463–75.

McDowell, R.E., Camoens, J.K., Van Vleck, L.D., Christensen, E. and Cabello Frias, E. (1976a) Factors affecting performance of Holsteins in subtropical regions of Mexico. *J.Dairy Sci.*, **59**, 722–9.

McDowell, R.E. and Hernandez-Urdaneta, A. (1975) Intensive systems for beef production in the tropics. *J.Anim.Sci.*, **41**(4), 1228–37.

McDowell, R.E., Hooven, N.W. and Camoens, J.K. (1976b) Effect of climate on performance of Holsteins in first lactation. *J.Dairy Sci.*, **59**, 965–73.

McDowell, R.E., Lee, D.H.K., Fohrman, M.H., Sykes, J.F. and Anderson, R.A. (1955) Rectal temperature and respiratory responses of Jersey and Sindhi-Jersey F_1 crossbred females to a standard hot atmosphere. *J.Dairy Sci.*, **38**, 1037–45.

McDowell, R.E., Wiggans, G.R., Camoens, J.K., Van Vleck, L.D. and St. Loius, D.G. (1976c) Sire comparisons for Holsteins in Mexico versus the United States and Canada. *J.Dairy Sci.*, **59**, 298–304.

MacFarlane, W.V. (1964) Terrestrial animals in dry heat. *In*: D.B. Dill (ed.) *Adaptation to the Environment*, pp. 509–33. American Physiological Society: Washington, DC.

McGinnies, W.G. (1968) Appraisal of research on vegetation of desert environments. *In*: W.G. McGinnies, B.J. Goldman and P. Paylore (eds) *Deserts of the World*, pp. 381–568. University of Arizona Press: Tucson, AZ.

McGinnies, W.G. (1979) Arid land ecosystems, common features throughout the world. *In*: D.W. Goodall, R.A. Perry and K.M.W. Howes (eds) *Arid land Ecosystems: Structure, Functioning and Management*. Cambridge University Press: Cambridge.

McGinnies, W.G. (1988) Climatic and biological conditions of arid lands: a comparison. *In*: E.E. Whitehead, C.F. Hutchinson, B.N. Timmermann and R.G. Vardy (eds) *Arid Lands Today and Tomorrow*, Westview Press: Boulder, CO.

McIlvain, E.H. and Shoop, M.C. (1971) Shade for improving cattle gains and rangeland use. *J.Range Mangt.*, **24**, 181–4.

McIlveen, R. (1992) *Fundamentals of Weather and Climate*. Chapman & Hall: London.

McKell, C.M. (ed.) (1989) *The Biology and Utilisation of Shrubs*. Academic Press: San Diego, CA.

Mackintosh, W.L.S. (1938) *Some Notes on the Abahirna and the Cattle Industry of Ankole*. Government Printer: Entebbe, Uganda.

McLean, J.A. (1963) The regional distribution of cutaneous moisture vaporisation in the Ayrshire calf. *J.Agric.Sci.*, **61**, 275–80.

McNaughton, S.J. (1987) Adaptation of herbivores to season changes in nutrient supply. *In*: J.B. Hacker and T.H. Ternouth (eds) *Nutrition of Herbivores*, pp. 391–408. Academic Press: Sydney.

MAFF/DAFS/DANI (1975) *Energy Allowances and Feeding Systems for Ruminants*. Tech.Bull.33. HMSO: London.

Mahadevan, P. (1970) Breeding methods. *In*: *Cattle Production in the Tropics. Vol.1, Breeds and Breeding*. Longman: London.

Maher, C. (1945) The goat: friend or foe? *E.Afr. Agric.For.J.*, **11**, 115–21.

Maiti, R.K. and Bidinger, F.R. (1981) Growth and development of the pearl millet plant. *ICRISAT Research Bulletin 6*. International Crops Research Institute for the Semi-Arid Tropics: Hyderabad.

Makkar, H.P.S. (1993) Anti-nutritional factors in foods for livestock. *In*: M. Gill, E. Owen, G.E. Pollott and T.L.J. Lawrence (eds) *Animal Production in Developing Countries*, pp. 69–85. Brit.Soc. Anim.Prod.Occ.Pub.16. British Society of Animal Production: Edinburgh.

Makkar, H.P.S., Singh, B. and Dawra, R.K. (1987) Tannin-nutrient interactions – a review. *Int.J.Anim. Sci.*, **2**, 127–40.

Makkar, H.P.S., Singh, B. and Negi, S.S. (1990) Tannin levels and their degree of polymerisation and

specific activity in some agro-industrial by-products. *Biol.Wastes*, **31**, 137–44.
Malaisse, F., Freson, R., Goffinet, G. and Malaisse-Mouset, M. (1975) Litterfall and litter breakdown in Miombo. *In*: F.B. Colley and E. Medina (eds) *Tropical Ecological Systems*, pp. 137–52. Springer-Verlag: New York, NY.
Malaysia (1967) *Land Capability Classification in West Malaysia – An Explanatory Handbook*. Economic Planning Unit, Department of the Prime Minister, Malaysia: Kuala Lumpur.
Maltby, E. (1989) Development, agricultural conversion and environmental investigations of wet soils and peat. *In*: E. Maltby and T. Wollersen (eds) *Soils and their Management: A Sino-European Perspective*, pp. 339–89. Elsevier Applied Science: London.
Malthus, T.R. (1798) *An Essay on the Principle of Population as it Affects the Future Improvement of Society with Remarks on the Speculation of Mr Goodwin, M. Concorcet and Other Writers*. J. Johnson in St. Paul's Churchyard: London.
Mannetje, L. 't and Nicholls, D.F. (1974) Beef production from pastures on granitic soils. *Ann.Rep. CSIRO Trop.Crops and Pastures Div., 1973–1974*. CSIRO: Canberra.
Marsh, T.D. and Dawson, V. (1948) Animal husbandry in Malaya. 2. The buffalo in Malaya. *Malay Agric.J.*, **31**, 102–14.
Martin, F.W. and Jones, A. (1986) Breeding sweet potatoes. *Plant Breed. Rev.*, **4**, 313–40.
Martin, G. and Fourrier, P. (1965) Trace elements in the cultivation of groundnuts in north Senegal. *Oleagineaux*, **20**, 287–91.
Martin, J.H., Leonard, W.H. and Stamp, D.L. (1976) *Principles of Field Crop Production*. Collier Macmillan: London.
Masefield, G.B. (1944) Some recent observations on the plaintain crop in Buganda. *E.Afr.Agric.For.J.*, **10**, 12–17.
Mason, I.L. (1975) Beefalo: much ado about nothing. *Wrld.Rev.Anim.Prod.*, **11**(4), 18–22.
Mason, I.L. (1976) Factors influencing the world distribution of beef cattle. *In*: A.J. Smith (ed.) *Beef Cattle Production in Developing Countries*, pp. 29–42. Centre for Tropical Veterinary Medicine, University of Edinburgh: Edinburgh.
Mason, I.L. (1981) Breeds. *In*: C. Gall (ed.) *Goat Production*, pp. 57–110. Academic Press: London.
Mason, I.L. (ed.) (1984) *Evolution of Domesticated Animals*. Longman: London.
Mason, I.L. and Maule, J.P. (1960) *The Indigenous Livestock of Eastern and South Africa*. Tech.Comm.No.14. CAB: Farnham Royal.
Mason, S.J. (1995) Sea surface temperatures and South African rainfall associations. *Int.J. Climatology*, **15**, 119–35.
Matharu, B.S. (1966) Camel care. *Ind.Fmg.*, **16**, 19–22.

Matthews, G.A. (1984) Herbicide application technology. *In*: T.J. Davis (ed.) *Proc. 4th Agriculture Sector Symposium*, 9–13 January 1984, Washington, DC, pp. 200–17. World Bank: Washington, DC.
Matthews, G.A. (1992) *Pesticide Application Methods*, 2nd edn. Longman: London.
Matthews, J. (1990) Mechanisation for the small farmer – lessons learnt and the way ahead. *In*: A. Speedy (ed.) *Developing World Agriculture*, pp. 140–48. Grosvenor Press International: London.
Matthews, R.B. and Stephens, W. (1998a) The role of photoperiod in determining seasonal yield variations in tea (*Camellia sinensis* L.). *Exp. Agric*. (in press).
Matthews, R.B. and Stephens, W. (1998b) CUPPA-Tea: A simulation model describing seasonal yield variation and potential production of tea (*Camellia sinensis* L.). I. Potential production. *Exp. Agric*. (in press).
Maule, J.P. (1953–54) Livestock breeds. *Brit.Agric. Bull. No.6*, 244–52.
Maule, J.P. (1990) *Cattle of the Tropics*. Centre for Tropical Veterinary Medicine, University of Edinburgh: Edinburgh.
Maust, L.E., McDowell, R.E. and Hooven, N.W. (1972) Effect of summer weather on performance of Holstein cows in three stages of lactation. *J.Dairy Sci.*, **55**, 1133–9.
von Maydell, H.-J. (1986) *Trees and Shrubs of the Sahel: Their Characteristics and Uses*. German Technical Co-operation (GTZ): Eschborn.
Meadows, M.E. (1996) Biogeography. *In*: W. Adams, A.S. Goudie and A.R. Orme (eds) *The Physical Geography of Africa*, pp. 161–72. Oxford University Press: Oxford.
Medina-Filho, H.P., Carvalho, A., Sondahl, M.R., Fazuoli, L.C. and Costa, W.M. (1984) Coffee breeding. *Plant Breed.Rev.*, **2**, 157–94.
van de Meerburg, R. (1992) The world trade in tea. *In*: K.C. Wilson and M.C. Clifford (eds) *Tea: Cultivation to Consumption*, pp. 649–86. Chapman & Hall: London.
Meganck, R.A. and Goebel, M. (1979) Shifting cultivation problems for parks in Latin America. *Parks*, **4**(2), 4–8.
Mehansho, H., Butler, L.G. and Carlson, D.M. (1987) Dietary tannins and salivary proline-rich proteins: interactions, induction and defense mechanisms. *Am.Rev.Nutr.*, **7**, 423–40.
Meiklejohn, J. (1953) The microbial aspects of soil nitrification. *E.Afr.Agric.For.J.*, **19**, 54–6.
Meiklejohn, J. (1955) Nitrogen problems in tropical soils. *Soils and Fertilizers*, **18**, 459–63.
Metcalfe, D.S. and Elkins, D.M. (1980) *Crop Production Principles and Practices*. Collier Macmillan: London.
Metianu, A. (1992) Soil preparation. *In*: *Tools for*

Agriculture. *A Guide to Appropriate Equipment for Smallholder Farmers*, pp. 23–8. Intermediate Technology Publications: London.

Miller, I.L. and Nobbs, R.C. (1976) Early wet season fertilisation of Para grass for use as saved fodder in the Northern Territory, Australia. *Trop. Agric.(Trin.)*, **53**, 217–24.

Miller, R. (1993) The most profitable cow – does Index indentify her? *Brit.Cattle Breeders' Club Digest*, **48**, 45–51.

Miller, S.F. (1982) Economics of weed control in the tropics and sub-tropics. *In: Proc.Brit.Crop Protection Conf., Weeds* 22–25 November 1982, Brighton, Vol. 2, pp. 679–87. British Crop Protection Council: Croydon.

Millet, M.A., Baker, A.J. and Satler, L.D. (1974) Pretreatment to enhance chemical, enzymatic and microbiological attack of cellulosic materials. *In*: C.R. Wilk (ed.) *Cellulose as a Chemical Energy Source*, pp. 193–221. Symp.No.5 Biotech.and Bioeng.Inter-Sci.Publ. John Wiley & Sons: New York, NY.

Mills, W.R. (1953) Nitrate accumulation in Uganda soils. *E.Afr.Agric.For.J.*, **19**, 53–4.

Milne, A.H. (1955) The humps of East African cattle. *Emp.J.Exp.Agric.*, **23**, 234–9.

Milne, G. (1936) A provisional soil map of East Africa. *Amani Memoirs*, **4**. Crown Agents for the Colonies: London.

Mitchell, G., Smith, C., Makower, M. and Bird, P.J.W.N. (1982) An economic appraisal of pig improvement. 1. Genetic and production aspects. *Anim.Prod.*, **35**, 215–24.

Mokwunye, A.Z., Chien, S.H. and Rhodes, E.R. (1986) Reactions of phosphate with tropical African soils. *In*: A.Z. Mokwunya and P.L.G. Vlek (eds) *Management of Nitrogen and Phosphorus Fertilizers in Sub-Saharan Africa*, pp. 253–81. Martinus Nijhoff: Dordrecht.

Monageng, K., Persaud, N., Carter, D.C., Gakale, L., Malapong, K., Heinrich, G., Siebert, J. and Mokete, N. (1990) Tillage and organic matter management for dryland farming in Botswana. *In*: E. Pushparajah and M. Latham (eds) *Organic Matter Management and Tillage in Humid and Subhumid Africa*, pp. 50–74. International Board for Soil Research and Management: Bangkok.

Monteith, J.L. (1986) How do plants manipulate the supply and demand of water? *Phil.Trans.Roy.Met.Soc. (London)*, **A316**, 245–59.

Monteith, J.L. (1991) Weather and water in the Sudano-Sahelian zone. *In*: M.V.K. Sivakumar, J.S. Wallace, C. Renard and C. Giroux (eds) Proceedings of Workshop on *Soil Water Balance in the Sudano-Sahelian Zone*, 18–23 February, pp. 11–29. IAHS Publication 199. International Association of Hydrological Sciences: Wallingford.

Monteith, J.L. and Unsworth, M.H. (1990) *Principles of Environmental Physics*. Edward Arnold: London.

Moody, K. (1974) Weeds and shifting cultivation. *FAO Soils Bull.No.24* pp. 155–66. FAO: Rome.

Moore, A.W. (1966) Non-symbiotic nitrogen fixation in soil and plant systems. *Soils and Fertilizers*, **29**, 113–28.

Moore, A.W. (1962) The influence of a legume on soil fertility under a grazed tropical pasture. *Emp.J.Expl. Agric.* **30**, 239–48.

Moorhouse, W. (1975) *Farm Management Accounting Service*. Rep.No.15. Economics Service Branch, Queensland Department of Agriculture: Queensland.

Moorman, F.R. and van Breemen, N. (1978) *Rice: Soil, Water, Land*. International Rice Research Institute: Los Banos, Philippines.

Morris, J. (1983) Smallholder mechanisation: man, animal or engine? *Outlook on Agric.*, **12**, 28–33.

Morris, T.R. (1975) Personal communication quoted by Sykes, A.H. (1977) Nutrition-environment interactions in poultry. *In*: W. Haresign, H. Swan and D. Lewis (eds) *Nutrition and the Climatic Environment*. pp. 17–29. Butterworths: London.

Motooka, P.S., Plucknett, D.L., Saiki, D.F. and Younge, O.R. (1967) *Pasture Establishment on Tropical Bushlands by Aerial Herbicide and Seeding Treatments in Kuai*. Hawaii Agric.Exp.Sta.Tch.Prog.Rep. No.165. Hawaii Agricultural Experimental Station: Hawaii.

Mueller-Harvey, L. Juo, A.S.R. and Wild, A. (1985) Soil organic C, N, S and P after forest clearance in Nigeria: Mineralization rates and spatial variability. *J.Soil Sci.*, **36**(4), 585–91.

Muinga, R.W., Thorpe, W. and Topps, J.H. (1992) Voluntary food intake, live-weight change and lactation performance of crossbred dairy cows given *ad libitum Pennisetum purpureum* (Napier grass var. Bana) supplemented with leucaenia forage in the lowland semi-humid tropics. *Anim.Prod.*, **55**(3), 331–7.

Muller, L. (1958) Advances in coffee production technology: mineral nutrition, detection and control of minor element deficiencies. *Coffee and Tea Industries*, **81**, 71–7.

Muller, Z.O. (1980) *Feed from Animal Wastes*. State of Knowledge. FAO: Rome.

Mullick, D.N. (1960) Effect of humidity and exposure to sun on the pulse rate, respiration rate, rectal temperature and haemoglobin level in different sexes of cattle and buffalo. *J.Agric.Sci.*, **54**, 391–4.

Mullins, C.E., Macleod, D.A., Northcote, K.H., Tisdell, J.M. and Young, I.M. (1990) Hard setting soils: behaviour, occurrence and management. *Adv. Soil Sci.*, **11**, 37–108.

Munro, J.M. (1987) *Cotton*, 2nd edn. Longman: London.

Munzinger, P. (ed.) (1982) *Animal Traction in Africa*. Schriftenreihe der GTZ, Deutsche Gesellschaft für Technische Zusammenarbeit (GTZ): No. 120. Eschborn, Germany.

Murdiati, T.B. and Mahyudin, P. (1985) The residual tannin and crude protein of *Calliandra callothyrus* and *Albizia falcataria* following incubation in heated and unheated rumen fluid. *Proc. 3rd Australasian Assn. for Anim.Prod.Anim.Sci.Cono*, Seoul, Korea, pp. 814–16.

Murray, M. (1975) *In*: G.M. Urquhart and J. Armour (eds) *Helminth Diseases of Cattle, Sheep and Horses in Europe*, pp. 92–6. University Press: Glasgow.

Murthy, R.S. and Pandey, S. (1981) Shifting cultivation in North-eastern India. *In*: T. Tingsanchali and H. Eggers (eds) *South-East Asian Regional Symposium on Problems of Soil Erosion and Sedimentation*. Asian Institute of Technology: Bangkok.

Myers, N. (1989) *Deforestation Rates in Tropical Forest and their Climatic Implications*. Friends of the Earth: London.

Nagarcenkar, R. (1974) Buffalo management. *Ind.Dairyman*, **26**, 305–306.

Nagarcenkar, R. (1975) Buffalo as a dairy animal. *Proc.All Ind.Conf.of Anim.Sci.and Livestock Breeders*. Punjab Agricultural University: Ludhiana, India.

Nagy, S. and Shaw, P.E. (1980) *Tropical and Subtropical Fruits*. Avi Press: Westport, CT.

Nair, P.E.R. (1990) *The Prospects and Promise of Agroforestry in the Tropics*. World Bank: Washington, DC.

Nair, P.K.R. (ed.) (1989) *Agroforestry Systems in the Tropics*. Kluwer: Dordrecht.

Nair, P.K.R. (1993) *An Introduction to Agroforestry*. Kluwer: Dordrecht.

Nakano, K. (1978) An ecological study of a village in northern Thailand. *Southeast Asian Studies*, **16**, 411–46.

Nakano, K. (1980) An ecological view of a subsistence economy based mainly on the production of rice in swiddens and in irrigated fields in a hilly region of northern Thailand. *Southeast Asian Studies*, **18**, 39–67.

Nakayama, H., Hidaka, R. and Ashizawa, F. (1991) Effects of testosterone injection on the semen quality of boars during high ambient temperature. *Anim.Reprod.Sci*., **25**, 73–82.

National Academy of Science (1975) *World Food and Nutrition Study: Interim Report*. National Academy of Science: Washington, DC.

National Research Council (eds) (1976) *Underexploited Tropical Plants with Promising Economic Value*. National Research Council: Washington, DC.

National Research Council (1978) *Nutrient Requirements of Domestic Animals. No.3. Dairy Cattle*. National Academy of Science: Washington, DC.

National Research Council (eds) (1979) *Tropical Legumes. Resources for the Future*. National Research Council: Washington, DC.

National Research Council (eds) (1983) *Amaranth: Modern Prospects for an Ancient Crop*. National Academy Press: Washington, DC.

National Research Council (eds) (1989) *Lost Crops of the Incas*. National Academy Press: Washington, DC.

National Research Council (1993) *Vetiver Grass: A Thin Green Line Against Erosion*. Board on Science and Technology for International Development, National Research Council. National Academy Press: Washington, DC.

Nawito, M.F., Shalash, M.R., Hoppe, R. and Rakha, A.M. (1967) Reproduction in the female camel. *Bull.Anim.Sci.Res.Inst.(Cairo) No.2*., Cairo.

Nene, Y.L., Hall, S.D. and Sheila, V.K. (1990) *The Pigeon Pea*. CAB International: Wallingford.

Nepstad, D.C., de Carvalho, C.R., Davidson, E.A., Jupp, P.H., Lefebure, P.A., Negreiros, G.H., da Silva, E.D., Store, T.A., Trumbore, S.E. and Vielra, S. (1994) The role of deep roots in the hydrological and carbon cycles of Amazonian forests and pastures. *Nature*, **372**, 666–9.

Nestel, B.L. and Creek, M.J. (1964) Animal production studies in Jamaica. V Liveweight production from Pangola pastures used in rearing and fattening beef cattle. *J.Agric.Sci*., **62**, 151–5.

Neue, H.U. and Scharpenseel, H.W. (1984) Gaseous products of the decomposition of organic matter in submerged soils. *In: Organic Matter and Rice*, pp. 311–28. International Rice Research Institute: Los Banos, Philippines.

Newton, J.D. and Said, A. (1957) Molybdenum deficiency in Latosols of Java. *Nature*, **180**, 1485–6.

Niamir, M. (1990) *Community Forestry – Herders Decision Making in Natural Resources Management in Arid and Semi-arid Africa*. FAO: Rome.

Niamir, M. (1991) Traditional African range management techniques: implications for rangeland management. *Pastoral Dev.Network Paper 31d*, pp. 1–9. Overseas Development Institute: London.

Nicholas, F.W. (1987) *Veterinary Genetics*. Clarendon Press: Oxford.

Nicou, R. and Chopart, J.L. (1979) Root growth and development in sandy and sandy-clay soils of Senegal. *In*: R. Lal and D.J. Greenland (eds) *Soil Physical Properties and Crop Production in the Tropics*, pp. 375–84. John Wiley & Sons: New York, NY.

Nieuwolt, S. (1977) *Tropical Climatology in Low Latitudes*. John Wiley & Sons: London.

Nir, D. (1974) *The Semi Arid World*. Longman: London.

Nolle, J. (1986) *Machines Modernes à Traction Animal*. Editions L'Harmattan: Paris.

Noorani, H. (1976) Cave paintings of Madhya Pradesh. *The Times of India*, 31 October 1976.
van Noordwijk, M., Lawson, G., Soumare, A., Groot, J.J.R. and Hairiah, K. (1996) Root distribution of trees and crops: competition or complementarity. In: C.K. Ong and P. Huxley (eds) *Tree-crop Interactions: A Physiological Approach*, pp. 319–64. CAB International/International Centre for Research in Agroforestry: Wallingford/Nairobi.
Norman, M.J.T. (1974) Beef production from tropical pastures. *Austr.Meat.Res.Com.Rev.*, **16**, 1–23 and **17**, 1–18.
Norman, M.J.T., Pearson, C.J. and Searle, P.G.E. (1984) *The Ecology of Tropical Food Crops*. Cambridge University Press: Cambridge.
Norton, A.J. (1989) Tillage and tillage implements in relation to water harvesting on Vertisols in semi-arid Zimbabwe. In: P. Ahn and C.E. Elliott (eds) *Vertisol Management in Africa*, pp. 143–56. Proc. No. 9. International Board for Soil Research and Management: Bangkok.
Nyamudeza, P., Mandiringana, O.T., Busangavanye, T. and Jones, E. (1991) The development of sustainable management or Vertisols in the lowveld of south east Zimbabwe. *IBSRAM Newsletter*, **20**, 6–9.
Nye, P.H. and Bertheux, M.N. (1957) The distribution of phosphorus in forest and savanna soils of the Gold Coast. *J.Agric.Sci.*, **49**, 145–59.
Nye, P.H. and Greenland, D.J. (1960) *The Soil under Shifting Cultivation*. Tech.Comm.No.51. Commonwealth Bureau of Soils: Harpenden.
Nye, P.H. and Stephens, D. (1962) Soil fertility. In: J.B. Wills (ed.) *Agriculture and Land Use in Ghana*, pp. 127–43. Oxford University Press: Oxford.
Ojeniyi, S.O. (1991) Comparison of row-tillage with no-tillage and manual methods: Effect on soil properties and cowpea. *Proc.12th Int.Conf.*, *Int.Soil Tillage Res.Organ*, Soil Tillage and Agricultural Sustainability, 8–12 July 1991, Ibadan, Nigeria, pp. 141–6. Ohio State University: Columbus, OH.
Ojha, T.P. (1987) Draft animals vis-à-vis tractor power on Indian farms. In: N.S.L. Srivastava and T.P. Ojha (eds) *Utilisation and Economics of Draft Animal Power*, pp. 244–59. Central Institute of Agricultural Engineering: Bhopal, India.
Okali, C., Sumberg, J. and Farrington, J. (1994) *Farmer Participatory Research: Rhetoric and Reality*. Intermediate Technology Publications: London.
Oke, T.R. (1978) *Boundary Layer Climates*. Methuen: London.
Okigbo, B.N. (1989) *Development of Sustainable Agricultural Production Systems in Africa*. International Institute of Tropical Agriculture: Ibadan, Nigeria.
Okigbo, B.N. and Greenland, D.J. (1976) Intercropping systems in tropical Africa. In: *Multiple Cropping*, pp. 63–101. American Society of Agronomy: Madison, WI.
Oldeman, L.R. (1994) The global extent of soil degradation. In: D.J. Greenland and I. Szabolics (eds) *Soil Resilience and Sustainable Land Use*, pp. 99–118. CAB International: Wallingford.
Oldeman, L.R., Hakkeling, R.T.A. and Sombroek, W.G. (1990) *World Map of the Status of Human-Induced Soil Degradation*. Revised Explanatory Note and Mapping at 1:10 million. International Soil Reference and Information Centre: Wageningen and UNEP: Nairobi.
Oloufa, M.M. (1968) Some aspects of reproductive efficiency in Egyptian cattle and buffaloes. *Egypt.Vet.Med.J.*, **15**, 173–85.
Ollerenshaw, C.B. (1971) Forecasting liver-fluke disease in England and Wales 1958–1968 with a comment on the influence of climate on the incidence of disease in some other countries. *Vet.Med.Rev.*, **18**, 289–312.
Olstead, J. (1993) Global warming in the dock. *Geog. Mag.*, **LXV** (9), 12–17.
Ong, C. (1994) Alley cropping – ecological pie in the sky? *Agroforestry Today*, **6**, 8–10.
Ong, C. (1996) A framework for quantifying the various effects of tree-crop interactions. In: C. Ong and P.A. Huxley (eds) *Tree-crop Interactions: A Physiological Approach*, pp. 1–23. CAB International/International Centre for Research in Agroforestry: Wallingford/Nairobi.
Ong, C. and Huxley, P.A. (1996) *Tree-crop Interactions: A Physiological Approach*. CAB International/International Centre for Research in Agroforestry: Wallingford/Nairobi.
Ong, C.K. and Black, C.R. (1994) Complementarity of resource use in intercropping and agroforestry systems. In: J.L. Monteith, R.K. Scott and M.H. Unsworth (eds) *Resource Capture by Crops, Proceedings of 52nd University of Nottingham Easter School*, pp. 255–278e. Nottingham University Press: Loughborough.
Ong, C.K., Black, C.R., Marshall, F. and Corlett, J. (1996) Principles of resource capture and utilisation of light and water. In: C. Ong and P.A. Huxley (eds) *Tree-crop Interactions: A Physiological Approach*, pp. 73–158. CAB International/International Centre for Research in Agroforestry. Wallingford/Nairobi.
Ong, C.K., Singh, R.P., Khan, A.A.H. and Osman, M. (1990) Recent advances in measuring water loss through trees. *Agroforestry Today*, **2**(3), 7–9.
Opara-Nadi, O.A. and Lal, R. (1987) Effects of land clearing and tillage methods on soil properties and maize root growth. *Field Crops Res.*, **15**, 193–206.
O'Riordan, T. (1995) *Environmental Science for Environmental Management*. Longman Scientific and Technical: Harlow.
Orskov, E.R. (1976) The effect of processing on digestion and utilisation of cereals by ruminants. *Proc.Nut.Soc.*, **35**, 245–52.

Osiru, D.S.O. and Willey, R.W. (1972) Studies on mixtures of dwarf sorghum and beans (*Phaseolus vulgaris*) with particular reference to plant population. *J.Agric.Sci.*, **79**, 531–40.

Othieno, C.O., Stephens, William and Carr, M.K.V. (1992) Yield variability at the Tea Research Foundation of Kenya. *Agric.For.Met.*, **61**, 237–52.

Ou, J.H. (1973) *A Handbook of Rice Diseases in the Tropics*. International Rice Research Institute: Los Banos, Philippines.

Ou, J.H. (1985) *Rice Diseases*, 2nd edn. CAB International: Wallingford.

Owen, E. (1976) Farm wastes: straw and other fibrous materials. *In*: A.N. Duckham, J.G.W. Jones and E.H. Roberts (eds) *Food Production and Consumption: the Efficiency of Human Food Chains and Nutrient Cycles*, pp. 299–318. North-Holland: Amsterdam.

Owen, E. and Jayasuriya, M.C.N. (1989) Use of crop residues as animal feeds in developing countries. *Res.Dev.in Agric.*, **6**, 129–38.

Owen, J.A. and Folland, C.K. (1988) Modelling the influence of sea surface temperatures on tropical rainfall. *In*: S. Gregory (ed.) *Recent Climatic Change*, pp. 141–53. Belhaven Press: London.

Oya, K. and Tokashiki, Y. (1984) Soil fertility in a shifting cultivation on Iriomote Island of Okinawa: Nutrients stored in the forest and their input to the soil. *Japan.J.Trop.Agric.*, **28**(4), 218–22.

Oyenuga, V.A. (1959) Effect of frequency of cutting on the yields and composition of some fodder grasses in Nigeria. *J.Agric.Sci.*, **53**, 25–33.

Pacey, A. and Cullis, A. (1986) *Rainwater Harvesting: The Collection of Rainfall and Runoff in Rural Areas*. Intermediate Technology Publications: London.

Padwick, G. Watts (1983) Fifty years of experimental agriculture. II, The maintenance of soil fertility in tropical Africa: A review. *Exp.Agric.*, **19**, 293–310.

Paling, R.W. (1990) *A Contribution to the Understanding of the Epidemiology and Control of Livestock Diseases in Africa*. Proefschrift: Utrecht.

Palmer, W.C. (1965) *Meteorological Drought*. US Weather Bureau Research Paper 45. US Weather Bureau: Washington, DC.

Pan, Y.S., Donegan, S.M. and Hayman, R.H. (1969) Sweating rate at different body regions in cattle and its correlation with some quantitative components of sweat gland volume for a given area of skin. *Austr.J.Agric.Res.*, **20**, 395–403.

Pandey, H.N., Nivarkar, A.E., Jana, D.N., Joshi, H.C. and Nawtiyal, L.P. (1989) Drinking water requirements of lactating cross bred cows during summer under free choice feeding systems. *Ind.J.Anim. Prod.Mangmt.*, **5**, 61–6.

Panin, A. and Ellis-Jones, J. (1994) Increasing the profitability of draft animal power. *In*: P. Starkey, E. Mwenya and J. Stares (eds) *Improving Animal Traction Technology, Proc.1st Workshop of the Animal Traction Network for Eastern and Southern Africa*, 18–23 January 1992, Lusaka, Zambia, pp. 94–103. Technical Centre for Agricultural and Rural Cooperation: Wageningen.

Panjarathinam, R. and Laxminarayana, H. (1974) A comparative study of microbial counts in the rumen liquor of cows and buffaloes. *Ind.Vet.J.*, **51**, 522–6.

Panse, V.G. and Khanna, R.C. (1964) Response of some important Indian food crops to fertilizers and factors influencing the response. *Ind.J.Agric.Sci.*, **34**, 172–202.

Parker, C. (1984) Herbicide use in small scale farming. *In*: T.J. Davis (ed.) *Proc.4th Agriculture Sector Symposium*, 9–13 January 1984, Washington, DC., pp. 218–30. World Bank: Washington, DC.

Parry, M. (1990) *Climate Change and World Agriculture*. Earthscan: London.

Partridge, I.J. (1979) Improvement of Nadi Blue grass (*Dicanthium caricosum*) pastures on hill land in Fiji with superphosphates and Sinatro: effects of stocking rate on beef production and botanical composition. *Trop.Grassl.*, **13**, 157–64.

Paterson, D.D. (1936) The cropping qualities of certain tropical fodder grasses. *Emp.J.Exp.Agric.*, **4**, 6–16.

Paterson, D.D. (1938) Further experiments with cultivated tropical fodder crops. *Emp.J.Exp.Agric.*, **6**, 323–40.

Paterson, D.D. (1939) The cultivation of perennial fodder grasses in Trinidad. *Trop.Agric.(Trin.)*, **16**, 55–7.

Payne, J.M. (1975) The Compton metabolic profile test. *Proc.Roy.Soc.Med.*, **65**, 181–3.

Payne, W.J.A. (1963) Relation of animal husbandry to human nutritional needs in East Africa. *E.Afr. Agric.For.J.*, **29**, 17–25.

Payne, W.J.A. (1970) *Cattle Production in the Tropics. Vol.1, Breeds and Breeding*. Longman: London.

Payne, W.J.A. (1976) Systems of beef production in developing countries. *In*: A.J. Smith (ed.) *Beef Production in Developing Countries*, pp. 242–57. Centre for Tropical Veterinary Medicine, University of Edinburgh: Edinburgh.

Payne, W.J.A. (1985) A review of the possibilities for integrating cattle and tree crop production systems in the tropics. *Forest Ecol. and Mangmt.*, **12**, 1–36.

Payne, W.J.A. (1990) *An Introduction to Animal Husbandry in the Tropics*, 4th. edn. Longman: London.

Payne, W.J.A. and Hancock, J. (1957) The direct effect of tropical climate on the performance of European-type cattle. II. Production. *Emp.J.Exp.Agric.*, **25**, 321–9.

Payne, W.J.A., Laing, W.I. and Raivoka, E.N. (1952) Breeding studies. *Agric.J.Fiji*, **23**, 9–13.

Pearce, F. (1995a) Fiddling while earth warms. *New Scientist*, **145**(1970), 14–15.

Pearce, F. (1995b) Global warming the jury delivers guilty verdict. *New Scientist*, **148**(2007), 6.

Pearce, F. (1996) Tropical smogs rival big city smoke. *New Scientist*, **150**(2030), 14.

Pearce, F. (1997) Global warming chills out over the Pacific. *New Scientist*, **153**(2070), 16.

Penman, H.L. (1948) Natural evaporation from open water, bare soil and grass. *Proc.Roy.Soc.A*, **193**, 120–45.

Peters K.J. (1993) Selection and breeding strategies for sustainable livestock production in developing countries with particular reference to dairy cattle production. *In*: M. Gill, E. Owen, G.E. Pollott and T.L.J. Lawrence (eds) *Animal Production in Developing Countries*, pp. 119–28. Brit.Soc.Anim.Prod. Occ.Pub.16, British Society of Animal Production: Edinburgh.

Peters, K.J. and Thorpe, W. (1988) Current status and trends in on-farm performance testing of cattle and sheep in Africa. *Proc.3rd Wrld. Cong. on Sheep and Cattle Breeding*, **1**, pp. 275–93.

Peters, R.L. (1991) Consequences of global warming for biological diversity. *In*: R.L. Wyman (ed.) *Global Climate Change and Life on Earth*, pp. 99–118. Routledge: London.

Peters, W.J. and Neuenschwander, L.F. (1988) *Slash and Burn: Farming in the Third World Forest*. University of Idaho Press: Moscow, ID.

Phillips, T.A. (1956) *An Agricultural Notebook*. Longman: London.

Pierce, F.J., Larson, W.E., Dowdy, R.H. and Graham, W.A.P. (1983) Productivity of soils: assessing long-term changes due to erosion. *Journal of Soil and Water Conservation*, **38**, 39–51.

Pieri, C.J.M.G. (1992) *Fertility of Soils: A Future for Farming in the West African Savannah*. Springer-Verlag: Berlin.

Pimentel, D. (ed.) (1993) *World Soil Erosion and Conservation*. Cambridge University Press: Cambridge.

Pingali, P., Bigot, Y. and Binswanger, H.P. (1987) *Agricultural Mechanisation and the Evolution of Farming Systems in Sub-Saharan Africa*. John Hopkins University Press: Baltimore.

Pinkas, A. and Hirstov, V. (1972) Some comparative biochemical and histological studies on calf and buffalo calf muscles. *Anim.Sci.*, **9**(7), 85–94.

Plucknett, D.L. and Smith, N.J.H. (1986) Historical perspectives on multiple cropping. *In*: C.A. Francis (ed.) *Multiple Cropping Systems*, pp. 20–39. Macmillan: New York, NY.

Polge, C. (1991) A project for the multiplication of bovine embryos. *Brit.Cattle Breeders' Club Digest*, **46**, 51–3.

Polikhronov, D.S. (1969) Comparison of productive characters in Murrah and Bulgarian buffaloes. *Zhivot.Nank.*, **6**(1), 95. (*ABA*, **37**, 3390).

Polikhronov, D.S., Panayotov, P. and Ognyanovid, A. (1971) Fattening of buffalo calves with complete ration mixtures. *Anim.Sci.*, **8**(6), 33–40.

Ponnamperuma, F.N. (1972) The chemistry of submerged soils. *Advances in Agron.*, **24**, 29–96.

Pons, L.J. (1989) Survey, reclamation and lowland agricultural management of acid sulphate soils. *In*: E. Maltby and T. Wollersen (eds) *Soils and their Management: A Sino-European Perspective*, pp. 295–304. Elsevier Applied Science: London.

Ponte, L. (1976) *The Cooling*. Prentice-Hall: Englewood Cliffs, NJ.

Popenoe, H. (1959) The influence of the shifting cycle on soil properties in Central America. *Proc. 9th Pacific Sci.Cong.*, **7**, 148–60.

Premkumar, P.D. (1994) *Farmers are Engineers: Indigenous Soil and Water Conservation Practices in a Participatory Watershed Development Programme*. PIDOW-MYRADA: Gulbarga, Karnataka, India, and copublished with Swiss Development Cooperation Field Office: Bangalore.

Prentice, A.N. (1972) *Cotton with Special Reference to Africa*. Longman: London.

Prentice, J.C., Cramer, W., Harrison, S.P., Leemans, R., Monserud, R.A. and Solomon, A.M. (1992) A global biome model based on plant physiology and dominance soil properties and climate. *J. of Biogeography*, **19**, 117–34.

Preston, T.R. (1974) Unpublished data quoted in 'Prospects for the intensification of cattle production in developing countries'. *In*: A.J. Smith (ed.) (1976) *Beef Production in Developing Countries*, pp. 242–57. Centre for Tropical Veterinary Medicine, University of Edinburgh: Edinburgh.

Preston, T.R. (1976) Prospects for the intensification of cattle production in developing countries. *In*: A.J. Smith (ed.) *Beef Cattle Production in Developing Countries*, pp. 242–57. Centre for Tropical Veterinary Medicine, University of Edinburgh: Edinburgh.

Preston, T.R. (1989) *The Development of Milk Production Systems in the Tropics*. Centre for Tropical Agriculture: Wageningen, Netherlands.

Preston, T.R. and Willis, M.B. (1969) Sugar cane as an energy source for the production of meat. *Outlook in Agric.*, **6**, 29–35.

Pretty, J. (1995) *Regenerating Agriculture: Policies and Practice for Sustainability and Self-Reliance*. Earthscan: London.

Probert, M.E. and Samosir, S. (1983) Sulphur in non-flooded tropical soils. *In*: G.J. Blair and A.R. Till (eds) *Sulphur in South-East Asian and South Pacific Agriculture*, pp. 15–27. University of New England: Armidale, NSW.

Proctor, J. (ed.) (1989) *Mineral Nutrients in Tropical Forest and Savannah Ecosystems. Special Publication No.9 for the British Ecological Society*. Blackwell Science: Oxford.

Protsch, R. and Berger, R. (1973) Earliest radio-carbon dates for domestic animals. *Science, USA*, **179**, 235–9.

Provenza, F.D., Burritt, E.A., Clausen, T.P., Bryant, J.P., Reichardt, P.B. and Distel, R.A. (1990) Conditioned flavour aversion: a mechanism for goats to avoid tannins in blackbrush. *Amer.Naturalist*, **136**, 810–28.

Purseglove, J.W. (1981) *Tropical Crops. Dicotyledons*, 2nd edn. Longman: London.

Purseglove, J.W. (1987) *Tropical Crops. Monocotyledons*, 2nd edn. Longman: London.

Purseglove, J.W., Brown, E.G., Green, C.L. and Robbins, S.R.J. (1981) *Spices*. Longman: London.

Putt, S.N.H. and Hanks, J.D. (1993) The identification and evaluation of disease constraints for extensive livestock production systems. *In*: M. Gill, E. Owen, G.E. Pollott and T.L.J. Lawrence (eds) *Animal Production in Developing Countries*, pp. 93–100. Brit.Soc.Anim.Prod.Occ.Pub.16. British Society of Animal Production: Edinburgh.

Putt, S.N.H., Shaw, A.P.M., Woods, A.J., Tyler, L. and James, A.D. (1987) *Veterinary Epidemiology and Economics in Africa*. ILCA Manual No.3. International Livestock Centre for Arid Regions: Nairobi.

Radley, B. (1992) Pest control and operator safety. *In: Tools for Agriculture. A Guide to Appropriate Equipment for Smallholder Farmers*, pp. 65–74. Intermediate Technology Publications: London.

Ragland, J.L. and Coleman, N.T. (1959) The effect of soil solution aluminium and calcium on root growth. *Soil Sci.Soc.of Amer.Proc.*, **23**, 355–7.

Rahmstorf, S. (1997) Ice cold in Paris. *New Scientist*, **153** (2068), 26–30.

Raintree, J.B. (ed.) (1987a) *D & D User's Manual: An Introduction to Agroforestry Diagnosis and Design*. International Centre for Research in Agroforestry: Nairobi.

Raintree, J.B. (1987b) The state of the art of agroforestry diagnosis and design. *Agroforestry Systems*, **5**, 219–50.

Raintree, J.B. (1991) *Socioeconomic Attributes of Trees and Tree Planting Practices*. Community Forestry Note No.9. FAO: Rome.

Raintree, J.B. and Warner, K. (1986) Agroforestry pathways for the intensification of shifting cultivation. *Agroforestry Systems*, **4**, 39–54.

Randhawa, N.S., Sinha, M.K. and Takkar, P.N. (1978) *Micronutrients. In: Soils and Rice*. International Rice Research Institute: Los Banos, Philippines.

Rao, M.K. and Nagarcenkar, R. (1977) Potentialities of the buffalo. *Wrld.Rev.Anim.Prod.*, **13**, 53–62.

Rao, M.R. and Willey, R.W. (1980) Evaluation of yield stability in inter-cropping studies on sorghum/pigeon pea. *Exp.Agric.*, **16**, 105–16.

Rasmussen, E.M. (1987) Global climate change and variability: effects on drought and desertification in Africa. *In*: M. Glantz (ed.) *Drought and Hunger in Africa*, pp. 3–22. Cambridge University Press: Cambridge.

Rattray, A. and Ellis, B.S. (1953) Maize and green manuring in S. Rhodesia. *Rhod.Agric.J.*, **49**, 188–99.

Rattray, J.M. (1957) The grasses and grass associations of S. Rhodesia. *Rhod.Agric.J.*, **54**, 197–234.

Reading, A., Thompson, R.D. and Millington, A.C. (1995) *Humid Tropical Environments*. Basil Blackwell: Oxford.

Reddy, V. and Butler, L.G. (1989) Incorporation of ^{14}C from (^{14}C) phenylalanine into condensed tannin of sorghum grain. *J.Agric.Food.Chem.*, **37**, 383–4.

Reed, J.D., Soller, H. and Woodward, A. (1990) Fodder tree and straw diets for sheep intake, growth, digestibility and the effects of phenolics on nitrogen utilisation. *Anim.Feed Sci.and Tech.*, **30**, 39–50.

Rees, M.C. and Minson, D.J. (1976) Fertilizer calcium as a factor affecting the voluntary intake, digestibility and retention time of Pangola grass (*Digitaria decumbens*) by sheep. *Brit.J.Nutr.*, **36**, 179–87.

Rege, J.E.O., Aboagye, G.S., Akah, S. and Ahunu, B.K. (1994) Crossbreeding Jersey with Ghana Shorthorn and Sokoto Gudali cattle in a tropical environment: additive and heterotic effects for milk production, reproduction and calf growth traits. *Anim.Prod.*, **59**, 21–9.

Reij, C., Mulder, P. and Begemann, L. (1988) *Water Harvesting for Plant Production*. Technical Paper No.91. World Bank: Washington, DC.

Reijntjes, C., Haverkort, B. and Waters-Bayer, A. (1992) *Farming for the Future. An Introduction to Low-External-Input and Sustainable Agriculture*. Macmillan: London.

Reiter, R.J. (1976) Effects of light on animals. (d) Effects on mammalian reproduction. *In*: H.D. Johnson (ed.) *Progress in Biometeorology*, Div.B. Vol.1. Pt.11. Swets and Zitlinger: Amsterdam.

Rendel, J. (1960) A study on relationships between blood groups and production characters in cattle. *Rep.Vl Int.Bloodgroup Cong. Munich (1959)*, pp. 8–23.

Rhoad, A.O. (1944) The Iberia Heat Tolerance Test for cattle. *Trop.Agric.(Trin.)*, **21**, 162–4.

Rhoades, J.D. (1974) Drainage for salinity control. *In*: J. van Schilfgaarde (ed.) *Drainage for Agriculture*, pp. 433–68. American Society of Agronomy: Madison, WI.

Rhodes, E.R. (1988) Africa – how much fertilizer needed? Case study of Sierra Leone. *Fertilizer Res.*, **17**(2), 101–18.

Ricaud, C., Egan, B.T., Gillaspie, A.G. and Hughes, G.C. (eds) (1989) *Diseases of Sugarcane*. Elsevier: Amsterdam.

Richards, B.M. and Voight, G.K. (1964) Role of mycorrhiza in nitrogen fixation. *Nature*, **201**, 310–11.

Richards, L.A. (ed.) (1954) *Diagnosis and Improvements of Saline and Alkaline Soils*. Handbook No.60. United States Department of Agriculture: Washington, DC.

Rife, C.D. (1962) Color and horn variation in water buffaloes. *J.Hered.*, **53**, 239–46.

Rinaudo, G., Alazard, D. and Moudiongui, A. (1988) Stem-nodulating legumes as green manure for rice in West Africa. *In*: *Green Manure for Rice Farming*, pp. 97–109. International Rice Research Institute: Los Banos, Philippines.

Risi, C. and Galwey, N.W. (1984) The *Chenopodium* grains of the Andes. *Adv.in Appl.Biol.*, **10**, 145–216.

Ristig, M. and McIntyre, I. (eds) (1981) *Diseases of Cattle in the Tropics*. Martinus Nijhoff: The Hague.

Ritchie, G.A. (ed.) (1979) *New Agricultural Crops*. Westview Press: Boulder, CO.

Ritzema, H.P. (ed.) (1994) *Drainage Principles and Applications*. International Institute for Land Reclamation: Wageningen.

Roache, K.L., Wellington, K.E. and Mahadevan, P. (1970) The extent of selection for milk yield among cows of the Jamaica Hope breed. *J.Agric.Sci.*, **74**, 469–71.

Roberts, N. (1994) The global environmental future. *In*: N. Roberts (ed.) *The Changing Global Environment*, pp. 3–21. Basil Blackwell: Oxford.

Robertshaw, D. (1982) Thermo-regulation of the goat. *Proc.3rd Int.Conf. Goat Prod. and Disease*, pp. 395–7. Tucson, AZ.

Robertshaw, D. and Taylor, C.R. (1969) A comparison of sweat gland activity in eight species of East African Bovids. *J.Physiol., London*, **203**, 135–43.

Robertson, A.W. and Frankignoul, C. (1990) The tropical circulation: a simple model versus as general model. *Qrtly.J.Roy.Met.Soc.*, **116**, 69–87.

Robinet, A.H. (1967) La chèvre Rousse de Maradi: Son exploitation et sa place dans l'economie et l'élevage de la Republique du Niger. *Revue Elév.Méd. vét.Pays.Trop. N.S.*, **20**, 129–86.

Robinson, J.B.D. (1957) The critical relationship between soil moisture content in the region of wilting point and the mineralization of natural soil nitrogen. *J.Agric.Sci.*, **49**, 100–105.

Robinson, J.B.D. (1959a) Camber bed cultivation of ground water (vlei) soils. 1. Experimental crop yields. *E.Afr.Agric.For.J.*, **24**, 184–91.

Robinson, J.B.D. (1959b) Camber bed cultivation of ground water (vlei) soils. 2. Modifications of the system. *E.Afr.Agric.For.J.*, **24**, 192–5.

Robinson, J.B.D., Brook, T.R. and de Vink, H.H.J. (1955) A cultivation system for ground water (vlei) soils. *E.Afr.Agric.For.J.*, **21**, 69–76.

Robinson, J.B.D. and Gacoka, P. (1962) Evidence of upward movement of nitrate during the dry season in the Kikuyu red loam coffee soil. *J.Soil Sci.*, **13**, 133–9.

Robison, D.M. and McKean, S.J. (1992) *Shifting Cultivation and Alternatives: An Annotated Bibliography*. CAB International: Wallingford.

Rocheleau, D., Weber, F. and Field-Juma, A. (1988) *Agroforestry in Dryland Africa*. ICRAF Science and Practice of Agroforestry Series No.3. International Centre for Research in Agroforestry: Nairobi.

Rodriquez, L.A., Mekannan, G., Wilcox, C.J., Martin, F.G. and Krienke, W.A. (1985) Effects of relative humidity, maximum and minimum temperatures, pregnancy and stage of lactation on milk composition and yield. *J.Dairy Sci.*, **68**(4), 973–8.

Rodriguez, M. (1992) Agricultural eguipment: The maintenance myth. *In: Tools for Agriculture. A Guide to Appropriate Equipment for Smallholder Farmers*, pp. 5–8. Intermediate Technology Publications: London.

Roger, P.A. and Kulasooriya, S.A. (1980) *Blue-green Algae and Rice*. International Rice Research Institute: Los Banos, Philippines.

Roger, P.A. and Watanabe, I. (1986) Technologies for utilizing biological nitrogen fixation in wetland rice: potentialities, current usage and limiting factors. *Fertilizer Res.*, **9**, 39–77.

Rollinson, D.H.L. (1955) Oestrus in Zebu cattle in Uganda. *Nature*, **176**, 352–3.

Rosenberg, M., Herz, Z., Davidson, M. and Folman, Y. (1977) Seasonal variations in post-partum plasma progesterone levels and conception in primiparous and multiparous dairy cows. *J.Reprod.Fertil.*, **51**, 363–74.

Rosenzweig, C. and Parry, M.C. (1993) Potential impacts of climate change on world food supply. *In*: H.M. Kaiser and T.E. Drennen (eds) *Agricultural Dimensions of Global Climate Change*. St. Lucie Press: Delray Beach, FL.

Rosenzweig, C. and Parry, M.C. (1994) Potential impacts of climate change on world food supply. *Nature*, **367**, 133–8.

Rowe, P. (1984) Breeding bananas and plantains. *Plant Breed.Rev.*, **2**, 135–56.

Rowe, P.R. and Rosales, F.E. (1996) Bananas and plantains. *In* J. Janick and J.N. Moore (eds) *Fruit Breeding. I: Trees and Tropical Fruit*, John Wiley & Sons, New York.

Rowell, D. (1981) Oxidation and reduction. *In*: D.J. Greenland and M.H.B. Hayes (eds) *The Chemistry of Soil Processes*, pp. 401–62. John Wiley & Sons: Chichester.

Roy, J.H.B., Balch, C.C., Miller, E.L., Ørskov, E.R. and Smith, R.H. (1977) Calculation of the N-requirement for ruminants from nitrogen metabolism studies. *In: Proc.of 2nd.Int.Symp.on Protein Metabolism and Nutr.*, pp. 126–9. Centre for Agricultural Publishing and Documentation: Wageningen.

Roy-Noël, J. (1979) Termites and soil properties. *In*: H.O. Mongi and P.A. Huxley (eds) *Soils Research and Agroforestry*, Proceedings of an Expert Consultation, pp. 271–95. International Centre for Research in Agroforestry: Nairobi.

RRI (1967) Division Report, 1966–71. Rubber Research Institute: Kuala Lumpur.

Ruddle, K., Johnson, D., Townsend, P.K. and Rees, J.D. (1978) *Palm Sago, a Tropical Starch from Marginal Lands*. University Press: Honolulu, HI.

Russell, D. (1993) *A Review of Research on Resource Management Systems of Cameroon's Forest Zone*. Resource, and Crop Mangt.Res.Mono. No.14. International Institute for Tropical Agriculture: Ibadan, Nigeria.

Russell, E.W. (1958) *The Fertility of Tropical Soils*. Rep.of Conf.of Directors and Sen.Agric.Officers of Overseas Depts. of Agric. Colonial Office, No.531. HMSO: London

Russell, E.W. (1971) Soil structure: its maintenance and improvement. *J.Soil Sci.*, **22**, 137–51.

Ruthenberg, H. (1972) *Farming Systems in the Tropics*. Clarendon Press: Oxford.

Ruthenberg, H. (1976) *Farming Systems in the Tropics*, 2nd edn. Clarendon Press: Oxford.

Ruthenberg, H. (1980) *Farming Systems in the Tropics*, 3rd edn. Clarendon Press, Oxford.

SADCC (1987) *The History of Soil Conservation in the SADCC Region*. Report No. 8. Soil and Water Conservation and Land Utilisation Coordination Unit, Southern African Development Coordination Conference: Maseru, Lesotho.

Salah, M.S., El-Nouty, F.D. and Al-Hajri, M.R. (1992) Effect of water sprinkling during the hot-dry summer season on semen quality of Holstein bulls in Saudi Arabia. *Anim.Prod.*, **55**, 59–63.

Sanchez, P.A. (1976) *Properties and management of soils in the tropics*. John Wiley & Sons, New York, 618 pp.

Sanchez, P.A. (1982) Nitrogen in shifting cultivation systems of Latin America. *Plant and Soil*, **67**, 91–103.

Sanchez, P.A. (1995) Science in agroforestry. *Agroforestry Systems*, **30**, 5–55.

Sanchez, P.A., Palm, C.A., Davey, C.B., Szott, L.T. and Russell, C.E. (1985) Trees as soil improvers in the humid tropics? *In*: M.G.R. Cannell and J.E. Jackson (eds) *Attributes of Trees as Crop Plants*, pp. 327–58. Institute of Terrestrial Ecology: Abbots Ripton.

Sanchez, P.A. and Salinas, J.G. (1981) Low input technology for managing Oxisols and Ultisols in tropical America. *Advances in Agron.*, **34**, 279–406.

Sands, E.B., Thomas, D.B., Knight, J. and Pratt, D.J. (1970) Preliminary selection of pasture plants for the semi-arid areas of Kenya. *E.Afr.Agric.For.J.*, **36**, 49–57.

Sarmiento, G. (1984) *The Ecology of Neotropical Savannas*. Harvard University Press: Cambridge, MA.

Saul, G.R. and Flinn, P.C. (1985) Effects of saline drinking water on growth and water and feed in-take of weaner heifers. *Austr.J.Exp.Agric.*, **25**, 734–8.

Scaife, M.A. (1968) Maize fertilizer trials in Western Tanzania. *J.Agric.Sci.*, **70**, 209–22.

Schalitchev, A. and Polikhronov, D.S. (1969) Investigations on the content and quality of milk in buffaloes of Murrah Breed imported from India. *Anim.Sci.*, **6**(3), 57–63.

Schemenauer, R.S. and Cereceda, P. (1992) The quality of fog water collected for domestic and agricultural use in Chile. *Amer.Met.Soc.*, **31**, 275–90.

Scherr, S.J., Barbier, B., Jackson, L.A. and Yadav, S. (1995) *Land Degradation in the Developing World: Implications for Food, Agriculture, and Environment to the Year 2020*. Food, Agriculture and Environment Discussion Paper. International Food Policy Research Institute: Washington, DC.

Schmida, A., Evenari, M. and Noy-Meir, I. (1986) Hot desert ecosystems. *In*: M. Evenari, I. Noy-Meir and D.W. Goodall (eds) *Ecosystems of the World: Hot Deserts and Arid Shrublands*, pp. 379–88. Elsevier: Oxford.

Schmidt-Nielsen, K., Schmidt-Nielsen, B., Houpt, T.R. and Jarnum, S.A. (1957) Body temperature of the camel and its relation to water economy. *Amer.J.Physiol.*, **188**, 103–12.

Schoettle, A. and Fahey, T.J. (1994) Foliage and fine root longevity in pines. *Ecol.Bull.*, **43**, 136–53.

Schoonhaven, A. van and Voysert, O. (1991) *Common Beans: Research for Crop Improvement*. CAB International: Wallingford.

Schwab, G., Elliott, W., Fangmeier, D. and Frevert, R.K. (1993) *Soil and Water Conservation Engineering*, 4th edn. John Wiley & Sons: New York, NY.

Schwerdtfeger, W. (ed.) (1976) *Climates of the World. Vol. 12: Climates of Central and South America*. Elsevier: London.

Scoones, I. (1987) Economic and biological carrying capacity implications for livestock development in the dryland communal areas of Zimbabwe. *Proc.Int.Semin.Dept.Biol.Sci.Univ.Zimbabwe, Paper 27b*, pp. 1–24. Overseas Development Institute: London.

Scoones, I. (ed.) (1994) *Living with Uncertainty: New Directions in Pastoral Development in Africa*. Intermediate Technology Publications: London.

Scoones, I. (1995) Policies for pastoralists: new directions for pastoral development in Africa. *In*: A. Binns (ed.) *People and Environment in Africa*, pp. 23–30. John Wiley & Sons: Chichester.

Seebeck, R.M. (1973) Sources of variation in the fertility of a herd of Zebu, British and Zebu × British cattle in N. Australia. *J.Agric.Sci.*, **81**, 253–62.

Seetharam, A., Riley, K.W. and Harinarayana, G. (eds) (1990) *Small Millets in Global Agriculture*. Aspect: London.

Seidel, G.E. (1984) Applications of embryo transfer and related technologies to cattle. *J.Dairy Sci.*, **67**(11), 2786–96.

Seif el Din, A.G. (1981) Agroforestry practices in dry regions. *In*: L. Buck (ed.) *Proceedings of the Kenya National Seminar on Agroforestry, 12–22 November 1980, Nairobi*, pp. 419–34. International Centre for Research in Agroforestry: Nairobi.

Seifert, G.S. (1971) Ecto- and endo-parasitic effects on the growth rates of zebu crossbred and British cattle in the field. *Austr.J.Agric.Res.*, **22**, 839–48.

Sellers, W. (1965) *Physical Climatology*. University of Chicago Press: Chicago, MI.

Semb, G. and Robinson, J.B.D. (1969) The natural nitrogen flush in different arable soils and climates in East Africa. *E.Afr.Agric.For.J.*, **34**, 350–70.

Semeniuk, V. (1994) Predicting the effect of sea level rise on mangroves in North-Western Australia. *J.of Coastal Research*, **10**(4), 1050–76.

Semple, J.A., Grieve, C.M. and Osbourn, D.F. (1966) The preparation and feeding value of Pangola grass silage. *Trop.Agric.(Trin.)*, **43**, 251–5.

Sessay, M.F. and Stocking, M.A. (1995) Soil productivity and fertility maintainenance of a degraded Oxisol in Sierra Leone. *In*: F. Ganry and B. Campbell (eds) *Sustainable Land Management in African Semi-arid and Sub-humid Zones*. Proceedings of the SCOPE (Scientific Committee on Problems of the Environment) Workshop, Dakar, Senegal, 1993, pp. 189–201. Centre de Cooperation Internationale en Recherche Agronomique Pour le Developpement (CIRAD): Montpellier.

Sethi, R.K. and Nagarcenkar, R. (1981) Inheritance of heat tolerance in buffaloes. *Int.J.Anim.Sci.*, **51**, 591–5.

Sewell, M.M.H. (1976) The role of management in the control of helminth diseases. *In*: A.J. Smith (ed.) *Beef Cattle Production in Developing Countries*, pp. 138–49. Centre for Tropical Veterinary Medicine, University of Edinburgh: Edinburgh.

Shafie, M.M. and El-Tannikhy, A.M. (1970) Comparative study of skin structure in Egyptian and imported cattle breeds and their crosses in relation to heat tolerance. *J.Anim.Prod.,UAR.*, **10**, 115–31.

Shahkhalili, Y., Finot, P.A., Hurrell, R. and Fern, E. (1990). Effects of food rich in polyphenols on nitrogen excretion in rats. *J.Nutr.*, **120**, 346–52.

Sharma, A. and Nagarcenkar, R. (1981) Studies on inheritance of sweat gland characteristics in Murrah buffaloes. *Wrld.Rev.Anim.Prod.*, **17**, 35–9.

Sharma, A.K. (1991) Agricultural mechanisation in Rajasthan, India. *Agric.Mech.in Asia, Africa and Latin America*, **20**, 69–72.

Sharma, P.K., De Datta, S.K. and Redulla, C.A. (1989) Effect of percolation rate on nutrient kinetics and rice yield in tropical soils. *Plant and Soil*, **119**, 111–26.

Sharon, D. (1972) The spottiness of rainfall in a desert area. *J.of Hydrology*, **17**, 161–75.

Sharon, D. (1981) The distribution in space of local rainfall in the Namib desert. *J.of Climatology*, **1**, 69–75.

Shaw, E. (1994) *Hydrology in Practice*. Chapman & Hall: London.

Shaxson, T.F., Hudson, N.W., Sanders, D.W., Roose, E. and Moldenhauer, W.C. (1989) *Land Husbandry: A Framework for Soil and Water Conservation*. Soil and Water Conservation Society: Ankeny, IA.

Shen, Jin-Hua (1980) Rice breeding in China. *In: Rice Improvement in China and Other Asian Countries*, pp. 9–30. International Rice Research Institute: Los Banos, Philippines.

Sheng, T.C. (1986) *Watershed Conservation: A Collection of Papers for Developing Countries*. Chinese Soil and Water Conservation Society: Taipei, Taiwan and Colorado State University: Fort Collins, CO.

Sheng, T.C. (1990) *Watershed Conservation II: A Collection of Papers for Developing Countries*. Chinese Soil and Water Conservation Society: Taipei, Taiwan and Colorado State University: Fort Collins, CO.

Shepard, B.M. (1990) Integrated pest management in rice: present status and future prospects in Southeast Asia. *In*: B.T. Grayson, M.B. Green and L.G. Copping (eds) *Pest Management in Rice*, pp. 258–68. Elsevier Applied Science: London.

Shetty, S.V.R. (1989) Design and evaluation of alternative production systems: Pearl millet/maize and cereal/groundnut systems in Mali. *In: Soil, Crop and Water Management Systems for Rainfed Agriculture in the Sudano-Sahelian Zone*, Proceedings of an International Workshop, 11–16 January 1987, ICRISAT Sahelian Centre, Niamey, pp. 291–301. International Crops Research Institute for the Semi-Arid Tropics: Patancheru, India.

Shorrocks, V.M. (1964) *Mineral Deficiencies in Hevea and Associated Cover Crops*. Rubber Research Institute of Malaysia: Kuala Lumpur.

Shuttleworth, W.J. (1979) *Evaporation*. Institute of Hydrology Report 56. Institute of Hydrology: Wallingford.

Siebert, B.D. and Macfarlane, W.V. (1971) Water turnover and renal function of dromedaries in the desert. *Physiol.Zool.*, **44**, 225–40.

Siegel, H.S. (1968) Adaptation of poultry. *In: Adaptation of Domestic Animals*, pp. 292–309. Lea and Febiger: Philadelphia, PA.

Silver, W.S. (1969) Biology and ecology of nitrogen fixation by symbiotic associations of non-leguminous plants. *Proc.Roy.Soc.B.*, **172**, 389–400.

Silvery, M.W., Coaldrake, J.E., Haydeck, F.P.,

Radcliff, D. and Smith, C.A. (1978) Beef cow performance from tropical pastures on semi-arid brigalow lands under intermittant drought. *Austr.J.Exp.Agric.Anim.Husb.*, **18**, 618–28.

Simm, G., Conington, J. and Bishop, S.C. (1994) Opportunities for genetic improvement of sheep and cattle in the hills and uplands. *In*: T.L.J. Lawrence, D.S. Parker and P. Rowlinson (eds) *Livestock Production and Land Use in Hills and Uplands*, pp. 51–66. Brit.Soc.Anim.Prod.Occ.Pub. 18. British Society of Animal Production: Edinburgh.

Simmonds, N.W. (1958) *Ensete* cultivation in the southern Highlands of Ethiopia: a review. *Trop.Agric.(Trin.)*, **35**, 302–307.

Simmonds, N.W. (1976) *The Evolution of Crop Plants*. Longman: London.

Simmonds, N.W. (1979) *Principles of Crop Improvement*. Longman: London.

Simmonds, N.W. (1981) Genotype (G), Environment (E) and GE components of crop yields. *Exp.Agric.* **17**, 355–62.

Simmonds, N.W. (1990) The social context of plant breeding. *Plant Breed.Abstr.*, **60**, 337–41.

Simmonds, N.W. (1991a) Genetics of horizontal resistance to diseases of crops. *Biol.Rev.*, **66**, 189–241.

Simmonds, N.W. (1991b) Selection for local adaptation in a plant breeding programme. *Theoretical and Applied Genetics*, **82**, 363–7.

Simmonds, N.W. (1993) Introgression and incorporation strategies for the use of crop genetic resources. *Biol.Rev.*, **68**, 539–62.

Simmonds, N.W. (1994) Diseases of tropical crops: problems and controls. *Botanical Journal of Scotland*, **47**, 129–37.

Simmonds, N.W. (1995a) Food crops: 500 years of travels, *CSSA Special Publication*, **23**, 31–45.

Simmonds, N.W. (1995b) Tree biology, forestry and agriculture, especially in the tropics. *Botanical Journal of Scotland*, **47**, 211–27.

Simmonds, N.W. (1996) Family selection strategy in plant breeding. *Euphytica*, **90**, 201–8.

Simoons, F.J. (1984) Gayal or mithan. *In*: I.L. Mason (ed.) *Evolution of Domesticated Animals*. Longman: London.

Simplicio, A.A., Reira, G.S., Nelson, A. and Pant, K.P. (1982) Seasonal variation in seminal and testicular characteristics of Brazilian Somali rams in the hot semi-arid climate of tropical northeast Brazil. *J.Reprod.Fertil.*, **66**, 735–8.

Simpson, J.R. (1960) The mechanism of surface nitrate accumulation on a bare fallow soil in Uganda. *J.Soil Sci.*, **11**, 45–60.

Sims, B.G., Johnston, B.F., Olmstead, A.L. and Maldonado, S.J. (1990) Animal-drawn implements for small farms in Mexico. *In*: A.W. Speedy (ed.) *Developing World Agriculture*, pp. 157–63. Grosvenor Press International: London.

Singh, R.N. (1961) *Role of Blue-green Algae in Nitrogen Economy of Indian Agriculture*. Indian Council of Agricultural Research: New Delhi.

Singh, S.B. and Desai, R.N. (1962) Production characters of Bhadawari buffalo cows. *Ind.Vet.J.*, **39**, 332–43.

Singh, S.P. (1992) Common bean improvement in the tropics. *Plant Breeding Reviews*, **10**, 199–270.

Singh, S.R. and Rachie, K.O. (eds) (1985) *Cowpea Research, Production and Utilization*. John Wiley & Sons: Chichester.

Singleton, V.L. (1981) Naturally occurring food toxicants: phenolic substances of plant origin common in foods. *Advances in Fd.Res.*, **27**, 149–242.

Sinha, S.K. (1991) Impacts of climate change on agriculture. *In*: J. Jager and H.L. Ferguson (eds) *Climate Change: Science, Impacts and Policy*, pp. 99–107. Cambridge University Press: Cambridge.

Siota, C.M., Castillo, A.P., Moog, F.A. and Javier, E.Q. (1978) Beef production on native, native/Stylo and native/Centro pastures. *Philippines J.Anim. Ind.*, **32**, 25–34.

Skerman, P.J. (1977) *Tropical Forage Legumes*. FAO: Rome.

Skinner, J.D. (1973) Technological aspects of domestication and harvesting of certain species of game in South Africa. *Proc.3rd Wildlife Conf.Anim.Prod. Melbourne*, **2**, 119–25.

SMA (1963) (Sugar Manufacturers' Association of Jamaica Ltd and Alcon Jamaica Ltd) The production costs and returns of eleven beef farms. Sugar Manufacturers' Association: Kingston, Jamaica.

Smart, P. and Herbertson, J.G. (eds) (1991) *Drainage Design*. Blackie & Son: Glasgow.

Smartt, J (1976) *Tropical Pulses*. Longman: London.

Smartt, J. (1990) *Grain Legumes. Evolution and Genetic Resources*. Cambridge University Press: Cambridge.

Smartt, J. and Simmonds, N.W. (eds) (1995) *Evolution of Crop Plants*, 2nd edn. Longman: London.

Smith, A.J. (1971) The productivity of laying pullets at high environmental temperatures. *Feedstuffs*, **17**, 26–7.

Smith, A.J. (1973) Some effects of high environmental temperatures on the productivity of laying hens (a review). *Trop.Anim.Hlth.Prod.*, **5**, 259–71.

Smith, A.J. (1974) Changes in the average egg weight and shell thickness of eggs produced by hens exposed to high environmental temperatures. *Trop.Anim.Hlth.Prod.*, **6**, 237–44.

Smith, A.J. (1984) Work animals in developing countries: Possibilities for improvement. *Span*, **27**, 27–9.

Smith, C. (1988) Applications of embryo transfer in animal breeding. *Theriogenology*, **29**, 203–12.

Smith, C.A. (1961) The utilisation of *Hyparrhenia* veld for the nutrition of cattle in the dry season. 1. The effects of nitrogenous fertilizers and mowing regimes on herbage yields. *J.Agric.Sci.*, **57**, 305–10.

Smith, C.A. (1963) Oversowing pasture legumes into the *Hyparrhenia* grassland of N. Rhodesia. *Nature*, **200**, 811–12.

Smith, C.A. (1965) Studies on the *Hyparrhenia* veld. VI. The fertilizer value of cattle excreta. *J.Agric.Sci.*, **64**, 403–406.

Smith, L.G. (1993) *Impact Assessment and Sustainable Resource Management*. Longman Scientific and Technical: Harlow.

Smith, M. (1990) *CROPWAT – A Computer Program for Irrigation Planning and Management*. Irrigation and Drainage Paper 46. FAO: Rome.

Smith, M. (1992) *Expert Consultation on Revision of FAO Methodolgies for Crop Water Requirements*. FAO: Rome.

Smithson, J.B. (1985) Breeding advances in chickpea at ICRISAT. *In*: G.E. Russell (ed.) *Progress in Plant Breeding*, pp. 223–38. Butterworth: London.

Soane, B.D. and Van Ouwerkerk, C. (eds) (1994) *Soil Compaction in Crop Production*. Elsevier: Amsterdam.

Soh, A.C. (1990) Oil palm breeding – breeding into the 21st century. *Plant Breed.Abstr.*, **60**, 1437–44.

Soil Survey Staff (1960) *Soil Classification – A Comprehensive System – 7th Approximation*. US Department of Agriculture: Washington, D.C.

Speirs, M. and Olsen, O. (1992) *Indigenous Integrated Farming Systems in the Sahel*. Wrld.Bank Tech.Paper No. 179. World Bank: Washington, DC.

Spier, S.J., Smith, B.P., Seawright, B.A., Norman, B.B., Ostrowski, S.R. and Oliver, M.N. (1987) Oak toxicosis in cattle in N. California: clinical and pathological findings. *J.Amer.Vet.Med.Assn.*, **191**, 958–64.

Spooner, R.L., Mazumber, N.K., Griffin, T.K., Kingwell, R.G., Wijeratne, W.V.S. and Wilson, C.D. (1973) Apparent heterozygote excess at the Amylase 1 locus in cattle. *Anim.Prod.*, **16**, 209–14.

Sprent, J.I. and Sutherland, J.M. (1990) Nitrogen fixing woody legumes. *Nitrogen Fixing Tree Research Reports*, **8**, 17–23.

Squire, G.R. (1990) *The Physiology of Tropical Crop Production*. CAB International: Wallingford.

Squires, V.R. (1988) Water and its functions, regulation and comparative use by ruminant livestock. *In*: D.C. Church (ed.) *The Ruminant Animal, Digestive Physiology and Nutrition*, pp. 217–26. Prentice-Hall: Englewood Cliffs, NJ.

Sreedharan, S. (1985) *Inheritance of body size measurements in relation to production efficiency in buffaloes*. PhD thesis, Punjab University: Chandigarh, India.

Srivastava, K.L. and Jangawad, L.S. (1988) Water balance and erosion rates of Vertisol watersheds under differential management. *In: Small Watershed Hydrology*, Proc.Workshop. International Crops Research Institute for the Semi-Arid Tropics: Patancheru, India.

Srivastava, V.K., Raizada, B.C. and Kulkarni, V.A. (1968) Carcass quality of Barberi and Jumna Pari type goats. *Ind.Vet.J.*, **45**, 219–25.

Stanley-Price, M.R., al-Harthy, H. and Whitcombe, R.P. (1986) *Fog Moisture and its Ecological Effects in the Oman Salalah*. Planning Committee for Development and Environment in the Southern Region: Sultanate of Oman.

Stanton, W.R. and Flach, M. (eds) (1980) *Sago. The Equatorial Swamp as a Natural Resource*. Martinus Nijhoff: The Hague.

Staples, R.R., Hornby, H.E. and Hornby, R.M. (1942) A study on the comparative effects of goats and cattle on a mixed grass-bush pasture. *E.Afr.Agric.For.J.*, **8**, 62–70.

Starkey, P. (1988) *Animal-drawn Wheeled Tool Carriers: Perfected Yet Rejected*. Deutsche Gesellschaft für Tech. Zusammenarbeit (GTZ) Eschborn/Vieweg: Braunschweig.

Starkey, P. and Faye, A. (1990) *Animal Traction for Agricultural Development*. Technical Centre for Agricultural and Rural Cooperation: Ede-Wageningen.

Stephens, D. (1962) The upward movement of nitrate in a bare soil in Uganda. *J.Soil.Sci.*, **13**, 52–9.

Stephens, D. (1970) Soil fertility. *In*: J.D. Jameson (ed.) *Agriculture in Uganda*, 2nd edn. Oxford University Press: London.

Stephens, William and Carr, M.K.V. (1991) Responses of tea (*Camellia sinensis*) to irrigation and fertiliser: II. Water use. *Exp.Agric.*, **27**, 193–210.

Steppler, H.A. and Nair, P.K.R. (eds) (1987) *Agroforestry: A Decade of Development*. International Centre for Research in Agroforestry: Nairobi.

Stewart, W.D.P. (1966) *Nitrogen in Plants*. Athlone Press, University of London: London.

Stiles, W. (1961) *Trace Elements in Plants*, 3rd edn. Cambridge University Press: Cambridge.

Stobbs, T.H. (1969) The use of liveweight-gain trials for pasture evaluation in the tropics. 3. The measurement of large pasture differences. *J.Brit.Grassl.Soc.*, **24**, 177–83.

Stocking, M.A. (1984) *Erosion and Productivity: A Review*. Soil Conservation Programme, Land and Water Development Division. FAO: Rome.

Stocking, M.A. (1985) Soil conservation policy in colonial Africa. *Agric.Hist.*, **59**, 148–61.

Stocking, M.A. (1993) Soil and water conservation for resource-poor farmers: designing acceptable technologies for rainfed conditions in Eastern India. *In*:

E. Baum, P. Wolff and M. Zobisch (eds) *Acceptance of Soil and Water Conservation: Strategies and Technologies, Vol. 3*, pp. 291–305. Topics in Applied Resource Management in the Tropics. German Institute for Tropical and Subtropical Agriculture: Witzenhausen.

Stocking, M.A. (1995) Soil erosion. *In*: W. Adams, A. Goudie and A. Orme (eds) *The Physical Geography of Africa*, pp. 253–65. Oxford University Press: Oxford.

Storey, H.H. and Leach, R. (1933) A sulphur deficiency of the tea bush. *Annals Appl.Biol.*, **20**, 23.

Stott, P. (1994) Savanna landscapes and global environmental change. *In*: N. Roberts (ed.) *The Changing Global Environment*, pp. 287–303. Basil Blackwell: Oxford.

Stover, R.H. and Simmonds, N.W. (1987) *Bananas*, 3rd edn. Longman: London.

Strahler, A.N. (1975) *Physical Geography*, 4th edn. John Wiley & Sons: New York, NY.

Stuber, C.W. (1994) Heterosis in plant breeding. *Plant Breeding Reviews*, **12**, 227–52.

Sumar, J. (1983) *Studies on Reproductive Physiology in Alpacas*. MS thesis, Coll.of Vet.Med., Univ. of Agric.Sci., Uppsala, Sweden.

Summerfield, R.J. and Bunting, A.H. (eds) (1980) *Advances in Legume Science*. Royal Botanical Gardens/MAFF, HMSO: London.

Summerfield, R.J. and Roberts, E.H. (eds) (1985) *Grain Legume Crops*. Collins: London.

Summer, G. (1988) *Precipitation: Process and Analysis*. John Wiley & Sons: Chichester.

Sutrisno, J.A.M., Janssen and Alkasuma (1990) Classification of acid sulphate soils: a proposal for improvements of the soil taxonomy system. *In*: *Workshop on Acid Sulphate Soils in the Humid Tropics*, pp. 71–80. Agency for Agricultural Research and Development: Jakarta.

Suttie, J.M. and Moore, C.E.M. (1966) *Desmodium uncinatum. Kenya Fmr.*, **116**, 18.

Swaminathan, M.S. (1985) IRRI's research and training agenda. *In*: *Impact of Science on Rice*, pp. 41–60. International Rice Research Institute: Los Banos, Philippines.

Swan, H. (1974) Forum on the use of ground straw in cattle rations. *Proc.Brit.Soc.Anim.Prod. New Series No. 3*, 116–49.

Sweet, R.J. (1991) The communal grazing cell experience in Botswana. *In*: C. Oxby (ed.) *Assisting African Livestock Keepers: The Experience of Four Projects*, pp. 16–30. Agric.Admin.Unit Occ.Paper No.12. Overseas Development Institute: London.

Swift, J. (1973) Disaster and a Sahelian nomad economy. *In*: *Drought in Africa*, Symposium. Centre for African Studies, University of London: London.

Swift, J. (1975) Pastoral nomadism as a form of land use: Twareg of the Adrar N. Iforus. *In*: T. Monod (ed.) *Pastoral Nomadism in Tropical Africa*. Oxford University Press: London.

Swift, M., Russell-Smith, J.A. and Perfect, T.J. (1981) .Decomposition and mineral-nutrient dynamics of plant litter in a regenerating bush-fallow in sub-humid tropical Nigeria. *J.Ecol.*, **69**(3), 981–95.

Swift, M.J., Heal, J.W. and Anderson, J.M. (1979) *Decomposition in Terrestrial Ecosystems*. University of California Press: Berkeley, CA.

Sykes, A.H. (1977) Nutrition-environment interactions in poultry. *In*: W. Haresign, H. Swan and D. Lewis (eds) *Nutrition and the Climatic Environment*, pp. 17–29. Butterworths: London.

Syrstad, O. (1985) Dairy merits of various *Bos taurus* × *Bos indicus* crosses. 36th.Ann.Mtng. Eur.Assn. Anim.Prod., Kallithea, Greece. EAAP: Rome.

Syrstad, O. (1990) Dairy cattle crossbreeding in the tropics: the importance of genotype × environment interactions. *Livestock Prod.Sci.*, **24**, 109–18.

Sys, C. (1959) Le classification des sols Congolese. *Proc.3rd.Int.Afric.Soils Conf.*, **1**, 303–12.

Sys, C. (1960) Principles of soil classification in the Belgian Congo. *Trans.7th Int.Cong.Soil Sci.*, **4**, 112–18.

Szott, L.T., Fernandes, E.C.M. and Sanchez, P.A. (1991) Soil-plant interactions in agroforestry systems. *In*: P.G. Jarvis (ed.) *Agroforestry: Principles and Practice*, pp. 127–52. Elsevier: Amsterdam.

Takahashi, K. and Arakawa, H. (eds) (1981) *Climates of the World. Vol. 9: Climates of Southern and Western Asia*. Elsevier: London.

Taneja, G.C. (1959) Sweating in cattle. V. Sweat prints. *J.Agric.Sci.*, **52**, 168–9.

Tattersfield, J.R. (1982) The role of research in increasing food crop potential in Zimbabwe. *The Zimbabwe Sci.News*, **16**, 6–10.

Taylor, C.R. and Lyman, C.P. (1967) A comparative study of the environmental physiology of an East African antelope, the eland and the Hereford steer. *Physiol.Zool.*, **40**, 280–95.

Tegene, B. (1992) *Erosion: Its Effect on the Properties and Productivity of Eutric Nitosols in Gununo Area, Southern Ethiopia, and Some Techniques of Its Control*. African Studies Series A9. University of Berne: Berne.

Tejwani, K.G. (1994) *Agroforestry in India*. Centre for Natural Resources and Environment Management: New Delhi.

Temple, R.S. and Reh, I. (1984) Livestock populations and factors affecting them. *In*: B.L. Nestel (ed.) *Development of Animal Production Systems*, pp. 33–62. Elsevier: Amsterdam.

Tergas, L.E., Blue, W.G. and Moore, J.E. (1971) Nutritive value of fertilized Jaragua grass

(*Hyparrhenia rufa*) in the wet-dry Pacific region of Costa Rica. *Trop.Agric.(Trin.)*, **48**, 1–8.

Terra, G.J.A. (1958) Farm systems in south-east Asia. *Netherlands J.Agric.Sci.*, **6**, 157–82.

Terry, P.J. (1984) *A Guide to Weed Control in East African Crops*. Kenya Literature Bureau: Nairobi.

Tewari, D.N. (1995) *Agroforestry for Increased Productivity, Sustainability and Poverty Alleviation*. International Book Distributors: Dehra Dun, India.

Thatcher, W.W. (1974) Effects of season, climate and temperature on reproduction and lactation. *J.Dairy Sci.*, **57**, 360–73.

Thom, H. (1958) A note on the gamma distribution. *Monthly Weather Review*, **86**(4), 117–22.

Thomas, D.S.G. and Middleton, N. (1994) *Desertification: Exploding the Myth*. John Wiley & Sons: Chichester.

Thomas, G.W. (1988) Elephant grass for soil erosion control and livestock feed. *In*: W.C. Moldenhauer and N.W. Hudson (eds) *Conservation Farming on Steep Slopes*, pp. 188–93. Soil and Water Conservation Society: Ankeny, IA.

Thornwaite, C.W. (1948) An approach towards a rational classification of climate. *Geog.Rev.*, **38**, 55–94.

Thornthwaite, C.W. (1954) The determination of potential evapotranspiration. *Publications in Climatology*, **7**(1).

Thornton, D.D. (1970) A stocking rate trial on rough grazing in Buganda. Part III. Liveweight gains in the last two years. *E.Afr.Agric.For.J.*, **35**, 331–5.

Thorpe, J. and Smith, G.D. (1949) Higher categories in soil classification: Order, sub-order and great soil groups. *Soil Sci.*, **67**, 117–26.

Thwaites, C.J. (1985) Physiological responses and productivity in sheep. *In*: M.K. Yousef (ed.) *Stress Physiology in Livestock. Vol.II, Ungulates*. CRC Press: Boca Raton, FL.

Tiffen, M. (1982) *Economic, Social and Institutional Aspects of Shifting Cultivation in Humid and Semi-humid Africa*. FAO: Rome.

Tiffen, M. (1993) Productivity and environmental conservation and a rapid population growth: a case study of Machakos district, Kenya. *J.Int.Dev.*, **5**(2), 207–23.

Tiffen, M. and Mortimore, M. (1990) *Theory and Practice in Plantation Agriculture*. Overseas Development Institute: London.

Tiffen, M., Mortimore, M. and Gichuki, F. (1994) *More People, Less Erosion: Environmental Recovery in Kenya*. John Wiley & Sons: Chichester.

Tindall, H.D. (1983) *Vegetables in the Tropics*. Macmillan: London.

Tinker, P.B.H. (1989) Monitoring, reclaiming and increasing soil fertility. *In*: E. Maltby and T. Wollersen (eds) *Soils and their Management: A Sino-European Perspective*, pp. 225–32. Elsevier Applied Science: London.

Tomar, N.S. and Mittal, K.K. (1960) Significance of the calving season in Hariana cows. *Ind.Vet.J.*, **37**, 367–70.

Tomar, S.P.S. and Desai, R.N. (1965) A study of growth rate in buffaloes maintained on military farms. *Ind.Vet.J.*, **42**, 116–25.

Torquebiau, E. (1992) Are tropical home gardens sustainable? *Agric.Ecosystems and Envir.*, **41**, 189–209.

Torres, F. (1983) Role of woody perennials in animal agroforestry. *Agroforestry Systems*, **1**, 131–63.

Tothill, J.C. and Mott, J.J. (eds) (1985) *Ecology and Management of the World's Savannas*. Australian Academy of Science: Canberra.

Tothill, J.D. (1940) *Agriculture in Uganda*. Oxford University Press: Oxford.

Trail, J.C.M. (1981) Merits and demerits of importing exotic cattle compared with the improvement of local breeds. *In*: A.J. Smith and R.C. Gunn (eds) *Intensive Animal Production in Developing Countries*, pp. 191–231. Brit.Soc.Anim.Prod.Occ.Pub.4. British Society of Animal Production: Edinburgh.

Trail, J.C.M., Feron, A., Pelo, M., Colardelle, C., Ordner, G., d'Ieteren, G., Durkin, J., Maehl, H. and Thorpe, W. (1988) Selection in trypano-tolerant cattle breeds in Africa. *Proc.3rd Wrld.Cong.on Sheep and Beef Cattle Breed.*, pp. 613–24.

Trail, J.C.M., d'Ieteren, G.D.M., Feron, A., Kakiese, O., Mulungo, M. and Pelo, M. (1991) Effect of trypanosome infection, control of parasitaemia and control of anaemia development on productivity of N'Dama cattle. *Acta Tropica*, **48**, 37–45.

Trewartha, G.T. (1981) *The Earth's Problem Climates*. University of Wisconsin Press: Madison, WI.

Tshibaka, T.B. (1992) *Labour in the Rural Household Economy of the Zairian Basin*. Res.Rep. No.90. International Food Policy Research Institute: Washington, DC.

Tubiana, M.J. and Tubiana, J. (1975) Tradition et developpement au Soudan oriental: l'exemple Zaghawa, *In*: T. Monod (ed.) *Pastoral Nomadism in Tropical Africa*. Oxford University Press: London.

Tulaphitak, T., Pairintra, C. and Kyuma, K. (1985) Changes in soil fertility and tilth under shifting cultivation. II. Changes in soil nutrient status. *Soil and Plant Nutr.*, **31**(2), 239–49.

Turenne, E.J.F. and Rapair, J.L. (1979) Culture itinerante et jachere forestiere: Mesures d'activite specifique de carbone de fractions de matiere organiques appliques a l'etude de renouvellement du stock organique en milieu forestier equatorial. *IAEA-Sm-235/35*. pp. 333–4. International Atomic Energy Authority: Vienna.

Turner, D.J. (1966) An investigation into the causes of low yield in late planted maize. *E.Afr.Agric.For.J.*, **31**, 249–60.

Turner, H.G. and Schleger, A.V. (1960) The significance of coat type in cattle. *Austr.J.Agric.Res.*, **11**, 645–63.

Turner, H.G. and Schleger, A.V. (1970) An analysis of growth processes in cattle coats and their relation to coat type and body weight gain. *Austr.J.Biol.Sci.*, **23**, 201–18.

Tyler, L. and Lee, R.P. (1990) Maintenance of health. *In*: W.J.A. Payne (ed.) *Introduction to Animal Husbandry in the Tropics*, pp. 35–86. Longman: London.

Uekermann, L., Joubert, D.M. and Steyn, G.Van D. (1974) The milking capacity of Boer goat does. *Wrld.Rev.Anim.Prod.*, **10**(4), 73–83.

UNESCO (1977a) *Development of Arid and Semi-arid Lands: Obstacles and Prospects.* Man and the Biosphere Technical. Note 6. UNESCO: Paris.

UNESCO (1977b) *Map of the World Distribution of Arid Regions.* Man and the Biosphere Technical Note 7. UNESCO: Paris.

UNESCO (1978) *Management of Natural Resources in Africa – Traditional Strategies and Modern Decision Making.* Man and the Biosphere Technical Note 9. UNESCO: Paris.

UNESCO/UNEP/FAO (1979) *Tropical Grazing Land Ecosystems.* Nat.Resources Res. No.16. UNESCO: Paris.

Unger, P.W. (1984) *Tillage Systems for Soil and Water Conservation.* FAO Soils Bull. No.54. FAO: Rome.

Unger, P.W. (1990) Conservation tillage systems. *Adv.Soil Sci.*, **13**, 27–68.

Urquhart, G.M., Armour, J., Duncan, J.L., Dunn, A.M. and Jennings, F.W. (1987) *Veterinary Parasitology.* Longman: London.

USDA (1975) *Soil Taxonomy.* Agric.Handbook No. 436. Soil Conservation Service, United States Department of Agriculture: Washington, DC.

USDA (1982) *Amendments to Soil Taxonomy.* Nat.Soil Taxon. Handbook 430-VI-Issue No.1. Soil Conservation Service, United States Department of Agriculture: Washington, DC.

USDA/ARS (1994) *Predicting Soil Erosion by Water: A Guide to Conservation Planning with the Revised Universal Soil Loss Equation.* Agric. Handbook No.703. Agricultural Research Service, United States Department of Agriculture: Washington, DC.

Vaccaro de Pearson, L. (1990) Survival of European dairy breeds and their crosses with Zebus in the tropics. *Anim.Breed.Abstr.*, **58**, 475–94.

Vallee, G. and Vuong, H.H. (1978) Floating rice in Mali. *In*: I.W. Buddenhagen and G.J. Persley (eds) *Rice in Africa*, pp. 243–8. Academic Press: London.

Vance, P.N., George, S. and Wohuinangu, J. (1983) Sulphur in the agriculture of Papua New Guinea. *In*: G.J. Blair and A.R. Till (eds) *Sulphur in South-East Asian and South Pacific Agriculture*, pp. 180–90. University of New England: Armidale. NSW.

Vandermaele, F.P. (1977) The role of animal production in world agriculture. *Wrld.Anim.Prod.Rev.*, **21**, 2–5.

Vandermeer, J. (1989) *The Ecology of Intercropping.* Cambridge University Press: Cambridge.

Van Es, A.J.H., Van Aggelen, D., Nijkamp. H.J., Vogt, J.E. and Scheele, C.W. (1973) *Z. Tierphysiol.Tierernahrgu.Futtermittelkde*, **32**, 121.

Van Hoorn, J.W. and Van Alphen, J.G. (1994) Salinity control. *In*: H.P. Ritzema (ed.) *Drainage Principles and Applications*, pp. 533–600. International Institute for Land Reclamation: Wageningen.

Van Hoven, W. (1984) Tannins and digestibility in greater kudu. *Can.J.Anim.Sci.*, **64**, 177–8.

Van Rensberg, P.J.J. (1956) Comparative values of fodder plants in Tanganyika. *E.Afr.Agric.For.J.*, **22**, 14–19.

Van Schilfgaarde, J. (ed.) (1974) *Drainage for Agriculture.* American Society of Agronomy: Madison, WI.

Venkateswarlu, N. and Rao, A.V. (1983) Response of pearl millet to inoculation with different strains of *Azospirillium brasiliensis*. *Plant and Soil*, **74**, 379–86.

Venter, H.A.W., Bonsma, J.C. and Skinner, J.D. (1973) The influence of climate on the reproduction of cattle. *Int.J.Biometeor.*, **17**, 147–51.

Vialard-Gordon and Richard, C. (1956) Etude Pluviometrique et physicometrique et économique des eaux de pluie a Saigon. *Agronomie Tropicale*, (11), 74.

Vicente-Chandler, J., Silva, S. and Figarella, J. (1959) The effect of nitrogen fertilization and frequency of cutting on the yield and composition of three tropical grasses. *Agron.J.*, **51**, 202–206.

Vilela, H., de Oliveira, S., Nascimento, C.H.F. and Gontijo, R.M. (1977) Efeito de pastagens de graminea e leguminosas sobre o ganho em peso de novilhos. l. época da 'seca'. *Arq.Esc.Vet.UFMG.*, **29**, 11–17.

Vine, H. (1953) Experiments on the maintenance of soil fertility at Ibadan, Nigeria. *Emp.J.Exp.Agric.*, **21**, 65–85.

Vogel, H. (1991) Conservation tillage for small-scale farming in Zimbabwe. *Proc.12th.Int.Conf., Int. Soil Tillage Res.Organ.*, Soil Tillage and Agricultural Sustainability. 8–12 July 1991, Ibadan, Nigeria, pp. 417–26. Ohio State University: Columbus, OH.

Vogel, H. (1993) Tillage effects on maize yield, rooting depth and soil water content on sandy soils in Zimbabwe. *Field Crops Res.*, **33**, 367–84.

Vogt, J.B.M. (1966) Responses to sulphur fertilization in N. Rhodesia. *Agrochimica*, **10**, 105–13.

Vries, C.A.de, Ferwerda, J.D. and Flach, M. (1967) Choice of food crops in relation to actual and potential production in the tropics. *Netherlands J.Agric.Sci.*, **15**, 241–8.

Walker, B.H. (1987) A general model of savanna structure and function. *In*: B.H. Walker (ed.) *Determinants of Tropical Savannas*, pp. 1–12. IRL Press: Oxford.

Walker, C.A. (1960) The population, morphology and evolutionary trends of apocrine sweat glands of African indigenous cattle. *J.Agric.Sci.*, **65**, 119–26.

Walker, D.H., Sinclair, F.L., Kendon, G., Robertson, D., Muetzelfeldt, D., Haggith, M. and Turner, G.S. (1994) *Agroforestry Knowledge Toolkit: Methodological Guidelines, Computer Software and Manual for AKT1 and AKT2, Supporting the Use of a Knowledge-based Systems Approach in Agroforestry Research and Extension*. School of Agriculture and Forest Sciences, University of Wales: Bangor.

Walker, G. (1996) Slash and grow. *New Scientist*, **151**(2048), 28–33.

Wallace, J.S. (1995) Towards a coupled light partitioning and transpiration model for use in intercrops and agroforestry. *In*: H. Sinoquet and P. Cruz (eds) *Ecophysiology of Tropical Intercropping*, pp. 153–62. INRA Editions: Paris.

Walter, H. (1972) *Ecology of Tropical and Subtropical Vegetation*. Oliver & Boyd: London.

Walton, P.D. (1962a) The effects of ridging on the cotton crop in the Eastern province of Uganda. *Emp.J.Exp.Agric.*, **30**, 63–76.

Walton, P.D. (1962b) Estimates of the water use by cotton crops at Serere, Uganda. *Emp.Cotton Grow.Rev.*, **39**, 241–51.

Wanous, M.K. (1990) Origin, taxonomy and ploidy of the millets and minor cereals. *Plant Varieties and Seeds*, **3**, 99–112.

Ward, H.K. (1968) Supplementation of beef cows grazing on veld. *Rhod.J.Agric.Res.*, **6**, 93–101.

Ward, H.K. (1987) *Crossbreeding: The Matapos Experiment*. Matapos Res. Sta.Bull. Matapos Research Station: Butawayo, Zimbabwe.

Ward, R.C. (1975) *Principles of Hydrology*. McGraw-Hill: London.

Warner, K. (1991) *Shifting Cultivators: Local Technical Knowledge and Natural Resource Management in the Humid Tropics*. Community Forestry Note No.8. FAO: Rome.

Warren, A. (1995) Changing understanding of African pastoralsim and the nature of environmental paradigms. *Trans.Inst. British Geographers*, **20**(2), 193–203.

Warren, A. and Khogali, M. (1992) *Assessment of Desertification and Drought in the Sudano-Sahelian Region 1985–91*. United Nations Sudano-Sahelian Office (UNSO): New York, NY.

Warrick, R.A., Barrow, E. and Wigley, T. (eds) (1993) *Climate and Sea Level Change: Observations, Projections and Implications*. Cambridge University Press: Cambridge.

Warrick, R.A., Shugars, H.H., Antionovsky, M.Ja., Tarrant, J.R. and Tucker, C.J. (1991) The effects of increased CO_2 and climate change on terrestrial ecosystems. *In*: B. Bolin, B.R. Doos, J. Jager and R.A. Warrick (eds) *The Greenhouse Effect, Climate Change and Ecosystems*, pp. 363–92. John Wiley & Sons: Chichester.

Watabe, T. (1981) *Report of the Scientific Survey on Traditional Cropping Systems in Tropical Asia, Parts 1 and 2*. Center for Southeast Asian Studies. Kyoto University: Kyoto, Japan.

Watson, G.A. (1983) Development of mixed tree and food crop systems in the humid tropics: a response to population pressure and deforestation. *Exp.Agric.*, **19**, 311–32.

Watson, G.A. (1990) Tree crops and farming systems in the humid tropics. *Exp.Agric.*, **26**, 143–60.

Weatherstone, J. (1992) Historical introduction. *In*: K.C. Wilson and M.C. Clifford (eds) *Tea: Cultivation to Consumption*, pp. 1–23. Chapman & Hall: London.

Webb, B.H. and Johnson, A.H. (1971) *Fundamentals of Dairy Chemistry*. AVI Publishing Co: Westport, CT.

Webster, A.J.F. (1992) The metabolisable protein system for ruminants. *In*: P.C. Garnsworthy, W. Haresign and D.J.A. Cole (eds) *Recent Advances in Animal Nutrition*, pp. 93–110. Butterworth-Heinemann: Oxford.

Webster, C.C. (1938) Experiments on the maintenance of soil fertility by green manuring. *Proc.3rd W.Afr.Agric.Conf.*, pp. 229–321. Colonial Office: London.

Webster, C.C. (1950) The improvement of yield in the tung oil tree (*Aleurites montana*). *Trop. Agric.(Trin.)*, **21**, 179–220.

Webster, C.C. and Baulkwill, W.J. (eds) (1989) *Rubber*. Longman: London.

Webster, C.C. and Wilson, P.N. (1980) *Agriculture in the Tropics*, 2nd edn. Longman: London.

Weerakoon, W.L. (1996) Issues in agroforestry development in Sri Lanka. *In*: P.A. Huxley and H. Ranasinghe (eds) *Agroforestry For Sustainable Development in Sri Lanka*. Proceedings of a 3-Day Participatory Training Course, 9–11 September, 1994, Colombo, Sri Lanka, pp. 29–59. University of Wales/University of Sri Jayewardenapura: Bangor/Colombo.

Weerakoon, W.L. and Seneviratne, A.M. (1982) Managing a sustainable farming system in Sri Lanka. *In*: *Proc.Brit.Crop ProtectionConf., Weeds*, 22–25 November 1982, Brighton, Vol.2, pp. 689–96. British Crop Protection Council: Croydon.

Weir, K.L., MacRae, I.C. and Allan, J. (1979) *Nitrogen*

Fixation Associated with the Root System of Tropical Grasses. CSIRO Trop.Crops and Pastures Div.Rep. 1977–8. CSIRO: Canberra.

Weiss, E.A. (1983) *Tropical Oil Seed Crops*. Longman: London.

Wellington, K.E., Mahadevan, P. and Roache, K.L. (1970) Production characteristics of the Jamaica Hope breed of dairy cattle. *J.Agric.Sci.*, **74**, 463–8.

Wesley-Smith, R.N. (1972) Liveweight gains of shorthorn steers on native and improved pastures at Adelaide River, Northern Territory. *Austr.J.Exp. Agric.Anim.Husb.*, **12**, 566–72.

Western, S. (1978) *Soil Survey Contracts and Quality Control*. Clarendon Press: Oxford.

Wetselaar, R. (1962) Nitrate distribution in tropical soils. II. Downward movement and accumulation of nitrates in the subsoil. *Plant and Soil*, **16**, 19–31.

White, J.M., Vinson, W.E. and Pearson, R.E. (1981) Dairy cattle improvement and genetics. *J.Dairy Sci.*, **64**, 1305–17.

Whiteman, P.C. (1977) Pastures in plantation agriculture. In: *Reg.Seminar Pasture Res.Dev.Solomon Is. and Pacific Region Proc.*, Honiara, Solomon Islands, pp. 144–53.

Whiteman, P.C. (1980) *Tropical Pasture Science*. Oxford University Press: Oxford.

Whitmore, T.C. (1989) Tropical forest nutrients, where do we stand? In: J. Proctor (ed.) *Mineral Nutrients in Tropical Forest and Savannah Ecosystems. Special Publication No. 9 of the British Ecological Society*, pp. 1–13. Blackwell Science: Oxford.

Whittaker, R.H. (1975) *Communities and Ecosystems*. Macmillan: New York, NY.

Whittemore, C.T. and Elsley, F.W.H. (1976) *Practical Pig Nutrition*. Farming Press: London.

Whittow, G.C. (1971) Ungulates. In: G.C. Whittow (ed.) *Comparative Physiology of Thermoregulation. Vol.2, Mammals*, pp. 191–280. Academic Press: New York, NY.

Wide, G.V. (1987) Practical utilisation of agro-industrial by-products in a feedlot in Malawai. In: *Utilization of Agricultural By-products as Livestock Feeds in Africa*, Proc.Afr.Res.Network for Agr. By-products (ARNAB). International Livestock Centre for Arid Regions: Addis Ababa.

Wiersum, K.F.P., Anspach, C.L., Boerboom, J.H.A., de Rouer, A. and Veer, C.P. (1985) *Changes in Shifting Cultivation in Africa*. FAO Forestry Paper 50/1. FAO: Rome.

Wight, W. (1958) *Annual Report of Tocklai Experiment Station, 1958*. Tocklai Experiment Station: Tocklai, Assam.

Wigley, T.M.L. and Raper, S.C.B. (1987) The global expansion of sea water associated with global warming. *Nature*, **330**, 127–31.

Wild, A. (1950) The retention of P by soil: A review. *J.Soil Sci.*, **1**, 221–38.

Wild, A. (ed) (1988) *Russell's Soil Conditions and Plant Growth*, 11th edn. Longman: Harlow.

Wilding, L.P. and Hossner, L.R. (1989) Causes and effects of acidity in Sahelian soils. In: *Soil, Crop and Water Management Systems for Rainfed Agriculture in the Sudano-Sahelian Zone*, Proceedings of an International Workshop, 11–16 January 1987, ICRISAT Sahelian Centre, Niamey, pp. 215–27. International Crops Research Institute for the Semi-Arid Tropics: Patancheru, India.

Willcocks, T.J. (1969) *Animal-drawn toolbar*. NIAE Overseas Liaison Unit Tech.Bull. No.2. National Institute of Agricultural Engineering: Silsoe, UK.

Willcocks, T.J. (1984) Tillage requirement in relation to soil type in semi-arid rainfed agriculture. *J.Agric.Eng.Res.*, **30**, 327–36.

Willcocks, T.J. and Twomlow, S.J. (1992) An evaluation of sustainable cultural practices for rainfed sorghum production on Vertisols in east Sudan. *Soil Tillage Res.*, **24**, 183–98.

Willey, R.W. and Osiru, D.S.O. (1972) Studies on mixtures of maize and beans (*Phaseolus vulgaris*) with reference to plant population. *J.Agric.Sci.*, **79**, 517–29.

Williams, E. (1970) Factors affecting the availability of soil phosphates and the efficiency of P fertilizer. *Proc.Anglo-Soviet Symp.Agro-Chem.Res.*

Williams, J.R., Renard, K.G. and Dyke, P.T. (1983) A new method for assessing the effect of erosion on productivity – the EPIC Model. *J.Soil and Water Conservation*, **38**, 381–3.

Williams, W.A. (1969) Effects of nitrogen from legumes and crop residues on soil fertility. *Proc.Conf.Biol.Ecol. of Nitrogen*. National Academy of Science: Washington, DC.

Williamson, G. and Payne, W.J.A. (1965) *An Introduction to Animal Husbandry in the Tropics*. Longman: London.

Willis, M.B. and Preston, T.R. (1970) Performance testing for beef: inter-relationships among traits in bulls tested from an early age. *Anim.Prod.*, **12**, 451–6.

Wills, J. B. (1962) The general pattern of land use. In: J.B. Wills (ed.) *Agriculture and Land Use in Ghana*, pp. 201–28. Oxford University Press: Oxford.

Wilmut, I. (1970) Embryo transfer in cattle breeding. *Wrld.Anim.Rev.*, **35**, 30–35.

Wilson, G.F., Adeeb, N.N. and Campling, R.C. (1973) Apparent digestibility of maize grain when given in various physical forms to adult sheep and cattle. *J.Agric.Sci.*, **80**, 259–67.

Wilson, J.R. and Ford, C.W. (1973) Temperature influences on the *in vitro* digestibility and soluble carbohydrate accumulation of tropical and temperate grasses. *Austr.J.Agric.Res.*, **24**, 187–95.

Wilson, K.C. and Clifford, M.C. (1992) *Tea: Cultivation to Consumption*. Chapman & Hall: London.

Wilson, P.N. (1957) Studies on the browsing and reproductive behaviour of the East African dwarf goat. *E.Afr.Agric.For.J.*, **23**, 138–47.

Wilson, P.N. (1958a) The effect of plane of nutrition on the carcass development and composition of the East African dwarf goat. I. Effect on the liveweight gains and the external measurements of the kids. *J.Agric.Sci.*, **50**, 198–210.

Wilson, P.N. (1958b) The effect of plane of nutrition on the growth and development of the East African dwarf goat. II. Age changes in the carcass composition of female kids. *J.Agric.Sci.*, **51**, 4–21.

Wilson, P.N. (1960) Effect of plane of nutrition on the growth and development of the East African dwarf goat. III. The effect of plane of nutrition and sex on the carcass composition of the kid at two stages of growth. *J.Agric.Sci.*, **54**, 104–30.

Wilson, P.N. (1961a) Observations on the grazing behaviour of cross-bred Zebu Holstein cattle managed on Pangola pastures in Trinidad. *Turrialba*, **11**, 55–71.

Wilson, P.N. (1961b) Palatability of water buffalo meat. *J.Agric.Soc. Trin.and Tob.*, **61**, 461–4.

Wilson, P.N. (1961c) The grazing behaviour and free-water intake of East African shorthorned Zebu heifers at Serere, Uganda. *J.Agric.Sci.*, **56**, 351–64.

Wilson, P.N. (1973) Livestock physiology and nutrition. *Phil.Trans.Roy.Soc.London B.*, **267**, 101–12.

Wilson, P.N. (1977) Inter-relationship between nutrients, feed, milk and money. *In*: *Preventive Medicine in Bovine Practice*, Proceedings of British Cattle Veterinary Association Conference, pp. 24–6. British Veterinary Association: London.

Wilson, P.N. (1979) Concentrates. *In*: W.H. Broster and H. Swan (eds) *Feeding Strategy for the High Yielding Dairy Cow*, pp. 374–97. Grenada Publishing: London.

Wilson, P.N., Barratt, M.A. and Butterworth, M.H. (1962) The water intake of milking cows grazing Pangola pastures under wet and dry season conditions in Trinidad. *J.Agric.Sci.*, **58**, 257–64.

Wilson, P.N. and Brigstocke, T.D.A. (1977) The commercial straw process. *Process Biochem.*, **12**(7), 17–21.

Wilson, P.N. and Osbourn, D.F. (1960) Compensatory growth after under nutrition in mammals and birds. *Bio.Rev.*, **35**, 324–59.

Wilson, R.T. (1984) *The Camel*. Longman: London.

Wilson, R.T. (1989) *Ecophysiology of the Camelidae and other Desert Ruminants*. Springer-Verlag: London.

Winter, W.H. (1987) Using fire and supplements to improve cattle production from monsoon tallgrass pastures. *Trop.Grassl.*, **21**, 71–81.

Winter, W.H., Edye, L.A. and Williams, W.T. (1977) Effects of fertilizer and stocking rate on pasture and beef production from sown pastures in Northern Cape York Peninsular. 2. Beef production and its relation to blood, faecal and pasture measurements. *Austr.J.Exp.Agric.Anim.Husb.*, **17**, 187–96.

Wischmeier, W.H. and Smith, D.D. (1978) *Predicting Rainfall Erosion Losses*. Agric. Handbook No. 537. USDA: Washington, DC.

WMO (1974) *Guide to Hydrological Practices*, 3rd ed. World Meteorological Organization No.168. World Meteorological Organization: Geneva.

Wong, P.W. (1964) Evidence for the presence of growth inhibiting substances in *Mikania cordata* (Burmf). *J.Rub.Res.Inst.Malaysia*, **18**, 231–42.

Wood, A.W. (1979) The effects of shifting cultivation on soil properties: An example from the Karimin Bomai Plateaus, Simbu Province, Papua New Guinea. *Papua New Guinea Agric.J.*, **30**, 1–14.

Wood, G.A.R. and Lass, R.A. (1985) *Cocoa*, 4th edn. Longman: London.

Woodhead, S. and Cooperdriver, G. (1979) Phenolic acids and resistance to insect attack in *Sorghum bicolar*. *Biochem.Systematics and Ecol.*, **7**, 309–10.

Woods, A.J. (1985) Sampling in animal health surveys. *Proc.Soc.Vet.Epid. and Prevent.Med.*, Reading.

Woolfe, J.A. (1992) *Sweet Potato*. Cambridge University Press: Cambridge.

Woomer, P. and Swift, M.J. (eds) (1994) *The Biological Management of Tropical Soil Fertility*. John Wiley & Sons: Chichester.

Woomer, P.L. and Swift, M.J. (1995) *The Biology and Fertility of Tropical Soils: Report of the Tropical Soil Biology and Fertility Programme, 1994*. Tropical Soil Biology and Fertility Programme: Nairobi.

World Bank (1984) *Land, Food and People*. FAO: Rome.

World Bank (1989) *Sub-Saharan Africa; From Crisis to Sustainable Growth*. World Bank: Washington, DC.

Wray, N.R., Simm, G., Thompson, R. and Bryan, J. (1991) Application of BLUP in beef breeding programmes. *Brit.Cattle Breeders' Club Digest*, **46**, 29–39.

Wricke, G. and Weber, W.E. (1988) *Quantitative Genetics and Selection in Plant Breeding*. De Gruyter: Berlin.

Wright, J.L. (1981) Crop coefficients for estimates of daily crop evapotranspiration. *In*: *Proceedings of Irrigation Scheduling Conference*, pp. 18–26. Chicago. American Society of Agricultural Engineers: St. Joseph, MI.

Wright, J.L. (1982) New evapotranspiration crop coefficients. *Proc.Am.Soc.Civ.Eng., Irrign.Drain.Div.*, **108**, 57–74.

Wrigley, G. (1988) *Coffee*. Longman: London.

Wrigley, T.M.L. and Raper, S.C.B. (1987) The global expansion of sea water associated with global warming. *Nature*, **330**, 127–31.

Wuethrich, B. (1995) El-Niño goes critical. *New Scientist*, **145**(1963), 32–5.

WWF (1993) *Some Like it Hot*. World Wildlife Fund: Gland.

Wyman, R.L. (1991) Multiple threats to wildlife. *In*: R.L. Wyman (ed.) *Global Climate Change and Life on Earth*, pp. 134–55. Routledge: London.

Yadav, B.C. and Suryanto, H. (1990) Comparative study of agriculture and mechanization in India and Indonesia. *Agric.Mech.in Asia, Africa and Latin America*, **21**, 59–66.

Yagil, R. (1985) The desert camel, Comparative physiological adaptation. *In: Comparative Animal Nutrition, Vol.5*. Karger: Basel.

Yamoah, C.F., Ay, P. and Agboola, A.A. (1986) The use of *Gliricidia sepium* for alley cropping in the Southern Guinea savanna zone of Nigeria. *Int.Tree Crops J.*, **3**(4), 267–79.

Yoshida, S. (1981) *Fundamentals of Rice Crop Science*. International Rice Research Institute: Los Banos, Philippines.

Youdeowei, A., Ezedinma, F.O.C. and Onazi, O.C. (1986) *Introduction to Tropical Agriculture*. Longman: London.

Young, A. (1989) *Agroforestry for Soil Conservation*. CAB International: Wallingford.

Young, A. (1990) Agroforestry, environment and sustainability. *Outlook on Agric.*, **19**, 155–60.

Young, A. (1997) *Agroforestry for Soil Management*, 2nd edn. CAB International: Wallingford.

Young, H.M. (1973) No-tillage farming in the United States – its profit and potential. *Outlook on Agric.*, **7**, 143–8.

Younge, O.R. and Plucknett, D.L. (1966) Quenching the high phosphorus fixation of Hawaiian Latosols. *Soil Sci.Soc.Amer.Proc.*, **30**, 653–5.

Yousef, M.K. (ed.) (1985) *Stress Physiology in Livestock, Vol.II, Ungulates*. CRC Press: Boca Raton, FL.

Yudelman, M., Coulter, J., Goffin, P., McCune, D. and Ocloo, E. (1992) An evaluation of the Sasakawa–Global 2000 project in Ghana. *In*: N.C. Russell and G.R. Dowswell (eds) *Africa's Agicultural Development in the 1990s: Can it be Sustained?*, pp. 45–55. Centre for Applied Studies in International Negotiations/Sasakawa Africa Association/Global 2000: Mexico, DF.

Zake, J.Y.K. (1987) Soil fertility and fertilizer research in Uganda. *In*: H. Ssali and L.B. Williams (eds) *Proc. East and Southeast African Fertilizer Management and Evaluation Network Workshop*, pp. 147–60. International Fertilizer Development Center: Muscle Shoals, AL.

Zoschke, A. (1990) Yield losses in tropical rice as influenced by the composition of the weed flora and the timing of its elimination. *In*: B.T. Grayson, M.B. Green and L.G. Copping (eds) *Pest Management in Rice*, pp. 300–13. Elsevier: London.

INDEX

Note: Primary entries are indicated in bold.

Abahima **391–3**
Acacia albida 101
Acacia mellifera 319
Acacia nilotica 101, 230, 311
Acacia pennata 321
Acacia raddiana 109
Acacia seyal 311
Acacia spp. 166, 236–7
Acacia themeda 314, 321
acclimatisation 390
acetate 309, 421
acid deposition **18**
acid treatment (of straw) 307
acrisols 9, 71
adaptation **390**
aflatoxin 269
African buffalo (*Syncerus caffer*) 340, 375, 377
Africander **see cattle**
agrochemicals **18**
agroforestry **160–66, 222–58**
agrosylviculture 223
artificial insemination (AI) 405, 462, **464–6**, 468, 472, 480–81, 484
akee (*Blighia sapida*) 259
albumin 459
alfalfa **see lucerne**
alfisols 53, 69–71, 153
alkali treatment (of straw) 307–8
alkaloids 268, 277–8, 310
   caffeine 277
   nicotine 278
   theobromine 277
alley cropping 99, 166–7, 176, 228, 230

alpacas 316, **351–7**, 463
aluminium 49, 53, 60, 67–8, 90, 168, 174, 240
   phosphate 61
   toxicity 64
amaranthus 263
amino acid 304, 308, 311, 456
   balance 428
   composition **425–6**
   requirements 445
ammonia 192, 303, 426
Anabaena 59
Ananus comosus **see pineapple**
Andropogon gayanus (Blue grass) 100, 161, 313, 326
animal behaviour – effect of heat on **383–5**
   by-products **303–4**, 307
   draught/power **see draught animals**
   feed **294–312**
   protein factor (APF) 431
Ankole cattle **see cattle**
anthrax 304, 451, 469
antibiotics 450
antibodies 435, 452
antigen 452
anti-nutritional factors **309–12**
antioxidant 430
Arachis hypogoea **see groundnut**
Aristida kewensis 314, 321
arrowroot (*Maranta arundinacea*) 258
ash 295–7, 354, 365
   nutritional value 407, 428

aubergine (Egg plant) (*Solanum melongena*) 259
avocado (*Persea americana*) 258, **268**, 287
Azadierachta indica (Neem) 101, 109
Azotobacter 326

bacteria 421, 460
bacterial count 412
   disease **451–2**
Bagasse 273, 303, 305
   pith 303
Bambara groundnut (*Voandzeia subterranea*) 258
bamboo **see Dendrocalmus spp.**
banana 42, 139, 200, **266–7**, 282, 289
   clones 287, 291
   food value 16, 159, 272
   intercropping 151
   plantations 137
   taxonomy 258
   weed control 138
Banteng (*Bibos javanicus*) 339
barley as animal feed 355
   amino acid composition **425**
   protein degradability 427
beans (*Phaseolus spp.*) 66, 136, 152, 165, 171, 227, 258, **268**, 283
beefalo 336
Berseem (*Ttrifolium alexandrinum*) 193, 295
Betel nut (*Areca catechu*) 259, 278

biological control 211, 252
  requirement 433
biomass 36–7, 147–8, 211, 240–41, 243, 245, 249, 251, 253, 282, 287
biomes 22, 24, 47
birthweight 387, 471
bison (*Bison bonasus*) 336, 340
blister blight 275
blackarm disease 279
blood 394, 458–9, 466
  meal 330
blue green algae 58, 192–3
best linear unbiased prediction (BLUP) 486
Bonsmara see cattle
bore holes 335, 401–2
boron 65–6
Bos bubalis bubalis **see water buffalo**
Bos indicus **see cattle: zebu**
Bos spp.
  B. bibos gaurus (Guar) 339, 340
  B. bibos frontalis (Gayal) 339
  B. javanicus (Banteng) 339
  B. nomadicus 339
  B. opisthonamus 339
  B. primigenius 339
  B. sauveli (Kouprey) 339
bovine spongiform encephalopathy (BSE) 304, 431
Brachiaria
  B. brizantha 410
  B. decumbens 331
  B. humidicola 100
  B. latifolia 100
  B. multiformis 410
  B. mutica (Para grass) 100, 295, 297–300, 324, 326, 330
Brahman see cattle, breeds
bread fruit (*Artocarpus altilis*) 258, **268**, 272
breeding – animal
  fractional 476
  index 486
  line 479
  seasonal 348, 387, 464
  value 480
breeding – plant **286**
breed societies 467, 479, **480–81**

bride price 391–2
broiler (chicken) 309, 336, 418
browse **319**, 354, 409
brucellosis 469
buffalo **see water buffalo**
  other species 340, 375–6
bunds 103–7, 109–10, 186–7
bulrush millet **see millet**
burning 63, 113–4, 125, 145, **148–50**, 153, 311, 320–21, 323, 325, 332
bush 320
  clearing 333
  control 320, 322–3
butter 398
  fat content 344, 350, 360, 365, 424, 470–71, 478, 485
  yield 386
butyrate 309
butyric acid 421

Cajanus Cajan **see Pigeon pea**
calcium
  in animal nutrition 310, 354, 366, 442, 444, 458
  in soil nutrition **62–5**, 68, 73, 75, 147–8, 155, 174, 193
  nutritional requirements for 429–30, 432, 436–7, 445
  phosphate 61
calf
  at foot 411
  mortality 436
  nutrition 435
  rearing 410
Calliandra spp. 166, 233
Calopogonium mucunoides 102, 225, 260
calorie/calorific value 421
calving
  index 387, 402
  interval 348, 463, 476
  percentage 329, 396
  seasonal 464
cambered beds 115–6
cambisols 9, 72
camels 44, 130, **351–7**, 377, 463
  Arabian (*Camelus dromedarius*) 336, 351, 353, 371, 375, 394
  Bactrian (*Camelus bactrianus*) 336, 351

nutrition 430, 435
  poll gland, of 353
  water requirements 435
Camellia sinensis **see Tea**
Canavalia ensiformis (Jack bean) 100, 164, 260
carbohydrate 311, 317–8, **419–24**
carbon 188, 250
carbon dioxide 35–6, 39, 42, 45–6, 48, 53, 59, 188–90, 193
  solid 480
carotene 354, 430
carrying capacity 322, 324, 327–30, 332, 392, 400, 403
casein 365, 426
cashew (*Anacardium occidentale*) 64, 258, 274
cashmere 364, 368
cassava (*Manihot esculenta*) 42–3, 64, 125, 151–3, 155, 157, 174, 183, 186, 227, 242, 246, 258, **264–5**, 282, 287, 291, 302–3, 306, 414, 420
Cassia siamea 101
Cassia tora 311
castor (*Ricinus communis*) 121, 258, 272
Casuarina papuena 58, 101, 239
catena 50, 79
cattelo 336, 340
cattle *Bos spp.* **337–42**
  Africander 387, 397, 480
  Ankole 391, 484
  Banteng 339
  Beefmaster 480
  Bibovine sub group 339
  Bisontine sub group 340
  Blood groups 466
  Bonsmara 480
  Brahman 381, 479
  Brangus 480
  breeding 348, 387, 466, 470–71, 474, 476, 479–80, 486
  Brown Swiss 327
  Bubaline sub group 340
  Cattalo 336, 340
  Charbray 480
  Criollo 350, 396
  crossbred 3, 305, 337, 375, 385, 397, 403, 411, 413, 415, 419, 468, 474–5, 476–7, 479–80
  crush 401

dip 397, 401
European (*Bos taurus*) **336–43**, 345, 347, 350, 365, 415, 429, 479, 485
evolution, of **338**
fat deposition, in 424
fattening, nutrition of **417–46**
feed, supplementary 400, 402, **406–7**
feeding standards **433–45**
genetic improvement, of **362–86**
grazing behaviour, of 384–5, 408
grazing, mixed 323
grazing, seasonal, of 394
growth rate 399, **436–7**
hair, of 357, 371, 376–7, 380–81, 390
heat tolerance, of 372, 375–6, 386–9
housing **407–9**, 451
Jamaica Black 480
Jamaica Hope 477–8, 480
Jamaica Red 480
Kenana 341
liveweight gains, of 433, 449, 457
management **391–416**
milk production 396–7, 406, 417, 422, 457, 469–71, 483
milk quality 308–9, 365, 386, 427
milk substitute 435–6
milk yields 343, 349, 386, 397, 413, 437, 469–70, 475–8, 483–5
milking parlours 407, 413
milking procedures **410–14**
Nellthrop 480
Nilotic 341
nutrition 350, 418, 459
nutritional requirements, of **434–8**
pastoralism **327–9**
ranching 330, 335, 340, 393, **396–403**, 450
reproduction 463
Sahiwal 327, 329, 341, 469, 473–4, 477
Sanga 342
Santa Gertrudis 379, 397, 473, 477, 479–80, 482
skin **381–2**

species see *Bos*
stocking rate 325, 327, **330–33**, 357, 394, 397, 410, 416
sweat 373–7
Taurine sub group 339
water requirements, of 388, 419, 435
water supplies, for 332, 334, **443**
working, nutrition of **437–8**
Zebu (*Bos indicus*) 305, **336–42**, 347, 350, 355, 365, 367, 379, 397, 415, 469, 485
cellulose 303, 308, 420
Cenchrus ciliaris 100, 314, 334
Centrosema pubescens 100, 102, 160, 162, 260, 300–301, 326, 410
cereals **302**, 307
cereal straws 56, 303, 308, 354, 421, 430, 435
cereal by-products 302, 441
cestodes 455
Chick pea (*Cicer arietinum*) 162–3, 258, 270, 290
chlorine 429
chloride 435
Chloris gayana (Rhodes grass) 100, 331, 334, 361
choline 442, 444
chromium 66, 429
Chromolaena odorata 146, 153, 157
cinnamon (*Cinnamomum zeylanicum*) 259, 278
citrus fruits (*Citrus spp.*) 64–5, 115, 236, 259, 267, 274, 287
Creutzfeldt-Jacob Disease (CJD) 431
clay 9, 49, 52–5, 59, 61–2, 64, 72, 76, 104, 115, 119, 122–3, 125, 139, 182, 186, 188
clearing **149**
climate **20–47**, **49–50**
climate classification **20–24**
climatic change **18**, 35, **45–7**
clone 204, 218, 285–7, 289, **291**
clove (*Eugenia caryophyllus*) 278
coat
 colour 380
 score 381
 texture 380
cobalt 58, 429

coca (*Erythroxylon coca*) 259, 278
cocoa/Cacao (*Theobroma cacao*) 42, 66, 74, 115, 136, 201, 203–4, 206, 224, 259, 267, **277–8**, 287, 289, 291, 311, 414
cocoa butter 277
coconut (*Cocos nucifera*) 42, 63, 74, 115, 200, 202–4, 224, 258, **267**, **271–2**, 277, 282, 286–7, 291, 406, 410
Cocoyam (*Xanthosoma sagittifolium*) 42, 258
coffees (*Coffea arabica*; *C. robusta*) 43, 63–6, 74, 139, 159, 200–201, 204, 206, 222, 224, 236, 241, 251, 259, **275–7**, 282, 286–7, 290–92, 414
coffee berry disease 278
cola nut (*Cola nitida*) 259, 278
Colocasia (Taro) **266**
colostrum 356, 435, 452
colostrum test 436
comfort zone 343, 371–2
Commiphora – Acacia 314–5
compost **170–71**, 451
concentrates 294, **305–9**, 411
concentrated feed 399
conception rate 386–7, 464
conduction 374, 377
contemporary comparison test 484
continuous cropping 191
continuous stocking 322
convection 374, 377
copper 65–6, 326, 442, 444
 deficiency 74, 428
 in animal nutrition 429, 431, 458
cystine **425**
cotton (*Gossypium spp.*) 43, 63–4, 66, 73, 102, 121, 127–8, 136, 139, 153, 157–9, 171–2, 174–5, 259, **279–80**, 286, 290, 292, 303
cottonseed 354, 402
 cake 305
 husk 305
 meal 406, 427
cover crops 99, 215, 260
Cowpea (*Vigna sinensis*; *V. unguiculata*) 90, 122, 124, 161–3, 174, 193, 224, 258, 290, 300
critical growth periods 318–9, 322

crop
  by products **304**, 399, 406, 414
  establishment 182
  residues **304**
  yield 88
cross breeding 474–5, 477 **see also cattle, crossbred**
Crotalaria juncea 162–3, 193, 260
  paulina 164
crude fibre (CF) 295–7, 300, 305, 350, 407
cucumber (*Cucumis sativus*) 35, 259, 274
Cymbopogon citratus 99
  nardus 99, 108
Cynodon dacrylon (Bermuda grass) 32, 100, 326, 330–31, 334
  plectostachyum 321, 326

dairy cows, nutritional requirements of **436–7**
dams 95, 103, 106, 110, 335, 401
Dasheen (*Colocasia esculenta*) 151, 250, 257
date palm (*Phoenix dactylifera*) 274, 287
day length 210, 387–8
defoliation **317**
Dendrocalmus spp. (bamboo) 101
derris (*Derris elliptica*) 260, 282
desert 22, 24, 43–4
desertification 17, 86
Desmodium spp. 325
  introtum 300–301
  uncinatum 300
digestibility 307–8, 311, 318–9, 324
digestion 306, 309–10
Digitaria abysinnica 321
  ciliaris 196
  decumbens (Pangola grass) 100, 152, 301, 318, 324–6, 331, 406–7, 413
Dioscorea spp. **see yams**
dipping 401, 455
disease
  control 335
  general **447–61**
  production 458
  resistance 287, 485
  tolerance 475
  vectors 447, 457

desoxyribonucleic acid (DNA) 469, 485
Dolichos lablab (or *Lablab purpureus*) 100, 270, 299–300
drainage **113–43**
drains, tile 114
draught animals 121, **130–33**, 340, 347, 404, 412
  nutritional requirements, of 437
  power 131, **132–3**, 357
dressing percentage 347, 406, 471
drought 35–6, 185, 399
dry season feeding **406–7**
durian (*Durio zibethynus*) 258

earthworms 53, 123
East coast fever (*Theileriosis*) 449, 451, 454
Eddo (*Colocasia esculenta*) 151, 257
egg 309, 382, 418
  production 382, 417, 422, 444, 470, 486
  weight 382, 445, 471
egg plant (aubergine) (*Solanum melongena*) 259
Elaeis guineensis **see oil palm**
elephant 336
elephant grass **see *pennisetum purpureum***
Eleusine corrcana **see Finger Millet**
  indica 196
  jaegeri 315
embryo transfer (ET) 466, 480
energy
  density 445
  digestible (DE) 421–3, 441
  gross (GE) 421–2
  metabolisable (ME) 305, 382, 422–3, **436–8**
  net (NE) 422–3
  requirements 355, 437–9
entisols 69–70, 72, 182
enzyme 306, 423, 426
equines 130
Eragrostis spp. 314
  curvula 100, 324
  superba 334
erosion
  gully 85, 105, 333
  hazard 95

by overgrazing 327, 334–5, 357, 368–9
sheet 85, 127
soil 17, 55, 80, 82–3, 91–3, **154–5**, 160, 187, 224
of soil nutrients 148
water 84, 86–7, 122
water, control of 102–3, 114, 118, 122, 126, 156–7, 224, 236, 317
wind 82, 85, 122
Erythrina fusca 225, 236, 277
ether extract (EE) 295, 354, 407
evaporation 25, 37, 49, 376–7
evaporation rate 401
evaporative cooling **373–4**
evaporative losses 419
evapotranspiration **38–41**, 49, 123, 210, 239

fallow 54–5, 63, 65–6, 68, 144–5, 148–9, 151, 153, 156–7, 159–60, 174–5, 186, 222, 228, 230, 295
farmyard manure **168–71**
Fasciola spp. 456, 460
fat 347, 355–6, 419–20, **423–4**
  abdominal 424
  deposition 424
  intramuscular 424
fatty acids 366, 423
fast degradable protein (FDP) **see protein**
feed conversion 470
  efficiency 417
  formulation 306
  intake **308–11**, 385, 442
  lots 355, 400, 414–5, 479
  specification 306
feeding
  stall 169
  standards **432–45**
  value 318
fences 333, 397, 401, 404–5
  electric 405
fencing 322, 332
ferrasols 9, 71
fertiliser 8, 13, 16–18, 46, 55, 58, 60–61, 63, 65, 88, 91, 137, 144, 158, 160, 164, 168–9, **171**, 173–6, 188, 211, 219–20, 222–3, 247, 253, 298, 323–4, 414, 430

use 14
fertility
  cattle 385, 387
  poultry 445
  sheep and goats 388
fibre 319, 420, 441
Finger millet (*Eleusine coracana*)
  see millet
fire 156, 246, 314, 320–22, 328, 410
fish 418
fish – meal
  amino acid content **425**
  nutritional value 305, 428, 440
  ponds 73
flies 408, 451–3, 455
fluorine 310, 458
fluorosis 429
fodder legumes **299**
  crops 420
foggage 406, 419
foot-and-mouth disease 449, 457
forage crops **294–301, 309**
forest **147–8**, 156
formaldehyde treatment (of straw) 307–8
fruit **274**
Fulani 328
fungal diseases **457–8**
fungi 288, 421
fungicides 457

Garcinnia indica 311
  mangostana 258
Guar (*Bos bibos gaurus*) 339–40
Gayal (*Bos bibos frontalis*) 339
genes 469
genetic improvement 417, **469–81**
genome map 485
gestation period 348, 354, 368, 463
ginger (*Zingiber officinale*) 259, 278
gleys/gleysols 9, 188
Gliricidia sepium 101, 166–7, 225, 233, 277
global warming **18**
glucose 421, 459
glucosinolate 310
glycine **425**
goat **357–70**
  African type 360–66
  Alpine 559, 364
  Anglo-Nubian 362, 364–5, 473

Angora 359, 366
Asiatic type 360, 362, 364, 366
Barbari 364, 366
Boer 361–2, 364, 366
breeds **346–7**
browsing, by 319–21, 418
bush control, by 323
classification **358–62**
effect of erosion 369
European type **358–9**
fertility 386–7, 466
Galla 361
gestation period 368
heat tolerance 373, 375, 377
Jumna Pari 360, 362, 364, 366
Kashmir 360, 366, 368
Malabari 364, 366
Marota 362
milk composition 365
milk yields **365–7**, 483
mixed stocking, with 331–3, 393–5
Nubian 359, 362, 366
nutrition 430
Oriental type 359–60
reproduction 463, 466
Saanen 358, 364, 368
skin 368
Small East African Dwarf 361
Sokota 368
Togenberg 358, 364
water requirements, of **435**
West African Dwarf 366
Golden berry (*Physalis peruviana*) 259
gossypol 310
grass 260, 294
  strips 106
  legume mixtures **299**, 416
  rains 121, 320
grazing 385
  alternate 455
  behaviour 408
  communal 333
  continuous 323
  management **322–5**
  mixed 323
  pattern 384
  period 384
  rotational 319, 322–3, 327, 332–3, 335, 394, 396, 404, 455
  strip 319

zero 319, 322, 327, 405, 456
green manure 163–4, 176, 193, 260
ground cover 317
groundnut (*Arachis hypogoea*) 43, 121, 126–7, 158, 174, 245, 258, **269–70**, 302, 402
  amino acid composition, of **425**
  Bambara 227, 258
  breeding 290
  fertiliser requirements, of 63, 66, 171–2
  intercropping, with 151–2
  nitrogen fixation, by 162–3
  protein degradability 427
  weed control, in 136, 139
  yield 165
growth rate 385–7, 470–71, 477
  compensatory 399
  hormone 485
Guatemala grass **see *Tripsacum laxum***
Guava (*Psidium guajhava*) 259

Haemonchosis/Haemonchus 456
hair 357, 371, 377
  coat 380–81
  colour 390
  fibre 376, 380
  follicle 380
harrow 119, 334
hay 295, 306, 324, 400, 421
  leguminous 300, 308
  nutritional quality of 319, 427, 435
  standing 316, 322, 325, 400, 406, 419
  vitamin content of 432
hedgerow intercropping 228–9
health, of livestock **447–61**
heat
  increment 422
  radiant 451
  stress 372, 389
  tolerance 359, **377–80**, 389, 478, 485
  tolerance index **377–80**
helminths **455–7**
hemicellulose 308, 420
hen 336, 384, 418
  battery 384
  laying 336
  nutrition, of 428

herbicide 103, 118, 123, 138, 140, **142–3**, 183, 187, 195, 211, 220, 309
herd book 467, 479, 481
   elite 467
heritability 284, 348–9, 470–72, 483–4
Heteropogon contortus 324, 326, 331
heterosis
   in animals 477
   in plants 284–5, 287
Hevea brasiliensis 311
HI (Harvest Index) 203–4
histidine **425**, **440**
histosols 9, 69–70
hoe 123–4, 126, 129, 137, 151–2, 334
homeothermy 371, 374, 377–8
hormone 310, 424
horns
   buffalo 344, 346, 351
   European cattle 343
   goat 359
   type 477, 481, 484
   Zebu cattle 341, 343, 391–2, 484
horses 130, 336–7, 375, 463
housing **407–9**, **451**
human health/nutrition **15–6**
   power **129–30**
humidity 38, 388, 401, 409
hump 341–3, 351, 355, 392
humus 51, 53, 66–8, 72, 146, **154**, 164
hurricanes 31
hybrid
   seed 197
   varieties 287, **290**
   vigour 477
hydrolysis 48, 308
hygiene, of livestock **447–61**
Hyparrhenia dissoluta 324
   filipendula 324
   rufa 324, 331
   spp. 314, 323, 325–6

Imperata cylindrica 146, 149, 152–3, 157, 313, 330–31
inbred lines (IBL) 286, **290**
insecticide 183, 196, 211, 261, 270, 278, 282, 334
inceptisols 72, 182

intercropping 151–2, 158, 165
iodine, animal requirements, for 429, 442, 444
Ipomea batatus **see sweet potato**
Iron
   animal nutrition, in 429–30, 442, 444
   compounds 60
   ferrous 181, 194
   oxides 50, 53, 174
   phosphate 61
   pyrites 85
   reduction 190
   silicates 49
   soil deficiency, of 65–6
irrigation **10**, 13–14, 84, 95–6, 116–18, 139, 179, 185, 217, 222–3, 247, 316, 335
ITCZ (Intertropical Convergence Zone) 28–32, 37–8

jackfruit (*Artocarpus heterophyllus*) 258
Jamaica Hope **see cattle**
joule 421, 441
jute (*Corchorus capsularis*) 259, 278

Kapok (*Ceiba pentandra*) 259, 280
Kikuyu grass (*Pennisetum clandestinum*) 315, 317
killing out percentage 360, 363
Kraal 392, 395, 401–2, 407, 448

lactation
   length 354, 396, 413, 478
   yield 360
lactose 354, 365
land
   capability classification 79–80
   clearing **113–43**
   evaluation **79–81**
   preparation 182
   rotation 145
   soaking 182
   suitability classification **79–81**
   tenure **2–4**, 109, 230
   use **13–5**
Lantana camera 321
latex 281, 203–4
leaching 118, 124, 148–9, 155–6, 253

leaf area index (LAI) 34, 202
leaf
   blight 261, 281
   canopy 202
   hopper 195, 197
   spot 267, 270, 279
legumes 15, 58, 63–4, 107, 160, 162, 164, 166, 176, 260, 270, 286, 290, 300, **302**, **316**, **325–7**
legume cover crops 102
Lens esculentum (Lentil) 163, 258, 270, 290
Leucaena leucocephala 58, 101, 166–8, 224, 232–3, 239, 247–8, 251, 300, 305, 327, 420
leucine **425**, **440**
ley 160
   farming 404, 418
lignin 303, 308, 420
liming 65
lipids **see fats and oils**
liquid feed **304**, 307
litchi (*Nephelium lappaceum*) 259
litter size 470–71, 483, 485
liver 426, 430, 432, 458
liver fluke 456, 460
livestock adaptation **371–90**
   breeding **462–86**
   classes **336–70**
   feeding **417–46**
   fertility **462–9**
   growth **285–6**
   improvement **417–46**
   records 484, 486
   reproduction **386–7**, **462–9**
liveweight gain 433, 449, 457
llama 316, 336, **351–7**, 463
Llanos 316
longevity 474, 476
Loudetia simplex 315, 331
lucerne (*Medicago sativa*) 295, 299, 230
Lupinus luteus (yellow lupin) 100
luvisols 9, 71, 120
lysine 425, 428

Macadamia nut (*Macadama spp.*) 259
Macroptilium atropupurpureum 101, 300–301
Macuna spp. 101–2, 163–4, 260, 295, 300

magnesium 49, **62–5**, 68, 73, 75, 147–8, 193, 241
  activity 62
  animal nutrition, in 429, 435, 442, 444
maize (*Zea mays*) 34–5, 74, 105, 127, 159, 249, 257, **261**, 283, 305
  amino acid composition, of **425**, 428
  bran 355
  breeding 286–7, 290
  disease control, in 153
  fertiliser requirements, of 74, 167, 171–4
  high lysine 302
  human food, for 15
  intercropping, with 151–2, 166, 183, 186, 229, 238, 243
  kernels 306
  kibbled 306
  mineral toxicity, of soil 64, 66
  protein degradability, of 427
  silage 309
  stover 303
  water deficit, of 36, 43
  water use 40
  yellow 424
  yield 58, 63, 90, 121, 124, 163–8, 175–6, 187
manganese
  in animal nutrition 429, 442, 444
  in soil 65–6, 194
mange 450
mango (*Mangifier induce*) 64, 258, 274, 282, 287, 311
mangosteen (*Garcenia mangostana*) 258
Manihot esculenta **see cassava**
Manila hemp (*Musa textilis*) 259, 278
manure 102, 298
marketing 4, 98, 398, 414
mastitis 412
meat
  and bone meal 330, 432, 440
  beef 338, 406
  dromedary, of 356–7
  goat, of 361, **363–5**, 369
  protein composition **425**, 427
  production 385, 440, 470, 481–2
mechanisation **129**, 184

Medicago sativa **see lucerne**
Mellotus philippensis 420
melon
  musk (*Cucumis melo*) 258
  water (*Citrullus lanatus*) 105, 258, 274
metabolic
  heat 377–8, 382
  profile 459–60
  rate 372
metabolisable energy (ME) **see energy**
metabolisable protein (MP) **see protein**
methane 18, 188, 193, 421–2
methionine 304, **425**, **440**
microflora/micro-organisms 421, 426, 452
micronisation 307–8
milk
  camel 355
  composition 309, 386, 427
  dromedary 354
  fat content 308–9
  goat **363–9**
  pink 456
  production 338, 357, 385, 394, 396–7, 406, 417–18, 422, 457, 469–71, 481, 483
  skim 411, 435–6
  substitute 435–6
  vein 481
  yield 343–4, 348–9, 354, 359, 365–6, 386, 397, 406, 413, 437, 469–70, 475–8, 483–5
milking
  cows 459
  once a day 411
  parlour 407, 413
  procedures **410–14**
  shed 408–9
millet 15, 34, 40, 43, 55, 127–8, 151–3, 155, 161, 166, 171, 186, 226, 249, 305, 354, 394
  Bulrush (*Pennisetum americanum*) 65, 106, 163, 165, 257, 262–4, 287, 290
  Finger (*Eleusine coracana*) 227, **263**
minerals 305, 311, 382, 401, 419
  deficiencies 335, 458–9
  mixture 341

nutritional value, of **428–30**
Miombo 50, 227, 314
mixed
  cropping 150, **164–6**
  farming 295, **403–12**
  stocking 332, 369
mohair 359, 364
molasses 301, 304–5, 354, 401, 405
molybdenum
  in animal nutrition 429–30
  in soil 65–6, 162, 326
monsoon 24, **29–30**, 37, 41, 43, 46, 49, 185
montane 41, 261, 315–16, 335
mosaic virus 264
mounds 125, 127–8, 151–2
mulch/mulching 54, 57, 99, **102–3**, 109, 115, 122–4, 130, 159, **171**, 219, 253, 267
multiple cropping 185, 222
multipurpose trees (MPTs) 230, 232–3, 239, 241–2, 246–7, 251–3, 255
Mung bean (*Phaseolus aureus*) 165–6, 193
Musa spp. **see banana**

N'Dama **see cattle**
nematode 153, 197, 267, 455
Nephelium lappaceum (litchi) 259
Nephotettix spp. (leaf hoppers) 195
net assimilation rate 37
net energy (NE) **see energy**
(NGO) non governmental organisation 97, 230
nickel 58, 66
nicotine 278
nicotinic acid 442, 444
nitrate 56–7
Nitrobacter 56
nitrogen 55, 57–9, 89, 124, **154**, 156, 161, 166, 170, 188, 193, 241, 323, 386, 424
  availability 151
  deficiency 190, 323
  endogenous 310
  fertiliser 18, 55, 164, 169, 171–5, 192, 197, 298
  fixation 56, 58–9, 102, 146–7, 161, 167, 239, 242, 300, 325–6
  fixing bacteria 192

flux 148
free extract (NFE) 295–7, 407, 421
liquid 480
loss, in soil 67, 149–50, 301
loss, by volatisation 183
transformation 191
urinary 324, 350
Nitrosomonas 56
nodules 58, 162, 326
nomad 393, 397, 404
nomadic pastoralism **393–6**
non-protein nitrogen (NPN) 303, **424–8**
no-tillage systems 122–3
nutmeg/mace (*Myristica fragrans*) 259, 278
nutritional requirements 417, 423, 433
  of cattle **433–8**
  of poultry **441–5**
  of sheep and goats **438**

Oca (*Oxalis tuberosa*) 258
oestrus 348, 368, 386, 462–3, 465–6
oils and fats **304**
oil palm (*Elaeis guineensis*) 42, 200, **214–21**, 258, **270–71**, 292, 410
  breeding 286–7, 289, 291
  plantation 202, 222
  pollinating weevil 219
  production 203–4, 282
  soil nutrient defficiency 64, 66, 74
  weed control 136
oil seed residues **302**, 307
okra (*Abelmoschus esculentus*) 151, 259, 274
open-pollinated population (OPP) 286, **290**
organic matter 60, 65–8, 74, 90, 125, 176, 184, 193
  complexes 55
  decomposition, of 48, 146, 150, 170, 182
  depletion, of 84–5, **88–90**, 104, 153
  mineralisation 56, 103, 155
  soil, in 236, 239–40, 250, 252
  supplies 147, 161, 169, 175
Oryza glaberrima 181

O. indica 197
O. javanica 197
O.sativa see rice
O.sinica 197
outbreeders **290**
overstocking 327–8, 330, 335, 394–5, 401, 403
ovulation 462–3
oxen 130, **437–8**
oxisols 55, 62, 69–71, 153, 182

paddy fields 179, 189, 193
palatability 298, 306, **308**, 347
palm oil 201, 214–16, 220
Panama disease 267
Pangola grass see *Digitaria decumbens*
Panicum maximum (Guinea grass) 100, 295–6, **298–9**, 326, 330–31, 334, 354, 410
  miliaceum 311
pantothenic acid 442, 444
papaya (pawpaw) (*Carica papaya*) 258, 274
Para grass see *Brachiaria mutica*
Paspalum dilatum 331
  distichum 196
  notatum (Bahia grass) 100, 326
  scrobiculatum 331
passion fruits (*Passiflora edulis*) 259, 274
pastorilism 44, **327–9**
pasture 222
  management **317–21**
  permanent 14
  productivity **416**
pawpaw (*Carica papaya*) 258, 274
peanut see groundnut
pearl millet see millet
pedigree 481–2
pelleting 307–8, 334
Pennisetum americanum see millet
  clandestinum (Kikuyu grass) 100, 315, 317, 326, 331
  pediculatum (Dinanath) 99–100
  purpureum (Elephant grass) 100, 108, 294, **296–301**, 313, 324–7, 330–31, 469
pepper (black) (*Piper nigrum*) 259, 278
peppers (sweet and hot) (*Capsicum spp.*) 151, 259, 274

performance testing 466, 481–2, 486
pesticide 16, 143, 247, 253
Phaseolus aureus (Mung bean) 193
  trilobus (*Pillipesera*) 193
  vulgaris 162
phenol 310
phenothiazine 456
phosphate
  fertiliser 60, 163, 169, **171–5**, 298, 324–6
  fixation 60
  requirements, of animals **428–30**
  requirements, of soil 61
  rock 174, 334
  soil 114, 123, 150, 162, 167, 190
Phosphorus
  animal nutrition, in 429–30, 432
  content of grass 318
  content of milk 354, 366
  deficiency 458
  livestock requirements, for **428–30**, **436–7**, **442**
  soil, in 18, **59–61**, 72, 89–90, 146–9, **154–5**, 188, 193
photoperiodism 34, 388
photosensitisation 456
photosynthesis 33–4, 36–7, 163, 192, 203, 209, 217, 250
photosynthetic active radiation (PAR) 249–50
pica 458
pig 294, 336–7
  genetic improvement 470, 483, 485–6
  growth 441
  heat regulation 386
  lactating 441
  nutrition 410, 424, 427, 429, 440
  porker 418
  pregnant sows 441, 463
  water loss 315
  water requirement 440
pigeon pea (*Cajanus cajan*) 100, 151–2, 158, 160–63, 165–6, 242, 258, 270, 290
pimento (all spice) (*Pimenta dioica*) 259, 278
pineapple (*Ananas comosus*) 64, 66, 74, 124, 200, 258, 274

pituitary gland 485
plaintain (*Musa spp.*) 258
plantation crops **200–21**
plant growth **33–41**
  hopper 195
plough 119–20, 356
pneumonia 395, 449, 451
population **5–6**, 156–9, 176, 329, 403
potash 171, 325
potassium
  activity 62
  animal nutrition, in **429**, **442**, **444**
  fertiliser 18, 169, 174, 298
  soil, in 49, **61–2**, 68, 123, 147–8, 155, 167, 241
  sweat, in 377
potato (*Solanum tuberosum*) 125, 258, **266**, 287
potential evapotranspiration (PEE) 39
poultry 336–7
  breeding 468, 470, 483, 486
  diet 294
  farms 309, 450
  heat tolerance, of **382–3**
  houses 388
  manure **304**, 310
  nutritional requirements, of **441–5**
  rations 431
  vitamin deficiency, of 432
  water requirements, of 420, 441
progeny testing 286, 466, 483–4, 486
propionate 309
propionic acid 421
Prosopis spp. 233, 236
  cineraria 101, 311
  juliflora 230, 311
protein 369, **424–8**, 471
  content 298, 302, 318, 354, 365, 386, 419, 471
  crude (CP) 295–7, 300, 305, 311, 324, 407, **425**, 427
  degradability 426
  digestibility 306, **308**, 319, 406, **427**
  digestible crude (DCP) 300, 325, 407, 437–9
  metabolisible (MP) 319, 437–9
  requirements 355, 382, 445
  rumen degradable (RDP) 436–9
  undegradable (UDP) 436, 438–9
protozoa 421, 451, **453–4**
pruning 251
Psidium guajava 259, 321
puddling 182, 185, 346
Pueraria phaseloides (Tropical Kudzu) 100, 102, 260, 300
pulse rate 348, 373
pulses 40, 43, 286, **302**, 307
pyrethrum (*Chrysanthemum cineranifolium*) 259, 281
pyridoxin 442, 444

Queensland arrowroot (*Canna edulis*) 258
quinine (Peruvian bark) (*Cinchona spp.*) 259, 278
quinoa 263

radiation 374, 377, 432, 451, 456
rainfall 20, **26–7**, **31–3**, 57
rainforest 22, 41–2, 94
rains 24, 29–30, 35
  grass 121, 320
ranching 330, 335, 340, 393, **396–403**, 450
range 402–3
rapeseed
  amino acid composition, of **425**
  protein degradability, of 427
ratoons 273
RDP (rumen degradable protein) **see protein**
relative humidity **see humidity**
repeatability 349, 470
reproduction efficiency 476
  performance 478
respiration rate 37, 348, 356, 373–4, 380, 382
rhinoceros beetle 272
rhizobium 58, 163, 194, 239
rhizosphere 59
riboflavin 442, 444
rice (*Oryza spp.*) 34, 74, 105, 128, 133, 144, 175, **178–99**, 202, 257, **260–61**, 282, 290, 295, 346
  amino acid composition, of **425**
  blast disease 196

bran 302, 305
breeding, of 197, 285–6, 288, 292
ecosystems 179
human food, as 15
husk 305
irrigated 73
paddy 140
soil requirements 63–4
straw 124, 350
upland 146, 151, 155, 159, 172
water use, of 40
Ricinus communis **see castor**
ridge and furrow 115–16, 151
rinderpest 395, 450
ring barking 321
ringworm 451, 457
root crops 53, **302–3**
rotary cultivator 135, 138
rotation 152–3, 158, **164–6**, 174, 176, 222
rubber (*Hevea brasilensis*) 259, **280–81**, 292
  breeding and selection 286–7, 289, 291
  grazing, under 410
  nutrient requirements, of 64–5, 174
  plantations 200–201, 215, 222
  production 203–4
  tapping 157
  weed control, of 136
  wind damage, to 74
rumen 308–9
  degradable protein (RDP) **see protein**
  fermentation 307, 311
  metabolism 310
  microflora 421, 436
rumination 294, 385, 436
runoff 33, 80, 118, 149, 154, 187, 236, 317, 334, 451
rust disease 261, 270, 276

Saccharum munja 100–101
  officinale (*cultivars*) **see sugar cane**
Sago palm (*Metroxylon spp.*) 258, **267**
Sahiwal **see cattle**
sainfoin 311
salinity **75–6**, 86, 195, 435

saliva/salivary glands 310, 436, 452
Salmonella 304, 310, 450, 460
salt 394, 401, 435
   rock 328, 418
Sanga **see cattle**
Santa Gertrudis **see cattle**
saturation deficit 209, 217
savanna 17, 42, 136, 146, **148–9**, 216, 398
   clearing 144, 151
   ecology 22, 24, 47, 314
   erosion 154
   fertility 145, 147, 153, 155, 174
   grazing 316, 397
   vegetation 51, 113
   weed control 152
scrapie 304, 431
screw worm (*Callitroga hominivorax*) 457
seasonal breeding 348, 387, 464
   day light variation 388
seasonality **387–9**, 399, 416
sebaceous gland 376
selection **285**
   index 350, 486
   mass 470, 479
   pressure 482
selenium 310, 429, 432, 442
semen 387, 450, 465–7, 480, 484
senna 167, 229, 236, 239, 243, 247–8
septicaemia 469
sesame (*Sesamum indicum*) 121, 160, 258, 272
Sesbania spp. 101, 167, 193, 233, 260
Setaria sphacelata 295–6, 299, 301, 325, 330–31
   splendida 295
shade 277, 334
shea butter tree 272
sheep 44, 332, 336, 369, 386–7, 410, 418, 463
   domestication, of 340
   grazing, by 316, **323**, 418
   mortality 395
   nomadic pastoralism, with 393–4
   reproduction 463, 466
   water loss, by 375
   requirements, of **435**
shelter 333, 369, 383, 451

shifting cultivation 16–17, **144–57**, 186, 228, 404
sib testing 482–3
silage 295, 301, 305–6, 322, 324, 354, 400, 413, 421
   vitamin content, of 430
silica 49, 194
Sim sim (*Sesamum indicum*) **see sesame**
simultaneous cropping **224–8**, 244
sisal (*Agave sisalana*) 66, 200, 259, 280, 291
skin pigmentation **381–2**
   thickness **382**
sleeping sickness (*Trypanosomiasis*) 15
slowly degraded protein (SDP) **see protein**
sodication/Sodicity **75–6**, 86
sodium 75, 310, 318, 377, 442, 444, 458
   animal nutrition, in 428–9
   hydroxide 75, 303
soil 8–9, **48–112**, 153
   acidity 64, 240
   classification **68–75**
   compaction 55, 118, 122
   conservation **82–112**, 115, 333
   degradation **83–95**
   erosion **see erosion**
   fertility **51**, 88, 91, 239
   mapping 68, 78, 80
   moisture 36
   organic matter (SOM) **see organic matter**
   profile **52**, 236, 241
   structure **52**, 53–4
   temperature 54, 209
   toxicity 75, 86, 174, 186
solar
   energy 25, 409
   radiation 31, 202, 209, 217, 344, 380, 387–8, 409
solids not fat (SNF) 354, 365, 386
sorghum (*Sorghum bicolor*) 40, 43, 158, **262**, 283
   amino acid composition, of **425**
   breeding 282, 286–7, 290
   human food, as 15
   intercropping 165
   irrigation, of 73

   livestock feed, as 294, 302, 305, 308, 394, 413
   nutrient requirements, of 155, 174
   production systems 127, 151–3
   silage 301
   soil toxicity, effect of 64, 66
   yields 121, 128, 161, 163, 166, 172
sour sop (*Annona muricata*) 258
sour veld 314, 321
sows **see pig**
soya bean (*Glycine max*) 35–6, 121, 127, 187, 272
   animal feed, as 295, 300, 309
   fertiliser requirements, of 172
   marketing 94
   meal 302
   seed inoculation 163
   soil toxicity, for 66
spaying 403
spice 278
spinach 151, 274
sprayers **140–42**, 219, 454
squash (*Cucurbita spp.*) 259, 274
starch 306, 420
   equivalent (SE) 422
stem borer 195, 261
stocking density 397
   rate 325, 327, **330–33**, 357, 394, 397, 410, 416
Stomoxys 327, 392, 453
storms 31–2
straight (feeds) 294, **301–2**
stratification 398
straw 56, 308, 354, 421
   quality 435
   vitamin content, of 430
Striga spp. 138, 152, 262
Stylosanthes guianensis (Stylo) 100, 300–301, 325, 331, 334
   humilis 334
subcutaneous fat 372, 424
subsoiling 116
sugar 420 (**see also sugar cane**)
sugar beet pulp 355
sugar cane (*Saccharum officinale*) 34, 36, 66, 115–16, 125, 258, **273–4**, 414
   animal feed, as 303, 405
   breeding 287, 289, 291–2
   by products 305

irrigation, of 73
nitrogen fixation, by 59, 326
nutrient requirements, of 74
plantations 200
soil toxicity, for 64, 66
weed control, of 139
sugar palm (*Arenga saccharifera*) 258
sulphur 63, 149–50, **154–5**, 167, 193–4
   animal nutrition, in 429, 435
   deficiency, in soil 63
sunlight 34, 432, 451
Sunn Hemp (*Crotalaria ochroleuca*) 136, 193
supplements 294, **305–9**
supplementary feed 400, 402, **406–7**
sustainability **17–18**, 88, 146, 241, 328
sustainable production **252–5**
sweat 374, 428, 485
sweat gland 345, 348, **374–7**, 485
   apocrine 372–3, 375–6
   ecrine 373
sweet potato (*Ipomoea batatas*) 125, 165, 258, **265**, 291
   breeding 287, 291
   fertiliser requirement, of 174
   weed control, in 136
sweet sop **see sour sop**
sweet veld 314
sylvopastoralism 223
synthetic nutrients **304**

taboo 1, 393
tallow 304
Tannia (*Xanthosoma spp.*) 42, 151, 258, **266**, 420
tannin **310**
   treatment 307–8
Taro (*Colocasia esculenta*) 125, 257
   giant (*Alocasia indica*) 257
TCIs (tree crop interfaces) 241, 246
TDN (total digestible nutrients) 423, 436–9
tea (*Camellia sinensis*) 200, 222, 241, 259, **275**
   breeding 291–2
   fertiliser requirement, of 174

   herbicide use, by 139
   plantations 201, **205–14**
   plucking 275
   soil nutrient deficiency, of 63–4
   yield 203–4
TDS (total dissolved solids) 435
Teff (*Eragrostis teff*) 102, 257, **263**
temperature
   ambient 372, 388, **443**, 445
   effects on crops 25, **37–8**, 45
   environmental 356, 383
   rectal 373, 380–81
Tephrosia candida 100, 102, 160
   vogelli 160
termite 53, 67–8, 79, 123–4, 252, 405
terrace 94, **107–8**, 179–80, 233
   bench 91, 96, 99, 104, 106
   intermittent 104
   orchard 104–5
   stone 106
testes 387
testosterone 387
Theileriosis 449
Themeda triandra 314–15, 321, 391
   australis 331
thermoregulation 385
thiamine 442, 444
threonine 425, **440**
ticks 328, 343, 411, 449, 452–3, 455, 457, 469, 473
tied ridging 83, 105, 120, **125–7**
tillage 53, **113–43**
tillering 186, 317
tilth 118
tobacco (*Nicotiana tabacum*) 36, 43, 159, 259, 290
   breeding 278, 286, 290
   pests and diseases, of 153
   soil conditions, for 64, 102, 174
tomato (*Lycospericon esculentum*) 35, 259, 274
tool carriers 122, 132–3
topography **50–51**
town dairies 350, 409
toxicity **310–11**, 459
trace elements/minerals 65–6, 306, 429, 434
tractor **133–6**, 338, 404
trade winds 25, 31, 38
transhumance 394–5, 404

transpiration
   animal 372
   plant 36
tree
   canopy 233
   nursery 235
   roots 243
   tomato (*Cyphomandra betacea*) 259
trematodes 455, 457
Trifolium alexandrinum (Berseem) 193
   semiphilosum 315
Tripsacum laxum (Guatemala grass) 100, 297–300
troughs 401–2
trypsin inhibitor 268
tryptophan **425**, **440**
Trypanosomiasis/
   trypanosome 389, 448–9, 451, 453–4
tsetse fly (*Glossina spp.*) 169, 234, 315, 327, 389, 447–9, 453, 457
Tung (*Aleuriles spp.*) 66
turmeric (*Curcuma longa*) 259, 278
tyrosine **425**

ultisols 55, 69–71, 153, 182
undegradable protein (UDP) **see protein**
urea 173, 305, 309–10, 405
urine 356, 421–2, 451, 456

vaccination/vaccines 395, 401, 452–3, 469
valine 425, **440**
vanilla (*Vanilla fragrans*) 259, 278
vegetables **274–5**, 287
velvet bean (*Mucuna spp.*) 102, 164, 295, 300
ventilation 409
vertisols 9, 69–70, 73, 123, 125, 127, 182
Vetiveria zizanioides (Vetiver grass) 99–101
vicuna 351–2
Vigna sinensis/*V.unguiculata* (Cowpea) 90, 101, 122, 124, 161–3, 174, 193, 224, 258, 290, 300

virus diseases
  animal **451–2**
  bunchy top 267
  mosaic 264
  rosette 270
vitamins 295, 305–6, 311, 382, 419, **430–32**
  A 366, **432–3**, 442, 444
  B 366, **431**, 442, 444
  C 354, 366, **431–2**
  D **432**, 442, 444
  E **442**, 442, 444
  K **431**, 444
  requirements, for pigs 440
volatile fatty acids (VFAs) 310, 421

water **419–20**
  availability 210, 217
  conservation **82–112**, 396
  consumption 389
  deprivation 355–6, 401, 419
  drinking 377, 383, 394, 419, 429, 435, 439
  erosion **see erosion**
  flow 86
  harvesting **105–7**
  intake 388, 420
  logging 114, 125
  loss 382, 420
  management 183
  metabolic 419
  pollution 160
  requirement 388, 419, 435, 439, 441, **443**
  supplies 327–8, 332–4, 385, **401**, 408, **443**
  table 114, 116–17, 186, 188, 210, 234
  use efficiency (WUE) **6–8**, 40–41
water buffalo (*Bos bubalis bubalis*) 182, 220, 336, **342–51**, 404, 418
  body fat 424
  breeds 346, 349
  classification 340
  draught 130
  milk 365
  river 345, 463
  swamp 345, 463
weed control **136–43**
well 117–18
wheat 15, 34, 40, 43, 73, 138, 163, 202, 286, 290
  protein degradability, of 427
whey 304, 431
wilt disease 267, 271
wind
  break 109, 166, 248
  speed 38
wilting point (WP) 36, 57
winged bean (*Psophocarpus tetragonolobus*) 258, 270
wool 357, 359, 371, 422

Yak (*Poephagus mutus*) 340
Yam (*Dioscorea spp.*) 118, 124–5, 151–2, 157, 160, 171, 174, 246, 258, **65–6**, 283, 291, 420
Yam bean (*Pachyrhizus tuberosus*) 258

Zea mays **see maize**
Zebu (*Bos indicus*) **see cattle, Zebu**
zig-zag breeding 475, 477
zinc 65–6, 194, 326
  deficiency 74
  animal nutrition, in 429, 442, 444